Chance Encounters

A First Course in Data Analysis and Inference

Christopher J. Wild
George A. F. Seber

University of Auckland

John Wiley & Sons, Inc.
New York Chichester Weinheim Brisbane Singapore Toronto

Editor	*Angela Battle*
Assistant Editor	*Mary O'Sullivan*
Senior Production Manager	*Lucille Buonocore*
Senior Production Editor	*Monique Calello*
Text Designer	*Nancy Field*
Cover Designer	*Karin Kincheloe*
Cover Art	*Roy Wiemann*
Illustration Editor	*Sigmund Malinowski*

Illustrations courtesy of Dr. Christopher J. Wild

"Dramatic Figures" created by JAK Graphics

This book was set in 10/12 Garamond by Publication Services and printed and bound by Quebecor Printing, Kingsport. The cover was printed by Phoenix Color Corporation.

This book is printed on acid-free paper. ∞

The paper in this book was manufactured by a mill whose forest management programs include sustained yield harvesting of its timberlands. Sustained yield harvesting principles ensure that the numbers of trees cut each year does not exceed the amount of new growth.

To order books or for customer service please, call 1(800)-CALL-WILEY (225-5945).

Library of Congress Cataloging in Publication Data:

Wild, C. J. (Christopher John), 1952-

Chance encounters: a first course in data analysis and inference / Christopher J. Wild, George A. F. Seber

p. cm.

Includes index.

ISBN 978-0-471-32936-7 (cl. : alk. paper)

1. Statistics. I. Seber, G. A. F. (George Arthur Frederick), 1938- II. Title.

QA276.12.W554 1999

519.5—dc21 99-042979

Printed in the United States of America

10 9

Preface

It is with some trepidation that we launch yet another elementary statistics book into a crowded market! What is distinctive about our book? It is an intuitive, data-oriented, graphical, and computer-oriented introduction to making sense of the world through statistics. We shall expand on each of these features in this preface.

Our Book Is Data Oriented

The main focus is on methods of summarizing and analyzing simple data structures. Our aim is to try to understand the processes producing the data so that we can come to conclusions and make decisions in the face of uncertainty. We emphasize the use of graphs, as they play a major part in the analysis and understanding of data structures. One of our maxims is "Don't do any formal statistical analysis without first graphing your data."

Our approach to data uses modern exploratory ideas. As the chapters unfold, we encourage the reader to see "exploratory" and "confirmatory" as part of the whole. For example, once we have explored the data using a variety of graphics, we choose an appropriate statistical tool to confirm what we see there.

Our examples and exercises use real data sets culled from many disciplines. Applications include the social, physical, biological, and medical sciences; and marketing, finance, and economics. They demonstrate that the subject of Statistics, with a capital S, is relevant to all aspects of our lives. This also allows us to foster the statistical thinking that occurs in the interplay between statistics and the context of a real investigation.

Our Book Is Computer Oriented

Computers have revolutionized the practice of statistics. Once the data have been entered, it is easy to call up a whole range of sophisticated tools and perform what used to be complicated statistical analyses with a few clicks on a keyboard. This has meant a shift in both teaching and authoring away from the mechanics of the calculations to a greater emphasis on interpreting the results of the computations.

The problem now for writers of general texts, however, is how to make this shift without turning the text into a tutorial manual for some particular software package. To avoid this problem, we have tried to give a broad picture of computational facilities using a selection of packages. In particular, we have opted for the popular Minitab package, originally available as a command language and now menu driven; the Excel spreadsheet; and the function language Splus and its freeware equivalent, R, available from Statlib. In addition to providing indicative computer code throughout, we provide some computer printouts to give readers a feel for what they can expect from a software package, and to show just how easy they are to get! In addition, our many detailed graphical displays highlight the sorts of things you can plot and draw electronically with the current technology.

Our Book Is Intuitive

We have endeavored to write a book that is expositional, chatty, and friendly. We try to explain concepts with words and graphics as much as possible. It is bad enough that students often arrive with an expectation that statistics will be dull without having a dull textbook as well! Students should find this book useful for self-study as well as for college and university courses. Sparsely worded statistics books can, at best, teach mechanical manipulations and mathematical theory. In contrast, we are interested in the big picture and the large statistical ideas. Since such statistical thinking can be demanding for the reader, an emphasis on understanding concepts sometimes demands extensive discussion. Snee [1993] pointed out the great variety in the ways people best learn and think. We have tried to make our presentation work on several levels and to cater to those who think best in verbal or pictorial ways as well as those who are more comfortable with algebra.

An example of this is our intuitive and data-oriented approach to probability in Chapter 4. We use two-way tables to motivate many of the ideas relating to probability and conditional probability. In particular, the instructor will be relieved to find that our approach avoids the need for Bayes' theorem and some of the more complex probability rules. We encourage the reader to be equally at home with percentages, proportions, and probabilities. When we changed wholesale to the two-way table approach, our course administrator found that the panic and overload on help facilities during the probability part of the course largely disappeared, with over half the class being able to successfully answer questions that were formerly answered using Bayes' theorem. We hope this approach will help to arrest the current decline in the teaching of probability that is occurring in elementary statistics courses. Some instructors may not wish to cover all this material, and we mention later how it can be reduced.

Valid Conclusions Serve as a Basis for Action

Statistics is about coming to conclusions in the face of uncertainty. We begin with questions and try to find answers. How the data are obtained—for example, how an experiment is designed—is critical to the quality and validity of the answers, and it also determines which methods of analysis are appropriate. We need to be able to check whether the assumptions underlying the methods are valid. Numerical summaries, plots, and various graphic aids are brought into play to trigger insights and to check the underlying assumptions. The interplay between graphics and methods is an important feature of our book, particularly in the later chapters.

Most of the material in this book has been extensively class tested. One of the authors (GAFS) initiated precalculus courses in statistics at Auckland 25 years ago. (The calculus-based courses are much older.) For some time, our first-year program has been successfully run on total quality management lines using a team approach (see Wild [1995] and Wild et al. [1997] for details). This is necessary as we now have nearly 3000 students taking the program each year. We are very grateful to the many people who have contributed and are still contributing to this program. Have we been successful? Two facts speak for themselves. First, our course reaches almost 50% of all new entrants and second, the retention to second-year applied statistics is now approaching 40%. It should be borne in mind that the majority of our students have no intention of proceeding further in statistics! We teach a broad spectrum of students with a wide variety of backgrounds and abilities, so our priority is to try to convey statistical ideas and statistical thinking rather than the mathematics.

Some Special Features

We have included a number of special features to help the reader focus on essentials.

- This book is written as a one-semester text. Time constraints, however, will mean that instructors may wish to select just some of the topics. We suggest ways of doing this in the following section. Instructors may also wish to incorporate some of the additional material that is available on our web site.
- We assume no prior knowledge of statistics. This is a precalculus text with very few mathematical requirements. We assume familiarity with arithmetic skills, simple algebra, some of the most basic concepts from set theory, the summation sign (in a very limited way), averages, square roots, and inequalities. There are occasional "extras," including references, that take the student beyond the general level of the material, which we have starred or relegated to footnotes. These additions should not interfere with the general flow.
- Our exercises come in two varieties: reasonably straightforward "practice" examples that generally concentrate on just one idea and immediately follow the textual development in each section, and more open-ended questions grouped as review exercises at the end of each chapter. In elementary texts there is a tendency for one problem to ask one question, so that knowledge becomes compartmentalized, whereas in real life we get a lot of information thrown at us that we have to sift through and that sometimes requires us to ask our own questions. We often use the context of the investigation underlying a real data set to stimulate statistical thinking.
- Most chapter sections end with a short quiz. Readers can use these to see if they have grasped and retained the key points. Instructors can also use them to generate class discussion.
- Case studies are given in some chapters. These studies expose the reader to some deeper statistical issues.
- Guides to the important learning themes include boxes and detailed summaries at the ends of the chapters.

Using the Material in a One-Semester Course

Although the layout of our book and the table of contents may be deceptively traditional in appearance (but beware of what is between the covers!), our approach uses modern exploratory ideas. Since the expositional and graphic approach to data analysis and inference moves more slowly than a purely methods approach, we don't expect most instructors to get through all the material. Some selection of topics, and even chapters, will be needed. To assist the reader, we have starred some sections that are not required for later use. Rather than simply giving a list of the topics in the chapters, we now wish to go through the chapters, indicating how an instructor might reduce the size of the course.

Chapter 1 is a broad-brush chapter that concentrates on themes and ideas. Many of these themes are picked up and embroidered later in the book. For example, random sampling is an important concept, and it is linked up with surveys and opinion polls. We need to distinguish carefully between experiments and observational studies, especially where the question of causality is concerned. This chapter may also be regarded as a public relations exercise! We explain why statistics is such an important subject and why it pervades our society. Because of the general nature of this chapter, a few key ideas can be summarized and much of the chapter can be assigned to students for reading in their own time. It is, however, an important chapter, as it underpins much of what we are trying to do in this book.

We regard Chapter 2 as a key "nuts and bolts" chapter describing numerical and graphical methods of summarizing data sets. Two features of this chapter are its emphasis on the best ways of displaying tables of data, and both the appeal and drawbacks of some popular types of plots. The chapter is comprehensive, with plenty of figures, and takes the reader slowly through each method. It is computer oriented, and with a computer students can learn these methods quite quickly with the chapter functioning as a manual. Some of the later topics (e.g., the Pareto diagram and segmented bar charts) can be omitted. Chapter 3 is also data oriented, except that our focus here is on a pair of random variables and the graphical methods that best convey their relationship. Regression is introduced informally here, and some instructors may consider the coverage of this topic in Chapter 3 sufficient. The more technical Chapter 12, on regression, can then be omitted. In both Chapters 2 and 3 we focus on the exploratory part of data analysis, whereas in Chapter 12 we emphasize the confirmatory aspects.

Chapter 4 is about probability, and it is here that many students lose their cool! We have already mentioned briefly what we regard as our "user-friendly" approach via two-way tables. Much of the chapter, however, is not needed for later chapters, so that, with the tough time constraints that most instructors work under, substantial trimming is possible, including the topic of conditional probability (Section 4.6) and much of statistical independence (Section 4.7).

The concept of random variable is an important one, and we find it helpful to treat discrete variables (Chapter 5) and continuous variables (Chapter 6) separately. In Chapter 5 the Binomial distribution has a prominent place, and the emphasis is on the model and underlying assumptions rather than on the mathematics. However, the supportive material on general probability functions in Section 5.2 can be reduced or omitted. The notion of expected value is important, but the motivation using data in Section 5.4 could be omitted or relegated to a computer simulation exercise. The idea of a probability density function is introduced intuitively using data in Chapter 5, and the Normal distribution plays a prominent role. Sums and differences of random variables are then discussed in Section 6.4 in anticipation of Chapter 7. Much of the approach in Section 6.4 can be incorporated into a computer exercise. Chapter 7, on ideas of estimation, is a pivotal chapter as it links the probability chapters (4 to 6) with the remaining inferential chapters (8 to 12). In Chapter 7 we discuss the important ideas of a sampling distribution and approximate Normality. As a first step toward confidence intervals and significance tests, we introduce the idea of a two-standard-error interval in Section 7.5, preparing the way for Chapters 8 and 9. However, much of Section 7.5 could be omitted, and the introduction to the t-distribution in Section 7.6 could be postponed for briefer mention later.

Chapter 8 deals directly with confidence intervals as an idea and uses confidence intervals for means, proportions, and differences between them as examples. Two features that can be omitted (much of Section 8.5) include a discussion of three different sampling situations that arise in media reports where one wishes to compare proportions, and the use and abuse of the margins of error published with opinion polls. Chapter 9 is about hypothesis testing, with much discussion on various practical issues such as the p-value, fixed-significance-level testing, tests versus confidence intervals, and practical versus statistical significance. This is an important chapter, as students tend to have trouble with this topic. Sections 9.4 and 9.6 can be omitted. In Chapters 8 and 9 we have achieved considerable simplification and unification via the repeated explicit use of T = [(estimate − parameter)/standard error] in working with tests and confidence intervals in the many situations where this idea can be applied. A simplifying trade-off has been to use the same form of standard error for both tests and intervals when these are first encountered. Standard usage is inconsistent here (e.g., in the differences between the treatments of means and proportions), and the differences encountered here are too minor to justify the additional layer of obscuring complexity.

In Chapter 10 we take another look at continuous variables. This time we bring all the tools out of our toolkit, including plots, confidence intervals, and hypothesis tests, and we show how these tools interact. We also discuss the important practical issues of assumptions and robustness. In addition to comparing two means, we introduce the analysis-of-variance technique for comparing more than two means. Design issues are looked at more closely, and paired comparisons and randomized designs are discussed. Nonparametric methods are also described, and the sign test is explained in some detail. Welch's approximate procedure, which is now the "default" in many computer packages for comparing two means, is preferred to the usual pooled two-sample test based on the assumption of equal standard deviations. Welch's test is more robust and works well. We regard this chapter as an important one in which data methods are brought together. Instructors, however, who cannot afford this level of detail may omit much of it. In Chapter 11 we try to do the same sorts of things for discrete data as we did for continuous data in Chapter 10. Goodness-of-fit tests and two-way tables are our main focus, and Simpson's paradox is discussed. Chapter 11 is a rather specialized chapter and can be omitted.

As already noted, Chapter 12, which supplements the first part of Chapter 3, discusses the importance of regression as a tool. Formal inference is presented only for the simple linear model along with some residual plots for diagnostic purposes. Correlation is also discussed. If only some of this chapter is to be used, Sections 12.1.3, 12.4.3, 12.4.4, and 12.5 could be omitted. Two optional chapters are available from our web site: Chapter 13 is about control charts; it covers the main types of control charts, and why and how they are used in practice. Chapter 14 is an introduction to time series using simple ideas of smoothing and deseasonalizing data. The discussion on index numbers includes how price indexes are constructed using survey data, something not generally found in other textbooks.

There are two final features. Although we use the computer to work for us, we have included in an appendix some tables that may be useful to the reader. Also, we have provided at the back of the book abbreviated answers to all the chapter exercises and the odd-numbered review exercises. No graphs are included there. Detailed solutions to all exercises, along with the accompanying graphics, are given in the solutions manual.

We have omitted some topics that are often found in elementary statistics books. In particular, combinatorial problems do not appear. However, instructors can easily add this topic to their courses. The term *variance* is seldom used in the main text. We have tried to follow Cobb's [1987] advice of not introducing two concepts when only one (standard deviation as a measure of spread) is really necessary. Normal-theory inferences about variances are not covered, although their extreme sensitivity to non-Normality is stressed and some robust alternatives are mentioned. The software package R previously mentioned was initially developed by fellow Aucklanders Ross Ihaka and Robert Gentleman, and is now burgeoning with a strong international team of collaborators. This package is similar in format to Splus; runs on Windows, Macintosh, and Unix platforms; and is available from Statlib free of charge. Note that Statlib can be reached (under "Links and Resources") from the American Statistical Association homepage at http://www.amstat.org.

Supplementary Materials for Instructors

- A complementary solutions manual is available to adopters of the book from the publisher.
- The Wiley web site, www.wiley.com/college/wild, provides the following resources:
  - Data disk
  - Additional data and exercises
  - PowerPoint slides
  - Two additional chapters
  - Additional topics
  - Teachers' forum

These are available for free download to adopters of the text. The two further detailed chapters on control charts and time series, alluded to earlier, may be particularly suitable for business-type courses. Initial material for additional topics was previously written for earlier drafts of the book when our local courses were larger than they are now. We do not intend to give a detailed list of this material here, but we do refer to some of what is available throughout the book. To give some idea of the sorts of topics, we list just a few: random numbers and their uses; the Poisson, Hypergeometric, Geometric, and Exponential distributions; Normal approximations; and a much more detailed treatment of the Wilcoxon nonparametric test for comparing two samples. We envisage the scope of web site growing to include a broader range of teaching materials and pedagogical ideas.

We welcome any comments or suggestions for improving this book and its supplementary materials. Please feel free to write to or contact us by fax (+64-9-3737108) or e-mail (ws@stat.auckland.ac.nz).

Acknowledgments

Our thanks go to the legion of statistical friends and colleagues who have influenced our thinking, and particularly to the many instructors at Auckland University who have used the drafts of this manuscript over the years and provided helpful feedback. Many of the ideas in this book, and some of the exercises, come from them. We would also like to sincerely thank the following people who reviewed a previous version of the manuscript and provided some very helpful comments:

Aidan Sudbury	*Monash University*
Gary Beus	*Brigham Young University*
Geoffrey McLachlan	*University of Queensland*
Howard Christensen	*Brigham Young University*
James Helmreich	*Marist College*
James Schott	*University of Central Florida*
John Spurrier	*University of South Carolina*
Linda Tappin	*Montclair State University*
Lisa Sullivan	*Boston University*
Maria Ripol	*University of Florida*
Mark Ecker	*University of Northern Iowa*
Mark Irwin	*Ohio State University*
Michael Petersons	*Macquarie University*
Vasant Waikar	*Miami University*
Ian Wright	*Curtin University*

We also owe a considerable debt to the editor, Brad Wiley II, for his important contribution to this book. His insight and penetrating comments combined with the reviewers' suggestions stirred us to undertake a major rewrite of the manuscript.

In conclusion, we wish to thank owners of copyright for permission to reproduce the following tables and figures: Table 3.2.1 (copyright by the American Society for Quality); Table 3.1.3 (reprinted with permission from *Technometrics,* copyright 1980 by the American Statistical Association, all rights reserved); Table 8 of Review Exercises 12 (reprinted with permission from *The Journal of the American Statistical Association,* copyright 1982 by the American Statistical Association, all rights reserved) and Appendix 3 (reprinted with permission from *The American Statistician,* copyright 1992 by the American Statistical Association, all rights reserved); Table 10.3.3 (copyright by the Institute of Mathematical Statistics); Table 1 of Review Exercises 8 (copyright by *The International Journal of Management*); Table 8.4.1 (copyright by the Royal Society of Chemistry); Table 4 of Review Exercises 8 (copyright by Oxford University Press); Table 7 of Review Exercises 11 (copyright by the American Marketing Association); Table 7 of Review Exercises 7 (copyright by the American Psychological Association); Table 3 of Review Exercises 9 (copyright by Guilford Publications); Table 6 of Review Exercises 10 (copyright by the Advertising Research Foundation); Figure 2 of Review Exercises 3 (copyright by Wadsworth Publishing Co.); and Table 2.5.2 (copyright by Taylor and Francis Publishers). Further acknowledgments are given in the text.

C. J. WILD

G. A. F. SEBER

Auckland, New Zealand
May 1999

Contents

CHAPTER 6
Continuous Random Variables 231

CHAPTER 7
Sampling Distributions of Estimates 277

CHAPTER 11
Tables of Counts 467

CHAPTER 12
Relationships between Quantitative Variables: Regression and Correlation 503

*CHAPTER 13
Control Charts *See web site*

*CHAPTER 14
Time Series *See web site*

APPENDIXES
Statistical Tables 557

CHAPTER 1

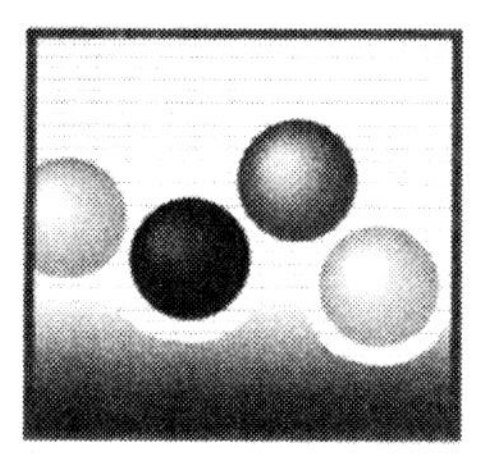

What Is Statistics?

Chapter Overview

Has anyone ever said "life is a gamble" to you? Such a statement reflects the feeling that our lives are surrounded by unpredictable, or "random," events. Some philosophers might argue that if we had perfect knowledge we might be able to predict everything. Others may disagree and argue that randomness is a basic feature of existence. Whatever our philosophical leanings, our knowledge is imperfect, so that from our vantage point events appear to be random. Everywhere there is variation caused by small, and sometimes large, unpredictable changes. For example, not only are we all different, but we all vary physically from day to day and even from minute to minute. Physical measurements such as blood pressure, red cell counts, and weight go up and down. Underlying the variation around us, however, are broad principles that make up our scientific knowledge. For example, we observe that taller people tend to be heavier, though not in every case. Our gas laws in chemistry (e.g., "the pressure exerted by the gas is proportional to its temperature") tell us about the average behavior of gas molecules in a container but not about individual molecules. Psychology can tell us about the behavior we can generally expect to get from a certain type of person (e.g., extrovert as opposed to introvert), but it cannot predict with certainty the behavior of any individual.

In addition to all the variation present in everything, we have problems when we try to take measurements. We have unpredictable experimental errors and measurement errors that will add still more variation to our data. Even in a subject like physics, which used to be called an "exact science," we cannot measure anything exactly. No matter how hard we try to calculate the speed of sound under controlled experimental conditions, we get a slightly different answer each time we carry out the experiment. In fact we cannot actually repeat an experiment under exactly the same experimental conditions. The conditions will always change, even if ever so slightly. Each time we repeat the experiment to confirm our answer, we are really performing a slightly different experiment. However, in spite of all this variation, we are aware that there are useful relationships, structures, and theories that can be uncovered by looking at the data. We try to see our way through the mass of information given by the data (the leaves of the tree) to find underlying principles (the branches). This is what the subject of statistics is about.

In this chapter we look at three fundamental kinds of statistical investigation: surveys, experiments, and observational studies. Each of these provides a different method for collecting data and making inferences about the uncertainty and variation we see. We begin in Section 1.1 by describing in detail the uses and misuses of surveys, and we introduce ideas of random sampling. A careful distinction is then made between a planned experiment (Section 1.2) and an observational study (Section 1.3) particularly with regard to investigating cause and effect. Section 1.4 discusses statistical investigation generally. All chapters conclude with a summary section (Section 1.5 here). Throughout, commonly used statistical terminology is introduced.

1.1 POLLS AND SURVEYS

As we have noted, we are all very different, both physically and mentally. Nobody is the same as me. Even identical twins are different. And nobody seems to hold the same sensible opinions that I hold! What's more, my circumstances and opinions can change from day to day. This makes the task of studying a population a difficult one. Usually it is impossible, or impossibly expensive, to investigate every single person. We must therefore settle for studying just part of the population. That is, we study a *sample*. Our goal is to have a sample that reflects the same sort of variation that the whole population exhibits, but on a smaller scale. Such a sample is called a survey. When the survey is aimed at finding out opinions on various issues, it is usually called an opinion poll, or *poll* for short. We shall see later that to get a representative survey we need to select the sample using random methods so that everyone in the population has the same chance of ending up in the sample.

⌘ EXAMPLE 1.1.1 *Survey on Female-Male Relationships*

American sexologist Shere Hite is known internationally for her best-selling writings on the relationships between men and women. *Time* magazine has called her "the doyenne of sex polls." Her research has been communicated through best-selling books, beginning with *The Hite Report: A National Study on Female Sexuality* in 1976, *The Hite Report on Male Sexuality* in 1981, and *Women and Love, a Cultural Revolution in Progress* in 1987. Like her earlier books, *Women and Love* was greeted by a furor in the American media and attracted worldwide attention: the book was the subject of a cover story in *Time* (12 October 1987) and even the *NZ Herald* (1 December 1987)[1] devoted almost a full page to discussion of the book and its critics. Hite's next book (*The Hite Report on the Family*, published 1994) ignited similar passions.

[1]The authors live in the city of Auckland, New Zealand. Our morning daily newspaper is the *New Zealand Herald*, abbreviated as *NZ Herald*.

Women in Love, a Cultural Revolution was based on the answers to a 127-item questionnaire from 4500 American women aged between 14 and 85 and painted a gloomy picture of relationships between men and women in the United States, particularly of marriage. Among the findings:

- 70% who were married for five years or more were having affairs—usually more for "emotional closeness" than for sex.
- 76% did not feel guilty about their infidelities.
- 87% had a woman friend that they felt closer to emotionally than they did to their husbands.
- 98% wanted "basic changes" to their love relationships.
- Only 13% married for more than two years were still "in love" with their husbands.
- 84% were emotionally unsatisfied with their relationships.
- 95% reported forms of "emotional and psychological harassment" from their men.

The critics were savage. Pulitzer Prize-winning columnist Ellen Goodman was reported as saying, "She [Hite] goes in with a prejudice and comes out with a statistic." The *Time* cover story stated, "as with Hite Reports I and II, the survey often seems merely to provide an occasion for the author's own male-bashing diatribes." The *NZ Herald* article said that "Hite uses statistics to bolster her opinion that American women are justifiably fed up with American men." Underlying the emotionalism were real disputes about what the survey findings said about American society. Shere Hite clearly believed that they were generally representative. Some critics believed they were not. Others simply believed that the research methodology was sufficiently flawed that one could not tell whether they were representative or not.

A survey is sometimes used to determine the views of a well-defined group of people. If Shere Hite's 4500 women[2] consisted of most or all "women chief executives of major corporations," we would be interested in the views of the group for their own sake. How did Hite get her women? She sent out 100,000 questionnaires to a variety of women's groups ranging from feminist organizations to church groups and garden clubs.

Basically, the 4500 were just those who replied and thus form a group too ill-defined to be of any interest in itself. Were the 4500 representative of American women in general, as Hite appeared to believe? First she sent out her questionnaires to women's groups. As *Time* stated, this strategy "means she was getting mostly one kind of person—'joiners'." People who are unhappy or unsatisfied may be more likely to be "joiners," perhaps because they need more companionship outside of the marriage. We call this type of problem, in which the population being sampled is an unrepresentative subgroup of the population of interest, *selection bias*. An even more important problem with the Hite study, however, is the fact that only 4.5% of those surveyed responded. *Time* quoted Regina Herzog, of the University of Michigan's Institute for Social Research, as saying, "Five percent could be any oddballs. We get pretty nervous if respondents in our own surveys go under 70%." Respondents to surveys differ from nonrespondents in one important way: they go to the trouble of filling

---

[2]With survey returns, round numbers like this are unusual; our guess is that it has been rounded to the nearest hundred. (This is an observation, not a criticism.)

out what in this case was a very long, complicated, and personal questionnaire. They may well differ with respect to the issues under study as well. This type of problem is called *nonresponse bias. Time* quoted pollster Hal Quinley as saying, "If sex was not very important then the woman wouldn't answer. If it was a burning issue, she would."

The sampling design of the Hite study was more complicated than portrayed in our description of it to date. After the first 1500 responses, Hite made a comparison between her respondents and the general U.S. female population and then tried to fill the gaps to ensure a sample that was fairly representative by age, geographic location, education levels, religion, and economic status. For example, the proportions of Roman Catholics and Protestants in the sample were roughly the same as the proportions in the general population. Does this make us feel happier about the results? Not really. What matters is that we want a sample whose members are representative with respect to their level of satisfaction with their love relationships. Every subgroup of the population has dissatisfied members. If we are still tending to get disproportionate numbers of dissatisfied people from each subgroup, then obtaining a sample that is representative with respect to these other demographic variables or characteristics has not helped us.

By drawing on other polls, the critics argued that Shere Hite's women were not representative. Some of the polls were conducted by reputable polling agencies while others were conducted by magazines. For example, a *Woman's Day* survey of 60,000 women and a *New Woman* survey of 34,000 women were quoted by *Time.* We have to be careful here. Magazine and newspaper polls are sometimes readership polls in which the survey questions are printed in the publication and interested readers send in the completed questionnaires (*self-selection*). Such polls are plagued by precisely the same problems as those of the Hite study—a population sampled that is not the population of interest, low response rates, and atypical respondents.[3] *Time* quoted Hite as admitting she didn't conduct a truly scientific study. "It's 4,500 people. That's enough for me." Having a large number of respondents is of no help. If the sample isn't representative, we are obviously no better off with a sample of 60,000 or 100,000, or even 1 million respondents, than we are with 100.

The biases mentioned above would ensure that a certain portion of the female population ruled themselves out of being surveyed by the method used. This group of women *could not* be represented in the final sample. Clearly any sensible sampling method must, at the very least, allow any woman from this group to have the same chance of being selected as any woman not in the group. Applying this to every possible group, what is needed is a random sample, namely one in which every woman in the population has the same chance of being selected. If the randomly selected sample was too small, then certain groups of women may not be represented simply by chance. However, if the survey was large enough, the sample would be representative in the sense that the makeup of the sample would reflect the makeup of the population.

---

[3]Self-selection biases can be huge. In 1990, the U.S. military deposed then Panamanian leader General Noriega. *Time* magazine (22 January 1990) noted that although the public opinion polls showed strong support for the intervention, letters to the editor in *Time* (a self-selection process) were running at more than 3 to 1 against.

✻ Example 1.1.2 *The Wrong Way to Take a Poll— Lessons from History*

Political polling first got started in a big way in the United States. Prominent in early polling was a respected magazine called the *Literary Digest.* It had correctly predicted the winner in every presidential election since 1916. In 1936 the *Digest* predicted that the incumbent, Roosevelt (Democrat), would get only 44% of the vote and his rival, Landon (Republican), would win. On election day Roosevelt won by a landslide with 62% of the vote. How did they manage to get it so wrong? The *Digest* was mailed to 10 million people whose names and addresses came from telephone directories and club memberships. It received only 2.4 million replies. The selection procedure was biased against the poor since, at the time, there were only 11 million residential telephones, and obviously these tended to be in wealthier homes. The United States was struggling to fight its way out of the Great Depression, and unemployment was high. It seems that prior to 1936, rich and poor voted along similar lines, but in 1936 this changed and the *Literary Digest* was caught unaware. Soon after, the *Literary Digest* went bankrupt! This story illustrates just how important it is to sample the correct population, that is, the actual population of interest. The technical term for the population of interest is the *target population.*

The next major error made by the American polls was in the 1948 presidential election, when all the major polls favored Dewey (Republican) over Truman (Democrat). Truman was elected President. During this period the major polls were using so-called *quota sampling* (each interviewer is assigned a fixed quota of subjects to interview with certain numbers to be made up, for example, by district, sex, race, and economic status). Within these quotas the interviewers could choose anyone they liked. They chose too many Republicans. Perhaps Republicans tended to be slightly easier to interview or were overrepresented in safer neighborhoods. Anyway, the Gallup poll chose too many Republican voters in each of the four presidential elections from 1936 to 1948. This reinforces our earlier point about the Hite study. Having a sample that is representative with respect to a long list of demographic factors does not ensure that the sample will be representative with respect to the factors under study.

Even though his major poll was using quota sampling in 1936, George Gallup carried out an experiment in which just 3000 people were chosen at random from the lists used by the *Literary Digest* and predicted the *Digest's* figures with an error of only 1%. Up to 1948 the Gallup poll used quota samples of 50,000. After 1948 Gallup and the other major polling organizations began to use random or chance sampling methods to choose their samples. Since this time they have used samples that may be only a tenth the size of their old quota samples, but their errors have been smaller.

We conclude this subsection with definitions of some of the jargon that you will often meet when reading descriptions of surveys and other forms of statistical investigation.

Target population: Complete set of individuals, objects, or *units* that we want information about.

Study population: Complete set of units that might possibly be included in the study. Ideally, the same as the target population, but often different (e.g., subset who can be contacted by telephone).

Sampling frame: List of units in the study population from which the sample will be drawn. (Not always applicable.)

Sampling protocol: Procedure used to select the sample.

Sample: Subset of units in the study population about which we actually collect information. We want to use information from the sampled units to tell us what is happening in the population as a whole.

Census: Attempt to sample the whole population.

Variable: A characteristic of each unit that we measure. For a human population, we may be interested in "age," "income," "amount of tax paid," and "gender." Each of these quantities is a variable.

Parameter: A numerical characteristic of the *population,* for example, average income in the whole population. Statistical investigations such as surveys are often used to estimate the unknown values of parameters of interest.

Statistic: A numerical characteristic of the *sample.* A (sample) statistic is often used to estimate a (population) parameter, for example, using the average income for the sample to estimate average income for the whole population.

1.1.1 Random Sampling

You must have seen market researchers interviewing people on street corners. Some of the biases in that process are easy to imagine. Very busy people walk by quickly and won't receive your glance. It is easiest to sample the tourists, window shoppers, and perhaps the unemployed—people who have nothing pressing to do. One would be unwilling to approach anyone who looked threatening in any way, such as someone who was obviously a member of a street gang. The types of people you react to in that way are under- or even unrepresented. In the interests of fairness you might think, "Well, I've just had a European male and a Hispanic female, but I haven't had anyone Chinese recently, so I'll approach that person over there." Thinking this way makes it very easy for minorities to become overrepresented. What we need is a method of choosing survey samples that avoids the (usually unintentional) biases to which people are prone. That method is random sampling.

The idea of taking a random sample is simple. We take a sampling frame, that is, a list of names or labels of every object in the population that we wish to sample. Suppose there are N objects in our frame (e.g., approximately $N = 30{,}000$ people in a New Zealand electoral district). We allot number 1 to the first object, 2 to the second, $3, 4, \ldots,$ going down the list. The final object has number N. We then need a mechanism to randomly choose a number between 1 and N, such as a variant of the lottery method. This fairly clumsy physical mechanism is feasible only if N is not too large. Suppose $N = 500$. Then we take 500 balls, paint each with a number from 1 to 500, and put them in a barrel. We rotate the barrel many times to thoroughly mix up the balls, blindly reach in, and pull out a ball. The number on the first ball chosen is the number of the first object in our sample. We continue to reach in and pull out more balls, each of which gives us the number of a new member of our sample. This is called *sampling without replacement* and gives rise to a ***simple random sample.*** If we want a sample of size $n = 20$, we draw 20 balls in this way. The important point is that at any draw, the balls are so thoroughly mixed up that any one of them is equally likely to be chosen.

An alternative to using the lottery method for choosing labeled objects in a random order is to use a table of random numbers. Most often the list of objects is stored in a computer, and the random numbers[4] are computer generated. To apply this methodology to the problem of drawing a sample from an electoral district, we could obtain a computer listing of everyone on the electoral roll for the particular electoral district, including both names and addresses. Having generated a sequence of random numbers, say 12,647, 32,678, 213, 53,231, ..., it is a simple matter to then print out the names and addresses of the 12,647th, 32,678th, 213th, 53,231st, ..., individuals on the list. Most statistical packages select a simple random sample of size n for you once you have the list in the computer and have specified n. In Minitab, look in the menus under `Calc` → `Random Data` → `Sample From Columns`, and choose sampling without replacement. In Splus and R, this can be done using the command `sample(x,n, replace=F)`. Excel also performs random sampling; look under `Tools` → `Data Analysis`, and choose `Sampling`. The sampling is with replacement, so you will have to eliminate repetitions and sample extras to make up the numbers. All of the packages we describe have extensive on-line help files that provide details of syntax.

Clearly, errors result when we use sample information to predict population values. We would get 100% accuracy only if we sampled the *whole* population. Also, different samples lead to different errors, as we see from the following example.

✘ Example 1.1.3 *Variation Due to Sampling*

Our colleagues Dr. Patricia Metcalf and Dr. Robert Scragg were principal investigators for a large workforce study that studied approximately 5700 people. We shall use the study group as a relatively small population from which to sample. One of the many issues the study addressed was body image: 69% of the people felt that they were overweight. Suppose we wanted to use a small sample to estimate this population figure. We took a random sample of these people of size 20 (sampling from the computer records) and observed that 45% of the sample felt they were overweight. This is quite a long way from the population value of 69%. We tried again and found that 85% of our next sample of 20 people thought they were overweight. For the third sample the figure was 65%. We were closer that time! Maybe we are getting better at this? In all, we took 10 samples each of size 20 from this population. Here are the percentages of people from each of the samples who thought they were overweight.

45, 85, 65, 75, 65, 90, 65, 55, 55, 75

The population percentage is 69%, but these sample percentages are scattered all the way from 45% to 90%. The process we are using is not producing reliable estimates. The most positive thing we can say is that we seem to have a roughly equal chance of getting an answer that is too big as of getting one that is too small.

You may well have heard that results from small samples are unreliable, so let us collect more information. This time we shall take random samples of 500 people. The first time we did this, 72.6% of the sample felt they were overweight. For the next

[4]Technically, these numbers are called *pseudorandom* numbers because of the way they are generated. For further details and applications see our web site. A table of random numbers is given in Appendix A1.

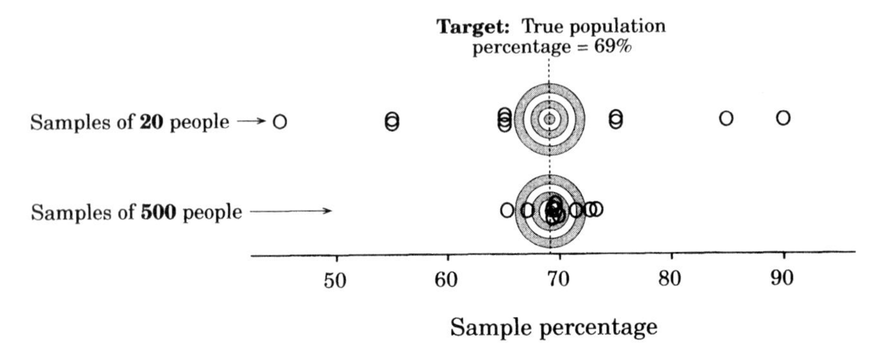

FIGURE 1.1.1 Comparing percentages from 10 different surveys each of 20 people with those from 10 surveys each of 500 people (all surveys from same population).

sample it was 69.8%. The results from 10 samples of size 500 follow. We have also displayed them graphically for comparative purposes in Fig. 1.1.1.

72.6, 69.8, 69.2, 73.2, 65.2, 69.6, 69.6, 69.2, 71.4, 67.0

None of the answers is exactly right, but we are clearly doing much better taking samples of size 500 than we were with 20. All of our answers are reasonably close to the population value of 69%. Taking an estimate from a sample of 500 people seems to be a reasonably reliable way of getting an estimate of the population value. Roughly half the sample values are too big, and roughly half are too small, so in this sense the sampling procedure is unbiased. The difference between the sample value and the true population value (69% here) we call a *random* or *chance error.* The bigger the sample, the smaller the chance error tends to be,[5] as illustrated in Fig. 1.1.1. Hence, for a large sample, we would usually have approximately the correct percentage of people who believe they are overweight. Similarly, we would also have approximately the correct percentages for a whole host of factors, such as left- and right-handedness, blood type, race, and many others. For large enough samples, therefore, characteristics of individuals in the population will be present in the sample in very similar percentages for almost every sample taken. We shall occasionally get a nonrepresentative sample, however, and unfortunately, we shall not know that this has occurred.

Besides avoiding biases, a second important advantage of random sampling is that the principles of equal likelihood that underlie such sampling can be used as a basis for formulating a mathematical theory of chance errors. The theory enables us to calculate the likely size of the error in our estimates. This cannot be done for the more subjective or systematic sampling procedures, and it is not particularly useful to have an estimate if you have no idea how accurate it is. We shall learn about the sampling variability of sample proportions and percentages, the likely sizes of the chance errors, and how to calculate sample sizes to keep the chance errors within reasonable bounds.

[5]We find out more about this later in the book. We must be careful not to quote a percentage without mentioning the sample size as well. Clearly, 50% expressing a certain view does not tell us much if our sample size is two!

The picture with random sampling is not completely rosy. For large populations, in particular, it will often be impossible to get a list of the entire population from which to sample. Even if you could, taking a simple random sample from a whole country for face-to-face interviews is an absurdly expensive process, especially if the population is widely dispersed. As a first step toward simplifying the process, we could use *cluster* sampling in which a household or "cluster" of individuals is used as our sampling unit rather than individuals. Although we still have to select a random sample of household units, we can do this by taking the cluster concept a step further by working with clusters of clusters. For example, one procedure would be to randomly sample some towns from a list of towns, then randomly sample some streets from within a town, and finally, randomly sample some households within those streets. Here randomness is still used to prevent bias, but the sampling scheme is much more complicated than simple random sampling. This scheme, which is called multi-stage cluster sampling, is certainly much more cost effective as interviewers can now generally get to all of their subjects within easy driving distance.

Another alternative to taking a simple random sample from a whole country is to use *stratified sampling*. We might decide that the states or provinces (called strata in statistical jargon) are sufficiently different to warrant taking a separate sample from each of the strata. Each provincial sample could then be chosen by some random mechanism such as multi-stage cluster sampling.

Example 1.1.4 *Sampling Plots of Land*

Although the preceding discussion focuses on surveys of people, the sampling methods apply to all kinds of populations. For example, suppose we have a large area of land that can be divided up into $N = 1000$ quadrants (square plots) of equal area. We would like to estimate the total number of certain objects (e.g., animals, plants, diseased trees, rabbit burrows) on the population area. One way often used to do this is to take a random sample of, say, $n = 100$ of the plots (10% of the population area) and count the number of objects on these sampled plots. Suppose we counted 324 objects. Since random sampling ensures that the sampled plots are generally representative of the whole population of plots, we could estimate the total number of objects by multiplying our sample total by 10 to get 3240 objects altogether. With many such populations the density of objects can vary substantially throughout the population area. If we know roughly where high and low concentrations are, we may be able to divide our population into strata, with the density being fairly constant within a particular stratum. We can then take a simple random sample of plots within each stratum.

1.1.2 Errors in Surveys

We have seen that the random sampling process introduces chance errors into our estimates simply because we use only a sample and not the whole population. The "margins of error" that accompany poll results in media (radio, TV, newspaper, etc.) reports refer only to this type of error. There are many other errors that creep into survey results that can be far more devastating than the chance errors resulting from sampling at random. The reported errors thus understate the real level of error. The survey literature makes a distinction between ***sampling errors,*** which arise from the decision to take a sample rather than trying to survey the whole population (which is what a *census* tries to do), and ***nonsampling errors*** which would be present even in a census. Except for self-selection, all the sources of error that we shall now describe are nonsampling errors. In fact many of them will be considerably larger in a census than in a comparable survey. Why? Because a survey, being a smaller operation, is easier to run well. For example, it is easier to hire only very good interviewers and ensure that they are all properly trained.

Selection Bias

We have already given examples of selection biases in which the population being sampled is an unrepresentative subgroup of the population in which the researcher is really interested.

Nonresponse Bias

A second major problem in sample surveys, discussed at the beginning of this chapter, is that of nonresponse. People may not return their questionnaires, or they may be out when the interviewer calls. As this part of the population may be different from the rest of the population with respect to the questions asked, appreciable levels of nonresponse can cause substantial bias in the results. It is important to try harder to obtain responses from these people. This may mean several letters or calling back several times to catch the person at home. In a study by Ward et al. [1985], people who were harder to contact tended to have higher incomes. Paying people to take part in a survey improves the response rate. This obviously makes a study more expensive, but it is sometimes done in marketing research studies that take quite a bit of each participant's time, for example, keeping a diary of buying habits. For mailed questionnaires, actually telephoning people and asking them to take part increases the response rate. Again, there is an associated cost.

Self-Selection

Self-selection occurs when people themselves decide whether or not to be surveyed; for example, questionnaires are made available to large numbers of people (e.g., the readers of a magazine) who may decide to fill them in and return them. Usually, very few do. The results tend to be akin to severe nonresponse. The method is frequently used as simply a fact-finding exercise, however. For example, hotels will often have a questionnaire available in the hotel room, and such questionnaires may tend to be filled in by disgruntled patrons only. Negative information, even if not representative, can still be useful (see Section 1.1.3). Similar comments apply to suggestion boxes.

Question Effects

You may be aware that questions can be slanted or loaded to help obtain the desired results. It is less well known that even fairly subtle variations in the wording of questions can have a measurable effect on the way people respond. Conducting surveys, especially if opinions rather than factual information is desired, is a very complex area. If you are actually going to be involved in constructing a survey, Kalton and Schuman [1982] is essential reading.

✖ Example 1.1.5 *Slightly Different Questions Lead to Very Different Answers*

A survey was carried out at Auckland University to study two types of question effects. Two versions of the questionnaire were prepared that were identical except for two questions. Version 1 asked:

1. "Are you in favor of giving special priority to buses in the rush hour?"
2. "Do you think that the cost of catching a bus into university is too high?"

Version 2 asked:

1. "Are you in favor of giving special priority to buses in the rush hour, or should cars have just as much priority as buses?"
2. "Taking into account the problems and costs associated with parking, do you think that the cost of catching a bus into university is too high?"

The two versions were each randomly allocated to students in a number of statistics classes. The total sample size was 1154, with 585 responding to version 1, and 569 to version 2. For question 1, 68% replied yes to version 1 and 47% to version 2. Clearly, there is a difference in response when one presents one or both sides of the case. For question 2, 75% said yes to version 1, and 63% said yes to version 2. Different wordings lead to different response patterns.

Survey-Format Effects

The survey format can affect the results. Moore [1979, p. 22] describes a study in which 44% of people surveyed by personal interview answered yes when asked if they favored contraceptives being made freely available to unmarried women, in contrast to 75% of those questioned by telephone or mail. Extreme care has to be taken with questions such as this, which people may be sensitive about answering honestly.

The responses to self-completion questionnaires have been shown to be affected by a host of seemingly inconsequential factors, including the location of a question on the questionnaire, the placement of the instructions and the general layout, and the colors of print for questions and instructions (see Kalton and Schuman [1982]). The order in which questions are asked can affect the way people reply in all types of surveys. So can the length of the survey.

In a commonly used form of telephone survey, households are telephoned randomly, and then someone within the household is asked to complete the survey. Some randomizing device may then be used, such as asking for the person whose birthday falls next. However, a person who lives alone is guaranteed to be asked when phoned, whereas a person in a family of five has only one chance in five of being asked. Thus people living singly or in pairs are overrepresented, while those living in large families are underrepresented. As young people tend to live in families, they tend to be underrepresented. Older people are more likely to live in pairs or alone (the children leave home, a partner may die), and they are overrepresented. Now suppose that a marketing research survey on media use, for example, is conducted this way and that the percentages of the audience using the various stations are reported directly from the survey (as often happens). Media catering to a young market will be reported as having a smaller audience than they really do and conversely for media appealing to older audiences. Sophisticated agencies, however, use reweighting techniques to try to correct for sample imbalances of this sort. Moore [1991, pp. 36-37] gives many more interesting facts about phone surveys.

Interviewer Effects

Different interviewers using the same questions can obtain different results; for example, the race of the interviewer can have a large effect on responses to questions with a racial dimension (see Moore [1991, p. 40]). In a large medical study with which we were associated, one question was treated quite differently by different interviewers, and the results had to be scrapped. Such possible effects can be checked on when a survey is given a suitable preliminary trial (called a pilot study in Section 1.1.4). Unfortunately, in our medical survey the problem was not picked up even in the pilot study, as it did not involve a broad enough spectrum of interviewers.

Asking the Right Question

In phrasing survey questions, and in fact for any statistical investigation, it is very important to be very clear about what you want to find out and then to be sure to phrase it unambiguously. The following illustration concerns the question of race in a New Zealand setting.

✖ Example 1.1.6 *Name Your Race*

The original inhabitants of New Zealand were the Maori, who first landed circa 900 A.D. from the Polynesian Islands to the north. In the national five-yearly census people say that they are Maori because they identify culturally with the Maori race rather than, say, the European race, even though, ancestrally, they are less than one quarter Maori.[6] Such responses are then useful for determining the numbers of people who are interested in issues specifically relating to the Maori people. For demographic statistics such as birth or death rates, however, the more usual definition of a Maori—namely, "half or more Maori blood"—is more appropriate (see Poole and Pole [1987]).[7] Clearly, the choice of question here depends on the purpose of the census.

Behavioral Considerations

Using surveys, we can only estimate the proportion of people who answer a question in a certain way, and this may not reflect their actual behavior. Some authors speak of *social desirability bias*. In post-election polls, more people claim to have voted than actually did vote, and more people claim to have voted for the winning party than

[6]The same sort of thing happens in the United States. In recent U.S. censuses there has been a big upsurge in the census counts of American Indians that could not be explained by birth and death statistics. The cause? Large numbers of people with small amounts of Native American blood who had previously self-identified as white are now self-identifying as American Indians.

[7]The census form has now been changed to reflect only cultural identity.

actually did so. Social desirability bias may not work the way you would expect, as shown by the following example.

✖ Example 1.1.7 *Smoking Marijuana—a Sensitive Question*

In 1987 a survey of 25,000 secondary school students was undertaken to estimate the extent of drug use among New Zealand high school children. Although the proportions claiming to use alcohol (51%) and marijuana (20%) were credible, large numbers claimed to have used hard drugs (e.g., 503 using cocaine, more than 200 using heroin), and this was at a time when the drug traffic was believed to be largely under control. There was a consensus among the police, education officials, and social workers, however, that the hard drug response rate was greatly overstated and that most of these students were simply claiming to have used hard drugs to impress their peers.

The data we want are the number actually using heroin, not the number claiming to have done so. With foresight the study designers may well have been able to overcome this problem. Perhaps there could have been some backup questions asking about the appearances and preparation of substances, which could have been helpful in weeding out many of the false returns. Additionally, it would be better if students could fill out their questionnaires in isolation rather than in a classroom with friends looking over one another's shoulders.

✖ Example 1.1.8 *What Is Your Real Age?*

When people are asked their age in surveys, they tend to give certain preferred ages, for example, 21, 25, 29, 35, 39. There are usually far more people with these ages in surveys than in the populations from which they were drawn. We can only conclude that some have been 39 for a good many years and that others turned 21 very soon after they were 17! Old people tend to *overstate* their age, advanced age being a source of status. Predictably, people also tend to lie, or refuse to answer, in response to sensitive or threatening questions, such as, "Have you driven while drunk?" or "Did you neglect to declare income on your tax return?" Atwood [1956] discussed the way hunters report their kills in wildlife surveys in the United States and detected several types of bias. Hunters tend to exaggerate their daily catches when they are low, or they can be genuinely forgetful. Poor memory combined with overstatement can lead to a catch being reported as a number ending in zero or five and somewhat above the true figure.

Transferring Findings

Very often people want to take data from one population and transfer the results to another. If the populations are very similar, it may be reasonable to transfer results. For example, New Zealanders may be interested in the Hite Report because they believe that their society is similar enough to American society for the patterns of sexual activity in the United States to give them some idea about patterns in New Zealand. Whether or not this is reasonable is usually a subject-matter decision, not a statistical one. Still, statistical arguments are often used to show that populations are *not* similar enough for such transference to be reasonable. Carrying across conclusions from one population to another can be very dangerous. Statistics allows us to make statements only about the population actually sampled. This may not be the population of interest.

EXAMPLE 1.1.9 *Transferring Child Abuse Statistics*

For many years, the Broadcasting Corporation of New Zealand ran telethons. For 24 hours one television channel, which reached the whole country, was given over to fund-raising for a particular cause. There was a lot of associated entertainment, and it was a major event, a sort of nationwide party. In 1988 the recipients of the funds were the survivors of domestic violence and sexual assault. Devastatingly effective ads were beamed night after night in the lead-up and were reproduced in the print media. One photograph of four cute babies bore the legend, "One of these children will be scarred for life." Another image was a normal scene of a father coming into a darkened room to check on his daughter before going to bed. This one was boldly titled, "It is not the dark she's afraid of." One of the figures constantly propagated was that one in eight children would be sexually abused by their own fathers. Nobody argued with these statements at the time, but investigative journalists for *Metro* (October 1988, pp. 134–149) and the *NZ Listener* (August 13–19, 1988, pp. 16–18) subsequently did some digging.

It was fairly clear that the preceeding figures, quoted mainly from overseas, shouldn't simply be transplanted into the New Zealand scene. There is also the question of how reliable the data is. Getting reliable data on anything as sensitive as sexual abuse is very difficult. These facts obviously worry a statistician, but did it really matter?[8] Trying to look at it from the point of view of the people involved in helping the survivors of domestic violence, our initial reaction was that maybe it didn't matter. The likely effect would simply be to make people more likely to support what everyone would agree was a good cause. But on further reflection, it isn't that simple. Take the father–daughter incest figures. Something as devastating as that happening to one girl in eight is frightening. It could be enough to damage the feelings of trust that most daughters have for their fathers. Yet if it really happens to only one girl in a hundred, it is a moderately rare occurrence, and the inflated figures may have led to very real and unnecessary harm. On another level, there is only so much money available for charity. If one cause is oversold, it benefits at the expense of other causes. Other people are then affected. Moreover, the cause itself cannot benefit in the long term, as people will become suspicious of its proponents. Given the suspect nature of the data then, should we ignore the data? Clearly the investigation is an important one, so perhaps the data should be used cautiously, with warnings attached. We need to be both statistically and socially responsible in the use of such data.

What is the point of all of these stories? Survey statisticians work in a world that is complex in human terms. They need to be aware of these complexities and be sensitive to them, so that they can foresee problems before they occur. Knowledge of human behavioral patterns is at least as important as technical statistical knowledge.

[8]These television ads subsequently won the Mobius Award for *outstanding creativity* (our italics), one of the top U.S. awards for radio and television advertising.

1.1.3 Unrepresentative Information Can Be Useful

Because of our desire to establish the importance of taking large samples at random and with high response rates to ensure representative samples that give reliable information, we may have left you with the misconception that information that does not meet these high standards is automatically useless. Information need not always be representative to be useful. It all depends on the purposes for which you want to use that information. For example, it was previously noted that questionnaires left in hotel rooms may provide useful information about the service provided.

EXAMPLE 1.1.10 ***Customer Satisfaction***

In any business operation, data obtained from customer complaints notoriously understates the true levels of customer dissatisfaction. It has been said that as few as 2% of unsatisfied customers complain but that as many as 35% may not come back (Davidow and Uttal [1989]). The data from individual complaints, however, can help management to identify faults and failures in their products or service procedures, which they may be able to correct or improve. In addition, *changes* in the overall level of complaints can signal whether customer dissatisfaction levels are increasing or decreasing. Knowing whether the situation is getting better or worse may be sufficient for a management team that is trying to make changes to improve the product or service they deliver.

1.1.4 Preparing for a Survey

Before conducting a full-scale sample survey, a great deal of preliminary work is needed, as demonstrated by the following example.

EXAMPLE 1.1.11 ***What Do You Think of McDonald's?***

Suppose we have the problem of measuring customer perception of the products and services provided by a McDonald's restaurant. There is variation in the level of service people receive at a restaurant. Also, different customers react to the same service in different ways. Therefore, before undertaking a scientific survey to determine what people think, we have to decide what to measure, that is, what sorts of things we should ask questions about and how we should phrase them. The factors we would most want to ask questions about in the survey are those factors that have the most influence on determining whether the customer will give the restaurant continued custom or decide not to return. The restaurant's management may have some strong opinions about what some of these factors are, such as speed of service, friendliness of personnel, cleanliness, and pleasantness of surroundings. They would be silly just to trust their own instincts, however. They may have overlooked an area that is very important to customers and thus to the continued health of their business.

To try to identify some of these factors, they will have to communicate with customers in a very unstructured way and try to elicit from customers what *really* is important to them, rather than what *management believes* is important to them. This part of the process is very time-consuming, so only a small group, possibly consisting of easily accessible people, can be talked to in this way. The information cannot be expected to be reliable or representative; we just hope that through open-ended discussion we are able to unearth the important issues.

Having done the preliminary investigation and written trial survey questions, we must now investigate any problems that might arise with the survey. This can be done by using just a small sample called a ***pilot survey.*** There are two types of pilot surveys. The first is aimed at checking whether the questions are clear and unambiguous. For example, are all the possible responses listed for each question? This checking can be done by giving the questionnaire to a small sample of people. Such a sample does not have to be random; in fact it would probably provide more information if the people selected were as diverse as possible. In some situations we may carry out several such pilot surveys until we are happy with the finished questionnaire. The need to foresee problems such as ambiguities, interviewer bias, deliberate lying, nonresponse, and so on is highlighted in a story told by N. Webb, a researcher for Gallup in Britain, in the discussion of Kalton and Schuman [1982]. In the early 1960s he did pilot work on one of his own surveys. The first question asked was whether the person was watching television at the time the doorbell rang, which seems straightforward enough. However, he had seen one woman who said yes working in her kitchen as he approached the house. When asked for clarification at the end of the interview, she said she *had been* watching TV but had just slipped out to wash a few cups! (A primary purpose of the survey was to gauge audience reaction to commercials.)

An important question to ask before carrying out the full survey is, "How many people should we sample?" Clearly, the bigger the sample, the more accurate the sample information but also the more costly the survey. These two aspects are conflicting, and the conflict can be resolved only if we can obtain rough estimates of the population quantities we are interested in and then use appropriate statistical theory (discussed in later chapters) to determine the size of the survey. If rough estimates are not available, and they usually aren't, we can obtain them by taking a second type of pilot survey, namely, one that must be randomly selected. We carry out a greatly scaled-down version of the full survey.

1.1.5 Who Uses Survey Methods?

Surveys and questionnaires are very much a fact of modern life. We come into the world via a questionnaire (a birth certificate), go out the same way (a death certificate), and seem to spend a good deal of the time in between filling them out! Surveys come in all shapes and sizes, ranging from a single question—such as the popular "What is the most important problem facing the country today?"—to a lengthy document that requires either an extensive interview or the collection of data over a period of time (e.g., all the television programs watched over the ratings period). We are particularly interested in *sample* surveys, those in which only a portion of the population under study is investigated.

A common form of survey is the opinion poll, in which people are asked a variety of political or social questions. Market surveys studying consumer responses to various products are also popular. Ratings for TV or radio audiences are based on surveying the viewing or listening habits of people. Social surveys that ask questions about, for example, local facilities (e.g., libraries) are needed for making resource decisions.

✖ EXAMPLE 1.1.12 *Distributing Profits Among Airlines*

Often when you buy tickets from one airline, particularly for international flights, some legs of the journey are on another carrier. The airlines find it too expensive to settle individual accounts. Instead they work out how much they owe each other on the basis of a sample of about 10% of the tickets (Neter [1989]).

✖ EXAMPLE 1.1.13 *Have You Paid Your Taxes?*

There is one form of sampling that might be called "sampling for intimidation"! In many countries a sample of people have their tax returns audited in detail. This tends to be in addition to a regular rotation in which everyone is audited, say, every five years. The sampling of tax returns is done in large part in the hope that the fear of being sampled will keep people honest. The insurance industry, particularly in the United States, uses a very similar strategy by only checking a sample of claims. Of course, the sampling also produces useful information on the extent of tax or insurance fraud. This can be used to fine-tune the trade-off between the cost of checking claims and the cost of fraud.

These are just a few examples of the many applications of surveys and who uses them. For further examples see our web site.

QUIZ FOR SECTION 1.1

1. In surveys, what is selection bias? (Section 1.1 introduction)
2. What is nonresponse bias, and how does it arise?
3. Why are many magazine surveys suspect?
4. What is quota sampling, and why is it unreliable?
5. What is meant by each of the following terms: target population, study population, sampling frame, sampling protocol, sample, census, variable, parameter, and statistic?
6. Why do we sample at random? (Section 1.1.1)
7. How could you implement the lottery method to randomly sample 10 students from a class of 250?
8. Complete the following: "The chance errors in survey estimates are generally smaller if we take ____ samples."
9. What is meant by saying that using the proportion of left-handers from a random sample to estimate the corresponding proportion in the population is an unbiased procedure?
10. What is cluster sampling, and why is it used?
11. What are some of the nonsampling errors that plague surveys? (Section 1.1.2)
12. If we take a random sample from a population, can we apply the results of our survey to other populations? Are sampling households at random and interviewing people at random on the street valid ways of sampling people from an urban population?

13. How do the following lead to biases or cause differences in response: nonresponse, self-selection, question effects, survey-format effects, interviewer effects, transferring findings?
14. Give an example where nonrepresentative information from a survey may be useful. (Section 1.1.3)
15. Why is it important to take a pilot survey? (Section 1.1.4)
16. Give an example of an unsatisfactory question in a questionnaire.
17. Can you think of some interesting applications of surveys not described here? (Section 1.1.5)

EXERCISES FOR SECTION 1.1

1. Suppose that you want to set up a nationwide survey about immigration issues. We want you to think as precisely as you can about the target population that *you* would be interested in. Who would you want included? Who would you want excluded? Can you define some rules to characterize your target population?
2. List some questions that people might be sensitive about answering in a survey interview.
3. What are some types of surveys that you have seen that almost certainly have large selection biases?
4. One important source of nonsampling error that we did not discuss was processing errors. Can you think of some ways in which errors could creep in during the data-recording and storage-stages of a survey?
5. A report called "No Place Like Home—Women's Experiences of Violence" was published in New Zealand in late 1989. It was based on questionnaires completed by 705 women who found the questionnaire in two magazines or at women's self-defense classes. Among many other findings were the following. Of the respondents, 91% were European compared with 81% in the population. (The other main racial groups were Maori and Pacific Islanders.) The report found that Europeans perpetrated 80% of all physical attacks but 92% of incest incidences. The *NZ Herald* headlined its story about the report, "Europeans high in sex abuse findings." Would you trust the figures published in the report as representative of women's experience in New Zealand? Why or why not? Are the figures indicative of a higher incest rate among Europeans?
6. Using a computer package or another method, take a random sample of 10 countries from those listed in Table 9 in Review Exercises 2. Print out the selected countries and their birthrates.

1.2 EXPERIMENTATION

We observed at the beginning of Section 1.1 that we are all very different and that we react differently on different occasions. For example, no two people react in exactly the same way to the same dose of a given drug, and a person's reaction will probably vary from day to day as well. If we want to compare the effects of two different drugs on two groups of people, then we have the problem that any drug difference gets obscured by the personal variation. It becomes hard to see whether the drugs really have different effects. This between-individual and within-individual variability is found with all biological organisms. Variability can make it very difficult to decide

"What is the best form of treatment?" and "In what quantities should it be given?" Some insects are killed by small doses of insecticide, while others survive comparatively massive doses. Some plants grow well with a certain type of fertilizer, while others fare badly. There is variability in human-made objects as well. No two items coming out of any manufacturing process are ever exactly identical, not even very simple objects such as ball bearings. The diameter of a wire changes over its length. The thickness of a board is always slightly different at different positions on its surface, and the average thicknesses of any two boards off the same production line are always different. If we want to examine the effects of two different treatments on our product, we have the same problem that any treatment difference is obscured by individual variation. We need to remember the following point at all times:

> In the real world, ***variability is everywhere and in everything.***

Because of this problem of variability, in each situation we need to design a specific experiment that allows us to disentangle the natural variation from any variation imposed by our treatment method. We now discuss the nature of statistical experiments and introduce some basic principles of experimental design.

EXAMPLE 1.2.1 *Comparing Two Varieties of Tomatoes*

Suppose we wished to compare two varieties of tomato. The purpose of the comparison is to find the variety that produces the greater quantity of marketable-quality fruit from a given area for large-scale commercial planting. What should we do? A simple approach would be to plant a block of land in each variety and measure the total weight of marketable fruit produced. There are some obvious difficulties, however, due to the variation in growing conditions. The variety that cropped most heavily may have done so simply because it was growing in better soil. Although we are not professional gardeners, we can think of a number of factors that affect growth: soil fertility, soil acidity, irrigation and drainage, wind exposure, exposure to sunlight (e.g., shading, north-facing or south-facing hillside). Unfortunately, no one knows exactly to what extent changes in these factors affect growth. Therefore, unless the two blocks of land are comparable with respect to all of these features, we won't be able to conclude that the more heavily producing variety is better, as it may just be planted in a block that is better suited to growth. If it was possible (and it never will be) to find two tracts of land that were identical in these respects, using just those two blocks for comparison would result in a fair comparison, but the differences found might be so particular to that particular combination of growing conditions that the results obtained would not be a good guide to full-scale agricultural production anyway.[9]

Let us think about it another way. Suppose we took a large block of land and subdivided it into smaller plots by laying down a rectangular grid. By using some sort of systematic design to decide what variety to plant in each plot, we may come unstuck if there is a feature of the land such as an unknown fertility gradient. We may still end up giving one variety better plots on average. Instead, let's do it randomly by numbering the plots and randomly choosing half of them to receive the first variety.

[9]In fact it would obviously be advantageous to do trials at a sample of sites chosen at random from around the target area to ensure that the results were applicable.

The rest receive the second variety. We might expect the random assignment to ensure that both varieties were planted in roughly the same numbers of high-fertility and low-fertility plots, high-pH and low-pH plots, well-drained and poorly-drained plots, and so on. In that sense we might expect the comparison of yields to be fair. Moreover, although we have thought of some factors affecting growth, there will be many more that we, and even the specialist, have not thought of, and we can expect the random assignment of treatments to ensure some rough balancing of those as well.

The same sorts of considerations apply to any experimental setup in which we are interested in the differences in the effects of treatments when applied to nonhomogeneous experimental units. Whether we are interested in comparing the effects of drugs on people, the speed of cooling on the strength of welded joints, different welfare regimes on employment patterns, dietary additives on the speed of cancer development in rats, or baking temperatures on the taste of cakes (here the experimental unit is the cake mixture), the simplest experimental design is to randomly allocate treatments to all your experimental units, usually in such a way that every treatment (e.g., variety, drug or baking temperature) is applied to the same number of units. For example, suppose there are nine units and three treatments. We could then number nine cards representing the units, shuffle the cards, and give the top three numbers treatment 1, the next three treatment 2, and the remaining three treatment 3. An example of this is given below.

2	3	1	2	1	3	3	2	1

This is called a ***completely randomized design,*** as the randomization (i.e., the shuffling of the cards) is carried out over all the units at once.

In the previous section we noted that random sampling tends to give representative samples *if our sample sizes are big enough*. Unfortunately, we can expect the random allocation of treatments to lead to representative samples (e.g., a fair division of the more- and less-fertile plots) only if we have a large number of experimental units to randomize. In many experiments this is not true (e.g., using six plots to compare two varieties), so in any particular experiment there may well be a lack of balance on some important factor. Random assignment still leads to fair or unbiased comparisons, but only in the sense of being fair or unbiased when *averaged* over a whole sequence of experiments. This is one of the reasons there is such an emphasis in science on results being repeatable. There are statistical techniques, however, that, except on rare occasions, protect one against claiming to have found treatment differences when the apparent differences are just caused by having unrepresentative samples (see Chapter 10).

Partly because random assignment of treatments does not necessarily ensure a fair comparison when the number of experimental units is small, more-complicated experimental designs are available to ensure fairness with respect to those factors that we believe to be very important. In our tomato example suppose that because of the small variation in the fertility of the land we were using, the only thing that we thought mattered greatly was drainage. We could then try to divide the land into two blocks, one well drained and one badly drained. These would then be subdivided into smaller plots, say ten plots per block. Then in each block, five plots would be assigned at random to the first variety, and the remaining five plots to the second variety. We would then compare the two varieties only within each block, so that well-drained plots are compared only with well-drained plots, and similarly for badly drained plots. This idea is called *blocking*. By allocating varieties to plots within a block at random, we would provide some protection against other extraneous factors.

Another application of this idea is in the comparison of the effects of two pain-relief drugs on people. If the situation allowed, we would like to try both drugs on each person and look at the differential effects within people. To protect ourselves against biases caused by a drift in day-to-day pain levels or other physical factors within a person that might affect the response, the order in which each person received the drugs would be randomly chosen. Here the block is the person, and we have two treatments (drugs) per person, applied in a random order. The randomization could be done in such a way that the same number received each drug first so that the design is "balanced" with regard to order. In the words of Box, Hunter, and Hunter [1978, p. 103] the best general strategy for experimentation is ***"block what you can and randomize what you cannot."***

We have talked so far about comparing two experimental treatments. An important special case of this is where one of the "treatments" is in fact no treatment at all. For example, we can compare the effect of treating soil with fertilizer with the effect of doing nothing. The group of experimental units given no (active) treatment is called the *control group*. By comparing a treatment group to a control group, we get an estimate of the effect of the treatment. Experiments that use people involve further difficulties, which we shall now illustrate.

✖ EXAMPLE 1.2.2 *The Psychological Effects of Taking a Pill*

There are psychological effects that complicate the effects of treatments on people. In medicine there is a tendency for people to get better because they think that something is being done about their complaint. Therefore, when you give someone a pain-relief drug, there are two effects at work: a chemical effect of the drug and a psychological boost. To give you some idea of the strength of these psychological effects, approximately 35% of patients have been shown to respond favorably to a placebo (or inert dummy treatment) for a wide range of conditions, including pain, depression, high blood pressure, infantile asthma, angina, and gastrointestinal problems (Grimshaw and Jaffe [1982]).

In evaluating the effect of the drug, we want to filter out the real effect of the drug from the effect of the patient's own psychology. The standard method is to compare what happens to patients who are given the drug (the treatment group) with what happens to patients given an inert dummy treatment called a placebo (the control group). This will work only if the patients do not know whether they are getting the real drug or the placebo. Similarly, when we are comparing two or more treatments, the subjects (patients) should not know which treatment they are receiving, if at all possible. This idea is called *blinding* the subjects. The results then are not contaminated by any preconceived ideas about the relative effectiveness of the treatments. One of the authors was involved in a comparative study of two asthma treatments, one in liquid form and one in powder form. To blind the subjects, each subject had to receive both a powder and a liquid. One was a real treatment, the other a placebo. Blinding, to whatever extent possible, tends to be desirable in any experiment involving human responses (e.g., medical, psychological, and educational experiments).

Example 1.2.3 *Who Gets the Free Milk?*

An experiment was performed in England to evaluate the effect of providing free milk to school children. There was a random allocation of children to the group who received milk (the treatment group) and the control group, which received no milk. Because the study designers were afraid that random assignment may not necessarily have balanced the groups on general health and social background at the classroom level, however, teachers were allowed to switch children between treatment and control to equalize the groups. It is easy to imagine the effect this must have had. Most teachers are caring people who would be unhappy to watch affluent children drinking free school milk while malnourished children go without. We expect that the sort of interchanging of students between groups that went on would result in too many malnourished children receiving the milk, thus biasing the study, perhaps severely.

To protect against the preceding sort of bias, medical studies are made *double blind* whenever possible. In addition to the subjects not knowing what treatment they are getting, the people administering the treatments don't know either. The people evaluating the results (e.g., deciding whether tissue samples are cancerous or not) should also be "blinded," called *triple blinding.*

Example 1.2.4 *Do You Have ESP?*

Later we shall see the results of a famous experiment in extrasensory perception (ESP). It involved recording whether the shape guessed by the "recipient" matched the shape supposedly projected mentally by the "sender." This was done 60,000 times. The results of each individual trial were recorded. Did the shapes match, or did they not? Now in any boring, repetitive process like this, there will inevitably be quite a few mistakes. Many people are in ESP research because they believe passionately that there is something to it, so it is quite conceivable that, even unconsciously, the mistakes would tend to run in favor of the matches, thus creating a small apparent effect of ESP. To protect oneself against such objections (or actual bias), one would want to blind the recording process, perhaps by making both the checking for matches and the recording of the results electronic.

Example 1.2.5 *The Famous Salk Vaccine Trial*

Many of the preceding issues arise in the famous story of the 1954 Salk vaccine trial. Polio plagued many parts of the world in the first half of this century. It struck in epidemics, causing deformity and death, particularly to children. In 1954 the U.S. Public Health service began a field trial of a vaccine developed by Jonas Salk. Two million children were involved, and of these, roughly half a million were vaccinated, half a million were left unvaccinated, and the parents of a million refused the vaccination. The trial was conducted on the most vulnerable group (in grades 1, 2, and 3) in some of the highest-risk areas of the country. The National Foundation for Infantile Paralysis (NFIP) put up a design in which all grade 2 children whose parents would consent were vaccinated, while all grades 1 and 3 children then formed the controls. What flaws are apparent here?

To begin with, polio is highly infectious, passed on by contact. It could therefore easily sweep through grade 2 in some districts, thus biasing the study against the vaccine, or it could appear in much greater numbers in grade 1 or grade 3, making the vaccine look better than it is. Also, the treatment group consisted only of children whose parents agreed to the vaccination, whereas no consent was required for the control group. At that time, well-educated, higher-income parents were more likely to agree to vaccination, so that treatment and control groups were unbalanced on social status. Such an imbalance could be important because higher-income children tended to live in more hygienic surroundings, and, surprisingly, this *increases* the incidence of polio. Children from less hygienic surroundings were more likely to contract mild cases when still young enough to be protected by antibodies from their mothers. They then generated their own antibodies, which protected them later on. We might expect this effect to bias the study against the vaccine.

Some school districts saw these problems with the NFIP design and decided to do it a better way—you guessed it—a double-blind, controlled, randomized study. The placebo was an injection of slightly salty water; the treatment and control groups were obtained by randomly assigning the vaccine or placebo to only those children whose parents consented to the vaccination. Whose results would you believe?

We have touched on only a few aspects of experimentation, ignoring many, including the important problems of experimental ethics. (Actually, the first people on whom Jonas Salk tried his vaccine were himself and his family. That was rather brave, but is it really more ethical to try out potentially dangerous substances on your own children? What about their rights?)

QUIZ FOR SECTION 1.2

1. In what sense does random allocation make for comparisons that are fair or unbiased?
2. What is a completely randomized design?
3. What is blocking?
4. Why do we use blocking in designing experiments?
5. What is a control group, and why are control groups used?
6. What is a placebo effect?
7. Why should we try to "blind" the investigator in an experiment?
8. Why should we try to "blind" human experimental subjects?
9. Complete the basic rule of experimentation: "____ what you can and ____ what you cannot."

EXERCISES FOR SECTION 1.2

1. Suppose you have 20 rats to use in an experiment to determine whether rewards of food or punishment by electric shock is most effective in terms of the time it takes for rats to learn a new task. How could you go about it? Why not just take the first 10 rats you catch to form the reward group and the rest to form the punishment group?

2. Consider an experiment designed to test two hardening processes for steel connecting rods. In the application the rods are to be 25 cm long. The steel comes in lengths of 1 m, and you are provided with four lengths for your experiment. Would it be adequate to take two lengths to form eight rods for testing the first process and the remainder to make eight rods for testing the second process? Can you suggest a better way of proceeding?

3. A December 1989 item in the *NZ Herald* said that in 1990 British school children would be used as guinea pigs in an experiment to determine whether vitamin and mineral pills improve intelligence. Some would be given different dosages of iron, calcium, zinc, and vitamins. Think about how you might go about conducting such an experiment. How are you going to measure intelligence? Are you interested in some measure of general intelligence or particular facilities and skills?

1.3 OBSERVATIONAL STUDIES

In the mid-1960s 51% of adult American males were regular cigarette smokers, and 70% had been regular smokers at some stage in their lives. Today smoking is a minority activity (e.g., only 30% of American males are regular smokers), and in many places around the world smoking is banned in workplaces, planes, restaurants, and other public places. Indeed, such bans now exist in growing numbers of countries. As you probably know, the reason for all this is the growing body of evidence implicating even secondhand (or sidestream) cigarette smoke as an important cause of many devastating health problems, most notably lung cancer and coronary heart disease.

Ever since the introduction of tobacco to Europe by people such as Sir Walter Raleigh, there has been vociferous opposition to smoking in some quarters. Much of this was on the grounds of filth and sinfulness, but some was on the grounds of health. The systematic gathering of evidence to support the view that smoking is a health hazard is relatively recent, however. Perhaps the first figures were from two doctors, Arkin and Wagner, who reported in 1936 that 90% of 135 male lung cancer victims were chronic smokers. (Remember, though, that most men smoked at that time.) In the next few years some retrospective (case-control) studies were performed. The basic idea here is to take a sample of lung cancer sufferers ("cases") and a control sample of unaffected people and compare their smoking histories. Lung cancer victims tended to have smoked more heavily on average than unaffected people.

EXAMPLE 1.3.1 *Effects of Smoking*

The first prospective study of smoking and disease outcomes was begun by Sir Richard Doll and Sir Bradford Hill in England in 1951. In this type of study the smoking habits of the study subjects are established at the onset of the study, in this case by a questionnaire sent to 60,000 British doctors. The study subjects were the 34,440 who replied. The participants were then followed through time, and deaths from cancers and some other causes were recorded, the deaths being obtained via the UK Registrars General. Over a period of 11 years, Doll and Hill found 7 lung cancer deaths per 100,000 men per year among nonsmokers, but 166 per year among heavy smokers. This is 24 times as many! For heart attacks the corresponding figures were 599 for heavy smokers and 422 for nonsmokers. While these figures do not look nearly as dramatic, the effect on heart disease is perhaps more important. Apparently there were $166 - 7 = 159$ excess deaths per 100,000 per year from lung cancer and $599 - 422 = 177$ excess deaths per 100,000 per year from heart attacks that might be ascribed to smoking.

While the evidence from the preceding example should have been enough to start smokers worrying, it did not constitute proof that smoking caused the excess deaths. Putting it another way, it did not constitute proof that the lung cancer rate of the "heavy smokers" group would have been as low as that of the "nonsmokers" had they never smoked. This was an *observational study,* not a controlled randomized experiment. The treatment group consisted just of those people who wanted to smoke; the control group consisted of those who decided not to. What was missing was a mechanism for ensuring that the groups were essentially identical in every other way except for their smoking habits. If the groups are not identical but differ in various ways, then one of these other differences could have been the real culprit; we shall discuss some possibilities later.

The essential ***differences between an experiment and an observational study*** for comparing the effects of treatments:

- In an ***experiment,*** the experimenter *determines* which experimental units receive which treatments.
- In an ***observational study,*** we simply compare units *that happen to have received* each of the treatments.
- An ***observational study*** is often useful for identifying possible causes of effects, but it ***cannot reliably establish causation.***
- Only properly designed and executed experiments can reliably demonstrate causation.

To perform a randomized experiment we would need unreasonable conditions. For example, we would need to find a group of people who would consent to smoke or not to smoke as directed. (If we wanted to gauge the effects of different levels of smoking, they would also have to agree to smoke an unknown prescribed number of cigarettes per day as well.) Then we could randomize them to the smoking group or to the control (nonsmoking) group. As in the nonrandomized prospective study, the participants would be followed through time, and any lung cancer deaths recorded. We could then compare lung cancer death rates. A substantially larger lung cancer death rate among smokers than among nonsmokers in a few large studies, or repeatedly in a number of smaller studies, could be interpreted as providing convincing evidence that smoking was *causing* additional cancers. The reason is that randomization evens everything else out so that smoking is the only remaining explanation. With observational studies we can never have this assurance.

Observational studies are necessary whenever the investigator cannot use controlled randomized experiments. Smoking is a case in point. Not only would it be unethical to purposely expose people to a potentially hazardous substance without powerful compensating benefits, it would be very hard to find anyone who would be prepared to smoke or not to smoke as directed on the toss of a coin. Even where experimentation is possible, the things we want to test in our experiments are usually first noticed in observational data. In addition, observational studies are ideal if you are interested in describing relationships between characteristics in a population (e.g., patterns of employment by geographic district and race, changes in the exchange rate over time). Their weakness is in *establishing* cause and effect.

Smoking is not only an activity that is carried out by hundreds of millions of people, but also a politically influential industry that indirectly affects the livelihoods of millions. It cannot realistically be banned on weak circumstantial evidence. Unfortunately, we can't do the clinical experiments that would establish once and for all whether the relationship is causal. Less direct approaches are needed. If smoking isn't the culprit, the real cause must be something that is associated both with smoking and with cancer, and we must try to think of such alternative mechanisms. Take stress for example. It may be that stress tends to make people both smoke more and be more susceptible to cancer.[10] Using observational data we could somehow divide our sample up according to the levels of stress that people experience. We would then look to see whether there are still higher rates of lung cancer deaths in smokers than in nonsmokers among people experiencing similar levels of stress. We call this process ***controlling*** for a variable; in this case that variable is stress.

We can take the idea of control further and control for both diet and stress at the same time. In other words, we compare lung cancer death rates between smokers and nonsmokers, but only by comparing individuals who have similar stress levels and similar diets. As you may have noticed, by continuing to subdivide in this way, we end up comparing smaller and smaller groups of people, and the method must eventually break down.[11] More sophisticated statistical methods are available to control or adjust for other characteristics that might be important. The intent of these methods is the same, however: We make comparisons only between individuals who are the same or very similar with respect to the factors under study. If there is still more lung cancer among smokers than nonsmokers, even when we have controlled for each or all of the other plausible explanations we can think of, we shall have a much stronger feeling that smoking really is a *cause* of lung cancer.

In 1957 R. A. Fisher, who, more than anyone else, was responsible for the development of the central ideas of randomized experimental design, was prominent in declaring that the case against tobacco smoking was not proven. Among Fisher's arguments were those we have given. The possible competing explanation he gave, namely, that there may be a genetic factor that predisposes people both to smoke and to lung cancer, is much harder to deal with, however. Where do we find people who are genetically identical to compare? The answer, of course, is identical twins. And there *is* evidence of a genetic component to smoking. Fisher showed that among pairs of identical twins of whom one smokes, the other twin is also more likely to smoke than is the case for fraternal twins (who are no closer genetically than other brothers and sisters), thus implying that smokers and nonsmokers have different genotypes. To control for a genetic effect we therefore look for pairs of identical twins in which one smokes and the other doesn't. Does the smoking twin tend to die of lung cancer before the nonsmoker? In fact they do, but not to as great an extent as in fraternal twins (Holt and Prentice [1974]). Unfortunately, using identical twins allows only for the genetic effect. Because, again, the data is observational and not from a randomized experiment, the observed difference could be explained by stress, diet, and so on. We still have to control for these other factors. Reid [1989] discusses the use of twin studies.

Observational studies are not the only way of approaching the question of whether smoking causes lung cancer and other health problems in humans. Researchers have also done proper randomized experiments using animals and found, for example, that smoking caused lesions, which were similar to lung cancers among human smokers, to appear in the lungs of dogs (Brown [1972, p. 49]). Issues arise here of whether the experimental conditions mimic human smoking sufficiently closely

[10] A factor that is related to both the response (here lung cancer) and the risk factor of interest (here smoking) is called a *confounding* variable. In the example we are worried that the effect of smoking may be *confounded* with the effect of stress.

[11] This idea is akin to using blocking in experiments, which we discussed prior to Example 1.2.2.

and whether the animal's metabolism is sufficiently close to that of a human, as far as the effects of tobacco smoke are concerned, for the results to be transferable. We can look for specific physical damage caused by smoking. Animal experiments have shown that smoking damages the lungs and blood vessels, and similar damage has been found in people who smoke heavily. On top of all the other evidence, there is a dose-response relationship, since the cancer rates go up with the amount of tobacco smoke consumed. Although all the observational studies and animal studies do not constitute completely watertight proof that smoking causes lung cancer, the overwhelming weight of evidence from many different sources is so strong that it is only prudent to act as if they do.

Epidemiologists are people who use statistical methods to try to find causes or risk factors for diseases. Very often, as in the case of smoking, they cannot use proper experiments to establish a causal effect and have to be content with a lower standard of "proof." The Surgeon General of the United States has laid down the following criteria for the establishment of a cause and effect relationship: (1) presence of a strong relationship (e.g., something such as the proportion of people contracting the disease being at least four times as high among people exposed to the possible hazard as among those not so exposed), (2) strong research design, (3) temporal relationship (a cause should precede an effect!), (4) dose-response relationship (the proportions contracting the disease should be higher at high exposures of the supposed hazard than at lower exposures), (5) reversible association (if the "cause" is removed—e.g., people stop smoking—the proportions contracting the disease should reduce), (6) consistency (the evidence of causality is stronger if many studies conducted in different places at different times produce similar results), (7) biological plausibility (the assertion that the potential hazard is a cause should not conflict with current biological understanding of the mechanism of the disease), (8) coherence with known facts. It is clear that each of these ideas strengthens evidence of causality, and many health agencies use similar criteria. But even if all the criteria are met, we still lack the certainty provided by proper randomized experiments.

QUIZ FOR SECTION 1.3

1. What is the difference between a designed experiment and an observational study?
2. Can you conclude causation from an observational study? Why or why not?
3. How do we try to investigate causation questions using observational studies?
4. What is the idea of *controlling for* a variable, and why is it used?

EXERCISES FOR SECTION 1.3

1. A 1991 Reuters wire service report stated that women in Indonesian villages with electric power are less likely to become pregnant than women in villages without electricity (i.e., the birthrate is lower in villages with electricity). Suggest an explanation other than the obvious one, "There is nothing much else to do."

2. A *University of Auckland News* item, titled "Smoking lowers marks," stated that nonsmokers taking Hebrew in Israel's University Entrance Examinations performed better than smokers. Is the inference that smoking lowers marks warranted? Why or why not?

3. It is often said that most car accidents occur within about 15 kilometers (10 miles) of home. Does this mean that you are safer on a long trip? What other explanation is possible?

4. A 1990 *Time* magazine story stated that male alcoholics have only about half the levels of the stomach enzyme alcohol dehydrogenase as healthy males.[12] Does this mean that low levels of alcohol dehydrogenase cause alcoholism? What other explanation is possible?

5. In New Zealand more than 10 men are sent to prison for every woman imprisoned. Does this show that female criminals are smarter? What else might it show?

6. Recently, members of our department investigated differences between the performances of girls and boys in mathematics on the national examination taken at the end of the final year of high school. Boys did better than girls on average on pure mathematics questions, while there was essentially no difference on statistics questions. One explanation ventured was that large numbers of boys take physics as well as mathematics, while comparatively few girls take physics. The suggestion was that common skills used in pure mathematics and physics tend to reinforce each other. How could you investigate whether boys tend to do better in pure mathematics simply because they are also taking physics? Can you think of any weaknesses with your method? Is a randomized experiment feasible?

1.4 STATISTICS: WHAT IS IT, AND WHO USES IT?

Statistics is concerned with one of the most basic of human needs, the need to find out more about the world and how it operates. Its focus is on extracting meaningful patterns from the variation that is everywhere.

> ***The subject matter of statistics is the process of finding out more about the real world by collecting and then making sense of data.***

Theoretical statisticians are largely concerned with developing methods for solving the problems involved in such a process, for example, finding new methods for analyzing (making sense of) types of data that existing methods cannot handle. Applied statisticians collaborate with specialists in other fields in applying existing methodologies to real-world problems. In fact, most statisticians are involved in both of these activities to a greater or lesser extent, and researchers in most quantitative fields of inquiry spend a great deal of their time doing applied statistics. Some even contribute new methods for collecting and analyzing data.

Let us begin by looking at the inquiry process that is the subject matter of statistics. This process is depicted in Fig. 1.4.1. It begins either with the need to solve a real-world problem or with simple curiosity.

(a) Problems

People try to solve an enormous diversity of problems using statistical thinking. We have already considered some of them in Section 1.1.5 on surveys. The following are just a few more:[13] agriculture (which varieties grow best?), economics (how are

[12] Alcohol dehydrogenase reduces the amount of pure alcohol that enters the bloodstream. Women tend to have lower levels and hence are affected more by alcoholic drinks than men.

[13] See our web site for further examples.

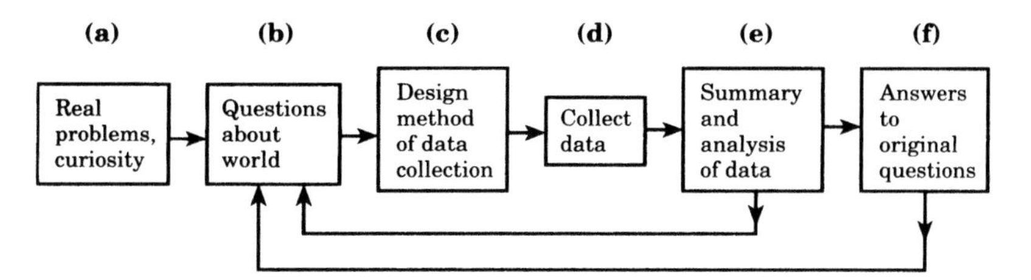

FIGURE 1.4.1 The investigative process.

living standards changing?), education (what is the best way to help students learn this subject?), environmental studies (do strong electric or magnetic fields induce higher cancer rates among people living close to them?), fisheries (are the current fish quotas imposed on the fishing industry appropriate?), literature (how does style change with the age of the author?), medicine (which drug is best?), meteorology (is global warming a reality?), psychology (how are shyness and loneliness related?), management (what is causing the problems my business is experiencing?) and zoology (is this animal population declining?).

Clearly the answer to the question of who uses statistics is that a high proportion of research people use statistics at some stage in their work—in fact, whenever they measure something. These days people try to measure or quantify almost anything!

(b) Questions

The investigative process begins with the asking of questions. These should initially be posed in nontechnical and, in particular, nonstatistical terms. Then we look for a form of investigation that will help us to answer them. Unfortunately, any single survey, experiment, or observational study can address only a limited number of precisely formulated questions. Suppose a company wanted to improve the quality of the paint surface on their refrigerators. The need may have arisen because of complaints or simply from a commitment to the production of quality goods. We look around to see what changes they could make in the painting process that might affect the wear, and this in turn may prompt us to ask a specific question such as, "What combination of paint thickness, baking temperature, and quantity of thinning agent will give the most hard-wearing surface?" The generation of such questions draws on subject-matter knowledge, in this case engineering knowledge.

Although questions should initially be posed in nonstatistical terms to keep subject matter priorities at the forefront, refining those questions and deciding how to answer them is intimately bound up with issues of "What should I measure?" and "How should I measure it?" The preceding paint surface question is not specific enough. We have to know the range of possible paint surface thicknesses, operating temperatures, and so on that we want to try and how we are going to measure "hard-wearingness." Will we measure it by effective lifetime? What do we mean by the "effective lifetime"? Are we going to measure the effective lifetimes of each surface under extreme conditions of heat and humidity so that the experiment won't take forever?[14] Vast amounts of time and money are wasted in unfocused studies because the investigators have never really thought through exactly what it was that they wanted to investigate. Statisticians can often do a very valuable job in helping investigators to formulate their questions concisely and in such a way that they might be answered from the sources of information available.

[14]This idea is called *accelerated (life) testing*.

(c) Design of Data-Collection Mechanism

Collection of data usually involves setting up an experiment, survey, or observational study. (The paint surface example obviously calls for an experiment.) Many statisticians specialize in the design of these forms of study. A statistician is aware of the assumptions required by existing methods of analysis, and he or she can help the investigator turn these assumptions into practical experimental requirements. For example, how do you take a sample of leaves from a tree to study leaf-shape relationships between various species? Or what sampling procedure do you use to estimate the number of insects (e.g., aphids) on a tree?

In discussing the planning stage it is often a good idea for the statistician to go to the "scene of the crime" and look at an experimental site. This gives the statistician a better feel for what is possible and what sort of randomness can be achieved. There is not much point in advocating a particular experimental method with good statistical properties if it is too difficult to carry out in practice. "What idiot made us use 4-meter plots when the tractor boom is only 3 meters wide?" Perhaps the most important consideration at the planning stage is whether the planned measurements are well suited to answering the original questions. This is not as simple as it may sound.

As noted in Section 1.1.4 on surveys, sample sizes or numbers of experimental observations required should be determined in advance. Too few observations may lead to information that is unrepresentative and imprecise. Generally, the more data we collect, the more reliable are any extrapolations from samples to populations or underlying mechanisms.

(d) Data Collection

Issues involved in the design phase that greatly influence the quality of the data collected include the training of interviewers, procedures to ensure accurate measurements, procedures to try to ensure that data is not corrupted when transcribed from one storage format to another, and the best form of computer or other storage to ensure quick and convenient retrieval for the analysis of part or all of the data. Well-laid-out worksheets mean that fewer mistakes are made in the recording process and in the transfer of the data onto a computer. We must consider the possibility that data will sometimes be recorded out in the field in poor light and pouring rain! Nowadays it is possible for the data-recording and data-storage steps of many experiments to be a largely automated operation using measuring devices containing microprocessors that are linked to a computer.

(e) Summary and Analysis

The main focus of this book is on methods of summary and analysis for some *simple* data structures. This is, after all, only an introduction to the subject. Graphs (Chapters 2 and 3) are very useful for finding trends and relationships in data. Then comes the attempt to confirm whether the patterns we have found in our small batches or samples of data have a wider validity. What can we say about what is going on in a whole population or in the process that generates our experimental data? In exploring our data we often become aware of aspects of the problem that we hadn't thought of before we started, and this generates new questions. They may be at least partially answerable from the data in hand, but often they will prompt the collection of new data.

(f) Answers

Finally, all going well, we can produce useful answers. We should check whether the answers we have really do answer our original, nonstatistical questions. The answers we get will often provoke further research. Applied statistics is not a technical activity that takes place in a back room. It is part of a process of arriving at and communicating understandings. Applied statisticians work as part of a team or for a client. They have to be able to explain the important messages that the data give to nonstatisticians in terms that the client or reader can understand. A careful, sophisticated analysis is quite useless if the results cannot be communicated to the client, so it will be important for you to practice explaining the results of the analyses you do in plain English.

Feedback

In Fig. 1.4.1 we have drawn the two most important places where the investigative process feeds back on itself in the sense of generating more questions and thus perpetuating the process. Unexpected features or inadequacies in the data collected often generate new questions, and even the answers to the original questions generate more questions. In a very real sense questions are more important than answers in the investigative process. It is often feasible to hire a specialist to help you answer a question, but if the question is never posed, then nothing will ever be learned.

1.5 SUMMARY

Statistics is concerned with one of the most basic of human needs, the need to find out more about the world and how it operates in the face of variation and uncertainty. A statistician needs to be more than just a narrow specialist. Common sense and an awareness of the subject area are vitally important. It is also essential to acquire the mental habit of curiosity, of constantly being on the lookout for anything (e.g., in the data) that looks interesting or strange and unexpected.

Although these qualities are absolutely vital to success as an investigator, they are almost impossible to teach in a statistics course. We have tried hard to address these issues in this book. Successful investigators gain a good deal of their practical good sense from experience and from stories about the experiences of others. In this chapter we have begun your education in this area with a selection of such stories. Through the course of the book you will read more stories. In many places throughout the book where review exercises contain descriptions of studies, we will ask you to criticize the study design and think of ways of improving it as a way of developing your practical awareness and intuition. We will also ask you to think about follow-up questions that the investigators should perhaps pursue as a way of developing the habit of curiosity.

Our primary aim in writing this chapter was to broaden your awareness of practical matters and introduce you to some of the ideas that are most basic to statistical studies, without expecting you to remember the details. You will often be referred back to parts of this chapter. Some of the ideas we particularly want you to remember from Chapter 1 follow.

The Subject of Statistics

Statistics is concerned with the process of finding out about the real world by collecting and then making sense of data (Section 1.4). An important feature is the quantification of uncertainty so that we can make firm decisions and yet know how likely we are to be right.

Experiments and Observational Studies

When we are interested in exploring questions of ***cause and effect*** we distinguish between observational studies and experiments.

The essential difference between an experiment and an observational study for comparing the effects of two or more treatments (e.g., smoking and not smoking) is that in an ***experiment*** the experimenter *determines* which subjects (experimental units) receive which treatments, whereas in an ***observational study*** we simply compare subjects *that happen to have received* each of the treatments.

Observational studies are widely used for identifying possible causes of effects but *cannot* reliably establish ***causation*** (Section 1.3). Only properly designed and executed experiments (Section 1.2) can reliably demonstrate causation.

The Role of Randomization

1. Well designed statistical studies employ randomization to avoid subjective and other biases.
   (i) Surveys and observational studies should use some form of random sampling to obtain representative samples.
   (ii) Experiments should use some form of random assignment of experimental subjects to treatment groups to try to ensure that comparisons are fair in the sense that treatment groups are as similar as possible in every way except for the treatment being used.
   (iii) In well designed experiments blocking is used to ensure fair comparisons with respect to factors that the experimenter knows are important, and randomization is used to try to obtain comparability with respect to unknown factors. "Block what you can and randomize what you cannot."
2. Randomization also allows the calculation of how much the estimates made from the study data are likely to be in error.

Blocking and Stratification

The word ***blocking*** is commonly used in an experimental design context, whereas the word ***stratification*** commonly is used in a survey or observational study context. Both expressions refer to the idea of making comparisons only within relatively similar groups of subjects (individuals or experimental units).

Sources of Error in Surveys

Random sampling leads to ***sampling errors.***
Nonsampling errors can be much larger than the sampling errors.

Possible sources of nonsampling errors include the following:

Selection bias: This arises when the population sampled is not exactly the population of interest.

Self-selection: Here the people themselves decide whether or not to be surveyed. The results tend to be akin to severe nonresponse.

Nonresponse bias: Nonrespondents tend to behave or think differently from respondents so that low response rates can lead to huge biases.

Question-wording effects: Even slight differences in question wording can produce measurable differences in how people respond.

Interviewer effects: Different interviewers asking the same questions can tend to obtain different answers.

Survey-format effects: A variety of factors, such as question order, questionnaire layout, self-administered questionnaire or interviewer, can affect the results.

Dealing with errors: Statistical methods are available for estimating the likely size of sampling errors. All we can do with nonsampling errors is to try to minimize them at the study-design stage.

Pilot survey: A pilot survey tests a survey on a relatively small group of people to try to identify any problems with the survey design before conducting the survey proper.

Some Jargon Used to Describe Experiments

Control group: This group of experimental units is given no treatment. The effect of a treatment can then be estimated by comparing what happened to each treatment group with what happened to the control group.

Blinding: This refers to the practice of preventing various people involved in the experiment from knowing which experimental subjects have received which treatment. One may be able to blind the subjects themselves, the people administering the treatments, and the people measuring the results.

Double blind: A term used in experiments on humans meaning that both the subjects and those administering the treatments have been blinded.

Placebo: An inert dummy treatment.

Placebo effect: The response caused in human subjects by the idea that they are being treated.

REVIEW EXERCISES 1

Although the following exercises can be attempted by individuals, many work particularly well as a basis for class discussion. Further exercises are given on our web site.

1. There can be very difficult measurement problems involved in seemingly simple situations. Take unemployment, for example:
   (a) Try to form your own definition of what it means to be "unemployed" (there are many types of people who are not in paid employment).

(b) What definition of "unemployment" is used in the collection of your country's national statistics? Would it be fairer if it was changed in some way?

(c) Why can simple comparisons of unemployment statistics between different countries be misleading?

2. Many hotels leave customer satisfaction surveys in hotel rooms for guests to fill out. Do you expect the views expressed on the forms to be representative? Why or why not? In what direction would you expect the results to be biased? What useful information might you as a hotel manager hope to extract from this source?

3. Most people base a good many of their opinions and impressions about the world on *anecdotal evidence,* that is, on the few cases or instances of something that have just happened to come to their attention (e.g., the four or five people who have told me about having homeopathic treatments for their ailments). Why is anecdotal evidence unreliable? How does it relate to the ideas in this chapter?

4. In the United States in 1990, one in every four black males between 20 and 29 was in prison or on parole. New Zealand Maori make up only 10% of the population, but nearly 50% of the prison population. Australian aborigines are 15 times more likely to be in jail than the rest of the population. Canadian Indians are heavily overrepresented in Canadian prisons. For some people, statistics of this type reinforce ethnic stereotypes. In all of these cases, the ethnic groups referred to have much larger proportions in the lower socioeconomic groups. How would you go about finding out whether these numbers can be explained simply by poverty?

There is another aspect to imprisonment. Not only do you have to be convicted of a criminal offense, you also have to be sentenced to prison. Many people have a strong suspicion that the sentencing process itself is often racist. Consider what issues might be involved in answering the question "Is sentencing racist?" and the types of information one might want to try to resolve them.

5. According to a *Sunday Star Times* article (13 June 1994), those who defend Shere Hite's use of statistics do so on three grounds: (i) her methods are fine as long as the sample sizes are big enough, and hers are; (ii) she is pioneering a new area in qualitative rather than quantitative data; and (iii) a good deal of social science research is not based on random samples, and most psychological studies use college students and generalize the results. What do you think about these arguments and the issues they raise?

6. An experiment was carried out locally to measure the effect of growth hormone on girls affected by a growth disorder called Turner's syndrome. All 34 girls in the study were given the growth hormone. Their heights were measured at the time the hormone was first administered and again one year later. What are some problems with such a study?

7. In research investigating whether Celts are more prone to alcoholism, schizophrenia, and other psychiatric disorders than other Britons, a psychiatrist investigated 1048 psychiatric patients at a London hospital (Press Association, July 1992). They were divided into 870 non-Celts and 177 Celts (determined by having names of Irish, Scottish, or Welsh origin). He found a greater incidence of alcoholism among Celts (35%) than among non-Celts (12.7%), but nothing to suggest higher incidences of other disorders.

(a) What problem can you see with the method used to determine ethnic origin?

(b) Suppose that there actually is a greater tendency for people of predominantly Celtic origin to become alcoholic than for non-Celts. Would the problem identified in part (a) tend to make the observed sample difference in alcoholism rates bigger or smaller than the true difference?

(c) To what population do the preceding results apply?

(d) Can you think of any factors that might limit the wider applicability of the results? Have you any other questions about this study?

8. As part of a study reported by Kaufmann [1992], 10 subjects with a history of developing headaches after drinking red wine each consumed 90 mL of a Californian Cabernet Sauvignon, and each developed a headache. One week later they each consumed 180 mL of the same type of wine, but this time one hour after taking an aspirin-like drug. Under these conditions only 2 of the 10 subjects developed a headache.

(a) Is this an observational study or an experiment?

(b) What is the population of interest?

(c) What question is the study actually investigating? What is the wider question of interest? Discuss how the results of the current study relate to that wider question.

(d) Comment on some possible deficiencies of the study and features that you would like to have seen incorporated in the way the study was designed or carried out.

(e) If you were doing research in this area, what would you want to look at next?

9. The Auckland Harbor Bridge links the North Shore to the rest of the city and is probably the most traveled piece of road in New Zealand. In 1989 there were about six fatal accidents on the bridge, and soon after long-called-for, mobile safety barriers were finally erected. An argument repeatedly raised against this sort of expenditure was that in terms of fatalities per kilometer traveled per car, the bridge was extraordinarily safe. Do you think this is a sound argument? If you wanted to maximize the number of lives saved given a fixed budget, what other measure of accident rate would be reasonable?

10. Is the current generation of people in their 20s generally taller than their parents were? This sounds like a simple question, but what are some of the issues of definition, measurement, and so on that must be resolved in order to answer it?

11. You have to lay out an experiment to compare the effects of three fertilizers on tree growth using a newly planted forest block on the side of a hill. How would you go about it? What are some things you would have to bear in mind?

12. The population density of a country is obtained by dividing the country's population by its land area. Why may comparisons of population densities between countries fail to compare how close together most of the people actually live?

13. Definitions of rape vary from forced sexual intercourse by being physically overpowered to succumbing to seduction or emotional pressure. A 1991 news story reported that the U.S. incidence of rape was 4 times that of Germany, 13 times that of Britain, and 20 times that of Japan. Why can we not take these figures at face value? In a similar vein, figures are often quoted in the media about trends over time in crime and other statistics by people with political points to make. What are the dangers here?

14. Transcendental meditation (TM) is a method of inducing deep relaxation which is supposed to be practiced twice a day for 20 minutes. A graph in a TM advertising brochure, said to be based on a five-year study of U.S. health insurance statistics, shows dramatically reduced hospital admissions for practitioners of TM compared with other people for problems with the nervous system; lung, throat, and nose; heart; injuries; tumors; and bone and muscle. The comparisons were between TM practitioners and nonpractitioners. Why is such a simple comparison unconvincing? What are some questions you would like to ask if you were evaluating the effectiveness of TM?

15. In Section 1.4 we discussed experimenting to find the combination of paint thickness, baking temperature, and quantity of thinning agent to produce the most long-lasting paint surface for a refrigerator. We raised the possibility of measuring the effective lifetimes of each surface under extreme conditions of heat and humidity (accelerated testing) to shorten the time the experiment would take. Can you foresee any potential problems with this approach?

16. A survey of New Zealand high school principals taken after widespread changes to the school system revealed that one in five were under medication for stress and almost half had visited doctors because of the pressure they were under. These figures came from the 250 questionnaires returned from 2500 sent out. How reliable do you think the results are and why?

CHAPTER 2

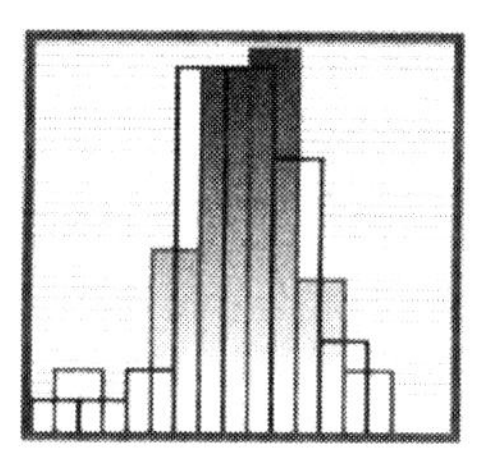

Tools for Exploring Univariate Data

Chapter Overview

A picture is worth more than a thousand words! In this chapter we shall look at methods of displaying observations on a single variable so that we can quickly see the main features of the data. In the previous chapter we discussed various methods of collecting data to help us answer questions about the real world. Once we have collected the data, we then wish to have a general look at it so as to get a feel for its main messages and any surprising features, before attempting to answer any formal questions. This phase of the investigation is generally known as the *exploratory stage*. The formal part, in which we try to make valid inferences from the data about what is happening in some wider population or process, is called the *confirmatory stage*.

We begin in Section 2.1 with a discussion of the types of variables that we encounter in statistical investigations and lay down some guidelines in Section 2.2 for the way the data should be presented, particularly in tables. Depending on the number of observations, we can use dot plots, stem-and-leaf plots, histograms, and box plots to help us get a feel for the "shape" of the data. These are discussed in detail in Section 2.3. We ask the question, "What can such plots tell us?" These plots can be supplemented by various numerical measures, described in Section 2.4, which focus on where the data are mainly located and how spread out are the observations. We introduce the mean, median, the so-called five-number summary, and the standard deviation as measures of location and spread. In Section 2.5 we discuss two related topics: repeated data that come from discrete observations and grouped data that arise when observations are grouped into intervals. Qualitative variables defining group membership are considered in Section 2.6, along with various graphical tools for pictorial representation.[1]

[1]The effects of changing the scale and adding or subtracting constants on these measures are discussed on our web site.

2.1 INTRODUCTION

The data in Table 2.1.1 are a subset of the data from a study on a series of male patients admitted to Greenlane Hospital[2] in Auckland after a heart attack. Each row of the data corresponds to measurements on a single patient. Every patient has measurements recorded for the following *variables.*[3]

- ID: A patient identifier instead of a name to protect a patient's privacy.
- EJEC: Ejection fraction, the percentage of blood in the left ventricle of the heart ejected in one beat.
- SYSVOL: End-systolic volume, a measure of the size of the heart. This is calculated from a two-dimensional silhouette picture of the left ventricle at its smallest part.
- DIAVOL: End-diastolic volume. The same as SYSVOL, except that the largest silhouette is used.
- OCCLU: Occlusion score, or percentage of the myocardium of the left ventricle supplied by arteries that are totally blocked.
- STEN: Stenosis score, or percentage supplied by the arteries that are significantly narrowed but not completely blocked.
- TIME: Time in months from when patient was admitted until OUTCOME (see following).
- OUTCOME: Coded variable with the following levels: 0 = alive at last follow up, 1 = sudden cardiac death, 2 = death within 30 days of heart attack, 3 = death from heart failure, 4 = death during or after coronary surgery, 5 = noncardiac death.
- AGE: The age in years of the patient at admission.
- SMOKE: Whether the patient continued to smoke; 1 = yes, 2 = no.
- BETA: Whether the patient was taking drugs called beta blockers; 1 = yes, 2 = no.
- CHOL: Blood cholesterol measured in mmoles per liter.
- SURG: Whether the patient had surgery, including reason for surgery. 0 = no surgery, 1 = surgery as part of trial, 2 = surgery for symptoms within one year, 3 = surgery for symptoms within one to five years, 4 = surgery for symptoms after five years.

How do we read the data table? The first patient listed had EJEC = 72, SYSVOL = 36, ..., AGE = 49, ..., SURG = 0; the second had EJEC = 52, SYSVOL = 74, ..., AGE = 54, ..., SURG = 1; and so on.

The object of the study was to determine which of the measured variables, AGE, EJEC, and so on, would be useful in predicting how long the patient will live after the heart attack and to find a way of making such predictions. There was also interest in the relationships between some of the variables in heart attack victims. For instance,

---

[2]Data courtesy of Dr. R. M. Norris.

[3]We often use capital letters for the name of a variable because it makes the name stand out in a sentence.

TABLE 2.1.1 Data on Male Heart Attack Patients

ID	EJEC	SYS-VOL	DIA-VOL	OCCLU	STEN	TIME	OUT-COME	AGE	SMOKE	BETA	CHOL[a]	SURG
390	72	36	131	0	0	143	0	49	2	2	59	0
279	52	74	155	37	63	143	0	54	2	2	68	1
391	62	52	137	33	47	16	2	56	2	2	52	0
201	50	165	329	33	30	143	0	42	2	2	39	0
202	50	47	95	0	100	143	0	46	2	2	74	1
69	27	124	170	77	23	143	0	57	2	2	*NA*	2
310	60	86	215	7	50	40	0	51	2	2	58	0
392	72	37	132	40	10	9	5	56	2	2	75	0
311	60	65	163	0	40	142	0	45	2	2	72	0
393	63	52	140	0	10	142	0	46	2	2	90	0
70	29	117	164	50	0	142	0	48	2	2	72	0
203	48	69	133	0	27	142	0	54	2	2	*NA*	0
394	59	54	133	30	13	142	0	39	2	1	*NA*	0
204	50	67	135	37	63	141	0	49	2	2	86	2
280	53	65	138	0	33	140	0	58	2	1	49	0
55	17	184	221	57	13	5	1	50	2	2	70	2
79	37	88	140	37	47	118	5	58	2	2	*NA*	0
205	45	106	193	33	43	140	0	47	1	1	38	1
206	43	85	150	0	50	23	5	51	2	2	61	0
312	60	59	149	7	37	139	0	43	2	1	56	0
80	38	103	168	47	43	100	1	55	2	2	62	1
281	57	53	124	0	57	140	0	58	2	1	93	0
207	44	68	121	27	60	139	0	55	2	2	63	1
282	51	53	109	0	77	139	0	41	2	2	45	4
396	63	58	157	0	73	139	0	51	2	2	60	0
208	49	81	157	13	13	139	0	49	2	2	60	0
209	48	58	112	0	0	72	1	56	2	2	57	0
283	58	71	167	27	0	138	0	45	2	1	46	0
210	42	92	159	0	0	139	0	57	2	2	58	0
397	68	50	156	0	100	138	0	51	2	1	*NA*	0
211	43	146	259	47	33	3	1	56	2	2	70	0
398	67	43	130	0	70	138	0	49	2	2	*NA*	3
284	52	70	146	0	23	137	0	47	1	2	*NA*	0
399	63	73	195	27	0	136	0	36	1	1	61	0
285	54	62	133	33	23	137	0	38	2	2	*NA*	0
71	37	93	148	47	0	137	0	59	2	2	*NA*	0
286	51	65	133	43	7	136	0	54	2	2	*NA*	0
212	42	95	163	40	10	109	3	57	2	2	*NA*	4
400	66	49	144	10	50	65	1	52	2	2	55	0
287	54	66	145	7	40	136	0	47	2	2	62	0
81	39	144	237	13	87	136	0	39	2	2	56	3
813	63	52	141	0	47	43	3	48	2	2	*NA*	0
68	30	219	314	33	45	76	1	53	1	2	*NA*	0
288	59	39	94	0	0	135	0	47	1	2	63	0
407	67	39	117	0	73	53	1	57	2	2	62	2

[a]*NA* = Not available (missing data code).

is there a relationship between SMOKE and OUTCOME? Notice how much variability there is in some of the columns of Table 2.1.1. It is this variation that we always encounter in data that makes it difficult to see any patterns. The "noise," in this case the natural variation, makes it difficult to determine the "signal," or message that the data are trying to convey to us.

What you see here is just a fraction of the data collected. Many more variables were measured than are represented here, and there were 616 patients in total. To present all 616 would take many pages. How can we even begin to find answers to the questions that the researchers were interested in among this indigestible mass of information? To make any progress at all, we need some tools for reducing the masses of figures to simple entities that we *can* understand. Pictures in the form of plots (graphs) provide the most effective means for assimilating numerical information. Together with numerical summaries (e.g., averages), plots provide us with the basic tools we shall need to understand data. We shall use the plots and data summaries introduced in this chapter throughout the book. But before we proceed, we shall use the data set in Table 2.1.1 to learn something about the nature of variables themselves.

2.1.1 Types of Variables

The major distinction we need to make is between quantitative and qualitative variables. ***Quantitative*** variables include all the things we usually think of as being measurements (e.g., distances, weights, angles) and also counts (e.g., numbers of objects). ***Qualitative*** variables, also sometimes called *factors* and *class variables,* define classes (or group membership). One example of a qualitative variable is MARITAL STATUS defining three groups (or having three *levels*) "never married," "married," and "previously married." These might be represented in a data set by the numerical codes 1, 2, and 3, respectively. Another example is BLOOD TYPE with levels A, B, AB, and O being the four possible blood types (phenotypes) under the ABO system. These four phenotypes might be recorded in a data set using the numerical codes 1, 2, 3, and 4, respectively. We are just classifying individuals into groups of people who share a certain characteristic, however. A value 3 recorded for a person's blood type simply means that he or she has type AB blood. The English word *measurement* implies some idea of distance between values. Not only does the idea of a distance between blood types make no sense at all, but the blood types cannot even be ordered from biggest to smallest. We could write them down in any order.

> ***Quantitative*** variables are ***measurements*** and counts.
> ***Qualitative*** variables describe ***group membership.***

There is a slight complication here in that sometimes we want to use an obviously quantitative variable to define groups. For example, we may want to compare the incomes of people broken down into 10-year age groups (e.g., the incomes of people aged 10–20, the incomes of those aged 21–30, the incomes of those aged 31–40, etc.). Here the quantitative variable AGE is being used to define group membership or, to put it another way, to define a qualitative variable AGEGROUP.

The preceding distinctions are very important, as the type of variable we have determines how we display and analyze data on that variable. We now look at quantitative and qualitative variables in more detail.

Quantitative Variables

Quantitative variables can either be discrete or continuous. A ***discrete*** random variable has gaps between each of the values it can take. Most of the examples of discrete variables we shall see in this book are counts, such as numbers of faulty parts, numbers of telephone calls, numbers of bacterial colonies, or numbers of people. For example, given a random sample of houses in a certain area, these are the numbers of people living in each of the first 10 houses sampled:

2 1 5 2 3 7 2 6 2 3

Continuous variables are things such as heights, weights, and temperatures. If measured sufficiently accurately, there would be no gaps between possible values. For instance, the following measurements, in millimeters, were recorded for the lengths of the first seven live lampshells of a random sample taken from the sea floor in New Zealand:

22.34 18.72 17.64 16.84 20.04 20.25 22.06

Because of the limitations of measuring devices, however, all measurements are essentially discrete. For example, our temperature gauge may only measure temperature to the nearest tenth of a degree. When the gaps between possible values are small, we treat the variables as continuous.

In analyzing data, the main criterion for deciding whether to treat a variable as discrete or continuous is whether the data on that variable contain a large number of different values that are seldom repeated or a relatively small number of distinct values that keep reappearing.

> Variables with ***few*** *repeated values* are treated as ***continuous.***
> Variables with ***many*** *repeated values* are treated as ***discrete.***

For example, bank balances such as $564.34, $1038.25, ... can be treated as continuous even though the amount in cents is actually discrete. Conversely, AGE, which is rounded to the nearest year in Table 2.1.1, could in some circumstances be treated as a discrete variable, even though aging is a continuous process.

Qualitative Variables

These variables classify objects or individuals into groups, such as by marital status. The categories can be *ordered,* such as income classified as high, medium, or low, or *unordered,* such as blood type or religion. Unordered qualitative variables are called ***categorical*** variables, whereas ordered ones are termed ***ordinal.*** We shall not stress this distinction, as we shall make very little use of it in this book. Values of a

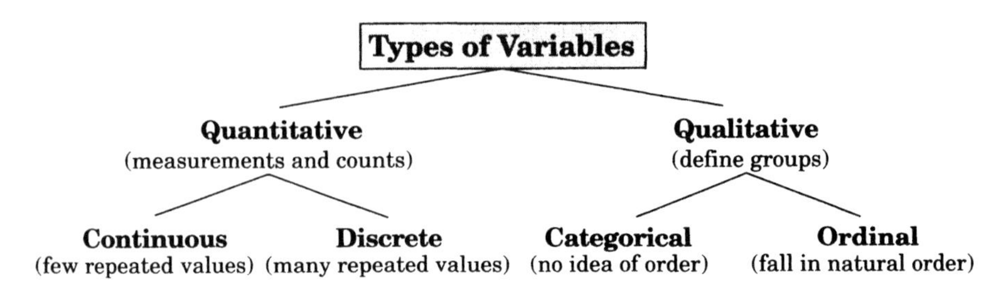

FIGURE 2.1.1 Tree diagram of types of variables.

qualitative variable can be represented by character strings (e.g., "male" and "female") or numerical codes (e.g., 1 and 2), as in Table 2.1.1. Codes are often used to increase the speed of entering data into a computer.

The distinctions made between types of variable in this discussion are summarized pictorially in tree-diagram form in Fig. 2.1.1.

To reinforce the points we have been making, we shall return to the data set in Table 2.1.1 that started this discussion and see how the variables there fit into this classification system.

The variables EJEC, SYSVOL, DIAVOL, TIME, AGE, and CHOL can be treated as continuous quantitative variables. Appreciable rounding is shown in the variables OCCLU and STEN which take values 0, 3, 7, 10, 13, 17, 20, 23, and so on. They were determined from looking at photographs, so the laboratory staff could not be too precise. They are therefore discrete quantitative variables, but since the gaps between the possible values (approximately 3) are small compared with the range of values (0–100), we could treat them as continuous variables. The variables BETA, SMOKE, SURG, and OUTCOME are all qualitative variables. BETA and SMOKE each have two levels, while SURG and OUTCOME have more.

QUIZ FOR SECTION 2.1

1. What is the difference between quantitative and qualitative variables?
2. What is the difference between a discrete variable and a continuous variable?
3. Name two ways in which observations on qualitative variables can be stored on a computer.
4. When would you treat a discrete random variable as though it were a continuous random variable? Can you give an example?

EXERCISES FOR SECTION 2.1

1. Try to reproduce the diagram of types of variables (Fig. 2.1.1) from memory.
2. Classify the following variables to be "measured" for a sample of 1000 employed people: annual family income; religion; ethnicity (race); distance traveled to work; usual mode of transportation; number of people in the household; smoking, classified as "none," "light," or "heavy"; number of cigarettes per day; "the political party you would be most likely to vote for;" and percentage of annual income saved.

Exercises 3 to 6 concern the data in Table 2.1.1.

3. Did the patient with ID = 69 continue to smoke? Did he have surgery, and if so, for what reason? Was he dead or alive at last follow-up?

4. What was the age at admission of the patient with ID = 201? What was his end-diastolic volume? Was he taking beta blockers? Did he have a cholesterol score recorded? If so, what was it?
5. What was the age at admission of the patient with ID = 203? What was his occlusion score? Was he taking beta blockers? Did he have a cholesterol score recorded? If so, what was it?
6. How many patients in the table continued to smoke? How many had surgery for symptoms after five years? How many were still alive at last follow-up?

2.2 PRESENTATION OF DATA

2.2.1 The Roles of Numbers

Although people are not very good at looking at and understanding a page of numbers, small tables of numbers in a written report can be understood if they are intelligently presented. Before we give rules about presentation, however, it is important to distinguish between two contrasting roles that numbers, and especially tables of numbers, have in written reports.

These are the two roles:

1. ***To convey information*** about the main features of the situation to the reader quickly. In particular, the reader should be able to make visual comparisons and see trends in the numbers themselves ***quickly and easily.***
2. To make the data available to the reader for ***detailed checking, analysis,*** or both.

In general, only tables of numbers designed for the first purpose should be in the body of a written report. The table should be there only to advance the arguments being made in the report. Tables designed for the second purpose should be included in reports only as appendixes.

It is seldom possible for a single table to fulfill both purposes.

For example, in Section 2.2.2 we shall recommend that numbers used for purpose 1 be severely rounded. In stark contrast, full accuracy should always be maintained in numbers used for purpose 2.

It is ***dangerous to round*** numbers that will be ***used in calculations.***

In our experience, students like to round the numbers obtained in the intermediate steps of a calculation. Never do this. It is amazing how fast these rounding errors can accumulate and falsify an answer. Subtractions, and to a lesser extent divisions, are particularly dangerous in this regard. When using a calculator, store the results of intermediate steps of a calculation in the memory.

- ***Round numbers for presentation***
- ***Maintain complete accuracy in numbers to be used in calculations.***

2.2.2 Guidelines for Tabular Presentation

The five guidelines to follow, adapted from Ehrenberg [1975, 1977, 1981], apply only to tables intended for purpose 1 above. (Recall that data tables used for detailed checking or analysis should maintain full accuracy.)

1. Round drastically.
2. Arrange the numbers you want compared in columns, not rows.
3. Order the columns by size.
4. Use row and column averages as a focus.
5. Provide verbal summaries.

Example 2.2.1 *Simplifying a Table of Gold Reserves*

Table 2.2.1 gives the gold reserves (in millions of fine troy ounces) held by central banks and governments for International Monetary Fund members with reported gold holdings from 1970 to 1990. The first thing we notice about the table is that it is hard to digest all the numbers. Looking at the first two entries in the 1970 column, we note that it would be much easier to compare 42 and 23 than it is to compare 42.01 and 22.59. In accordance with the first guideline, Ehrenberg recommends keeping only two of what he calls "busy" digits in the presented numbers. Much of the time this just means using two significant figures.[4] We have done this to get Table 2.2.2. Note how much easier this table is to digest.

Suppose we are interested in comparing the countries within each year. Guideline 2 tells us to list the numbers under the year columns, as we have done in Table 2.2.2. This guideline follows from the observation that numbers are easier to read down than across. For example, it is easier to make comparisons down column 1 than across the row. Therefore, if the primary objective was to show changes over time and comparing countries was secondary, we would have YEAR running vertically and COUNTRY running horizontally.

If we are interested in using Table 2.2.1 like a list to look up particular countries, then our alphabetical ordering is fine. If we want to make comparisons, however, we can apply guideline 3 and order the rows in order of decreasing magnitude. This has been done in Table 2.2.2 by the most recent (1990) gold holdings. It is immediately clear where the largest and smallest gold reserves are held.

[4] If the numbers were, for example, 2119.23, 2136.32, 2129.87, and 2145.16, we would present 2119, 2136, 2130, and 2145. Only the second two digits are busy.

TABLE 2.2.1 Gold Reserves[a] of Gold-Holding IMF Countries

Country	1970	1975	1980	1985	1990
Belgium	42.01	42.17	34.18	34.18	30.23
Canada	22.59	21.95	20.98	20.11	14.76
France	100.91	100.93	81.85	81.85	81.85
Italy	82.48	82.48	66.67	66.67	66.67
Japan	15.22	21.11	24.23	24.33	24.23
Netherlands	51.06	54.33	43.94	43.94	43.94
Switzerland	78.03	83.20	83.28	83.28	83.28
UK	38.52	21.03	18.84	19.03	18.94
U.S.	316.34	274.71	264.32	262.65	261.91

[a]Units: millions of troy ounces.

Source: Reprinted with permission from *The World Almanac and Book of Facts*. (1988, p. 103; 1993, p. 140).

TABLE 2.2.2 Simplified Table of Gold Reserves[a] of IMF Countries

Country	1970	1975	1980	1985	1990	Average
U.S.	320	270	260	260	260	280
Switzerland	78	83	83	83	83	82
France	100	100	82	82	82	89
Italy	82	82	67	67	67	73
Netherlands	51	54	44	44	44	47
Belgium	42	42	34	34	30	37
Japan	15	21	24	24	24	22
UK	39	21	19	19	19	23
Canada	23	22	21	20	15	20
Average	83	78	71	71	70	

[a]Units: millions of troy ounces.

Following guideline 4, averages[5] can be helpful for indicating overall patterns in the table and guiding our intuition when looking at the entries within the table. Even when the patterns differ between rows and columns, the averages provide useful reference points for determining the nature of the differences. In Table 2.2.2 the row averages[6] give an overview of reserves held by each country over the period. The column averages have declined over time, particularly from 1970 to 1975 and from 1975 to 1980. (What other patterns can you detect?)

Finally it is important to provide a verbal summary that highlights the important features of the table that you want the reader to absorb. For example, it could be stated that all the countries except Switzerland and Japan reduced their gold reserves from 1970 to 1990.

[5]We can use totals if they are more appropriate to the situation than averages.

[6]The averages quoted here are rounded versions of the true averages, not averages of the rounded numbers.

In this book our main focus is on the analysis of data for understanding, rather than on presenting numerical information to a reader. The importance of presentation cannot be overstated, however. A good understanding of a situation or a data set may well be useless if you cannot communicate that understanding to others. For a fuller discussion we recommend that you read Ehrenberg [1981]. We conclude by noting that computer spreadsheet programs (e.g., Excel or Lotus 123) are excellent for looking at tabular data because they facilitate quick and easy sorting by any row or column and the rounding of numbers.

QUIZ FOR SECTION 2.2

1. For what two purposes are tables of numbers presented?
2. When should you round numbers, and when should you preserve full accuracy?
3. How should you arrange the numbers you are most interested in comparing?
4. Should a table be left to tell its own story?

EXERCISES FOR SECTION 2.2

1. (a) Simplify Table 2.2.3 and use the simplified table to comment on the relative importance of the various gold producers in 1991. Comment on any trends in production you see over time and any exceptions to these trends.

TABLE 2.2.3 World Gold Production
(in millions of troy ounces)

		Africa			N. & S. Amer.				Others			
Year	World prod.	S. Afr.	Ghana	Zaire	U.S.	Can.	Mexico	Colmb.	Aust.	China	Phlp.	USSR
1972	44.84	29.25	.724	.141	1.449	2.078	.146	0.19	0.75	—	0.61	—
1974	40.12	24.39	.614	.131	1.126	1.698	.134	0.27	0.51	—	0.54	—
1975	38.48	22.94	.524	.103	1.052	1.654	.145	0.31	0.53	—	0.50	—
1976	39.02	22.94	.532	.091	1.048	1.692	.163	0.30	0.50	—	0.50	—
1977	38.91	22.50	.481	.080	1.100	1.734	.213	0.26	0.62	—	0.56	—
1978	38.98	22.65	.402	.076	0.998	1.735	.202	0.25	0.65	—	0.59	—
1979	38.77	22.62	.362	.070	0.964	1.644	.190	0.27	0.60	—	0.54	—
1980	39.20	21.67	.353	.040	0.970	1.627	.196	0.51	0.55	—	0.75	8.43
1982	43.08	21.36	.331	.062	1.466	2.081	.214	0.47	0.87	1.80	0.83	8.55
1983	45.00	21.85	.287	.193	2.002	2.363	.198	0.44	0.98	1.85	0.82	8.60
1984	46.48	21.86	.287	.117	2.084	2.683	.271	0.80	1.30	1.90	0.83	8.65
1985	49.28	21.57	.299	.063	2.427	2.815	.266	1.14	1.88	1.95	1.06	8.70
1986	51.53	20.51	.287	.168	3.739	3.365	.251	1.29	2.41	2.10	1.30	8.85
1987	53.03	19.18	.328	.141	4.947	3.724	.257	0.85	3.56	2.30	1.05	8.85
1988	58.45	19.88	.373	.140	6.460	4.110	.297	0.93	4.89	2.50	1.13	9.00
1989	63.50	19.53	.429	.113	8.536	5.093	.267	0.87	6.52	2.57	1.13	9.16
1990	68.00	19.38	.541	.136	9.330	5.381	.268	0.94	7.85	3.22	0.79	9.71
1991	67.26	19.32	.846	.145	9.320	5.553	.270	0.96	7.53	3.86	0.80	7.72

Source: Reprinted with permission from *The World Almanac and Book of Facts*. (1988, p. 103; 1993, p. 140).

(b) What proportion of world gold production did South Africa have in 1972? in 1980? in 1986? (In each case divide South Africa's production for the year by world production for that year.)

(c) What *percentage* of world gold production did the United States have in 1972? in 1980? in 1986? in 1991?

(d) What country was the second largest producer in the 1970s? in the 1980s? in the 1990s?

2. Find a table of interest to you (perhaps in a yearbook of your national statistics), and try to simplify it to make the important features more obvious.

2.3 SIMPLE PLOTS FOR CONTINUOUS VARIABLES

This section discusses plots that are useful to see how our data on a single continuous variable are behaving.

2.3.1 Dot Plots

The simplest type of plot we can do is to plot a batch of numbers on a scale. For example, let us plot the numbers

$$3 \quad 4 \quad 4.5 \quad 4.5 \quad 6 \quad 8$$

as in Fig. 2.3.1. Note that we have displaced the second 4.5 vertically above the first so that it is not obscured. This displacement of points is often called ***stacking.*** Dot plots are good for small to moderate batches of data.

Because they display the distances between individual points, dot plots are good for showing features such as clusters of points, gaps, and outliers, as in Fig. 2.3.2. Outliers are very atypical observations, which in this case means observations that seem too far from the bulk of the data. Outliers should always be labeled on a plot (e.g., using the name of the person or object measured). The implications of outliers are discussed further in Section 2.3.4 and in later chapters.

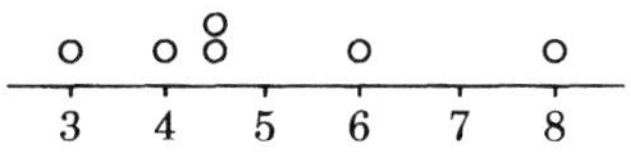

3 4 5 6 7 8

FIGURE 2.3.1 Dot plot.

FIGURE 2.3.2 Dot plot showing special features.

EXAMPLE 2.3.1 *Using a Dot Plot*

University teachers exploit the features of dot plots with respect to clusters and gaps in allocating grades for small classes. For example, in Fig. 2.3.3 we have the marks for one of our small third-year classes together with the final grades.

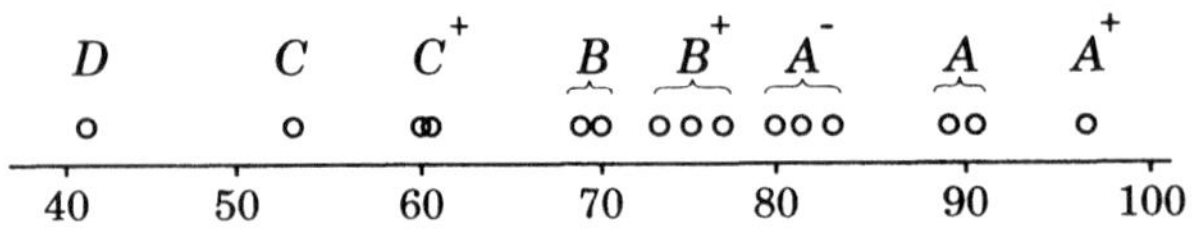

D C C^+ B B^+ A^- A A^+

40 50 60 70 80 90 100

FIGURE 2.3.3 Grading of a university course.

Dot plots are available in Minitab under the menu items `Graph` → `Dotplot`. Splus and R use `stripplot` and employ the term `dotplot` for an entirely different type of graph.

Scale Breaks

Figure 2.3.4(a) does a really good job of displaying just how far away the outlier at 61 is from the rest of the data. The rest of the data, however, is so bunched up that it is hard to see what is happening there. In Fig. 2.3.4(b) we have broken the scale to the right of the majority of the data. This has enabled us to stretch out that portion of the plot that tells us what is happening in the bulk of the data. The scale break marks, //, visually signal that part of the scale is missing, and then we see the outlier. Thus Fig. 2.3.4(b) is very good at showing us what is happening in the bulk of the data. We can still see the outlier but cannot see how discrepant it is. Scale breaks are sometimes seen in other types of plots as well. We shall not discuss them further.

Attaching Labels to Dot Plots

If there are not too many data points, we can actually label each of the points, as is done in Fig. 2.3.5. Bar graphs (see Section 2.6) can also be used for presenting data sets like this where we want to associate labels with observations.

The plots to follow, namely stem-and-leaf plots and histograms, work better than dot plots with larger sets of data, say, more than about 20 or 30 observations. Dot plots become cluttered at this stage, whereas stem-and-leaf plots and histograms are good at showing the relative densities of observations on the scale. In this and all that follows, however, there are no hard and fast rules. It is best to look at several types of

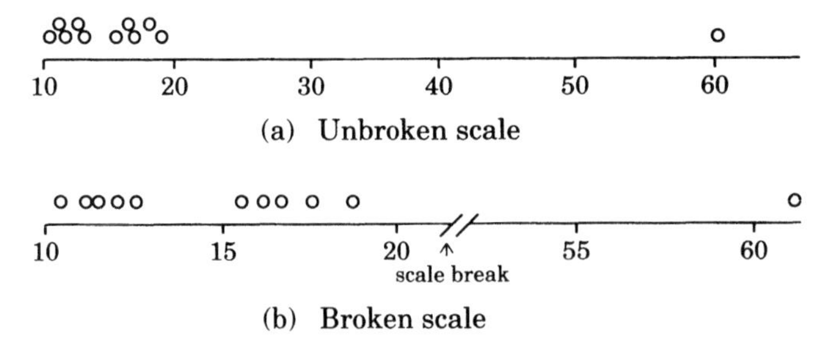

FIGURE 2.3.4 Dot plot with and without a scale break.

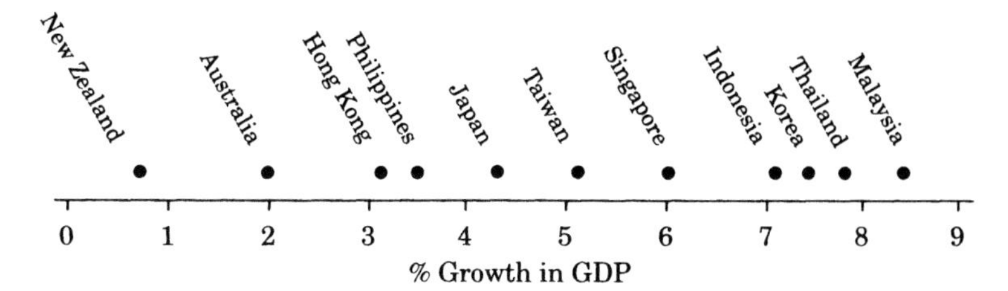

FIGURE 2.3.5 Forecast of percent growth in GDP for 1990 for some Southeast Asian and Pacific countries.

plots and then ask yourself the question, "Which of these plots seems best to display the features I see in this data?"

QUIZ FOR SECTION 2.3.1

1. What is an outlier?
2. What features of data are dot plots good at revealing?
3. When should we use dot plots?

EXERCISES FOR SECTION 2.3.1

Following are the 1995 unemployment rates (%), reported by the International Monetary Fund, of those countries classified as industrial:
Members of European Economic Community (EEC): 13 (Belgium), 10 (Denmark), 11.6 (France), 9.4 (Germany), 9.5 (Greece), 13.2 (Ireland), 12 (Italy), 8.4 (Netherlands), 7.2 (Portugal), 22.9 (Spain), 8.2 (UK).
Nonmembers of the EEC: 8.5 (Australia), 4.6 (Austria), 9.5 (Canada), 17.2 (Finland), 5 (Iceland), 3.1 (Japan), 5 (Norway), 6.4 (New Zealand), 7.5 (Sweden), 4.2 (Switzerland), 5.6 (United States).

1. Construct a dot plot of unemployment rates for EEC members. Attach labels by hand, and comment on what you see.
2. Construct a dot plot of unemployment rates for EEC nonmembers. Attach labels by hand, and comment on what you see.
3. Have you any ideas about how you might construct a single plot for comparing the unemployment rates of EEC members with those of nonmembers of the EEC?

2.3.2 Stem-and-Leaf Plots

Below is a *stem-and-leaf* plot (also known as a *stem plot*) of the 45 observations on the variable EJEC given in Table 2.1.1. The values were 72, 52, ..., 67. Stem-and-leaf plots are closely related to dot plots but with the data grouped in a way that retains much, and often all, of the numerical information. They are very good for plotting batches of numbers containing about 15 to 150 data points. The stem-and-leaf plot is built up from the numbers themselves.

Units: 7 | 2 = 72

```
1 | 7
2 | 7 9
3 | 0 7 7 8 9
4 | 2 2 3 3 4 5 8 8 9
5 | 0 0 0 1 1 2 2 3 4 4 7 8 9 9
6 | 0 0 0 2 3 3 3 3 6 7 7 8
7 | 2 2
```

We split each number into two parts:

$$\begin{array}{cc} a & | \ b \\ \uparrow & \uparrow \\ \text{stem} & \text{leaf} \end{array}$$

where the leaf is a *single* digit. In the preceding plot, 72 has been split into 7 | 2, and 52 into 5 | 2. In general, where we make the split depends on the range of the data. We arrange the stems in a vertical array and draw a spacer, for example, a vertical line, to the right of each stem. We then add the leaves to the plot.

To show how this is done, we use the following data, which represent the breaking strengths in kilograms of 11 samples of wire:[7]

220 214 222 218 223 210 223 210 227 225 212

Construction of the Plot

Suppose we decide to split each number after the second digit so that 220 becomes

$$\begin{array}{cc} 22 & | \ 0 \\ \uparrow & \uparrow \\ \text{stem} & \text{leaf} \end{array}$$

There are only two different stems in the data, namely 21 and 22. The step by step procedure is now as follows:

1. Plot stems

```
21 |
22 |
```

2. Add 220 to plot

```
21 |
22 | 0
      ↑
  leaf of 220
```

3. Add 214 to plot

```
21 | 4 ←
22 | 0
```

4. Add 222 to plot

```
21 | 4
22 | 0 2
        ↑
    leaf of 222
```

[7] We are using only these few numbers to show how a plot is constructed, not to exemplify a good plot.

5. Add 218 to plot

21	4 8 ←
22	0 2

Having plotted all the numbers we get

21	4 8 0 0 2
22	0 2 3 3 7 5

Finally we ***order the leaves*** to obtain

21	0 0 2 4 8
22	0 2 3 3 5 7

When the plot is constructed this way, the first row will contain all numbers falling in the 210s, and the second row will contain all numbers falling in the 220s. We say that we have ***class intervals*** of size 10 kg.

Lengthening the Plot

In the preceding wire-data plot, all the data are stacked up together on two lines. This will give us no information at all except the comparative numbers of observations in the 210–219 class interval and the 220–229 class interval. We could spread out the preceding plot by putting leaves 0, 1, 2, 3, 4 on one line and leaves 5, 6, 7, 8, 9 on the next. This leads to

21	0 0 2 4
21	8
22	0 2 3 3
22	5 7

This plot has 5-kg class intervals.

We can spread the plot out still further by using five lines per stem. This gives separate lines for leaves (0, 1), (2, 3), (4, 5), (6, 7), (8, 9). The stem-and-leaf plot is shown in Fig. 2.3.6, with 2-kg class intervals. It is so spread out, however, that it has become almost a dot plot. As an exercise you might try to reproduce the preceding two stem-and-leaf plots for yourself.

21	0 0
21	2
21	4
21	
21	8
22	0
22	2 3 3
22	5
22	7

FIGURE 2.3.6 Stem-and-leaf plot for breaking strengths.

Rounding Numbers (Shortening the Plot)

Suppose we wished to plot numbers with a lot of digits, such as

6459 2142 3127 786 7659 2341 3217 876

Since each leaf is a single digit, we would end up plotting 645 | 9, 214 | 2, and so on. This would give us an impossibly long plot with at least $765 - 78 = 683$ stems! We can effectively shorten the plot by rounding each number to the nearest hundred, giving the numbers

6500 2100 3100 800 7700 2300 3200 900

Dropping off the two zeros from each number, we then end up plotting

6 | 5 2 | 1 3 | 1 0 | 8 7 | 7 2 | 3 3 | 2 0 | 9

Rather than construct a plot, we have simply indicated how to select the numbers for plotting together with the position of the split. We would arrive at the same stem-and-leaf plot if the numbers happened to be 64.59, 21.42, . . . , or 645.9, 214.2, ... In conclusion we emphasize again that each leaf is a ***single*** digit.

The method of rounding the numbers is essentially a way of shortening a plot. This raises the question of how many stems should be used. A stem-and-leaf plot with fewer than five active stems is fairly uninformative. We would not usually use more than of the order of 10–15 active stems unless the data set is larger than about 50. As we shall see, however, hard and fast rules are unproductive, as plots of different lengths can convey different information.

The Units Statement

To help a reader reconstruct the numbers from the plot, we always put in a ***units*** statement at the top of the plot. For example,

Units: 8 | 3 = 83,000

tells the reader that the split is coming between the tens of thousands and the thousands. Thus a number with stem 9 and leaf 7 is read as 97,000. In contrast,

Units: 8 | 3 = 0.083

tells the reader that a number with stem 9 and leaf 7 is read as 0.097. Statistical software packages that produce a stem-and-leaf plot each have their own form of units statement.

Computing

Although stem-and-leaf plots were developed as a quick way of obtaining an informative plot by hand, most statistical software packages can automatically produce them. For example, they are available in Minitab under `Graph` → `Stem-and-Leaf`, or in Splus and R using the command `stem`. The programs allow the "scaling" (i.e., stretching or contracting) of plots (see on-line help for syntax).

EXAMPLE 2.3.2 *Stem-and-Leaf for Traffic Death Rates*

The data in Table 2.3.1 give the traffic death rates (per 100,000 of population) for 30 countries during the mid-1980s. A straightforward stem-and-leaf plot of this data is given in Fig. 2.3.7(a). To construct this plot we placed the split between the stem and the leaf at the decimal point. For example, Australia has a stem of 17 and a leaf of 4. The plot has 22 active stems and looks rather stretched out. We can also produce a plot with 12 stems by putting the split between the tens and the units and using five lines per stem. All the numbers then have to be rounded to the nearest unit to do this. The observation for Australia now becomes 17 with a stem of 1 and a leaf of 7. The resulting plot, given in Fig. 2.3.7(b), is similar in shape to the previous one. However, in using it, some of the initial information is lost because of the rounding.

TABLE 2.3.1 Traffic Death Rates (per 100,000 Population) for 30 Countries[a]

17.4 Australia	20.1 Austria	19.9 Belgium	12.5 Bulgaria	15.8 Canada
10.1 Czechoslovakia	13.0 Denmark	11.6 Finland	20.0 France	12.0 E. Germany
13.1 W. Germany	21.1 Greece	5.4 Hong Kong	17.1 Hungary	15.3 Ireland
10.3 Israel	10.4 Japan	26.8 Kuwait	11.3 Netherlands	20.1 New Zealand
10.5 Norway	14.6 Poland	25.6 Portugal	12.6 Singapore	9.8 Sweden
15.7 Switzerland	18.6 United States	12.1 N. Ireland	12.0 Scotland	10.1 England & Wales

[a]Data for 1983, 1984, or 1985, depending on the country (prior to reunification of Germany).
Source: Hutchinson [1987, p. 3].

(a) Units: 17 | 4 = 17.4 deaths per 100,000

Stem	Leaf
5	4
6	
7	
8	
9	8
10	1 1 3 4 5
11	3 6
12	0 0 1 5 6
13	0 1
14	6
15	3 7 8
16	
17	1 4
18	6
19	9
20	0 1 1
21	1
22	
23	
24	
25	6
26	8

Collapse to → 12 stems

(b) Units: 1 | 7 = 17 deaths per 100,000

Stem	Leaf
0	5
0	
0	
1	0 0 0 0 0 1 1
1	2 2 2 2 3 3 3 3
1	5 5
1	6 6 7 7
1	9
2	0 0 0 0 1
2	
2	
2	6 7

FIGURE 2.3.7 Two stem-and-leaf plots for the traffic deaths data.

What do we see in Fig. 2.3.7? (We shall suggest what to look for in these plots in an organized fashion in Section 2.3.4.) What stands out most is the abnormally small value at 5.4 and the abnormally large values at 25.6 and 26.8. Going back to Table 2.3.1, we see that the 5.4 belongs to Hong Kong. It is perhaps not surprising that Hong Kong has such a low rate because it has a large population living in a very small land area and so the use of private cars is probably low. This possible explanation is something we can check out. Of the very large values, the 25.6 belongs to Portugal, and the 26.8 belongs to Kuwait. We have no idea why these countries have such high death rates. This suggests an area for further investigation. We also see a division of the central mass of the data into two groups in Fig. 2.3.7(a). There is a noticeable gap between 15.8 and 17.1. The upper group from 17.1 to 21.1 consists of Hungary, Australia, United States, Belgium, France, Austria, New Zealand, and Greece. Is there anything that these countries have in common that sets them apart from the other countries? Greater use of motor vehicles? Less motorway/freeway driving? Worse roads? The analysis of data is an interplay between what you see in the data and what you know about the subject under investigation. And as we stressed in Chapter 1, the data themselves often trigger more questions about the world and keep the investigative cycle going.

Which of the two plots in Fig. 2.3.7 should we use? The conventional wisdom, as stated previously concerning the recommended number of stems, says Fig. 2.3.7(b). As is often the case with plots, however, we learn from both. Gaps in such plots can be an artifact of the way the stems are selected, so it is sometimes a good idea to try using different stems.

EXERCISES FOR SECTION 2.3.2

1. How do we read a number with (i) stem 15 and leaf 4 and (ii) stem 6 and leaf 7 if the units statement reads
   (a) $12 \mid 9 = 1.29$? (b) $12 \mid 9 = 129$? (c) $12 \mid 9 = 0.00129$?
2. Determine the appropriate stems and leaves for the following data extracts. In each case give the units statement.
   (a) Diameter in centimeters: 1.63, 2.12, 1.80, 2.02, 1.74, 1.81, 2.08, 1.85, 2.14.
   (b) Depth in meters: 543, 625, 651, 728, 659, 417, 832, 746, 528, 664.
   (c) Length in meters: 16.321, 17.144, 16.492, 17.526, 16.114, 17.633, 15.835, 16.463, 17.237, 16.871.
   (d) Breaking strength in kilograms: 1663, 1688, 1656, 1664, 1614, 1668, 1659, 1638, 1653.
3. Reconstruct the data in Table 2.3.1 from the stem-and-leaf in Fig. 2.3.7(a) by reattaching each leaf to its stem.
4. Construct a stem-and-leaf plot of the data on the variable DIAVOL in Table 2.1.1 by hand: (a) using about 6 lines; and (b) using about 14 lines. Comment on what you see. Use a computer to construct the plot.
5. Use a computer to obtain a stem-and-leaf plot of the traffic accident data in Table 2.3.1. Compare your plot with Fig. 2.3.7 to see how many stems the computer has used.
6. The impact strengths of gear teeth (in lb.-ft) at position 7 on 33 gear blanks are as follows:[8]

2287	2275	1946	2150	2228	1695	2000	2006	1945	2006	2209
2216	1934	1904	1958	1964	2066	2222	2066	1964	2150	2114
2125	2210	1588	2234	2210	2156	2204	1641	2263	2120	2156

[8]The story is described in Example 3.2.2.

Construct a stem-and-leaf plot. Comment on what you see. (Keep this data set on file for future exercises.)

2.3.3 Histograms

For illustrative purposes, we shall use the data given on the lengths of *female* coyotes given in Table 2.3.2. A histogram of these data is given in Fig. 2.3.8(a). As you can see, a histogram is basically a set of boxes. Its shape is determined by three features, the number, width, and height of the boxes. We shall see that they are very closely related to stem-and-leaf plots and that the shapes of both can be interpreted in the same way. Computers will produce any histograms that you need. Knowing how they are constructed by hand, however, is an important first step toward understanding their meaning and use.

We begin constructing a histogram by picking out the maximum and minimum values and marking these points on the horizontal axis of our graph. In the simplest type of histogram, we then split the range between these two extreme values into a convenient number of equal-sized class intervals and count the number of observations falling in each of these intervals. If possible, we like the length of the class intervals and the class interval boundaries to be nice round numbers. Between 5 and 15 intervals is a common rule of thumb, with fewer intervals if the number of observations is small or more if it is large. Finally, above each interval we draw a rectangle with height equal to the number of observations falling in the interval (also known as the *frequency* of observations in that interval).

Looking at Table 2.3.2, we see that all female coyote lengths fall between 71.0 cm and 102.5 cm. A convenient choice for class intervals might be to take seven intervals each of length 5 cm. The counts of the numbers of observations falling into each class interval (called *frequencies*) for each of the seven intervals are given in Table 2.3.3. Here the interval $70\text{–}75^-$ represents all lengths from 70.0 to 74.9. The height of the rectangle above a class interval is equal to the frequency (i.e., counted number) of observations in that interval. For example, the height for the interval $85\text{–}90^-$ is 12. The resulting histogram is shown in Fig. 2.3.8(a).

As can be seen from Table 2.3.3 and Fig. 2.3.8, stem-and-leaf plots and histograms are very similar.[9] A stem-and-leaf plot is like a histogram turned on its side. Histograms are perhaps more attractive to look at, though stem-and-leaf plots have a number of advantages for medium-sized data sets. For example, they return information on the individual values, and the data are sorted. They are quick to produce by hand and retain information on the individual values. They also provide a mechanism for ordering (sorting) the data from smallest to biggest,[10] which will be very useful to us in the future. The choice of where to split the numbers to obtain the stem and how many lines to use per stem corresponds roughly to the choice of class intervals in a histogram. In both types of plot, we probably won't see the patterns in the data if there are too many or too few class intervals. If a stem-and-leaf plot or histogram uses a very stretched-out scale, it behaves like a dot plot rather than a display of relative densities of data. At the other extreme, using too few class intervals can result in smoothing over and obscuring interesting features of the data, for example, separation into clusters and abnormal values.

---

[9]The reader may have noticed that the counts in the table are not the same as the number of leaves in the stem-and-leaf plot. The reason is that 84.5 rounds to 85 and goes into the stem-and-leaf interval 85–90 rather than the 80–84 interval.

[10]Though data are easily sorted by computers using a command like SORT.

TABLE 2.3.2 Coyote Lengths Data[a] (cm)

Females											
93.0	97.0	92.0	101.6	93.0	84.5	102.5	97.8	91.0	98.0	93.5	91.7
90.2	91.5	80.0	86.4	91.4	83.5	88.0	71.0	81.3	88.5	86.5	90.0
84.0	89.5	84.0	85.0	87.0	88.0	86.5	96.0	87.0	93.5	93.5	90.0
85.0	97.0	86.0	73.7								
Males											
97.0	95.0	96.0	91.0	95.0	84.5	88.0	96.0	96.0	87.0	95.0	100.0
101.0	96.0	93.0	92.5	95.0	98.5	88.0	81.3	91.4	88.9	86.4	101.6
83.8	104.1	88.9	92.0	91.0	90.0	85.0	93.5	78.0	100.5	103.0	91.0
105.0	86.0	95.5	86.5	90.5	80.0	80.0					

[a]Coyote captured in Nova Scotia, Canada. Data courtesy of Dr. Vera Eastwood.

TABLE 2.3.3 Frequency Table for Female Coyote Lengths

Class Interval	Tally	Frequency	Stem-and-leaf plot	
70 – 75⁻	ll	2	7	1 4
75 – 80⁻		0	7	
80 – 85⁻	~~llll~~ l	6	8	0 1 4 4 4
85 – 90⁻	~~llll~~ ~~llll~~ ll	12	8	5 5 5 6 6 7 7 7 7 8 8 9
90 – 95⁻	~~llll~~ ~~llll~~ lll	13	9	0 0 0 0 1 1 2 2 2 3 3 4 4 4
95 – 100⁻	~~llll~~	5	9	6 7 7 8 8
100 – 105⁻	ll	2	10	2 3
Total		40		

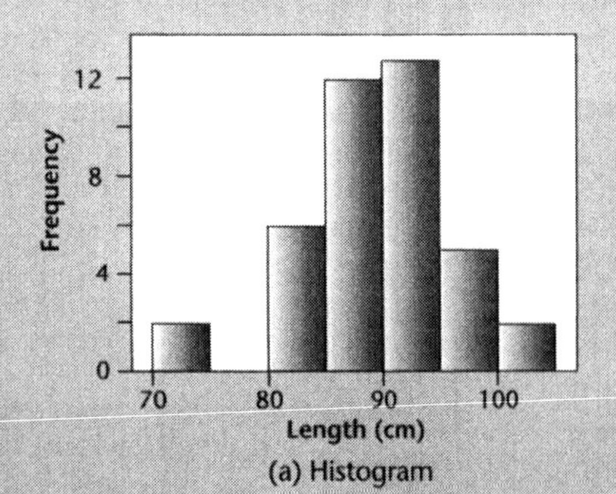

(a) Histogram

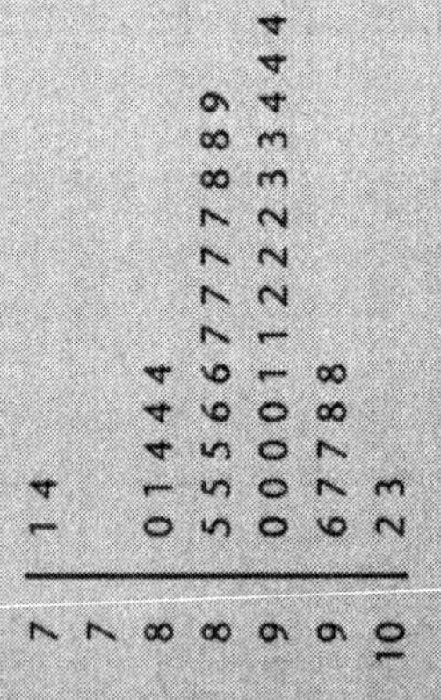

(b) Stem-and-leaf plot rotated

FIGURE 2.3.8 Histogram of the female coyote-lengths data.

It is sometimes convenient to use a ***relative frequency histogram.*** Here the height of each rectangle is made equal to the *relative frequency* (i.e., proportion) of the observations in the class interval. For example, the height of the interval 70–75^- would now be $2/40 = 0.05$ or 5%. The vertical scale now goes from 0 to 1 or, if expressed in percentages rather than proportions, 0% to 100%. This relative frequency histogram has exactly the same shape as before; only the vertical scale is changed. When we want to compare histograms based on different numbers of individuals (as with male and female coyotes), we would almost always want to use relative frequency histograms.

The key features that the eye picks up from histograms are the overall shape and the areas of the rectangles. With equal class intervals these areas are proportional to the frequencies. Figure 2.3.9(a) through (c) shows how the shape of the histogram depends (1) on the width you choose for your class intervals and (2) on where you place the ***boundaries*** of the class intervals. Ideally, you should look at several histograms with different choices of class-interval width and boundary positions to get an impression of shape.

Histograms are available from general purpose statistical packages, for example, under `Graph` → `Histogram` in Minitab and using `hist(varname)` in Splus and R. In Excel you go from `Tools` → `Data Analysis` and choose `Histogram`. This produces a frequency table from which a histogram can be produced using a chart wizard. (Note that there are problems with labeling the x-axis.) The packages have default settings for class intervals that can be changed (see on-line help files for syntax).

Some packages produce a "smooth" version of the histogram, which avoids the problem of where to put the class-interval boundaries. The simplest of these is called the ***density trace,*** as in Fig. 2.3.9(d). (A description of this technique is given on our web site.)

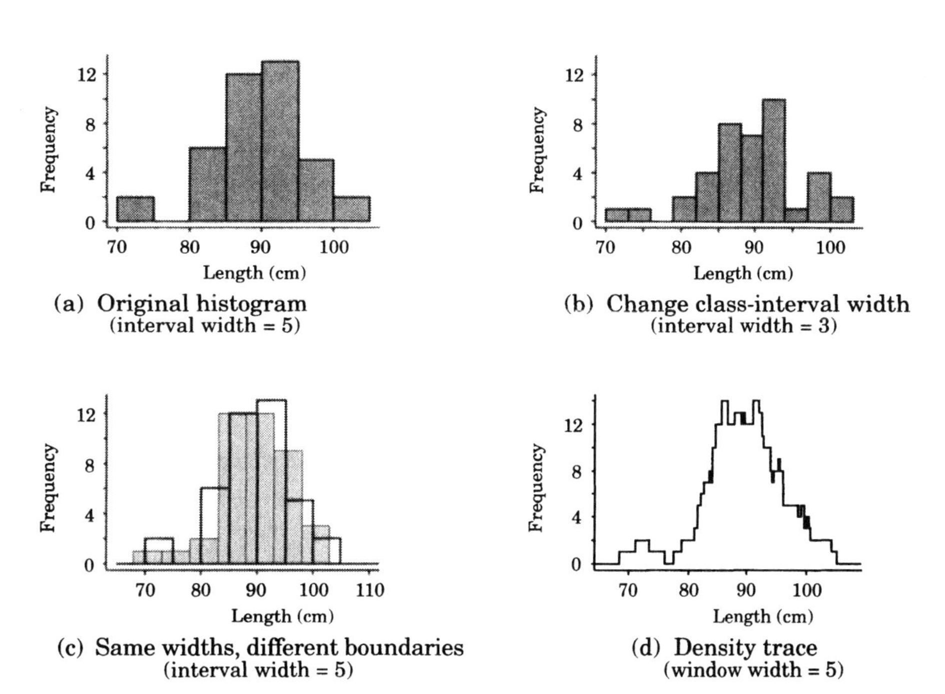

FIGURE 2.3.9 Histograms and density trace of female coyote-lengths data.

QUIZ FOR SECTION 2.3.3

1. What advantages does a stem-and-leaf plot have over a histogram? (Three are mentioned in the text.)
2. The shape of a histogram can be quite drastically altered by choosing different class-interval boundaries. What type of plot does not have this problem? What other factor affects the shape of a histogram?

EXERCISES FOR SECTION 2.3.3

1. Using two-unit class intervals, construct a frequency table and corresponding histogram of the traffic accident data in Table 2.3.1. Compare your results with the stem-and-leaf plot in Fig. 2.3.7(b).
2. Construct a frequency table and corresponding histogram for the lengths of *male* coyotes (data in Table 2.3.2). You now have a histogram for males and another for females [Fig. 2.3.8(a)]. Comment on the shapes of the two histograms.
3. From Fig. 2.3.8 and Table 2.3.3, what percentage of female coyotes have a length less than 90 cm?

2.3.4 Interpreting Stem-and-Leaf Plots and Histograms

Figure 2.3.10 shows some of the features that we might look for in histograms and stem-and-leaf plots. We do not react to the minor ups and downs in such plots but tend to look for a smoothed general impression. Hence, most of the curves in Fig. 2.3.10 are drawn as smooth. The graphs are oriented in the same way as histograms; to relate these shapes to stem-and-leafs you will have to mentally (or physically) rotate the pictures through 90 degrees.

Outliers

First we look for discrepant observations. Provided we have a sufficient number of class intervals, stem-and-leaf plots are very useful for showing any observations that are well away from the main body of the data [Fig. 2.3.10(k)]. We should look more closely at such observations to see why they are different. Are they mistakes, or is something unusual and interesting going on here? Can we explain what caused it? To check whether or not the outliers are mistakes, we go back to the original source of the data (e.g., the hand-recorded observations, patient notes). We can ask subject experts about their plausibility. It is not always possible to determine whether an outlier is a mistake or a real observation.

We next turn our attention to the body of the data.

The Existence of More Than One Peak

Figure 2.3.10(b) shows two peaks, and (c) shows three. These peaks are called ***modes.*** When there is a single peak, the mode represents the most popular value (or the most popular class interval). The presence of several modes usually indicates that there are several distinct groups in the data. For example, if the data representing the diameters of trees exhibited two modes (i.e., was ***bimodal***) then we would tend to suspect two different varieties of trees, two different planting dates, or trees from two different

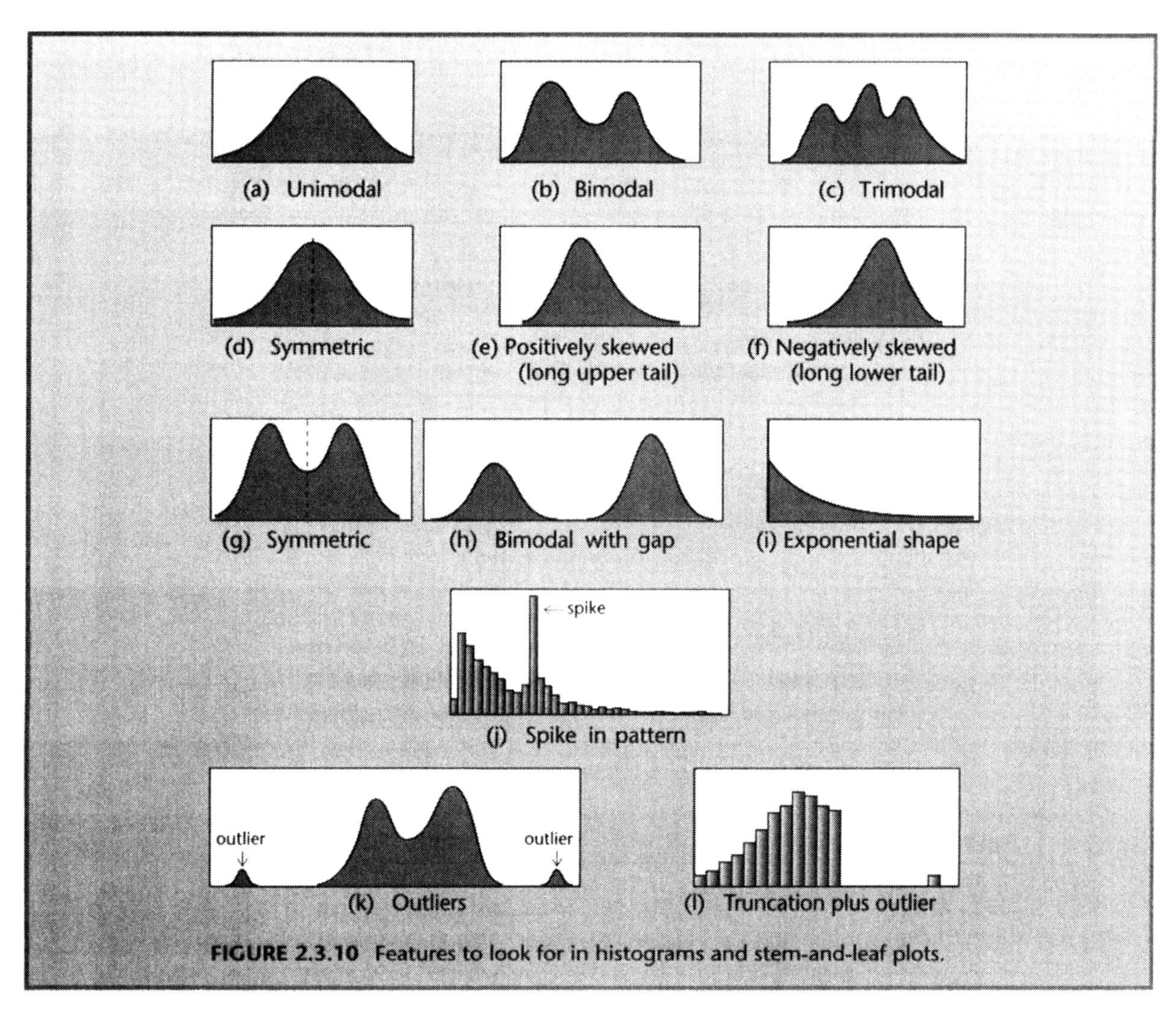

FIGURE 2.3.10 Features to look for in histograms and stem-and-leaf plots.

locations. We would probably want to investigate the reason for the two modes. If you look at the heights of people without separating out males and females, you get a bimodal distribution, as shown in Fig. 2.3.11 in which the histogram is constructed from the very people whose heights are being graphed. Figure 2.3.10(h) shows a more extreme version of bimodality where there appears to be complete separation between two groups.

Shape of the Distribution

After noting modality, we look at shape in more detail. Does the plot appear to be close to symmetry, as in Fig. 2.3.10(d) and (g)? Or does it show moderate skewness, as in Fig. 2.3.10(e) and (f), or extreme positive skewness, as in (i)? We look to see whether the envelope of the graph is roughly bell-shaped, as in Fig. 2.3.10(a) and (d), or whether it has some other shape, such as the *exponential* shape shown in Fig. 2.3.10(i).

We shall see later that when applying formal statistical techniques to data, statisticians usually prefer to work with a *symmetric* bell-shaped histogram like Fig. 2.3.10(d). Unfortunately, nature isn't always that obliging, although we can often find ways of re-expressing (or *transforming*) the data to make it conform.

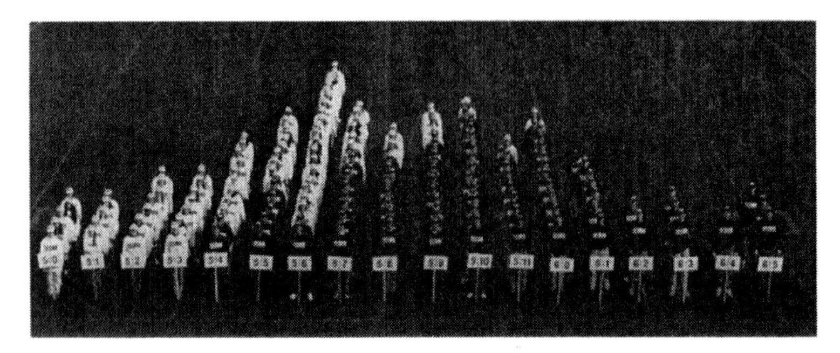

FIGURE 2.3.11 Histogram of heights constructed using the people. Photograph by Peter Morenus in conjunction with Prof. Linda Strausberg, University of Connecticut. Subjects are University of Connecticut genetics students, females in white tops, males in dark tops.

A stem-and-leaf plot or histogram can be helpful if we are interested in constructing a statistical model for the process that describes the data. The shape of the histogram may suggest a mathematical function whose curve fits the histogram closely (see Chapter 6).

Central Values and Spread

We note where the data appear to be centered and how spread out they are. If there is more than one mode, we note the positions of the modes.

Be Suspicious of Abrupt Changes

A doctoral student of the authors was screening the data from a big diabetes study (6000 subjects) prior to a complicated analysis involving many variables. One variable was a blood measure determined in the laboratory. The data on that variable are depicted in the histogram in Fig. 2.3.10(j). Standing out from the general smooth pattern[11] is a big spike, which occurred at 10 units. She made inquiries at the lab and found that, for the first six months, the lab had been reporting any value below 10 as 10. This caused the spike. Fortunately, it was possible to go back and correct these early values. If she hadn't looked at the histogram, the measurements on that variable would have included a lot of nonsense values, and the results she reported would have been misleading.

Figure 2.3.10(l) shows another abrupt change. This pattern was evident in some manufacturing data shown to us by a colleague who is a quality-improvement consultant. We see much of a bell-shaped curve, and then it suddenly stops, except for a single large outlier. The observations were on the hardness of samples of steel, which, as it turned out, had been subject to 100% inspection. Any items above a specified upper limit had been rejected as "out of specifications." The outlier was an item that had somehow slipped through the inspection net.

If you see an abrupt change, try to establish the reason for it.

QUIZ FOR SECTION 2.3.4

1. What was another reason given for plotting data on a variable, apart from interest in how the data on that variable behaves?

[11]Even though there are a large number of class intervals, the histogram looks reasonably smooth because the data set is very large.

2. What does it mean for a histogram or stem-and-leaf plot to be bimodal? What do we suspect when we see a bimodal plot?
3. What are outliers, and how do they show up in these plots? What should we try to do when we see them?
4. What do we mean by symmetry and positive and negative skewness?
5. What shape do we call exponential?
6. You were told to be suspicious of something. What was it?

EXERCISES FOR SECTION 2.3.4

1. Use a computer to construct a histogram of the data in the EJEC column in Table 2.1.1. Comment on the shape.
2. Do the same for OCCLU. Comment on the shape of the histogram. Are there any outliers?
3. Do the same for DIAVOL. Construct a stem-and-leaf plot and compare it with your histogram. What do you see in these plots?

2.4 NUMERICAL SUMMARIES FOR CONTINUOUS VARIABLES

2.4.1 Locating the Center of the Data

In addition to graphic displays, it is often important to have numeric measures of the main features of the data, particularly if we want to make precise comparisons of many different batches of data. Following is the *minimal* set of summary statistics automatically produced for the variable AGE in Table 2.1.1 by Minitab from `Basic Statistics → Display Descriptive Statistics`. Some of the items in this printout will be discussed later. For the present we will look at just the items labeled "mean" and "median."

Standard deviation (pointing to StDev)

Descriptive Statistics

Variable	N	Mean	Median	TrMean	StDev	SE Mean
age	45	50.133	51.000	50.366	6.092	0.908
Variable	**Minimum**	**Maximum**	**Q1**	**Q3**		
age	36.000	59.000	46.500	56.000		

Lower quartile (pointing to Q1) *Upper quartile* (pointing to Q3)

When trying to summarize data numerically, the most basic feature is the "center" of the data, which is a measure of the "typical size" of the observations. There are several ways of measuring this; the two most commonly used are the *sample mean* and the *sample median*.

The sample mean is the usual average[12] of the observations.

$$\text{The } \textbf{\textit{sample mean}} = \frac{\textbf{\textit{Sum}} \text{ of the observations}}{\textbf{\textit{Number}} \text{ of observations}}$$

The sample median is simply the value in the middle position when all the observations are written in order of size from smallest to largest. For example, given the seven numbers

$$\begin{array}{ccccccc} 1 & 4 & 7 & 9 & 10 & 12 & 14 \\ & & & \uparrow & & & \\ & & & \text{Median} & & & \end{array}$$

we see that the sample median is 9. We note that the sample median is 4 [= (7 + 1)/2] observations in from either end.

Now consider the eight numbers

$$\begin{array}{cccccccc} 11 & 13 & 15 & 16 & 19 & 21 & 22 & 25 \\ & & & \uparrow & & & & \\ & & & \text{Median} & & & & \end{array}$$

From the idea of the sample median falling in the middle position, it is clear that we want the sample median to be at the position of the arrow, which is midway between the fourth and fifth observations, that is, $4\frac{1}{2}$[= (8 + 1)/2] observations in from either end. When the sample median falls between two observations, we get its value by averaging the observations on either side, which here gives a sample median of (16 + 19)/2 = 17.5. We can write two rules for the sample median that work whether the sample size n (i.e., the number of data points) is odd or even.

> The ***sample median*** is the $\frac{n+1}{2}$th largest observation.

> If $\frac{n+1}{2}$ is not a whole number, the median is the average of the two observations on either side.

Example 2.4.1 *Bowling Average*

The numbers of wickets taken by a well-known bowler in a series of nine cricket tests were 3, 4, 2, 3, 1, 5, 4, 5, 2. Thus $n = 9$, and the sample mean is

$$\frac{3+4+2+3+1+5+4+5+2}{9} = 3.22$$

[12]We use the term the *sample mean* rather than simply the *mean* (similarly median) to distinguish between the average of a batch of numbers and the *mean* of something known as a probability distribution, which we shall first meet in Chapter 5.

The ordered observations are 1, 2, 2, 3, 3, 4, 4, 5, 5, and $(n + 1)/2 = (9 + 1)/2 = 5$. The fifth observation is 3, so the sample median is 3. Here the mean and the median are quite close.

It is helpful to use symbols to represent the sample mean and the sample median. Let $x_1, x_2, \ldots, x_n$ denote the n observations in the sample. In Example 2.4.1, $x_1 = 3$, $x_2 = 4$, and so on, until $x_9 = 2$. We denote the sample mean by $\bar{x}$, which is read as "x-bar" and calculated using

(Sample mean) $$\bar{x} = \frac{x_1 + x_2 + \cdots + x_n}{n} \left(= \frac{\sum x_i}{n} \right)$$

- ***The sample mean is denoted by*** $\bar{x}$.
- ***The sample median will be denoted by* Med.**

EXAMPLE 2.4.2 *Average and Median Survival Times*

The survival times (in days) of the first six patients in the Stanford heart transplant program were 15, 3, 46, 632, 126, 64. The mean survival time is

$$\bar{x} = \frac{15 + 3 + 46 + 632 + 126 + 64}{6} = 147.7 \text{ days}$$

The ordered observations are 3, 15, 46, 64, 126, 623. Since $n = 6$, $(n + 1)/2 = (6 + 1)/2 = 3\frac{1}{2}$, so the sample median is the $3\frac{1}{2}$th observation, or the average of the third and fourth observations. Hence,

$$\text{Med} = \frac{46 + 64}{2} = 55$$

As this example shows, the sample mean and sample median can differ substantially.

Although we can imagine the sample median as simply the middle value, it is not so easy to visualize the sample mean. One helpful way of doing this is to imagine that the dot plot of the data consists of equal-sized steel balls attached to a very light, rigid rod. Then the mean is simply the center of gravity of the system, that is, the point at which the construction balances. For example, in Fig. 2.4.1 only (a) is balanced; in (b) and (c) the rod will rotate as shown.[13]

The sample mean is where the dot plot balances.

[13]How does the sample mean of the original observations relate to the histogram? It is approximately the point of balance for the histogram (see Section 2.5.2 for a numerical example). As we know, however, minor changes to the class intervals can change the shape of these plots, so we cannot make any exact statement. How does the median relate to a histogram? It is close to the point that has half the area under the histogram to the right of it and half to the left.

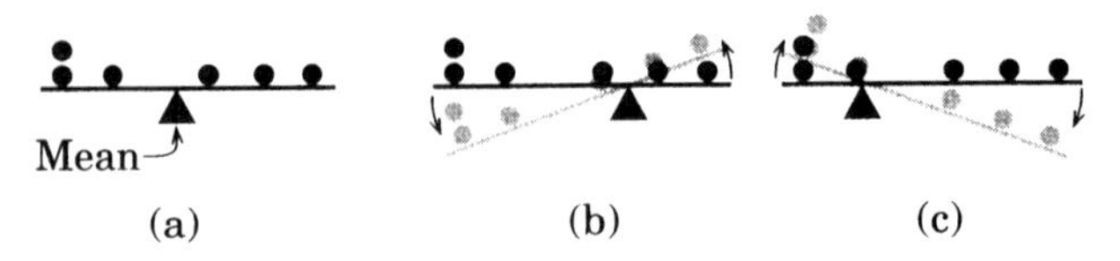

FIGURE 2.4.1 Mechanical construction representing a dot plot: (a) shows a balanced rod, while (b) and (c) show unbalanced rods.

Relationship between the Mean and Median

If the dot plot is perfectly symmetric about some point P, as in Fig. 2.4.2(a), then the sample mean and the median are identical (our beaded rod would balance at P, which is also the median). Similarly, if our stem-and-leaf plot or histogram is roughly symmetrical, then the mean and median will be very close together.

As the distribution of the data becomes skewed, the sample mean moves away from the sample median in the direction of the longer tail. This is illustrated with a dot plot in Fig. 2.4.2(b). Here we make the two rightmost values from Fig. 2.4.2(a) much larger so that the two corresponding dots move further to the right. The sample mean (i.e., the point of balance) also moves to the right, but the sample median remains the same. Thus the sample mean is affected by extreme values, but the sample median is not and tends to stay with the main body of the data.

Although the sample mean is by far the most commonly quoted measure of location, the sample median is a more robust measure, as it is not affected by a few incorrect extreme values (outliers), so it can be used for "dirty" data. Also, for very skewed data sets the sample median is likely to be a more sensible measure of center than the mean. Summaries of income data for a group of people, for example, generally quote the median income rather than the mean income. This is because the median, being the halfway point, is better at reflecting the common experience than the mean. The latter can be badly inflated by the presence of one or two multimillionaires.

In any real situation, which of these two measures (the sample mean or median) should we use? According to Canadian statistician Jock Mackay, there is a single correct answer to this and every other statistical question, "It all depends"! Although we usually say medians are a better measure for skewed distributions than means, this must be tempered by common sense.

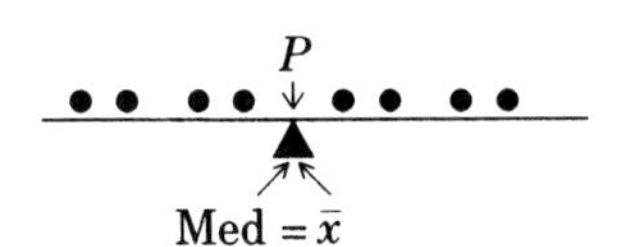

(a) Data symmetric about P

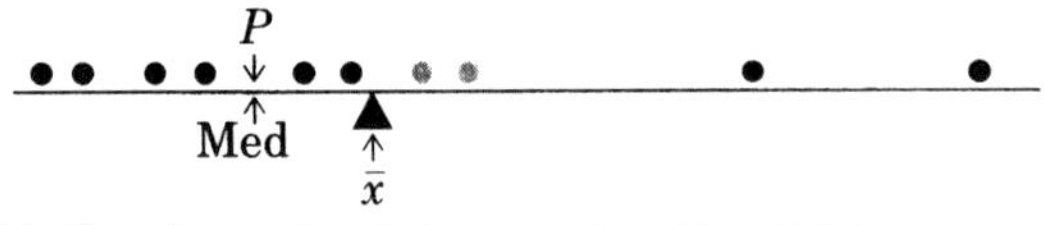

(b) Two largest points moved to the right

FIGURE 2.4.2 The mean and the median. [Grey disks in (b) are the "ghosts" of the points that were moved.]

✖ Example 2.4.3 *Mean or Median?*

A *Time* magazine story (24 February 1986) described a dispute between the American Medical Association (AMA) and the American Bar Association (ABA) about the rising costs of malpractice insurance for doctors. The AMA used *means* to show that the average amount awarded in malpractice lawsuits rose by over 50% from 1980 to 1984. The lawyers, however, quoted the *medians* for these two periods as both being the same, that is, no rise at all! Which figures are more relevant? Clearly, in this case the means tell us more because they are proportional to the total cost of insurance claims, and it is this total cost that is passed on in insurance premiums. Although the medians hadn't changed, there had been an increase in large settlements (e.g., as much as $14 million to David Berg, a university student who suffered brain damage during minor elective surgery, of which his lawyer took $5.3 million). There had been some inflation in the United States over the interim years so that an increase in dollar amounts may not mean an increase in real costs. Perhaps the most relevant figures therefore are those based on the *percentage* of a doctor's income paid out in insurance premiums. These were almost constant over 1976–1984, even though the dollar payments nearly doubled.

We conclude this subsection with two warnings. The first is this:

Quoting a single center for the data may not be sensible.

For example, if the histogram of the data shows two distinct peaks, as in Fig. 2.3.10(b), this may indicate that the underlying sampled population contains two different populations (e.g., males and females) who tend to score differently on this variable. It is then desirable, if at all possible, to come up with a well-defined rule of classification that will divide the data into two batches. This is called ***stratification.*** Separate measures of location or center can then be found for each batch, or ***stratum.***

Our second warning is this:

Beware of inappropriate averaging.

Some people seem to have a philosophy, "If they're numbers, average 'em." *You* probably would not do anything as crazy as in the cartoon, but some people come very close. It makes sense only to average observations on the *same variable* (i.e., same measurement scale). Moreover, an average or median of a *qualitative* variable is totally meaningless.

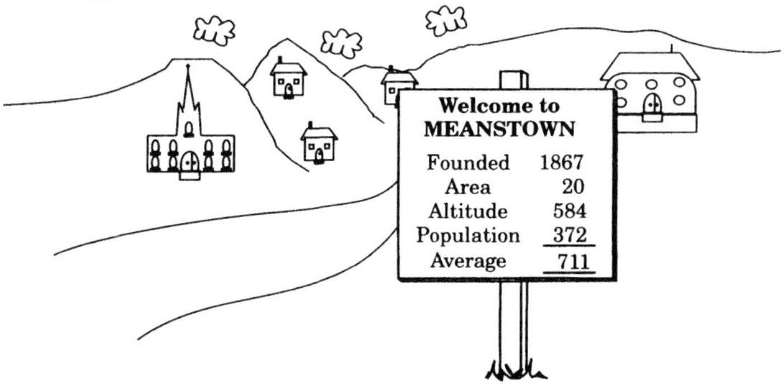

Suggested by a 1977 cartoon in *The New Yorker* magazine by Dana Fradon.

QUIZ FOR SECTION 2.4.1

1. How is the sample mean related to the dot plot?
2. If $(n + 1)/2$ is not a whole number, such as 23.5, how do we obtain the sample median?
3. Why is the sample median usually preferred to the sample mean for skewed data? Why is it preferred for "dirty" data?
4. Under what circumstances may quoting a single "center" (be it a mean or median) not make sense?
5. What can we say about the sample mean of a qualitative variable?

EXERCISES FOR SECTION 2.4.1

1. Find the median of the following data sets:
   (a) 1 3 6 9 10 14
   (b) 1 1 2 4 5 6 8
   (c) 3.0 4.0 2.1 3.5 2.6 3.0 4.8 2.5
2. Use a computer to find the mean and the median lengths of the female Nova Scotia coyotes in Table 2.3.2. You can also read off the median from the stem-and-leaf plot to the right of Table 2.3.3. Why are the two answers different? Does it matter?
3. Find the mean and the median lengths of the male Nova Scotia coyotes given in Table 2.3.2. Compare your answers with those for the female coyotes. What can we say about the average lengths of female and male coyotes?
4. Find the mean age of the patients in Table 2.1.1 who were alive at last follow-up and did not have surgery. There are 23 such patients, and you will need to look at the variables OUTCOME and SURG to identify them. Construct a stem-and-leaf of the ages of these patients (which orders the data) and read off the median. (Keep your plot for an exercise in the next section.)

2.4.2 The Five-Number Summary

Having introduced the idea of the center, or location, of a batch of data, we now ask, "When does it make sense to summarize a data set by a *single* number, such as the sample median or the sample mean?" The answer is, "Very seldom."

A colleague of ours visited an organization that sells and services copiers. They were proud of their service record. The average time to respond to a fault (i.e., to have a service person on the site) was under 4 hours. Their clients, however, were not as impressed by this as the company thought they should be. All that the clients seemed to be able to remember was the odd occasion when they had to wait an excessive time (e.g., 24 hours) for service. Clearly an average call-out time of 4 hours, with all calls being between 3 hours and 5 hours, is very different as far as a customer is concerned from an average call-out time of 4 hours with times ranging from 30 minutes to 24 hours. The issue here is the amount of variability in the call-out times. The customer wants as little unpredictability as possible for planning purposes; for example, should they just do other work and wait while the copier is being fixed or send an urgent copying job elsewhere? In statistical terms this means that the customer wants the service times to vary as little as possible, as well as to have a low mean or median service time.

We now augment the sample median as a measure of center or location of the data with other indicators of where the data is to give the *five-number summary.* This includes the minimum, the maximum, and the quartiles, which we shall now introduce. The maximum and the minimum are sometimes called the *extremes.*

The Lower and Upper Quartiles (Q_1 and Q_3)

Just as the sample median divides the ordered data into two halves, each with the same number of observations, the *quartiles* are those points that divide the data into quarters. The lower or first quartile (denoted Q_1) divides the bottom half of the data in two, and the upper or third quartile (denoted Q_3) divides the top half of the data in two. The median is also known as the second quartile (Q_2). The following batch of numbers illustrates what we mean.

```
3   4   7   8   11   13   21   29
      ↑        ↑         ↑
      Q1      Med        Q3
```

Before giving precise definitions for the quartiles, we should note that several definitions are used in the literature and by different software packages.[14] Fortunately, the different definitions give very similar answers for any batch of numbers that is large enough for the computation of quartiles to make much sense. The minor differences are then of no practical importance.

The rule we shall adopt is as follows:

> The first quartile (Q_1) is the median of all the observations whose *position* is strictly below the *position* of the median, and the third quartile (Q_3) is the median of those above.

✱ EXAMPLE 2.4.4 *Computing the Median and Quartiles*

Consider the ordered set of numbers:

2 4 5 6 6 8 10 10 12

Here $n = 9$, so $(n + 1)/2 = (9 + 1)/2 = 5$. The fifth position gives the median, so Med = 6. Removing this position, we require the median of the numbers 2, 4, 5, and 6. Thus,

$$Q_1 = \frac{4+5}{2} = 4.5$$

Similarly,

$$Q_3 = \frac{10+10}{2} = 10$$

[14] See, for example, Freund and Perles [1987] and Frigge et al. [1989].

Hence, we have

$$\begin{array}{ccccccccc} 2 & 4 & 5 & 6 & 6 & 8 & 10 & 10 & 12 \\ & \uparrow & & & \uparrow & & & \uparrow & \\ & Q_1 & & & \text{Med} & & & Q_3 & \end{array}$$

EXAMPLE 2.4.5 *Computing the Median and Quartiles*

Consider the ordered data set:

2 5 7 11 12 14

Here $n = 6$, so $(n + 1)/2 = (6 + 1)/2 = 3.5$. Hence the median is at position 3.5, so

$$\text{Med} = \frac{7 + 11}{2} = 9$$

The median of the numbers 2, 5, and 7 is 5, and the median of the numbers 11, 12, and 14 is 12. Hence $Q_1 = 5$ and $Q_3 = 12$, giving the following:

$$\begin{array}{cccccc} 2 & 5 & 7 & 11 & 12 & 14 \\ & \uparrow & \multicolumn{2}{c}{\uparrow} & \uparrow & \\ & Q_1 & \multicolumn{2}{c}{\text{Med}} & Q_3 & \end{array}$$

For working by hand we can read the median and quartiles for a large data set off a stem-and-leaf plot, and indeed, we sometimes improve the plot by marking the positions of these values.

EXAMPLE 2.4.6 *Quartiles for Traffic Death Rates*

Let's look again at the traffic death rates data, or more specifically the stem-and-leaf plot in Fig. 2.3.7(a). Here there are $n = 30$ observations. Thus $(30 + 1)/2 = 15.5$, so we average the 15th observation (13.0) and the 16th observation (13.1) to obtain a median of $(13.0 + 13.1)/2 = 13.05$. There are 15 observations below the median position, and $(15 + 1)/2 = 8$. Hence the lower quartile is the eighth observation from the bottom, giving $Q_1 = 11.3$. Also, the upper quartile is the eighth largest observation from the top, and $Q_3 = 18.6$. (The unit for all these numbers is deaths per 100,000 of population.)

Defining the Five-Number Summary

We now have all the ingredients we need for the five-number summary highlighted in the following box:

The five-number summary = (Min, Q_1, Med, Q_3, Max)

Here Min and Max are the smallest and largest observations. The preceding five numbers all give us information about the location of the data in a summary form. The quartiles have the added feature of telling us where the middle 50% of the data lie (i.e., between the quartiles) and therefore give us some idea as to how spread out the data points are. We can expect these numbers to be useful for plotting purposes, as we find in the next section. We see from the following printout that all five numbers plus several others are available from Minitab by using `Basic Statistics → Display Descriptive Statistics`. Splus and R use `summary` to do the same thing. Excel's `Descriptive Statistics`, found under `Tools → Data Analysis`, do not include quartiles. These can be found using `QUARTILE` (see help files for syntax).

Standard deviation

Descriptive Statistics

Variable	N	Mean	Median	TrMean	StDev	SE Mean
age	45	50.133	51.000	50.366	6.092	0.908
Variable	Minimum	Maximum	Q1	Q3		
age	36.000	59.000	46.500	56.000		

Lower quartile *Upper quartile*

Percentiles

The lower quartile, median, and upper quartile can be described as the three quartiles that divide the data into approximately four equal parts (approximately, as some quartiles lie on the boundaries). If we have a large set of data, we may wish to divide the ordered data into roughly 100 equal parts using 99 *percentiles*. For example, about 1% of the observations lie below the 1st percentile, and about 1% lie above the 99th percentile. The lower quartile is the 25th percentile, as 25% of observations lie below this value. Similarly, the median is the 50th percentile, and the upper quartile is the 75th percentile. The observations up to, say, the 20th percentile represent the lower 20% of the data.

✖ Example 2.4.6 (cont.) *Five-Number Summary*

We have just found Q_1, Med, and Q_3 for the traffic deaths data. Returning to Fig. 2.3.7(a), we find that the minimum and maximum are respectively 5.4 and 26.8. Thus we have the five-number summary (in deaths per 100,000 of population):

$$\text{Min} = 5.4 \quad Q_1 = 11.3 \quad \text{Med} = 13.05 \quad Q_3 = 18.6 \quad \text{Max} = 26.8$$

QUIZ FOR SECTION 2.4.2

1. What is the five-number summary?
2. What are the quartiles intended to do?
3. How are the quartiles defined?
4. What type of plot provides a way of ordering the data from smallest to biggest by hand?

EXERCISES FOR SECTION 2.4.2

1. Find the five-number summaries for the following data:

   (a) 1 2 2 5 8 8 11 12 14

   (b) 1 3 5 9 9 11 13 13 18 20

2. A sample of 20 men aged under 20 from the workforce study conducted by Dr. Patricia Metcalf and Dr. Robert Scragg yielded the following waist-to-hip ratios. Find the five-number summary for the data.

1.04	0.85	0.86	0.83	0.88	0.89	0.91	0.88	0.87	0.87
0.68	0.84	0.83	0.87	0.93	0.75	0.93	0.99	0.93	0.88

Data courtesy of Dr. P. Metcalf and Dr. R. Scragg, U. of Auckland.

3. In problem 4 of Exercises 2.4.1 you constructed a stem-and-leaf plot of the ages of the patients in Table 2.1.1 who were alive at last follow-up and did not have surgery. Use your plot to compute the five-number summary for the ages of these patients.

4. If you were told that, for your age and sex, you were between the 76th and 77th percentile for your height and between the 45th and 46th percentile for your weight, what sort of body shape would you have?

2.4.3 Measuring the "Spread" of the Data

The Range

The *range* is the simplest measure of how spread out the data are, being simply the difference between the largest and smallest sample values.

$$\text{Range} = \text{Largest observation} - \text{Smallest observation}$$

Although the range is easy to compute, it is severely affected by the occasional extreme observation (outlier). We do not use it except in control charts that are designed for monitoring the variability of processes over time. In this application simplicity and computational speed are sometimes a great advantage (see Chapter 13 on our web site).

The Interquartile Range

A compromise that retains some of the simplicity and interpretability of the range, but is much less sensitive to the presence of outliers, is the interquartile range (or midspread). This is defined as the difference between the upper and lower quartiles:

$$\text{Interquartile range} = \text{Upper quartile} - \text{Lower quartile}$$

or

$$\boxed{\text{IQR} = Q_3 - Q_1}$$

Thus the interquartile range is the range spanned by the central half of the data. As with the median, IQR is not disturbed by the presence of a few very large or very small observations (outliers).

EXAMPLE 2.4.6 (cont.) *Finding the Quartiles and Interquartile Range*

We found that $Q_1 = 11.3$ and $Q_3 = 18.6$ for the traffic deaths data. Thus,

$$\text{IQR} = 18.6 - 11.3 = 7.3 \text{ deaths per } 100{,}000$$

This tells us that the middle half of the observations has a range of 7.3 deaths per 100,000 of population.

The Sample Standard Deviation

The sample standard deviation, denoted s_X, is probably the most popular measure of spread. Although this measure is not easy to interpret, it plays a fundamental role in the statistical inference and decision making used later in this book. You therefore need to become familiar with it. The formula for the sample standard deviation is[15]

$$s_X = \sqrt{\frac{(x_1 - \overline{x})^2 + (x_2 - \overline{x})^2 + \cdots + (x_n - \overline{x})^2}{n-1}} = \sqrt{\frac{1}{n-1}\sum(x_i - \overline{x})^2}$$

To see how the formula works, we consider the batch of five numbers:

5 2 3 4 8

These numbers have sample mean $\overline{x} = 22/5 = 4.4$, so

$$s_X^2 = \frac{1}{4}\left[(5-4.4)^2 + (2-4.4)^2 + (3-4.4)^2 + (4-4.4)^2 + (8-4.4)^2\right] = 5.3$$

Thus $s_X = \sqrt{5.3} = 2.30$.

Of course, these days we don't need to use the preceding formula to calculate sample standard deviations, as computers and many hand calculators do it for us automatically.

Interpreting the Sample Standard Deviation

We stated earlier that the sample standard deviation is not particularly easy to interpret. We want you simply to think of it as a measure of spread and use the following results to obtain a feeling for its size. The standard deviation is zero only if there is no spread (i.e., all observations are identical). Also, for many unimodal, moderately symmetric sets of data, experience has taught us that the following rules usually give reasonably close approximations.[16]

[15]Some books use a divisor of n instead of $n - 1$ in some situations. Hand calculators usually allow for both divisors. Most software packages use $n - 1$. It makes little practical difference for $n > 10$, however.

[16]In addition to the 68%–95% rule for "well-behaved samples," there is a weaker rule that follows from a mathematical result called Chebyshev's inequality and tells us that *for any sample, at least 75%* of the observations will lie within two standard deviations of the mean and *at least 89%* will lie within three standard deviations. (See also Section 13.1.2 on our web site.)

68%–95% Rule

Approximately,

- **68%** of the data lie within **1** sd of $\overline{x}$, that is, between $\overline{x} - s_X$ and $\overline{x} + s_X$.
- **95%** of the data lie within **2** sd's of $\overline{x}$, that is, between $\overline{x} - 2s_X$ and $\overline{x} + 2s_X$.

  (Note: Here *sd* is used to abbreviate *standard deviation.*)

For the lengths of the 40 female coyotes (Table 2.3.2), $\overline{x} = 89.24$ cm and $s_X = 6.548196$ cm, so the rule would lead us to expect that roughly 68% of observations would lie between $\overline{x} - s_X \approx 82.7$ cm and $\overline{x} + s_X \approx 95.8$ cm. In fact, 29 of the 40 observations, or 72.5%, lie within these limits. Similarly, we would expect that roughly 95% would lie between $\overline{x} - 2s_X \approx 76.1$ cm and $\overline{x} + 2s_X \approx 102.3$ cm; the actual percentage is 92.5%.

Not only is the standard deviation difficult to interpret directly, it is also sensitive to extreme observations. (Can you see why?) You might ask why so many people use such an "unfriendly" measure. The reason is bound up in the development of statistical theory based on a very important symmetric, bell-shaped curve for finding probabilities called the Normal distribution. We consider this distribution in Chapter 6.

Many textbooks emphasize the *square* of the sample standard deviation, s_X^2, which is called the ***sample variance.*** For large n, $n - 1 \approx n$, and s_X^2 is approximately $\frac{1}{n}\sum(x_i - \overline{x})^2$, which is the average of the squared distances of the observations from the mean. However, a measure of spread must be in the same units as the observations (and $\overline{x}$), so we must take the square root of the variance.

QUIZ FOR SECTION 2.4.3

1. What feature of the data is each of the following summary statistics trying to measure: the interquartile range? the range? the standard deviation?
2. How is each of the quantities in (1) defined?
3. What does the 68%–95% rule tell you about the standard deviation as a measure of spread?

EXERCISES FOR SECTION 2.4.3

1. Find the sample mean and standard deviation for the end-diastolic volumes of the patients in Table 2.1.1 who were taking beta blockers. (You will need to use the variable BETA to identify them.)
2. Use the data on the lengths of *male* coyotes given in Table 2.3.2.
   (a) Find the sample mean and sample standard deviation of these data. How do these values compare to those for the female coyotes given in the text? Which data set is more variable? If you randomly selected a female and a male coyote from each of the two data sets, what could you say about their lengths?
   (b) A rule of thumb in the text says that often about 68% of observations lie within one standard deviation of the mean and 95% within two standard deviations. What is the true percentage of these data that lie within one standard deviation of the mean? within two standard deviations?
   (c) Find the range and the interquartile range. Another rough rule of thumb says that the sample standard deviation is often about three quarters (75%) of the interquartile range. How well does this approximation work here?

2.4.4 Plotting Data Summaries: The Box Plot

The five-number summary can be represented pictorially using a box plot. In Fig. 2.4.3 we have a box plot produced from the data set SYSVOL in Table 2.1.1.

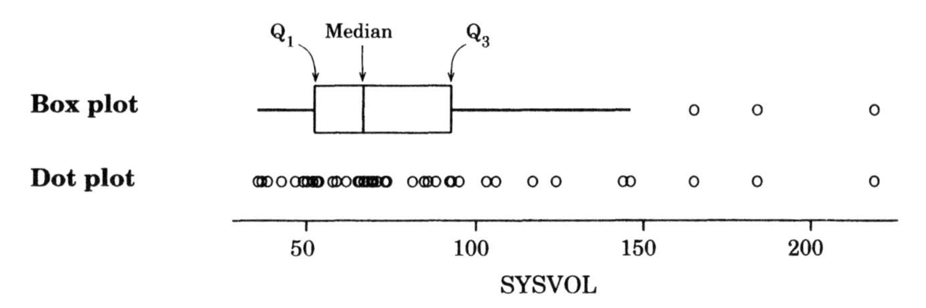

FIGURE 2.4.3 Box plot for SYSVOL.

We now describe how such a plot is constructed using Fig. 2.4.4. We briefly give details for drawing a box plot by hand to give you a better idea of what the various elements represent. To draw a box plot, we start with a horizontal or vertical scale that encompasses the range of the data.[17] On or by that scale we carry out the following:

1. Draw a box extending from the lower quartile to the upper quartile, with a line across marking the position of the median.
2. Calculate $Q_1 - 1\frac{1}{2} \times \text{IQR}$. Find the smallest observation that is no smaller than this value. Mark its position with a small cross and draw a line connecting the cross to the edge of the box. This gives the left-hand "whisker."[18]
3. Similarly, to construct the right-hand whisker, find the largest observation that is within $1\frac{1}{2}$ interquartile ranges of the upper quartile, mark its position with a cross, and connect the cross to the box.
4. Any observations that fall outside the whiskers are plotted individually. These observations are called ***outside values*** and should be labeled for further investigation or checking. They are often *but not necessarily* outliers. The box plot is designed so that approximately 99.5% of observations fall inside the whiskers for data from a Normal distribution. (Normal distributions are discussed in Chapter 6.)

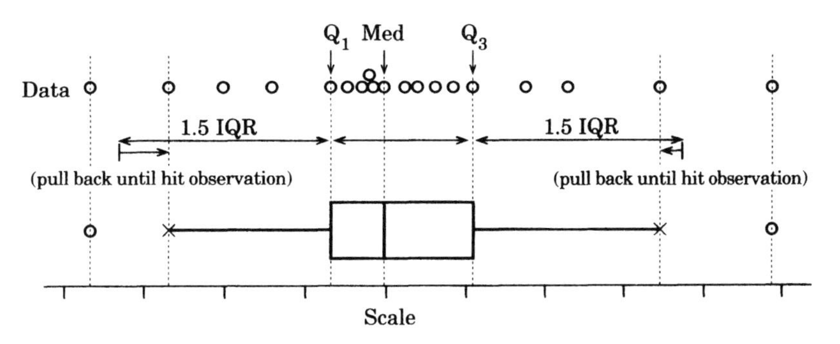

FIGURE 2.4.4 Construction of a box plot.

[17] Different statistical computer packages construct box plots differently. The main variation occurs in the definition of the quartile. See Frigge et al. [1989].

[18] The positions where the whiskers end are sometimes called the ***adjacent values.***

✱ Example 2.4.7 *Plots for the Breaking Strengths of Gear Teeth*

The breaking strengths of gear teeth in certain positions of a gear are given as a stem-and-leaf plot, a box plot, and a dot plot in Fig. 2.4.5. This is a portion of a larger data set given later in Example 3.2.2 and relates to strength measurements taken on teeth in positions 4 and 10. From the stem-and-leaf plot we notice the gap in the plot between 2040 and 2100. Also, all the plots reveal two large outliers at 2520 and 2590. Clearly, a box plot has much less information than a stem-and-leaf plot. For example, the separation of the data in the stem-and-leaf into a lower block from 1980 to 2040 and a much bigger block from 2100 to 2340 has been lost in the box plot. The lower outside value in the box plot is not an outlier and is close to the whisker, whereas the two large outside values appear to be outliers and stand out clearly as such.

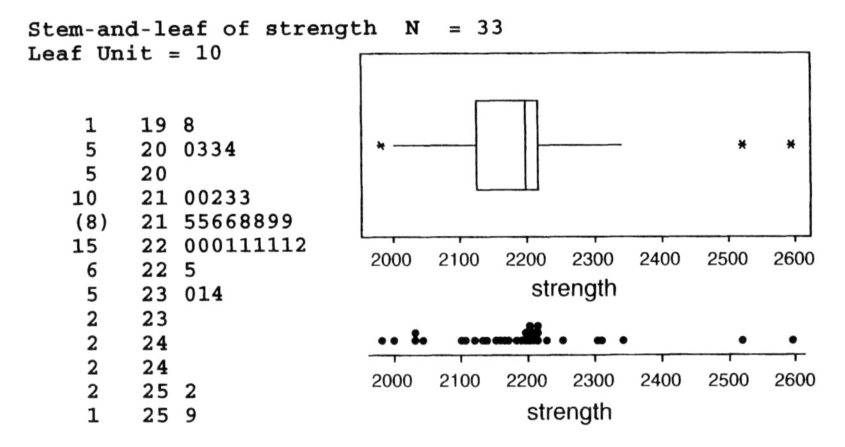

FIGURE 2.4.5 Three graphs of the breaking-strength data for gear teeth in positions 4 and 10 (Minitab output).

The real point in drawing box plots is to provide a quick visual summary for comparing *several* sets of data, as we do in Section 3.2. When comparing a very large number of batches, we tend to concentrate on the most prominent features, such as location and spread, which the box plot shows quite well. Box plots also show outliers (as we have seen) and skewness, as in Fig. 2.4.6(a). However, a wide box

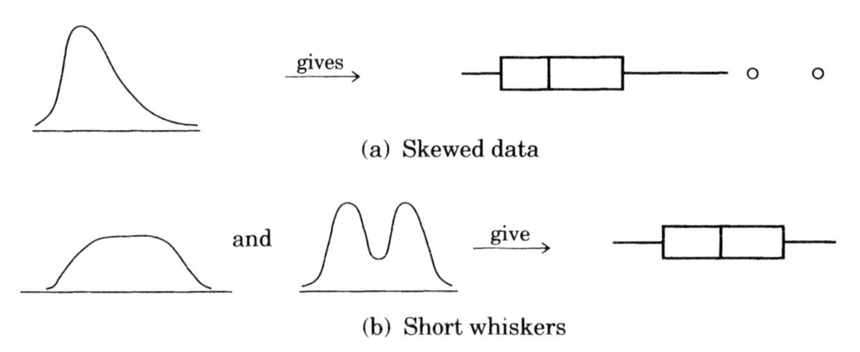

FIGURE 2.4.6 Relating the stem-and-leaf to the box plot.

with short whiskers could be coming from a bimodal distribution or a very short-tailed distribution, as shown in Fig. 2.4.6(b).

A slight modification of the box plot that we like for data sets of about 15 to 40 observations is shown later in Fig. 3.2.7, where we have superimposed box plot information on top of the dot plot to give the best of both worlds. We shall call this type of plot a ***dot–box plot.***

QUIZ FOR SECTION 2.4.4

1. What features of the data are summarized by the central box in the box plot?
2. How far out from each end of the box do you go to plot the whiskers?
3. What are outside values, and are they necessarily outliers?

EXERCISES FOR SECTION 2.4.4

1. The dietary intake of carbohydrate (in mg per day) for a random sample of 20 women aged under 40 taken from the workforce study of Drs. Metcalf and Scragg is given below.

199	162	327	145	149	351	453	374	287	151	201	375	223	230	193
229	206	144	152	164	121	190	158	145	129	168	173	189	589	247

Data courtesy of Dr. P. Metcalf and Dr. R. Scragg, U. of Auckland.

Construct a box plot of the data. What do you conclude?

2. The following numbers are the times (in seconds) taken for a computer to complete a file lookup task using a large database.

2.0	2.4	1.6	4.0	1.3	1.4	3.4	2.1	5.1	5.4	3.3	2.9
8.5	2.9	1.7	3.8	0.8	0.7	1.4	1.5	1.8	2.2	1.6	

Data courtesy of E. Wild.

Construct a dot plot and a box plot for this set of numbers. What do the plots tell you?

2.5 REPEATED AND GROUPED DATA

2.5.1 Repeated Data (Discrete Variables)

As stated in Section 2.1, a quantitative variable is treated as discrete (rather than continuous) for the purposes of data analysis if it takes a relatively small number of distinct values, with some or all of them often repeated. Such data are typically presented using a *frequency table* and depicted using a *bar graph*. We shall show you how to compute a sample mean and standard deviation from a frequency table.

EXAMPLE 2.5.1 *Mean Word Length*

Consider the set of 100 data points given in Table 2.5.1. These are the word lengths of the first 100 words on a randomly chosen page from a book written by one of the authors.[19]

TABLE 2.5.1 Word Lengths for the First 100 Words on a Randomly Chosen Page

3	2	2	4	4	4	3	9	9	3	6	2	3	2	3	4	6	5	3	4
2	3	4	5	2	9	5	8	3	2	4	5	2	4	1	4	2	5	2	5
3	6	9	6	3	2	3	4	4	4	2	2	4	2	3	7	4	2	6	4
2	5	9	2	3	7	11	2	3	6	4	4	7	6	6	10	4	3	5	7
7	7	5	10	3	2	3	9	4	5	5	4	4	3	5	2	5	2	4	2

There are only 11 distinct numbers in this data set, and most of them occur repeatedly. This type of behavior is quite common with data consisting of counts. We shall now summarize this data using the following frequency table obtained from Minitab:

Value u_j	1	2	3	4	5	6	7	8	9	10	11
Frequency f_j	1	22	18	22	13	8	6	1	6	2	1

In other words, 1 appears once in the data, 2 appears 22 times, 3 appears 18 times, and so on. The sample mean (average) of the numbers in the table is

$$\begin{aligned}\bar{x} &= \frac{1}{n}(x_1 + x_2 + \cdots + x_{100}) \\ &= \frac{1}{100}(1 + \overbrace{2 + 2 + \cdots + 2}^{22\text{ times}} + \overbrace{3 + 3 + \cdots + 3}^{18\text{ times}} + \cdots + 11) \\ &= \frac{1}{100}(1 \times 1 + 2 \times 22 + 3 \times 18 + \cdots + 11 \times 1) \\ &= 4.35\end{aligned}$$

Using this information to infer a general pattern, we get the following formula for calculating sample means from repeated data:[20]

$$\boxed{\bar{x} = \frac{1}{n}\text{Sum of (value} \times \text{frequency of occurrence)}}$$

The standard deviation is more complicated.[21] Many hand calculators allow you to enter values and frequencies and will automatically calculate means and standard

[19] Data such as word lengths, sentence lengths, occurrences of words such as *and* and *the*, and so on can be used to quantify that elusive quality of writing called *style*.

[20] Mathematically, we have $\bar{x} = \frac{1}{n}(u_1 f_1 + u_2 f_2 + \cdots + u_k f_k) = \frac{1}{n}\sum_{j=1}^{k} u_j f_j$. Here k is the number of distinct values in the data ($k = 11$ in the example), the u_j's are the distinct values of the x's, and the frequency f_j is the number of times u_j occurs in the data. We note that $\sum_j f_j = n$.

[21] For the interested reader, the standard deviation is given by $s_X = \sqrt{\frac{1}{n-1}\sum_{j=1}^{k} f_j(u_j - \bar{x})^2}$.

deviations from data in this format. We note that if we add the frequencies of all the distinct values, we get the total number of observations in the data set.

The frequencies add to n.

The preceding formulas in the boxes (and footnotes) are useful only for computing a mean and standard deviation using an unsophisticated hand calculator when the data is available in the form of a frequency table. For example, it is faster to compute

$$\frac{1}{100}(1 \times 1 + 2 \times 22 + \cdots + 11 \times 1)$$

than it is to to add up all 100 observations individually.

Actually, we rarely use the formula for the mean explicitly. We mention it here because we use it to motivate a theoretical idea called the *expected value* in Chapter 5. In practice, once you have the file of data, it does not matter whether you have repeated data or not when it comes to finding such statistics as the mean and standard deviation.

Frequency Tables and Bar Graphs

We now use another example to discuss the use of frequency tables and bar graphs in more detail.

EXAMPLE 2.5.2 *Frequency Table for the Depths of Deep-Living Fish*

Table 2.5.2 is a frequency table taken from Haedrich and Merrett [1988], who investigated species of deep-living fish in the North Atlantic Ocean. Regions of the ocean were divided into 45 zonal strata. Table 2.5.2 tells us that 117 species were found in just a single stratum, 61 species were found in two strata, 37 in three, and so on. There were 330 species altogether. We have depicted the information in Table 2.5.2 in a bar graph in Fig. 2.5.1.

The frequency table. In the frequency table, Table 2.5.2, we have four columns:

- **Column 1** gives the u_j's, the distinct values taken by the variables. (In Example 2.5.2 the variable being used is the number of strata containing a given species.)
- **Column 2** gives the f_j's, that is, the frequencies, or the number of times u_j is observed. (In Example 2.5.2 this is the number of species found in u_j strata.)
- **Column 3** gives the relative frequencies f_j/n expressed as a *percentage*.[22] It tells us the percentage of the sample for which the variable takes the value u_j. If we just used the relative frequencies without converting them to percentages, that would tell us about the *proportion* of the sample for which the variable takes the value u_j.

[22] Multiply the relative frequency by 100%.

TABLE 2.5.2 Frequency Table for the Occurrence of Fish Species in Ocean Strata

No. of strata in which species occur (u_j)	Frequency (No. of species) (f_j)	Percentage of species $\left(\frac{f_j}{n} \times 100\right)$	Cumulative[a] percentage
1	117	35.5	35.5
2	61	18.5	53.9
3	37	11.2	65.2
4	24	7.3	72.4
5	23	7.0	79.4
6	12	3.6	83.0
7	14	4.2	87.3
8	10	3.0	90.3
9	9	2.7	93.0
10+	23	7.0	100.0
	$n = 330$	100.0	

[a]The cumulative percentage column gives the true cumulative percentages rounded to one decimal place, which sometimes gives an answer slightly different from that obtained by adding the rounded numbers in the percentage column.

Source: Haedrich and Merrett [1988]

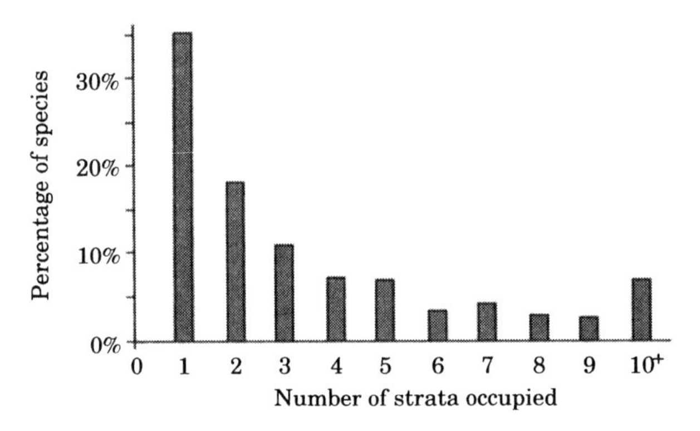

FIGURE 2.5.1 Bar graph for species data.

- **Column** 4 gives the *cumulative percentage*. It tells us the percentage of the data for which the variable takes a value which is u_j or less. (e.g., from Table 2.5.2, 53.9% of species are found in two or fewer strata, and 72.4% of species are found in no more than four strata.)

This is a fairly common layout for a frequency table for a discrete variable. Although not all published tables include all four columns, this is what we prefer. Notice that we have combined the strata 10, 11, 12, ... for ease of presentation. If one species, for example, happened to be found in 20 strata, then our table would need to be twice as long to show all the data, and there would probably be a lot of zero entries in the frequency column (f_j). However, is it a good idea to combine data in this way? Clearly, you cannot use Table 2.5.2 to calculate a sample mean, as some of the u_j have been "lost" by pooling. This again emphasizes the need to keep all the data for calculations rather than for display. Also, a scientist would be particularly interested in these other values, so they should be presented somewhere in the report.

The bar graph. A bar graph or bar chart, as in Fig. 2.5.1, differs from a histogram in that the rectangles are not joined up. This visually emphasizes the fact that the values the variable takes are distinct and that each rectangle represents a single value. In contrast, rectangles that are joined up, as in a histogram, visually suggest that the variable can take values anywhere along the axis. It is bad practice to use joined-up rectangles with data that is clearly discrete.

In Fig. 2.5.1 we used the percentage column of Table 2.5.2 as our vertical scale. We might instead have used the frequencies (column 2). It makes no difference to the shape of the plot. The choice of scale depends on whether we want the reader to take note of the actual counts, or just to see how the values are distributed across the sample. If we were thinking in terms of the data being representative of a larger population and it was that larger population that was of real interest, then we would always use percentages (or proportions) as our vertical scale. We could also use the percentages from the sample to give us an idea of the corresponding percentages for the whole population.

Computing. Bar graphs are produced by all general-purpose statistical packages, for example, under `Graph` → `Chart` in Minitab, using the `barplot` command in Splus and R, and using the chart wizard in Excel. Frequency tables are also produced from raw data, for example, in Minitab under `Stat` → `Tables` → `Tally`, by the `table` command in Splus and R, and in Excel under `Tools` → `Data Analysis` then choosing `Histograms`.

QUIZ FOR SECTION 2.5.1

1. What sort of table and graph do you use for summarizing repeated discrete data?
2. In what way does your graph of the previous question differ from a histogram? Why the difference?
3. Describe what is meant by a cumulative percentage.

EXERCISES FOR SECTION 2.5.1

1. In a 1910 study of the emission of alpha particles from a polonium source, Rutherford and Geiger counted the number of particles striking a screen in each of 2608 time intervals of length one eighth of a minute. In Table 2.5.3 we have the number of time intervals in which 0, 1, 2, 3, ... particles had been observed.
   (a) Assuming that all six observations in the category 11+ are actually 11s, compute the sample mean of the number of particles observed in a time interval.
   (b) Construct a bar graph of these data.
   (c) Add a percentage and cumulative percentage column to the table.
   (d) In what percentage of the time intervals were exactly four particles observed?
   (e) In what percentage of the time intervals were four or fewer particles observed?
   (f) What is the smallest number u of particles for which we can say that for at least 90% of time intervals, there were no more than u particles observed?
2. Does it make sense to compute the median, quartiles, and interquartile range for discrete data? Justify your answer by referring to Table 2.5.2.

TABLE 2.5.3 Rutherford and Geiger's Data

Number of particles u_j	Observed frequency f_j	Observed proportion f_j/n
0	57	0.022
1	203	0.078
2	383	0.147
3	525	0.201
4	532	0.204
5	408	0.156
6	273	0.105
7	139	0.053
8	45	0.017
9	27	0.010
10	10	0.004
11+	6	0.002
	2608	

2.5.2 Grouped Data (Continuous Variables)

With many practical studies, the actual value of an observation is not recorded, only which class interval it is in. For example, in questionnaires the age classes (such as 20–24 years) rather than individual ages are used. Even if individual values are recorded, official agencies often publish such data in summarized form as a frequency table. It is therefore useful to be able to estimate means and standard deviations using such tables.

Table 2.5.4 is a frequency table for the female coyote lengths given in Table 2.3.2. We can estimate the sample mean and standard deviation of the original data by treating each observation in a class interval as if it fell at the midpoint of the interval and using the method for obtaining the mean and standard deviation of repeated data from Section 2.5.1. Applying this idea to Table 2.5.4, we get a sample mean of 89.625 and a sample standard deviation of 6.59. These may be compared with the real (ungrouped)

TABLE 2.5.4 Frequency Table for Female Coyote Lengths

Class interval	Midpoint (u_j)	Frequency (f_j)
$70\text{–}75^-$	72.5	2
$75\text{–}80^-$	77.5	0
$80\text{–}85^-$	82.5	6
$85\text{–}90^-$	87.5	12
$90\text{–}95^-$	92.5	13
$95\text{–}100^-$	97.5	5
$100\text{–}105^-$	102.5	2
Total		40

values of 89.24 and 6.55 respectively. These estimates aren't bad, considering how coarsely the data in the frequency table has been grouped.

EXERCISES FOR SECTION 2.5.2

The dietary intakes of carbohydrate (in milligrams per day) for each of a sample of 62 women aged under 40 were recorded, and the frequency table, Table 2.5.5, was constructed.

1. Use your frequency table to estimate the sample mean (and standard deviation if you can) of the original numbers. Compare your answers from this grouped data method with the actual sample mean and standard deviation of 229.73 and 94.02, respectively, computed from the original 62 observations.
2. Add a column of relative frequencies, and draw a relative frequency histogram. What shape does it have? What proportion of the females in the sample have a daily intake:
   (a) of less than 300 mg/day?
   (b) between 300 and 450 mg/day?
   (c) of 450 mg/day or more?
3. What would happen to your relative frequency histogram if you combined the highest four class intervals into a single interval 400–600?

TABLE 2.5.5 Female Daily Carbohydrate Intake (mg/day)

Class interval	Midpoint (u_j)	Frequency (f_j)
100–150$^-$	125	14
150–200$^-$	175	14
200–250$^-$	225	13
250–300$^-$	275	10
300–350$^-$	325	5
350–400$^-$	375	3
400–450$^-$	425	1
450–500$^-$	475	1
500–550$^-$	525	0
550–600$^-$	575	1
Total		62

Data courtesy of Dr. Patrica Metcalf.

2.6 QUALITATIVE VARIABLES

In the data on heart attack patients that we used to begin this chapter, there were three qualitative variables. One of those variables, SURG, recorded whether the patient had surgery including the reason for surgery. It had levels coded as 0 = no surgery performed, 1 = surgery as part of trial, 2 = surgery for symptoms within one year, 3 = surgery for symptoms within one to five years, 4 = surgery for symptoms after five years.

How do we investigate data on a single qualitative variable?

We simply count how many individuals (or objects) in our sample fall into each class. We recommend presenting the results in the form of a ***frequency table*** and depicting them using a ***bar graph***. When presenting such tables and graphs in a report, however, it is best to try to avoid using the numeric codes that may have been used to record the data. If at all possible, we should use informative verbal descriptions of the categories or classes that form the levels of the variable. Packages generally handle qualitative variables in the same way as we described for discrete variables earlier. Most handle character values (names) as well as numeric codes.

A frequency table is given for the variable SURG in Table 2.6.1 and a bar graph is given in Fig. 2.6.1. Although we have used the numeric code for SURG in the graph, we have also added labels. We emphasize that the table given relates to the whole data set, not just the portion given in Table 2.1.1. Note also that we have used *percentage* for the vertical scale of the bar graph rather than *frequency* because the actual frequencies are just a reflection of the sample size, whereas percentages give us an idea of the proportions of patients who are likely to fall into each class in the future.

Ordering Categories by Size

There was a natural ordering in the levels of the variable SURG that we did not feel we could ignore. We would have wanted to keep "under one year," "one to five years," and "after five years" in that order regardless of what happened to the percentages. Just as with the tables in Section 2.2, however, if there is no compelling order for the categories, we should order them by size, with the most important ones (i.e., the biggest percentages) coming first.

TABLE 2.6.1 Frequency Table for the Variable SURG

	SURG	Frequency	Percentage	Cumulative percentage
No surgery performed	0	409	66.4	66.4
Surg. as part of trial	1	89	14.4	80.8
Surg. for sympt. within 1 year	2	72	11.7	92.5
Surg. for sympt. 1 to 5 years	3	29	4.7	97.2
Surg. for sympt. >5 years	4	17	2.8	100.0
		616	100.0	

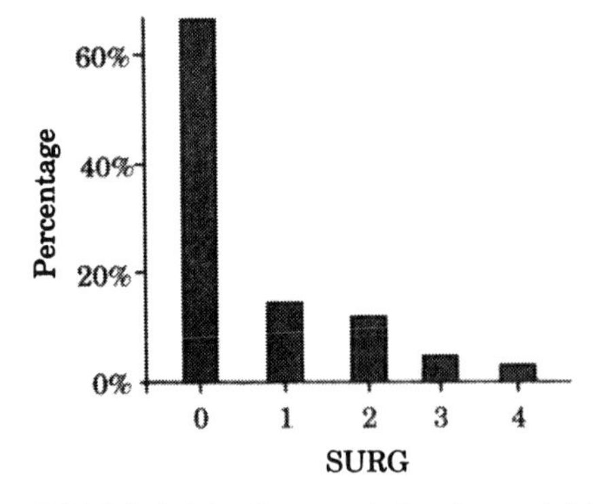

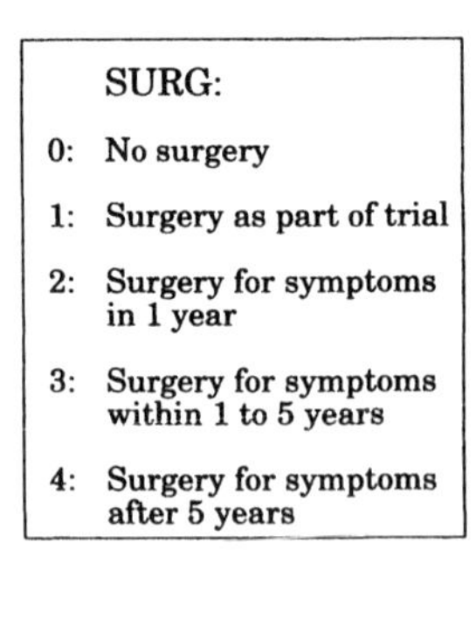

FIGURE 2.6.1 Bar graph for the variable SURG.

To date we have used bar graphs only for representing frequency tables from a discrete or a qualitative variable. Besides displaying frequencies (counts), they can be used to display amounts of money, time, or any other quantitative variable where the intent of the presenter is to relate labels to relative importance or relative size.

> ***Use bar graphs to relate labels to relative size.***

For example, Fig. 2.6.2 plots the cost of commercial rents in 24 major cities around the world.[23] Note how the labels are ordered according to the size of the variable *with the biggest on the left.*

> Where possible, ***order items by size.***

When the plot is organized this way, our attention is automatically focused on the most important items. We see immediately where the most expensive (and cheaper) rents are in Fig. 2.6.2. Although the rents are given in New Zealand dollars, the relative heights are the same in any currency. To convert to any other currency, you simply change the numbers on the vertical scale of the graph.

This data could also have been represented using a labeled dot plot like Fig. 2.3.5. The dot plot is good at showing distances between points and therefore features such as gaps, clusters, and outliers. The bar graph is better for displaying multiplicative information. For example, we can see from Fig. 2.6.2 that rents in Tokyo and the West End of London were much higher than those elsewhere, with London rents being about 90% of Tokyo rents and more than twice as expensive as those in Paris.

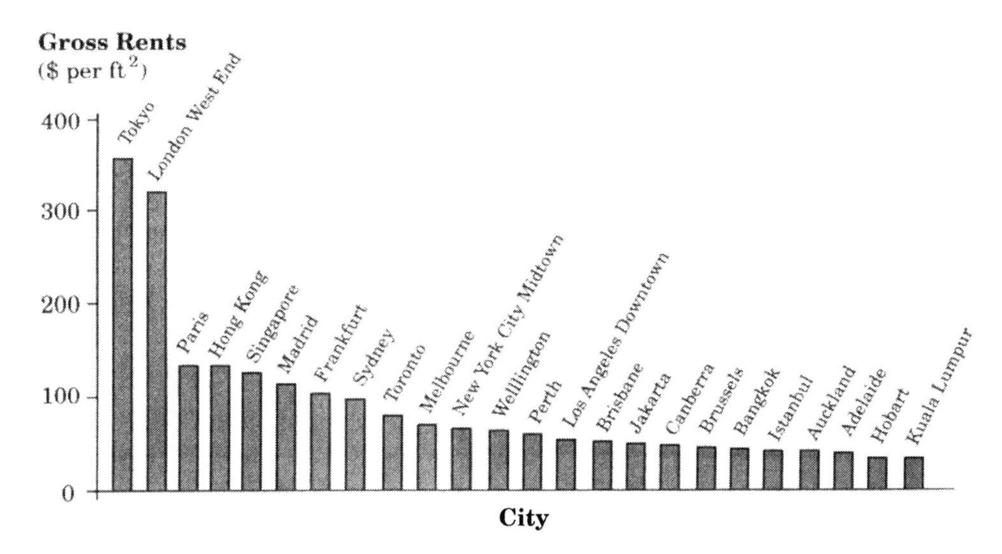

FIGURE 2.6.2 Cost of commercial rents around the world.

[23] Source: Redrawn from *New Zealand Herald* (20 Feb. 1991, Sect. 5, p. 2).

2.6.1 Other Forms of Graphs

The ***pie chart,*** cf. Fig. 2.6.3(b), is frequently used in reports for relating labels to relative size. It is relevant for displaying the "measurement" for each object (or individual) as a proportion of the total, that is, as a proportion of the sum of all the "measurements." The "measurements" can be counts from a frequency table or continuous variables, such as time, amount of money, distance, and so on, but the idea of *proportion of the whole* has to make sense.[24] The ***segmented bar*** graph, Fig. 2.6.3(c), is another form of graph that performs the same function as a pie chart. Here, the length of a bar is subdivided according to the proportions being plotted. Note that at the time of writing all of the types of graph in Figs. 2.6.3 and 2.6.4 are produced by word processors such as Microsoft Word as well as by spreadsheet software such as Excel and statistical packages.

EXAMPLE 2.6.1 *Various Graphs for the World Gold Production*

We return to the figures on world gold production given in Table 2.2.3 and ask, "Where is most of the gold being produced?" Table 2.6.2 gives the percentages of world gold being produced by various countries. Data are given for every second year from 1983 to 1991, but for the time being we shall just look at the situation in 1991 (later exercises will look for trends over time). Categories (in this case countries) have been ordered by the size of their percentage of world gold production in 1991 so that the

TABLE 2.6.2 Percentages of the World's Gold Production[a]

Country	1983	1985	1987	1989	1991
S. Africa	48.6	43.8	36.2	30.8	28.7
U.S.	4.4	5.0	9.3	13.4	13.9
USSR	19.1	17.7	16.7	14.4	11.5
Australia	2.2	3.8	6.7	10.3	11.2
Canada	5.3	5.7	7.0	8.0	8.3
China	4.1	4.0	4.3	4.0	5.7
Rest	16.3	20.2	19.7	19.0	20.8

[a]Derived from Table 2.2.3

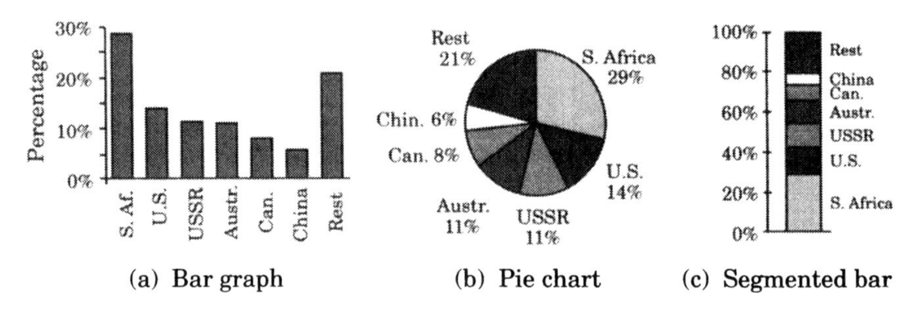

FIGURE 2.6.3 Percentages of the world's gold production in 1991.

[24]The angle in the pie for representing the ith "measurement" x_i is $\frac{x_i}{T} \times 360°$, where T is the total $\sum x_i$.

order of importance of the various producers is obvious to the reader. The "Rest" category has been placed last. There are no major producers incorporated in the 20.8% of world production ascribed to "Rest." It is made up of several countries each producing relatively small amounts. The 1991 percentages are displayed in Fig. 2.6.3 as (a) a bar graph, (b) a pie chart, and (c) a segmented bar. All three figures give a visual impression of the relative sizes of each country's gold production, which for most people communicates itself much more quickly to the brain than the numbers themselves.

Exercise: Table 2.6.2 was derived from the data in Table 2.2.3. Can you see how it was done?

2.6.2 Choosing between Types of Graphs

The pie chart and the segmented bar are appropriate only when you are trying to convey the idea of proportions of a single whole. The bar graph does not convey this idea at all. Research into how accurately people read graphic information, however, shows that subjective impressions of relative size are communicated more effectively by bar graphs than pie charts or segmented bars or rods. This can be seen to some extent by inspecting the graphs in Fig. 2.6.4, all of which depict a situation where group A makes up 22% of the whole and the other percentages are 13% (B), 23% (C), 7% (D), 25% (E), and 10% (F). The small differences between 22%, 23%, and 25% are readily visible in the bar graph Fig. 2.6.4(a) but are hard to detect in the pie chart Fig. 2.6.4(c).

The attempt to add three-dimensional perspective to graphs, as in the perspective bar graphs [Fig 2.6.4(b)] and perspective pie charts [Fig 2.6.4(d)], tends to make reading scales and judging relative sizes even more difficult. On a perspective pie chart, impressions of relative size are further influenced by such things as the placing around the pie and the colors used to distinguish the pieces of the pie.

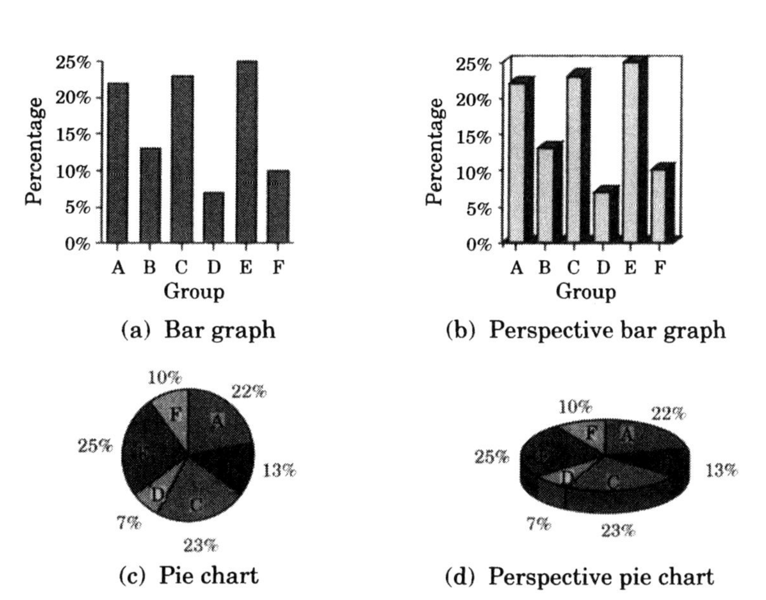

FIGURE 2.6.4 Comparisons of plots.

Although simple bar graphs convey relative size information more effectively than the other types of plot presented in Figs. 2.6.3 and 2.6.4, the editors of magazines, newspapers, and prospectuses love the "three-dimensional" shapes because they look so attractive on a page (even more so in color). This is an example of a general conflict between the desire to communicate information and the desire to produce visually appealing reports. Unfortunately, the variety of devices used in the press for turning simple graphs into interesting graphic designs (e.g., three-dimensional) almost invariably result in obscuring and even distorting the information being presented; see Tufte [1983, 1990] for examples and a broad discussion of graphic presentation.

2.6.3 Pareto Chart

EXAMPLE 2.6.2 ***Listing the Major Defects in Electric Fans***

The data given in Table 2.6.3 represent a list of all major defects found on electric fans after assembly using 100% inspection over 27 weeks. The number and proportion of each type of defect is listed.

TABLE 2.6.3 Major Defects Found in Electric Fans over 27 Weeks

Problem	Number	% of total[a]
Excess weld	3	2.86
Cell spring broken	2	1.90
Latch spring bent	1	0.95
Latch spring weld	2	1.90
Latch broken	4	3.81
Paint scratched	51	48.57
Paint on spring	5	4.76
Defective cell	10	9.52
Filter rail broken	5	4.76
Extra switch	2	1.90
Screen not seated	3	2.86
Screen rusted	1	0.95
Faulty capacitor	2	1.90
Faulty power pack	3	2.86
Flow lines in louver	9	8.57
Loose cell guides	2	1.90
Total	105	99.97

[a]Note that the total percentage is 99.97% rather than 100% because of rounding error.

When it comes to finding and fixing faults in a system like the one in Example 2.6.2, total quality management practitioners talk about "distinguishing the important few from the trivial many." To make a real improvement in the process, the important few causes of the problems have to be remedied. Time and money spent on the trivial many will have very little impact. The ***Pareto chart*** for the data in Table 2.6.3 is essentially a bar graph with the items ordered by size, as in Fig. 2.6.5. Ordering the fault types by size, as this Pareto chart does, focuses the attention of the improvement team on the most important types.

What data can we use in a Pareto chart? Having categorized the faults in a system, we collect data on all faults for a period of time, recording for each fault type data such as the number of faults, the cost to repair the faults caused, time that the assembly line has had to shut down because of the faults caused, or the monetary cost to the organization of the fault type. Any of these measures can be used for constructing the Pareto chart. It is important that the measure chosen reflects the real impact of each type of fault on the operation, however. Numbers or percentages of faults, as used in Table 2.6.3, make sense if the impact of each fault is roughly the same, regardless of its type.

✖ Example 2.6.2 (cont.)

Reordering the percentages in Table 2.6.3, we get Table 2.6.4. We note that almost half of the faults correspond to paint being scratched, so the top priority will be to try to find and fix whatever is causing the scratching. Next in importance come defective cells and flow lines. Together, the top three problems make up two thirds of all defects. Also, 11 types of defect generally contribute only a few percent each, though collectively they contribute 23.82% (see "All other types"). Table 2.6.4 is best displayed graphically as Fig. 2.6.5, which we have noted is called a Pareto diagram or chart.

TABLE 2.6.4 Major Defects in Order of Importance

Problem	% of total	Cumulative % of total
Paint scratched	48.57	48.57
Defective cell	9.52	58.09
Flow lines	8.57	66.66
Paint on spring	4.76	71.42
Filter rail broken	4.76	76.18
All other types (11)	23.82	100.00

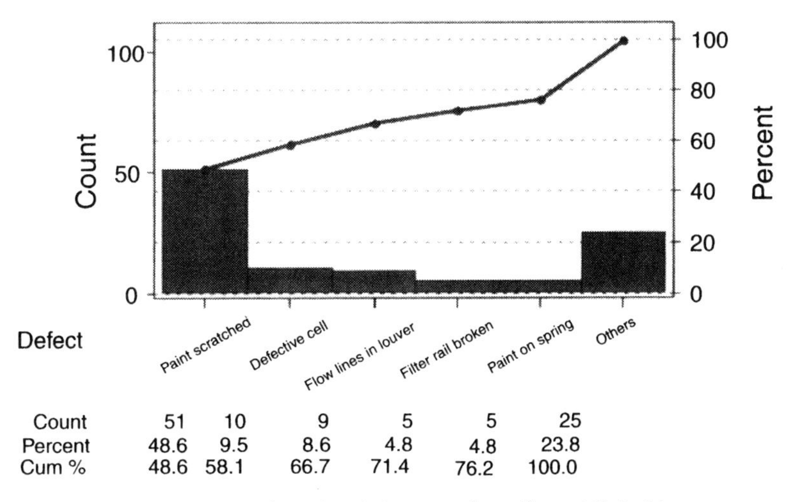

FIGURE 2.6.5 Pareto chart for defects on fans (from Minitab).

This figure was produced by Minitab (under `Stat → Quality Tools → Pareto Chart`). Using the vertical scale for the cumulative percentages on the right-hand side of the plot, a line is drawn on the chart that shows the cumulative percentage due to all problems listed up to that point.

QUIZ FOR SECTION 2.6

1. How do you obtain a frequency table for a qualitative variable?
2. How is this information best plotted?
3. In constructing a bar graph for comparing different categories, how would you order the categories?
4. Give reasons both for and against using a pie chart.
5. When should you use a three-dimensional pie chart? What about three-dimensional histograms?
6. What is a Pareto chart? What is its main purpose?

EXERCISES FOR SECTION 2.6

1. Construct a frequency table and bar graph for SURG for the portion of the heart attack data given in Table 2.1.1.
2. Using the data in Table 2.6.2, produce a bar graph and a pie chart for percent gold production in 1985. What changes do you notice between 1985 and 1991 (compare with Fig. 2.6.3)?
3. The following figures, estimated from a graph in *The Economist* (9 June 1990, p. 4) give the total media spending on advertising (in billions of U.S. dollars) in 1989 of the 10 biggest-spending countries:

   4 (Australia), 13 (Britain), 7 (Canada), 8 (France), 2 (Netherlands), 5.5 (Italy), 30 (Japan), 5.5 (Spain), 87 (U.S.), 10 (West Germany).

   Plot this data in a way that clearly indicates where most of the advertising dollars were spent in 1989.
4. In a factory, machines use a die cast to press out a particular manufacturing component. The machines don't operate all the time because of various incidents (time lost is called *downtime*). A particular machine was followed up for one week on a given shift, and the downtimes caused by various factors are listed in Table 2.6.5. Using the percent downtime as the measure of importance, construct a Pareto chart. What factors would you try to eliminate?

TABLE 2.6.5 Downtime Causes for Press No. 7, Shift No. 1

Reason	Minutes	Reason	Minutes
Die change	48	Start new coil	81
Nobody at machine	43	Inspection	20
Operator adjustment	32	Operator change	6
Maintenance	21	R & R die for repair	44
Die repair (in press)	548	Material handling	98
Die repair (in tool room)	224	Steel jammed	7

2.7 SUMMARY

This chapter introduced you to the most basic tools used for understanding, summarizing, and presenting data, namely, a variety of plots and summary statistics. Most of these will be used repeatedly throughout the remainder of the book. There are three basic types of information in the chapter:

(a) ***Construction:*** Details of how to construct particular plots or how to calculate particular types of summary statistics.

(b) ***Interpretation:*** (i) Advice about how to interpret what you see in a plot and (ii) material that tries to give you an intuitive understanding of what a summary statistic tells you about the data.

(c) ***Deciding what tools to use in a given situation:*** Since computer programs construct plots and calculate summary statistics automatically, the skills that are now most important in the real world are (b) and (c). Even using computers, you still have to know what tool you want the computer to use (c) and be able to interpret the results (b). The chapter described each tool in turn and the way it is used. This summary emphasizes the issue of what tool to use when.

Different tools are appropriate for different types of data. The first idea introduced was that of a ***variable.*** A variable is simply a type of measurement or, more generally, a type of information being collected. Typically, information is collected on several variables. We begin with tools for looking at data on a single variable. The tools used to analyze and summarize data depend on the type of variable one is interested in, as is laid out in Table 2.7.1.

The plot and summary statistic entries for continuous variables are too complicated to fit into Table 2.7.1. Table 2.7.2 lays out plotting tools for continuous variables, while Table 2.7.3 presents the various summary statistics in widespread use.

TABLE 2.7.1 Data on a Single Variable

Variable type		Summary		Plots
quantitative (measurements and counts)	↗ **continuous** (few repeated values)	See Table 2.7.3		See Table 2.7.2
	↘ **discrete** (many repeated values)	frequency table, sample mean, sample standard deviation	→	bar graph
qualitative (defines groups)		frequency table[a]	→	bar graph (or pie chart)

[a]Order categories by size (frequency) unless there is a compelling reason to do otherwise.

TABLE 2.7.2 Plots for Data on a Single Continuous Variable

Sample size		Plot type	Strengths of plot
small (e.g., $n \leq 20$)	↗	**dot plot**	–displays distances between points well and thus good at revealing clusters, gaps, and outliers –can attach labels
	↘	**bar graph**[a]	–relates relative sizes to labels well
moderate (e.g., $15 \leq n \leq 150$)	→	**stem-and-leaf**	–displays relative density of observations on the scale well and thus reveals distributional shape and outliers –can often get original data from plot
large (e.g., $n \geq 50$)	→	**histogram**	–displays relative density of observations on the scale well and thus reveals distributional shape

[a]Order categories by size unless there is a compelling reason to do otherwise.

TABLE 2.7.3 Summary Statistics for Data on a Single Continuous Variable

Feature of Data			Comments
center	↗	sample mean ($\overline{x}$)	–$\sum \frac{x_i}{n}$ –sensitive to outliers and dirty data
	↘	sample median (Med)	–position $= \frac{(n+1)}{2}$ –divides data in two –insensitive to outliers –usually more sensible for skewed data
quarters	↗	lower quartile (Q_1)	–defines boundary of lower quarter of data
	↘	upper quartile (Q_3)	–defines boundary of upper quarter of data
spread	↗	sample standard deviation (s_X)	–approx. 68% of data within ±1 s.d. of mean –sensitive to outliers
	↘	sample interquartile range (IQR)	–IQR $= Q_3 - Q_1$ –range of central 50% of data –insensitive to outliers

Note: Five-number summary = (Min, Q_1, Med, Q_3, Max)

2.7.1 Numbers and Tables of Data

The way these are used depends on the purpose for which they are being used:

(i) ***To convey understanding to people***
Principle: Strip away all unnecessary detail and organize to facilitate relevant comparisons.

(ii) ***For use in calculations***
Principle: Preserve full detail to avoid rounding errors.

(iii) ***For reference***
Principle: Preserve as much detail as possible and organize for ease of finding information.

2.7.2 Plotting

The way these are used also depends on the purpose for which they are being used:

(i) ***Exploration***
Principle: Look at the data in as many different ways as possible, searching for its important features.

(ii) ***Communication to others*** (this follows exploration)
Principle: Be selective. Choose the displays that *best show* to a reader the features you have observed.

GLOSSARY OF SELECTED TERMS

cluster: A group of data points that fall very close together.

grouped data: Data that have been summarized in terms of class intervals and the number (frequency) of data points falling into each class interval.

mode: In terms of distributional shape, a mode is a peak, and *the* mode is the highest peak. In terms of raw observations, the mode is the most commonly occurring observation. In terms of a frequency table or histogram, the modal class is the one with the highest frequency of observations in it.

bimodal: Used in terms of distributional shape to mean that the shape has two peaks. Also, trimodal (three peaks), and so on. See Fig. 2.3.10.

symmetry and skewness: See Fig. 2.3.10.

outlier: An observation that is unusually far from the bulk of the data.

relative frequency of the data with some property: Proportion of the data having that property.

repeated data: Data in which distinct values keep recurring.

REVIEW EXERCISES 2

1. The extract provided in Table 1 gives the fuel consumption at a steady 80 km/hr in liters per 100 km of all vehicles in the 1001-1350-cc category in tests conducted by the N.Z. Ministry of Energy in March 1983.
   (a) Construct a stem-and-leaf plot for the fuel consumption. Comment on any features.
   (b) Find the median and interquartile range of the fuel consumption.
   (c) Obtain the box plot. Compare it with your stem-and-leaf plot.
   (d) Are there any outliers? If so, can you give an explanation?

2. Simplify Table 2 and use your simplified table to comment on the relative importance of the various economic sectors as contributors to the American gross domestic product (GDP) in 1994. Comment too on any trends you see over time and any exceptions to these trends. Some of the growth in GDP over time will be attributable simply to inflation. What should you look at to track changes in the importance of the different economic sectors to the U.S. economy over time?

TABLE 1 Fuel Efficiency of 1000–1350-cc Cars (in liters per 100 km at 80 km/hr)

	Efficiency	Transmission[a] (Gearbox)
Ford		
Laser L hatchback	5.9	M4
Laser GL hatchback	5.9	M4
Mazda		
323 hatchback	5.8	M4
323 hatchback	5.7	M5
323 wagon	6.2	M4
Mitsubishi		
Mirage II GL 3-dr hatchback	5.6	M4
Mirage II saloon	5.7	M4
Mirage II GLX 3-dr hatchback	5.3	M4 × 2
Mirage II GLX 5-dr hatchback	5.4	M4 × 2
Mirage GLX saloon	5.4	M4 × 2
Mirage II Geneva 5-dr hatchback	5.4	M4 × 2
Mirage II Geneva saloon	5.4	M4 × 2
Nissan/Datsun		
Sunny wagon	6.4	M4
Nissan Sunny sedan	5.4	M4
Toyota		
Starlet DL 5-dr hatchback	5.6	M4
Corolla DX liftback	5.9	M5
Corolla DX sedan	5.9	M4
Corolla DX sedan	7.2	A
Corolla GL sedan	6.1	M5
Corolla GL sedan	7.2	A
Corolla wagon	5.9	M4

[a]Transmission codes: M4 = manual 4 speed; M5 = manual 5 speed; M4 × 2 = manual 4 speed with dual (high/low) ratios; A = automatic

Source: *Fuel Efficiency Guide,* N.Z. Ministry of Energy, March 1983.

3. The data in Table 3 came from the FA Carling Premier League Table of the professional soccer league in England. The data show how many goals each team in the premier division had scored in the 1994-1995 season as of 5 February 1995, in games played both at their own ground (home goals) and at their opposition's ground (away goals). The teams are ordered by their position in the league (i.e., Blackburn was leading the league at the time).
   (a) Find the sample mean and sample standard deviation for both the home goals data and the away goals data. What do you conclude?
   (b) What would be the best way of constructing stem-and-leaf plots to compare the two data sets? Construct such plots. Do they reinforce your conclusion in (a)?
   (c) Find the five-number summary for the number of home goals scored from the stem-and-leaf plot.
   (d) The five-number summary for the away goals is (4, 10, 15, 19, 22). Draw or obtain box plots for the two data sets. Comment on their shapes.
   (e) Bearing in mind that the two data sets don't come from independent sources (they come from the same source, namely, the football clubs), can you think of a better way of organizing the data?

TABLE 2 U.S. Gross Domestic Product by Sector (billions of dollars)

	1959	1967	1977	1982	1987	1992	1994
Agriculture, forestry, fisheries	20.3	24.9	54.3	77.1	88.6	112.4	117.8
Mining	12.5	15.2	54.1	149.5	88.3	92.2	90.1
Construction	23.7	39.5	93.8	129.8	217.0	229.7	269.2
Manufacturing durable goods	81.7	134.1	277.6	377.4	514.4	573.4	673.1
Manufacturing nondurable goods	58.6	86.7	184.7	272.3	374.6	490.2	524.0
Transportation & public utilities	45.0	70.5	179.5	293.2	420.7	528.8	606.4
Wholesale trade	36.1	57.8	142.3	219.6	300.3	406.5	461.9
Retail trade	49.1	78.2	190.2	288.1	436.5	544.3	609.9
Finance, insurance, real estate	69.0	117.4	283.7	504.2	830.3	1148.8	1273.7
Services	48.4	90.8	255.5	471.8	785.1	1200.8	1342.7
Federal government	35.0	56.9	120.0	188.5	255.2	321.4	327.1
State and local government	29.8	60.9	173.5	273.4	398.1	552.2	604.3
Statistical discrepancy	−2.0	0.7	17.7	−2.8	−16.8	43.7	31.2
Gross domestic product	507.2	833.6	2026.9	3242.1	4692.3	6244.4	6931.4

Source: U.S. Bureau of Economic Analysis.

TABLE 3 Premier League Table for the 1994–1995 Season

Team	Home goals	Away goals	Team	Home goals	Away goals
Blackburn Rovers	39	19	Chelsea	19	15
Manchester United	27	21	Manchester City	27	8
Newcastle United	28	16	Aston Villa	13	19
Liverpool	24	21	Southampton	18	19
Nottingham Forest	22	18	Crystal Palace	9	12
Tottenham Hotspur	22	22	Queen's Park Rangers	23	15
Leeds United	20	14	Everton	23	4
Sheffield Wednesday	18	18	West Ham United	14	10
Wimbledon	18	13	Coventry City	12	13
Norwich City	19	6	Ipswich Town	19	10
Arsenal	15	15	Leicester City	16	8

4. Table 4 shows the area of original rain forest, the area of the 1990 rain forest, and the 1990 annual rate of deforestation for various countries.
   (a) Rearrange the table following the guidelines given in this chapter for presenting tabular data, assuming that the rate of deforestation is the quantity of interest. What do you learn from the table?
   (b) Create a new column of data by calculating the actual amount of rain forest destroyed in 1990.
   (c) Which country contributed most to global deforestation in 1990?
   (d) Which country was destroying the biggest proportion of its rain forest in 1990?
   (e) Which country had destroyed the biggest proportion of its rain forest in the time up until 1990?

TABLE 4 Deforestation of Rain Forest

Country	Original rain forest (1000 km^2)	1990 rain forest (1000 km^2)	1990 annual rate of deforestation (%)
Bolivia	90	70	2.1
Brazil	2860	2200	2.3
Colombia	700	279	2.3
Guyana	500	410	0.1
Indonesia	1220	860	1.4
Ivory Coast	160	16	15.6
Malaysia	305	157	3.1
Nigeria	72	28	14.3
Peru	700	515	0.7
Philippines	250	50	5.4
Thailand	435	74	8.4
Venezuela	420	350	0.4

Source: World Wildlife Fund, *Special Report No. 6,* May 1991.

5. In an article titled "Women on the Fast Track in Professions . . . " in the French newspaper *Le Monde* in 1993, Marie-Claude Betbeder wrote about the increasing proportion of women in some of the professions. Her table of percentages for 1982 and 1990 is given in Table 5.
   (a) What percentage of dental surgeons in 1990 were women?
   (b) Construct a new column containing changes in female participation from 1982 to 1990. Construct a bar graph of the changes that shows where the biggest changes have occurred.
   (c) One might conjecture that it was easiest for professions with the smallest female participation in 1982 to show the biggest gains. Construct a plot that will enable you to investigate this conjecture.
   (d) Construct a suitable figure for comparing the percentages of women in the different professions both within each year and between the two years.

TABLE 5 Women in Professions

Profession	Percentage 1982	Percentage 1990
Medical specialists	22	31.5
General practitioners	13	27
Dental surgeons	26	31
Psychologists & psychotherapists	71	73.5
Vets	12	27
Pharmacists	49	52
Barristers	33	40
Notaries	4	12
Legal & tax advisers	28	24
Charted accountants & auditors	11.5	16
Architects	5	15
Bailiffs	15.5	25

Source: *Le Monde,* 17 February 1993.

6. Barnett [1978] gave a number of small data sets that may or may not contain outliers. Some of these follow. In each case, discuss whether you think that one or more outliers are present. (Some of the answers will surprise you!)

(a) On New Year's Eve 1961 in Wick, England, the following hourly temperature readings were obtained:

43 43 41 41 41 42 43 58 (midnight) 58 41 41

(b) In a study of poultry growth, the weights (in kilograms) of a particular bird on successive weighings were

1.20 1.60 1.90 1.55 2.20 2.25

(c) Ten dice were thrown 10 times; one record of the numbers of sixes had the form:

2 0 3 12 2 0 1 1 3

Note that there are nine numbers. What do you think happened?

(d) The capacities (in milliliters) for a sample of 17 skulls were as follows :

1230 1318 1380 1240 1630 1378 1348 1380 1470 1445 1360 1410
1540 1260 1364 1410 1545

7. A fishing experiment was carried out in the North Sea on a fisheries research ship to see what effect the mesh size had on trawl fishing. Two mesh sizes were used: 35 mm (A) and 87 mm (B). The boat used had two nets, which were identical in every way except for the mesh size of the cod ends (the rear, or aft, part of the net where the escape attempts occur). The design used for fishing was to trawl for one hour before swapping the nets. This was repeated, giving the pattern ABAB. In Table 6 we have the numbers of haddock caught for the two nets for each fish size category.

TABLE 6 Effect of Mesh Size on Trawl Fishing

	Frequency			Frequency	
Length (cm)	35-mm mesh	87-mm mesh	Length (cm)	35-mm mesh	87-mm mesh
$23.5–24.5^-$	1	0	$35.5–36.5^-$	68	78
$24.5–25.5^-$	1	0	$36.5–37.5^-$	37	52
$25.5–26.5^-$	3	0	$37.5–38.5^-$	33	40
$26.5–27.5^-$	14	1	$38.5–39.5^-$	12	17
$27.5–28.5^-$	30	5	$39.5–40.5^-$	5	17
$28.5–29.5^-$	49	19	$40.5–41.5^-$	6	14
$29.5–30.5^-$	60	29	$41.5–42.5^-$	10	10
$30.5–31.5^-$	49	51	$42.5–43.5^-$	1	4
$31.5–32.5^-$	70	91	$43.5–44.5^-$	6	6
$32.5–33.5^-$	108	120	$44.5–45.5^-$	2	2
$33.5–34.5^-$	88	118	$45.5–46.5^-$	1	5
$34.5–35.5^-$	84	107	$46.5–47.5^-$	0	1

Source: Data courtesy of Dr. Russell Millar.

(a) What do you think of the design for changing the mesh size, given that changing the net is not an easy operation? Can you think of other ways of conducting the experiment?

(b) How many fish are caught with each mesh size? Is the difference what you would expect? What do you think caused this effect?

(c) Calculate the mean size and standard deviation for the data from each mesh size. What do these numbers tell you?

(d) We wish to compare the histograms for the two mesh sizes. Explain why relative frequency rather than ordinary histograms should be used.

(e) Using a class interval of 1 cm, plot both relative frequency histograms on the same graph. What do you conclude? Is the difference in shapes as expected?

(f) Do the same thing as in (e), but using a 2-cm class interval. What effect does this extra grouping have on the comparison, if any?

8. A national census (carried out, for example, every 5 years in New Zealand and every 10 years in the United States) provides a great deal of information about such things as living conditions, employment, and so on. Table 7 gives the number of occupants of private dwellings taken from the N.Z. 1986 and 1991 censuses.

(a) Can you calculate the mean number of occupants per private dwelling for either census? Why or why not?

(b) For each year, calculate the percentages of houses with one occupant, two occupants, and so on. What changes have taken place with regard to occupancy from 1986 to 1991?

(c) Draw a separate bar graph for each year. What do you see? Can you think of a better way of displaying the change in occupancy patterns?

9. Pianka [1994] studied seven desert species of monitor lizards, or goannas, (genus *Varanus*) in the Great Victorian Desert of Australia. He was interested in comparing the ecologies of the seven species and in looking for possibly evolutionary trends. Table 8 is an extract from one of his tables of data giving details of the combined stomach contents of 82 animals from one species, *Varanus eremius.*

(a) Construct a bar graph for the numbers of each prey species. What do you conclude?

(b) Convert the frequencies for each prey species into percentages. If you were comparing Table 8 with a similar table for a different species, such as *Varanus gouldi,* based on the contents of 60 stomachs, would you use frequencies or percentages in constructing your bar graphs?

(c) The volumes in column 2 can also be converted to percentages. Explain whether or not a pie chart based on these percentages is a good idea.

TABLE 7 Occupants in Private Dwellings

Number of occupants	1986 census dwellings	1991 census dwellings
1	213,876	248,085
2	332,574	375,453
3	185,517	205,557
4	191,772	191,646
5	101,280	97,380
6	38,547	36,648
7	13,848	12,771
8 or more	11,187	10,122
Total	1,088,601	1,177,662

Source: *NZ Official Year Book 1995,* Statistics NZ

TABLE 8 Stomach Contents of 82 Lizards

Prey species	Number	Volume (ml)
Centipedes	2	1.0
Spiders	1	0.1
Scorpions	3	2.7
Grasshoppers	28	16.6
Cockroaches	3	3.0
Caterpillars	1	0.1
Unidentified insects	7	2.6
Lizards	53	83.6
Total	98	109.7

Source: Pianka [1994].

(d) If you were interested in studying diet preferences for this species, would you use the percentages from the numbers or the volumes?

(e) Food preference is one possible reason why so many lizards were eaten. What other factor might explain the pattern of consumption? How could you find out whether *Varanus eremius* eats lizards in preference to other prey species?

Exercises 10 to 15 concern the data presented in Table 9. These data are vital statistics for various countries. They are projections for 1996 based on the information in the U.S. Census Bureau's International Database at the time the request for data was made (late 1996). The second variable *Type* defines groups of countries.

TYPE = 1: Countries classified by United Nations agencies as industrialized (We have not included Japan as the United Nations customarily does.)

TYPE = 2: A selection of countries in Africa

TYPE = 3: Southeast Asian countries plus China and Japan

TYPE = 4: Central and South American countries

The word *both* in some column titles refers to both males and females considered as a single group.

10. Construct a labeled dot plot of the birth rates ("both" column) for the industrialized countries. Also construct a bar graph (use a sensible ordering). What, if anything, does the information in these graphs suggest to you?

11. Repeat the previous exercise for the life expectancies for the African countries.

12. (a) Compare male with female life expectancies for the industrialized countries using

(i) stem-and-leaf plots

(ii) dot plots

(iii) bar graphs

(b) Design modifications of each of these types of plot that would also enable you to compare countries. (Try to create something new here.) Compare your ideas with those of someone else.

13. It is clear from the data that female life expectancies tend to be greater than male life expectancies, and you might ask how much greater.

(a) Consider the difference *diff* = *female life expectancy* − *male life expectancy.* If a country has value 3 for *diff,* what does this tell us? (Explain in words.) The value 0 is special. Why? What do values greater and less than 0 correspond to?

(b) Consider the ratio *ratio* = *female life expectancy/male life expectancy*. If a country has value 1.2 for *ratio,* what does this tell us? (Explain in words.) The value 1 is special. Why? What do values greater and less than 1 correspond to?

14. Apply the following steps (a)–(f) to some of the following variables:

(i) infant mortality (both)

(ii) female infant mortality divided by male infant mortality

(iii) life expectancy (both)

(iv) the difference between female life expectancy and male life expectancy

(v) birthrate

(vi) death rate

(a) Construct appropriate plots that allow you to compare the groups of countries.

(b) Write down any interesting features that you see in the plots.

(c) Look for countries that stand out from their group and label the plots appropriately.

(d) Try to think of any possible explanations for what you see in (b) and (c).

TABLE 9 International Vital Statistics

Country	Type	Infant mortality rate			Life expectancy					
		Both	Male	Female	Both	Male	Female	Births[a]	Deaths[a]	Net migrants[a]
Australia	1	5.5	6.1	4.9	79.39	76.44	82.50	13.99	6.88	2.74
Austria	1	6.2	7.0	5.4	76.53	73.38	79.84	11.19	10.43	3.34
Canada	1	6.1	6.8	5.4	79.07	75.67	82.65	13.33	7.17	4.47
Denmark	1	4.8	5.6	4.1	77.30	73.78	81.01	12.24	10.42	2.00
Finland	1	4.9	4.7	5.1	75.47	73.82	77.18	11.32	10.92	0.58
France	1	6.2	7.1	5.2	78.36	74.47	82.46	10.86	9.03	1.18
Germany	1	6.0	6.6	5.3	75.95	72.80	79.27	9.66	11.20	8.25
Ireland	1	6.4	7.1	5.7	75.58	72.88	78.46	13.22	8.93	−6.46
Italy	1	6.9	7.6	6.2	78.06	74.85	81.48	9.87	9.82	1.25
Neth.	1	4.9	5.4	4.3	77.73	74.91	80.68	12.08	8.70	2.25
N.Z.	1	6.7	7.7	5.6	77.01	73.96	80.21	15.78	7.72	3.17
Norway	1	4.9	5.6	4.2	77.53	74.63	80.61	11.96	10.68	3.57
Spain	1	6.3	6.9	5.6	78.26	74.95	81.81	10.04	8.86	0.44
Sweden	1	4.5	4.9	4.1	78.06	75.62	80.63	11.55	11.43	5.48
Switz.	1	5.4	6.0	4.8	77.62	74.58	80.82	11.35	9.64	4.20
UK	1	6.4	7.2	5.7	76.41	73.78	79.17	13.12	11.24	0.30
U.S.	1	6.7	7.7	5.6	75.95	72.65	79.41	14.80	8.80	3.10
Algeria	2	48.7	51.2	46.1	68.31	67.22	69.46	28.51	5.90	−0.49
Angola	2	138.9	150.9	126.3	46.80	44.65	49.06	44.58	17.66	−0.14
Egypt	2	72.8	74.6	70.8	61.43	59.51	63.46	28.18	8.70	−0.35
Ethiopia	2	122.8	133.1	112.2	46.85	45.71	48.02	46.05	17.53	−1.36
Gambia	2	80.5	88.6	72.2	52.96	50.74	55.24	44.44	13.66	4.72
Kenya	2	55.3	58.3	52.2	55.61	55.53	55.69	33.38	10.30	−0.35
Libya	2	59.5	63.9	54.9	64.67	62.48	66.97	44.42	7.70	0.00
Malawi	2	139.9	147.4	132.2	36.16	35.87	36.46	41.56	24.48	0.00
Morocco	2	43.2	47.9	38.3	69.52	67.53	71.61	27.39	5.77	−1.08
Namibia	2	47.2	51.4	42.8	64.48	62.85	66.16	37.29	7.98	0.00
Nigeria	2	72.4	75.7	68.9	54.34	53.06	55.65	42.89	12.71	0.34
S. Africa	2	48.8	51.0	46.5	59.47	57.21	61.80	27.91	10.32	0.00
Tanzania	2	105.9	117.5	94.0	42.34	40.95	43.78	41.31	19.47	−10.36
Uganda	2	99.4	108.3	90.2	40.29	39.98	40.60	45.92	20.72	−2.80
Zaire	2	108.0	118.1	97.6	46.70	44.97	48.47	48.10	16.90	−14.56
Zambia	2	96.1	102.1	90.0	36.31	36.15	36.46	44.73	23.65	0.00
Zimbabwe	2	72.8	77.8	67.6	41.85	41.91	41.78	32.34	18.2	0.00
China	3	39.6	31.5	48.6	69.62	68.33	71.06	17.01	6.92	−0.34
Hong Kong	3	5.1	5.2	5.0	82.19	78.88	85.71	10.50	5.23	12.42
Indonesia	3	63.1	69.0	56.8	61.64	59.51	63.88	23.67	8.38	0.00
Japan	3	4.4	4.8	4.0	79.55	76.57	82.68	10.19	7.71	−0.4
Malaysia	3	24.0	28.5	19.1	69.75	66.82	72.89	26.20	5.49	0.00
Singapore	3	4.7	5.2	4.3	78.13	75.07	81.39	16.28	4.56	7.29
S. Korea	3	8.2	8.5	7.9	73.26	69.65	77.39	16.24	5.66	−0.35
Taiwan	3	7.0	7.5	6.5	76.02	73.43	78.82	15.01	5.52	−0.61
Thailand	3	33.4	36.3	30.3	68.60	64.89	72.49	17.29	7.00	0.00
Argentina	4	28.3	31.3	25.0	71.66	68.37	75.12	19.41	8.62	0.18
Bolivia	4	67.5	73.2	61.6	59.81	56.94	62.82	32.37	10.75	−3.41
Brazil	4	55.3	58.8	51.6	61.62	56.67	66.81	20.80	9.19	0.00
Chile	4	13.6	14.8	12.4	74.49	71.26	77.72	18.09	5.68	0.00
Colombia	4	25.8	29.0	22.5	72.81	69.97	75.73	21.34	4.65	−0.13
Ecuador	4	34.8	39.7	29.7	71.09	68.49	73.82	25.06	5.50	0.00
El Salv.	4	31.9	34.1	29.5	68.88	65.44	72.50	28.30	5.81	−4.4
Honduras	4	41.8	45.8	37.7	68.42	66.01	70.96	33.38	5.83	−1.53
Mexico	4	25.0	30.1	19.6	73.67	70.07	77.45	26.24	4.58	−2.97
Nicaragua	4	45.8	52.1	39.3	65.72	63.41	68.13	33.83	6.01	−1.17
Panama	4	29.7	31.4	27.9	73.92	71.19	76.75	23.20	5.42	−1.42
Paraguay	4	23.2	24.6	21.8	73.84	72.33	75.43	30.97	4.31	0.00
Peru	4	52.2	54.2	50.0	69.13	66.97	71.39	24.33	6.13	−0.76
Venez.	4	29.5	33.3	25.5	72.09	69.11	75.29	24.39	5.09	−0.36

[a]Per 1000 of population.

Source: U.S. Bureau of the Census International Database, 1996.

(e) What questions do the graphs prompt you to ask?

(f) What additional information would you need to obtain to answer your questions in (e)?

15. A number of miscellaneous questions follow:

(a) Birthrate is defined as the number of births per 1000 of population. What factors affect a country's birthrate?

(b) The death rates of the industrialized countries appear to be higher on average than those of the Asian and Central and South American countries. Can you suggest any reasons why this might be?

(c) In the industrialized group, the female-to-male ratio of infant mortalities for Finland is highly unusual. Is this because the individual infant mortality rates are highly unusual?

(d) Consider the mean of the birthrates for the Southeast Asian countries (including China and Japan). What meaning, if any, does this number have? Does it tell you about the average birthrate for the whole region? Why or why not? Do your answers change if we consider the median rather than the mean?

(e) A 1996 UN report stated that in 70 mostly third-world countries average incomes in the 1990s were lower than in 1980, whereas 15 third-world countries mostly in Asia had experienced sustained economic growth. The obvious inference is that globally things have been getting worse. A feature writer suggested that the 15 countries that had made progress included China and India, which alone account for nearly half of the human race, so the global picture was actually much more optimistic than it had been painted. What do you think is the more relevant measure here?

(f) *Migrants* refers to net migration, that is, those entering minus those leaving. How would you construct a variable *Increase* that gives the net increase in population size (per thousand) for the year for each country?

Further exercises, some involving large data sets, are given on our web site.

CHAPTER 3

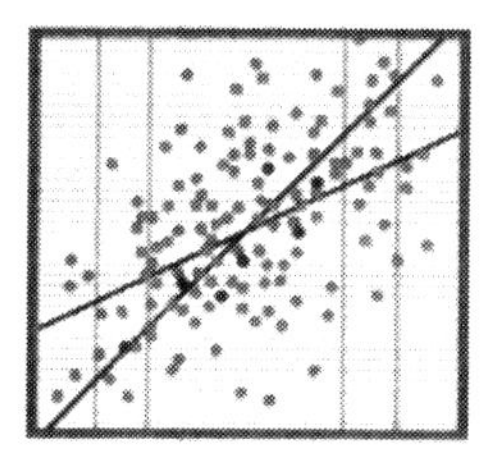

Exploratory Tools for Relationships

Chapter Overview

In the last chapter we focused on data from a single variable. We now turn our attention to the problem of studying data from two variables. Clearly, we can examine each variable on its own using the methods of Chapter 2. We may also wish to see if there is any relationship between the two variables, however. For example, knowledge of the relationship between *severity of injury* and *days in hospital* would be useful in allocating hospital beds. Knowledge of the relationship between *tax rate* and *tax take* would be useful in setting company tax rates. And knowledge of the relationship between level of *fertilizer applied* and *crop yields* would help farmers make sensible decisions. These examples are all applications in which we would use what is known about one variable to *predict* the behavior of another.

The chapter concentrates on relationships between two variables. As each variable is either quantitative or qualitative, there are three possible combinations to consider: quantitative versus quantitative, quantitative versus qualitative, and qualitative versus qualitative. As with single variables, we again try to use pictures to convey the main features of any relationship that might exist.

In the case of two quantitative variables the key graphical tool is the scatter plot, which we consider in some detail in Section 3.1. The theory for such plots is developed in Chapter 12. When one of the variables is qualitative, and therefore used for defining group membership, the groups can be compared visually using a set of box plots or, in some cases, a set of dot plots or histograms. This is done in Section 3.2. Looking for relationships between two qualitative variables is a little more difficult. The starting point is the two-way table of proportions, discussed in Section 3.3. Finally, in Section 3.4 we begin to ask whether the patterns we see in our data are real or whether they could be some sort of random artifact.

3.1 TWO QUANTITATIVE VARIABLES

3.1.1 Plotting the Data

If you look back at the heart attack data of Section 2.1 and Table 2.1.1, you will see that there are a number of continuous quantitative variables in the data set, among them EJEC, SYSVOL, DIAVOL, and TIME. How do we look for a relationship between two such variables, for example, between SYSVOL and DIAVOL? The primary tool for investigating relationships between quantitative variables is the scatter plot. Figure 3.1.1 presents a scatter plot of DIAVOL values versus SYSVOL values. Each point on the graph represents the diastolic heart volume and systolic heart volume for the same person.

> ***Use scatter plots*** to explore relationships between quantitative variables.

The most striking feature of the plot is the way in which DIAVOL values tend to increase as the corresponding SYSVOL values increase according to an approximately linear pattern. That such a relationship should exist is entirely understandable once we realize that these two variables are just different ways of trying to capture the idea of the size of the heart. The heart is expanding and contracting all the time, and the two measures are taken under different conditions. Still, some people have smaller hearts and tend to produce smaller values for both variables, and others have larger hearts and tend to produce larger values for both variables. The relationship we see here is not an exact relationship but a pattern or a trend. As is typical of the patterns we see in scatter plots, there is variation (or scatter) about the trend. Sometimes the variation is so great that it is hard to see whether there really is any trend; sometimes the variation about the trend is small, and the relationship is excellent for prediction. We shall find that there is considerably more that can be read from plots such as this as this section proceeds.

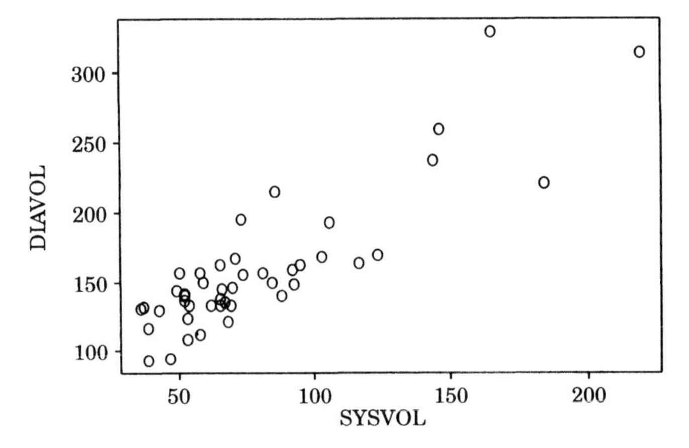

FIGURE 3.1.1 Scatter plot of SYSVOL versus DIAVOL for the heart attack data in Table 2.1.1.

Scatter-plot facilities are easy to find in the menus of most statistical packages, for example, under `Graph` → `Plot` in Minitab. They are available via Excel's chart wizard.[1] Almost all of the graphs in this chapter were produced using the freeware package R where, as in Splus, the basic scatter-plot command is `plot(x,y)`.

Overprinting is a common problem with scatter plots that occurs when two or more observations have to be plotted in the same position or sufficiently close to the same position that it appears visually that there is only one observation there. When this happens, the visual impression the scatter plot gives can be misleading because the overprinted data looks less dense than it really is. We generally plot points using open circles because these are distinguishable unless points coincide virtually exactly (see the lower-left corner of Fig. 3.1.1, where the data is densest). Solid shapes such as ● are much more prone to overprinting.[2]

Having established the basic idea of looking at relationships between quantitative variables by using scatter plots, we shall go on to a more detailed examination of ways of thinking about relationships and the patterns that can be seen in scatter plots. We start with a sequence of examples that we will use in the discussion.

EXAMPLE 3.1.1 *Chernobyl, Radioactivity, and Increased Death Rates*

According to an article in *The Economist* (30 January 1988), there was a noticeable increase in the number of Americans dying in the summer following the Chernobyl nuclear accident in Ukraine in 1986. A radioactive plume reached America 11 days after the reactor explosion. Measures of radioactivity in rain water went up over 1000-fold, and this was later picked up in milk samples (a routine indicator of radioactivity). Table 3.1.1 gives the average peak radioactivity in milk samples and the percentage increase in the death rate for nine regions of the United States. We shall call these variables RADIATION and PERCENT, respectively.

A scatter plot of the PERCENT versus RADIATION is given in Fig. 3.1.2. Here each point corresponds to a data pair for a given region. For example, the leftmost

TABLE 3.1.1 Deaths and Radiation in Milk after Chernobyl

Region	Peak radioactivity in milk (picocuries/L)	Percentage increase in death rate
Middle Atlantic	23	2.2
South Atlantic	20	2.4
New England	22	1.9
East North-Central	29	3.9
West North-Central	32	3.6
East Southern	21	2.6
Central Southern	16	0.0
Mountain	37	4.2
Pacific	44	5.0

Source: Estimated from a graph in *The Economist* (30 January 1988, p. 75).

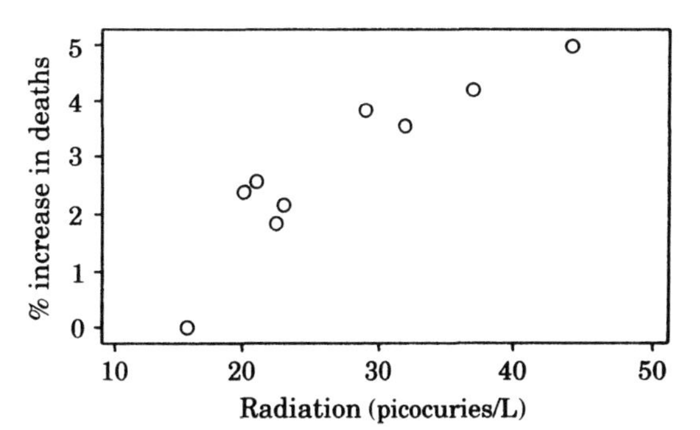

FIGURE 3.1.2 Chernobyl data.

[1] Excel refers to all plots as *charts*, so a scatter plot becomes a *scatter chart*.

[2] For discussion of approaches to the overprinting problem, see Chambers et al. [1983].

point corresponds to "Central Southern" and has coordinates $y = 0.0$ and $x = 16$. There is a striking pattern in this data as well. As the radiation dose in milk goes up, so does the death rate. The pattern follows a gentle curve. Did the increased levels of radiation from Chernobyl cause the larger number of deaths in the United States in the summer of 1986? This is observational data (in contrast with data from a randomized experiment), so the issue is not clear-cut. Not only did the death rates go up after the incident, however, but we seem to have observed a dose-response relationship (see Section 1.3). Once more we see variation (or scatter) about the trend.

Example 3.1.2 *Completion Times and Computer Workloads*

The data in Table 3.1.2 came from an investigation of the completion times experienced by users of a multiuser computer system at various levels of total workload for the system. Each data point corresponds to an experiment in which x terminals all initiated the same task. The variables measured are TIME PER TASK and the NUMBER OF TERMINALS running the task. Figure 3.1.3 plots time per task against the number of terminals. The plot shows a linear trend and scatter about the trend. Even repeat experiments using the same number of terminals give quite different values for time per task.

TABLE 3.1.2 Computer Timing Data

Number of terminals:	40	50	60	45	40	10	30	20
Time per task (secs):	9.9	17.8	18.4	16.5	11.9	5.5	11.0	8.1
Number of terminals:	50	30	65	40	65	65		
Time per task (secs):	15.1	13.3	21.8	13.8	18.6	19.8		

Source: Data courtesy of E. Wild.

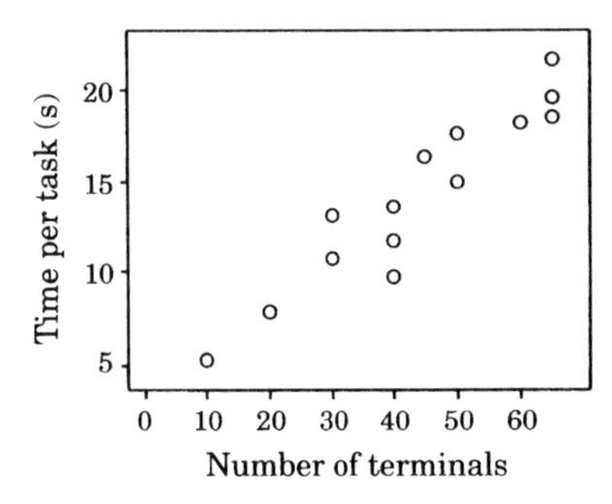

FIGURE 3.1.3 Computer timing data.

Example 3.1.3 *Pollutants in Exhaust Gas Emissions*

The data in Table 3.1.3 come from a study discussed by Lorenzen [1980] of the pollutants in the exhaust gases emitted by automobiles. Forty-six identical vehicles were used, and for each vehicle the amount of the following pollutants was measured (in grams per mile): hydrocarbons (HC), carbon monoxide (CO), and nitrogen oxides (NOX). To look at the relationship between NOX emissions and CO emissions, we

TABLE 3.1.3 Gaseous Emissions in Car Exhausts (grams per mile)

Car	HC	CO	NOX	Car	HC	CO	NOX	Car	HC	CO	NOX
1	0.50	5.01	1.28	17	0.83	15.13	0.49	32	0.52	4.29	2.94
2	0.65	14.67	0.72	18	0.57	5.04	1.49	33	0.56	5.36	1.26
3	0.46	8.60	1.17	19	0.34	3.95	1.38	34	0.70	14.83	1.16
4	0.41	4.42	1.31	20	0.41	3.38	1.33	35	0.51	5.69	1.73
5	0.41	4.95	1.16	21	0.37	4.12	1.20	36	0.52	6.35	1.45
6	0.39	7.24	1.45	22	1.02	23.53	0.86	37	0.57	6.02	1.31
7	0.44	7.51	1.08	23	0.87	19.00	0.78	38	0.51	5.79	1.51
8	0.55	12.30	1.22	24	1.10	22.92	0.57	39	0.36	2.03	1.80
9	0.72	14.59	0.60	25	0.65	11.20	0.95	40	0.48	4.62	1.47
10	0.64	7.98	1.32	26	0.43	3.81	1.79	41	0.52	6.78	1.15
11	0.83	11.53	1.32	27	0.48	3.45	2.20	42	0.61	8.43	1.06
12	0.38	4.10	1.47	28	0.41	1.85	2.27	43	0.58	6.02	0.97
13	0.38	5.21	1.24	29	0.51	4.10	1.78	44	0.46	3.99	2.01
14	0.50	12.10	1.44	30	0.41	2.26	1.87	45	0.47	5.22	1.12
15	0.60	9.62	0.71	31	0.47	4.74	1.83	46	0.55	7.47	1.39
16	0.73	14.97	0.51								

Source: Lorenzen [1980].

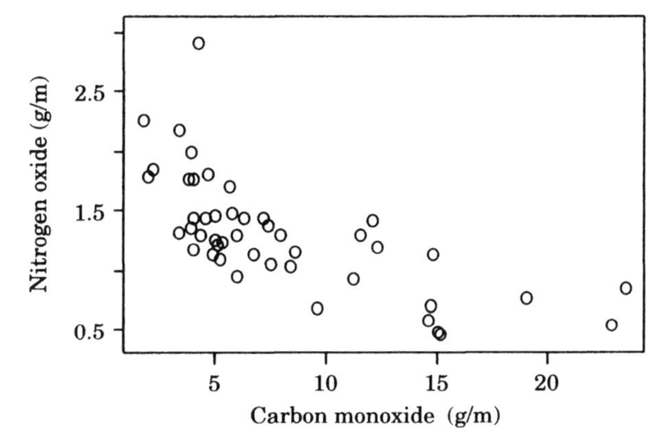

FIGURE 3.1.4 Gaseous emissions in car exhausts.

constructed a scatter plot of NOX level versus CO level, which is given in Fig. 3.1.4. Again there is a pattern. It isn't very strong, but we can see that vehicles that emit low levels of CO tend to emit relatively large levels of NOX. The pattern curves downward in a reverse J shape.

✖ Example 3.1.4 *Are We Running Faster?*

Table 3.1.4 gives the winning times in seconds for the men's 1500-meter race at the Olympic Games from 1900 to 1988.[3] A scatter plot of TIME versus YEAR for each

[3]There were no Olympic Games held in 1916, 1940, and 1944 because of the two World Wars. There was a boycott in 1980 by some countries including the United States. There was another boycott in 1984, this time by Warsaw Pact countries. These countries do not usually produce strong 1500-meter contenders. The 1992 time, which looks unusually slow, was a very tactical race.

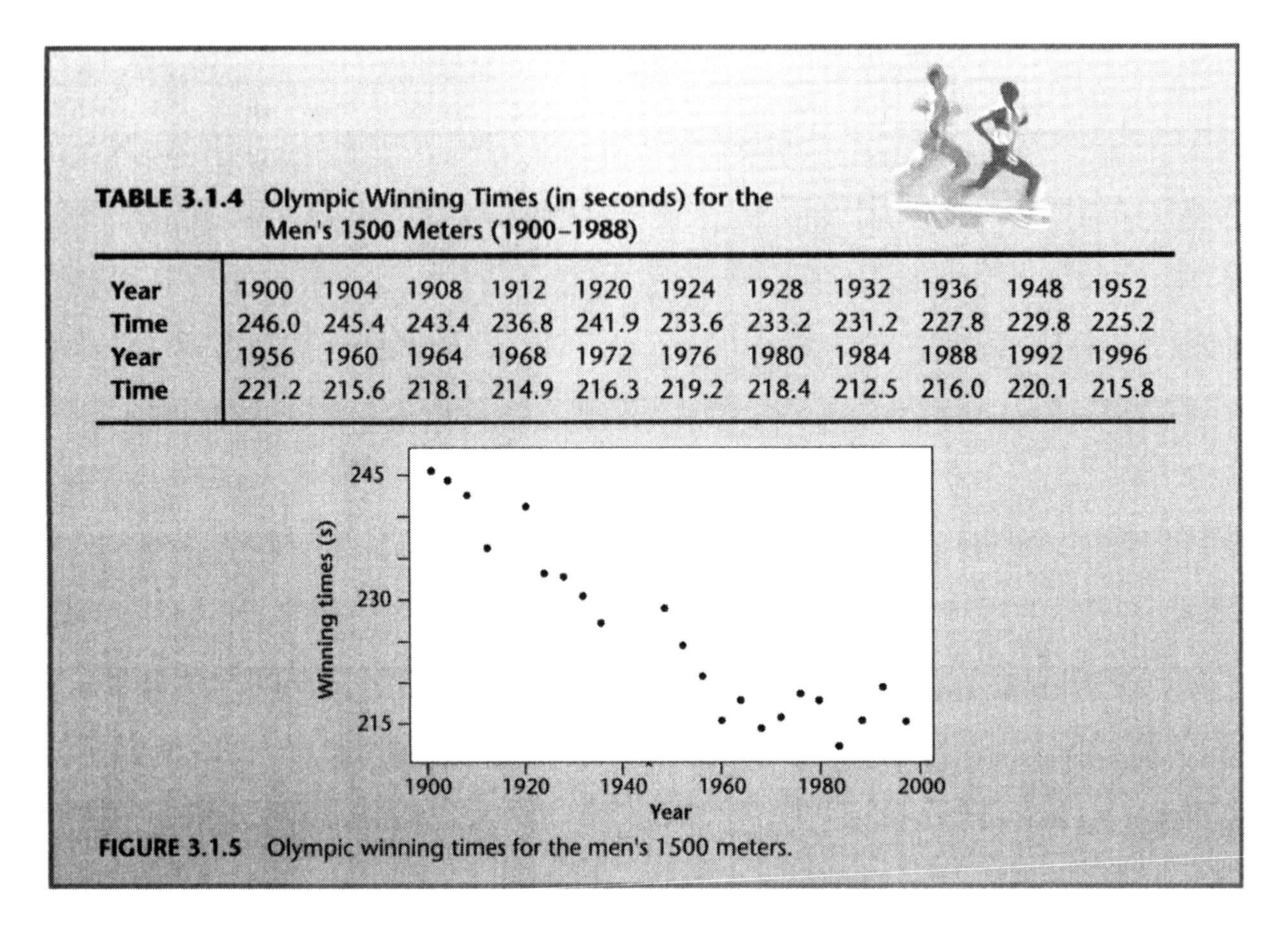

TABLE 3.1.4 Olympic Winning Times (in seconds) for the Men's 1500 Meters (1900–1988)

Year	1900	1904	1908	1912	1920	1924	1928	1932	1936	1948	1952
Time	246.0	245.4	243.4	236.8	241.9	233.6	233.2	231.2	227.8	229.8	225.2
Year	1956	1960	1964	1968	1972	1976	1980	1984	1988	1992	1996
Time	221.2	215.6	218.1	214.9	216.3	219.2	218.4	212.5	216.0	220.1	215.8

FIGURE 3.1.5 Olympic winning times for the men's 1500 meters.

Olympic Games is given in Fig. 3.1.5. Note that winning times are getting smaller as YEAR increases and that the downward pattern is linear until the 1970s, after which it appears to level off.

Random and Nonrandom Variables

In each of the data sets for the preceding four examples we have measurements on two or more characteristics of either each individual in our study, or on each repetition of an experiment. In Example 3.1.1 we sampled nine regions and measured two variables, RADIATION and PERCENT. Both variables are *random* in that we don't know the values of either until we examine each region. In Example 3.1.2, however, the NUMBER OF TERMINALS being used is nonrandom because it is controlled by the experimenter. Such variables are called *controlled* variables. The variable TIME PER TASK is random. We don't know what values it will take until we actually run the experiments. In Example 3.1.3 we sampled cars and then measured three types of emission: HC, CO, and NOX. All three of these variables are random. In Example 3.1.4 the variable TIME is random: we do not know the winning time until the experiment (the race) is over. Although the variable YEAR is not actually something we control, it is essentially predictable.[4] Therefore we classify it as a nonrandom variable. Similar examples of nonrandom variables include DEPTH in an observational study that

[4]It is not entirely predictable because sometimes the games are not held, for example, during the two World Wars.

determines the percentage of plankton present at various depths in the ocean, or AGE in a study that determines cancer death rates at various ages. In contrast, if one sampled 500 people and then measured the age of each person, AGE would be a random variable.

QUIZ FOR SECTION 3.1.1

1. What is a quantitative variable (Section 2.1)?
2. What basic tool is used for exploring relationships between quantitative variables?
3. What is a controlled variable?
4. What is the difference between a random and a nonrandom variable?

EXERCISES FOR SECTION 3.1.1

1. In each of the following pairs of variables X and Y, identify whether one variable is random and one is nonrandom or whether both are random.
   (a) X = amount of fertilizer applied, Y = crop yield at each of several sites with different levels of fertilizer application
   (b) X = decrease in cervical cancer mortality rate, Y = proportion of women undergoing cervical-smear tests (for each of 10 areas)
   (c) X = time after injection, Y = concentration of drug in blood taken every hour for 24 hours
   (d) X = year, Y = average retail sales to a household (for each of 11 years)
   (e) X = measurement on an alcohol-dependence scale, Y = result of an alcoholism screening test (on each of 25 different subjects)
   (f) X = number of workers, Y = number of supervisors (for each of 27 industrial establishments)
   (g) X = average body weight per litter, Y = average brain weight per litter (for each of 20 litters of mice)
   (h) X = age of tree, Y = volume of wood in the tree (for each tree in a forest stand)
   (i) X = age of a tree, Y = volume of wood in the tree (measured each year for 25 years)
2. To check your facility with producing scatter plots, plot the small data set in Table 3.1.1 both by hand and by using a computer package. Compare your plots to Fig. 3.1.2.

3.1.2 Modeling the Relationship

Regression

Regression is a way of studying relationships between variables that is useful when it makes sense to try to predict or explain the behavior of one variable in terms of the behavior of one or more other variables. The variable whose behavior we want to predict or explain is called the ***response*** variable and is conventionally denoted Y. For regression to make sense, Y must be a random variable. The other variables are of interest only insofar as they can help us to understand, explain, or predict the behavior of Y. They are called ***explanatory variables.*** For the time being we consider only a single explanatory variable[5] and denote it by X. In plotting it is conventional to use the

[5]There are many competing terms that are also widely used in various subject areas. The Y variable is variously called the *dependent* or *endogenous* variable, and X the *independent, exogenous, predictor, carrier,* or *regressor* variable or *covariate.*

vertical axis to represent the response variable Y and the horizontal axis to represent the explanatory variable X. (This is what is conventionally meant when we say that we plot "Y versus X.")

The regression way of thinking is particularly natural when Y is random and X is controlled or nonrandom. For example, we are interested in how Y = TIME PER TASK responds to changes as we change X = NUMBER OF TERMINALS in Example 3.1.2. We are interested in how Y = WINNING TIME changes with X = YEAR in Example 3.1.4. Regression can also be useful, however, when both variables are random. It would be useful in Example 3.1.3, for example, if we wanted to be able to predict the Y = NOX emission levels of cars from their X = CO emission levels.

When both variables are random, situations arise when we want to investigate the closeness of the relationship in a way that treats the variables symmetrically, without thinking about using one variable to predict the other. In such cases we can use a different approach based on the concept of correlation discussed in Chapter 12.

Trend and Scatter

There are two main components of a regression relationship: *trend* and *scatter*. The trend is the pattern we see when we look at the scatter plot. Figure 3.1.6 gives scatter plots for three data sets. The left-hand plot is simply the unadorned scatter plot. On the right-hand plot we have drawn a line or curve on the scatter plot by eye to summarize the trend that we see in the plot. In these examples the trend curve never summarizes the relationship perfectly; the data points do not lie exactly on the hand-drawn trend curve but appear to be scattered about it.

Regression relationship = trend + residual scatter

Note that in Fig. 3.1.6(a) there is a similar amount of scatter about the line regardless of the value of disposable income. In contrast, there is much less scatter about the curve in Fig. 3.1.6(c) when the value of GESTATIONAL AGE is small (left side of the scale) than when GESTATIONAL AGE is high. The data in Fig. 3.1.6(c) came from ultrasound measurements on babies taken before they were born. In less technical language we thus can say that there is much less variation in the liver lengths of young (and therefore small) babies than there is among older (and bigger) babies. Figure 3.1.7(a) gives a scatter plot of engine displacement versus weight for 74 models of automobile.[6] We have annotated this plot to form Fig. 3.1.7(b) by adding a trend curve (solid line) and dotted curves above and below the trend that capture the usual extent of the scatter. All of these curves were drawn by eye.

Trend curves by eye. Figure 3.1.8(a) shows a scatter plot with two possible summarizing trend lines superimposed. Which does the better job? Most people who have not looked at the right-hand plot pick the steeper line. Bear in mind, however, that when drawing trend curves, we are are trying to predict Y-values from their X-values. Look now at Fig. 3.1.8(b) and you will see that when the X-value is large, the actual Y-values are almost always smaller than would be predicted by the steeper line. In other words, the steeper line would give bad predictions when X is large. It also gives bad predictions when X is very small. For small X, virtually all of the Y-values

[6] Plotted from data given in Chambers et al. [1983] and available from `Statlib`.

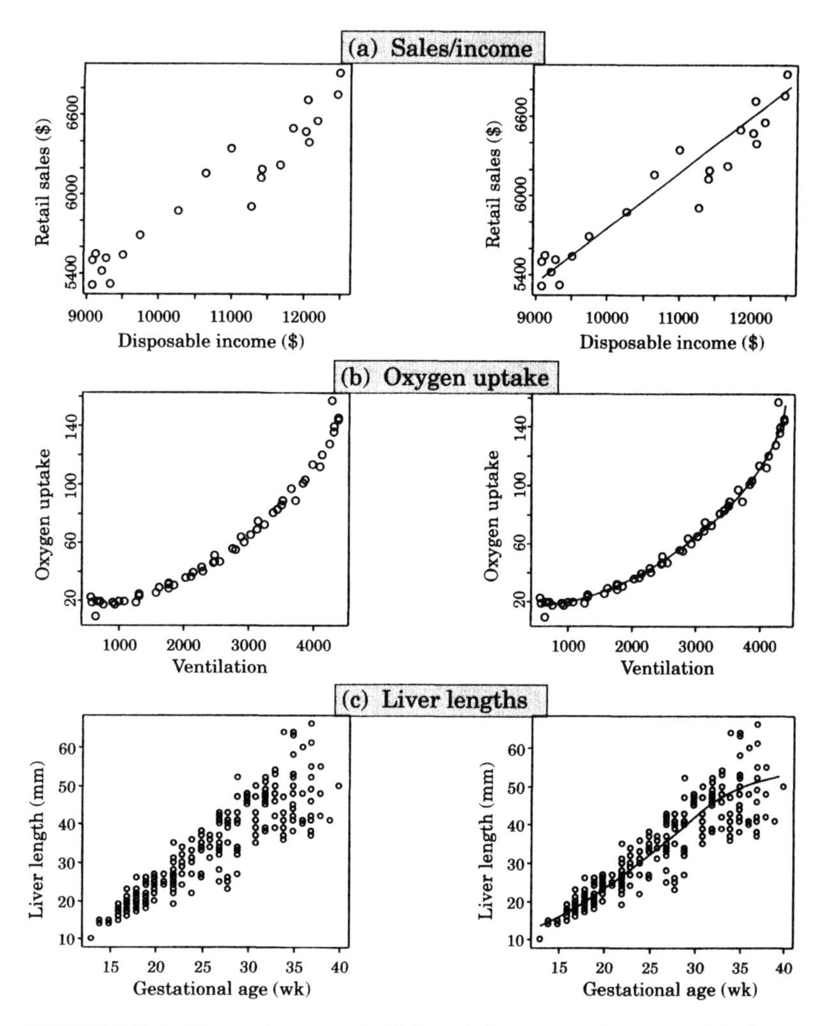

FIGURE 3.1.6 Three data sets (with hand-drawn trend curves added).

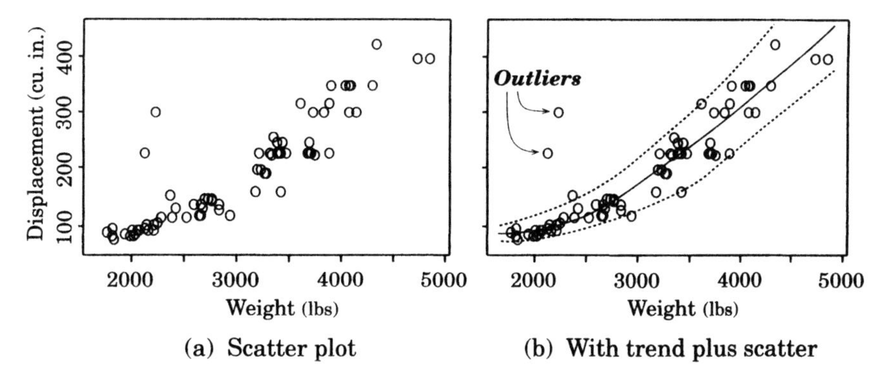

FIGURE 3.1.7 Displacement versus weight for 74 models of automobile.

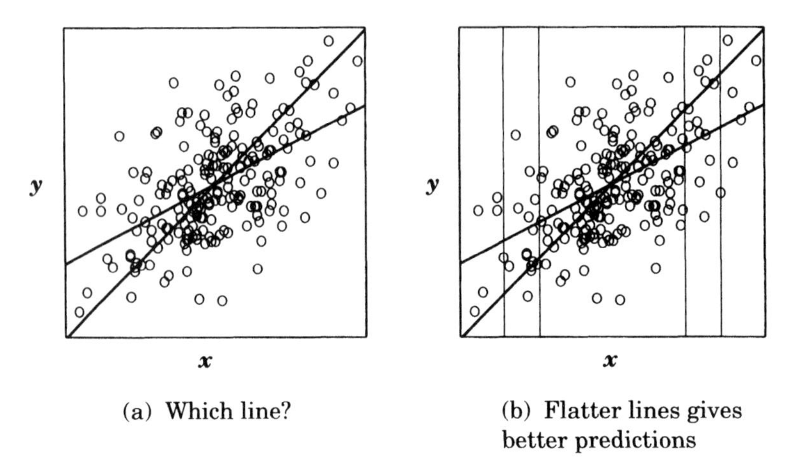

(a) Which line?

(b) Flatter lines gives better predictions

FIGURE 3.1.8 Educating the eye to look vertically.

are bigger than the steep line would predict. The sort of thing we have been doing when sketching trend curves is called ***smoothing.*** The eye has remarkable power for providing smoothing, but we need to educate it a little. We should mentally divide the data into vertical strips and ensure that our trend curve is going through near the middle of all the Y-values in the strip. The flatter line in Fig. 3.1.8(b) is behaving like this and is giving better predictions as a consequence. You will note that the curve in Fig. 3.1.6(c) is also behaving in this way. (In Chapter 12 we shall discuss formal methods for producing summarizing trend curves for scatter plots. The flatter line in Fig. 3.1.8 is the *least-squares regression line.*)

Outliers. Another important feature to look for in scatter plots is *outliers.* These are the "oddballs," the observations that look very different from the rest of the data. In the regression context, outliers are values that are unusually far from the trend curve. In other words, they are further away from the trend curve than one might expect from the usual level of scatter. There are two outliers identified in Fig. 3.1.7(b). Figure 3.1.4 shows an outlier at a carbon monoxide level of approximately 4.

One's usual *initial* suspicion about any outlier is that it is a mistake, for example, a typing or transcription error. Some studies suggest that we can expect something like 0.1% to even 10% of the data to actually be in error. If an observation is clearly atypical, every effort should be made to check its original source and correct the value if possible. Figure 3.1.9 plotting SYSVOL versus DIAVOL was produced from the full heart attack data set described in Section 2.1. The point labeled B is an outlier, a clear

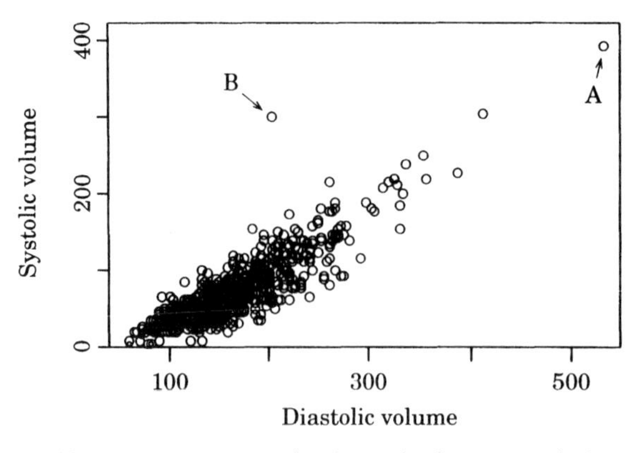

FIGURE 3.1.9 Scatter plot from the heart attack data.

oddball far from the standard relationship seen between these two variables. Point A does not look bad according to the scatter-plot trend, but it shows an unusually large value of the volume. It would stand out in a one-dimensional plot of either variable. When we checked the values for these patients in the original handwritten case notes, we found that B was in error, and we corrected it, whereas A seemed to be an actual value.[7]

If we knew that A was an error but couldn't find the true value, we would have just simply omitted this data point from the analysis. Outliers can have a big effect on the conclusions you reach when using some of the tools of formal analysis that we meet later in the book. We need to be very careful about omitting any points from a data set, however. If after checking original sources we are still in doubt, we should carry out our analysis with and without the outlier. If we are lucky, including the outlier may not affect the basic conclusions that we draw from the data. Sometimes an outlier is not a mistake, and spotting it alerts us to something unusual having happened. It indicates a breakdown of the general pattern, and it may be important (or commercially advantageous) to find out why. Again, outliers sometimes arise simply because large unexplained variations occur from time to time in the population or experiment we are studying.

Strong versus Weak Relationships

Figure 3.1.6(b)[8] depicts a very strong relationship. The observations all fall very close to the trend curve. The relationships in Fig. 3.1.6(a) and (c) are substantially weaker. Figure 3.1.10 depicts a relationship that is real but very weak.[9] It comes from 58 sexually abused children in a study conducted by Auckland psychiatrists Drs. Sally Merry and Leah Andrews. The children were rated on a measure of psychological disturbance by both their teacher and the nonabusing parent.

Why do some relationships look weak and others look strong? A relationship looks strong [as in Fig. 3.1.6(b)] if the residual scatter is small compared with the range of

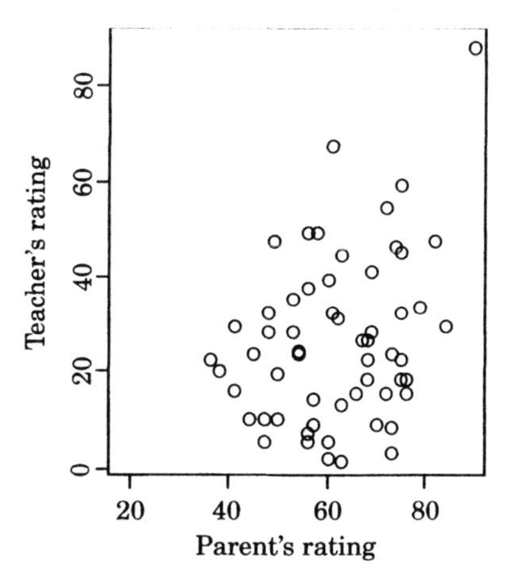

FIGURE 3.1.10 Parent's rating versus teacher's rating for abused children.

---

[7]The process of "cleaning up" data sets from very big studies before formal analysis can take months of hard work.

[8]Plotted from data from Prof. G. W. Bennett, University of Waterloo.

[9]A formal significance test shows that it is too strong to be plausibly explained in terms of chance alone.

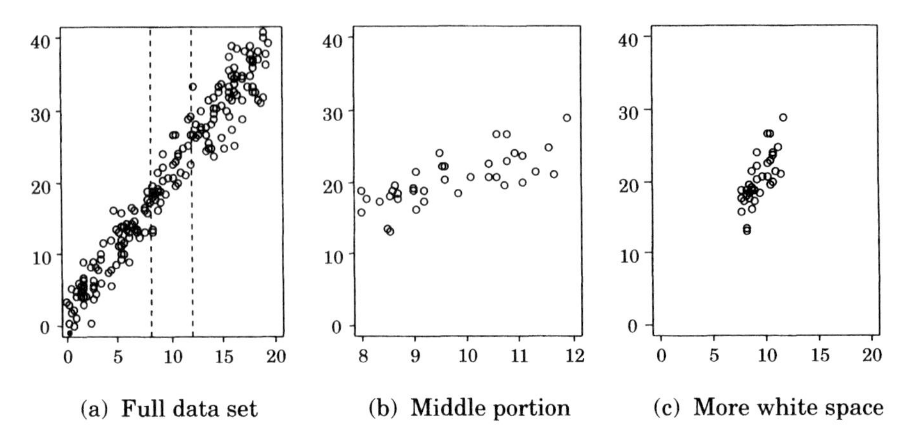

FIGURE 3.1.11 Visual impressions from scatter plots.

values taken on by the trend curve. The trend curve explains most of the differences we see between the Y-values. If the average size of the residual scatter is not small compared with the range of values taken on by the trend curve, the relationship looks weak. In this case there is still a lot of unexplained variation in the Y-values. There are two lessons we wish to convey to you about the distinction between strong and weak relationships.

(i) If you take a strong relationship, as in Fig. 3.1.11(a), and plot it only for a limited range of X-values, as in Fig. 3.1.11(b), it will look much weaker. In (b) we have plotted only the data that fall between the vertical dotted lines in (a).

(ii) Other factors affect our visual impressions of the strength of a relationship, as can be seen by comparing Fig. 3.1.11(b) and (c). Both graphs plot the same set of points, but the x-scale is more compact and there is more white space in Fig. 3.1.11(c). The result is that the relationship looks much stronger.

Item (ii) is simply a visual illusion, whereas item (i) is real. If a relationship between X and Y looks weak, it may just be because you are looking at a very restricted range of values of X.

Association between variables. When we see trend relationships in our plots as we did in the preceding examples, provided the patterns are too strong to be explained by chance alone, the two variables are said to be *associated.* If Y tends to increase with X, the variables are said to be *positively associated.* If Y tends to decrease as X gets bigger, the variables are said to be *negatively associated.*

QUIZ FOR SECTION 3.1.2

1. When people talk about plotting Y versus X, which variable is conventionally represented on the horizontal axis and which on the vertical axis?
2. What are the roles of the response variable and the explanatory variable in regression? On a scatter plot, which axis is conventionally used for the explanatory variable and which for the response?
3. What are the two main components of a regression relationship?
4. What do we call observations that are further from the trend curve than expected when compared with the usual level of scatter?
5. Should outliers simply be discarded when analyzing data? Justify your answer.

6. What should you immediately do when you identify an outlier?
7. What makes some relationships look weak and others look strong?
8. Under what circumstances can a strong relationship look weak in a scatter plot?
9. What do we mean by a positive association between two variables? a negative association?

EXERCISES FOR SECTION 3.1.2

1. Draw a trend curve (or line) by eye on each scatter plot in Section 3.1.1. Also add curves like those in Fig. 3.1.7(b) that capture the "usual level of scatter."
2. Take each part of problem 1 of the exercises on Section 3.1.1 in which both X and Y are random. Determine whether you use one of the variables to predict or explain the behavior of the other or whether you would just want to look at the closeness of any relationship in a symmetrical way.
3. We often talk about a person's "blood pressure" as though it is an inherent characteristic of that person. In fact, a person's blood pressure is different each time you measure it. One thing it reacts to is stress. The following table gives two systolic blood pressure readings for each of 20 people sampled from those participating in a large study. The first was taken five minutes after they came in for the interview, and the second some time later.

1st Reading:	116	122	136	132	128	124	110	110	128	126
2nd Reading:	114	120	134	126	128	118	112	102	126	124
1st Reading:	130	122	134	132	136	142	134	140	134	160
2nd Reading:	128	124	122	130	126	130	128	136	134	160

(a) Plot Y = 2nd Reading versus X = 1st Reading, and describe anything interesting you see in the plot.
(b) Add the $y = x$ line to the plot.
(c) Do you see anything else of interest? Suggest an explanation.

4. The literary style of different authors differs widely from one to another, and it is possible to measure these differences statistically. By style we mean those aspects of writing that might be independent of the subject matter, for example, the lengths of words and sentences or the frequencies of different words such as *the* and *to*. The table below gives the percentages of these two words on each page for a random sample of 18 pages from two books, *Pride and Prejudice* by Jane Austen and *Spy Hook* by Len Deighton.

Pride and Prejudice									
"to" (%):	2.03	2.24	2.37	1.80	1.79	1.58	1.70	2.55	1.68
"the" (%):	5.29	3.06	3.96	3.51	2.95	2.05	3.39	3.90	4.11
"to" (%):	3.73	1.24	3.07	2.77	1.91	2.08	1.13	2.07	3.42
"the" (%):	2.55	2.66	4.11	8.39	2.97	2.54	3.67	2.99	3.71
Spy Hook									
"to" (%):	3.78	1.46	1.81	2.56	4.27	2.92	3.85	2.88	1.95
"the" (%):	4.56	6.42	4.16	4.31	4.51	3.70	3.59	3.91	5.62
"to" (%):	1.76	2.92	2.48	2.75	2.73	2.53	2.97	1.36	3.28
"the" (%):	7.12	6.90	2.28	3.87	5.94	2.86	7.65	5.51	4.93

(a) Plot Y = percentage of *to* versus X = percentage of *the* for each book.
(b) Note any special features of your two plots.
(c) Plot the data from both books on the same scatter plot using a different plotting symbol (or color) for each book. Which is more informative, the two plots in (a) or the single plot?
(d) Is it possible to differentiate between *Pride and Prejudice* and *Spy Hook* using these data?

3.1.3 The Prediction Problem

One of the most important reasons for looking at relationships between a response variable Y and one or more explanatory variables is so that we can use the information about explanatory-variable values to predict the response. Figure 3.1.12 replots the data in Fig. 3.1.6(c). The data were taken on 258 normal fetuses. The response variable is the LIVER LENGTH (in millimeters) measured by ultrasound. The explanatory variable is the GESTATIONAL AGE (in weeks). The trend curve (solid line) was drawn by eye. It is easy to see from the plot that we can use this data to predict the liver length of a fetus at any given gestational age. We have used the graph to predict the liver length of a fetus with a gestational age of 25 weeks. From the graph we read off a ***predicted value*** of $\widehat{y} \approx 31$ mm.

Moreover, we can see from the graph that there is quite a lot of variability in the liver lengths of babies aged 25 weeks. We have also drawn in by eye two dashed curves to contain the usual level of scatter. By projecting from these dashed curves, we have obtained an interval of values that we think will be fairly likely to contain the liver length of a normal fetus from this population,[10] namely, the interval from about 22 mm to 42 mm. This provides an intuitive introduction to ***prediction intervals,*** discussed in Section 12.4.3. As should be clear from Fig. 3.1.12, we cannot predict a new response with any precision from a weak relationship. A large amount of scatter about the trend will lead to wide prediction intervals.

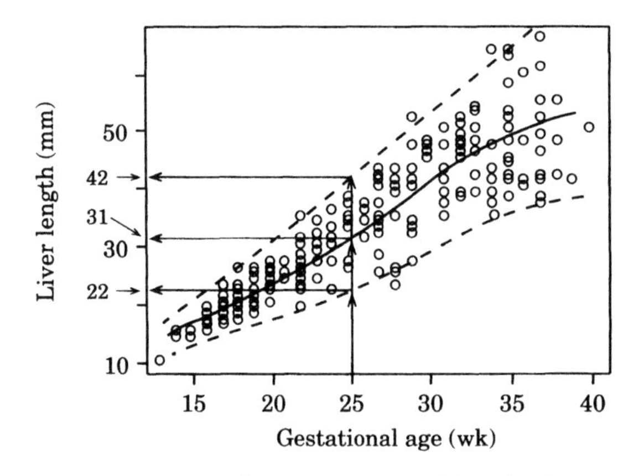

FIGURE 3.1.12 The prediction problem for liver length data.

[10]The researchers who collected the data depicted in Fig. 3.1.12 did so in order that they could construct intervals like these, which would enable them to screen fetuses for liver abnormalities. A fetus's liver would be suspect if its length fell outside the interval.

When can we make predictions with any degree of reliability?

Predicted values and prediction intervals can sensibly be calculated when X is controlled and Y is random or when both X and Y are random. Prediction can be used with confidence, however, only if the mechanism generating the new observation or new individual is the same as the mechanism that generated the past data in the scatter plot. For example, we can use Fig. 3.1.12 to predict the liver lengths only of new fetuses drawn from the same population as the population that gave rise to the figure. The relationship between gestational age and liver length may be different in a different population.

EXAMPLE 3.1.5 *Dangers of Prediction*

Margolin [1988] reports data from an experiment in which a drop of urine at a certain concentration is dropped on a petri dish (or plate) and some time later the number of visible bacterial colonies growing on the plate is counted. This was done for various concentrations with some replication (i.e., the same concentration being tried several times). Focus on Fig. 3.1.13(a). It shows the number of visible colonies increasing linearly with increasing concentration of urine.

How many colonies would one be likely to see on a plate at a concentration of 11.60 ml/plate? It would be quite natural to extrapolate the linear pattern seen in Fig. 3.1.13(a) to make a prediction. A summary line drawn on the plot by hand gave us a prediction of 35 colonies at a concentration of 11.6 (not shown). In fact, the experimenters collected data at this concentration, so we can check out our prediction. How good was it? Completely useless! The average number of colonies at concentration of 11.6 is actually only about 15 [see Fig. 3.1.13(b)]. The pattern is clearly nonlinear. At high concentrations the urine becomes toxic, and rather than continuing to rise, the numbers of colonies growing actually drop off.

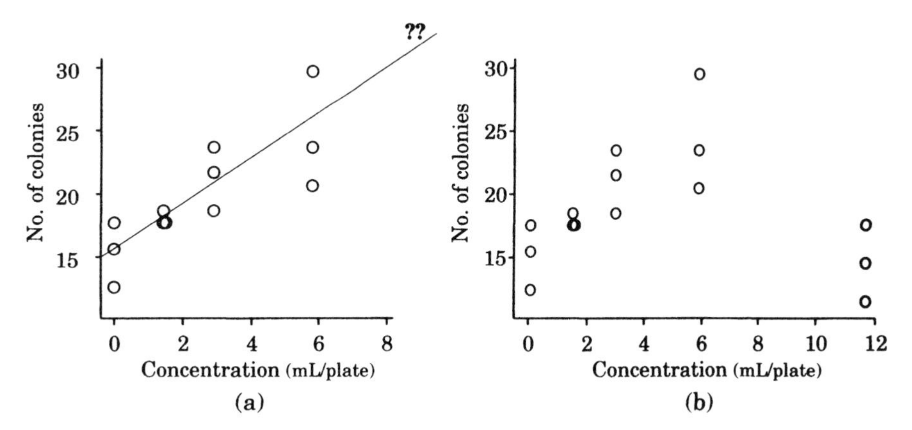

FIGURE 3.1.13 The dangers of predicting outside the range of the data. (Plotted from data in Margolin [1988].)

Example 3.1.5 illustrates a very important point. It can be dangerous to predict outside the range of X-values that we have observed. A relationship that fits the data well may not extend outside that range.

Be very cautious when predicting outside the range of the data.

Thus we can make predictions with any degree of reliability only within the range of the data[11] and for events that are being generated by the same processes that generated the data we have. The further we get from the data, the greater the opportunities for making really major errors. This is why predictions about what will happen in the future made on the basis of past data (here X is time) are so often wrong.

QUIZ FOR SECTION 3.1.3

1. Why can we not predict with any precision from a weak relationship?
2. Under what circumstances can prediction be used with any confidence?
3. Why is prediction outside of the range of the data dangerous?

EXERCISES FOR SECTION 3.1.3

The object of these exercises is to make a single-number prediction and an interval prediction, as in Fig. 3.1.12, but using other data sets.

1. From Fig. 3.1.1, predict the DIAVOL value for a new patient of this type whose SYSVOL value is 100.
2. From Fig. 3.1.4, predict the nitrogen oxide emissions of another car for which the carbon monoxide emission was 5 g/mi.
3. From Fig. 3.1.5, predict the winning time for the 1500 meters in the year 2000.

3.1.4 Other Patterns

Trend-plus-scatter patterns are not the only patterns we see in scatter plots. If you look hard at a scatter plot, you will sometimes see patterns that prompt the response, "That looks a bit odd! I wonder what is going on there?" This may lead to important insights, or it may just remind you of something you should have thought about but perhaps didn't. We shall now show you two examples in Fig. 3.1.14. In each case the left-hand plot is an unadorned scatter plot, whereas in the right-hand plot some feature of the plot has been emphasized. The data for both examples, together with a much more detailed discussion, is given in Chambers et al. [1983].

Figure 3.1.14(a) plots WEIGHT versus PRICE for 74 models of automobile on the U.S. market. If you look hard at it, there appear to be two strands to the plot, as emphasized in the right-hand plot. Is this pattern real? What caused it? If the raw data are inspected, it turns out that the upper strand corresponds to domestic U.S. cars and the lower strand to foreign cars. Thus the relationship between weight and price is different for the two different types of car (domestically produced and imported). We would probably want to analyze the relationships for the two types of car separately. If you had not thought to allow for a "country of origin" difference, the scatter plot may prompt you to do so.

Figure 3.1.14(b) plots the PETAL WIDTH versus PETAL LENGTH for 150 iris flower petals from the well-known Fisher's iris data. Notice the separation into two clusters. It turns out that the iris petals in the lower-left cluster came from a different variety of iris from those in the upper cluster. The upper cluster actually comes from a mixture

[11] The situation is different if we are not just using patterns seen in data but are instead basing predictions on well-understood and trusted theory.

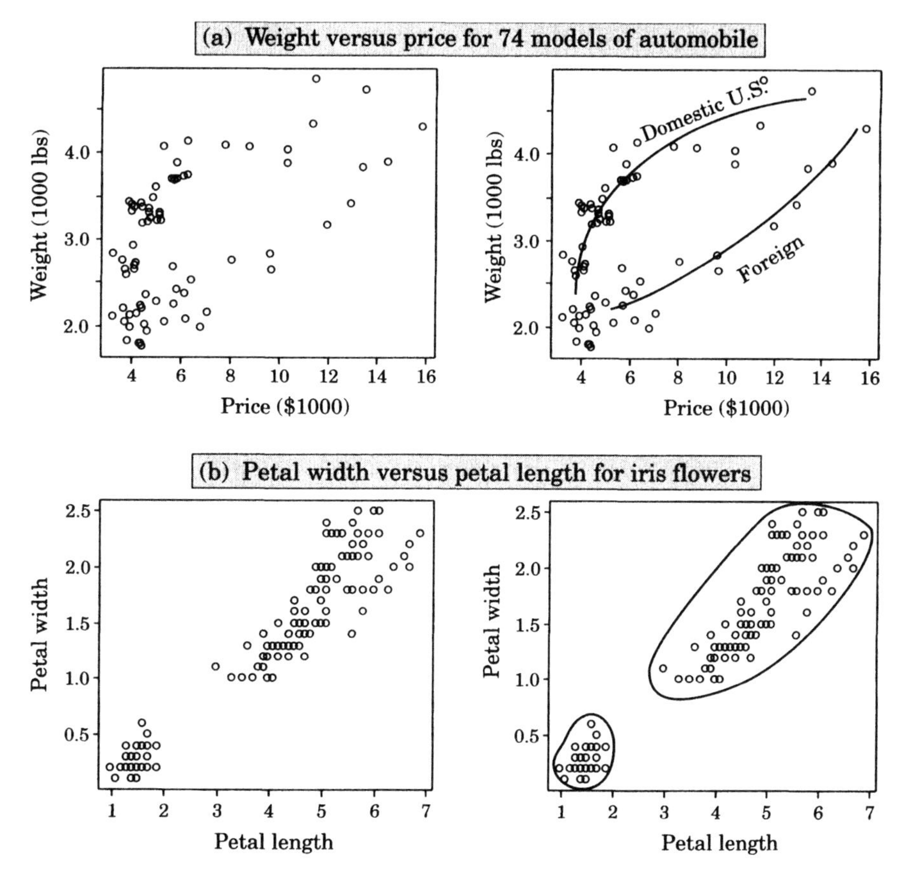

FIGURE 3.1.14 Two examples exhibiting interesting structure.

of two varieties. Scatter plots of other characteristics of the petals largely separate the two groups in this cluster.

When both X and Y are random, clusters in a scatter plot signal the possible presence of several groups, as do clusters in a dot plot or two or more modes in a histogram. It then becomes important to find out what defines these groups. A host of supplementary questions may follow, including, "Is the regression relationship different for different groups?" "In what way is it different?" and "Can the relationship enable us to classify new objects successfully with regard to group membership on the basis of X and Y measurements alone?"

Scientific inquiry is detective work. We should always be on the lookout for clues about what is really happening.

3.2 QUANTITATIVE VERSUS QUALITATIVE VARIABLES

EXAMPLE 3.2.1 *Female Life Expectancy versus Type of Country*

If you look back at the international vital statistics data given in Table 9 of Review Exercises 2, you will see that each row of the table contains the data for a particular

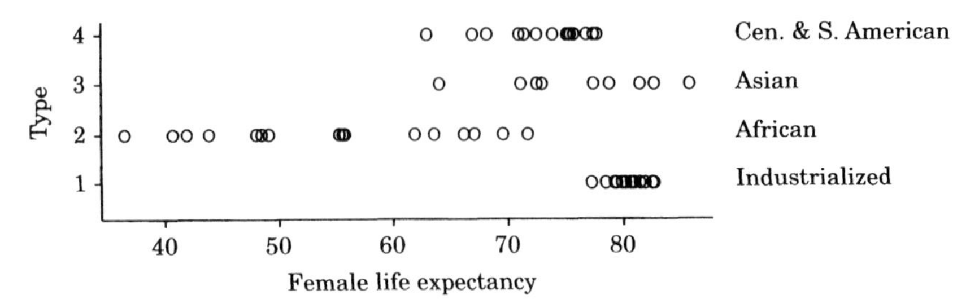

FIGURE 3.2.1 Relationship between FEMALE LIFE EXPECTANCY (quantitative) and TYPE of country (qualitative).

country. All but one of the variables are quantitative variables. The exception is the variable TYPE, which is qualitative (or categorical) and tells us which group the country belongs to. The groups, or types, of country are (1) industrialized, (2) African, (3) Asian, and (4) Central and South American. How do we look for a relationship between a quantitative variable such as FEMALE LIFE EXPECTANCY and a qualitative variable such as TYPE? We plot the values of the quantitative variable (life expectancy) for each group against the same scale, as in Fig. 3.2.1.

> To explore the relationship between a ***quantitative*** variable and a ***qualitative*** variable, ***plot the values*** of the quantitative variable ***for each group against the same scale.***

We shall find later in this section that there are other ways of plotting the values for each group against the same scale (e.g., with box plots), but the general idea is always the same. Figure 3.2.1 was constructed as a scatter plot of TYPE versus FEMALE LIFE EXPECTANCY. The group labels were added later.[12] If the variable defining type of country had contained character expressions (e.g., "African," "Asian," and so on) instead of numeric codes, we could have created a numeric-code equivalent like TYPE to use in constructing the plot.

A great deal of information can be read from Fig. 3.2.1. We see, for example, that industrialized countries tend to have the highest life expectancies and there is little country-to-country variation (small spread), whereas the African countries have the lowest average life expectancy but there is huge country-to-country variation (large spread). Some African countries (which ones?) have higher life expectancies than some of the Asian countries, and so on. The current purpose, however, is not an analysis of this particular data but establishing the connections between the data set, the types of variables, and an appropriate plot.

EXAMPLE 3.2.2 *Strength of Gear Teeth versus Position*

The data in Table 3.2.1 were taken from Gunter [1988]. They originated from the Ford Motor Company and concerned gear blanks that Ford bought. The teeth, Ford engineers found, were breaking at too low a stress. To help find out what was going

[12] In future plots of this type we have removed the vertical axis. We have done this simply because the vertical axis contains no useful information once group labels have been placed on the plot. Special programs also produce these plots, for example, the function called `stripplot` in Splus and R. In Minitab, look under `Graph` → `Dotplot`. For the `by variable` use the qualitative (grouping) variable. The Minitab plots look subtly different because Minitab puts a horizontal axis line under each group.

TABLE 3.2.1 Impact Strengths of Gear Teeth (in lb-ft)

Position 1	Positions 2 and 12	Positions 3 and 11	Positions 4 and 10	Positions 5 and 9	Positions 6 and 8	Position 7
1976	2425	2228	2186	2228	2431	2287
1916	2000	2347	2521	2180	2250	2275
2090	2251	2251	2156	2114	2311	1946
2000	2096	2222	2216	2365	2210	2150
2323	2132	1940	2593	2299	2329	2228
1904	1964	1904	2204	2072	2263	1695
2048	1750	1820	2228	2323	2353	2000
2222	2018	2012	2198	2449	2251	2006
2048	1766	2204	2150	2300	2275	1945
2174		2144	2311	2078	1958	2006
1976		2305	2102	2150	2185	2209
2138		2042	2138	2377		2216
2455		2120	1982	2108		1934
1886		2419	2042	2257		1904
2246		2162	2030	2383		1958
2287		2251	2216	2323		1964
2030		2222	2305	2246		2066
2210			2204	2251		2222
2084			2198	2156		2066
2383			2204	2419		1964
2132			2162	2329		2150
2210			2120	2198		2114
2222			2108	2269		2125
1766			2030	2287		2210
2078			2180	2330		1588
1994			2251	2329		2234
2198			2210	2228		2210
2162			2216			2156
1874			2168			2204
2132			2210			1641
2108			2341			2263
1892			2000			2120
1671			2132			2156

Source: Gunter [1988].

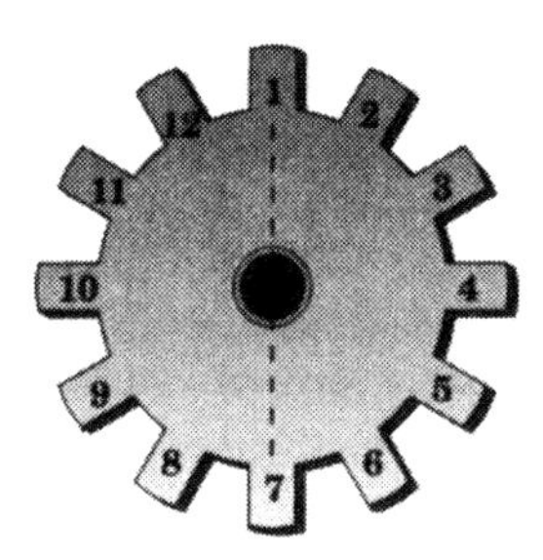

	A	B
1	Impact strength	position
2	1976	P 1
3	1916	P 1
4	2090	P 1
•••	•••	•••
•••	•••	•••
35	2425	P 2 & 12
36	2000	P 2 & 12
•••	•••	•••
•••	•••	•••
163	2120	P 7
164	2156	P 7

FIGURE 3.2.2 Gear and fragment of data spreadsheet.

wrong, the impact strengths required to break each tooth were obtained for teeth in each position in the diagram in Fig. 3.2.2. The positions of the teeth are important and are related to position in the mold. Observations on pairs of teeth in a symmetrical position about the vertical axis are grouped. Whether a tooth is a 2 or a 12 will depend on which face of the blank is uppermost. Positions 1 and 7 are distinguishable, however.

The data in Table 3.2.1 are presented in a different format from that used for the vital statistics data. We can represent them in a similar two-column format, as in the spreadsheet extract shown in Fig. 3.2.2. This type of format (see also Table 2.1.1) is more common for the machine storage of data. We want to have a look at the data in Table 3.2.1 and see where we find the weakest teeth. We are going to postpone this for the moment, however, and use parts of this data to illustrate various ways of making comparisons using the basic idea of *plotting each of the groups against the same scale.*

3.2.1 Using Dot Plots

For small samples we like to use dot plots. In Fig. 3.2.3 we have plotted our data along a horizontal scale (as we did in Fig. 3.2.1). We might equally well have arranged our plots vertically using a vertical scale. Figure 3.2.3 compares the impact strengths of teeth in positions 2 and 12, 3 and 11, and 6 and 8 (the smaller groups in the data set). The main thing we see is a slight increase in impact strengths as we move from the first group (positions 2 and 12) to the second group (positions 3 and 11). The shift in average level is fairly small, however, compared with the variability of impact strengths (from gear blank to gear blank). There is a much more dramatic increase in average impact strength moving from positions 3 and 11 to positions 6 and 8. Moreover, there is much less variability in impact strength at positions 6 and 8, with all the points being fairly tightly clustered except for one outlying low value.

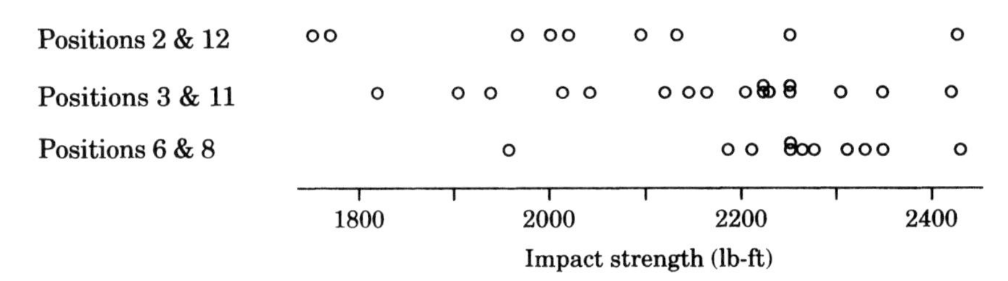

FIGURE 3.2.3 Using dot plots to compare three groups for the gear impact strength data.

3.2.2 Using Stem-and-Leaf Plots

We can also use stem-and-leaf plots for larger groups, as in Figs. 3.2.4 and 3.2.5. Figure 3.2.4 is interesting in that we have arranged the stem-and-leafs back to back, which makes comparisons particularly easy. Of course, this only works for two groups. For more than two, we have to use plots like Fig. 3.2.5. In Fig. 3.2.4 we see lower average impact strengths in position 7. Note that the straggling values for positions 4 and 10 are all large, whereas in position 7 they are all small. Position 7 shows signs of bimodality and a long lower tail.

Units 15 | 9 = 1590 lb–ft

Positions 4 and 10		Position 7
	15	9
	16	4
	16	
	17	0
	17	
	18	
	18	
	19	0 3
8	19	5 5 6 6 6
4 3 3 0	20	0 1 1
	20	7 7
4 3 2 1 0	21	1 2 3
9 8 7 6 6 5	21	5 5 6 6
3 2 2 2 1 1 0 0 0 0 0	22	0 1 1 1 2 2 3 3
5	22	6 8 9
4 1 1	23	
	23	
	24	
	24	
2	25	
9	25	

FIGURE 3.2.4 Using back-to-back stem-and-leafs to compare two groups for the gear impact strength data.

	Position 1		Positions 4 and 10		Position 7
15		15		15	9
16		16		16	4
16	7	16		16	
17		17		17	0
17	7	17		17	
18		18		18	
18	7 9 9	18		18	
19	0 2	19		19	0 3
19	8 8 9	19	8	19	5 5 6 6 6
20	0 3	20	0 3 3 4	20	0 1 1
20	5 5 8 8 9	20		20	7 7
21	1 3 3 4	21	0 1 2 3 4	21	1 2 3
21	6 7	21	5 6 6 7 8 9	21	5 5 6 6
22	0 1 1 2 2	22	0 0 0 0 0 1 1 2 2 2 3	22	0 1 1 1 2 2 3 3
22	5 9	22	5	22	6 8 9
23	2	23	1 1 4	23	
23	8	23		23	
24		24		24	
24	6	24		24	
25		25	2	25	
25		25	9	25	

FIGURE 3.2.5 Using stem-and-leaf plots to compare three groups for the gear impact strength data.

3.2.3 Using Box Plots

Figure 3.2.6 takes the same three groups as Fig. 3.2.5 and compares them using box plots.[13] Finally, in Fig. 3.2.7 we compare all the groups. Since some groups are small and some are large, we have used a compromise plot, namely, dot plots with the box plot information—minus the whiskers—superimposed.

There is a nice postscript to this example. If you look at the plots, you will see that, apart from position 7 itself, the further the teeth get away from position 7, the weaker they tend to get. When Ford's statistician showed the supplier of the blanks a plot like this,[14] the supplier immediately saw what was happening. The gear blanks were made using a powdered metallurgical molding process in which powder was blown into the blanks through an orifice at position 7. Position 1 was furthest from that orifice. Gunter states that the decreased strength and consistency the further from

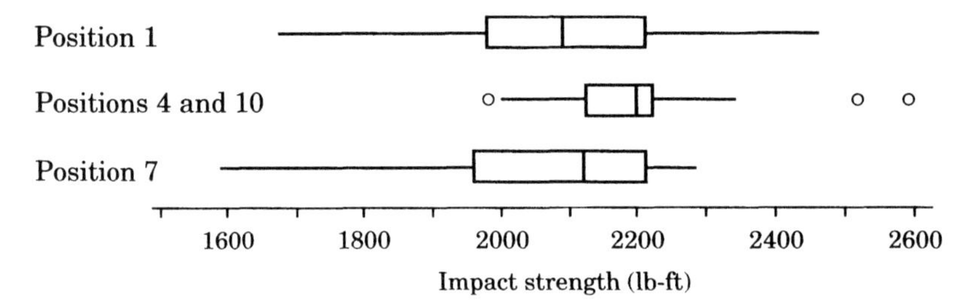

FIGURE 3.2.6 Using box plots to compare three groups for the gear impact strength data.

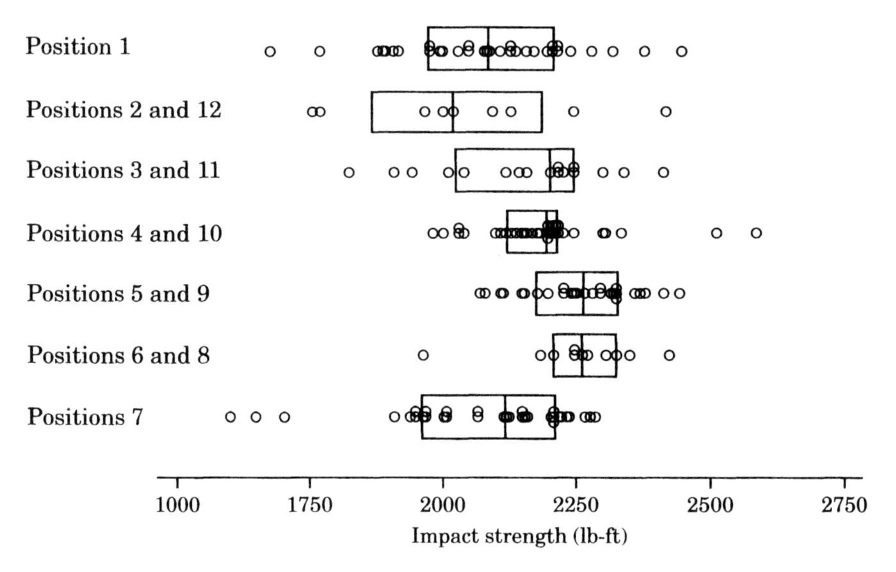

FIGURE 3.2.7 Using dot–box plots to compare all of the groups for the gear impact strength data.

[13] Comparative box plots are available in almost all statistics packages, for example, under `Graph` → `BoxPlot` in Minitab, where the grouping variable is entered as the X-variable. See also the `boxplot` function in Splus and R.

[14] Actually, he just used box plots.

position 7 was due to the fact that the powder was not packing tightly enough into the corners of the mold, resulting in insufficient density and tooth strength on firing. Identifying the problem is the first step toward fixing it.

3.2.4 An Organized Approach to Comparing Groups

As previously noted, the essential requirement for comparing groups is that we plot each group against the same scale. Any of the types of plot that we have seen can be used (dot plots, stem-and-leaf plots, histograms, box plots, dot-box plots). The type of plot to use for all the groups depends largely on the sample sizes of the groups. The guidelines given for single variables apply here as well, but with one difference. It is difficult to compare more than about two or three histograms or stem-and-leaf plots. However, box plots are very useful for comparing larger numbers of groups with moderate to large samples. (Our favorite set of comparative box plots, given in Fig. 2 of Review Exercises 3, compares 36 groups.)

When presented with such plots, we like to begin looking at gross features and finish up looking at details. We would thus look for the following features in the order given:

(i) *Changes in average level from plot to plot*
How big are the shifts in the mean compared with the usual level of variability? Are they small? Is there complete separation?

(ii) *Differences in variability* (spread)

(iii) *Differences in gross distributional shape*
(e.g., differences in type of skewness or in number of modes)

(iv) *Details of individual groups* (including identifying any extreme or unusual points)

Having read the information in the plot, we should then start thinking about *why* the data is behaving as it is (as in the gear-mold story).

QUIZ FOR SECTION 3.2

1. What general principle governs the construction of plots to explore the relationship between a quantitative variable and a qualitative variable?

2. Which plots work best when there is a large number of groups to be compared?

3. What type of plot works best when there is a small number of observations in each group?

4. What general principles govern the way in which we should look at plots that compare groups?

EXERCISES FOR SECTION 3.2

1. A sample of 27 children aged about 8–9 years who had an inborn error of metabolism known as transient neonatal tyrosinemia (TNT) were compared with a closely matched sample of 27 normal children (the control group) by their scores on the Illinois Test of Psycholingual Ability. This test gives scores on 10 variables. One of these is the auditory reception score.

The data for the two groups on this score are as follows:

Control Group													
40	35	30	22	21	39	39	22	44	34	30	26	44	36
30	18	27	30	33	26	31	29	34	36	42	32	38	
TNT Group													
26	31	28	19	31	35	37	41	35	29	18	38	31	26
33	27	22	37	24	11	24	28	17	31	34	41	23	

Data courtesy of Peter Mullins, University of Auckland

(a) Compare the control group and TNT group using: (i) dot plots, (ii) back-to-back stem-and-leaf plots, and (iii) box plots.

(b) What do you conclude from the plots?

(c) Does one type of plot show you anything the others do not, or are all showing the same features of the data here?

2. Waist-to-hip ratios of men and women sampled from the workforce-study database are as follows:

Men under 45									
0.84	0.99	0.78	0.91	1.06	0.98	0.86	0.89	0.87	0.92
0.90	0.89	0.86	0.92	0.87	0.80	0.91	0.94	0.84	0.98
Men over 45									
0.93	0.90	1.01	0.88	0.84	0.92	0.91	1.02	0.85	0.98
0.81	0.99	0.92	0.87	0.92	0.97	0.93	0.91	0.84	0.88
Women under 45									
0.92	0.76	0.64	0.80	0.71	0.82	0.75	0.80	0.87	0.84
0.70	0.73	0.83	0.80	0.74	0.80	0.84	0.69	0.77	0.77
Women over 45									
0.84	0.89	0.87	0.80	0.76	0.92	0.73	0.76	0.81	0.77
0.84	0.86	0.75	0.79	0.83	0.86	0.85	0.81	0.87	0.80

Data courtesy of of Dr. P. Metcalf and Dr. R. Scragg, University of Auckland

(a) Compare the four groups by plotting box plots.

(b) What do you conclude from your plots? Are there any age differences for each sex?

3.3 TWO QUALITATIVE VARIABLES

For a single qualitative variable we constructed a frequency table by counting the number of observations falling into each level (group, or category) of the variable and depicted the results in a bar graph (see Section 2.6). When we want to look at the relationship between two categorical variables, we first count the number of observations falling into each *combination* of levels of the two variables and lay them out in a ***two-way table of counts.*** This process is known as ***cross-tabulation*** or *cross-classification.*

✖ Example 3.3.1 *Cultural Differences in Body Image*

We shall use the workforce-study database of Drs. Metcalf and Scragg to investigate whether there are ethnic differences in women's perceptions of the extent to which they are overweight. Figure 3.3.1(a) contains a fragment of the data. (There are well over 100 variables in the database.) We restricted our attention to 246 women who had very similar values of a measure of overweight called body-mass index. All of these women were a little lighter than average for their height. We are interested in the relationship between the two *qualitative* or categorical variables, namely, BODY IMAGE and ETHNICITY. Here BODY IMAGE is each woman's own rating of her size (`slight.uw` is "slightly underweight," `right` is "about right," `slight.ow` is "slightly overweight," and so on). The women were classified into four ethnic groups: Asian, European, Maori, and Pacific Islander.

The first step toward investigating the relationship between BODY IMAGE and ETHNICITY is to *cross-tabulate* these two variables and construct the two-way table of counts which is shown in Fig. 3.3.1(b).[15] It may be helpful to think of a table like this as corresponding to a set of pigeonholes or letter boxes. We slot each individual into the correct pigeonhole (determined here by the woman's ethnicity and her body image), and once we are finished, we count the number of individuals in each of the pigeonholes. In technical terms, the pigeonholes are called the ***cells** of the table.*

The totals at the bottom of the table are ***column totals.*** If you add all the numbers in the second column, you get 60, telling you that 60 of the women thought their weight was "about right." The totals in the right-hand column are ***row totals.*** If you add all the numbers in the first row, you get 50, which tells you that 50 of the women in the data set were Asian. Both the row totals and the column totals add to give the ***grand total*** in the lower right-hand corner. This tells us that we are dealing with a total of 246 women. What does looking at the table tell us about the relationship between ethnicity and body image? With experience you may learn to see patterns in such tables. We find it more generally helpful, however, to calculate and plot appropriate proportions to get a picture of the relationship. Before we can do this, we need to know a little about the calculation of proportions from a two-way table of counts.

First, we shall discuss *whole-table* proportions, or the proportions of the whole set of women falling into each cell of the table. This turns out to be equivalent to dividing each cell count by the grand total (or 246). For example, of the 246 women, 21 were both Asian and considered themselves "slightly overweight," a proportion of $21/246 = 0.085$ (or 8.5%). The proportion of women who were both European and considered themselves "very overweight" is $3/246 = 0.012$ (or 1.2%).

The row and column totals can be helpful in calculating some proportions. The proportion of all women who were Pacific Islander is 31/246. Here we got the total number of Pacific Island women for the numerator from the fourth row total. Similarly, the proportion of all women who thought they were "slightly underweight" is 10/246. All of the entries in Fig. 3.3.1(c) were computed in this way. [In fact all the computations in Fig. 3.3.1(c) and (d) were done on a spreadsheet, with the resulting tables then copied and pasted into the figure.]

---

[15]To perform a cross-tabulation in Minitab, use the menu item `Stat → Tables → Cross Tabulate`. The Splus function is called `crosstabs`. Cross-tabulation can be performed in Excel using `PivotTable` under the `Data` menu.

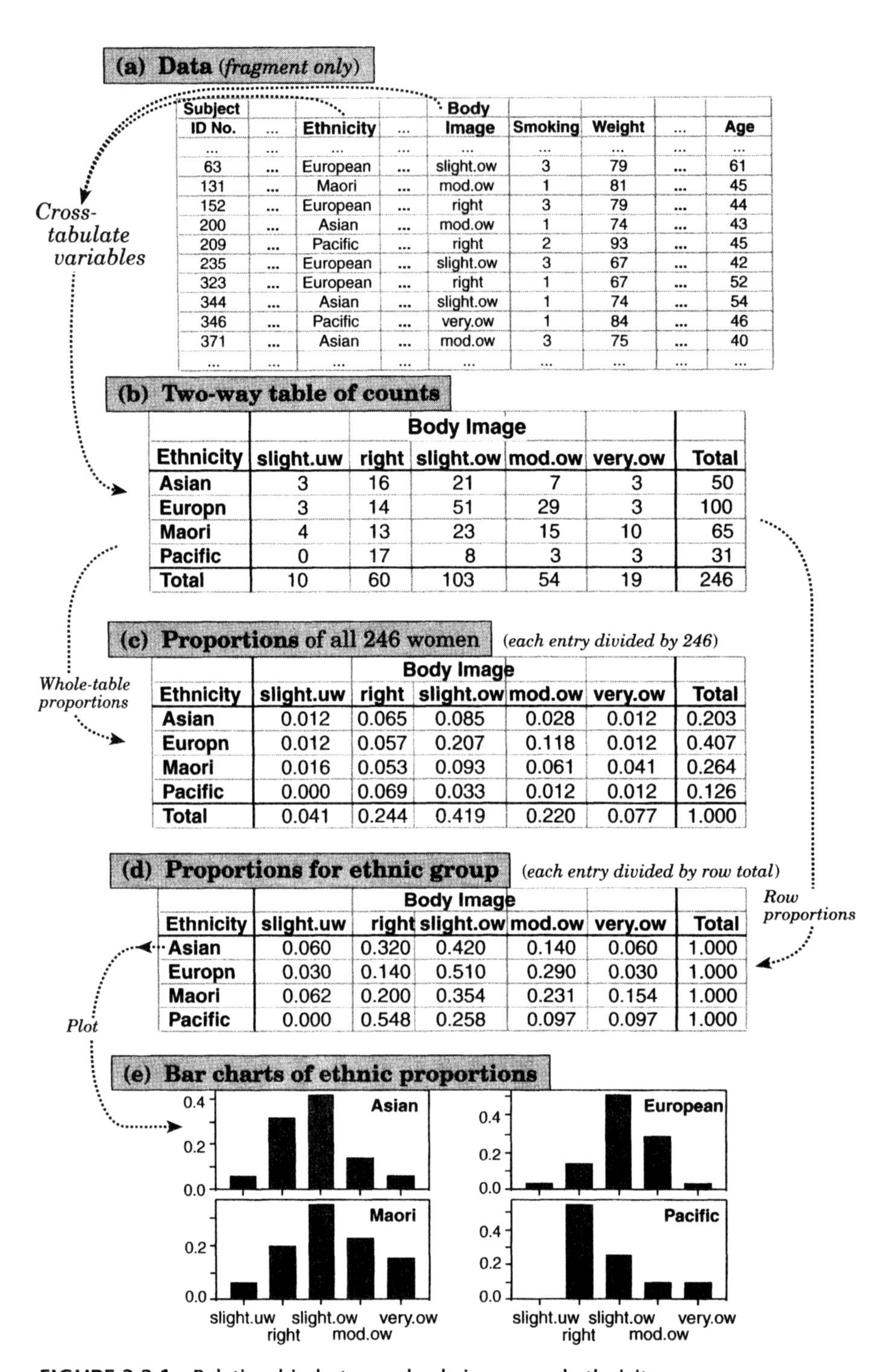

(a) Data *(fragment only)*

Subject ID No.	...	Ethnicity	...	Body Image	Smoking	Weight	...	Age
...	...	...	...	...	...	...	...	...
63	...	European	...	slight.ow	3	79	...	61
131	...	Maori	...	mod.ow	1	81	...	45
152	...	European	...	right	3	79	...	44
200	...	Asian	...	mod.ow	1	74	...	43
209	...	Pacific	...	right	2	93	...	45
235	...	European	...	slight.ow	3	67	...	42
323	...	European	...	right	1	67	...	52
344	...	Asian	...	slight.ow	1	74	...	54
346	...	Pacific	...	very.ow	1	84	...	46
371	...	Asian	...	mod.ow	3	75	...	40
...	...	...	...	...	...	...	...	...

(b) Two-way table of counts

	Body Image					
Ethnicity	slight.uw	right	slight.ow	mod.ow	very.ow	Total
Asian	3	16	21	7	3	50
Europn	3	14	51	29	3	100
Maori	4	13	23	15	10	65
Pacific	0	17	8	3	3	31
Total	10	60	103	54	19	246

(c) Proportions of all 246 women *(each entry divided by 246)*

	Body Image					
Ethnicity	slight.uw	right	slight.ow	mod.ow	very.ow	Total
Asian	0.012	0.065	0.085	0.028	0.012	0.203
Europn	0.012	0.057	0.207	0.118	0.012	0.407
Maori	0.016	0.053	0.093	0.061	0.041	0.264
Pacific	0.000	0.069	0.033	0.012	0.012	0.126
Total	0.041	0.244	0.419	0.220	0.077	1.000

(d) Proportions for ethnic group *(each entry divided by row total)*

	Body Image					
Ethnicity	slight.uw	right	slight.ow	mod.ow	very.ow	Total
Asian	0.060	0.320	0.420	0.140	0.060	1.000
Europn	0.030	0.140	0.510	0.290	0.030	1.000
Maori	0.062	0.200	0.354	0.231	0.154	1.000
Pacific	0.000	0.548	0.258	0.097	0.097	1.000

FIGURE 3.3.1 Relationship between body image and ethnicity.

Sometimes we want to concentrate on a single row or column and (temporarily) ignore the rest of the table. For example, the first row contains all our information about Asian women. To obtain the proportion of Asian women who thought they were "slightly underweight," we see that there are 50 Asian women and 3 of these thought themselves "slightly underweight"—a proportion of 3/50. The proportion of Asian women who thought they were about "right" is 16/50. Each entry in the first row of the table in Fig. 3.3.1(d) is obtained by dividing the cell count by 50. This gives us the proportions of Asian women falling into each one of the body-image classes. Similarly, the second row relates entirely to European women. When we divide each cell count in the second row by the row total of 100, we get the proportions of European women falling into each body-image class.

For the body-image problem, it is these *row proportions* that are interesting. Each row represents the behavior of a particular ethnic group. Figure 3.3.1(e) presents a separate bar chart for each set of row proportions in the table of Fig. 3.3.1(d). The most striking features emerging from the plot are that European women seem to be more likely to place themselves in the overweight categories than is the case for the other ethnic groups, and Pacific Island women seem more likely to be relaxed about their weight (in the sense of classifying themselves as "right"). Remember that we are talking about differences in the way the women think about themselves and that the actual degree of overweight for all of these women is very similar.

The two important things to take away from this section are the idea of cross-tabulating (or cross-classifying) two qualitative variables to produce a two-way table of counts and the ability to read the information in such a table. We study the calculation of various proportions of interest from two-way tables in Chapter 4. The plotting and analysis of count data is the subject of Chapter 11.

QUIZ FOR SECTION 3.3

1. What are the two words commonly used to describe the process of translating individual-level data on two qualitative variables into a two-way table of counts?
2. How do we form whole-table proportions? row proportions? column proportions?

EXERCISES FOR SECTION 3.3

1. Construct a two-way table of counts relating BETA and SURG for the heart attack patients in Table 2.1.1.
2. Table 3.3.1 reports statistics concerning all Japanese children having accidents (including fatalities) on the roads in 1984. It is a two-way table relating SCHOOL (the level of school a child was attending) and ACTIVITY (what the child was doing at the time of the accident). Compute the proportion of
   (a) all children in junior high and traveling to or from school;
   (b) all children in kindergarten and in the "other" level of ACTIVITY;
   (c) all children who were going to or from school;
   (d) all children who were in kindergarten;
   (e) kindergarten children who were going to or from school;
   (f) first- to third-grade children who were playing;
   (g) children in the "other" level of ACTIVITY who were infants.

TABLE 3.3.1 Two-Way Table Relating School and Activity

	School					
Activity	Infant	Kinder.	1st–3rd Gr.	4th–6th Gr.	Jr. high	Total
To/from School	0	1,205	3,163	1,213	1,151	6,732
Playing	2,184	2,300	2,373	719	182	7,758
Other	6,552	7,447	7,642	2,560	1,168	25,369
Total	8,736	10,952	13,178	4,492	2,501	39,859

Source: *Statistics '84. Road Accidents Japan.* Tokyo: International Association of Traffic and Safety Sciences. Counts reconstructed from percentages by the authors.

3.4 ARE THE PATTERNS YOU SEE REAL OR RANDOM?

(A Role for Probability)

We have spent some time looking at batches of data to find patterns (trends) and exceptions. Often our data has been sampled from a wider population (or several populations). Usually it is that wider population that we want to find out about, not the few individuals who have fallen by chance into our sample. Or we may be dealing with an experiment with unpredictable outcomes. To what extent do the patterns we see in our plots represent real features of the wider population(s) or some underlying mechanism generating the data, and to what extent are they accidental by-products of random sampling? In posing such questions we are coming out of the *exploratory* phase of our analysis and entering the *confirmatory* phase, where we now want to confirm that any features we see are real. To make any progress in the confirmatory phase, we have to study how the sampling process behaves. We shall find that the confirmatory phase requires the calculation of the chances, or probabilities, of certain risks. Two examples follow.

✻ EXAMPLE 3.4.1 *Drugs and Their Effect on Blood Pressure*

Suppose two drugs for reducing blood pressure were tried on 20 hypertensive males of similar age and build. The 20 males were split randomly into two groups of 10 using, say, the lottery method, with the first group receiving drug A and the second group drug B. The reduction in diastolic blood pressure, in millimeters of mercury (mm Hg), was recorded for each man. Ignoring cost differentials and differences in side effects, how would we determine which drug was more effective?

Suppose we draw dot plots of the data and obtain Fig. 3.4.1(a). We first note that the drugs act differently on different men, with reductions ranging from about 4 to 10 mm Hg. Some men using A have done better than others using B. It is quite conceivable that A will work better than B for some men, while for others the reverse will be true. If you were a hypertensive male of similar age and build, which drug would you use? Since the two patterns in Fig. 3.4.1(a) are almost identical, you would feel intuitively that there is no difference in the drugs. On average, the reduction by A

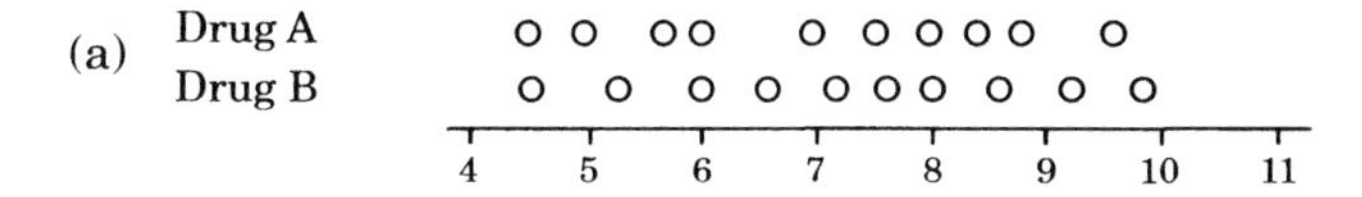

FIGURE 3.4.1 Reduction in Blood Pressure (in mm Hg). Patterns are (a) similar, (b) non-overlapping, (c) overlapping.

would be about the same as that by B. Looking at the problem another way, suppose we chose a reduction of, say, 7 mm Hg. The chance (probability) of doing better than this (as estimated by the proportion of people with bigger reductions) is about the same for both drugs.

If the data gave us Fig. 3.4.1(b) instead of Fig. 3.4.1(a), then using arguments similar to those preceding, we would conclude that A is more effective than B as there is no overlap of the two patterns. What often happens in practice, however, is that we get Fig. 3.4.1(c), in which the two patterns overlap. Here we would tend to choose A, but would we be confident about our choice? Diastolic blood pressure varies, not only from person to person, but within an individual from hour to hour, depending on level of activity, food intake, and so on. It may have happened that the men labeled (i) and (ii) in Fig. 3.4.1(c) had big reductions with drug A because of chance environmental effects. If these two observations were moved substantially to the left, the two patterns would not look so different. The problem here is to determine when an observed difference in performance is due to an actual difference in the drugs or simply due to random variation. To deal with this problem we need to develop a theory of probability that will enable us to attach probabilities to random observed differences when, in fact, there is no difference in the overall performance of the two drugs.

Example 3.4.2 *Estimating the Size of an Animal Population*

An important problem in statistical ecology is how to determine the size of an animal population, such as the number of fish in a lake. A large number of techniques for providing an answer are available (see Seber [1982]), but the best-known one is the capture-recapture method.

We begin with an enclosed population of size N. A random sample of M (say, 100) individuals are captured, tagged or marked in some way, and then released back into the population. After allowing time for the marked and unmarked to mix sufficiently, a second simple random sample of size n $(= 50$, say) is taken (without replacement) and m $(= 10$, say) are observed to be marked. Then, if the second sample is exactly representative of the whole population, the proportion of marked individuals in the sample will be the same as the proportion of marked individuals in the population.

Hence

$$\frac{m}{n} = \frac{M}{N} \quad \text{or} \quad \frac{10}{50} = \frac{100}{N}$$

which we can solve for N and get $N = 500$.

There are a number of practical assumptions underlying this method of estimation. Perhaps you can think of some of them, for example, that animals do not lose their marks (tags). In addition to whether appropriate assumptions are satisfied, we also have to take into account sampling variation. The sample proportion of tagged individuals will never be exactly equal to the population proportion, so that equating the two will lead to just an estimate of N and not the true value. For example, if m was 9 or 11 in our above example, then

$$N = \frac{Mn}{m} = \frac{100 \times 50}{m} = \frac{5000}{m}$$

has values, to the nearest whole number, of 556 and 455, respectively.

Since the estimate of N can vary each time the experiment is repeated, we would like to make a statement about the extent of the variation. For example, if we knew that our estimate of N would be within 5% of the true value of N for 99 out of 100 repetitions of the experiment, we would then have reasonable confidence in our method. If we were planning to do such an experiment, we would know that we would have to be extremely unlucky to get an estimate that was not within 5% of the true value; the chances of this happening are only 1/100. Once the experiment was done, this fact would give us a fair amount of confidence in the answer we obtained. Therefore, to assess the accuracy of any estimation procedure, we need to be able to calculate "chances," or probabilities, of certain risks. For this reason we need to spend time in this book developing ideas of probability. We do this in the next chapter.

3.5 SUMMARY

In this chapter we have concentrated on investigating relationships between two variables. We make the following distinctions between variables:

- A ***quantitative variable*** is a measurement or a count.
- A ***qualitative variable*** (also called a *categorical* variable or grouping *factor*) tells us to which of a set of groups an observation belongs.

As each of our two variables can be either quantitative or qualitative, we have three possible combinations, as in Table 3.5.1.

We look for relationships between two variables as follows:

1. ***Quantitative versus quantitative***
   (a) The chief tool for comparing two quantitative variables is the ***scatter plot.***
   (b) Predictable variables are said to be ***nonrandom.*** Unpredictable variables are said to be ***random.***

TABLE 3.5.1 Looking at Relationships between Variables

Variable types	Tools in Chapter 3
Both quantitative	Scatter plot
One quantitative One qualitative	Plotting: – Plot data for each group against same scale. – Use dot plots, box plots ($n \geq 15$), stem-and-leaf plots, or histograms, depending on batch sizes. – Use dot plots and box plots if number of groups is large
Both qualitative	Cross-tabulate to form two-way tables of counts. Plot appropriate proportions.

(c) ***Controlled variables*** are nonrandom variables whose values are manipulated by the experimenter.

(d) In ***regression*** we have a variable whose behavior we want to predict or explain and another variable that we want to use in forming our predictions or explanations.

- The ***response variable*** (denoted Y) is the variable whose behavior we want to predict or explain.
- Y must be random and is usually continuous.
- Y is plotted on the vertical axis.
- The ***explanatory variable*** (denoted X) is used to form our predictions or explain the behavior of the response.
- X is quantitative (discrete or continuous). It may be random or nonrandom.
- A ***regression relationship*** = ***trend*** + ***scatter.***
- ***Outliers*** are values that are unusually far from the trend curve.

(e) ***Strong or weak relationships.*** A relationship may look weak because it has been plotted over only a limited range of x-values.

(f) ***Prediction:***

- It is very dangerous to predict outside the range of the data.
- We cannot make precise predictions from a weak relationship.

2. ***Quantitative versus qualitative***
We plot the values of the quantitative variable for ***each group against the same scale*** and then compare groups. We can use any of the following: dot plots, stem-and-leaf plots, box plots, or histograms. Box plots are particularly good when we have many groups and more than about 20 observations in each group.

3. ***Qualitative versus qualitative***
We ***cross-tabulate*** the variables to form a ***two-way table of counts*** (augmented by row and column sums). Relationships can be explored by plotting appropriate proportions calculated from the table.

REVIEW EXERCISES 3

Note that the following exercises also make use of the tools given in Chapter 2.

1. Following are the building consent fees (in $) that would be charged for a $55,000 addition of a bedroom and bathroom to an existing house by each of 29 N.Z. cities. The corresponding number in parentheses is the average number of inspections that would be carried out during the building process.

541 (5), 743 (6), 1193 (5), 790 (5.5), 575 (5), 538 (6.5), 412 (3), 556 (8), 787 (5), 701 (10), 555 (4), 708 (6), 684 (7), 727 (8), 647 (7), 786 (6), 773 (6), 574 (4), 840 (5), 463 (3), 491 (4), 673 (5), 609 (6), 540 (5), 789 (4), 925 (3), 742 (7), 1068 (4), 716 (4).

(a) Construct a stem-and-leaf plot of the data and compute the five-number summary.

(b) How many times more expensive is the most expensive city than the cheapest city?

(c) There is clearly a great deal of variability between cities in the amount they charge. One might wonder whether the more expensive cities work harder for their fees. Construct a plot that displays the relationship between the number of inspection visits and the amount charged. What do you see?

2. Lampshells are of particular interest because they have fossil records going back a very long time. Although they are rare worldwide, they are quite abundant in parts of New Zealand. In 1979, two scuba divers associated with the New Zealand Oceanographic Institute collected samples of the shells from the sea floor of Paterson Inlet, Stewart Island. The aim of the biologists was to see what differences existed between the live and dead shellfish. The divers placed three square frames fairly close to each other and then proceeded to scoop up all the shells and sediments from inside each square. This material was then shoveled into three plastic bags, which were sealed and taken to the surface. On shore, shells from each square were sorted into piles of live and dead animals. The length of each shell was measured with dial calipers accurate to 0.01 mm. In Table 1 we give a sample of the data collected from square 1.

(a) Given the aim of the biologists for collecting the data, do you think the sampling design is satisfactory? Would your answer be the same if you were interested in estimating the density of shells per unit area on the sea floor? If not, why not?

(b) Construct back-to-back stem-and-leaf plots for the shells from live and dead animals. What do you conclude? Do either of the plots have any special features?

(c) Construct side-by-side box plots using the same horizontal scale. Do these plots confirm your conclusions in (b)? Have you lost or gained any information by using box plots?

(d) What proportion of live animals have a shell length shorter than 15 mm? What is the proportion for dead animals?

3. Some years ago, a friend of one of the authors wanted to buy a particular model of GMC truck somewhere in the range 1979–1981. He started watching the newspaper advertisements and recording YEAR and PRICE. He obtained the data in Table 2. In fact, he kept a running scatter plot. In other words, each time he saw another truck advertised, he added its YEAR and PRICE to the plot.

(a) Construct a scatter plot of the first 23 observations. What can you say about the relationship between YEAR and PRICE?

TABLE 1 Lengths of Lampshells

Live animals							
22.34	18.72	17.64	16.84	20.04	20.05	22.06	25.18
10.28	13.08	21.48	6.18	13.06	16.92	19.96	18.08
5.76	5.90	12.98	4.18	19.02	15.54	20.68	6.38
13.08	21.26	21.86	21.06	9.06	23.82	15.64	4.12
8.62	15.02	7.22	15.23	4.84	4.13	7.76	24.48
Dead animals							
24.93	16.56	16.82	21.35	23.34	19.95	23.38	18.45
24.74	16.22	12.37	18.27	10.83	19.86	22.62	24.56
21.38	24.32	22.71	20.06	17.70	20.12	22.52	25.93
19.12	19.95	20.22	13.75	20.49	21.55		

Source: *Statistics at Work*, N.Z. Statistical Association (1982).

TABLE 2 Truck Prices[a]

Obs	Year	Price	Obs	Year	Price	Obs	Year	Price	Obs	Year	Price
1	80	6995	7	84	14500	13	81	8950	19	78	4750
2	84	12500	8	78	2400	14	81	8500	20	84	12500
3	75	4000	9	80	7995	15	82	8995	21	75	1400
4	81	8950	10	85	12900	16	77	3000	22	85	9700
5	77	4500	11	80	4995	17	82	9700	23	81	8995
6	78	4800	12	88	16000	18	84	8000	24	80	????

[a] Prices in Canadian dollars.

(b) Most people seeking a 1979–1981 model would only follow the prices of trucks in those years. Our friend recorded everything he saw from 1975 to 1988. Why was it a good idea to include the additional years? (Look at your plot.)

(c) Draw a summarizing trend line on your graph by eye. What would you estimate the average price of a 1980 truck to be? (Hint: Use your trend line and read the 1980 value off your graph.)

(d) There is obviously quite a lot variability in prices of trucks even from the same year. We want to take this into account. Draw a summarizing envelope around the points by eye. What range of values would you expect prices for a 1980 truck to be in?

(e) The 24th truck was a 1980 at $2000. Add it to the plot. The truck was in good condition. What would you do?

(f) What other variables apart from YEAR might affect the price of a car or truck? How would you "measure" each of these quantities? Can you think of ways in which you could add any of this additional information to the graph?

4. Consider Fig. 3.1.9 and its description. Use the scatter plot and your own intuition to answer the following:

(a) What systolic volume would you expect a man of this type to have if his diastolic volume was 200? (Give a single-number estimate.)

(b) Into what range of values would you be fairly sure that systolic volume would fall if the diastolic volume was 200?

(c) Repeat (a) and (b) for a diastolic volume of 100.

5. In 1989 *Time* magazine ran a 27-page article giving photographs of everybody who had been killed by a gun in the United States in a single week in May (464 people) and a little summary information for each. Professor Winston Cherry of the University of Waterloo compiled the stem-and-leaf plots of Fig. 1 by categorizing the deaths into four classes and plotting the age at which each person had died (Units 2 | 3 = 23 years). Four people were not represented, as their ages were missing.

(a) What can you see about the ages at which people die from each of the four modes of death by looking at the stem-and-leaf plots?

(b) If we had more space, we would have arranged the plots differently on the page. What would be a better arrangement?

(c) Find five-number summaries for each of the groups, and construct box plots to compare the groups. What important information in Fig. 1 is hidden entirely when we look at box plots? Is the median gun-suicide age a particularly meaningful summary statistic? Justify your answer.

(d) Guess the shape of a stem-and-leaf plot of the ages for all gun deaths (i.e., treated as a single, combined group). You may wish to construct the plot to check your answer.

(e) Would you expect the patterns here to apply to other western countries, for example, United Kingdom, Canada, Australia, New Zealand? Justify your answer.

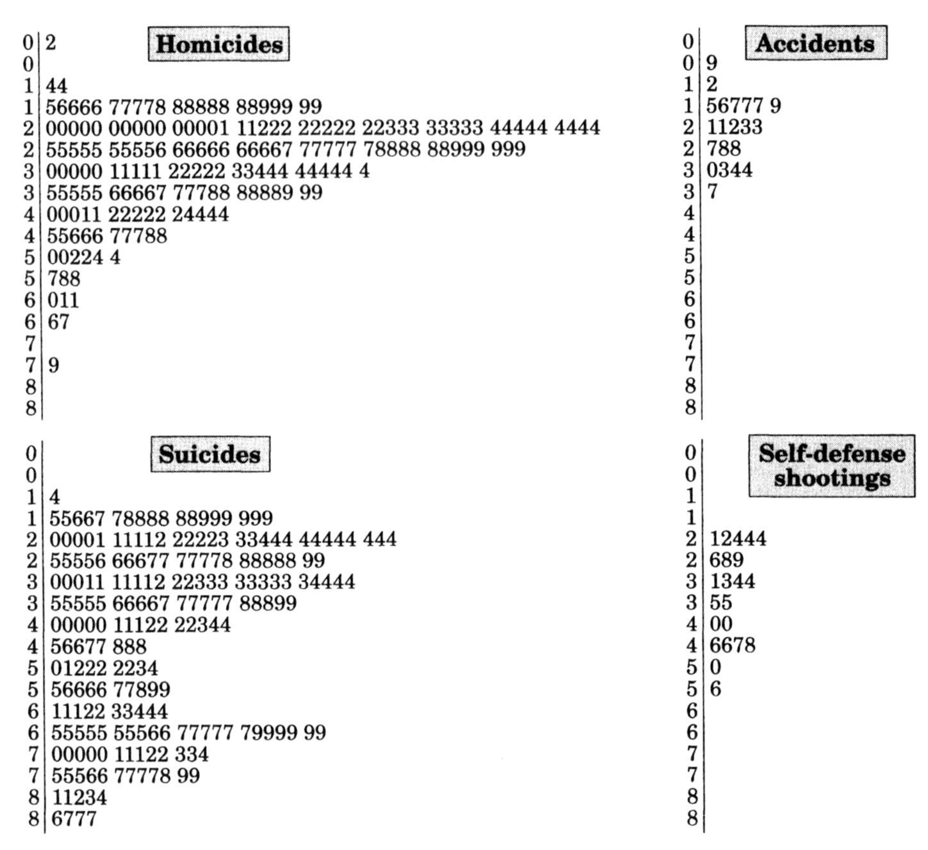

FIGURE 1 Ages of all U.S. gun deaths during one week in May 1989. Source: Compiled by Professor Winston Cherry from an article in *Time,* 17 July 1989.

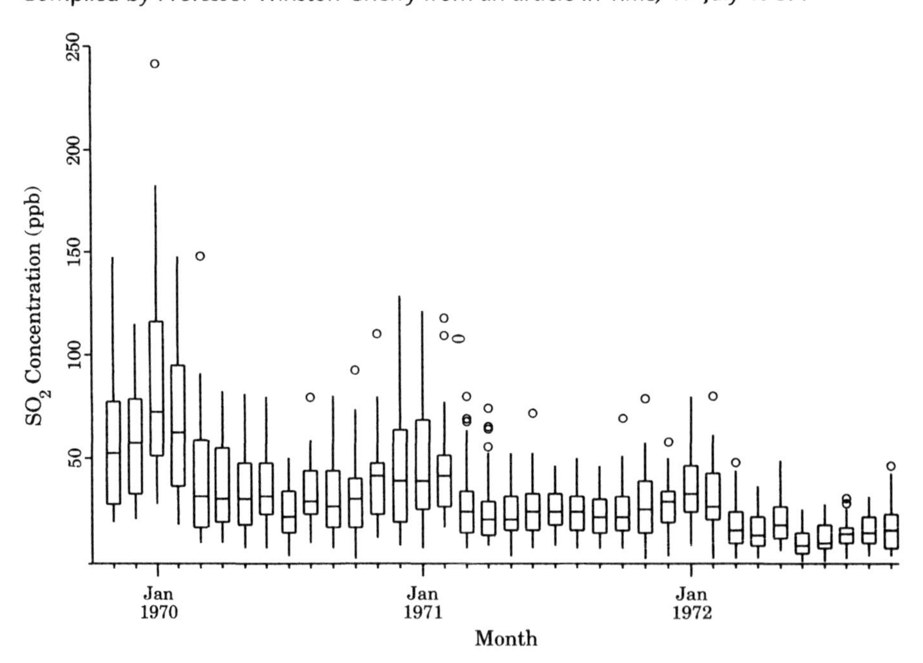

FIGURE 2 Box plots of daily maximum atmospheric SO_2 concentrations in Bayonne, New Jersey, for the months from November 1969 to October 1972. Source: Redrawn with permission from Chambers et al. [1983, p. 59].

6. Figure 2, redrawn from Chambers et al. [1983], contains our favorite set of box plots. We like it because there are many different features in this data that are revealed by the graph. The data plotted are daily maximum atmospheric SO_2 concentrations in Bayonne, New Jersey, from November 1969 to October 1972. Each box plot summarizes the daily readings for a particular month. Discuss any features you see in this data.

7. Figure 3 plots data on the dimensions of three species of fish (bream, pike, and perch) caught in Lake Laengelmavesi near Tampere in Finland. The data[16] are a subset of the fish-catch data in the *Journal of Statistics Education*'s data archive submitted by Juha Puranen. Several variables were measured for each fish caught. The variables we have used in the plots are (maximal) length, (maximal) height, (maximal) width, and weight. In the scatter plots of weight versus length, width versus length, and height versus length, different plotting symbols are used to distinguish between types of fish (+ for bream, * for pike, and o for perch).
   (a) This set of plots reveals many interesting features in these data. What features can you detect? Begin by summarizing the main features of each figure.
   (b) Which pair of variables gives the best discrimination between species?

8. Figure 4 plots the percentages of world gold production given in Table 2.6.2 for the years 1983, 1987, and 1991. The types of graph are commonly available (e.g., at the time of writing, in Microsoft Excel, Microsoft Word, and other spreadsheet programs and word processors) and widely used. The purpose of this exercise is to take what you have learned so far about graphics and think a little bit further, with slightly more complicated data.
   (a) Work out how each type of plot is constructed.
   (b) What can you see in the plots about trends in world gold production over the years?
   (c) Comment on the strengths and weaknesses of each type of plot. Are there ways in which you could improve some of them?
   (d) Which type of plot best displays the features *that you see* in the data? If you were to use two types to display this data, what would your second choice be? Would your first choice on its own be adequate?

   You will probably have noticed how South Africa's share of world production has declined over time. An obvious question is whether this is because South Africa is producing less or other people are producing more. Suppose we want to plot actual gold production figures, not percentages. All the types of plots in Fig. 4 can be adapted, with a little thought, to displaying actual production figures rather than percentages.

   (e) How would you modify each type of graph in Fig. 4 so that it displayed actual production figures?
   (f) If you have suitable computer software, make the modification, replot the data, and comment on what you see.

9. Apply suitable graphs suggested by the GDP data in problem 2 and Table 2 in Review Exercises 2 to look for trends over time in the contributions of the various economic sectors to American GDP.

10. Scatter plots are produced by major word processors (you do not need a statistical package). How can you create a simple dot plot for comparing several groups [e.g., Fig. 3(b) without the boxes] by just using the scatter-plot facility of your computer? How can you create a dot plot for a single group? Try your ideas out.

Further exercises, some involving large data sets, are given on our web site.

---

[16]We have changed some labeling anomalies, for example, the weights are in column 3.

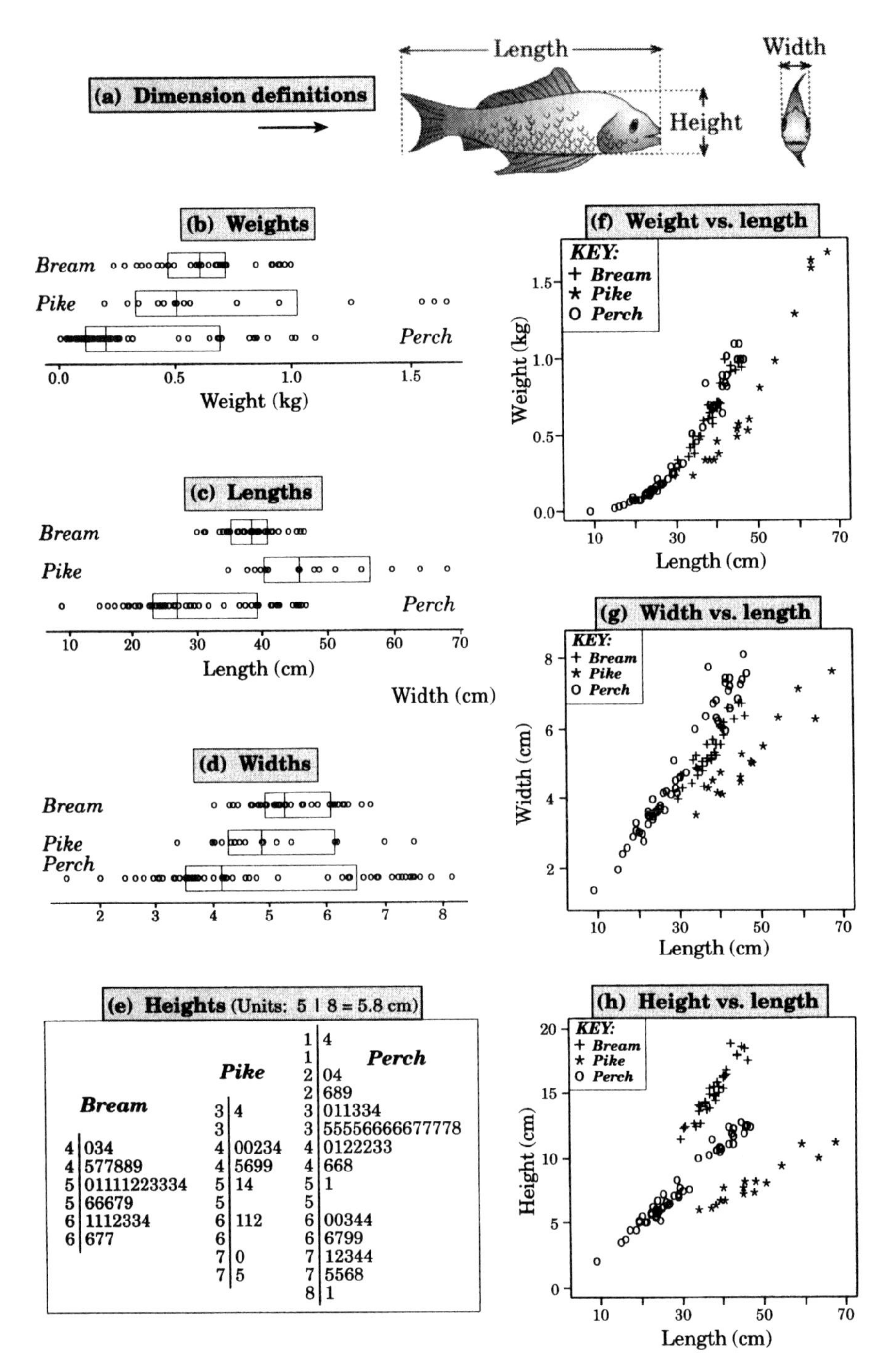

FIGURE 3 Dimensions for three species of fish (bream, pike, and perch).

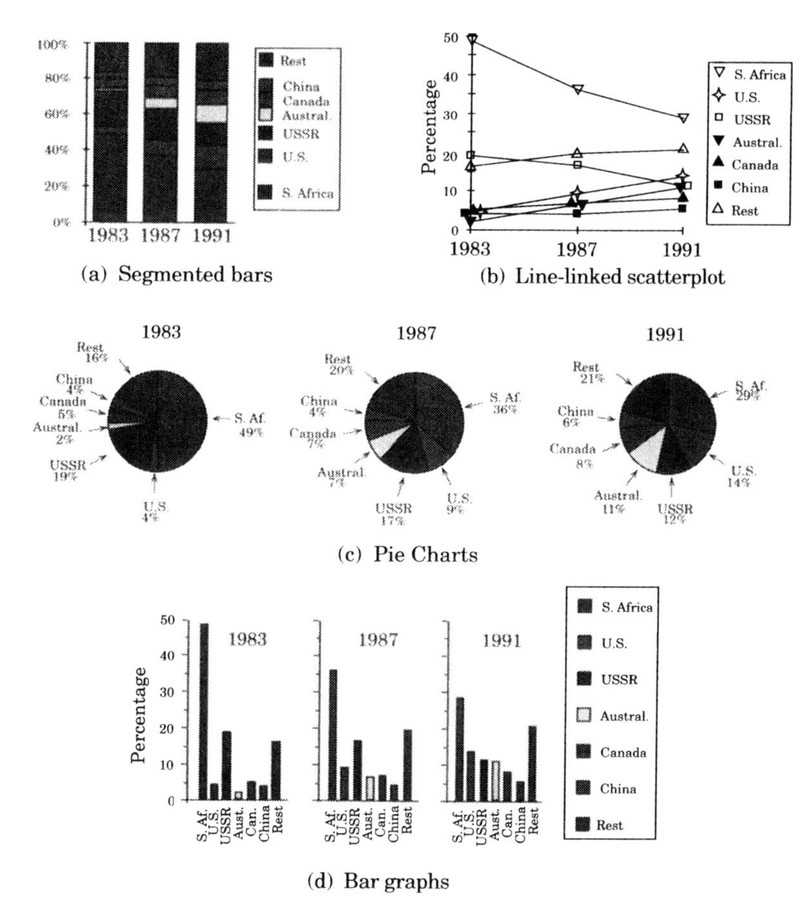

(a) Segmented bars

(b) Line-linked scatterplot

(c) Pie Charts

(d) Bar graphs

FIGURE 4 Percentages of world gold production over time.

CHAPTER 4

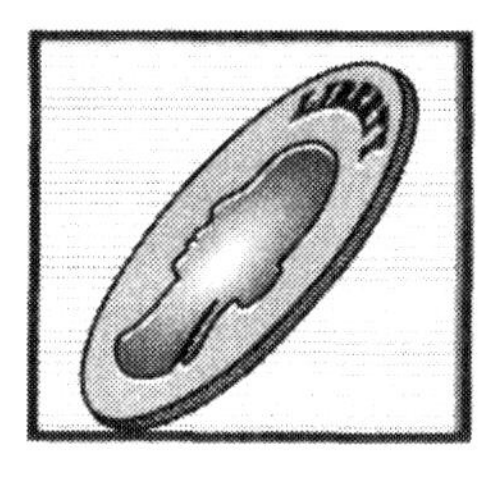

Probabilities and Proportions

Chapter Overview

While the graphic and numeric methods of Chapters 2 and 3 provide us with tools for summarizing data, probability theory, the subject of this chapter, provides a foundation for developing statistical theory. Most people have an intuitive feeling for probability, but care is needed as intuition can lead you astray if it does not have a sure foundation.

There are different ways of thinking about what probability means, and three such ways are discussed in Sections 4.1 through 4.3. After talking about the meaning of events and their probabilities in Section 4.4, we introduce a number of rules for calculating probabilities in Section 4.5. Then follow two important ideas: conditional probability in Section 4.6 and independent events in Section 4.7. We shall exploit two aids in developing these ideas. The first is the two-way table of counts or proportions, which enables us to understand in a simple way how many probabilities encountered in real life are computed. The second is the tree diagram. This is often useful in clarifying our thinking about sequences of events. A number of case studies will highlight some practical features of probability and its possible pitfalls. We make linkages between probabilities and proportions of a finite population and point out that both can be manipulated using the same rules. An important function of this chapter is to give the reader a facility for dealing with data encountered in the form of counts and proportions, particularly in two-way tables.

4.1 INTRODUCTION

"If I toss a coin, what is the probability that it will turn up heads?" It's a rather silly question isn't it? Everyone knows the answer: "a half," "one chance in two," or "fifty-fifty." But let us look a bit more deeply behind this response that everyone makes.

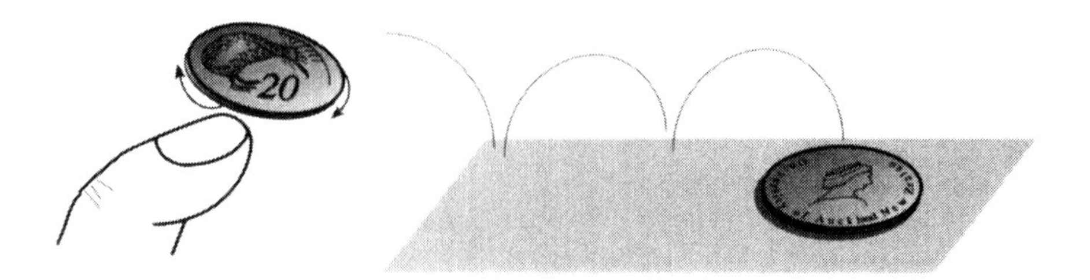

First, why is the probability one half? When you ask a class, a dialogue like the one following often develops. "The probability is one half because the coin is equally likely to come down heads or tails." Well, it could conceivably land on its edge, but we can fix that by tossing again. So why are the two outcomes, heads and tails, equally likely? "Because it's a fair coin." Sounds as if somebody has taken statistics before, but that's just jargon, isn't it? What does it mean? "Well, it's symmetrical." Have you ever seen a symmetrical coin? They are always different on both sides, with different bumps and indentations. These might influence the chances that the coin comes down heads. How could we investigate this? "We could toss the coin lots of times and see what happens." This leads us to an intuitively attractive approach to probabilities for repeatable "experiments" such as coin tossing or die rolling: Probabilities are described in terms of *long-run relative frequencies* from repeated trials. Assuming these relative frequencies become stable after a large enough number of trials, the probability could be defined as the limiting relative frequency. Well, it turns out that several people have tried this.

English mathematician John Kerrich was lecturing at the University of Copenhagen when World War II broke out. He was arrested by the Germans and spent the war interned in a camp in Jutland. To help pass the time he performed some experiments in probability. One of these involved tossing a coin 10,000 times and recording the results.[1] Figure 4.1.1 graphs the proportion of heads Kerrich obtained, up to and including toss numbers 10, 20, ..., 100, 200, ..., 900, 1,000, 2,000, ..., 9,000, 10,000. The data, from Kerrich [1964], are also given in Freedman et al. [1991, Table 1, p. 248].

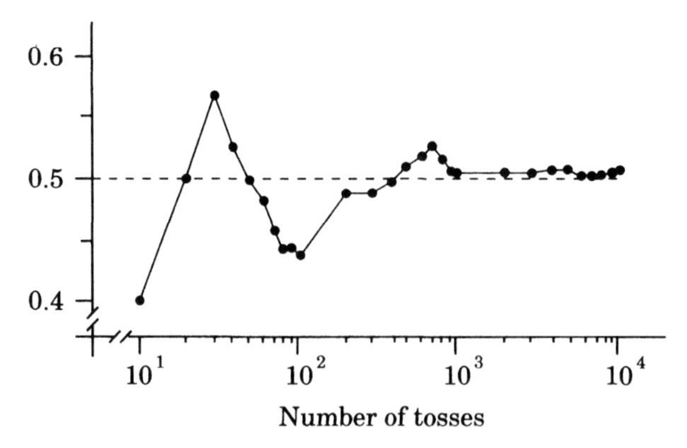

FIGURE 4.1.1 Proportion of heads versus number of tosses for John Kerrich's coin-tossing experiment.

[1]John Kerrich hasn't been the only person with time on his hands. In the eighteenth century, French naturalist Comte de Buffon performed 4,040 tosses for 2,048 heads; around 1900, Karl Pearson, one of the pioneers of modern statistics, made 24,000 tosses for 12,012 heads.

For the coin Kerrich was using, the proportion of heads certainly seems to be settling close to $\frac{1}{2}$. This is an *empirical* or experimental approach to probability. We never know the exact probability this way, but we can get a pretty good estimate. Kerrich's data give us an example of the frequently observed fact that in a long sequence of independent repetitions of a random phenomenon, the proportion of times in which an outcome occurs gets closer and closer to a fixed number, which we can call the probability that the outcome occurs. This is the ***relative frequency*** definition of probability. The long-run predictability of relative frequencies follows from a mathematical result called the law of large numbers.

But let's not abandon the first approach via symmetry quite so soon. The answer it gave us agreed well with Kerrich's experiment. It was based on an artificial idealization of reality—what we call a ***model.*** The coin is imagined as being completely symmetrical, so there should be no preference for landing heads or tails. Each should occur half the time. No coin is exactly symmetrical, however, nor in all likelihood has a probability of landing heads of exactly $\frac{1}{2}$. Yet experience has told us that the answer the model gives is close enough for all practical purposes. Although real coins are not symmetrical, they are close enough to being symmetrical for the model to work well.

4.2 COIN TOSSING AND PROBABILITY MODELS

At a University of London seminar series, N.Z. statistician Brian Dawkins asked the very question we used to open this discussion. He then left the stage and came back with an enormous coin, almost as big as himself.

The best he could manage was a single half turn. There was no element of chance at all. Some people can even make the tossing of an ordinary coin predictable. Program 15 of the *Against All Odds* video series (COMAP [1989]) shows Harvard probabilist Persi Diaconis tossing a fair coin so that it lands heads every time.[2] It is largely a matter of being able to repeat the same action. By just trying to make the action very similar we could probably shift the chances from 50:50, but most of us don't have that degree of interest or muscular control. A model in which coins turn up heads half the time and tails the other half in a totally unpredictable order is an excellent description of what we see when we toss a coin. Our physical model of a symmetrical coin gives rise to a ***probability model*** for the experiment. A probability model has two essential components: the *sample space,* which is simply a list of all the outcomes the experiment can have, and a list of *probabilities,* where each probability is intended to be the probability that the corresponding outcome occurs (see Section 4.4.4).

[2]Diaconis was a performing magician before training as a mathematician.

In sport, coins are tossed to decide which end of the field a team is to defend or who is going to get the ball first. It is quite common to call the outcome after the coin has been tossed but before it has fallen. Even with a normal coin, the side that it lands on is virtually completely determined by a number of factors such as which side was up when it started, the degree of spin, the speed and angle with which it left the thumb, and how far it has to fall. If we knew all this, then with sufficient expertise in physics, we could write down some equations that are thought to govern the motion of the coin (a mathematical model like Newton's laws) and use these to work out which way up the coin should land. The mathematical model will give us precise predictions, but it has some drawbacks. It will be complicated, and it will require us to have some way of accurately measuring quantities such as speeds and angles. Our probability model, on the other hand, is simple and requires no such information. But we pay for this by not being able to predict specific occurrences (e.g., whether this toss will result in a head). We also need to assume that the various physical factors, such as those previously mentioned, vary in an unpredictable fashion.

Probability models can talk only about average behavior in the long run. In a very long sequence of coin tosses we shall see very nearly the same *proportions* (or relative frequencies) of heads and tails. This is an example of what is popularly referred to as the "law of averages." Many misconceptions have developed around the idea of the "law of averages." Even though the proportion (relative frequency) of heads becomes more and more stable as the number of tosses increases, inspection of Kerrich's data in Freedman et al. [1991, p. 248] reveals that the difference between the actual numbers of heads and tails becomes more and more variable. The law of averages applies to relative frequencies, not to absolute numbers. Many people believe that after a sequence of 10 straight heads the chances of getting a tail is much bigger than 50% "because the law of averages will be trying to even things up." The law of averages says nothing at all about short-run behavior. If you inspect very long sequences of coin tosses, find all instances of 10 heads in a row, and then look at what happens next each time, you will find that the next toss is still a tail about half of the time and a head about half of the time. Short-term behavior is utterly unpredictable. And, of course, these are precisely the reasons coins are used to start sports games—the results are "fair" but unpredictable.

Let us return to our mathematical model for coin tossing. Even if we could make all the measurements required by a mathematical model of coin tossing, it is unlikely that it would always give us the correct answers. There would be factors affecting the experiment for which we had not allowed. For example, a gust of wind or slight variations in the degree of bounce in the table could still render our answer wrong some of the time. Furthermore, measurements can never be made completely accurately. Measurement errors are often random in appearance and may be large enough to adversely affect the answers. In building probability models for real-world situations, we need to model both the predictable factors that we know about, using mathematics, and the unpredictable factors, using probability. Some of the unpredictable factors may or may not actually be random. This does not matter. The important thing is that they are unpredictable to us and look random. We therefore describe these unpredictable elements in terms of random events with associated probabilities. We saw in Chapter 3 that a model that consists of a predictable (deterministic) part and an unpredictable (stochastic) part is typically used when we fit a straight line to data. We model the predictable part, the pattern or trend, by a straight line, and we model the unpredictable part, the variation of the points about the line, using ideas of randomness and probability. This more complex type of model is developed further in Chapter 12. In the meantime, however, we shall confine ourselves to very simple situations.

Another Two-Outcome Probability Model

The sex of a child is a two-outcome "experiment" whose outcome still appears to be random. With current scientific knowledge it is still impossible, short of resorting to artificial insemination techniques, to *control* the sex of the conceived child. This is despite Aristotle's belief that boys tend to be produced if the father is highly excited and other venerable solutions such as waiting until the wind is in the north, keeping one's boots on, and eating raw eggs. To the best of our knowledge, none of them work reliably! Conception can be viewed as a race. The winner is the first sperm to reach the egg and still have enough energy to penetrate its wall. Does it carry an X chromosome resulting in a girl, or a Y, giving a boy? Looking around and seeing roughly equal numbers of males and females, we may decide that it is just like tossing a coin. Thus we could use the same probability model, but with boys and girls replacing heads and tails. The set of possible outcomes is {boy, girl}, and each outcome has a probability of a half. This model works reasonably well, but it has deficiencies. In most countries the birthrate for boys is slightly higher than that for girls. In New Zealand, for example, roughly 52% of births are boys.

Another deficiency in the analogy between tossing coins and having children becomes apparent when we come to have the second child. The chances of getting a girl should be the same whether or not the first child was a girl (after all, the coin doesn't know whether it came down heads or tails last time). This idea is called independence. There is some evidence, however, that people whose first child is a girl or a boy have a second child of the same sex slightly more frequently than one of the opposite sex. Having made these points, however, we note that the deficiencies of the coin-tossing model as a description of children's sex order in families are slight and that probabilities from the coin-tossing model are accurate enough for almost all practical purposes.

QUIZ ON SECTION 4.2

1. What does the law of averages say about the behavior of coin tosses?
2. What are two widely held misconceptions about what the law of averages says about coin tosses?
3. Describe two ways in which the coin-tossing model is inadequate for describing the sex order of children in families. Are the deficiencies big enough to be important?

4.3 WHERE DO PROBABILITIES COME FROM?

In the previous section, we stated that a probability model had two essential components, the sample space, which is a list of all the possible outcomes our experiment can have, and a list of probabilities, one for each outcome. But where do the probabilities that we meet in everyday life come from? We have seen some examples in the previous section. Here is another one.

In 1977 a PanAm jumbo jet and a KLM jumbo jet collided on an airport runway in the Canary Islands, killing 581 people. One jet was taxiing after landing, while the other was taking off. Soon after, well-known Australian statistician Terry Speed noticed the following wire service report in *The West Australian:*

> NEW YORK, Mon: Mr. Webster Todd, Chairman of the American National Transportation Safety Board, said today statistics showed that the chances of two jumbo jets colliding on the ground were about 6 million to one . . . .—AAP

Many people are frightened of flying, and major air disasters increase such fears. It seems clear from the report that the National Transportation Safety Board has responded with a scientific assessment based on hard data ("statistics showed . . . ") of the chances of such an accident occurring again so that people could put their fears into perspective. Terry Speed, who has strong research interests in probability, was intrigued by this and wondered how the board had calculated their figure, so Terry wrote to the chairman. He received the following reply from a high government official, which we reproduce with his permission:

> Dear Professor Speed,
>
> In response to your aerogram of April 5, 1977, the chairman's statement concerning the chances of two jumbo jets colliding (6 million to one) has no statistical validity nor was it intended to be a rigorous or precise probability statement. The statement was made to emphasize the intuitive feeling that such an occurrence indeed has a very remote but not impossible chance of happening.
>
> Thank you for your interest in this regard.

At best, the quoted probability was a subjective assessment. At worst, it was a vanishingly small number plucked out of thin air to reassure the public.

We have seen examples now of each of the three main ways that probabilities are assigned to events: (a) from models, (b) from data, and (c) subjectively. These different ways are now described in detail.

4.3.1 Probabilities from Models

We can sometimes think up a sufficiently simple model of a real experiment in which it is easy to determine a probability. The simplest cases of this occur when the model leads us to believe that the outcomes are equally likely. This is why we believe that the probability of getting a head when tossing a coin is $\frac{1}{2}$, that the chances of any particular outcome (say, a 4) on rolling a standard die is $\frac{1}{6}$, and that the chances of drawing any particular card (say, ace of hearts) from a standard deck of cards is 1 in 52. Unfortunately, the method tends to be limited to a few special cases that are simple enough to be treated in this way. Now we know that the probabilities that the model gives will only be approximately true for the real experiment. If the assumptions of the model are sufficiently wrong, the answers can be completely wrong. Let us think about card games. The probabilities given for card games depend critically on the cards being "well shuffled" so that their order is random. From experience, we know that children and beginners can't shuffle cards very well. Some clumps of cards don't get mixed up much. Experts are so clever at shuffling that it is hard for a layperson to know whether they are shuffling for randomness or shuffling to their own advantage. There are also many famous examples of supposedly random lottery draws that have exhibited behavior that is clearly nonrandom. Perhaps the most famous example is the U.S. military draft lottery of 1970 during the Vietnam War.

The military draft was based on the dates of birth of 18-year-old men and worked as follows. Three hundred and sixty-six identical cylinders were used, each containing a slip of paper with a day of the year written on it (1952, the men's birth year, was a leap year). The cylinders were poured into a two-foot bowl, supposedly randomly mixed. An official drew them out one by one. The order of drawing was the *draft number.* Thus if June 10 was drawn fifth, all 18-year-old men born on June 10 had the draft number 5. The actual draft was performed by conscripting all of the men who had a draft number lower than some limit, where the limit was set to fulfill the quota for soldiers for that year. Recall that most of these soldiers would end up fighting in the jungles of Vietnam, so randomness of draw was important in the interests of fairness. Everyone should have the same chance of being drafted. In 1970, however, reporters noted that men born later in the year tended to have lower draft numbers (and therefore a greater chance of being drafted) than those born earlier in the year (see Fig. 4.3.1). Statistical calculations revealed that such a strong association between draft number and birth date could be expected to occur less than once every thousand years with truly random lotteries. Moreover, from a close inspection of the mixing process devised by Colonel Fox and Captain Pascoe, one might have expected that December cylinders would tend to be closer to the top of the bowl than those for January, say.

The job of devising the lottery was subsequently taken from the military and given to statisticians at the U.S. National Bureau of Standards, who devised a scheme based on various levels of randomization. Dates were placed into date capsules in random order determined by a table of random numbers. The date cylinders were placed in a drum in random order (using another table). Draft numbers (1 to 365 unless a leap year) were placed in a second drum in random order (using a third table). Both drums were rotated for an hour. In front of TV cameras an official then pulled out cylinders in pairs, one from each drum, to choose the date of birth and its associated draft number. As Moore [1991] says, "It's awful, but it's random." For further details see Moore [1991, pp. 62-65], Fienberg [1971], and Rosenblatt and Filliben [1971].

4.3.2 Probabilities from Data

If we did not have a model for coin tossing that was so obviously a good one, a natural approximation for the probability of John Kerrich's coin coming up heads would be the proportion that Kerrich observed, namely, 5067/10,000. Moreover, most

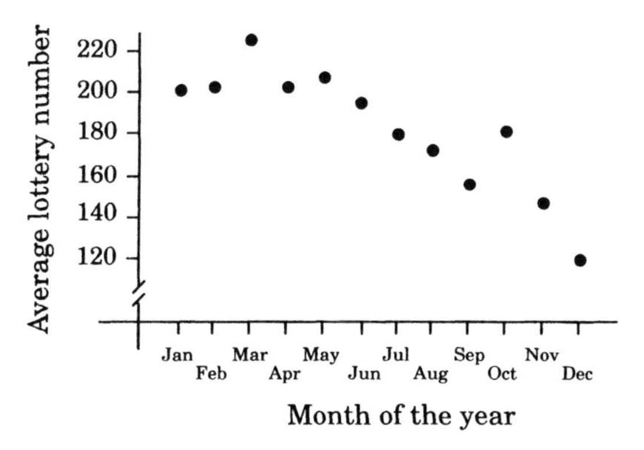

FIGURE 4.3.1 Average lottery numbers by month. Replotted from data in Fienberg [1971].

people would transfer this figure to their own coins because they could see no reason why their coin should behave differently from Kerrich's. Because the probability of an outcome is the long-run relative frequency if we can independently repeat the experiment over and over again, the bigger the sample observed, the more reliable the answer. We would have more faith in Kerrich's figure based on 10,000 tosses than a figure based on only 100 or even 1000 tosses.[3]

A major source of quoted probabilities for events is data on the relative frequencies of these same events in the past. In New Zealand over recent years, roughly 600 people from a population of about 3 million were killed on the roads. Most people would be fairly comfortable with a statement of the form, "The probability that a randomly selected New Zealander will die on the roads next year is about 600/3,000,000." There are two important considerations, however. First, we can make such statements only if we think that the underlying process is stable over time. For example, if we knew that the open-road speed limit would be increased before the end of next year, our estimate would have to be revised upward, and it is by no means clear how we should do this. Second, as previously noted, our relative frequencies have to be taken from large numbers for us to have much confidence in them as probabilities. A real success story of the use of historical relative frequencies to provide probabilities of similar events in the future is provided by the life insurance industry. Life insurance companies need good estimates of the chances that various types of people will die within a given period so that they can calculate what premiums have to be charged to cover the maximum probable level of claims. Referring back to the wire service report of Section 4.3, the U.S. National Transportation Safety Board could conceivably have quoted a relative frequency probability. Every day hundreds of commercial airliners take off and land all around the world. The Safety Board could have estimated the number of takeoffs and landings in the previous 10 years, say, counted the number of runway collisions, and quoted a figure such as one runway collision per 6 million takeoffs. Many newspaper readers probably interpreted the board's (purely subjective) statement in such terms.

There are two important relationships between probabilities from theoretical models and probabilities based on relative-frequency data. First, if the model is reasonable, then for any experiment that can be repeated over and over again, the probability of an event obtained from the model tells us the relative frequency with which that event will occur over the long run. Second, we may choose not to use the observed relative frequencies as our probabilities but simply use them to make us feel happier about our probabilities obtained from the model. (Otherwise we should throw away the model!) For example, suppose we observed 24 heads in 40 tosses of some particular coin. Few people would then use $\frac{24}{40}$ for the probability of getting a head for that coin. Most would observe that 24 in 40 was reasonably in line with the model-based value of $\frac{1}{2}$ and thus feel happier about using this value in future.

4.3.3 Subjective Probabilities

In 1998 the horse Indian Charlie was the favorite going into the Kentucky Derby. Suppose that an ordinary race-goer thought that the chances of Indian Charlie winning

[3]We shall see how the estimate of a probability taken from a relative frequency becomes more accurate as the number of repetitions increases in Chapter 7.

TABLE 4.3.1 Fatality Rates

Accident	Average chance of death per year per individual
Motor travel	1 in 3,000
Air travel	1 in 100,000
Reactor accidents (based on 100 reactors in the U.S.)	
(a) within a few weeks	1 in 300,000,000
(b) within about 20 weeks	1 in 16,000,000

Source: *Reactor Safety Study*, cited in Speed [1977].

the Derby were $\frac{3}{4}$. What would he or she mean? It doesn't make sense to think in terms of the 1998 Kentucky Derby being run many times and Indian Charlie winning three quarters of those races. The gambler doesn't make this assessment on the basis that Indian Charlie has won three quarters of its past races. The serious gambler makes an assessment from a subjective pooling of all relevant information of which he or she may be aware, including Indian Charlie's past record, the records of other horses in the race, the state of the track, and any information picked up about the current form of the horses. But basically it comes down to a numerical measure of the strength of that gambler's belief in the proposition that Indian Charlie will win. Another gambler exposed to the same sources of information would have a different strength of belief. Some gamblers may have a strong sense of belief and thus give the proposition a high probability for what may appear to most of us to be ludicrous reasons. (In the event, Indian Charlie was beaten by stable mate Real Quiet, and this was the 18th time the favorite had been beaten in 20 years!)

In contrast to subjective probabilities, most people can agree about relative-frequency probabilities and even about model probabilities if they are backed up by data. Unfortunately, subjective probabilities often masquerade as frequency probabilities. Table 4.3.1, taken from Speed [1977], was abridged from a table in the *Reactor Safety Study* (or Rasmussen Report), a major U.S. governmental study of the safety of nuclear power facilities. The first two probabilities are frequency probabilities based on plenty of hard data. One out of every 3,000 of the millions of people in the United States die on the roads per year. One in every 100,000 dies in an air accident in a year. The juxtaposition with these figures makes it appear that the following reactor accident values are also frequency probabilities based on data. In fact, they are based on calculations of the likelihoods of chains of events (see Section 4.7.3). Many of the individual probabilities in the chain were somebody's subjective assessment, so the result is really only a subjective probability.

4.3.4 Manipulation of Probabilities

While not all statisticians agree about what probabilities should be associated with a particular real-world event, all do agree on how probabilities should be combined and manipulated. This is the subject of the following section. It even applies to subjective probabilities. Subjective Bayesian statisticians, who believe that the process of statistical inference should be concerned with using data to refine one's subjective degree of belief in a theory or statement, still use the ordinary rules of probability for this refinement process.

QUIZ ON SECTION 4.3

1. What are the three types of probability we typically encounter?
2. Give examples of each from your own experience.
3. What assumption underlies probabilities given for card games?
4. When the relative frequency of an event in the past is used to estimate the probability that it will occur in the future, what assumption is being made?
5. What do all statisticians agree about with respect to probabilities? (Section 4.3.4)
6. When a weather forecaster says that there is a 70% chance of rain tomorrow, what do you think this statement means?

EXERCISES FOR SECTION 4.3

1. Suppose we make a spinner as shown in the picture. The experiment is to spin the pointer vigorously and see what color it stops on. How would you obtain a relative frequency probability for the probability that it stops on grey?

   Can you calculate a model-based probability of stopping on grey? If so, how? What assumptions do you need to make?

2. Consider shaking a thumb tack in a cup and tossing the tack out onto a table. It can land one of two ways (see picture). How would you construct a relative frequency probability for the probability that the tack lands point down?

   Can you construct a model-based probability? Justify your answer.

3. A random-number table is constructed from a sequence of digits. Each new digit in the sequence is obtained by choosing a digit at random from the 10 digits 0, 1, . . . , 9. Say whether each of the following statements is true or false and why.
   (a) Each column should have the same number of 9s in it.
   (b) Each column should have a similar number of 4s as 5s.
   (c) After three 5s in a row, the next number is less likely to be a 5.
   (d) We are less likely to see the sequence 1, 2, 3, 4, 5 than to see the sequence 2, 7, 4, 9, 3.

4.4 SIMPLE PROBABILITY MODELS

4.4.1 Sample Spaces

We begin with the idea of a *random experiment*, that is, an experiment whose outcome cannot be predicted. The term *experiment* is used in its widest sense. It can mean either a naturally occurring phenomenon (e.g., measuring the height of high tide on a given day, counting the number of aphids on a leaf), a scientific experiment (e.g., measuring the speed of sound or the blood pressure of a patient), or a sampling experiment (e.g., choosing a person at random from a class of students using the lottery method and recording some characteristic of the person).

> A ***sample space***, S, for a random experiment is the set of all possible outcomes of the experiment.

In simple examples we can represent the sample space simply as a list. With more complicated examples some mathematical representation may be necessary. There are two important considerations in the way we list the outcomes. First, every outcome must be represented. Second, to avoid ambiguity, no outcome can be represented twice. This means that any outcome gives rise to one and only one member of the list.

You will find that, apart from examples based on data tables and case studies, a number of the examples in this chapter are very simple and not very "realistic." For example, you will often see examples about tossing a coin or sampling colored balls from a barrel. Just as tossing a coin can serve as a useful model for sex outcomes when having children, however, we shall find in Chapter 5 that these very simple physical experiments become the basis of models for a vast array of real applications. In the meantime, we shall just concentrate on using them to enable us to explore how probabilities and probability models behave.

✖ Example 4.4.1 *Sample Spaces*

(a) If we toss a coin twice, we can represent the four possible outcomes in terms of heads (H) and tails (T) as

$$S = \{HH, HT, TH, TT\}$$

where HT, for example, indicates a head followed by a tail.

(b) Similarly, three tosses give

$$S = \{HHH, HHT, HTH, THH, HTT, THT, TTH, TTT\}$$

(c) If we roll two dice and record the numbers facing uppermost on each die, we could use

$$S = \{(1, 1), (1, 2), \ldots, (1, 6), (2, 1), \ldots, (2, 6), (3, 1), \ldots, (6, 6)\}$$

where we have represented each of the 36 possible outcomes as a pair. For example, (1, 6) represents rolling a 1 with the first die and a 6 with the second.

(d) Suppose we interview a person at random and record their religious preference (if any). A sample space for the outcome of the interview might be

$$S = \{\text{Buddhism, Christianity, Hinduism, Islam, Other, None}\}$$

Here the category "Other" would be used to capture all of those who adhere to a religion not listed. Those under "None" would consist of people who adhere to no religion. Have we got a sample space? We have accounted for all religions. We have not listed all answers people will give us, however. We need a further category, which we might call "Nonresponse," to include, for example, those who refuse to answer. Also, as with most classification systems, there are problems with category definition. Consider Christianity. There may be some sects that you would be unsure whether to list under "Christianity" or under "Other."

(e) Suppose that in contrast to (a) and (b), we now toss a coin until the first tail appears. Then

$$S = \{T, HT, HHT, HHHT, \ldots\}$$

The row of dots here means that the pattern keeps on repeating forever. There is no limit to the number of heads we could conceivably throw before our first tail.

(f) Suppose the experiment is to measure tomorrow's rainfall. A possible S is the set of all numbers greater than or equal to zero, which we could write as the interval $[0, \infty]$. If we were prepared to believe that there was no way that the day's rainfall would be greater than 30 mm, we could restrict S to all numbers from 0 mm to 30 mm, which we could write as $[0, 30]$.

More than one sample space can be used to describe an experiment. This is why we talk about *a* sample space and not *the* sample space for an experiment. In (b) we could count the number of heads in the three tosses and use S_1 = {0 heads, 1 head, 2 heads, 3 heads}. Every outcome of the experiment is represented and represented only once, as required. Which sample space we use depends on the type of question we wish to answer. For example, S_1 lets us talk about the number of heads but does not let us distinguish between the order in which heads and tails fall, while S lets us address both issues.

4.4.2 Events

We often want to talk about a collection of outcomes that share some characteristic, for example, the outcomes resulting in at least one head if we toss a coin twice. This leads to the following definition of an event:

> An ***event*** is a collection of outcomes.
> An event ***occurs*** if any outcome making up that event occurs.

The sample space itself is an event. An event may contain only a single outcome.[4]

[4]Technically, an event is a subset of the sample space.

✖ EXAMPLE 4.4.2 *Events*

(a) In Example 4.4.1(a), we toss a coin twice, giving $S = \{HH, HT, TH, TT\}$. The event A = "at least one head" is given by $A = \{HH, HT, TH\}$. If any one of these three outcomes occurs when we toss the coin twice, then event A has occurred.

(b) In Example 4.4.1(d), the event B = "has a religious preference that is not Buddhism or Christianity" is given by B = {Hinduism, Islam, Other}.

> The ***complement*** of an event A, denoted $\overline{A}$, occurs if A does not occur.

The complement of A, denoted $\overline{A}$, contains all outcomes not in A. It is sometimes helpful to read $\overline{A}$ as "not A."

✖ EXAMPLE 4.4.3 *Complementary Events*

(a) The complement of A in Example 4.4.2(a) is $\overline{A} = \{TT\}$. Verbally, the complementary event to "at least one head" is "no heads," which in this case is the same as "two tails."

(b) In Example 4.4.1(b) where we toss a coin three times, the event B = "at least two heads" $= \{HHH, HHT, HTH, THH\}$ has complement $\overline{B} = \{HTT, THT, TTH, TTT\}$. Verbally, the complementary event to "at least two heads" is "at most one head" or, equivalently here, "at least two tails."

It is useful to represent events diagrammatically. These are called *Venn diagrams*. We tend to represent the sample space S as a rectangular box. Events inside S are represented by the contents of a closed shape. Any shape will do, although we shall usually use a circle, as in Fig. 4.4.1(a). In Fig. 4.4.1(b) we have shaded the contents of A (which we think of as representing all of the outcomes in A), whereas in Fig. 4.4.1(c) we have shaded the contents of $\overline{A}$.

EXERCISES FOR SECTION 4.4.2

1. In Example 4.4.1(c) let A = "sum of the faces uppermost is 4." List the outcomes in A.

2. In Example 4.4.1(e) let A = "even number of tosses before the tail." What outcomes are in A? Describe $\overline{A}$ both verbally and by listing its outcomes.

3. Write down a sample space for tossing a coin until we have two tails, three heads, or a maximum of four tosses. What outcomes are in the event A = "three tosses made"?

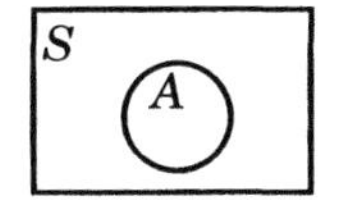

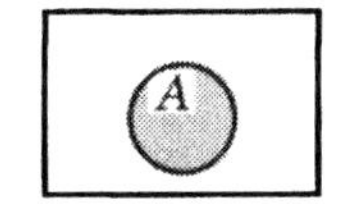

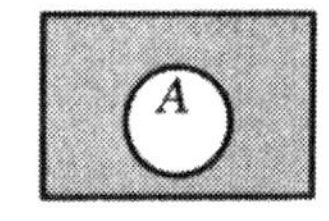

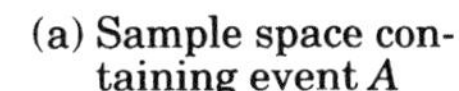

(a) Sample space containing event A (b) Event A shaded (c) $\overline{A}$ shaded

FIGURE 4.4.1 An event A in the sample space S.

4.4.3 Combining Events

We now look at two ways of combining two events A and B: "A or B occurs" (where *or* is used in the inclusive sense of A or B or both[5]) and "both A and B occur."

> ***A* or *B*** contains all outcomes in A **or** B (or both).
> ***A* and *B*** contains all outcomes which are in **both** A and B.

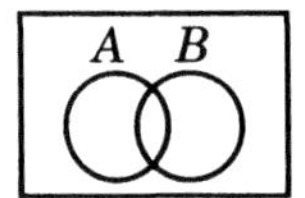

(a) Events A and B

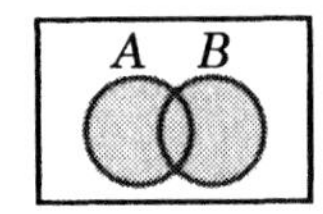

(b) "A or B" shaded

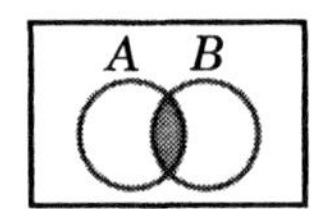

(c) "A and B" shaded

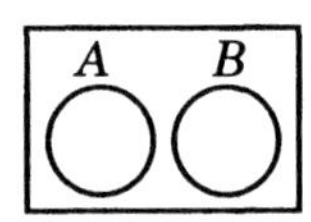

(d) Mutually exclusive events

FIGURE 4.4.2 Two events.

✖ Example 4.4.4 *Combining Events*

In Example 4.4.1(a) we have $S = \{HH, HT, TH, TT\}$. Let A be the event "at least one head" and B be the event "at least one tail." Then

$$A = \{HH, HT, TH\} \quad B = \{HT, TH, TT\}$$
$$A \text{ and } B = \{HT, TH\} = \text{"exactly one head"}$$
$$A \text{ or } B = \{HH, HT, TH, TT\} = S$$

Two events A and B that have no outcomes in common are said to be ***mutually exclusive.***

> ***Mutually exclusive*** events cannot occur at the same time.

You may find the phrase "A excludes B" a useful memory aid for the meaning of "mutually exclusive." Any event A and its complement $\overline{A}$ are mutually exclusive. We usually represent mutually exclusive events diagrammatically by nonoverlapping shapes, as in Fig. 4.4.2(d).

✖ Example 4.4.5 *Mutually Exclusive Events*

(a) When tossing a coin twice [Examples 4.4.1(a), 4.4.2(a)], the events "head first toss" $= \{HH, HT\}$ and "tail first toss" $= \{TH, TT\}$ are mutually exclusive. The

---

[5]This is sometimes written as "and/or." Other phrases we shall use are the "union of A and B" and "at least one of A and B occurs."

two events "head first toss" = $\{HH, HT\}$ and "head second toss" = $\{HH, TH\}$ are not mutually exclusive, however. Both of these events occur if we observe two heads.

(b) Consider rolling a die twice [Example 4.4.1(c)]. Let A = "sum from the two faces is 4," B = "3 on first roll," and C = "sum from the two faces is 7." Then $A = \{(1,3),(2,2),(3,1)\}$, $B = \{(3,1),(3,2),\ldots,(3,6)\}$, and $C = \{(1,6),(2,5),(3,4),(4,3),(5,2),(6,1)\}$. Now A and C are mutually exclusive. A and B can occur together, however, because the outcome (3, 1) is in both A and B. Also, B and C can occur together because the outcome (3, 4) is in both B and C.

EXERCISES FOR SECTION 4.4.3

1. An experiment consists of tossing a coin and rolling a die. Give a sample space for the experiment. Let A = "die scores 3" and B = "coin is heads." By listing outcomes, write down expressions for (i) A, (ii) B, (iii) A and B, and (iv) A or B.

2. For the ABO blood system, a person can be one of the four phenotypes A, B, O, or AB. Two people are chosen at random. Give a sample space for the pair of phenotype outcomes. List outcomes for the following events:
   (a) C = "both people have the same phenotypes"
   (b) D = "at least one person has phenotype A"
   (c) C and D

4.4.4 Probability Distributions

Traditional usage dictates that probabilities are numbers scaled to lie between 0 and 1 (or 0% and 100%) and that outcomes with probability 0 cannot occur. In addition, we say that events with probability one or 100% are certain to occur. We now go on to define the term *probability distribution,* but only for models with finite sample spaces or with infinite sample spaces that can be represented as a list, for example, $S = \{H, TH, TTH, TTTH, \ldots\}$.

Suppose $S = \{s_1, s_2, s_3, \ldots\}$ is such a sample space. A list of numbers $p_1, p_2, \ldots$ is a ***probability distribution*** for S, provided the p_i's satisfy both

(i) the p_i's lie between 0 and 1, **($0 \le p_i \le 1$)**, and

(ii) the sum of all the p_i's is 1. **($p_1 + p_2 + \cdots = 1$)**.

According to the probability model, p_i is the probability that outcome s_i occurs. We write $p_i = \text{pr}(s_i)$.

> Probabilities lie between 0 and 1, and they add to 1.

In practice, our aim is not just to specify a mathematically valid probability model, namely, one that satisfies the preceding conditions (i) and (ii), but also to specify a model in which the stated probabilities give a good approximation of the actual behavior of the experiment.

EXAMPLE 4.4.6 *Probability Distributions*

(a) Consider tossing a coin twice so that $S = \{HH, HT, TH, TT\}$. For a fair coin each outcome should be equally likely, so the probabilities for the four outcomes, p_1, p_2, p_3, and p_4, should be identical. If we want them to add to 1, each value must therefore be $\frac{1}{4}$, that is,

$$\text{pr}(HH) = \tfrac{1}{4} \quad \text{pr}(HT) = \tfrac{1}{4} \quad \text{pr}(TH) = \tfrac{1}{4} \quad \text{pr}(TT) = \tfrac{1}{4}$$

These probabilities constitute a probability distribution for S.

(b) Consider choosing a three-child family at random and looking at the sexes of the children from first born to last born.

$$S = \{GGG, GGB, GBG, BGG, GBB, BGB, BBG, BBB\}$$

Let us assume that looking at the sexes of a randomly chosen three-child family is like tossing a fair coin three times and that each of the 8 ($= 2^3$) outcomes in S is equally likely to occur. Since the probabilities must add to 1, each outcome has probability $\frac{1}{8}$.

(c) To vary this a little, consider a couple having children who will stop when they have a child of each sex or when they have three children. For this "experiment" the sample space is

$$S = \{GGG, GGB, GB, BG, BBG, BBB\}$$

By comparing (a) and (b), a reasonable probability distribution might be

$$\text{pr}(GGG) = \text{pr}(GGB) = \text{pr}(BBG) = \text{pr}(BBB) = \tfrac{1}{8}$$

and

$$\text{pr}(GB) = \text{pr}(BG) = \tfrac{1}{4}$$

These values are between 0 and 1 and add to 1, so they qualify as a probability distribution.[6]

Probabilities of Events

The ***probability of event A*** can be obtained by adding up the probabilities of all the outcomes in A.

EXAMPLE 4.4.7 *Probabilities of Events*

(a) In Example 4.4.6(a), let the event A = "at least one head" = $\{HH, HT, TH\}$. By adding the probabilities of the three outcomes in A, we find

$$\text{pr}(A) = \text{pr}(HH) + \text{pr}(HT) + \text{pr}(TH) = \tfrac{1}{4} + \tfrac{1}{4} + \tfrac{1}{4} = \tfrac{3}{4}$$

[6]After reading Section 4.7 you will be able to derive these probabilities as following from the assumptions that $\text{pr}(G) = \text{pr}(B) = \frac{1}{2}$ and the sexes of different children are statistically independent.

(b) Similarly, for Example 4.4.6(c) with C being the event "first child is a girl," we have $C = \{GGG, GGB, GB\}$ and

$$\text{pr}(C) = \tfrac{1}{8} + \tfrac{1}{8} + \tfrac{1}{4} = \tfrac{1}{2}$$

Equally Likely Outcomes

If S consists of 10 equally likely outcomes, each with probability p, then since the probabilities add to 1, we have $10p = 1$ or $p = \frac{1}{10}$. Thus each outcome has a probability of $\frac{1}{10}$. If A has four outcomes in it, then using our addition rule, we have $\text{pr}(A) = \frac{1}{10} + \frac{1}{10} + \frac{1}{10} + \frac{1}{10} = \frac{4}{10}$. More generally, for any finite sample space with equally likely outcomes,

$$\mathbf{pr}(A) = \frac{\textbf{Number of outcomes in } A}{\textbf{Total number of outcomes in } S}$$

We note that pr(A) is the *proportion* of the list of outcomes that correspond to the event A.

Example 4.4.8 *Equally Likely Outcomes*

In Example 4.4.7(a), three out of four outcomes are in A, so that $\text{pr}(A) = \frac{3}{4}$.

Example 4.4.9 *Cell Proportions for a Two-Way Table*

Table 4.4.1, called a ***two-way table of counts*** or *contingency table* (introduced briefly in Section 3.3), cross-classifies job losses in the United States over a three-year period. Job losses are broken down by the sex of the person who lost the job and the reason given for losing it. The entries in the table represent the number of job losses (in thousands) by people of the particular sex for a particular reason. There were 5,584,000 jobs lost (to the nearest thousand) and of these, 1,703,000 were lost by males because the workplace moved or closed down.

Suppose we decided to choose a lost job at random so that we could investigate the circumstances. The sample space for this experiment consists of all the 5,584,000

TABLE 4.4.1 Job Losses in the United States (in thousands) for 1987 to 1991

	Reason for job loss			
	Workplace moved/closed	Slack work	Position abolished	Total
Male	1,703	1,196	548	3,447
Female	1,210	564	363	2,137
Total	2,913	1,760	911	5,584

Source: Constructed, with permission, from data in *The World Almanac* [1993, p. 157].

lost jobs. Since these are all equally likely to be chosen, we can obtain probabilities by counting numbers of outcomes. The outcomes making up the event "lost by female because workplace moved/closed down" are all job losses with this property, and there are 1,210,000 of them. Thus

$$\text{pr}(\textit{lost by female because workplace moved/closed down}) = \frac{1{,}210{,}000}{5{,}584{,}000} = 0.217$$

to three decimal places. There were 2,137,000 jobs lost by females, so that

$$\text{pr}(\textit{lost by female}) = \frac{2{,}137{,}000}{5{,}584{,}000} = 0.383$$

The event "lost by male for slack work" has 1,196,000 outcomes, so that

$$\text{pr}(\textit{lost by male for slack work}) = \frac{1{,}196{,}000}{5{,}584{,}000} = 0.214$$

We can convert these to percentages by multiplying by 100%, so the last answer becomes 21.4%.

The set of six combinations of classes in the table ("lost by male because workspace moved/closed," . . . , "lost by female because position abolished") forms an alternative sample space for this experiment because all eventualities are allowed for and no outcomes are represented twice. The sample space consisting of all 5,584,000 individual job losses was more useful, however, because its outcomes are all equally likely and we thus could obtain any relevant probabilities almost immediately. Note that the so-called cell probabilities relating to entries in the table turn out to be the corresponding proportions of the population of job losses. In fact, we could have converted the entries in the original table from counts to proportions by simply dividing each entry by 5584 to get the two-way table of *proportions,* Table 4.4.2. Such tables were introduced in Section 3.3.

We note that the table of proportions, Table 4.4.2, has the property that the entries in each row add up to the corresponding row total and the entries in each column add up to the corresponding column total. The grand total is 1.000. We can check that we have done our sums right by seeing that the entries in the last column add to 1.000 and the entries in the last row also add to 1.000, that is, the table "checks down and across." In tables that follow, we shall find that the data are sometimes presented as counts (frequencies) and sometimes as proportions. In the latter case it is important that the total frequency, 5584 in this example, is also given. We then have some idea of the original individual frequencies by multiplying each proportion by the

TABLE 4.4.2 Proportions of Job Losses from Table 4.4.1

	Reason for job loss			
	Workplace moved/closed	Slack work	Position abolished	Row totals
Male	.305	.214	.098	.617
Female	.217	.101	.065	.383
Column totals	.522	.315	.163	1.000

total frequency. We may not recover them exactly, however, because of rounding off the proportions, in this case to three decimal places.

EXERCISES FOR SECTION 4.4.4

1. Two dice are rolled. What is the probability that the sum of the two faces uppermost is (i) 9? (ii) even?
2. Refer to Example 4.4.9 for the following questions.
   (a) What is the probability that a randomly chosen job loss was (i) by a female whose position was abolished? (ii) caused by the position being abolished?
   (b) Take the six classes in the table and represent them as a sample space with an associated probability distribution.
   (c) Do the same thing as (b) but using the three categories of job loss as a sample space.
   (d) In Example 4.4.9 why did we employ all job losses as our sample space and not the sample space in (b)?
3. The following questions require background thinking about the data in Table 4.4.1.
   (a) In Example 4.4.9 we were careful not to equate the 5,584,000 job losses over the three-year period with 5,584,000 people. Why are the two not equivalent?
   (b) There were fewer job losses by females over this period. Does this demonstrate that women are more reliable workers?
   (c) If you wanted to compare the relative chance of a random male losing his job with that of a random female losing her job, what measure would you use?

4.4.5 Probabilities and Proportions

When we have a real (finite) population of units (e.g., people), the theory of equally likely outcomes tells us that, when we choose a unit at random, the probability that a unit with property A is chosen is numerically identical to the proportion of units in the population with property A. For example, if 10% of the population is left-handed, the chances that a randomly chosen person is left-handed is also 10%.

The concepts of a proportion and a probability are quite distinct. A proportion is a partial description of a real population—a form of summary. Probabilities tell us about the chances of something happening in a random experiment. The fact that proportions are numerically identical to probabilities for a real population under the experiment "choose a unit at random," however, means that we can use the probability notation and any formulas derived for manipulating probabilities to solve problems involving proportions as well. At times, where it makes good practical sense, we shall express a real problem in terms of proportions of a population rather than probabilities. We prefer to do this rather than introducing the artificial "choose a unit at random" that would enable us to write everything in terms of probabilities.

QUIZ ON SECTION 4.4

1. What is a sample space? What are the two essential criteria that must be satisfied by a possible sample space? (Section 4.4.1)
2. What is an event? (Section 4.4.2)
3. If A is an event, what do we mean by its complement, $\overline{A}$? When does $\overline{A}$ occur?

4. If A and B are events, when does "A or B" occur? When does "A and B" occur? (Section 4.4.3)

5. What are the two essential properties of a probability distribution $p_1, p_2, \ldots, p_n$? (Section 4.4.4)

6. How do we get the probability of an event from the probabilities of outcomes that make up that event?

7. If all outcomes are equally likely, how do we calculate pr(A)?

8. How do the concepts of a proportion and a probability differ? Under what circumstances are they numerically identical? What does this imply about using probability formulas to manipulate proportions of a population? (Section 4.4.5)

4.5 PROBABILITY RULES

The following rules are clear for finite sample spaces[7] in which we obtain the probability of an event by adding the p_i's for all outcomes in that event. The outcomes do not need to be equally likely.

- **Rule 1:** The sample space is certain to occur.[8]

$$\mathbf{pr}(S) = 1$$

- **Rule 2:** pr(A does ***not*** occur) = 1 − pr(A does occur). [Alternatively, pr(an event occurs) = 1 − pr(it doesn't occur).]

$$\mathbf{pr}(\overline{A}) = 1 - \mathbf{pr}(A)$$

- **Rule 3:** Addition rule for mutually exclusive (non-overlapping) events: pr(A or B occurs) = pr(A occurs) + pr(B occurs)

$$\mathbf{pr}(A \text{ or } B) = \mathbf{pr}(A) + \mathbf{pr}(B)$$

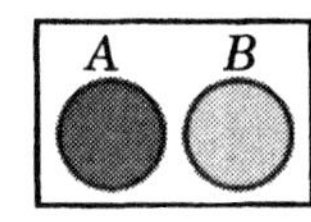

[*Aside: What happens if A and B overlap? In adding pr(A) + pr(B), we use the p_i's relating to outcomes in A and B twice. Thus we adjust by subtracting pr(A and B). This gives us

$$\text{pr}(A \text{ or } B \text{ occurs}) = \text{pr}(A \text{ occurs}) + \text{pr}(B \text{ occurs}) - \text{pr}(A \text{ and } B \text{ occur})]$$

[7] The rules apply generally, not just to finite sample spaces.

[8] The sample space contains all possible outcomes.

EXAMPLE 4.5.1 *Random Numbers*

A random number from 1 to 10 is selected from a table of random numbers. Let A be the event that "the number selected is 9 or less." As all 10 possible outcomes are equally likely, $\text{pr}(A) = \frac{9}{10} = 0.9$. The complement of event A has only one outcome, so $\text{pr}(\overline{A}) = \frac{1}{10} = 0.1$. Note that

$$\text{pr}(A) = 1 - \text{pr}(\overline{A}) = 1 - 0.1 = 0.9$$

This formula is very useful in any situation where $\text{pr}(\overline{A})$ is easier to obtain than $\text{pr}(A)$. It is used in this way in Example 4.5.4 and frequently thereafter.

EXAMPLE 4.5.2 *Probability Rules and a Two-Way Table*

Suppose that between the hours of 9.00 A.M. and 5.30 P.M., Dr. Wild is available for student help 70% of the time, Dr. Seber is available 60% of the time, and both are available (simultaneously) 50% of the time. A student comes for help at some random time in these hours. Let A = "Wild in" and B = "Seber in." Then A and B = "both in." If a time is chosen at random, then $\text{pr}(A) = 0.7$, $\text{pr}(B) = 0.6$ and $\text{pr}(A \text{ and } B) = 0.5$. What is the probability, $\text{pr}(A \text{ or } B)$, that at least one of Wild and Seber are in?

We can obtain this probability, and many other probabilities, without using any algebra if we put our information into a two-way table, as in Fig. 4.5.1. The left-hand (descriptive) table categorizes the possible outcomes of Wild being in or out and Seber being in or out. Each interior cell in the table is to contain the probability that a particular combination occurs. At present we have information on only one combination, namely, pr(*Wild in* and *Seber in*) = 0.5, and we have placed that in the table. The left column total tells us about Seber being in regardless of where Wild is, so that is where pr(*Seber in*) = 0.6 belongs. Similarly, pr(*Wild in*) = 0.7 belongs in the top row-total position. Everything else in the table is unknown at present. The right-hand (algebraic) table relates all this information to the algebraic events and notation we set up earlier.

Descriptive Table

		Seber In	Seber Out	Total
Wild	In	0.5	?	0.7
	Out	?	?	?
	Total	0.6	?	1.00

pr(*Wild in* **and** *Seber in*) → 0.5; pr(*Seber in*) → 0.6; pr(*Wild in*) → 0.7

Algebraic Table

	B	$\overline{B}$	Total
A	$\text{pr}(A \text{ and } B)$	$\text{pr}(A \text{ and } \overline{B})$	$\text{pr}(A)$
$\overline{A}$	$\text{pr}(\overline{A} \text{ and } B)$	$\text{pr}(\overline{A} \text{ and } \overline{B})$	$\text{pr}(\overline{A})$
Total	$\text{pr}(B)$	$\text{pr}(\overline{B})$	1.00

FIGURE 4.5.1 Putting Wild–Seber information into a two-way table.

The missing entries in the left-hand table can now be filled in using the simple idea that rows add to row totals and columns add to column totals.

.5	?	.7
?	?	?
.6	?	1.00

→

.5	?	.7
?	?	.3
.6	.4	1.00

→

.5	.2	.7
?	?	.3
.6	.4	1.00

→

.5	.2	.7
.1	.2	.3
.6	.4	1.00

TABLE 4.5.1
Completed Probability Table

	Seber		
Wild	In	Out	Total
In	.5	.2	.7
Out	.1	.2	.3
Total	.6	.4	1.0

Our final table is reproduced in Table 4.5.1. To find the probability that at least one person is in, namely pr(*A* or *B*), we now sum the probabilities in the (row 1, column 1), (row 1, column 2), and (row 2, column 1) positions:[9]

$$\text{pr}(A \text{ or } B) = 0.5 + 0.2 + 0.1 = 0.8$$

We can also find a number of other probabilities from the table, such as the following. (Algebraic expressions are appended to each line that follows, not because we want you to work with complicated algebraic expressions, but just so that you will appreciate the power and elegance of the table.)

pr(*Wild in* and *Seber out*) = 0.2 $[\text{pr}(A \text{ and } \overline{B})]$

pr(*At least one of Wild and Seber are out*) = 0.2 + 0.1 + 0.2 $[\text{pr}(\overline{A} \text{ or } \overline{B})]$

pr(*One of Wild and Seber is in* and *the other is out*) = 0.2 + 0.1 $[\text{pr}((A \text{ and } \overline{B}) \text{ or } (\overline{A} \text{ and } B))]$

pr(*Both Wild and Seber are in* or *both are out*) = 0.5 + 0.2 $[\text{pr}((A \text{ and } B) \text{ or } (\overline{A} \text{ and } \overline{B}))]$

The preceding example is typical of a number of situations where we have partial information about various combinations of events and we wish to find some new probability. To do this we simply fill in the appropriate two-way table. We have used some ideas here that seem natural in the context of the table, but we have not formalized them yet. One idea used was adding up the probabilities of mutually exclusive events.

4.5.1 Addition Rule for More Than Two Mutually Exclusive Events

Events $A_1, A_2, \ldots, A_k$ are all mutually exclusive if they have no overlap, that is, if no two of them can occur at the same time. If they are mutually exclusive, we can get the probability that at least one of them occurs (sometimes referred to as the union)

---

[9] We can also find this result using the equation in the aside following rule 3, namely,

$$\text{pr}(A \text{ or } B) = \text{pr}(A) + \text{pr}(B) - \text{pr}(A \text{ and } B) = 0.7 + 0.6 - 0.5 = 0.8$$

simply by adding:

$$\text{pr}(A_1 \text{ or } A_2 \text{ or } \ldots \text{ or } A_k) = \text{pr}(A_1) + \text{pr}(A_2) + \cdots + \text{pr}(A_k)$$

The addition law therefore applies to any number of mutually exclusive events.

✖ Example 4.5.3 *Equally Likely Random Numbers*

A number is drawn at random from 1 to 10 so that the sample space $S = \{1, 2, \ldots, 10\}$. Let A_1 = "an even number chosen," $A_2 = \{1, 3, 5\}$, and $A_3 = \{7\}$. If $B = A_1$ or A_2 or A_3, find pr(B).

Since the 10 outcomes are equally likely, $\text{pr}(A_1) = \frac{5}{10}$, $\text{pr}(A_2) = \frac{3}{10}$, and $\text{pr}(A_3) = \frac{1}{10}$. The three events A_i are mutually exclusive so that, by the addition law,

$$\text{pr}(B) = \text{pr}(A_1) + \text{pr}(A_2) + \text{pr}(A_3) = \frac{5}{10} + \frac{3}{10} + \frac{1}{10} = \frac{9}{10}$$

Alternatively, $B = \{1, 2, 3, 4, 5, 6, 7, 8, 10\}$ so that $\overline{B} = \{9\}$ and, by rule 2,

$$\text{pr}(B) = 1 - \text{pr}(\overline{B}) = 1 - \frac{1}{10} = \frac{9}{10}$$

✖ Example 4.5.4 *Events Relating to the Marital Status of U.S. Couples*

Table 4.5.2 is a two-way table of proportions that cross-classifies male-female couples in the United States who are not married to each other by the marital status of the partners. Each entry within the table is the proportion of couples with a given combination of marital statuses. Let us consider choosing a couple at random. We shall take all of the couples represented in the table as our sample space. The table proportions give us the probabilities of the events defined by the row and column titles (see Example 4.4.9 and Section 4.4.5). Thus the probability of getting a couple in which

TABLE 4.5.2 Proportions of Unmarried Male-Female Couples Sharing a Household in the United States, 1991

	Female				
Male	**Never married**	**Divorced**	**Widowed**	**Married to other**	**Total**
Never married	0.401	.111	.017	.025	.554
Divorced	.117	.195	.024	.017	.353
Widowed	.006	.008	.016	.001	.031
Married to other	.021	.022	.003	.016	.062
Total	.545	.336	.060	.059	1.000

Source: Constructed, with permission, from data in *The World Almanac* [1993, p. 942].

the male has never been married and the female is divorced is 0.111. Here "married to other" means living as a member of a couple with one person while married to someone else.

The 16 cells in the table, which relate to different combinations of the status of the male and the female partner, correspond to mutually exclusive events (no couple belongs to more than one cell). For this reason we can obtain probabilities of collections of cells by simply adding cell probabilities. Moreover, these 16 mutually exclusive events account for all the couples in the sample space so that their probabilities add to 1 (by rule 1).[10]

Each row total in the table tells us about the proportion of couples with males in the given category, regardless of the status of the females. Thus in 35.3% of couples the male is divorced, or equivalently, 0.353 is the probability that the male of a randomly chosen couple is divorced.

We shall continue to use Table 4.5.2 to further illustrate the use of the probability rules in the previous subsection. The complement of the event "at least one member of the couple has been married" is the event that both are in the "never married" category. Thus by rule 2,

$$\begin{aligned}\text{pr}(\textit{At least one has been married}) &= 1 - \text{pr}(\textit{Both never married}) \\ &= 1 - 0.401 = 0.599\end{aligned}$$

Note how much simpler it is in this case to use the complement than to calculate the desired probability directly by adding the probabilities of the 15 cells that fall into the event "at least one married."

QUIZ ON SECTION 4.5

1. If A and B are mutually exclusive, what is the probability that both occur? What is the probability that at least one occurs?
2. If we have two or more mutually exclusive events, how do we find the probability that at least one of them occurs?
3. Why is it sometimes easier to compute pr(A) from $1 - \text{pr}(\overline{A})$?

EXERCISES FOR SECTION 4.5

1. A house needs to be reroofed in the spring. To do this a dry, windless day is needed. The probability of getting a dry day is 0.7, a windy day is 0.4, and a wet and windy day is 0.2. What is the probability of getting
   (a) a wet day?
   (b) a day that is either wet, windy, or both?
   (c) a day when the house can be reroofed?
2. Using the data in Table 4.5.2, what is the probability that for a randomly chosen unmarried couple
   (a) the male is divorced or married to someone else?
   (b) both the male and the female are either divorced or married to someone else?
   (c) neither is married to anyone else?

[10] Such a collection of events is technically known as a partition of S.

(d) at least one is married to someone else?

(e) the male is married to someone else or the female is divorced or both?

(f) the female is divorced and the male is not divorced?

3. A young man with a barren love life feels tempted to become a contestant on the television game show "Blind Date." He decides to watch a few programs first to assess his chances of being paired with a suitable date, someone whom he finds attractive and who is no taller than he is (as he is self-conscious about his height). After watching 50 female contestants, he decides that he is not attracted to 8, that 12 are too tall, and that 9 are attractive but too tall. If these figures are typical, what is the probability of getting someone

   (a) who is both unattractive and too tall?

   (b) whom he likes, that is, is not unattractive or too tall?

   (c) who is not too tall but is unattractive?

4. For the data and situation in Example 4.4.9, what is the probability that a random job loss

   (a) was not by a male who lost it because the workplace moved?

   (b) was by a male or someone who lost it because the workplace moved?

   (c) was by a male but for some reason other than the workplace moving?

4.6 CONDITIONAL PROBABILITY

4.6.1 Definition

Our assessment of the chances that an event will occur can be very different depending on the information that we have. An estimate of the probability that your house will collapse tomorrow should clearly be much larger if a violent earthquake is expected than it would be if there were no reason to expect unusual seismic activity. The two examples that follow give a more concrete demonstration of how an assessment of the chances of an event A occurring may change radically if we are given information about whether event B has occurred or not.

EXAMPLE 4.6.1 *Conditional Probability of Zero*

Suppose we toss two fair coins and $S = \{HH, HT, TH, TT\}$. Let A = "two tails" = $\{TT\}$ and B = "at least one head" = $\{HH, HT, TH\}$. Since all four outcomes in S are equally likely, $P(A) = \frac{1}{4}$. If we know that B has occurred, however, then A cannot occur. Hence the *conditional* probability of A *given that* B *has occurred* is 0.

EXAMPLE 4.6.2 *Conditional Probabilities from a Two-Way Table of Frequencies*

Table 4.6.1 was obtained by cross-classifying 400 patients with a form of skin cancer called malignant melanoma with respect to the histological type[11] of their cancer and its location (site) on their bodies. We see, for example, that 33 patients have nodular

[11] Histological type means the type of abnormality observed in the cells that make up the cancer.

TABLE 4.6.1 Four Hundred Melanoma Patients by Type and Site

Type	Site: Head and neck	Site: Trunk	Site: Extremities	Row totals
Hutchinson's melanomic freckle	22	2	10	34
Superficial spreading melanoma	16	54	115	185
Nodular	19	33	73	125
Indeterminate	11	17	28	56
Column totals	68	106	226	400

Reproduced from Roberts et al. [1981].

melanoma on the trunk while a total of 226 have some form of melanoma on the extremities. Suppose we were to select one of the 400 patients at random.

Let A = "cancer is on the trunk." Clearly, pr(A) = 106/400. Let B be the event "cancer type is nodular." If we are given the information that a patient with nodular cancer was selected, the probability that the selected patient has cancer on the trunk may be different—this latter probability is pr(A given B). We are now concerned only with the patients who have nodular cancer, that is, the 125 patients in the third row of the table. Of this group 33 have cancer on the trunk, and each of them is equally likely to be chosen from the group with nodular cancer. Hence (where "#" is read "number of")

$$\text{pr}(A \text{ given } B) = \frac{33}{125} = \frac{\#\text{ nodular patients with cancer on trunk}}{\#\text{ nodular patients}}$$

and we denote this probability by pr($A \mid B$). In more general terms

$$\text{pr}(A \mid B) = \frac{\#\text{ outcomes in } A \text{ and } B}{\#\text{ outcomes in } B}$$

$$= \frac{\#\text{ outcomes in } A \text{ and } B/\#\text{ outcomes in } S}{\#\text{ outcomes in } B/\#\text{ outcomes in } S} = \frac{\text{pr}(A \text{ and } B)}{\text{pr}(B)}$$

What about sample spaces where the outcomes are not all equally likely? The preceding expression for pr($A \mid B$) can still be justified in much the same way when probabilities are regarded as long-run relative frequencies. We therefore use the expression as a general definition of pr($A \mid B$) for all situations.

> The ***conditional probability*** of A occurring *given* that B occurs is given by
>
> $$\text{pr}(A \mid B) = \frac{\text{pr}(A \text{ and } B)}{\text{pr}(B)}$$

Hence the probability that A occurs given B has occurred is the probability that both occur divided by the probability B occurs.[12] Similarly,

$$\text{pr}(B \mid A) = \frac{\text{pr}(A \text{ and } B)}{\text{pr}(A)}$$

Example 4.6.3 *Conditional Probabilities from a Two-Way Table of Proportions*

The data in Table 4.6.2 come from a poll of 900 New Zealanders, aged 15 or more years, carried out by the *New Zealand Listener* in 1996. The question asked was, "Do you approve of abortion when the child is not wanted?"

TABLE 4.6.2 Proportions of Females and Males and Their Responses to Abortion

	Yes	No	Don't know	Total
Females	.2489	.2178	.0733	.5400
Males	.1967	.2267	.0366	.4600
Total	.4456	.4445	.1099	1.0000

Source: The results of a poll of 900 New Zealanders from *New Zealand Listener* 1996.

Suppose that one of the people who responded was chosen at random and we want to calculate the conditional probability that a person favors an abortion, given that we know whether the person is female or male. We can work this out using the conditional probability formula as follows:

$$\text{pr}(yes \mid female) = \frac{\text{pr}(yes \text{ and } female)}{\text{pr}(female)}$$

$$= \frac{0.2489}{0.5400} = 0.4609$$

Similarly,

$$\text{pr}(yes \mid male) = \frac{0.1967}{0.4600} = 0.4276$$

We see that the probability is higher for females than males, which is perhaps not unexpected. Notice that both of the probabilities are calculated by dividing each entry in column 1 by the corresponding row total. For example, the first probability is simply the proportion of the first row total in the category "yes."

Our answers can also be interpreted in terms of proportions of the people surveyed rather than probabilities for a randomly selected respondent. Let us now compute pr($yes \mid female$), interpreted intuitively in terms of proportions, without the conditional probability rule. We are interested only in the 54% of the complete sample who are females. A subset of the females, constituting 24.89% *of the complete sample,* say yes to abortion. What proportion (fraction) *of females* does this constitute?[13] It

---

[12]We must have positive pr(B). It makes no intuitive sense anyway to compute the conditional probability of A given B when B cannot occur.

[13]Recall from basic arithmetic that when you want to know what fraction a is of b, you divide.

makes up 24.89%/54.00%, or .2489/.5400 = .4609 ≈ 46%. It may help some readers to go a step further and imagine that these proportions all relate to 10,000 people. We can therefore multiply all numbers in the table by 10,000 (chosen because it removes all decimal points) and think in terms of ratios of numbers of people (more generally, frequencies) rather than ratios of proportions. Thus we have 5400 females of whom 2489 (or 46%) say yes to abortion.

Instead of working with row totals, we can also work with column totals. For example,

$$\text{pr}(\textit{females} \mid \textit{yes}) = \frac{0.2489}{0.4456} = 0.5586$$

From this we can say that about 56% of people in the survey who favor abortion are females. In answering questions about two-way tables, you will be asked to provide answers in terms of probabilities, proportions, or percentages; the general public is often a lot happier with percentages than proportions or probabilities.

Finally, we might ask how useful such proportions are. If the people surveyed represent a random sample of the N.Z. population over 15 years of age, then we can use these proportions as estimates for the whole population.

EXERCISES FOR SECTION 4.6.1

1. Using Table 4.6.1, what is the probability that a randomly chosen patient has
   (a) nodular cancer, given they have cancer of the head and neck?
   (b) cancer of the head, neck, or trunk, given that their cancer is nodular?
2. In Example 4.5.2 Wild was available with probability 0.7, Seber with probability 0.6, and both with probability 0.5. What is the probability that Seber is available, given that Wild is available?

The conditional probability formula is not difficult to use. A more difficult skill is recognizing where a conditional probability is required. No one in real life will ever ask you, "What is the probability of A given B?" The problem arises in other guises, for example, "If A occurs, how likely is B?" or in terms of proportions of a population, "What proportion of those with property A also have property B?" In the exercises to follow, we have also included a few unconditional statements to keep you alert.

3. The following questions relate to Example 4.4.9 and Table 4.4.1.
   (a) (i) What is the probability that a random job loss was by a female and the reason given was slack work?
   (ii) Where a job was lost by a female, what is the probability that the reason given was slack work?
   (b) (i) What proportion of jobs were lost by males for slack work?
   (ii) What proportion of male job losses were for slack work?
   (iii) What proportion of job losses were due to slack work?
   (c) What proportion of all job losses was due to the position being abolished (i) for females? (ii) for males? (iii) in general?
4. The following questions relate to the data and story of Example 4.5.4 and Table 4.5.2. Parts (a) and (b) concern a randomly selected couple.
   (a) What is the probability that (i) the male is divorced? (ii) the female is divorced if the male is divorced? (iii) the male partner of a divorced female is divorced? (iv) the male is divorced and the female is divorced?

(b) (i) Consider all the relevant conditional probabilities. Where the female has been widowed, what is the marital status of the male most likely to be?

(ii) You can answer part (i) without doing the conditional probability calculations. Look at your calculations. Can you see why (and how) this can be done?

(iii) If the male is "never married," what is the female most likely to be? What is the probability that she will have this status?

(c) (i) For what proportion of couples in which the male is "never married" is the female also "never married"?

(ii) For what proportion of couples is the male "never married" and the female "never married"?

(d) For what proportion of couples in which the male is no longer married (i.e., divorced or widowed) is the female also no longer married?

5. From Table 4.6.2, find the proportion of males who did not answer yes.

4.6.2 Multiplication Rule

Sometimes the available information is conditional, and we have to use it to find the probabilities (or proportions) of other events. This often involves using the so-called multiplication rule, often as part of a larger calculation.

EXAMPLE 4.6.4 *Multiplication Rule for the Probability of a Joint Event*

In 1992 14% of the population of Israel was Arabic, and of those, 52% were described as living below the poverty line. What proportion of the population of Israel consisted of Arabic people living below the poverty line?

There are two basic events here: "Arabic" and "living below the poverty line," which we shall shorten to "poor." We see that pr(*Arabic*) = 0.14. The 52%, however, relates only to the Arabic subset of the population and is thus conditional. We have 0.52 = pr(*Poor* | *Arabic*). The quantity we want is pr(*Poor* and *Arabic*).

Multiplication rule:

$$\text{pr}(A \text{ and } B) = \text{pr}(A \mid B)\,\text{pr}(B) = \text{pr}(B \mid A)\,\text{pr}(A)$$

The multiplication rule follows directly from the definition of conditional probability.

EXAMPLE 4.6.4 (cont.) *Multiplication Rule for the Probability of a Joint Event*

The multiplication rule gives us

$$\text{pr}(Poor \text{ and } Arabic) = \text{pr}(Poor \mid Arabic)\text{pr}(Arabic)$$
$$= 0.52 \times 0.14 = 0.0728$$

or just over 7% of the population being Arabic living below the poverty line.

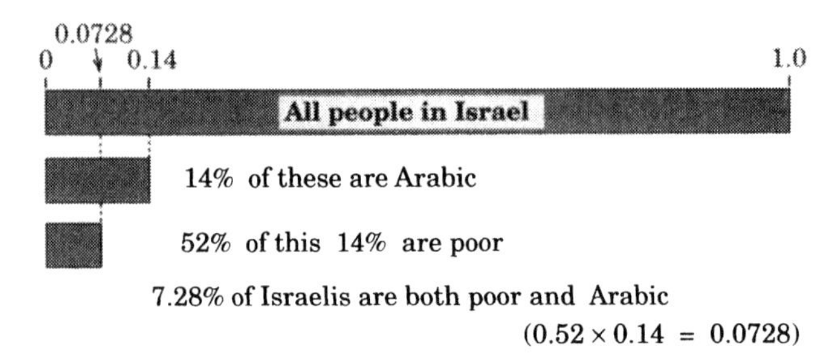

FIGURE 4.6.1 Illustration of the multiplication rule.

A more intuitive way of understanding the multiplication rule is as a "proportion of a proportion" or a "percentage of a percentage." We have 14% of the whole population being Arabic and 52% of the Arabic population being poor, so in terms of the whole population 52% of the 14% are both poor and Arabic. You may recall from basic arithmetic that when you want to find a fraction of a fraction, you multiply. The situation is illustrated in Fig. 4.6.1. Remember that a percentage is a shorthand way of expressing a fraction, for example, $14\% = \frac{14}{100}$. Thus 52% × 14% is really

$$52\% \times 14\% = \frac{52}{100} \times \frac{14}{100} = \frac{728}{10{,}000} = 7.28\%$$

EXAMPLE 4.6.5 *Multiplication Rule for a Chain of Events*

Two balls are drawn at random, one at a time without replacement, from a box of 4 white and 2 red balls. What is the probability that both balls are white?

Let W_1 and W_2 represent the events "first ball is white" and "second ball is white," respectively. Then

$$\text{pr}(W_1 \text{ and } W_2) = \text{pr}(W_1)\text{pr}(W_2 \mid W_1)$$

As each ball is equally likely to be selected as the first ball, $\text{pr}(W_1)$ is the number of white balls divided by the number of balls, that is, $\frac{4}{6}$. Once the first ball is chosen, there are 5 balls left, of which 3 are white. Hence $\text{pr}(W_2 \mid W_1) = \frac{3}{5}$ and

$$\text{pr}(W_1 \text{ and } W_2) = \frac{4}{6} \times \frac{3}{5} = \frac{2}{5}$$

EXERCISES FOR SECTION 4.6.2

1. In the United States, approximately 1% of the population is schizophrenic, 0.8% of people are homeless, and one-third of the homeless are schizophrenic. Using probability notation and the events "homeless" and "schizophrenic," write down the three pieces of information given here in terms of probability statements about events. (Note: the 1% is inaccurate—see Review Exercises 4, problem 10.)

2. According to a survey reported in the *Kitchener-Waterloo Record* (17 May 1989), 26% of residents of the Canadian province of British Columbia aged between 18 and 25 had used cocaine, and 77% of those who tried it once used it again. Based on these figures, what proportion of B.C. residents in this age group had used cocaine at least twice?

3. Referring to Example 4.6.5, what is the probability that both balls are red?

4.6.3 More-Complicated Calculations Using Conditional Probabilities

✱ Example 4.6.6 *Using Conditional Probabilities for a Sequence of Events*

Suppose we sample 2 balls at random, one at a time without replacement, from an urn containing 4 black balls and 3 white balls. We want to calculate the probability that the second ball is black. When we come to make the second draw, the chances of drawing a black ball depend on what ball was removed at the first draw because that determines the composition of the balls in the urn. We shall therefore have to use information that comes naturally in the form of conditional probabilities. We use the same notation as in Example 4.6.5, for example, B_2 denotes the event that the second ball sampled is black.

Tree Diagrams

One method that is sometimes used to tackle problems like the one in Example 4.6.6 involves a type of diagram called a *(probability) tree diagram.* These diagrams often provide a convenient way of organizing (and then using) conditional probability information. To aid the discussion, Fig. 4.6.2 gives a tree diagram for the situation in Example 4.6.6. Along the way, we shall state some general rules for constructing and using such trees.

The probability written beside each line segment in the tree is the probability that the right-hand event on the line segment occurs *given* the occurrence of all the events that have appeared along that path so far (reading from left to right). Each time a branching occurs in the tree, we want to cover all eventualities, so the probabilities beside any "fan" of line segments should add to unity.

Because the probability information on a line segment is conditional on what has gone before, the order in which the tree branches should reflect the type of information that is available. In Example 4.6.6 we have unconditional probability information about the first draw so the first set of branches of the tree (represented by B_1 and W_1) concern the first draw. The readily available probability information about the second draw depends on (i.e., is conditional on) what happened at the first draw and thus

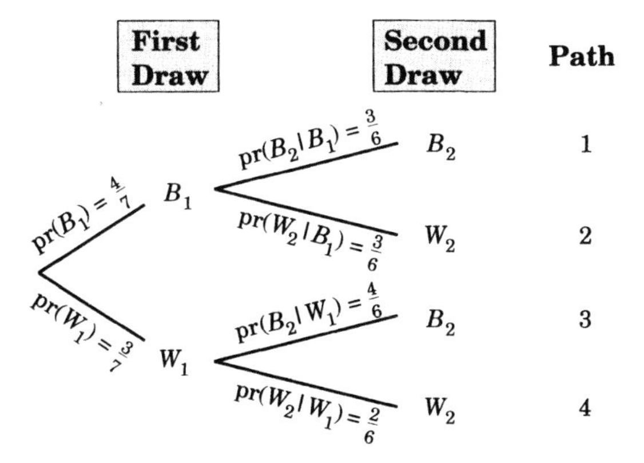

FIGURE 4.6.2 Tree diagram for a sampling problem.

forms the second set of branches. We draw the tree to represent all four possible outcomes: "B_1 and B_2," "B_1 and W_2," "W_1 and B_2," and "W_1 and W_2." These four outcomes (events) are mutually exclusive and give all the possibilities.

Rules for Use

(i) ***Multiply along a path*** to get the probability that all of the events on that path occur. (This uses the multiplication rule for conditional probabilities.)

(ii) ***Add*** the probabilities of ***all whole paths*** in which an event occurs to obtain the probability of that event occurring. (This uses the addition rule for mutually exclusive events.)

EXAMPLE 4.6.6 (cont.) *Using the Rules*

To get the probability of obtaining a black ball on the second draw, the rules tell us to multiply along paths and add whole paths containing a black ball on the second draw (namely, paths 1 and 3). This gives us

$$\text{pr}(B_2) = \frac{4}{7} \times \frac{3}{6} + \frac{3}{7} \times \frac{4}{6} = \frac{4}{7}$$

EXAMPLE 4.6.7 *Tree Diagram and Two-Way Table Methods*

We revisit the data in Example 4.6.4. A 1992 news report stated that 11% of Israel's Jewish population and 52% of its Arabic citizens lived below the poverty line. Arabic citizens were reported to make up 14% of the population of Israel. We shall assume that these two groups account for the whole population of Israel so that 86% of the population is Jewish. We shall determine (i) proportion of the Israel's population living below the poverty line, and (ii) the proportion of poor people in Israel who were Arabic. In probability notation, we can write the information we have been given as pr(*Poor* | *Jewish*) = 0.11, pr(*Poor* | *Arabic*) = 0.52, pr(*Arabic*) = 0.14 and pr(*Jewish*) = 0.86.

There are two factors at work here: ethnic group and whether or not someone is poor. We could use a tree to work on this problem splitting first on ethnicity, because we have unconditional information about this, and then on poverty because our information about poverty is conditional on what ethnic group is being discussed. This has been done in Fig. 4.6.3.

The event "being poor" corresponds to paths 1 and 3. Our rules tell us to obtain pr(*Poor*) by multiplying along paths and adding whole paths containing this event. Thus, pr(*Poor*) = $0.14 \times 0.52 + 0.86 \times 0.11 = 0.1674$. This tells us that approximately 17% of the population of Israel lives below the poverty line. We can now find the proportion of poor people who are Arabic by

$$\text{pr}(\textit{Arabic} \mid \textit{Poor}) = \frac{\text{pr}(\textit{Poor} \text{ and } \textit{Arabic})}{\text{pr}(\textit{Poor})} = \frac{0.14 \times 0.52}{0.1674} = 0.4349$$

We see that almost half (43%) of Israel's poor are Arabic.

This approach may look reasonably simple—once we have laid out the tree for you! We have found, however, that many of our students are better able to solve problems using the table method to follow. Our two factors, ethnicity and poverty, become

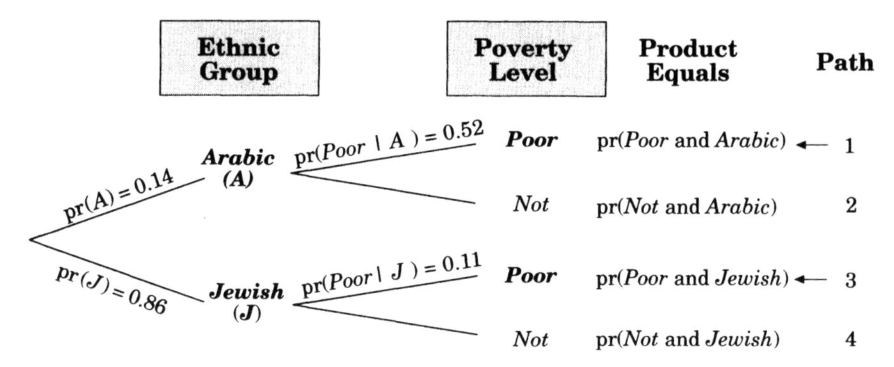

FIGURE 4.6.3 Poverty in Israel.

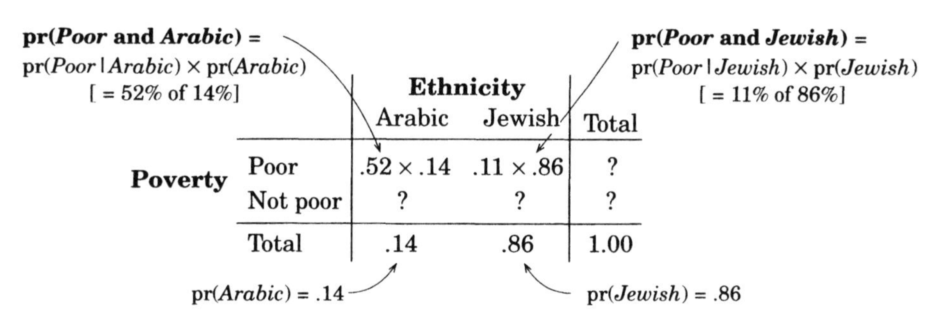

		Ethnicity		
		Arabic	Jewish	Total
Poverty	Poor	.52 × .14	.11 × .86	?
	Not poor	?	?	?
	Total	.14	.86	1.00

FIGURE 4.6.4 Proportions by ethnicity and poverty.

the two dimensions of a two-way table. Our information that 14% of the population are Arab and 86% are Jewish belong in the column totals of the table. The entries inside the two-way table of proportions need to be of the form pr(*A* and *B*), not conditional information. Thus our conditional information, for example, pr(*Poor* | *Arabic*) = 0.52, cannot be inserted as it stands. However, we can construct the right type of proportions using the multiplication rule pr(*A* and *B*) = pr(*B* | *A*) pr(*A*), as shown in Fig. 4.6.4.

We then fill in all the unknown entries in the table, using the simple idea that sums across rows have to give row totals and sums down columns must produce the column totals, as follows.

.0728	.0946	?
?	?	?
.14	.86	1.00

→

.0728	.0946	.1674
?	?	?
.14	.86	1.00

→

.0728	.0946	.1674
.0672	.7654	.8326
.14	.86	1.00

We got from the second to the third of the small tables above by making the columns add up. The third small table is our completed table, which we present with all the labels attached as Table 4.6.3. Many probabilities can be read off Table 4.6.3, including pr(*Poor*) = 0.1674. We can obtain any of the conditional probabilities in the normal way including

$$\text{pr}(Jewish \mid Poor) = \frac{\text{pr}(Poor \text{ and } Jewish)}{\text{pr}(Poor)} = \frac{0.0946}{0.1674} = 0.5651$$

TABLE 4.6.3 Proportions by Ethnicity and Poverty

		Ethnicity		
		Arabic	Jewish	Total
Poverty	Poor	.0728	.0946	.1674
	Not poor	.0672	.7654	.8326
	Total	.14	.86	1.00

We note that in pr(*Jewish* | *Poor*) the order of the conditioning is reversed from the order in the information given to us, which was of the form pr(*Poor* | *Jewish*). The table has allowed us to reverse the order of the conditioning.[14]

Many problems involving ***the reversal of order of conditional probabilities*** can be solved by constructing two-way tables.

Example 4.6.8 *Constructing a Two-Way Table*

From a 1990 study issued by the National Academy of Sciences in the United States it was found that, of American women using contraception, 38% are sterilized, 32% use oral contraceptives, 24% use barrier methods (diaphragm, condom, cervical caps), 3% use IUDs, and 3% rely on spermicides (foams, creams, jellies). If we define the failure rates of a method as the percentage of those who become pregnant during a year of use of the method, then the failure rates for each of these methods are approximately as follows: sterilization 0%, the oral contraceptive pill 5%, barrier methods 14%, IUDs 6%, and spermicides 26%. One question we would like to answer is, What percentage of women using contraception experience an unwanted pregnancy over the course of a year? Another is, If we look only at women who experienced contraceptive failure, what proportions were using each type of contraceptive method?

There are two factors at work in this problem as well. One is *Method*—the type of contraceptive method the woman was using. The other is *Outcome*—whether or not the woman experienced contraceptive failure. This suggests constructing a two-way table with dimensions *Method* and *Outcome*, as in Fig. 4.6.5. We are given information about the proportions of women using each method, pr(*Sterilized*) = 0.38, pr(*Oral*) = 0.32, ... This information can be inserted directly into the column totals, as in Fig. 4.6.5. Our other information is about the proportion of women experiencing failure conditional on the type of contraceptive used, for example, we see pr(*Fail* | *Sterilized*) = 0, pr(*Fail* | *Oral*) = 0.05, pr(*Fail* | *Barrier*) = 0.14, and so on. This cannot be inserted directly because the interior entries in the table must be of the form pr(*A* and *B*) rather than conditional probabilities. Once more, we construct these from the conditional probabilities using the multiplication rule, as shown in Fig. 4.6.5.

[14] The two-way table has avoided the use of the so-called Bayes' theorem, which is a more traditional way of solving these problems in other textbooks.

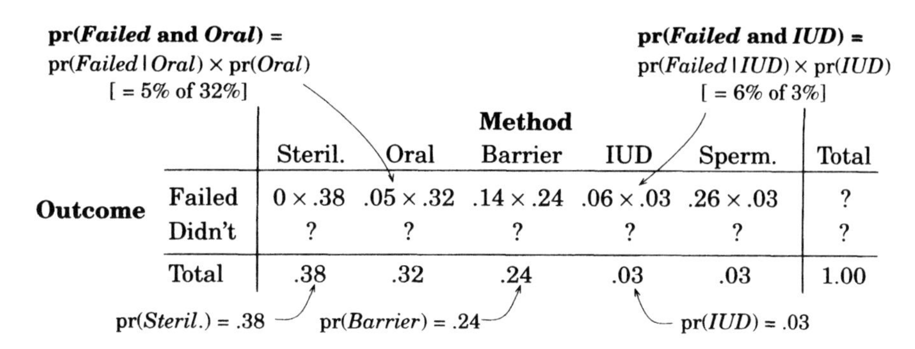

FIGURE 4.6.5 Proportions by outcome and method.

TABLE 4.6.4 Table Constructed from the Data in Example 4.6.8

		Method					
		Steril.	Oral	Barrier	IUD	Sperm.	Total
Outcome	Failed	0	.0160	.0336	.0018	.0078	.0592
	Didn't	.3800	.3040	.2064	.0282	.0222	.9408
	Total	.3800	.3200	.2400	.0300	.0300	1.0000

Knowing that the table should add up going across rows and down columns enables us to fill in the missing entries and obtain Table 4.6.4.

From the completed table, we are able to read off pr(*Failed*) = 0.0592, or approximately 6% of the women sampled had experienced contraceptive failure. Conditional probabilities such as pr(*Barrier* | *Failure*) are easily constructed from the information in the table:

$$\text{pr}(Barrier \mid Failure) = \frac{\text{pr}(Barrier \text{ and } Failure)}{\text{pr}(Failure)} = \frac{0.0336}{0.0592} = 0.568$$

This tells us that, of the women experiencing contraceptive failure, 57% were using IUDs. This sort of information is often quoted in the media. If you read those words, you would tend to think, "IUDs must be a particularly unreliable form of contraception." However, pr(*IUD* | *Failure*) is the wrong probability for deciding which method to use. The relevant probability is pr(*Failure* | *IUD*), that is, the failure rate among those using the method. For IUDs this is 6%, which is nearly as good as the 5% failure rate for oral contraceptives and much better than barrier methods or spermicides.

EXERCISES FOR SECTION 4.6.3

Solve the following problems by setting up an appropriate two-way table.

1. In New Zealand, 3.24% of Europeans and 1.77% of Maori have type AB blood. A blood bank in a district where the population is 85% European and 15% Maori wants to know how much AB blood to stock. What percentage of people in the district have AB blood? What percentage of the people in the AB blood group are Maori?

2. The chances of a child being left-handed are 1 in 2 if both parents are left-handed, 1 in 6 if one parent is left-handed, and 1 in 16 if neither parent is left-handed (*New Zealand Herald*, 5 January 1991). Suppose that, of couples having children, in 2% both father and mother are left-handed, in 20% one is left-handed, and in the rest neither is left-handed. What is the probability of a randomly chosen child being left-handed? What is the probability that neither parent of a left-handed child is left-handed?

3. University of Florida sociologist Michael Radelet believed that if you killed a white person in Florida, the chances of getting the death penalty were three times greater than if you had killed a black person (*Gainesville Sun*, 20 October 1986). In a study, Radelet classified 326 murderers by race of the victim and type of sentence given to the murderer. He found that 36 of the convicted murderers received the death sentence. Of this group, 30 had murdered a white person, whereas 184 of the group that did not receive the death sentence had murdered a white person. If a victim from this study was white, what is the probability that the murderer of this victim received the death sentence? Do you agree with Radelet?

QUIZ ON SECTION 4.6

1. In pr($A \mid B$), how should the symbol "|" be read? (Section 4.6.1)
2. Give an example where A and B are two events with pr(A) $\neq 0$ but pr($A \mid B$) $= 0$.
3. If event A always occurs when B occurs, what can you say about pr($A \mid B$)?
4. When drawing a probability tree for a particular problem, how do you know what events to use for the first fan of branches and which to use for the subsequent fans?
   What probability do you use to label a line segment?
   How do you find the probability that all events along a given branch occur?
   How do you find the probability that a particular event occurs? (Section 4.6.3)
5. What results do you use for two-way tables to fill in unknown entries?

CASE STUDY 4.6.1 *Testing for AIDS*

It is well known that AIDS is one of the most important public health problems facing the world today. The sense of extreme crisis has dulled in western countries as the 1990s end, but AIDS is rampant in parts of Africa. AIDS is believed to be caused by the human immunodeficiency virus (HIV), but many years can elapse between HIV infection and the development of AIDS. This case study is based on the situation in the early '90s when there were still widespread demands for screening whole populations for HIV infection. In 1990 the World Health Organization (WHO) projected between 25 and 30 million cases of HIV infection worldwide by the year 2000. The United States, where 200,000 AIDS cases had been reported by mid-1992, was the worst-affected western country, largely because the epidemic began earlier in the United States. By 1990 WHO estimated that one in every 75 males and one in every 700 females in the United States was infected with HIV.

The enzyme-linked immunosorbent assay (ELISA) test was the main test used to screen blood samples for antibodies to the HIV virus (rather than the virus itself). It gives a measured mean absorbance ratio for HIV (previously called HTLV) antibodies. Table 4.6.5 gives the absorbance ratio values for 297 healthy blood donors and 88 HIV patients. Healthy donors tend to give low ratios, but some are quite high, partly because the test also responds to some other types of antibody, such as human leucocyte antigen or HLA (Gastwirth [1987, p. 220]). HIV patients tend to have high ratios, but a few give lower values because they have not been able to mount a strong immune reaction.

To use this test in practice, we need a cutoff value so that those who fall below the value are deemed to have tested negatively and those above to have tested positively. Any such cutoff will involve misclassifying some people without HIV as having a positive HIV test (which will be a huge emotional shock), and some people with HIV as having a negative HIV test (with consequences to their own health, the health of people about them, the integrity of the

TABLE 4.6.5 Number of Individuals Having a Given Mean Absorbance Ratio (MAR) in the ELISA for HIV Antibodies

MAR	Healthy donor		HIV patients	
< 2	202	275	0	2
2– 2.99	73		2	
3– 3.99	15		7	
4– 4.99	3		7	
5– 5.99	2		15	
6–11.99	2		36	
12+	0		21	
Total	297		88	

Adapted from Weiss et al. [1985].

blood bank, etc.). Using a cutoff ratio of 3 we find that of the healthy people[15] in Table 4.6.5, 275/297 = 0.926 test negatively (22 false positives) and of HIV patients 86/88 = 0.98 test positively (2 false negatives). It should be noted that the false-negative rate may be an undercount.[16] Better results than these have been obtained with the multiple use of ELISA (Gastwirth [1987, p. 236]) and with modern commercial versions of the test. The proportions given above are only rough estimates from small samples. Nevertheless, making use of the numerical equivalence between proportions of a population and probabilities for a randomly chosen individual (Section 4.4.5), we shall use these proportions as if they were true probabilities. Hence

$$\text{pr}(\textit{Positive} \mid \textit{HIV}) = 0.98$$

and, rounding off to two decimal places (as the information is very approximate),

$$\text{pr}(\textit{Negative} \mid \textit{Not HIV}) = 0.93$$

We shall consider the effect of screening the whole U.S. population for AIDS in 1991. At that time, the proportion of Americans infected with HIV was about 1%. We are interested in the proportion of Americans who would test positively and the proportion of those testing positively who would actually have AIDS.

There are two factors of interest in this problem. First is *disease status*—a person either has HIV or does not. Second is *test result*—the person's test result is either positive or negative. We shall construct a two-way table in the usual way to form Fig. 4.6.6. We know that

$$\text{pr}(\textit{Positive} \mid \textit{HIV}) = 0.98$$

$$\text{pr}(\textit{Negative} \mid \textit{Not HIV}) = 0.93$$

$$\text{pr}(\textit{HIV}) = 0.01$$

and thus

$$\text{pr}(\textit{Not HIV}) = 0.99$$

We shall place this information into the table in the usual way, recalling that entries in the interior of the table have to be of the form pr(*A* and *B*). They cannot be conditional.

We now complete the table using the simple idea that the rows and columns of the table must add up to give the

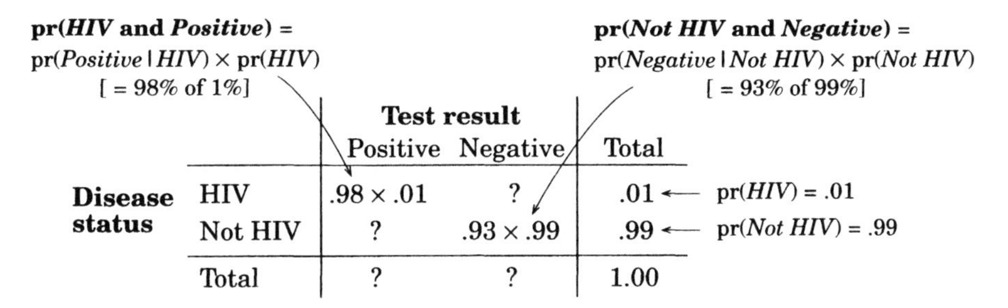

FIGURE 4.6.6 Putting HIV information into the table.

[15] In the medical and biostatistical literatures, the probability of correctly diagnosing a sick individual as "sick" is called the *sensitivity* of a test, while the probability of correctly diagnosing that a healthy individual does not have the condition of interest is called the *specificity* of that test.

[16] It appears that the virus takes 6 to 12 weeks to provoke antibody production (*Time*, 2 March 1987, p. 44). Also, *Time* (12 June 1989) reports cases of infected men who had not produced antibodies for up to three years.

totals as follows:

.0098	?	.01
?	.9207	.99
?	?	1.00

→

.0098	.0002	.01
.0693	.9207	.99
?	?	1.00

→

.0098	.0002	.01
.0693	.9207	.99
.0791	.9209	1.00

This gives us Table 4.6.6 as our completed table.

We can read pr(*Positive*) = 0.0791 off Table 4.6.6. This is a surprising result. Although only 1% of the population have HIV, 8% would test positively. The majority of the people in any sample who tested positively would not in fact have HIV.[17] This is despite the fact that the test seems reasonably good. After all, it correctly classifies 98% of people with HIV as having HIV and 93% of people without HIV are classified as being HIV free.

TABLE 4.6.6 Proportions by Disease Status and Test Result

		Test result		
		Positive	Negative	Total
Disease status	HIV	.0098	.0002	.01
	Not HIV	.0693	.9207	.99
	Total	.0791	.9209	1.00

So what proportion of people testing positively on ELISA would actually have HIV? We want pr(*HIV* | *Positive*). The conditioning here is in the reverse order to that in the original information supplied to us. We can get everything we need from Table 4.6.6, however.

$$\text{pr}(HIV \mid Positive) = \frac{\text{pr}(HIV \text{ and } Positive)}{\text{pr}(Positive)} = \frac{0.0098}{0.0791} = 0.124$$

As the previous figures suggested, if the whole U.S. population had been screened, only 1 person in 8 (12.4%) who tested positively would have had HIV. The other 7 out of every 8 would be false positives. The situation in other western countries would have been even more extreme, as Table 4.6.7 shows.

The values of pr(*HIV* | *Positive*) given for each of the countries were calculated exactly as above. The only thing that changed from country to country was the value used for pr(*HIV*), the proportion of the population with HIV. [Note that the American figures have changed slightly due to the use of pr(*HIV*) = .00864 rather than the rounded value of 0.01 used in the detailed calculations.]

The most extreme case in Table 4.6.7 is Ireland. If the Irish government had decided to screen the total population of 3.6 million people for HIV in 1990, from the figures above, roughly 250,000 (7%) would have tested positively, and of these, only about 1250 (0.5% of the positives) would

TABLE 4.6.7 Proportions Infected with HIV

Country	No. AIDS[a] cases	Population[b] (millions)	pr(*HIV*)[c]	pr(*HIV* | *Positive*)
United States	218,301	252.7	0.00864	0.109
Canada	6,116	26.7	0.00229	0.031
Australia	3,238	16.8	0.00193	0.026
New Zealand	323	3.4	0.00095	0.013
United Kingdom	5,451	57.3	0.00095	0.013
Ireland	142	3.6	0.00039	0.005

[a]Source: AIDS—New Zealand, November 1992.

[b]1991 estimates, except for Ireland, for which May 1990 figures are given.

[c]Proportion of population infected by HIV. These are very rough. We have assumed that the proportion of HIV-infected people is 10 times larger than the proportion of AIDS cases. This is the approximate relationship between the number of U.S. cases and the U.S. Centers for Disease Control's estimate of the number of HIV-infected Americans in 1990.

[17]People without HIV who test positively are called *false positives*.

have HIV. How do we tell these 1250 people apart from the rest of the 250,000? In the case of HIV there was another more expensive and more specific test, called the Western blot test, that could be used.[18] Thus any screening program would have to include funding for both ELISA tests for everyone and Western blot tests for a quarter of a million people.[19]

Although the value of pr(*HIV* | *Positive*) in Table 4.6.7 varies with the proportion of people with HIV in the population to some extent, all of the entries in the table are small. We don't want to leave you with the reverse misapprehension that pr(*HIV* | *Positive*) is always small. Among intravenous drug users in New York in 1988, it was estimated that 86% had HIV (*New Zealand Herald,* 17 November 1988). Using a value of pr(*HIV*) = 0.86 in the calculations produces pr(*Positive*) = 0.853 and pr(*HIV* | *Positive*) = 0.988. If all New York drug addicts had been screened, almost every person testing positively (98.8%) would have had HIV.

This sort of "good-but-imperfect-test" situation is widespread. It applies to large numbers of medical screening procedures (diabetes, cervical cancer, breast cancer, etc.).[20] It applies to polygraph lie detector tests (some people who are not lying show the physiological symptoms interpreted as a sign of lying, while some people who are lying do not). It applies to psychological and intellectual tests performed to judge the suitability of job applicants (some people who are capable of doing the job well will fail the tests, while some who are not will pass the tests). It can also apply to the testing of urine or blood samples to detect drug use. The type of problem we see in Table 4.6.7, where the majority of those who would test positive would be false positives, is very common in screening for relatively rare conditions. Similar behavior could be expected in testing for drug use among a population in which drug use is rare or using lie detector tests on a group of people in which the vast majority told the truth. An alternative strategy, as indicated by the results for New York drug addicts, is to try to identify high-risk subpopulations and screen only those. With medical screening, particularly in an area as sensitive as AIDS, this can be political dynamite.

So far, we have been using pr(*HIV* | *Positive*) to think about the proportion of those testing positive in a screened population who actually have HIV. But what does pr(*HIV* | *Positive*) mean for an individual?

Let's get personal and imagine that you have just tested positive. Clearly, this would be a major trauma for you. *Time* (2 March 1988) quoted a health professional as saying, "The test tends to rip people's lives apart." *The Economist* (4 July 1992) told a story of a young American who committed suicide on learning that he had tested positive for HIV: "He believed his chance of carrying the virus was 96%. It was 10%."[21] What is pr(*HIV* | *Positive*) for me, that is, what is the probability that *I* have HIV given that *I* have just tested positive on an ELISA test?

We have to think in terms of being a random representative of some population. We saw earlier that the value of pr(*HIV* | *Positive*) depends critically on the value of pr(*HIV*) for the population from which the individual is sampled. None of us can usefully be thought of as a randomly selected individual from our own country as far as HIV is concerned because we know that HIV is much more prevalent in some sections of the population than others. To obtain a value of pr(*HIV* | *Positive*) for yourself, a value for pr(*HIV*) is required that gives the proportion of people who have HIV *among people as much as possible like yourself* with respect to the known risk factors for AIDS. If you are a New York drug addict who shares needles, a positive ELISA test is fairly conclusive. If you have always lived in a monogamous sexual relationship, believe your partner to have done the same, don't share needles, and didn't have a blood transfusion prior to the testing of the blood supply, a positive ELISA test is almost certainly a false positive.

4.7 STATISTICAL INDEPENDENCE

4.7.1 Two Events

We have seen (e.g., Example 4.6.3) that our assessment of the chances that an event occurs can change drastically depending on the information we have about other

[18] In medical terms (see footnote 15) the Western blot is more specific but not as sensitive as ELISA.

[19] Such testing is not cheap! The state of Illinois introduced screening as a condition for a marriage license in 1988. In the first 11 months 150,000 people were screened at a cost of $5.5 million (23 were infected). Many other states now do similar screening.

[20] A screening test, however, is designed to identify a group at increased risk of a condition.

[21] It is surprising that someone was given the results of a positive result on a single test. In New Zealand, people are not told that they have tested positive unless they have also tested positive on a second ELISA test and on a Western blot test.

events. In general, pr($A \mid B$) and pr(A) are different. If there is no change, however, that is, pr($A \mid B$) = pr(A), then knowing whether B has occurred gives no new information about the chances of A occurring. We then say that A and B are ***statistically independent*** and we have the following definition:

> Events A and B are statistically independent if
>
> $$\text{pr}(A \mid B) = \text{pr}(A)$$

In this case the multiplication formula pr(A and B) = pr(B) pr($A \mid B$) becomes pr(A and B) = pr(A) pr(B), and we take this as our working rule.

> If A and B are statistically independent, then
>
> $$\text{pr}(A \text{ and } B) = \text{pr}(A)\,\text{pr}(B)$$

From the working rule we find that we also have pr($B \mid A$) = pr(B) so that it doesn't matter whether our definition of independence is given in terms of pr($A \mid B$) or pr($B \mid A$): If A is independent of B, then B is also independent of A. Our main use of the working rule is to calculate pr(A and B) when we know that A and B are independent. It can also be used to check for statistical independence in a probability model by checking whether the formula works.

Unfortunately, it is only possible to establish statistical independence theoretically in rather trivial models. To establish that events are exactly independent in a more complex setting by doing an experiment and collecting data would require an infinite amount of data. In most statistical modeling of the real world, independence is assumed when it seems reasonable to do so on the basis of subject-matter knowledge or intuition. The notion of *physical independence* gives one way of thinking about this important modeling issue. Events are *physically independent* if there is no physical way in which the outcome of one event can influence the outcome of the other. Under any sensible probability model, physically independent events will be statistically independent.

Note: If A and B are independent, then so are A and $\overline{B}$, $\overline{A}$ and B, $\overline{A}$ and $\overline{B}$.[22] This makes intuitive sense for physically independent experiments. Suppose A can occur in experiment 1 and B in experiment 2. If the two experiments are physically independent, then the occurrence or nonoccurrence of A will not affect the occurrence or nonoccurrence of B.

✖ Example 4.7.1 *Independent Blood Phenotypes*

According to one study, 30% of black Americans have type A blood, and 26% have the genetic marker PGM 1-2. If these characteristics appear independently, and there is

[22]These can be proved formally using the probability rules.

strong evidence that they do, then the probability that a black American has both is

$$\begin{aligned}\text{pr}(\textit{Type A} \text{ and } \textit{PGM 1-2}) &= \text{pr}(\textit{Type A}) \times \text{pr}(\textit{PGM 1-2}) \\ &= 0.3 \times 0.26 = 0.078\end{aligned}$$

The probability that a black American has type A blood but does not have PGM 1-2 is

$$\begin{aligned}\text{pr}(\textit{Type A} \text{ and } \overline{\textit{PGM 1-2}}) &= \text{pr}(\textit{Type A}) \times \text{pr}(\overline{\textit{PGM 1-2}}) \\ &= 0.3 \times 0.74 = 0.222\end{aligned}$$

Example 4.7.2 *Independent Tosses of a Coin*

Suppose we toss a coin twice, so that $S = \{HH, HT, TH, TT\}$. Let

$$\begin{aligned}A &= \text{"at least one head"} = \{HH, HT, TH\} \\ B &= \text{"at least one tail"} = \{HT, TH, TT\}\end{aligned}$$

Assuming equally likely outcomes, we have $\text{pr}(A) = \frac{3}{4}$ and $\text{pr}(B) = \frac{3}{4}$. However, the probability that both events occur is

$$\text{pr}(A \text{ and } B) = \text{pr}(\{HT, TH\}) = \tfrac{1}{2} \neq \text{pr}(A) \times \text{pr}(B)$$

Thus the events A and B are *not* independent.

Example 4.7.3 *Two-Way Table for Independent Blood Phenotypes*

There are a large number of genetically based blood group systems that have been used for typing blood. Two of these are the Rh system (with blood types Rh+ and Rh−) and the Kell system (with blood types K+ and K−). It is found that any person's blood type in any one system is independent of his or her blood type in any other. It is known that, for Europeans in New Zealand, about 81% are Rh+ and about 8% are K+. If a European New Zealander is chosen at random, what is the probability that he or she is either (Rh+, K+) or (Rh−, K−)?

We can now construct a two-way table. From the preceding information we have $\text{pr}(Rh+) = 0.81$, $\text{pr}(Rh-) = 0.19$, $\text{pr}(K+) = 0.08$, and $\text{pr}(K-) = 0.92$. These probabilities belong in the row and column totals respectively. How do we get the probabilities in the interior of the table? We use a multiplication rule, except that this time we use it for independent events and not for conditional probabilities. For example

$$\text{pr}(Rh+ \text{ and } K+) = \text{pr}(Rh+)\text{pr}(K+) = 0.81 \times 0.08 = 0.0648$$

In a similar manner we can calculate all the probabilities in Table 4.7.1.

To find the answer to our original question, we simply add the probabilities for the $(Rh+, K+)$ and $(Rh-, K-)$ cells to get $0.0648 + 0.1748 = 0.2396$.

Suppose now that a murder victim has blood type $(Rh+, K-)$ but has a blood stain on him with type $(Rh-, K+)$, presumably from the assailant. What is the

TABLE 4.7.1 Blood Type Data

.92 × .81   pr(*RH*+) = .81

.08 × .81

.08 × .19

	K+	*K*-	Total
Rh+	?	?	.81
Rh-	?	?	.19
Total	.08	.92	1.00

pr(*K*+) = .08   .92 × .19

→

	K+	*K*-	Total
Rh+	.0648	.7452	.81
Rh-	.0152	.1748	.19
Total	.08	.92	1.00

probability that a randomly selected person matches this type? The answer is read from Table 4.7.1, namely 0.0152, which is quite small. Suppose further that one of the suspects has this blood type. Either the suspect is innocent and a random event with probability 0.0152 has occurred, or the suspect is guilty. What do you think?

EXERCISES FOR SECTION 4.7.1

1. According to a study on 3433 women conducted by the Alan Guttmacher Institute in the United States (*Globe and Mail,* 7 August 1989), 6% of women on the contraceptive pill can expect to become pregnant in the first year compared with 14% of women who do not use the pill but whose partners use condoms. What are the chances of the woman becoming pregnant in the first year if she is on the pill *and* he uses condoms? (Assume independence. Is this a reasonable assumption?)

2. White North Americans in California have blood phenotypes A, B, O, and AB with probabilities 0.41, 0.11, 0.45, and 0.03, respectively. If two whites are chosen at random, what is the probability that they have the same phenotypes? Why may you assume that the two are independent?

4.7.2 Positive and Negative Association

In humans, independence of characteristics, as in Example 4.7.1, tends to be the exception rather than the rule. Some things we know tend to go together, for example blond hair and blue eyes. Someone with blond hair is much more likely to have blue eyes than someone with brown or black hair. We say the events "having blond hair" and "having blue eyes" are positively associated. Suppose we look at the population in which 30% have blond hair and 25% have blue eyes. If we assumed independence, we would say that the proportion with both is

$$\begin{aligned}\text{pr}(\textit{blond hair} \text{ and } \textit{blue eyes}) &= \text{pr}(\textit{blond hair})\,\text{pr}(\textit{blue eyes}) = 0.3 \times 0.25 \\ &= 0.075\end{aligned}$$

Since these events are not independent, we should have used

$$\begin{aligned}\text{pr}(\textit{blond hair} \text{ and } \textit{blue eyes}) &= \text{pr}(\textit{blond hair})\,\text{pr}(\textit{blue eyes} \mid \textit{blond hair}) \\ &= 0.3 \times \textbf{?}\end{aligned}$$

Among blond-haired people the proportion with blue eyes is high, probably much closer to 80% than 25%. The product 0.3×0.8 is then much larger than 0.075. Assuming independence when events are positively associated can lead to answers that are far too small.

Many human characteristics are negatively associated as well; that is, if you have one you are much less likely to have the other. Black hair and blue eyes is one example. Assuming independence when events are negatively associated leads to answers that are too big.

*4.7.3 Mutual Independence of More Than Two Events

The n events $A_1, A_2, \ldots, A_n$ are ***mutually independent*** if

$$\text{pr}(A_1 \text{ and } A_2 \text{ and } \ldots \text{ and } A_n) = \text{pr}(A_1)\,\text{pr}(A_2)\cdots\text{pr}(A_n)$$

and the same type of multiplication formula holds for *any* subcollection of the events.[23] We shall use this result often in Chapter 5 for events like tosses of a coin, which we know are physically independent. The preceding formula and the independence assumption are frequently abused, however. Often it is relatively easy to get information about individual probabilities, for example the proportions of individuals who own their own houses, who believe in abortion, who have a high intelligence, and who hold strong religious beliefs. To calculate the probability that all these conditions hold at the same time, we need a generalization of the multiplication rule of Section 4.6.2. This requires information about conditional probabilities, for example, the proportion of strongly religious people *among those* who have high intelligence and believe in abortion and own their own house and so on. Clearly, information on society broken down to this level is hard to find. What often happens is that, in the *absence of knowledge* of the appropriate conditional probabilities, people *assume* independence. From the discussion of the previous section, this can lead to answers that are grossly too small or grossly too large—and we won't know! The two case studies that follow the exercises for this section contain salutary tales.

EXERCISES FOR SECTION 4.7.3

1. Consider the following scenario. A woman in her early 40s wants to have a baby so that the child can serve as a bone marrow donor for her teenage daughter who has a virulent form of leukemia. To make things even more difficult, the husband had a vasectomy many years ago. What are the mother's chances of success? Now suppose the chance of successfully reversing such a vasectomy is 50%. At her age the chances of conception are about 75%. Suppose that the chance of siblings of the same parents having matched blood marrow is 25% and the chance of a bone-marrow transplant curing leukemia in this patient is estimated to be about 70%. To be successful, all of these things have to happen.
   (a) Assuming independence, what is the mother's chance of success?
   (b) Which criteria do you think are independent, and which are you doubtful about?
   [Note: Such an event actually took place! See *Time* (5 March 1990, p. 41, and 23 March 1992, p. 49).

2. In 1991 a power failure at an AT&T switching center triggered loss of telephone service to over 1 million people in New York City and caused havoc, especially with air traffic (*Time*, 30

---

[23]This is stronger than just requiring every pair of events to be independent; that is, $\text{pr}(A_i \text{ and } A_j) = \text{pr}(A_i)\,\text{pr}(A_j)$, all $i \neq j$. If $n = 4$, the product rule has to hold not only for all four events but also for any three events and for any two events.

September 1991). When AT&T switched to their own power generation equipment, a power surge tripped a battery-powered emergency backup system. This triggered battery-powered alarms to warn that the backup system had been activated. Unfortunately, the alarm system's sirens did not come on and the visual warnings were not noticed for over five hours. When the batteries ran down the resulting power failure immediately shut off three huge switches that route telephone calls. Which of the events "power failure," "sirens not working," "visual signals not noticed," "batteries ran down," and "routing switches shut off" do you think are likely to be independent? Which seem to be direct causes and effects? Which are likely to be related but do not appear to be causes and effects?

CASE STUDY 4.7.1 *People v. Collins*

One of the first occasions in which a conviction was obtained in an American court largely on statistical evidence was the case of *People v. Collins.* In 1964 Mrs. Juanita Brooks was knocked over while walking home with her shopping basket. When she got up, she saw a young woman running away and found that her purse was missing. The young woman was described as having blond hair in a pony tail and wearing something dark. Another witness, John Bass, saw such a woman get into a yellow car driven by a black male with a beard and mustache. Collins and his wife, Janet, fitted the description. Bass picked out Collins in a lineup, but there were problems with the identification. To help what may have been a weak identification, the prosecutor called on a college mathematics instructor. This witness explained the product rule for probabilities of mutually independent events. The prosecutor continued by having the mathematical witness apply the product rule to this case, which he proceeded to do. He assumed the following probabilities (really relative frequencies here) given in Table 4.7.2 for each of the characteristics. Using the product rule to obtain the chances that a random couple meets all the characteristics in the given description, he multiplied the individual probabilities to obtain $\frac{1}{10} \times \frac{1}{4} \times \cdots \times \frac{1}{1000} = 1$ chance in 12 million. The chances of finding such a couple was so overwhelmingly small that the possibility of the police finding another couple fitting the description probably never entered the jurors' heads. The jury was convinced, and the Collins couple were convicted.

In 1968 the California Supreme Court threw the verdict out. Some of the holes in the argument should be clear from our earlier discussions. Some of the preceding characteristics are clearly not independent, for example, man with a mustache and black man with a beard, since most men with beards also have mustaches. Furthermore, if you have a "girl with blond hair" and "black man with beard," the chance of having an "interracial couple" is close to 1, not $\frac{1}{1000}$, so that from this alone the answer is too small by a factor of about 1000. Also, the prosecution had presented no evidence to support the values chosen for their probabilities.

The defense also presented a much more subtle probability argument. The police found one couple fitting the description, so at least one such couple existed. The defense calculated the conditional probability that two or more such couples exist given that at least one couple exists. This probability turns out to be quite large (about 35%), even using the prosecution's 1 in 12 million figure. Thus reasonable doubt that the Collinses committed the crime has been created.[24]

TABLE 4.7.2 Frequencies Assumed by the Prosecution

Yellow car	$\frac{1}{10}$	Girl with blond hair	$\frac{1}{3}$
Man with mustache	$\frac{1}{4}$	Black man with beard	$\frac{1}{10}$
Girl with ponytail	$\frac{1}{10}$	Interracial couple in car	$\frac{1}{1000}$

CASE STUDY 4.7.2 *Nuclear reactor safety*

Speed [1977] reviewed the use of probability arguments in the Reactor Safety Study, a major U.S. government study of the safety of nuclear power (see U.S. Nuclear Regulatory Commission [1975]). This study had come out with estimates such as one chance in 20,000 per reactor per year of a core meltdown; one chance in a thousand million per

[24]For further information about this case, see Jonakait [1983], Fairley and Mosteller [1974].

reactor per year that containment would fail, releasing virtually all the volatile and gaseous fission products into the atmosphere; one chance in 16 million per year of an individual in the United States being killed by a reactor accident; and so on. How reliable were these figures? We won't go into technical details but will focus on some elementary issues.

The calculations were based on a *fault tree analysis*, which is a diagram linking all the things that might go wrong and cause disaster. Hundreds of probability statements were made about things such as the following:

(i) The pump will work when required.

(ii) The operator will turn on the switch when required.

(iii) The safety system is undergoing maintenance when required.

(iv) Under stated conditions a steam explosion will occur.

Among more technical criticisms, Speed [1977] condemned the Reactor Safety Study on three elementary grounds:

(a) Individual probabilities ascribed to events were based on little or no data and were sometimes purely subjective.

(b) There were unfounded assumptions of independence in chains of events that could cause gross underestimates of the probability.

(c) Fault tree analysis can consider the chances of failure only from an *anticipated* cause. It is possible that an unanticipated cause of failure may have a reasonably large probability of occurring.

While the study was in draft form, there was an incident in which the Browns Ferry plant in Alabama was closed down due to a large electrical fire in the control room. The draft had not even considered such contingencies and had to be modified to include a statement to this effect. The same incident helps illustrate point (b). Reasoning of the following form was used.

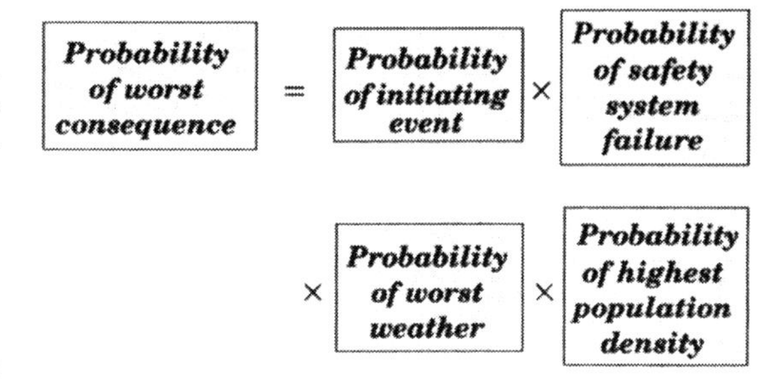

Note the assumed independence of "initiating event" and "safety system failure." Rather than being independent, however, some initiating events such as large-scale electrical fires can actually contribute to safety system failure. Other assumptions were made that people either noticed or did not notice warning signs on a meter *independently*. Yet sometimes an event that causes one person not to notice the dial, for example, a smoke-filled room and the panic of a fire, will cause others not to notice it either. There are therefore positive associations between events in both of these examples, and as we know from Section 4.7.2, assuming independence when there are positive associations can cause gross underestimates of probabilities. Speed's conclusions from all this were that "it would be nothing short of a miracle if the overall system probabilities ever bore any relation to reactor experience."[25]

The moral of this subsection is that we should be highly suspicious of any probabilities obtained under independence assumptions unless convincing reasons or data are given to justify the independence.

QUIZ ON SECTION 4.7

1. What does it mean for two events A and B to be statistically independent?
2. Why is the working rule under independence, pr(A and B) = pr(A)pr(B), just a special case of the multiplication rule of Section 4.6.2? What two uses do we make of the working rule? (Section 4.7.1)
3. What do we mean when we say two human characteristics are positively associated? negatively associated? (Section 4.7.2)
4. What happens to the calculated pr(A and B) if we treat positively associated events as independent? if we treat negatively associated events as independent?

[25]Speed [1985] considers the "Sizewell B Probabilistic Safety Study," a more recent study that avoids some of the very worst of the abuses of the earlier report. He finds that some of the basic criticisms still apply. Speed [1985] also references some of the literature on this subject.

5. Why do people often treat events as independent? When can we trust their answers? (Section 4.7.3)

6. What is an inherent difficulty with a fault tree for calculating risks? (Case Study 4.7.2)

4.8 SUMMARY

From Chapter 4, we hope you absorb these ideas:

(a) ***Some basic ideas about probabilities,*** such as how probabilities arise, the idea of a simple probability model, the important (but very different) notions of mutually exclusive events and statistically independent events, the idea that a probability depends on the information available, and a formalization of this through conditional probability.

(b) ***Some facility with manipulating probabilities and proportions.*** There are some standard types of probability manipulation that will be used repeatedly, particularly in the following chapter. The adding of probabilities of mutually exclusive events to get the probability that at least one of them occurs and multiplying the probabilities of independent events to get the probability that all of them occur, in particular, fall into this class. Also, some problems can best be solved by constructing two-way tables of counts or proportions.

You will find that many of the very simple examples that you have used and thought about in this chapter (e.g., tossing coins or sampling colored balls) will become models (analogies) that will enable you to solve some practically important problems in the following chapters. Conditional probabilities play no further part in the book.[26] They are included here because of their importance in thinking about a number of very important practical problems (as illustrated in the case studies) and to combat the widespread misuse of probability arguments, particularly those based on assumptions of independence.

This summary is divided into two sections, the first dealing with the main concepts or ideas about probability and the second dealing with formulas for calculating and manipulating probabilities.

4.8.1 Summary of Concepts

1. The ***probabilities*** people quote come from three main sources:
   (i) ***Models*** (idealizations such as the notion of equally likely outcomes, which suggest probabilities by symmetry)
   (ii) ***Data*** (e.g., relative frequencies with which the event has occurred in the past)
   (iii) ***Subjective feelings*** representing a degree of belief

2. A simple ***probability model*** consists of a *sample space* and a *probability distribution* (definitions to follow).

[26]They do provide an alternative method of answering many of the problems in the next chapter, however.

3. A ***sample space, S,*** for a random experiment is the set of all possible outcomes of the experiment.

4. A list of numbers $p_1, p_2, \ldots$ is a ***probability distribution*** for a discrete sample space $S = \{s_1, s_2, s_3, \ldots\}$, provided that (i) all of the p_i's lie between 0 and 1 and (ii) they add to 1.
   According to the probability model, p_i is the probability that outcome s_i occurs. We write $p_i = \text{pr}(s_i)$.

5. An ***event*** is a collection of outcomes. An event ***occurs*** if any outcome making up that event occurs.

6. The ***probability of event A*** can be obtained by adding up the probabilities of all the outcomes in A.

7. If all outcomes are equally likely,

$$\text{pr}(A) = \frac{\text{number of outcomes in } A}{\text{total number of outcomes}}$$

8. The ***complement*** of an event A, denoted $\overline{A}$, occurs if A does not occur.

9. It is useful to represent events diagrammatically using ***Venn diagrams.***

10. A ***union*** of events, ***A* or *B***, contains all outcomes in A or B (including those in both). It occurs if at least one of A or B occurs.

11. An ***intersection*** of events, ***A* and *B***, contains all outcomes that are in *both* A and B. It occurs only if *both* A and B occur.

12. ***Mutually exclusive*** events cannot occur at the same time.

13. The ***conditional probability*** of A occurring ***given*** that B occurs is given by

$$\text{pr}(A \mid B) = \frac{\text{pr}(A \text{ and } B)}{\text{pr}(B)}$$

14. Events A and B are ***statistically independent*** if knowing whether B has occurred gives no new information about the chances of A occurring, that is, if $\text{pr}(A \mid B) = \text{pr}(A)$.

15. If events are physically independent, then under any sensible probability model they are also statistically independent.

16. Assuming that events are independent when in reality they are not can often lead to answers that are grossly too big or grossly too small.

4.8.2 Summary of Useful Formulas

A. For discrete sample spaces, **pr(*A*)** can be obtained by adding the probabilities of all outcomes in A.

B. For equally likely outcomes in a finite sample space

$$\text{pr}(A) = \frac{\text{number of outcomes in } A}{\text{total number of outcomes}}$$

C. **General Probability Rules**

1. $\text{pr}(S) = 1$
2. $\text{pr}(\overline{A}) = 1 - \text{pr}(A)$.

3. If A and B are mutually exclusive events, then

$$\text{pr}(A \text{ or } B) = \text{pr}(A) + \text{pr}(B)$$

where *or* is used in the inclusive sense.

4. If $A_1, A_2, \ldots, A_k$ are mutually exclusive events, then

$$\text{pr}(A_1 \text{ or } A_2 \text{ or } \ldots \text{ or } A_k) = \text{pr}(A_1) + \text{pr}(A_2) + \cdots + \text{pr}(A_k)$$

D. **Conditional Probability**

1. Definition:

$$\text{pr}(A \mid B) = \frac{\text{pr}(A \text{ and } B)}{\text{pr}(B)}$$

2. Multiplication rule:

$$\text{pr}(A \text{ and } B) = \text{pr}(B \mid A)\text{pr}(A) = \text{pr}(A \mid B)\text{pr}(B)$$

E. **Independence**

1. If A and B are independent events, then

$$\text{pr}(A \text{ and } B) = \text{pr}(A)\text{pr}(B)$$

2. If $A_1, A_2, \ldots, A_n$ are mutually independent, it follows that

$$\text{pr}(A_1 \text{ and } A_2 \text{ and } \ldots \text{ and } A_n) = \text{pr}(A_1)\text{pr}(A_2)\cdots\text{pr}(A_n)$$

REVIEW EXERCISES 4

1. Give a suitable sample space S for each of the following random experiments. (Note that there may be more than one answer for S.)

(a) A lightbulb is chosen at random from a batch of bulbs. It either works or it doesn't.

(b) A student is selected at random from your class, and his or her number of siblings (brothers and sisters) is recorded.

(c) A person is interviewed on the street, and the number of his or her birth parents who are alive is noted as part of a questionnaire.

(d) You have a thermometer hanging at home, and you read the temperature at a given time every day.

(e) One lunchtime you count the number of people in line at the student cafeteria.

(f) A member of your class of 10 students is chosen at random, and his or her height is measured.

(g) A randomly selected student is interviewed, and the student is asked what form of transport he or she used to get to the university that day.

2. According to the 1991 N.Z. census, a randomly selected dwelling would have 1, 2, ... occupants with probabilities given by the following table:

No. of occupants	1	2	3	4	5	6 or more
Probability	.21	.32	.17	.16	—	.05

Source: Statistics NZ, 1994.

Obtain the missing probability corresponding to 5 occupants.

3. An experiment has two outcomes, A and $\overline{A}$. If A is three times as likely to occur as $\overline{A}$, what is pr(A)?

4. Suppose we roll a normal six-sided die and observe the score. The sample space for this random experiment is $S = \{1, 2, 3, 4, 5, 6\}$. Define the events $A = \{1, 4\}$, $B = \{6\}$, $C = \{1, 2, 3\}$, and $D = \{2, 3, 5\}$.
   (a) What is the event E that the score is an odd number?
   (b) What is the complement of D?
   (c) Which of the events A to E form a partition of S?
   (d) Which of the events B to E are mutually exclusive to A?
   (e) Find the following events: (i) A or C, (ii) C and D, (iii) C and $\overline{D}$.

5. One of the biggest problems with conducting a mail survey is the poor response rate. In an effort to reduce nonresponse, several different techniques for formatting questionnaires have been proposed. An experiment was conducted to study the effect of the questionnaire layout and page size on response in a mail survey. A group of students at a Dutch university were questioned about their attitudes toward suicide. Four different types of questionnaire formats were used. The results of the survey are shown in Table 1.
   (a) What proportion of the sample responded to the questionnaire?
   (b) What proportion of the sample received the typeset small-page version?
   (c) What proportion of those who received a typeset large-page version actually responded to the questionnaire?
   (d) What proportion of the sample received a typeset large-page questionnaire and responded?
   (e) What proportion of those who responded to the questionnaire actually received a typewritten large-page questionnaire?
   (f) By looking at the response rates for each of the four formats, what do you conclude from the study?

6. French gamblers in the seventeenth century used to bet that at least one 1 would turn up on four rolls of a die. A "1" is often called an ace. In a similar game they also bet on whether at least one double ace would occur on 24 rolls of a pair of dice. A French nobleman, the

TABLE 1 Questionnaire Formats

Format	Responses	Nonresponses	Total
Typewritten (small page)	86	57	143
Typewritten (large page)	191	97	288
Typeset (small page)	72	69	141
Typeset (large page)	192	92	284
Total	541	315	856

Reproduced with permission of author and publisher from Jansen, J.H. "Effect of questionnaire layout and size and issue-involvement on response rates in mail surveys." *Perceptual and Motor Skills*, 1985, **61**, 139–142. © Perceptual and Motor Skills, 1985.

Chevalier de Méré, reasoned that both events were equally likely, arguing essentially as follows:

GAME 1: There is 1 chance in 6 of getting an ace on one roll, so in four rolls the chances are four times that, or $4/6 = 2/3$.

GAME 2: There is a chance of 1/36 of getting a pair of aces from rolling a single pair, so for 24 pairs the chances are 24 times that, or 2/3 again.

From experience, however, de Méré doubted his arguments. It seemed to him that in practice the first event occurred more frequently. This situation came to be called the Paradox of Chevalier de Méré.

De Méré passed his problem on to Blaise Pascal (1623–1662). It was solved in a correspondence between two of the greatest mathematical minds of the day, Pascal himself and Pierre de Fermat (1601–1665). With your understanding of probability theory the problem should not be very difficult. Pascal and Fermat, however, were starting from scratch with very few of the probabilistic ideas that make thinking about such problems fairly easy for us. In fact, their correspondence was the starting point for the development of a mathematical theory of probability.

(a) What is wrong with Chevalier de Méré's arguments? (Hint: For game 1 apply the argument when you use only two rolls.)

(b) What is the true probability in either case?

[Note: Although the probabilities are different, they are very similar. Later we shall do some calculations to get some idea of the immense number of games one would have to watch to detect fairly reliably that these probabilities are in fact different.]

7. The University of Auckland has faculties as follows (with the percentage of the student body in each following in parentheses): Arts (30%), Commerce (19%), Science (18%), Engineering (7%), Law (7%), Education (6%), Medicine (4%), and other (9%), where "other" encompasses the remaining faculties, which are smaller than those listed. The percentages of female students within these faculties are: Arts (65%), Commerce (41%), Science (39%), Engineering (15%), Law (52%), Education (82%), Medicine (49%), and other (47%).

(a) Construct a two-way table showing the percentages of males and females in the various faculties.

(b) What percentage of Auckland's students are female?

(c) What percentage of Auckland's female students are in (i) Arts? (ii) Law? (iii) Engineering? (iv) Education?

8. In 1980 a U.S. Senate committee was investigating the feasibility of a national screening program to detect child abuse. A team of consultants came up with the following estimates: 1 child in 100 is abused, a physician can detect an abused child about 90% of the time, and a national screening program using physicians would incorrectly label about 3% of the nonabused children as abused.

(a) Using the preceding information, what is the probability that a child is actually abused, given that the screening program diagnoses the child as such?

(b) Do you think it is appropriate to apply the preceding information to both boys and girls?

(c) Would the information be relevant today?

9. There is a 40% to 60% chance that a pregnant woman with the HIV virus will pass it on to her child. Approximately 1% of all black teenage girls who bore children in New York City during 1988 were infected with HIV (*Time*, 2 July 1990). Taking the lower figure (40%), what proportion of the babies born to black teenage girls in 1988 were infected? Express your answer as a rate of infected babies per 1000 births.

10. Most of the figures to follow come from a *Time* cover story (6 July 1992) about effective new drugs for treating schizophrenia. All are U.S. figures, and all are approximate.

(a) One in four schizophrenics attempts suicide. Of those who attempt it, 1 in 10 succeeds. What proportion of schizophrenics actually commit suicide?

Approximately 1% of of the population is schizophrenic, 0.8% of people are homeless, and one-third of the homeless are schizophrenic.

(b) Using a two-way table, find the proportion of schizophrenics who are homeless.

A child has a 10% chance of becoming schizophrenic if one parent is schizophrenic and a 40% chance if both are. The chance of a child of a schizophrenic developing the condition is reduced, but only slightly, when raised by adoptive parents without the condition.

(c) Does this information suggest anything to you about hereditary and environmental effects for schizophrenia?

(d) It would be interesting to know what percentage of schizophrenics had both parents schizophrenic. What additional information would we need to resolve this?

(e) Assuming that people marry independently of their susceptibility to schizophrenia and that each couple has one child, answer the question raised in (d).

(f) Actually, "1% of Americans are schizophrenic" is not quite correct. The article really says that 1% of Americans *develop* schizophrenia. What can you therefore say about the percentage who *have* schizophrenia?

11. The final grade of students in a first-year course in statistics at Auckland was made up of 30% from assignments and tests while the course is in progress (the "course-work mark") and 70% from an examination at the end of the course. Table 2, showing the relationship between course-work mark and final grade, is taken from the results of 1270 students who completed the course in 1992. The entries in the interior of the table are the percentages of those students within the given range of course-work marks who got each of the final grades.

(a) Can you think of a reason for presenting the data in this way (or a use for it as presented)?

Suppose that we randomly select a student who completed the course.

(b) Describe in words a sample space for this experiment.

(c) What is the probability that the student got between 25 and 30 for course work?

(d) What is the probability that the student got between 25 and 30 for course work and an A as a final grade?

Suppose that the figures in Table 2 lead to reasonable probabilities for a randomly selected student currently taking the course.

(e) What is the probability the student will get an A as a final grade, given that he or she got between 25 and 30 for course work?

(f) What is the probability that a student with between 10 and 15 for course work will get a B final grade or better?

TABLE 2 Course-Work Marks and Final Grades

Percentage of class	Course-work mark	Final grade				Total
		Failing grade	C	B	A	
2.7	≤ 5	91.2%	8.8%	0	0	100%
4.6	5^{+}–10	79.7	20.3	0	0	100
18.1	10^{+}–15	53.9	35.7	10.0	0.4	100
33.9	15^{+}–20	13.3	45.1	34.9	6.7	100
30.6	20^{+}–25	0.3	12.6	60.3	26.8	100
10.1	25^{+}–30	0	0	14.7	85.3	100
100						

The remainder of this exercise talks about proportions of the 1992 class.

(g) What proportion of the class got between 15 and 20 on course work?

(h) What proportion of the class got between 15 and 20 on course work and failed?

(i) What proportion of the class with between 15 and 20 on course work failed?

(j) What proportion of the entire class failed? What proportion passed?

(k) What proportion of the class got A final grades?

(l) What proportion of those with A final grades got over 25 on course work?

12. In North America, as in Australasia, cancer is the second leading cause of death after heart diseases. Accidents account for only about a fifth as many deaths as cancer. Table 3 gives the incidence rates (as new cases per 100,000 of population) and the mortality rates (as deaths per 100,000) for seven leading cancer sites.

(a) Ignoring the "other"category, what cancer has most new cases in a year?

(b) Ignoring the "other"category, what cancer kills (i) the most people? (ii) the most males? (iii) the most females?

(c) What proportion of breast cancer deaths are males?

(d) What proportion of male cancer deaths are due to (i) lung cancer? (ii) colorectal cancer?

(e) What listed cancers affect (i) males and not females? (ii) females and not males?

The numbers of cancers contracted in a year and the numbers of deaths in a year from that cancer give us a rough estimate of the chances that a particular cancer will eventually kill someone who contracts it.

(f) Ignoring the "other" category, which listed cancer is most likely to end up being fatal (i) regardless of gender? (ii) for men? (iii) for women? In each case, give the probability of eventual death.

(g) What assumptions underlie the type of calculations done in (f)? When are these assumptions sensible?

TABLE 3 Cancer Incidence and Mortality Rates

	New Cases			Deaths		
Cancer sites	**Male**	**Female**	**Total**	**Male**	**Female**	**Total**
Oral	20.6	9.7	30.3	5.2	2.8	8.0
Colorectal	79.0	77.0	156.0	28.9	29.4	58.3
Lung	102.0	66.0	168.0	93.0	53.0	146.0
Skin[a]	17.0	15.0	32.0	4.1	2.6	6.7
Breast	1.0	180.0	181.0	0.3	46.0	46.3
Uterus	0	45.5	45.5	0	10.0	10.0
Prostate	99.0	0	99.0	28.0	0	28.0
Other	246.4	171.8	418.2	115.5	101.2	216.7
Total	565.0[b]	565.0[b]	1130.0	275.0	245.0	520.0

[a]Excludes nonmelanoma skin cancer.

[b]These are the same simply because of the rounding of the data.

Source: Constructed, with permission, from data in *The World Almanac and Book of Facts* [1993, p. 223] except for prostate figures, which are 1988 figures from American Cancer Society.

(h) Doctors are more likely to talk about things such as the five-year survival rate than about the chances of eventual death from a disease. What are some problems with "the chance that you will die from it" as a measure of the seriousness of a disease?

13. Reporting on an address by Professor William Doe, President of the Gastroenterological Society of Australia, *The Weekend Australian* (30 June 1990) stated that tests used widely throughout Australia to detect bowel cancer could lead to death because of widely inaccurate results. The tests, which used chemicals to detect the presence of blood in feces, were available over the counter at pharmacies and widely used for mass screening programs. According to Professor Doe, they delivered a negative result on 30% to 50% of cancers (we shall assume 40%), and "these people are tragically assured that they don't have cancer." A positive result is delivered to 40% of people who do not have bowel cancer or its early warning signs, some of these resulting from animal blood in the bowel from digested meat. There are about 8000 new cases of bowel cancer per year in Australia out of a population of approximately 16 million. We shall therefore estimate the proportion of undiagnosed cases of bowel cancer and its precursors in Australia at the time as approximately 0.0005. If a mass screening of all Australians was undertaken using this test,

(a) what proportion of bowel cancer cases would be overlooked?

(b) what proportion of Australians would give a positive test?

(c) what proportion of people testing positively would actually have the disease?

(d) how could you reduce the false-positive rate?

[Note: The good news is that more reliable tests are now available.]

14. In 1993 a team of scientists from Johns Hopkins University and the University of Helsinki reported in *Science* [1993, vol. 260, p. 751] their discovery of a genetic marker for so-called familial cancer of the colon. This accounts for about one in seven cases of cancer of the colon (hereafter CaCo), which is the second leading cause of cancer deaths in the world. The scientists estimated that one person in 200 carries the defective gene, that 95% of people with the gene will develop cancer, and that of those who get cancer, 60% will get cancer of the colon.

(a) From these figures, what percentage of people will develop CaCo from this mechanism?

Professor Vogelstein of Johns Hopkins predicted a diagnostic test based on the genetic discovery. (The existing screening test used worldwide misses more than 70% of tumors.) Between 5 and 10 million Americans are presently considered to be at increased risk of CaCo because of a strong family history of the disease. Vogelstein believes that 75% of these people will find that they do not have the implicated genetic marker and that these people bear only the average risk of CaCo, which is 1 chance in 20.

(b) What proportion of those with a "strong family history" will get CaCo? What proportion of those who will get CaCo carry the defective gene?

Detection of those carrying the marker is useful because their colons can then be scanned annually using a fiber-optic scope, a procedure known as colonoscopy, which costs about $1000. (The five-year survival rate for CaCo is 90% when detected early.)

(c) Taking the lower figure of 5 million Americans at increased risk, how much money could be spent annually on colonoscopies if Vogelstein's predictions are borne out?

15. Two tennis players play a three-set match (i.e., they keep playing until one player wins two sets). Suppose that the two players are evenly matched so that each player has a 50% chance of winning a set and that the outcomes of different sets are independent.

(a) Write down a sample space for the experiment.

Let A denote the event "player 1 wins the match" and B denote the event "match finishes in two sets."

(b) Find (i) pr(A), (ii) pr(B), (iii) pr(A and B), (iv) pr(A or B), (v) pr($B \mid A$).

(c) Are A and B independent? (Explain your answer.)

(d) Are A and B mutually exclusive?

16. Each day the price of a certain stock either moves up one cent or moves down one cent. It moves up with probability 1/3 and down with probability 2/3 independently of previous movements. We are interested in what the price will be in three days' time. Let A be the event that the price has increased at the end of three days, and let B be the event that the price drops on the first day. Write down a suitable sample space for this "experiment" and calculate the following:
(a) pr(A), (b) pr(B), (c) pr(A and B), (d) pr($A \mid B$). (e) Are A and B (i) independent? (ii) mutually exclusive? Explain your answer.

17. Sanderson Smith [Smith, 1990] told the following story. He received an unsolicited phone call from a marketing research firm asking him to keep records and receipts for his purchases of certain items over a period of three months. Some incentives were offered. The firm was to send him some journal booklets in which to record his purchases, one journal for each of the three months. If he returned the journal for a given month along with receipts to verify his stated purchases, he would go into that month's drawing. Each drawing would give him 1 chance in 10 of winning a prize of $25. Furthermore, if he returned all three journals and receipts on time, he would have 1 chance in 100 of winning a three-day trip to Las Vegas and 1 chance in 350 of independently winning a two-week holiday for two in Hawaii. Sanderson Smith did some calculations to figure out whether it was worth participating. Regarding the trips as the major prizes and the $25 prizes as minor and assuming that he fulfilled all the conditions, what is the probability of winning

(a) (i) none of the minor prizes? (ii) all three minor prizes? (iii) exactly one minor prize? (iv) exactly two minor prizes?

(b) (i) neither major prize? (ii) Las Vegas but not Hawaii? (iii) both major prizes?

(c) (i) no prizes at all? (ii) one or more minor prizes, but no major prize?

18. Writing in *Rolling Stone* in February 1991, humorist P. J. O'Rourke ruminated on his experiences covering the Gulf War from Saudi Arabia. He pondered his chances of receiving a direct hit by a Scud missile. He figured that a Scud, which carried about 113 kg of explosive, would have a blast area of 91 m in diameter at most and that missiles were being lobbed into an area of eastern Saudi Arabia that was roughly 80 km long and 48 km wide. Given that Scuds were inaccurately aimed, you can assume that missiles fall randomly on this area.

(a) What was his probability of being in the blast area of a missile?

(b) If 20 missiles fell onto the area, what is the probability of escaping all 20?

19. Suppose that an insurance company has classified its drivers into three classes using various criteria. Of drivers insured with them, 20% fall in the low-risk category, 70% into the medium-risk category, and 10% are high risk. From the historical records, 1% of low-risk drivers make a claim in any one year. Corresponding figures for medium-risk and high-risk drivers are 4% and 10%, respectively. Assume that there have been no changes in the company's business that would prevent the historical record from being a good guide to the immediate future.

(a) Construct a two-way table of probabilities.

(b) What is the probability that a randomly selected driver in the medium-risk category will have a claim?

(c) If we select a driver at random, what is the probability that the driver has a claim and is in the medium-risk category?

(d) What proportion of drivers can be expected to have a claim (regardless of risk category)?

(e) What proportion of claims can we expect to be made by drivers classified as high risk?

*(f) If a driver goes three years without a claim, what is the probability he or she had been classified into the low-risk group? (Treat different years as independent.)

20. In Chapter 1 of Brook et al. [1986], Barry Singer gives an entertaining and informative discussion of many of the ideas in this chapter and about finding a mate as well. This exercise revolves around your chances of finding an ideal mate.

(a) Make a list of the characteristics you consider absolutely essential in a partner (e.g., sufficiently intelligent, similar sense of humor, etc.).

(b) Estimate or guess, for each item on the list, the proportion of people you meet who satisfy the listed criterion (e.g., 30% are intelligent enough). Think of each of these numbers as being the probability that a new person you meet will satisfy the listed criterion.

(c) Assume independence of the listed criteria. How do you get from the individual probabilities in (b) to the probability that a new person you meet will satisfy every criterion on your list? Make the calculation.

(d) Which criteria on your list, if any, are obviously not independent? Which of these are positively associated? Which are negatively associated?

(e) Assume you meet 300 new people a year. From the calculation in (c), how many years would you expect it will take for you, on average, to find a person meeting all your essential criteria?

(f) Are you depressed? Are some of your criteria not so essential after all? Which items on the list are making the search so difficult? How do the associations affect the answer? If you are still interested in this problem, read Barry Singer's chapter.

21. Consider the life table given in Table 4. It gives the numbers of males and females surviving at different ages per 100,000 born and can thus be interpreted as a way of presenting the probability that a randomly chosen baby will still be alive at each age.

(a) Find the probability that (i) a boy who survives the first year of life will reach 20, (ii) a 20-year-old woman will reach 60.

(b) By evaluating the probabilities of dying within a five-year period given being alive at the start of the period, what is the second most dangerous five-year period up to age 40 (i) for a male? (ii) for a female? You may find the answer to (i) surprising. Can you think of an explanation?

TABLE 4 Numbers Surviving per 100,000 Born

Age (years)	Males	Females
0	100,000	100,000
1	98,826	99,035
5	98,565	98,827
10	98,414	98,692
15	98,216	98,569
20	97,473	98,304
25	96,594	97,998
30	95,851	97,684
35	95,180	97,317
40	94,407	96,792
60	82,690	89,099

Source: NZ Life Tables, 1985–1987.

(c) For a 20-year-old man and a 20-year-old woman, what is the probability that (i) both will reach 60? (ii) the woman will reach 60 but the man will die before 60? (iii) neither will reach 60?

(d) What assumption did you have to make about the lifetimes of the two people in part (c) in order to do the calculations? Would this assumption hold if the man and woman were a married couple? If you think the assumption would fail, would you expect the association between lifetimes to be positive or negative? Why?

(e) Assume that 20-year-old men and women exist in equal numbers. What is the probability that a random 20-year-old person will live until 40?

(f) What assumption is being made by applying the answers in (a)(ii), (c), and (e) to people who are currently 20 years old? How valid do you think the assumption is?

22. The distribution of blood types for the New Zealand European population is as follows:

40% type A   9% type B   49% type O   2% type AB

Suppose that the blood types of European married couples are independent and that both the husband and the wife follow this distribution of blood type. In the following questions we assume that the couple is randomly selected.

(a) If the wife has type B blood, what is the probability that the husband has type B blood?

(b) What is the probability that both the husband and wife have type B blood?

(c) What is the probability that at least one member of the couple has type B blood?

(d) What is the probability that both the husband and wife have the same type blood?

(e) An individual with type B blood can safely receive transfusions only from people with type B or type O blood. What is the probability that the husband of a woman with type B blood is an acceptable blood donor for her?

23. In her 9 September 1990 "Ask Marilyn" column in *Parade Magazine,* Marilyn vos Savant (reported to be the holder of the world's highest IQ) posed a problem that attracted thousands of replies and lots of the controversy. The problem is as follows. "Suppose you're on a game show and given a choice of three doors. Behind one is a car; behind the others are goats. You pick door No. 1, and the host, who knows what is behind them, opens No. 3, which has a goat. He then asks if you want to pick No. 2. Should you switch?" Although it is not in the original statement of the problem, it is clear from Marilyn's replies in subsequent columns that the host will always open a door that does not belong to the player and has a goat behind it.[27] What is the probability of winning the car in a situation like this if your strategy is always to switch? never to switch?

24. The following, a historic problem called Bertrand's Box Problem, is taken from R. G. Seyman's discussion of Morgan et al. [1991]. A box contains three drawers: one containing two gold coins, one containing two silver coins, and one containing one gold and one silver coin. A drawer is chosen at random, and a coin is randomly selected from that drawer. If the selected coin turns out to be gold, what is the probability that the chosen drawer is the one with the two gold coins?[28]

25. Digital data are transmitted as a sequence of signals that represent 0s and 1s. Suppose that such data are being transmitted to a satellite and then relayed to a distant site. Suppose

---

[27] It is possible, though more complicated, to solve the problem without knowing this—see Morgan et al. [1991].

[28] Another version presents this problem as a card con trick. There are three cards with their two sides colored respectively red/red, red/black, and black/black. A card is chosen at random. The player can see that the top side is red and has to guess the bottom side. Most people intuitively think that the bottom side is equally likely to be red or black. There is a potential for cheating gamblers out of money because the actual probability that the bottom side is red is quite different from 50%.

that, due to electrical interference in the atmosphere, there is a 1-in-1000 chance that a transmitted 0 will be reversed between the sender and satellite (i.e., distorted to the extent that the satellite's receiver interprets it as a 1) and a 2-in-1000 chance that a transmitted 1 will be reversed. Suppose that 40% of transmitted digits are 0s.

(a) What is the probability that a transmitted digit is correctly received by the satellite?

(b) Assuming independence, what is the probability that all digits are received correctly (i) if 1000 digits are transmitted? (ii) if 10,000 are transmitted?

(c) Suppose that between the satellite and the receiver, the chances of reversal are twice as large as they were between the sender and satellite. Assuming independence, what is the probability that a digit reaches the receiver as originally sent?

26. In reliability theory, components in a system are said to be *in parallel* if the system fails only if all the components in the system fail. For example, if we have an alarm warning system that gives warnings via a flashing light, a siren, and a digital display, the warning system fails only if all three devices fail to alert people to the problem.

(a) Suppose that the light fails with probability 0.01, the siren with probability 0.02, and the display with probability 0.08. Assuming failures occur independently, what is the probability that the system fails?

(b) Particularly where components fail independently, putting parallel components into a system can make the system dramatically more reliable than any of its components, as in (a). The results are not as good as the independence calculations show, however, if there are common causes of failure (resulting in positive association). Can you think of a possible common cause of failure for all three components?

(c) Can you think of anything from your own experience where components of a system act in parallel? (It need not be a mechanical or electronic system; it could be an administrative system.)

At the other extreme from a parallel system, a set of components is said to be *in series* when the system fails if any one of the components fails. As a simplified example, suppose you plan to travel to Los Angeles for a meeting by driving to the airport, catching a flight to Los Angeles, picking up a rental car, and driving to the site of the meeting. There are four "components" discussed here: the initial drive, the flight, picking up the rental car, and another drive (we could obviously break each of these down further), which are in series. Suppose that you have estimated the probabilities of something going wrong with each of these components that would stop your getting to the meeting on time to be 0.02, 0.05, 0.08, and 0.03, respectively.

(d) Assuming independent causes of failure, what is the probability of failing to make the meeting on time? (Hint: It is easier to work with succeeding than failing for series systems.)

Note from your answer how, in contrast to a parallel system, the series system is much less reliable than any of its components.

(e) What are some systems of series components that you use or have experienced?

***27.** A large study of college students by Professor Stanley Coren of the University of British Columbia and Dr. Diane Halpern of San Bernardino State University showed that left-handed people had an 89% greater risk of having a car accident than right-handed people (*New Zealand Herald,* 5 January 1991). Left-handed people make up 10% of drivers. What percentage of people having car accidents are left-handed?

(Hint: Work with p, where p denotes the probability that a right-handed person has a car accident. You do not actually need to know p to solve the problem, but if you are having trouble working with an unknown p, assume a value for it and see whether you can see why your answer does not depend on the value assumed.)

CHAPTER 5

Discrete Random Variables

Chapter Overview

This chapter is one of two chapters dealing with random variables. After introducing the notion of a random variable in Section 5.1, we discuss discrete random variables; continuous random variables are left to the next chapter. We recommend that you have another look at Section 2.1, where random variables are first introduced. Next on the menu, in Section 5.2, we learn about calculating simple probabilities using a probability function. Coin-tossing and urn-sampling models are introduced in Section 5.3 as analogies for practical applications. Emphasis is given to model recognition. The Binomial probability function is introduced as the distribution of the number of heads in n tosses of a biased coin and as an approximation for the urn-sampling model. The theoretical idea of expected value is introduced in Section 5.4. This leads to the notions of population mean and population standard deviation, population analogues of the sample mean and sample standard deviation that we met in Chapter 2. Finally, the mean and standard deviation are obtained for $aX + b$ in terms of those for X.[1]

5.1 RANDOM VARIABLES

In Section 4.4 we talked about the idea of a random experiment. Recall that we used the term *experiment* in its widest sense. It can refer to a naturally occurring phenomenon, a scientific experiment, or a sampling experiment in which an "object," such as a person or plot, is selected at random from a population of objects. In statistical applications we then want to measure, or observe, different aspects or characteristics of the outcome of our experiment. Such an outcome is unpredictable, and an associated variable is therefore called a random variable.

[1]Other topics, including the Geometric, Hypergeometric, and Poisson distributions, are discussed on our web site.

> A ***random variable*** is a type of measurement taken on the outcome of a random experiment.

We use uppercase letters X, Y, Z, and so on to represent random variables. If our experiment consists of sampling a person from some population, we may be interested in measuring their yearly income (X, say), accommodation costs (Y), blood pressure (Z), or some other characteristic.

We use the term *measurement* loosely. We may have a variable X = "marital status" with three categories: "never married," "married," and "previously married," with which we associate the numerical codes 1, 2, and 3, respectively. Then the X-measurement of an individual who is currently married is $X = 2$. In Section 2.1 after looking at heart attack data we distinguished between several types of variables, and now would be a good time to have another look at that section. We saw that discrete variables have gaps between the values they can take, and these values and gaps are typically whole numbers. In this chapter we concentrate on discrete random variables.

5.2 PROBABILITY FUNCTIONS

EXAMPLE 5.2.1 *Number of Heads Is a Discrete Random Variable*

Consider the experiment of tossing a coin twice, and define the variable X = "number of heads." A sample space for this experiment is given by $S = \{HH, HT, TH, TT\}$, and X can take values 0, 1, and 2. We observe that $X = 0$ if the outcome is TT, $X = 1$ if the outcome is HT or TH, and $X = 2$ if the outcome is HH.

We use small letters x, y, z, and so on to represent possible values that the corresponding random variables X, Y, Z, and so on can take. The statement $X = x$ defines an event consisting of all outcomes with X-measurement equal to x. In Example 5.2.1 $X = 1$ is the event $\{HT, TH\}$, while $X = 2$ consists of $\{HH\}$. Thus we can assign probabilities to events of the form $X = x$ as in Section 4.4.4 by adding the probabilities of all those outcomes that have X-measurement equal to x. This leads to the following definition.

> The ***probability function*** for a discrete random variable X gives
>
> $$\mathbf{pr}(X = x)$$
>
> for every value x that X can take.

Where there is no possibility of confusion between X and some other variable, $\text{pr}(X = x)$ is often abbreviated to $\mathbf{pr}(x)$. As with probability distributions, $0 \le \text{pr}(x) \le 1$, and the values of $\text{pr}(x)$ must add to 1. This provides a useful check on our calculations.

✖ EXAMPLE 5.2.2 *Probability Function for the Number of Heads*

Consider tossing a coin twice as in Example 5.2.1. If the coin is unbiased so that each of the four outcomes is equally likely, then pr(0) = pr(*TT*) = $\frac{1}{4}$, pr(1) = pr(*TH*, *HT*) = $\frac{2}{4}$, and pr(2) = pr(*HH*) = $\frac{1}{4}$. This probability function is conveniently represented as a table.

x	0	1	2
pr(x)	$\frac{1}{4}$	$\frac{1}{2}$	$\frac{1}{4}$

The probabilities add to 1 as required. Values of x not represented in the table have probability 0.

✖ EXAMPLE 5.2.3 *Bar Graph for a Probability Function*

This is continuation of Example 4.4.6(c), in which a couple has children until they have at least one of each sex or a maximum of three children. The sample space and probability distribution can be represented as

Outcome	GGG	GGB	GB	BG	BBG	BBB
Probability	$\frac{1}{8}$	$\frac{1}{8}$	$\frac{1}{4}$	$\frac{1}{4}$	$\frac{1}{8}$	$\frac{1}{8}$

Let X be the number of girls in the family. Then X takes values 0, 1, 2, 3 with probability function

x	0	1	2	3
pr(x)	$\frac{1}{8}$	$\frac{5}{8}$	$\frac{1}{8}$	$\frac{1}{8}$

The probabilities of the events $X = 0$, $X = 2$, and $X = 3$ are easy to get, as they correspond to single outcomes. However, pr($X = 1$) = pr(GB) + pr(BG) + pr(BBG) = $\frac{1}{4} + \frac{1}{4} + \frac{1}{8} = \frac{5}{8}$. We note that the probabilities add to 1 as required.

Probability functions are best represented pictorially as a bar graph. Figure 5.2.1 contains a bar graph of the preceding probability function.

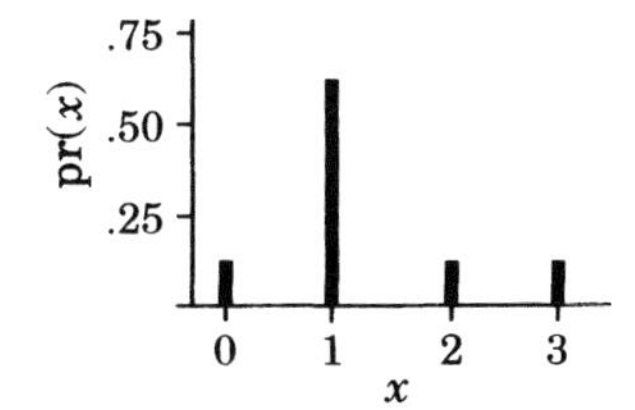

FIGURE 5.2.1 Bar graph of a probability function.

EXAMPLE 5.2.4 *Probability Function for Tossing a Biased Coin*

Let us complicate the "tossing a coin twice" example (Examples 5.2.1 and 5.2.2) by allowing a biased coin for which the probability of getting a head is p, where p is not necessarily $\frac{1}{2}$. As before, pr($X = 0$) = pr(TT). By TT we really mean "T_1 and T_2," or "tail on first toss" and "tail on second toss." Thus

$$\text{pr}(TT) = \text{pr}(T_1 \text{ and } T_2) = \text{pr}(T_1) \times \text{pr}(T_2) \quad \text{(because tosses are independent)}$$

$$= (1-p) \times (1-p) = (1-p)^2$$

Similarly, pr(HT) = $p(1-p)$, pr(TH) = $(1-p)p$ and pr(HH) = p^2. Thus pr($X = 0$) = pr(TT) = $(1-p)^2$, pr($X = 1$) = pr(HT) + pr(TH) = $2p(1-p)$, and pr($X = 2$) = pr(HH) = p^2. A table is again a convenient representation of the probability function:

x	0	1	2
pr(x)	$(1-p)^2$	$2p(1-p)$	p^2

The next example provides a probability function based on data. It is used to introduce several new ideas about using a probability function to calculate probabilities. The idea of *cumulative probabilities* is introduced, and we demonstrate the use of cumulative probabilities for calculating *upper-tail* probabilities and *interval probabilities.*

EXAMPLE 5.2.5 *Probabilities for Lengths of Hospital Stays*

People going into the hospital for surgery usually want to know how long they are likely to be in there. They do not expect an exact answer, knowing that different people require different lengths of time to recover. "What are the chances I'll be out in less than a week?" The best indication is the length of the hospital stays for previous patients who have been in a similar situation. The following spreadsheet fragment summarizes the length of hospital stays of 263 patients for a major operation as a frequency table. The second to last row gives the proportions (relative frequencies) to three decimal places. For example, the proportion of patients who stayed for seven days is 79/263 = 0.300 (or 30%). The last row gives the cumulative proportions. We see that a proportion of 0.882 (or 88.2%) stayed for seven days or less.

Days stayed	x	4	5	6	7	8	9	10	*Total*
Frequency		10	30	113	79	21	8	2	263
Proportion	**pr($X = x$)**	0.038	0.114	0.430	0.300	0.080	0.030	0.008	1.000
Cumulative Proportion	**pr($X \le x$)**	0.038	0.152	0.582	0.882	0.962	0.992	1.000	

We shall treat these proportions as giving the probability distribution for a random variable X, the number of days that a new patient will remain in hospital before discharge. Thus, pr($X = 7$) = 0.300 and pr($X = 10$) = 0.008.

Working with the Individual Probabilities [pr($X = x$)]

In the following, pr($X = x$) is abbreviated to pr(x).

The probability of staying more than a week (i.e., for 8 or more days) is

$$\text{pr}(X \geq 8) = \text{pr}(8) + \text{pr}(9) + \text{pr}(10) = 0.080 + 0.030 + 0.008 = 0.118$$

The probability of a stay of less than six days is

$$\text{pr}(X < 6) = \text{pr}(X \leq 5) = \text{pr}(4) + \text{pr}(5) = 0.038 + 0.114 = 0.152$$

The probability the stay will be between five and seven days inclusive is

$$\text{pr}(5 \leq X \leq 7) = \text{pr}(5) + \text{pr}(6) + \text{pr}(7) = 0.114 + 0.430 + 0.300 = 0.844$$

Taking Care with Language

It is fairly simple to add the probabilities of x-values, as we have done above. Where students often get into trouble is in deciding from a written passage which x-values should be included and, in particular, whether or not to include the numbers at either end of a sequence of x-values. Including or excluding a number can make a substantial difference in the answer you get, so it is important to be careful! When we are dealing with whole numbers, the expressions within the following sets have the same meaning ("≡" means "is equivalent to"):

- *Set 1:* "More than 7" ≡ "at least 8" ≡ "8 or more" ≡ "no fewer than 8"
  The first expression translates naturally to $X > 7$ and the last three to $X \geq 8$. In all cases, we are referring to the numbers 8, 9, ...
- *Set 2:* "Less than 7" ≡ "at most 6" ≡ "6 or fewer" ≡ "no more than 6"
  The first expression naturally translates to $X < 7$ and the last three to $X \leq 6$. In all cases, we are referring to the numbers up to and including 6.
- *Intervals:* In the phrase "between 4 and 8," the word *between* is ambiguous. Do we include the endpoints 4 and 8 or omit them? "Between 4 and 8 inclusive" means 4, 5, 6, 7, 8, whereas if the phrase "strictly between 4 and 8" is used, the reference is probably to 5, 6, 7. "Between 5 and 7 inclusive" would be less open to misinterpretation. Most often verbal descriptions of intervals use combinations of the expressions in sets 1 and 2, for example, "more than 5 days but no more than 8." We have to decipher such expressions and decide which endpoints are included.

Working with Cumulative Probabilities [pr($X \leq x$)]

A skill that will be very important when using the Binomial distribution in Section 5.3 is the ability to work with so-called ***cumulative*** (or *lower-tail*) probabilities, which are of the form pr($X \leq x$). The probabilities themselves will be obtained from computer programs, but here we discuss the way we work with them using the hospital probability table (repeated in simplified form for easy reference).

Days stayed	**x**	4	5	6	7	8	9	10	*Total*
Individual	**pr(X = x)**	0.038	0.114	0.430	0.300	0.080	0.030	0.008	1.000
Cumulative	**pr(X ≤ x)**	0.038	0.152	0.582	0.882	0.962	0.992	1.000	

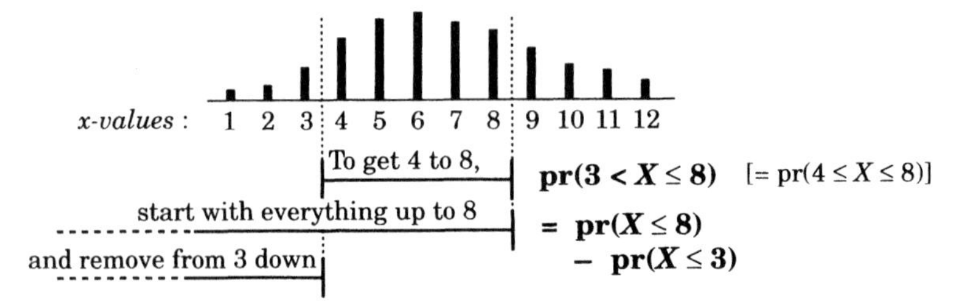

FIGURE 5.2.2 Interval probabilities from cumulative probabilities. (This figure represents an arbitrary distribution, not the hospital distribution.)

First we note that in the cumulative row, the probabilities are adding up (*accumulating*) as we go from left to right: $\text{pr}(X \leq 5) = \text{pr}(X \leq 4) + \text{pr}(5)$, $\text{pr}(X \leq 6) = \text{pr}(X \leq 5) + \text{pr}(6)$, and so on. If we want to find the probability that the new patient's stay in hospital is no longer than 6 days, $\text{pr}(X \leq 6)$, this is already a cumulative probability, and we can read it from the table. We see that $\text{pr}(X \leq 6) = 0.582$.

Suppose we want to find the probability that the stay is 9 days or more, $\text{pr}(X \geq 9)$. We call a probability of the form $\text{pr}(X \geq x)$ an ***upper-tail*** probability. When we need an upper-tail probability, we convert the problem into one involving a cumulative (or lower-tail) probability by using rule 2 of Section 4.5 for complementary events, namely,

$$\text{pr}(A \text{ occurs}) = 1 - \text{pr}(A \text{ does not occur})$$

If the number of days, X, is not "9 or greater" then it must be "8 or smaller." Thus,

$$\text{pr}(X \geq 9) = 1 - \text{pr}(X \leq 8)$$

The latter probability is a cumulative probability that we can read from the table. Then, $\text{pr}(X \geq 9) = 1 - \text{pr}(X \leq 8) = 1 - 0.962 = 0.038$. Similarly, we have $\text{pr}(X \geq 6) = 1 - \text{pr}(X \leq 5) = 1 - 0.152 = 0.848$.

Finally we must consider ***interval probabilities,*** that is, probabilities of the form $\text{pr}(a < X \leq b)$. The reasoning is depicted in Fig. 5.2.2, which shows the calculation of $\text{pr}(3 < X \leq 8)$. Generally,

$$\text{pr}(a < X \leq b) = \text{pr}(X \leq b) - \text{pr}(X \leq a)$$

Note that when we subtract $\text{pr}(X \leq a)$, we remove a from the interval, that is, a is not included.

We shall now apply this to the hospital probabilities. Rather than blindly following a formula, we advise that you think intuitively, as in Fig. 5.2.2, and ask yourself, "What is the first of the values below the desired interval that I need *to remove?*" The probability of a stay of between 6 and 8 days inclusive is given by

$$\text{pr}(X \leq 8) - \text{pr}(X \leq 5) = 0.962 - 0.152 = 0.810$$

Here we want to include 6, so we need to remove the numbers from 5 on down. The probability of strictly more than 7 days but no more than 9 is given by

$$\text{pr}(X \leq 9) - \text{pr}(X \leq 7) = 0.992 - 0.882 = 0.110$$

Here we do not want to include 7, so 7 is the first of the values we want to remove.

QUIZ ON SECTIONS 5.1 AND 5.2

1. What is a random variable? What is a *discrete* random variable? (Section 5.1)

2. What general principle is used for finding pr($X = x$)? (Section 5.2)

3. What two general properties must be satisfied by the probabilities making up a probability function?

4. What are the two names given to probabilities of the form pr($X \leq x$)?

5. How do we find an upper-tail probability from a cumulative probability?

6. When we use pr($X \leq 12$) − pr($X \leq 5$) to calculate the probability that X falls within an interval of values, what numbers are included in the interval? Why?

EXERCISES FOR SECTION 5.2

1. One number in the following table for the probability function of a random variable X is incorrect. Which is it, and what should the correct value be? Justify your answers.

x	1	2	3	4	5
pr($X = x$)	0.07	0.10	1.10	0.32	0.40

2. Of 200 adults, 176 own one TV set, 22 own two TV sets, and 2 own three TV sets. A person is chosen at random. What is the probability function of X, the number of TV sets owned by that person?

3. Consider Example 4.4.6(b) and construct a table for the probability function of X, the number of girls in a three-child family, where the probability of getting a girl is $\frac{1}{2}$.

4. From Example 5.2.5, what is the probability that a patient spends more than 4 days in the hospital but less than 8? Answer this question using both individual probabilities and cumulative probabilities.

5. Suppose a discrete random variable X has probability function given by

x	3	4	5	6	7	8	9	10	11	12	13
pr($X = x$)	.07	.01	.09	.01	.16	.25	.20	.03	.02	.11	.05

(a) Construct a row of cumulative probabilities for this table.

Using both the probabilities of individual values and cumulative probabilities, compute the probability that

(b) $X \leq 5$, (c) $X > 9$, (d) $X \geq 9$,
(e) $X < 12$, (f) $5 \leq X \leq 9$, (g) $4 < X < 11$.
(h) What is pr($X = 14$)? Why? (i) What is pr($X < 3$)? Why?

6. Read Case Study 5.2.1, which follows these exercises, and then answer the following questions. A woman decides to give up trying if the first three attempts are unsuccessful. What is the probability of this happening? Another woman wants to increase her chances of success to at least 50%. How many times should she try?

CASE STUDY 5.2.1 *Chances of success with artificial insemination*

A *New Zealand Herald* report quoted an Auckland obstetrician as stating that the chances of a successful pregnancy resulting from implanting a frozen embryo are about 1 in 10. Suppose a couple who are desperate to have children will continue to try this procedure until the woman becomes pregnant. We shall assume that the process is just like tossing a biased coin[2] until the first "head" occurs, with "heads" being analogous to "becoming pregnant." The probability of becoming pregnant at any "toss" is $p = 0.1$. Let X be the number of times the couple will try the procedure up to and including the successful attempt. Suppose the couple want to know how many times they should try this procedure to give themselves at least a 40% chance of success.

Now $X = 1$ means that the first attempt is successful, so $\text{pr}(X = 1) = p = 0.1$. When $X = 2$, the first attempt is unsuccessful, but the second attempt is successful. Assuming independence of these two events,

$$\text{pr}(X = 2) = 0.9 \times 0.1 = 0.09$$

When $X = 3$, there are two failures followed by a success, so that

$$\text{pr}(X = 3) = 0.9 \times 0.9 \times 0.1 = 0.9^2 \times 0.1 = 0.081$$

Similarly

$$\begin{aligned}\text{pr}(X = 4) &= 0.9 \times 0.9 \times 0.9 \times 0.1 \\ &= 0.9^3 \times 0.1 = 0.0729\end{aligned}$$

$$\text{pr}(X = 5) = 0.9^4 \times 0.1 = 0.06561$$

Can you see the pattern? Now

$$\text{pr}(X = 1) = 0.1$$

$$\begin{aligned}\text{pr}(X \le 2) &= \text{pr}(X = 1) + \text{pr}(X = 2) \\ &= 0.1 + 0.09 = 0.19\end{aligned}$$

$$\begin{aligned}\text{pr}(X \le 3) &= \text{pr}(X \le 2) + \text{pr}(X = 3) \\ &= 0.19 + 0.081 = 0.271\end{aligned}$$

$$\begin{aligned}\text{pr}(X \le 4) &= \text{pr}(X \le 3) + \text{pr}(X = 4) \\ &= 0.271 + 0.0729 = 0.3439\end{aligned}$$

$$\begin{aligned}\text{pr}(X \le 5) &= \text{pr}(X \le 4) + \text{pr}(X = 5) \\ &= 0.3439 + 0.06561 = 0.40951\end{aligned}$$

The last probability is greater than 0.40. The couple would need to keep trying up to and including their fifth attempt.

5.3 THE BINOMIAL DISTRIBUTION

5.3.1 Sampling from a Finite Population

Survey data is obtained by taking a sample from a finite population. When polling human populations, dichotomous (or binary) questions are often asked. These are questions with two possible answers, for example, "agree or disagree" or "yes or no." Countless surveys in the United States in 1998 have asked some variant of "Do you think President Bill Clinton should be impeached?" A 1998 press release from the Harvard School of Public Health reported on a study in which 474 surgery patients sampled from Pennsylvania were asked whether they were aware of the statewide *Consumer Guide* that contained such information as the mortality rates of every surgeon (only 57 patients said yes). When asked whether they knew the rating of their own surgeon or hospital, less than 1% said yes. Another 1998 Harvard press release concerned a study of violence in public schools, in which 1558 junior and senior U.S. school children were sampled. Of these children, 41% agreed with the statement, "If I am challenged, I am going to fight," and 66% reported witnessing or participating in fights in the past year. In acceptance sampling in industry we sample from a batch of industrial components and test each component to see whether or not the component is defective (see Case Study 5.3.1). If too many are defective, the batch is

[2]We discuss further what such an assumption entails in Section 5.3.2.

(a) ***Two-color urn-sampling model***

N balls in an urn, of which there are

M black balls

$N - M$ white balls

Sample n balls and count X = # black balls in sample

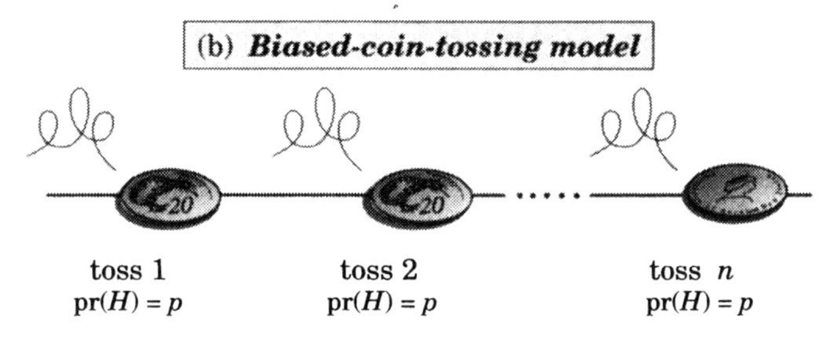

Perform n tosses and count X = # heads

FIGURE 5.3.1 The two-color urn model and the biased-coin model.

rejected. In biology we may sample an animal population and see how many animals have a specific genetic mutation.

In each of these situations we take a sample of n objects from a finite population, look at each sampled object in turn to see whether or not it has the particular characteristic of interest (e.g., knows the rating of their surgeon), and count X, the number of sampled objects that have this characteristic. Thus, in sampling terms, all of the preceding situations are essentially identical. We have a simple *physical model* or analogy for them: the two-color urn model depicted in Fig. 5.3.1(a). Our population is analogous to the N balls in the urn. The number of individuals in the population with the characteristic of interest is analogous to the M black balls. We sample n balls and count X, the number of balls in the sample that are black.

In almost all real sampling the sample size n is small compared to the population size N. Under these conditions the probability distribution of X is well approximated by a very famous probability distribution called the *Binomial distribution.* Programs that calculate these probabilities require the sample size n and the value of $p = M/N$, the proportion of "black balls" in the whole population (e.g., the proportion of all U.S. voters who wanted Clinton impeached). We shall learn more about the Binomial distribution shortly.

5.3.2 The Binomial Assumptions

Another way that data arises is by *observing a random process* that sequentially produces binary outcomes (i.e., outcomes of the yes/no or pass/fail type) in an unpredictable way as time goes on. An example from manufacturing concerns computer chips coming off the end of a production line. Each new chip produced by the line may be defective (fail a set of tests) or be acceptable. Over time we observe that some chips are defective and others are not defective, but the pattern of defectives and nondefectives looks random, and we have no way of knowing ahead of time which ones will be defective and which will not.

We now have another physical model that is sometimes applicable for binary observations on a random process, the model of tossing a biased coin, as in Fig. 5.3.1(b). We imagine a *biased* or unbalanced coin because we want to allow the probability of getting a "head" (e.g., the chip being defective) to take values other than 0.5. What properties should we look for in order to classify a real situation as being "just like tossing a coin"? The biased-coin-tossing model is a physical model for any random process that we can think of as being made up of a series of random "tosses" or ***trials,*** with the following essential features.

1. ***Two outcomes:*** Each trial ("toss") has only two outcomes, conventionally referred to as ***success*** ("heads") and ***failure*** ("tails").
2. ***Constant p:*** The probability p of getting a success has the same value, for each trial.
3. ***Independence:*** The results of the trials are independent of one another.

Assessing whether the coin-tossing model is a valid analogy for a real process, such as that producing our computer chips, comes down to an assessment of whether conditions 2 and 3 are reasonable approximations. For condition 2 to hold, the average rate at which the line produced defective items would have to be constant over time. This would not be true if the machines used in the manufacturing had started to drift out of adjustment and were steadily getting worse. For condition 3 to hold, the chance of a particular item being defective should not depend on what has happened to previous items. (In practice, this usually happens only if the items are sampled sufficiently far apart in the production sequence.)

The *Binomial distribution* is the distribution of X, the number of heads in n tosses of a biased coin (where n is fixed in advance). Consider a coin constructed so that the probability of obtaining a head is p. When we make n tosses, X (the number of heads) can take values $0, 1, 2, \ldots, n$. The probability distribution of X is called the ***Binomial distribution,***[3] and we write $\boldsymbol{X \sim}$ **Binomial($\boldsymbol{n, p}$)**. (More formally, the Binomial is the distribution of the number of successes in n trials for which the probability of success on any particular trial is p.)

> The distribution of the number of heads in $\boldsymbol{n}$ tosses of a biased coin is called the ***Binomial distribution.***

The values n and p are called *parameters* of the distribution. When $n = 5$ and $p = 0.2$, we are talking about the probability distribution for the number of heads X obtained when we use a biased coin constructed so that pr(H) $= 0.2$ for any given toss, and we make $n = 5$ tosses. (Alternatively, think of testing $n = 5$ computer

[3]The word Binomial comes from the Latin *Bi,* meaning two, and *nomen,* meaning name. Here we have two outcomes, heads and tails. Several other important distributions relate to the biased-coin-tossing model. Whereas the *Binomial* distribution is the distribution of the number of heads in a *fixed* number of tosses, the *Geometric* distribution is the distribution of the number of tosses up to and including the first head, and the *Negative Binomial* distribution is the number of tosses up to and including the kth (e.g., 4th) head.

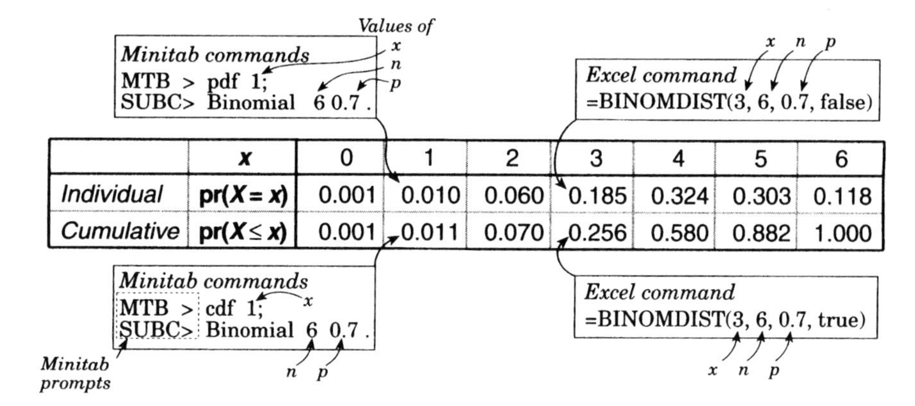

	x	0	1	2	3	4	5	6
Individual	pr(X = x)	0.001	0.010	0.060	0.185	0.324	0.303	0.118
Cumulative	pr(X ≤ x)	0.001	0.011	0.070	0.256	0.580	0.882	1.000

FIGURE 5.3.2 Binomial($n = 6, p = 0.7$) probabilities.
Table constructed in Excel and annotated with both the Minitab commands and the Excel commands that produce the probabilities.

chips where pr(*chip defective*) = 0.2.) The Binomial probability function for $n = 5$ and $p = 0.2$ can be shown to be (to 3 decimal places)

x	0	1	2	3	4	5
pr(x)	0.328	0.410	0.205	0.051	0.006	0.000

Under these circumstances we have a probability of 0.205 of obtaining 2 heads from our 5 tosses (or 2 defective chips among our 5 chips). If we make $n = 6$ tosses and pr(H) = p = 0.7, however, then the distribution of X is as given in Fig. 5.3.2. The probability that 6 tosses will result in 2 heads is 0.060. Figure 5.3.2 also adds a row containing the cumulative probabilities. We see that pr($X \leq 3$) = 0.256.

Although Fig. 5.3.2 gives the form of the Minitab commands for Binomial probabilities, we prefer to use the menus. Look under `Calc` → `Probability Distributions` → `Binomial`. This brings up a dialog box to be completed. Check `probability` for pr($X = x$) and `cumulative` for pr($X \leq x$). The form of the Excel command is `BINOMDIST(x,n,p,cumulative)` where `cumulative` is set to `false` for individual probabilities and to `true` for cumulative probabilities. The Splus and R commands are `dbinom(x,n,p)` for pr($X = x$) and `pbinom(x,n,p)` for pr($X \leq x$). Binomial probabilities are available from all general-purpose statistical packages and a number of calculators (e.g., graphics calculators). We provide a table of individual probabilities in Appendix A2. Working with a formula for pr($X = x$) is discussed briefly at the end of this section.

✖ EXAMPLE 5.3.1 *Binomial Probabilities for Numbers of Cars Caught Speeding*

On a particular stretch of highway, 30% of cars exceed the 100 km/hr speed limit. Fifteen cars pass a hidden police camera. We are concerned with X, the number of these cars that are speeding. We shall leave it to the reader to think about the conditions required for this to be like tossing a biased coin 15 times with p = pr(H) = pr(*speeding*) = 0.3. If these conditions are satisfied, the distribution of X is Binomial($n = 15, p = 0.3$). What is the probability that seven or more cars are speeding, pr($X \geq 7$)? This is an upper-tail probability so, as in Section 5.2, we convert it to a calculation involving a cumulative (or lower-tail) probability by using the complement. If X is not 7 or more, then it is 6 or less, so that (to three decimal places)

$$\text{pr}(X \geq 7) = 1 - \text{pr}(X \leq 6) = 1 - 0.8688574 = 0.131$$

If we add the individual probabilities pr(7) + pr(8) + $\cdots$ + pr(15) obtained from the $n = 15, p = 0.3$ entry in Appendix A2, we get $0.081 + 0.035 + \cdots = 0.132$. The small discrepancy is due to rounding error caused by only using three decimal places in the calculations.

We now find the probability that between 3 and 8 cars (inclusive) are speeding. We want the lower endpoint, 3, to remain in our interval, so we remove the values from 2 down.

$$\text{pr}(3 \leq X \leq 8) = \text{pr}(X \leq 8) - \text{pr}(X \leq 2) = 0.9847575 - 0.1268277 = .858$$

(to three decimal places). This time adding the individual terms pr(3) + pr(5) + $\cdots$ + pr(8) from Appendix A2 gives $0.170 + 0.219 + \cdots + 0.035 = 0.858$ again.

Example 5.3.2 *Binomial Probabilities to See if Hypnosis Enhances ESP*

ESP researchers often use a shuffled deck of so-called Zener cards. There are equal numbers of five types of card in the deck showing five very different shapes.

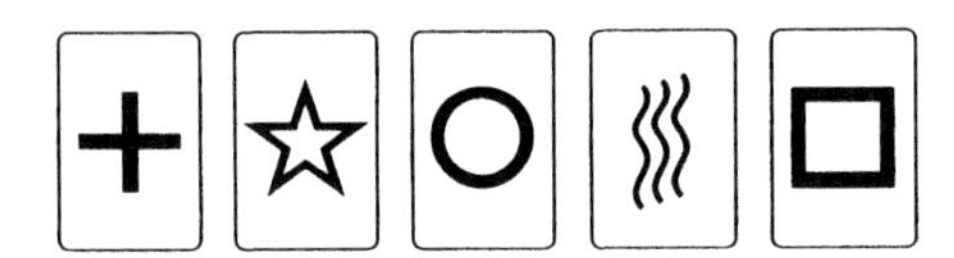

Some parapsychologists believe that hypnosis can enhance a person's ESP ability. Casler [1964] performed an experiment to test this hypothesis. A card was randomly selected from the deck. A hypnotized person concentrated on the card, and a hypnotized student tried to guess its identity. This was done 100 times by each student. Fifteen students were tested, giving a total of 1500 guesses. Of the 1500 guesses, 325 were correct. Now, if only random guessing was operating, the students would have 1 chance in 5 of correctly identifying each card. We would expect about one fifth of the 1500 guesses, or 300, would turn out to be correct. In the experiment there were 325 correct guesses, slightly more than the 300 we might have expected by chance. Is this small discrepancy enough to convince us that something other than chance was operating? How likely are we to get something like 325 correct just by random guessing?

The problem is like tossing a coin 1500 times, with a correct guess being analogous to the coin coming up heads. We have $n = 1500$ trials, and the probability of success on each trial is $p = 0.2$, so the distribution of X, the number of correct guesses, is Binomial($n = 1500, p = 0.2$). Splus's binomial command[4] `pbinom(324,1500,0.2)`

[4]The versions of Minitab and Excel running on our computers could not handle a value of n as large as 1500.

tells us that $\text{pr}(X \leq 324) = 0.94$ so that approximately 19 times out of 20, the number of correct random guesses in a set of 1500 is smaller than the 325 obtained in Casler's experiment. This suggests that something other than just guessing may have been going on. The results would have to be repeatable in other similar experiments, however, before we would be convinced.

✖ Example 5.3.3 *Binomial Probabilities and Number of Presidents' Sons*

A 1993 article in *The Economist* noted that the presidents of the United States have had, between them, 61 daughters and 90 sons. The context was a discussion of biological theories about organisms "wanting" to get as many of their genes into future generations as possible. A successful son can give you more grandchildren than a daughter. We shall check whether a sex imbalance like that seen in the presidents' children is likely to occur by chance alone. We shall once again assume that begetting children is just like tossing a coin, with sons and daughters being equally likely. Using this analogy, presidents "tossed the coin" 151 times. The distribution of the number of sons among 151 children is Binomial($n = 151, p = 0.5$). What is the probability of getting 90 or more sons?

$$\text{pr}(X \geq 90) = 1 - \text{pr}(X \leq 89) = 1 - 0.9888124 = 0.011$$

(to three decimal places). There is only about 1 chance in 100 of having 90 or more sons in 151 children. It seems implausible that an imbalance like that seen in the presidents' children would have arisen by chance alone. Or is it?

We are going to take time out for some statistical lessons that arise naturally out of the previous two examples. We start with Casler's ESP experiment. If only one such experiment had been done, this is a surprisingly large number of correct identifications. Sometimes lots of people do experiments that are similar, if not the same, however. Our calculations showed that we get 325 or more correct guesses for about 1 in every 20 sets of 1500 random guesses. If lots of people were doing such experiments, some will show discrepancies like Casler's by chance alone, even if there is no real ESP effect at all. If only unusual results get reported (publication bias), we shall get the mistaken impression that there is a real ESP effect. This is one of the reasons why science insists on results being repeatable.

We can read even less of significance into presidents having an unusual number of sons. Reports like this come about as a consequence of society's fascination with the unusual. There are many groups, such as chemists, musicians, mathematicians, and so on, for which things like excesses of boys or unusually large numbers of left-handers are reported. We notice unusual events and then try to find meaning in them. Important discoveries often occur in this way. But highly unusual events can also happen by chance alone. The preceding calculation tells us that in the long run, with chance alone operating, we get about 1 set of 151 children in every 100 such sets that has 90 or more sons. Suppose there are 1000 sets of 151 children whose parents are notable for some reason or other. We would expect about 10 of these sets of children to consist of 90 or more sons by chance alone. If someone noticed one of these imbalances, say in the children of Nobel prize winners, it would get reported. Then people would start thinking there must be a deep and significant reason why Nobel prize winners had so many sons. We note that different "identifiable groups of parents" will have different numbers of children (not all 151), so the calculation does not apply exactly, but the general idea of the effect of chance happenings combined with the reporting of "the unusual" is important.

5.3.3 Sampling from a Finite Population Revisited

We return to the situation where we have a finite population of size N in which M individuals have some characteristic of interest (e.g., disapprove of human cloning). We take a simple random sample of size n from the population and count X, the number of individuals in the sample with that characteristic. Our physical model for such situations is the urn model in Fig. 5.3.1(a). We stated that, when the sample size n is much smaller than the population size N, the distribution of X is approximately Binomial(n, p), where $p = M/N$ is the proportion of individuals in the population with the desired characteristic. Why is this?

Supposed we sampled from the urn at random with replacement. By reviewing the conditions given in Section 5.3.2, we see that sampling *with replacement* is just like tossing a biased coin. Each time we select a ball, there are two possible outcomes, a black ball or a white ball (condition 1), the probability of selecting a black ball is the proportion of black balls in the urn, $p = M/N$, which is always the same (condition 2) and does not depend in any way on previous selections (condition 3). Thus the distribution of X, the number of black balls in the sample, is exactly Binomial$(n, p = M/N)$.

All real sampling, however, is done *without* replacement. It is clearly inefficient to interview the same person twice or measure the same object twice. As we remove balls from the urn, the proportion of black balls in the urn changes, and thus the probability of selecting a black ball changes depending on how many black and white balls have been removed on past draws. This violates the independence condition 3. Thus, the Binomial distribution no longer applies exactly.[5]

When the sample size n is small compared to M and $N - M$ (respectively, the number of black balls and white balls in the urn), however, the proportion of black balls in the urn changes very little from $p = M/N$ as balls are randomly selected and removed from the urn. In these circumstances the Binomial$(n, p = M/N)$ distribution is a good approximation to the true distribution of X. This accounts for many, if not most, occasions in which the Binomial distribution is used in practice. The approximation works well enough for most practical purposes if the sample size is no bigger than about 10% of the population size ($n/N < 0.1$), and this is true even for large n.

Summary

If we take a sample of size n from a much larger population in which a proportion p have a characteristic of interest, the distribution of X, the number in the sample with that characteristic, is approximately Binomial(n, p).

(Operating rule: Approximation is adequate if $n/N < 0.1$.)

---

[5]The actual distribution for sampling without replacement in the two-color-urn model is called the *Hypergeometric distribution*. We shall not work with Hypergeometric probabilities here, but they can be obtained from many statistical packages and from Excel. Skills such as using cumulative probabilities to find upper-tail probabilities and interval probabilities apply to any discrete distribution, not just to the Binomial.

Practical applications of the urn model where the Binomial approximation does not work include many gambling games (e.g., lotto games, keno, and poker) and some acceptance-sampling applications (see Case Study 5.3.1).

✖ EXAMPLE 5.3.4 *Using the Binomial Approximation*

Suppose you need medical treatment for a particular problem. Although you do not know it, 70% of doctors would recommend drug treatment for this problem and 30% would recommend surgery. You want to make sure you get the right treatment, so, just to be on the safe side, you go to three doctors. Regard these doctors as a random sample from the population of doctors you might have approached. What are the chances that the majority opinion (i.e., two or more) favors surgery?

This is clearly an urn-sampling situation with black balls corresponding to doctors who favor surgery. Our sample size $n = 3$ is much smaller than the population size, and the population proportion is $p = 0.3$. The distribution of X, the number who would recommend surgery, is approximately Binomial($n = 3, p = 0.3$). Using this distribution, we find that $\text{pr}(X \geq 2) = 0.216$. Slightly more than one person in five who acted in this way would receive a majority opinion in favor of surgery.

The Formula for Binomial Probabilities

The following formula is derived on our web site. When $X \sim \text{Binomial}(n, p)$,

$$\text{pr}(X = x) = \binom{n}{x} p^x (1 - p)^{n-x} \qquad \text{for} \qquad x = 0, 1, 2, \ldots, n$$

The symbol $\binom{n}{x}$ represents the number of different ways of choosing x objects from a set of n objects. It can be calculated using the formula

$$\binom{n}{x} = \frac{n!}{x!(n-x)!}$$

This formula makes use of factorial ($n!$) notation. By 4! we mean $4 \times 3 \times 2 \times 1 = 24$. Similarly $5! = 5 \times 4 \times 3 \times 2 \times 1 = 120$. Generally, $n! = n \times (n-1) \times (n-2) \times \cdots \times 3 \times 2 \times 1$. By definition $0! = 1$. Almost all scientific calculators have a button for $n!$, and many even have a $\binom{n}{x}$ button (it could alternatively be labeled with a notation like ^nC_x or C_x^n). When $n = 5$ and $p = 0.3$, we have

$$\text{pr}(X = 2) = \binom{5}{2}(.3)^2(.7)^3 = \frac{5!}{2! \times 3!} \times (.3)^2 \times (.7)^3 = 0.3087$$

QUIZ ON SECTION 5.3

1. Why are models like the biased-coin-tossing model and the two-color urn-sampling model useful in practice?
2. For what types of situations is the urn-sampling model useful? Give examples.
3. For what types of situations is the biased-coin model useful? Give the three essential conditions for its applicability.

4. What is the distribution of the number of heads in n tosses of a biased coin? Identify the meaning of each of the parameters of the distribution.

5. Under what conditions does the Binomial distribution apply to samples taken without replacement from a finite population? What do n and p represent in this context?

EXERCISES FOR SECTION 5.3

1. The purpose of this exercise is to ensure that you can confidently use your computer to calculate Binomial probabilities.

   If $X \sim \text{Binomial}(n = 10, p = 0.3)$, what is the probability that [use cumulative probabilities in all cases but (a)]

   (a) pr($X = 3$)? (b) pr($X \geq 4$)? (c) pr($X > 6$)?
   (d) pr($X < 7$)? (e) pr($X \leq 1$)? (f) pr($3 < X < 8$)?
   (g) pr($4 \leq X < 8$)? (h) pr($X > 12$)? (i) pr($X < 11$)?

2. This problem is concerned with practicing model-recognition skills. For each of the following situations try to answer the following questions. Is the coin-tossing or the urn-sampling model more appropriate? What additional assumptions would be required for the physical model to hold? Give the values of the parameters. If the value of a parameter is unknown, write "unknown." Can you use the Binomial distribution? (Justify your answer.) Give the values of the Binomial parameters.

   (a) Over time, 20% of cars parking in a short-stay car park on Monday morning stay longer than the allowed 60 minutes. The first 10 cars parking in such spaces are watched one Monday morning, and X_1 is the number that stay longer than 60 minutes.
   (b) Of 10,000 cars parked in such car parks in the city that day, 50 cars are watched, and X_2 is the number of those watched that stay longer than 60 minutes.
   (c) Of 7400 subscribers to *Consumer* magazine, 2730 had home computers. Suppose a company doing market research was dialing these people at random. One interviewer dialed 50 people. X_3 is the number dialed who had a home computer.
   (d) Of people with home computers, 45% do not use them for playing games. X_4 is the number of people, out of a random sample of 100, who do not use their computer for games.
   (e) The English version of the word game Scrabble has 100 tiles. Of these, 12 have the letter *E* written on them. The tiles are mixed up, and then every player begins with 7 tiles. X_5 is the number of *E* tiles that a single player gets.
   (f) In a report of violent crimes in the United States, 64% of reported violent crimes did not involve weapons. Of 50 randomly chosen violent crime reports, X_6 is the number in which no weapon was involved.
   (g) For the violent crime reports from a particular small town, 89 involved weapons and 99 did not. Of 30 randomly chosen reports, X_7 is the number that did not involve weapons.
   (h) We roll a standard die 10 times and count X_8, the number of 6s.
   (i) We deal a 7-card hand from a standard deck of cards and count X_9, the number of aces.
   (j) It is known that 10% of the population are left-handed. We randomly sample 30 people and count X_{10}, the number of left-handers in the sample.
   (k) It is known that 30% of a brand of wheel bearing will function adequately in continuous use for a year. We test a sample of 20 bearings and count X_{11}, the number in the sample that function adequately.
   (l) Experience has shown that 30% of headache patients have their headaches cured by a placebo pill. Out of the next 50 headache patients whom a particular doctor sees, we count X_{12}, the number cured by the placebo.

CASE STUDY 5.3.1 *Acceptance sampling*

Manufacturers whose product incorporates a component or part received in batches from a supplier will often take a sample of each batch of parts they receive and send back the batch if there are too many defective parts in the sample. This is called *acceptance sampling*. Many such schemes in common use in industry were developed for the U.S. military and are encapsulated in military standards. This Case Study explores some of the issues involved.

Suppose that an incoming batch contains 3000 components. We shall make our rejection decision on the basis of a random sample of size 20 and will reject the batch if we see k or more defectives in our sample, where k is yet to be determined. This is clearly an urn-sampling situation, with black balls corresponding to defective components, $N = 3000$, M unknown, and $n = 20$. Our $n/N < 0.1$ rule for the adequacy of the Binomial approximation is easily satisfied, so the distribution of the number of defective components in the sample is approximately Binomial($n = 20, p$).

Suppose that we want to protect ourselves against accepting any batch that contains 10% or more defectives. (In practice, limits on what constitutes an acceptable batch are much smaller than this; 10% has been chosen for simplicity.) We shall set k so that when the proportion of defectives in the whole batch is $p = 0.1$, the probability of getting k or more defectives in the sample (and thus rejecting the batch) is high. Using a computer we found the following probabilities for the Binomial($n = 20, p = 0.1$).

k	0	1	2	3	4	5	...
pr($X \geq k$)	1.00	0.878	0.608	0.323	0.133	0.043	...

If we set $k = 2$ (i.e., reject if we see 2 or more defectives), we have a 61% chance of rejecting the batch, or a 39% chance of accepting a bad batch. This is clearly too high. If we set $k = 1$ (i.e., reject if we see even 1 defective), we have an 88% chance of rejecting the batch, or only a 12% chance of accepting it. Note that if the true proportion of defectives is higher than 10%, that is, $p > 0.1$, there is an even higher chance of rejecting the batch [e.g., when $p = 0.15$, pr($X \geq 1$) = 0.961].

This sounds much better from the purchaser's point of view. There is at worst a 12% chance of accepting a substandard batch. But let us now look at the problem from the supplier's point of view. What are the chances that this procedure will send back acceptable batches? What we need are values of pr($X \geq 1$) for values of p that are smaller than 0.1. For $n = 20$ we give the probabilities for the following three values of p:

Proportion defective in batch:	p	.01	.05	.1
Probability of rejecting batch:	pr($X \geq 1$)	0.182	0.642	0.878

The supplier can justifiably be quite upset because 64% of batches with 5% defective components would be sent back by this rule, and even 18% of batches with only 1% defectives would be sent back.

The problem we have experienced is in large part due to our having used $n = 20$, too small a sample size for the proportion of defectives in the sample to be a reliable estimate of the proportion in the population. What if we took a sample of 200 components? The distribution of the number of defectives is now Binomial($n = 200, p = 0.1$).

We made up a table of values of pr($X \geq k$) like the preceding one for $k = 0, 1, \ldots, 20$. (This can be done very quickly with software such as Minitab or Excel.) We found that pr($X \geq 14$) = 0.943, pr($X \geq 15$) = 0.907, pr($X \geq 16$) = 0.857. If we sent back any batch with 15 or more defectives, we would have just less than 1 chance in 10 of accepting a batch with 10% defectives or more. Addressing the problem from the supplier's viewpoint, we calculated the probability that a batch would be sent back, pr($X \geq 15$), for $n = 200$ and various values of p.

p	.01	.02	.03	.04	.05	.06	.07	.08	.09	.1
pr($X \geq 15$)	.000	.000	.001	.015	.078	.222	.430	.640	.804	.907

We see that the chances of sending back a batch with a very small proportion of defectives is now small, for example, essentially zero if the batch has 2% defectives ($p = 0.02$). A batch with 7% defectives ($p = 0.07$) has a 43% chance of being sent back.

Thinking exercise: What are the some of the costs that would have to be traded off if you wanted to come up with a cost-effective acceptance-sampling scheme?

5.4 EXPECTED VALUES

5.4.1 Formula and Terminology

We now introduce an important concept called the *expected value of a random variable,* which plays a fundamental role in the chapters to follow. Before defining expected values, we motivate our definition by a simple example.

EXAMPLE 5.4.1 *Expected Winnings from Gambling*

A friend has approached you with a request because they know you have been studying probability. The friend has been invited to take part in a game of chance that costs \$1.50 for each game played. The player wins \$1, \$2, or \$3 with respective probabilities of 0.6, 0.3, and 0.1. Is the game a rip-off?

Our first analysis of the situation is entirely intuitive. Suppose you played 100 games. With a 60% chance of winning \$1, you would expect to win \$1 for about 60

Prize (\$)	x	1	2	3	
Probability	pr(x)	0.6	0.3	0.1	
What we would "expect" from 100 games					*add across row*
Number of games won		0.6×100	0.3×100	0.1×100	↙
\$ won		$1 \times 0.6 \times 100$	$2 \times 0.3 \times 100$	$3 \times 0.1 \times 100$	Sum

Total prize money = Sum; ***Average prize money*** = Sum/100
= $1 \times 0.6 + 2 \times 0.3 + 3 \times 0.1$
= 1.5

FIGURE 5.4.1 Intuitive analysis of a game of chance.

of the 100 games. Similarly, you would expect to win \$2 for about 30 games and \$3 for about 10 games. This gives total prize money for 100 games of about \$60 + \$60 + \$30 = \$150, or average winnings of \$150/100, amounting to \$1.50 per game. The game looks fair in the sense that the expected average winnings per game is the same as the entry fee.

Figure 5.4.1 gives the probability function for this situation (first two rows) and then breaks down the steps used in the intuitive computation. We see that our intuitive calculation of "expected" prize money *per game* is equivalent to using the formula:

"Expected" prize money per game = Sum of (value × probability of value)

Not quite trusting our intuition completely, we next simulated the playing of this game a large number of times on a computer and observed what happened. We did this by programming the computer to produce random numbers, with a 1 being produced with probability 0.6, a 2 with probability 0.3, and a 3 with probability 0.1. We did this N times for various N and counted the numbers of 1s, 2s, and 3s that occurred. The results are given in Table 5.4.1.

TABLE 5.4.1 Average Winnings from a Game Conducted *N* Times

Number of games played (*N*)	Prize won in dollars (x) 1	2	3	Average winnings per game ($\overline{x}$)
	frequencies (*Relative frequencies*)			
100	64 (*.64*)	25 (*.25*)	11 (*.11*)	1.7
1,000	573 (*.573*)	316 (*.316*)	111 (*.111*)	1.538
10,000	5995 (*.5995*)	3015 (*.3015*)	990 (*.099*)	1.4995
20,000	11917 (*.5959*)	6080 (*.3040*)	2000 (*.1001*)	1.5042
30,000	17946 (*.5982*)	9049 (*.3016*)	3005 (*.1002*)	1.5020
∞	(*.6*)	(*.3*)	(*.1*)	1.5

For $N = 30{,}000$, we compute the average (sample mean) $\overline{x}$ using the repeated-data formula of Section 2.5.1:

$$\begin{aligned}
\overline{x} &= \frac{1}{N}\sum(\textit{value} \times \textit{frequency}) \\
&= \frac{1}{30{,}000}(1 \times 17{,}946 + 2 \times 9049 + 3 \times 3005) \\
&= 1 \times \frac{17{,}946}{30{,}000} + 2 \times \frac{9049}{30{,}000} + 3 \times \frac{3005}{30{,}000} \\
&= \sum(\textit{value} \times \textit{relative frequency}) \\
&= 1 \times 0.5982 + 2 \times 0.3016 + 3 \times 0.1002 \\
&= 1.5020
\end{aligned}$$

In Section 4.1 we saw how the relative frequency (or proportion) of heads in tossing an unbiased coin approached the probability of $\frac{1}{2}$ as the number of tosses increased. The same sort of thing is happening in Table 5.4.1. As N gets very big, the relative frequency of each value x of the random variable X gets very close to its probability pr(x) (for $x = 1, 2, 3$). This means that $\overline{x}$ gets very close to

$$\sum_{i=1}^{3} x\,\text{pr}(x) = 1 \times 0.6 + 2 \times 0.3 + 3 \times 0.1 = 1.5$$

This last equation is represented by the last entry, $N = \infty$, in Table 5.4.1, where $N = \infty$ really means "N extremely large." (In fact, if we use only two decimal places, we see that we have reached this state by $N = 10{,}000$.) We note that the long-term average result is the same as our intuitive result. The intuition many people have with respect to the word "expect" is connected with what should happen on average over many, many repetitions of the experiment.

The sum $\sum x\,\mathrm{pr}(x)$ in the preceding example is called the *expected value* of X and is denoted by $\mathrm{E}(X)$. Thus, for a general distribution of X we have

> The ***expected value:***
> $\mathrm{E}(X) = \sum \boldsymbol{x}\,\mathbf{pr}(\boldsymbol{x})$ = Sum of (*value* $\times$ *probability of value*)

Expected values tell us about long-run average behavior in many repetitions of an experiment. They are obviously an important guide for activities that are often repeated. The gambling game in Example 5.4.1 is fair in the sense that the long-term average winnings, or expected winnings per game, is the same as the entry fee. Of course, most gambling games are not fair because the promoter of the game is in business to make a profit. If you look at the expected returns on "investment," it is hard to see why people play state lotteries. The main purpose of such lotteries is to make money for government projects. New Zealand's lotto game returns about 55 cents to the players for every dollar they spend. If you choose your lotto numbers randomly, 55 cents is the expected prize money for an investment of $1.00. In other words, if you played an enormous number of games, the amount of money you won would be about 55% of what you had paid out for tickets.[6] In the very long run, you are guaranteed to lose money. The main prizes are very large, however, and the probabilities of winning them are tiny so that no one ever plays enough games for this long-run averaging behavior to become reliable. Many people are prepared to essentially write off the cost of playing as being small enough that they hardly notice it, against the slim hope of winning a very large amount that would make a big difference to their lives.[7]

However you rationalize it, though, not only are lottery players guaranteed to lose in the long run, the odds are that they will lose in the short run too. The Kansas instant lottery of Review Exercises 5, problem 14, returns about 47 cents on the dollar to the players. Wasserstein and Boyer [1990] worked out the chances of winning more than you lose after playing this game n times. The chances are best after 10 games, at which point roughly 1 person in 6 will be ahead of the game. But after 100 games, only 1 person in 50 will have made a profit.

EXAMPLE 5.4.2 *Expected Value Calculated*

In Example 5.2.3 we had a random variable with probability function

x	0	1	2	3
pr(x)	$\frac{1}{8}$	$\frac{5}{8}$	$\frac{1}{8}$	$\frac{1}{8}$

Here

$$\mathrm{E}(X) = \sum x\,\mathrm{pr}(x) = 0 \times \frac{1}{8} + 1 \times \frac{5}{8} + 2 \times \frac{1}{8} + 3 \times \frac{1}{8} = 1.25$$

---

[6] If you don't choose your numbers randomly, things are more complicated. The payout depends on how many people choose the same numbers as you and therefore share any prize. The expected returns would be better if you had some way of picking unpopular combinations of numbers.

[7] Some U.S. lotto payouts have exceeded $100 million. The idea that people's behavior with respect to profits and losses often cannot be expressed simply in terms of money is recognized by economists and decision theorists who work in terms of quantities they call ***utilities***.

✖ Example 5.4.3 *Expected Value for a Binomial Distribution*

If $X \sim \text{Binomial}(n = 3, p = 0.1)$, then X can take values 0, 1, 2, and 3. Using probabilities obtained from the table in Appendix A2, we find

$$\begin{aligned}\text{E}(X) &= \sum x\text{pr}(x) = 0 \times \text{pr}(0) + 1 \times \text{pr}(1) + 2 \times \text{pr}(2) + 3 \times \text{pr}(3)\\ &= 0 + 1 \times 0.243 + 2 \times 0.027 + 3 \times 0.001\\ &= 0.3\end{aligned}$$

We note that here $\text{E}(X) = 0.3 = 3 \times 0.1 = np$.

If we look at Example 5.2.2, we see that $X \sim \text{Binomial}(n = 2, p = 0.5)$. Then

$$\begin{aligned}\text{E}(X) &= \sum x\,\text{pr}(x)\\ &= 0 \times \frac{1}{4} + 1 \times \frac{1}{2} + 2 \times \frac{1}{4} = 1\end{aligned}$$

We note again that $\text{E}(X) = 1 = 2 \times 0.5 = np$.

Suppose we take a random sample of size 1000 from a large population in which 20% of people had been audited by tax authorities. We would intuitively expect about 20% of the people in the sample, that is, $1000 \times 0.2 = 200$ of them, to have been audited. Our intuitive expected value is $1000 \times .2 = n \times p$, as in the previous example. In fact, it can be shown theoretically that whenever $X \sim \text{Binomial}(n, p)$, $\text{E}(X) = \sum x\,\text{pr}(x) = np$.

For the Binomial distribution: $\quad \textbf{E}(\boldsymbol{X}) = \boldsymbol{np}$

It is conventional to call E(X) the *mean* of the distribution of X and denote it by the Greek symbol for "m," namely μ. If there are several random variables being discussed, we use the name of the random variable as a subscript to μ so that it is clear which random variable is being discussed. If there is no possibility of ambiguity, the subscript on μ is often omitted.

$\boldsymbol{\mu_X} = \textbf{E}(\boldsymbol{X})$ is called the ***mean*** of the distribution of $\boldsymbol{X}$.

✖ Example 5.4.4 *Expected Value for a Finite Population*

Suppose we consider a finite population of 200 students taking a particular university course. We know that 120 have one assignment due that week, 60 have two assigments due, and 20 have three assigments due. If a student is chosen at random from the population, then the probabilities that the student has one, two, or three assignments due are $\frac{120}{200} = 0.6$, $\frac{60}{200} = 0.3$, and $\frac{20}{200} = 0.1$, respectively. For this random experiment we have the same probability distribution as in Example 5.4.1, so the expected value of X is again 1.5. Now

$$\begin{aligned} E(X) &= 1 \times \frac{120}{200} + 2 \times \frac{60}{200} + 3 \times \frac{20}{200} \\ &= \frac{1}{200}(1 \times 120 + 2 \times 60 + 3 \times 20) \end{aligned}$$

This last line is also the average of all 200 values of the variable *number of assignments due* in this population (see the repeated-data formula for $\bar{x}$ in Section 2.5.1), that is, E(X) is numerically identical to the mean of the finite population.

Largely because E(X) is the ordinary average (or mean) of all the values for a finite population, like the one in the preceding example, it is often called the *population mean*. This name has stuck for E(X) generally and not just when applied to finite populations. It is also used in situations like the one described in Example 5.4.1, where the population is essentially *conceptual* in that it has to be generated by repeating the experiment a very large number of times. Thus in general we have that

> μ_X = **E(X)** is usually called the ***population mean.***

This is the terminology we shall use. It is shorter than calling $\mu = E(X)$ the "mean of the distribution of X" and serves to distinguish it from the "sample mean," which is the ordinary average of a batch of numbers.

There is one more connection we wish to make. Just as $\bar{x}$ is the point where a dot plot or histogram of a batch of numbers balances (Fig. 2.4.1 in Section 2.4.1), μ_X is the point where the bar graph of pr($X = x$) balances, as in Fig. 5.4.2.

> μ_X is the point where the bar graph of pr($X = x$) balances.

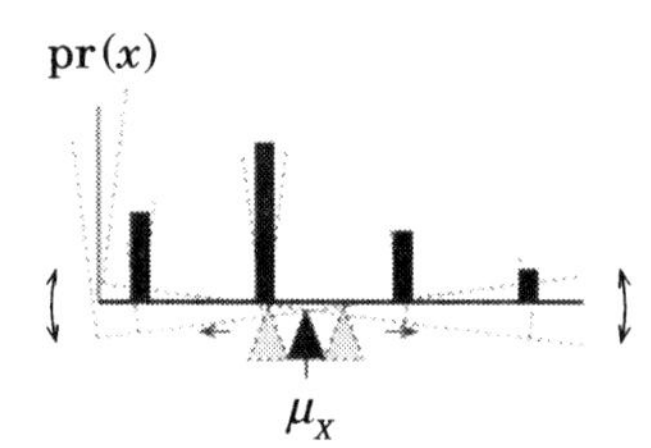

FIGURE 5.4.2 The mean μ_X is the balance point.

EXERCISES FOR SECTION 5.4.1

1. Suppose that a random variable X has probability function

x	2	3	5	7
pr(x)	0.2	0.1	0.3	0.4

   Find μ_X, the expected value of X.

2. Compute $\mu_X = E(X)$ where $X \sim$ Binomial($n = 2, p = 0.3$) and check that your answer equals $n \times p$.

5.4.2 Population Standard Deviation

We have discussed the idea of a mean of the distribution of X, $\mu = E(X)$. Now we want to be able to talk about the standard deviation of the distribution of X, which we shall denote[8] sd(X). We use the same intuitive idea as for the standard deviation of a batch of numbers. The standard deviation is the square root of the average squared distance of X from the mean μ, but for distributions we use E(.) to do the averaging.

> The ***population standard deviation*** is
>
> $$\text{sd}(X) = \sqrt{E[(X-\mu)^2]}$$

Just as with population mean and sample mean, we use the word *population* in the term *population standard deviation* to distinguish it from *sample* standard deviation, which is the standard deviation of a batch of numbers. The *population* standard deviation is the standard deviation of a distribution. It tells you about the spread of the distribution or, equivalently, about how much X varies. In Example 5.4.1 we saw that $\bar{x}$ approaches μ_X when N becomes very large. In a similar fashion we find that s_X, the (repeated-data) sample standard deviation, approaches sd(X) when n becomes large.

The square of the standard deviation, $E[(X-\mu)^2]$, is known as the ***variance*** of X. To compute sd(X), we need to be able to compute $E[(X-\mu)^2]$. How are we to do this? For discrete distributions, just as we compute $E(X)$ as $\sum x\,\text{pr}(x)$, we calculate $E[(X-\mu)^2]$ using

$$E[(X-\mu)^2] = \sum (x-\mu)^2\,\text{pr}(x)$$

We now illustrate the use of the formula.[9]

Example 5.4.5 *Standard Deviation for a Binomial Distribution*

If $X \sim \text{Binomial}(n = 3, p = 0.1)$ then X can take values 0, 1, 2, 3, and we calculated in Example 5.4.3 that $\mu = E(X) = 0.3$. Then

$$\begin{aligned} E[(X-\mu)^2] &= \sum (x-\mu)^2\text{pr}(x) \\ &= (0-0.3)^2\text{pr}(0) + (1-0.3)^2\text{pr}(1) + (2-0.3)^2\text{pr}(2) + (3-0.3)^2\text{pr}(3) \\ &= 0.27 \end{aligned}$$

(pr(0), pr(1), etc., are given in Example 5.4.3.) Thus $\text{sd}(X) = \sqrt{E[(X-\mu)^2]} = \sqrt{0.27} = 0.5196$.

Our answer, $\text{sd}(X) = \sqrt{0.27}$, for the Binomial distribution in Example 5.4.5 satisfies $\text{sd}(X) = \sqrt{np(1-p)}$. It can be shown mathematically that this formula holds for any Binomial(n, p) distribution.

---

[8]Some books use the notation σ_X instead of sd(X).

[9]Spreadsheets are excellent for calculations like this.

> ***For the Binomial distribution:***
>
> $\mathrm{E}(X) = np, \quad \mathrm{sd}(X) = \sqrt{np(1-p)}$

EXERCISES FOR SECTION 5.4.2

1. Compute sd(X) for Exercises for Section 5.4.1, problem 1.
2. Compute sd(X) for Example 5.4.2.
3. Compute sd(X) for Exercises for Section 5.4.1, problem 2.

5.4.3 The Effect of Rescaling Random Variables

Suppose that the length of a cell-phone toll call has a distribution with mean E(X) = 1.5 minutes and a standard deviation of sd(X) = 1.1 minutes. If the telephone company charges a fixed connection fee of 50 cents and then 40 cents per minute, what is the mean and standard deviation of the distribution of charges? Let Y be the charge. We have

$$Cost = 40 \times Time + 50 \qquad \text{that is, } Y = 40X + 50$$

which is an example of

$$Y = aX + b$$

where X is a random variable and a and b are constants (fixed numbers).

Before we answer our question, let us some look at some data. Suppose that three calls are made lasting 1.6, 1.2, and 2.0 minutes, respectively. That is, we have a sample of size 3, with $x_1 = 1.6$, $x_2 = 1.2$, and $x_3 = 2.0$. We then find that the sample mean and standard deviation are $\overline{x} = 1.6$ and $s_X = 0.4$, respectively. In terms of the y-values, the corresponding values are $y_1 = 114$, $y_2 = 98$, and $y_3 = 130$, with $\overline{y} = 114$ and $s_Y = 16$. We see that $\overline{y} = 40 \times \overline{x} + 50$ and $s_Y = 40 \times s_X$. This is not unexpected, as we are simply multiplying each x-value by 40 and adding 50; we would expect to do the same to the sample mean. Adding a fixed amount of 50 to every observation, however, would have no effect on the variability. But multiplying x-values by 40 will increase the distances between the values by 40 times, so the variability of the y-values would be expected to be 40 times the variability of the x-values. Means and standard deviations of distributions behave in exactly the same way, because they are either population or limiting versions of the same quantities. We therefore have

$$\mathrm{E}(aX + b) = a\,\mathrm{E}(X) + b \quad \text{and} \quad \mathrm{sd}(aX + b) = |a|\,\mathrm{sd}(X)$$

We use the *absolute value*[10] of a, or $|a|$, with the standard deviation because we want our measure of variability, sd, to be always positive and in some situations a may be

[10] To get the *absolute value* of a number, we take the size of the number and make the result positive, for example, $|-2| = |2| = 2$, $|-6.3| = |6.3| = 6.3$.

negative. Application of these results to the preceding example is straightforward. Intuitively, the expected cost is 40 cents (per minute) times the expected time taken (in minutes) plus the fixed cost of 50 cents. Thus

$$E(40X + 50) = 40\,E(X) + 50 = 40 \times 1.5 + 50 = 110 \text{ cents}$$

and

$$sd(40X + 50) = 40\,sd(X) = 40 \times 1.1 = 44 \text{ cents}$$

To see how the formulas work with a negative multiplier, we note that $E(-7X + 3) = -7\,E(X) + 3$, but $sd(-7X + 3) = |-7|\,sd(X) = 7\,sd(X)$.

It will also be helpful to have some facility with using these results algebraically, so we shall do one theoretical example. This example will be important in later chapters.

✱ Example 5.4.6 *Count and Proportion for the Binomial Distribution*

If Y is the number of heads in n tosses of a biased coin, then we know that $Y \sim$ Binomial(n, p). The *proportion* of heads is simply Y/n, which we denote by $\widehat{P}$. What is $E(\widehat{P})$ and $sd(\widehat{P})$?

$$E(\widehat{P}) = E\left(\frac{1}{n}Y\right) = \frac{1}{n}E(Y) = \frac{1}{n}np = p$$

$$sd(\widehat{P}) = sd\left(\frac{1}{n}Y\right) = \frac{1}{n}sd(Y) = \frac{1}{n}\sqrt{np(1-p)} = \sqrt{\frac{p(1-p)}{n}}$$

EXERCISES FOR SECTION 5.4.3

Suppose $E(X) = 3$ and $sd(X) = 2$. Compute $E(Y)$ and $sd(Y)$, where

(a) $Y = 2X$ (b) $Y = 4 + X$ (c) $Y = X + 4$
(d) $Y = 3X + 2$ (e) $Y = 4 + 5X$ (f) $Y = -5X$
(g) $Y = -5X + 4$ (h) $Y = 4 - 5X$ (i) $Y = -7X - 9$

QUIZ ON SECTION 5.4

1. What does the expected value of X tell you about? (Section 5.4.1)

2. Why is the expected value also called the population mean? (Section 5.4.1)

3. What is the relationship between the population mean and the bar graph of the probability function? (Section 5.4.1)

4. What are the mean and standard deviation of the Binomial distribution? (Sections 5.4.1 and 5.4.2)

5. Why is $sd(X + 10) = sd(X)$? Why is $sd(2X) = 2\,sd(X)$? (Section 5.4.3)

5.5 SUMMARY

Random variable

A type of measurement made on the outcome of a random experiment.

Probability function

$\text{pr}(X = x)$ for every value X can take, abbreviated $\text{pr}(x)$.

Expected value for a random variable X, denoted $E(X)$.

- Also called the ***population mean*** and denoted μ_X (abbreviated to μ).
- Is a measure of the long-run average of X-values in many repetitions of the experiment.
- Formula:

$$\mu_X = E(X) = \sum x\,\text{pr}(x) \qquad \text{(for a discrete random variable } X\text{)}$$

Standard deviation for a random variable X, denoted $\text{sd}(X)$.

- Also called the ***population standard deviation*** and denoted σ_X (abbreviated σ).
- Is a measure of the variability of X-values.
- Formula:

$$\sigma_X = \text{sd}(X) = \sqrt{E[(X - \mu)^2]}$$

$$\left[\text{for a discrete random variable } X, \quad E[(X - \mu)^2] = \sum (x - \mu)^2\,\text{pr}(x)\right]$$

Linear functions for any constants a and b,

$$E(aX + b) = a\,E(X) + b \quad \text{and} \quad \text{sd}(aX + b) = |a|\,\text{sd}(X)$$

Sampling from a finite population

The urn model is a physical model for situations in which we sample n individuals at random from a finite population and count X, the number of individuals with a characteristic of interest.

When $n/N < 0.1$, the distribution of X is approximately Binomial(n, p), where p is the population proportion with the characteristic of interest.

Observing a random process

The biased-coin-tossing model is a physical model for situations which can be characterized as a series of *trials* where:

1. each trial has only two outcomes: *success* and *failure;*
2. $p = \text{pr(success)}$ is the same for every trial; and
3. trials are independent.

The distribution of the number of successes in n trials (or the number of heads in n tosses) is Binomial(n, p). The Binomial distribution has

$$E(X) = \mu_X = np \qquad sd(X) = \sigma_X = \sqrt{np(1-p)}$$

REVIEW EXERCISES 5

1. The main aim of this question is simply to give practice in identifying an appropriate model and distribution from a description of a sampling situation. You will look for situations that look like coin tossing (Binomial for a fixed number of tosses) and situations that look like urn sampling. Some situations do not fit into either of these categories.

   For each of the following situations, write down the sampling model and identify the values of any parameters. In urn-sampling situations, consider whether a Binomial approximation can be used. If neither of the models is appropriate, write "not applicable."

   (a) In the long run, 80% of Ace light bulbs last for 1000 hours of continuous operation. You need to have 20 lights in constant operation in your attic for a small business enterprise, so you buy a batch of 20 Ace bulbs. Let X_1 be the number of these bulbs that have to be replaced by the time 1000 hours are up.

   (b) In (a), because of the continual need for replacement bulbs, you buy a batch of 1000 cheap bulbs. Of these, 100 have disconnected filaments. You start off by using 20 bulbs (which we assume are randomly chosen). Let X_2 be the number with disconnected filaments.

   (c) Suppose that telephone calls come into the university switchboard randomly at a rate of 100 per hour. Let X_3 be the number of calls in a one-hour period.

   (d) In (c), 60% of the callers know the extension number of the person they wish to call. Suppose 120 calls are received in a given hour. Let X_4 be the number of callers who know the extension number.

   (e) It so happened that of the 120 calls in (d), 70 callers knew the extension number and 50 did not. Assume calls go randomly to telephone operators. Suppose telephone operator A took 10 calls. Of the calls taken by operator A, let X_5 be the number made by callers who knew the extension number.

   (f) Suppose 20 patients, of whom 9 had flu, came to a doctor's office on a particular morning. The order of arrival was random as far as having flu was concerned. The doctor had time to see only 15 patients before lunch. Let X_6 be the number of flu patients seen before lunch.

   (g) Let X_7 be the number of 1s in 12 rolls of a fair die.

   (h) Let X_8 be the number of rolls if you keep rolling until you get the first 6.

   (i) Let X_9 be the number of double 1s on 12 rolls of a pair of fair dice.

   (j) A Scrabble set consists of 98 tiles, of which 44 are vowels. The game begins by selecting 7 tiles at random. Let X_{10} be the number of vowels selected.

2. Do you think that the Binomial distribution is appropriate for the following random variables? Give reasons for your answer.

   (a) A student takes a multiple-choice statistics test in which there are a 100 questions, each with 5 alternatives, of which only one is the correct answer. The student does not know much and guesses quite a few of the questions. Let X_1 be the number of questions he gets right.

   (b) The student takes the test home and gives it to his little brother, who does not know anything about statistics. His little brother gets X_2 right.

(c) A history professor proofreads a history book and counts X_3, the number of pages that are error free.

(d) A mathematics professor proofreads an advanced mathematics book and counts X_4, the number of pages that are error free.

(e) A couple continue to have children until they get a boy. The number of children they have is X_5.

(f) The government of a country would like see the monthly inflation rate kept below 0.2%. Let X_6 be the number of months in a 12-month period when this occurs.

3. The following simple model is sometimes used as a first approximation for studying an animal population. A population of 420 small animals of a particular species is scattered over an area 2 km by 2 km. This area is divided up into 400 square plots, each of size 100 m by 100 m. Twenty of these plots are selected at random. Using a team of observers, the total number of animals seen at a certain time on these 20 plots is recorded. It is assumed that the animals are independent of one another (i.e., no socializing!) and each animal moves randomly over the whole population area.

   (a) What is the probability that a given animal is seen on the sampled area?

   (b) Let X be the number of animals seen on these plots. Explain why we can use the Binomial model for X. Give the parameters of this distribution.

   (c) Let W be the number of animals found on a single plot. State the distribution and parameters of W.

   (d) Give two reasons why modeling the numbers of animals found on the sampled plots with a Binomial distribution may not be appropriate (i.e., which of our assumptions are not likely to be valid?).

4. A lake has 1000 fish in it. A scientist catches 50 of the fish and tags them. The fish are then released. After they have moved around for a while, the scientist catches 10 fish. Let X be the number of tagged fish in her sample.

   (a) What assumptions would need to be satisfied if the urn-sampling model is to apply?

   (b) What approximate distribution could you use for X?

   (c) Using this distribution, find the probability of obtaining at least one tagged fish in the sample.

5. Over 92% of the world's trade is carried by sea. There are nearly 80,000 merchant ships of over 500 metric tons operating around the world. Each year more than 120 of these ships are lost at sea. Assume that there are exactly 80,000 merchants ships at the start of the year and that 120 of these are lost during the year.

   (a) A large shipping corporation operates a fleet of 160 merchant ships. Let L be the number of ships the company will lose in a year. What distribution can you use for L? Identify any parameters.

   (b) What is the number of ships that the corporation expects to lose in a year?

   (c) What assumptions were made in (a)? Do you think that these assumptions are reasonable here?

6. According to an 1989 Toronto *Globe and Mail* report, a study on 3,433 women conducted by the Alan Guttmacher Institute in the United States found that approximately 11% of women using the cervical cap method of contraception can expect to become pregnant within the first year.[11] Suppose a family planning clinic fits 30 women with cervical caps in early January.

   (a) What is the probability that none will become pregnant before the end of the year?

   (b) What is the probability that no more than 2 will become pregnant before the end of the year?

   (c) What is the probability that none will be pregnant by the end of two years?

---

[11] Compared with 16% for the diaphragm, 14% whose partners use condoms, and 6% on the contraceptive pill.

7. Approximately 18% of lots of condoms tested by the U.S. Food and Drug Administration between 1987 and 1989 failed leakage tests.[12] It's hard to know if this applies to individual condoms, but let us suppose it does. In a packet of 12 condoms,
   (a) what is the distribution of the number that would fail the test?
   (b) what is the probability none would fail the test?

8. Suppose that experience has shown that only $\frac{1}{3}$ of all patients having a particular disease will recover if given the standard treatment. A new drug is to be tested on a group of 12 volunteers. If health regulations require that at least 7 of these patients should recover before the new drug can be licensed, what is the probability the drug will be discredited even if it increases the individual recovery rate to $\frac{1}{2}$?

9. In a nuclear reactor the fission process is controlled by inserting into the radioactive core a number of special rods whose purpose is to absorb the neutrons emitted by the critical mass. The effect is to slow down the nuclear chain reaction. When functioning properly, these rods serve as the first-line defense against a disastrous core meltdown.

   Suppose that a particular reactor has 10 of these control rods (in real life there would probably be more than 100), each operating independently and each having a 0.80 probability of being properly inserted in the event of an incident. Furthermore, suppose that a meltdown will be prevented if at least half the rods perform satisfactorily. What is the probability that the system will fail when needed?

10. Samuel Pepys, whose diaries so vividly chronicle life in seventeenth century England, was a friend of Sir Isaac Newton. His interest in gambling prompted him to ask Newton whether one is more likely to get
    (a) at least one 6 when six dice are rolled,
    (b) at least two 6s when 12 dice are rolled, or
    (c) at least three 6s when 18 dice are rolled?

    The pair exchanged several letters before Newton was able to convince Pepys that (a) was most probable. Compute these three probabilities.

11. Suppose X has a probability function given by the following table but *one* of the given probabilities is in error:

X	−3	0	1	3	8
pr(X = 3)	0.23	−0.39	0.18	0.17	0.13

    (a) Which one of the probabilities is in error? Give the correct value and use it in answering the following questions.
    (b) What is the probability that X is at least 1?
    (c) What is the probability that X is no more than 0?
    (d) Calculate the expected value and standard deviation of X.

12. A standard die has its faces painted different colors. Faces 3, 4, and 6 are red; faces 2 and 5 are black; and face 1 is white.
    (a) Find the probability that when the die is rolled, a black or even-numbered face shows uppermost.

    A game is played by rolling the die once. If any of the red faces show uppermost, the player wins the dollar amount showing, while if a black face shows uppermost, the player loses twice the amount showing. If a white face shows uppermost, he wins or loses nothing.
    (b) Find the probability function of the player's winnings X.
    (c) Find the expected amount won. Would you play this game?

[12]The failure rate in Canada was even worse, with 40% of lots failing Health and Welfare Dept. tests over this period!

TABLE 1 Almond Delight Prizes

"Free utilities"	$ value	Number of such vouchers distributed
Light home for week	1.61	246,130
Wash and dry clothes for month	7.85	4,500
Hot water for 2 months	25.90	900
Water lawn all summer	34.29	360
Air condition home for 2 months	224.13	65
Free utilities for 6 months	558.00	45

13. Almond Delight is a breakfast cereal sold in the United States. In 1988 the manufacturers ran a promotion with the come-on to the purchaser, "Instantly win up to six months of FREE UTILITIES" up to a value of $558. (Utilities are things such as electricity, gas, and water). A colleague bought a box of Almond Delight for $1.84 and sent us the box. The company had distributed 252,000 checks or vouchers among 3.78 million boxes of Almond Delight, with at most one going into any box as shown in Table 1. If a check was found in a box, the purchaser could redeem the prize by sending in a self-addressed, stamped envelope.

Let X be the value of vouchers found in the packet of Almond Delight bought.

(a) Write down the probability function of X in table form.

(b) Calculate the expected value of X.

(c) Calculate the standard deviation of X.

(d) Would you buy Almond Delight if you found another brand that you liked equally well at $1.60? Why or why not? (Allow 40 cents for the cost of envelopes and postage.)

14. An instant lottery played in Kansas called "Lucky Numbers" costs $1.00 per ticket. The prizes and chances of winning them are:

Prize	$0	Free ticket	$3	$7	$11	$21	$2100
Prob	by subtraction	$\frac{1}{10}$	$\frac{1}{30}$	$\frac{1}{100}$	$\frac{1}{100}$	$\frac{1}{150}$	$\frac{1}{300{,}000}$

We want to find the expected return on $1.00 spent on playing this game. Free tickets entitle the player to play again, so let us decide that every time we get a free ticket we shall play again until we get a proper outcome. We get the relevant probabilities, which ignore free tickets, by dividing each of the probabilities above by 0.9. (Why?) Now work out the expected return on a $1.00 ticket. What is the standard deviation?

15. Suppose that, to date, rivets that are outside the specification limits have been produced by a stable process at a consistent rate of 1 in every 100. The following (artificial) inspection process is operated. Periodically, a sample of 8 rivets is taken, and if 2 or more of the 8 are defective, the production line is halted and adjusted. What is the probability that

(a) production will be halted on the strength of the sample if the process has not changed?

(b) production will not be halted even though the process is now producing 2% of its rivets outside the specification limits?

***16.** In the *People v. Collins* case described in Case Study 4.7.1 (Section 4.7.3), we saw how the defendants were convicted largely on the basis of a probabilistic argument, which showed that the chances that a randomly selected couple fit the description of the couple involved were overwhelmingly small (1 in 12 million). At the appeal at the Supreme

Court of California, the defense attacked the prosecution's case by attacking the independence assumption. Some of the characteristics listed clearly tend to go together such as mustaches and beards and black men and interracial couples. Their most devastating argument, however, involved a conditional probability. The probability of finding a couple to fit the description was so small that it probably never entered the jurors' heads that there might be another couple fitting the description. The defense calculated the conditional probability that there were two or more couples fitting the description, given that at least one such couple existed. This probability was quite large, even using the prosecution's 1 in 12 million figure. Reasonable doubt about whether the convicted couple were guilty had clearly been established. We shall calculate some of these probabilities now.

Suppose there were n couples in Los Angeles (or maybe of the whole of California) and that the chances of any given couple matching the description is 1 in 12 million. Assuming independence of couples,

(a) what is the distribution of X, the number of couples who fit the description? Justify your answer.

(b) show that $\text{pr}(X \geq 2 \mid X \geq 1) = \dfrac{1 - \text{pr}(0) - \text{pr}(1)}{1 - \text{pr}(0)}$.

(c) evaluate the probability in (b) when $n = 1$ million, 4 million, and 10 million.

17. A *Time* article (13 March 1989, Medicine section) sounded warning bells about the fast-growing in vitro fertilization industry, which caters to infertile couples who are desperate to have children. At the time, U.S. in vitro programs charged in the vicinity of $7,000 per attempt. There is a lot of variability in the success rates both between clinics and within a clinic over time, but we will use an average rate of 1 success in every 10 attempts.

Suppose that 4 attempts ($28,000) is the maximum a couple feels they are prepared to pay and that they will try until they are successful up to that maximum. (See Case Study 5.2.1.)

(a) Write a probability function for the number of attempts made.

(b) Compute the expected number of attempts made. Also compute the standard deviation of the number of attempts made.

(c) What is the expected cost when embarking on this program?

(d) What is the probability of still being childless after paying $28,000?

The calculations you have made assume that the probability of success is always the same at 10% for every attempt. This will not be true. Quoted success rates will be averaged over both couples and attempts. Suppose the population of people attending the clinics is made up of three groups. Imagine that 30% is composed of those who will get pregnant comparatively easily, say pr(*success*) = 0.2 per attempt; 30% are average with pr(*success*) = 0.1 per attempt; and the third group of 40% have real difficulty and have pr(*success*) = 0.01.[13] After a couple are successful, they drop out of the program. Now start with a large number, say 100,000 people and perform the following calculations.

(e) Calculate the number in each group you expect to conceive on the first attempt and hence the number in each group who make a second attempt.

Note: It is possible to find the probabilities of conceiving on the first try, on the second given failure at the first try, and so on, using the methods of Chapter 4. You might even like to try it. The method you were led through here is much simpler, though informal.

(f) Find the number in each group getting pregnant on the second attempt and hence the number in each group who make a third attempt. Repeat for the third and fourth attempts.

---

[13] In practice there will be a continuum of difficulty to conceive, not several well-defined groups.

(g) Now find the proportions *of all couples who make the attempt* who conceive at each of the first, second, third, and fourth attempts. Observe how these proportions decrease.

Note: The decrease is largely due to the fact that, as the number of attempts increases, the "difficult" group makes up a bigger and bigger proportion of those making the attempt.

18. Suppose you are interested in the proportion p of defective items being produced by a manufacturing process. You suspect that p is fairly small. You inspect 50 items and find no defectives. Using 0 as an estimate of p is not particularly useful. It would be more useful if we could put an upper limit on p, for example, "We are fairly sure p is no bigger than 0.07." (Note: The actual number, 0.07 here, is irrelevant to the question.)

(a) What is the distribution of X, the number of defectives in a sample of 50?

(b) What assumptions are implicit in your answer to (a)? Write some circumstances in which they would not be valid.

(c) Plot $\text{pr}(X = 0)$ versus p for $p = 0, 0.02, 0.04, 0.06, 0.08$, and sketch a smooth curve through the points. (This should show you the shape of the curve.)

(d) Write the expression for $\text{pr}(X = 0)$ in terms of p. We want to find the value of p for which $\text{pr}(X = 0) = 0.1$. Solve the equation for p if you can. Otherwise read it off your graph in (c).

Note: Your answer to (d) is a reasonable upper limit for p since any larger value is unlikely to give 0 defectives (less than 1 chance in 10).

19. A 8 March 1989 report in the *Kitchener-Waterloo Record* (p. B11) discussed a successful method of using DNA probes for sexing embryo calves developed by the Salk Institute Biotechnology and Industrial Associates in California. The method correctly identified the sex of all but 1 of 91 calves. Imitate the previous problem to calculate a likely upper limit for the failure rate p of the method by calculating the probability of getting at most 1 failure for various values of p and using an approximate graphical solution [compare with problem 18(d)].

20. Everyone seems to love to hate their postal service and New Zealand is no exception. When N.Z. Post set up a new mail system, letters were to go by either "Standard Post" at a cost of 40 cents or "Fast Post" at a cost of 80 cents. Fast Post was promoted vigorously. According to N.Z. Post publicity, the chances of a Fast Post letter being delivered to the addressee the following day were "close to 100%." Very soon after the introduction of the new system, the *New Zealand Herald* conducted an experiment to check on the claims. Newspaper staff mailed 15 letters all over the country from the Otara Post Office in Auckland. According to the resulting story, only 11 had reached their destination by the next day. This doesn't look very "close to 100%."

(a) On the grounds that we would have been even more disturbed by fewer than 11 letters arriving, determine the probability of 11 or fewer letters being delivered in a day if each had a 95% chance of doing so.

(b) What assumptions have you made?

(c) From your answer to (a), do you believe that N.Z. Post's claims were valid, or was the new system suffering from what might be kindly referred to as "teething problems"?

Note: You need not consider this in your answer, but there are different ways we might be inclined to interpret the 95%. Should it mean 95% of letters? This should be relatively easy for N.Z. Post to arrange if most of the traffic flows between main centers with single mail connections. Or should it mean 95% of destinations? This would give much more confidence to customers sending a letter somewhere other than a main center. The *Herald* sent its letters to 15 different cities and towns.

21. A chromosome mutation linked with color blindness occurs in one in every 10,000 births on average. Approximately 20,000 babies will be born in Auckland this year. What is the probability that

(a) none will have the mutation?

(b) at least one will have the mutation?

(c) no more than three will have the mutation?

22. Brain cancer is a rare disease. In any year there are about 3.1 cases per 100,000 of population.[14] Suppose a small medical insurance company has 150,000 people on its books. How many claims stemming from brain cancer should the company expect in any year? What is the probability of getting more than 6 claims related to brain cancer in a year?

23. At a company's Christmas social function, attended by the sister of a colleague of ours, 10 bottles of champagne were raffled off to those present. There were about 50 people at the function. The 50 names were placed on cards in a box, the box was stirred, and the name of the winner was drawn. The winner's name was returned to the box for the next draw. The sister won 3 of the 10 bottles of champagne. By the third time she was getting a little embarrassed. She definitely doubted the fairness of the draw. Even worse, she doesn't drink, and so she didn't even appreciate the champagne!

Assuming a random draw,

(a) What is the distribution of X, the number of bottles won by a given person? Justify your answer and give the value(s) of the parameter(s).

(b) What is

(i) pr($X = 0$) and pr($X \leq 2$)?

(ii) the probability of winning 3 or more bottles?

(c) Do you think the cards in the box were well stirred?

But perhaps it isn't as bad as all that. Sure, the chances of a particular person winning so many bottles are tiny, but after all, there were 50 people in the room. Maybe the chances of it happening to somebody in the room are quite reasonable.

(d) The events E_i = "*i*th person wins 3 or more bottles" are not independent. Make an observation that establishes this.

(e) Even though the events are not independent, since pr(E_i) is so small the independence assumption probably won't hurt us too much. Assuming independence, what is the probability that nobody wins 3 or more bottles? What is the probability that at least one person wins 3 or more bottles?

(f) Does this change your opinion in (c)?

24. (Computer simulation exercise) Suppose that every day for a whole week the weather forecast quotes a 40% chance of rain. We can generate the outcome on a particular date using a Binomial($n = 1, p = 0.4$) random number. A 1 corresponds to a day with rain and a 0 to a day with no rain. Seven such observations give you the outcomes for a week, assuming it rains or not independently each day with probability 0.4.

Note: Alternatively, you could use a sequence of digits from the random number table in Appendix A1. Translate any digit in {0, 1, 2, 3} to a wet day and any other digit to a dry day.

(a) Generate 40 weeks of data. How many of your simulated weeks had more dry days than wet days? How many had more wet days than dry days? What is the longest run of wet days that you found in a single week? How about dry days? What proportion of the 280 days you generated were wet?

(b) For each of the 40 weeks you generated in (a), write down the number of wet days. What you are seeing are 40 random observations from a Binomial($n = 7, p = 0.4$) distribution. Why? Now use your computer package to generate 40 Binomial($n = 7, p = 0.4$) random numbers.

(c) Simulate the number of wet days in each of 200 weeks by generating 200 Binomial($n = 7, p = 0.4$) random numbers. Obtain a *relative*-frequency table and a bar graph (bar chart) of these numbers of wet days. Compare your relative frequencies with the

[14] U.S. figures from *Time* (24 December 1990, p. 41).

Binomial($n = 7, p = 0.4$) probabilities and your bar graph with a bar graph of the Binomial($n = 7, p = 0.4$) probabilities. What do you see?

(d) By inspecting your 40 weeks in (a), construct a frequency table of the longest run of dry days within a week. (For each entry x in the table, $x = 0, 1, 2, \ldots$, the corresponding frequency f_x is the number of weeks for which the longest run of dry days is x.) Add a column of *relative* frequencies.

(e) Your relative-frequency table can be thought of as an estimate of the probability distribution of the longest run of dry days in a week. Use the table to estimate the expected value of the longest run of dry days.

CHAPTER 6

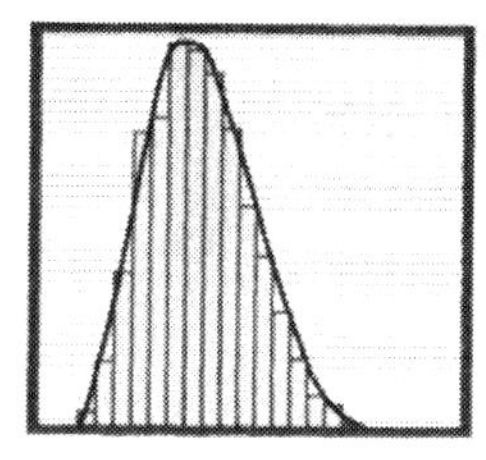

Continuous Random Variables and the Normal Distribution

Chapter Overview

This chapter introduces the density function as a means of summarizing the pattern of variation in a continuous random variable. It shows how probabilities are represented by areas under the density curve and the mean is the center of gravity of the curve. In Section 6.2 the Normal distribution is introduced and we work with probabilities as areas and with percentiles or quantiles using the Normal distribution. The advantages of measuring distance in terms of numbers of standard deviations are discussed via z-scores in Section 6.3. Section 6.4 explores the transmission of variability via the distributions of sums and differences of random quantities. Throughout the chapter, primary emphasis is on probabilities and quantiles obtained from computer programs, although the use of tables is also described in Section 6.3.

6.1 INTRODUCTION

6.1.1 Proportions as Areas: The Standardized Histogram

✖ EXAMPLE 6.1.1 *Dietary Carbohydrate in the Workforce*

We have used the workforce database before. It is information collected by Drs. Metcalf and Scragg in a study of 5929 people working in a wide variety of working environments, mainly from Auckland, N.Z. Recall that a variable is continuous if there are no gaps between the values it can take. Many variables collected in this study are

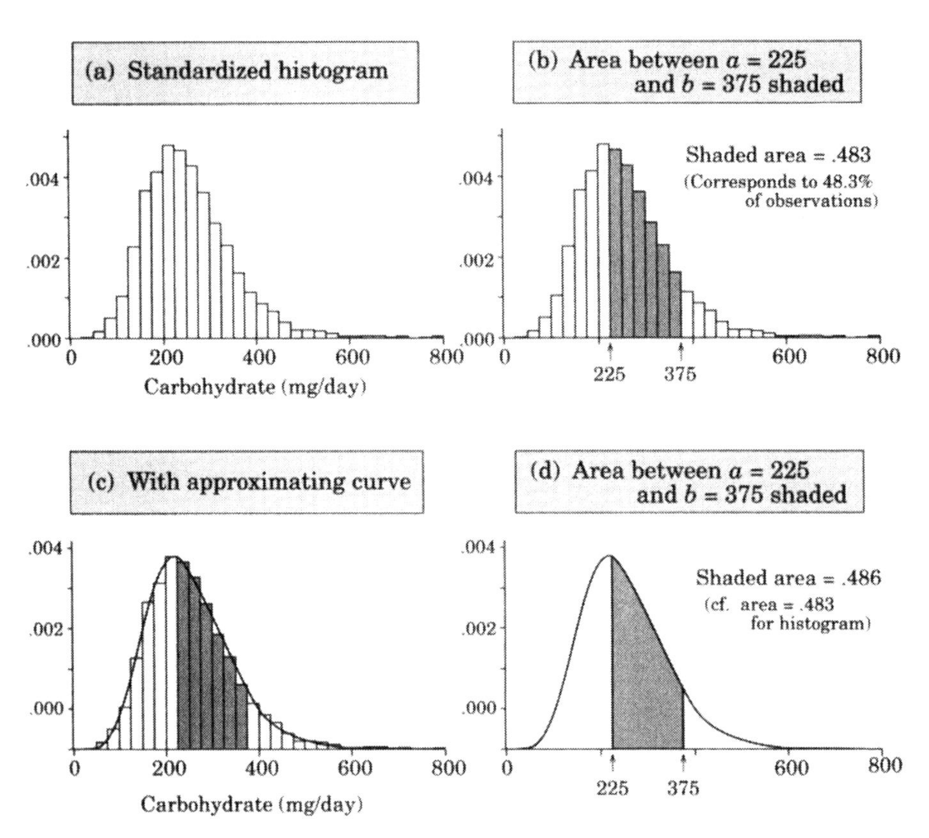

FIGURE 6.1.1 Dietary intake of carbohydrate (mg/day) for 5929 people in the workforce database.

continuous, for example physical dimensions such as heights, weights, hip and waist measurements, blood pressure measurements, concentrations of various chemicals in the blood. The variable we are using here is average daily intake of carbohydrate in the diet. A histogram of the data is given in Fig. 6.1.1(a). It is unimodal, with the modal interval extending from 200 mg/day to 225 mg/day and skewed toward large values. The histogram shows a huge amount of variability in dietary intake of carbohydrate from person to person, with the biggest consumers of carbohydrate ingesting more than 10 times as much carbohydrate as some of the smaller consumers.

Usually, as in Section 2.3.3, histograms are drawn with the height of the rectangle for the jth class interval being either the frequency (count) f_j or the relative frequency f_j/n of observations falling into that interval. If we use the relative frequency, the ***height*** of the rectangle tells us what proportion of observations fall into the jth interval, and this can be read off the vertical scale of the plot. For a ***standardized histogram,*** we adjust the height of the rectangle so that its *area* is equal to the relative frequency. This is accomplished by changing the height to $f_j/(nw)$, where w is the width of the interval. [The area (width $\times$ height) is then f_j/n.] So now it is the *area* of the jth rectangle that tells us what proportion of data points lie in the jth class interval.

In Fig. 6.1.1(a) the proportion of people in the database whose daily carbohydrate intake falls within any class interval is the area of the corresponding rectangle. We can get the proportion of people whose carbohydrate intake falls between the limits a = 225 mg/day and b = 375 mg/day by adding the areas of all the rectangles between these limits. Thus the proportion of people with carbohydrate intakes between 225 mg/day and 375 mg/day is the shaded area in the histogram in Fig. 6.1.1(b) between these limits, which turns out to be 0.483 or 48.3% of people in the database. This correspondence between area and proportions (later probabilities) is the foundation on which the theory of probability for continuous variables is built. Additionally, we note that the total area under the standardized histogram is $\sum\left(\frac{f_j}{nw} \times w\right) = \frac{\sum f_j}{n} = 1$. We now summarize what we have learned about standardized histograms.

For a ***standardized histogram:***

- The vertical scale is Relative frequency/interval width
- Total area under histogram = **1**
- ***The proportion*** of the data between a and b is the ***area*** under the histogram between a and b

In Fig. 6.1.1(c) we have superimposed an approximating smooth curve on the standardized histogram. Figure 6.1.1(d) contains just the approximating smooth curve. We have also shaded the area under the curve between the same limits as the histogram (225 to 375) and calculated the shaded area. It turned out to be 0.486, which is very close to the area of 0.483 under the histogram. Recall that areas under the histogram correspond to proportions of people in the database. It is obvious from the visual comparison with Fig. 6.1.1(c) that areas under the smooth curve between pairs of limits are in quite close agreement with the corresponding areas under the histogram and thus with proportions of people in the database.

6.1.2 Probabilities and Density Curves

When we look at enormous sets of data (much larger than the workforce database) on a continuous variable, histograms with very small class intervals tend to be well approximated by smooth curves. It is not a large step from this to imagining a model for a random process generating data on a continuous variable X in which the probability distribution is described by a smooth curve. The way in which the curve describes probabilities is that the probability that a random observation falls between limits a and b is given by *the area under the curve* between those limits, as in Fig. 6.1.1(d). We call the smooth curve describing the behavior of a continuous random variable X a ***density curve,*** or more formally, the graph of a *probability density function.*

Random observations may arise by sampling a member of a large population at random. They also arise from observing the output of a process that produces items that vary from item to item in random way. For example, the volumes of wine in 750-ml bottles produced by a bottling plant inevitably vary somewhat in an unpredictable way from bottle to bottle. Before going on to calculate probabilities as areas, we shall spend a little time looking at the behavior of random data. What would data look like if it was produced by a random process for which the probabilities are described by the smooth curve in Fig. 6.1.1(c and d)? We have investigated this using computer-generated random numbers. (You may find it helpful to think of each random observation we obtain as the dietary carbohydrate intake of a randomly sampled individual from some huge population, e.g., the population of the United States.) The first 15 random X-values we obtained were (to the nearest integer):

194 348 243 106 285 233 322 299 160 383 185 391 455 155 210

Figure 6.1.2(a) presents dot plots of six different sets of 15 random observations from this process. Each set makes quite a different pattern, although all observations we see fall between 100 and 600. The remainder of the figure gives histograms of much larger sets of data generated from the process. The density curve is superimposed each time to remind us of the shape of the probability distribution that is producing this data. Histograms in the left-hand column use 30 class intervals to cover the range from 0 to 650 mg/day. Those in the right-hand column use 70 class intervals. The top four histograms relate to samples of size 300. Histograms in the same row are plots of the same set of data. We note that our histograms for data sets of size 300 do not look very much like the density curve. We looked at large numbers of histograms of sets of 300 observations; some looked roughly like the density curve, but many did not. We note also that the shapes of histograms from different sets of 300 observations from this process can look quite different from one another. As the sample sizes become larger, however, the histograms begin to resemble the density curve more and more closely (shown) and they do so more and more reliably (not shown). We need much larger sample sizes before the histograms reliably mimic the basic shape of the density curve when using narrow class intervals than we do using wider class intervals. In turns out that the smooth density curve is reproduced as the intervals become very narrow (tend to zero) and the number of observations becomes enormous (tends to infinity).

Taking Stock

Let us summarize the lessons we have learned so far in this section:

- The distribution generating data on a continuous variable is described by a smooth curve called a *density curve.*
- Probabilities are represented by areas under the curve.
- Total area under the curve = 1.
- It takes very large sample sizes before the regularity of the underlying curve is reliably reflected in histograms of data from the distribution.

For a continuous X the ***probability*** a random observation falls between a and b equals the ***area*** under the density curve between a and b

6.1.3 Some Comparisons with the Discrete Case

Just as for the distribution of a discrete random variable, the distribution of a continuous random variable has a mean, which we refer to as the ***population mean*** denoted by E(X) or μ_X, and a standard deviation which we refer to as the ***population***

(a) Dot plots of 6 sets of 15 random observations

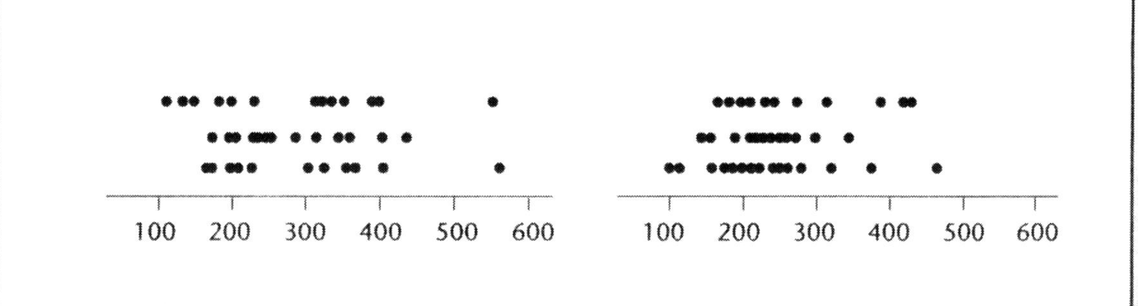

(b) Histograms with density curve superimposed

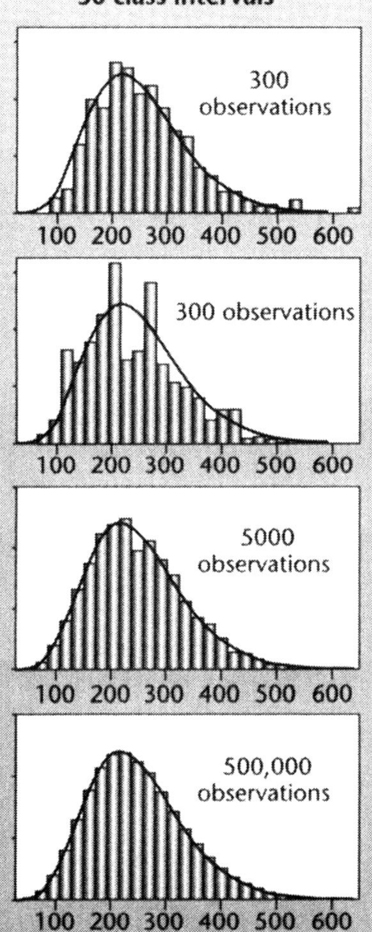

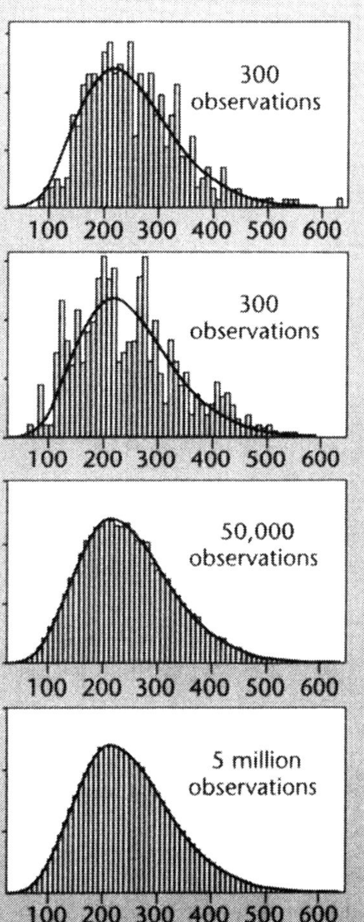

FIGURE 6.1.2 Behavior of data randomly generated from the smooth density curve in Fig. 6.1.1(d).

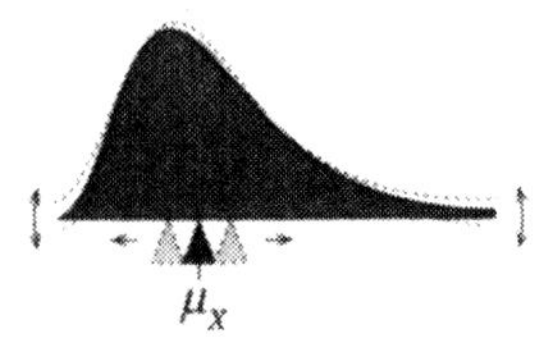

FIGURE 6.1.3 The mean is where the density curve balances.

standard deviation denoted by sd(X) or[1] σ_X. Recall that in the discrete case, a line graph of the probability function balances at the position of μ_X (see Fig. 5.4.2). In the continuous case μ_X is the center of gravity of the density curve. In other words, if we could make a cardboard cutout of the density curve, μ_X is the point on the x-axis where the cutout would balance, as in Fig. 6.1.3. As for the discrete case, when we multiply a random quantity X by a constant a and/or add a constant b:

$$\mathrm{E}(aX + b) = a\,\mathrm{E}(X) + b \qquad \text{and} \qquad \mathrm{sd}(aX + b) = |a|\,\mathrm{sd}(X)$$

Let us return to thinking in terms of the random process investigated in the previous subsection. One important way in which the continuous random variables differ from the discrete case is as follows. If we need to determine something such as the probability that the carbohydrate intake X falls between 150 and 200, then we find that $\mathrm{pr}(150 \le X \le 200) = \mathrm{pr}(150 \le X < 200) = \mathrm{pr}(150 < X \le 200) = \mathrm{pr}(150 < X < 200)$. This property only holds for continuous random variables and not for discrete random variables, where we have to be very careful about how we treat interval endpoints.

In calculations involving a continuous random variable, *we do not have to worry about whether interval endpoints are included or excluded.*

This statement is true because, for a continuous random variable X, the probability that X takes any particular value exactly is 0. For example, the probability that a carbohydrate intake is exactly 200, that is, $\mathrm{pr}(X = 200)$, is 0. Many people find this idea rather disconcerting. Every time we generate a random X from the preceeding process, it falls somewhere between 0 and about 650. Lots of observations fall in the vicinity of 200. How can $\mathrm{pr}(X = 200) = 0$? The probability of a carbohydrate intake being 200 to the nearest whole unit, namely $\mathrm{pr}(199.5 \le X \le 200.5)$, is given by the area between these endpoints and turns out to be about 0.0046. If we want to be within 0.1 of 200, the interval becomes one tenth as wide and the area becomes roughly one tenth as large giving $\mathrm{pr}(199.05 \le X \le 200.05) \approx 0.00046$. Every extra decimal place of accuracy we demand reduces the interval width, and hence the probability, by another factor of 10, for example, $\mathrm{pr}(199.00005 \le X \le 200.00005) \approx 0.00000046$ and so on. The more decimal places we demand, the closer to 0 the probability gets. This is not a problem in practice, however. Because we can never achieve infinitely many

[1] If there is no ambiguity as to which random variable we are talking about, we shall sometimes drop the subscript from μ_X and σ_X and simply write μ and σ.

decimal places of accuracy in a measurement, the *relevant* probability never reduces to 0. For example, it would be optimistic to think that we could measure a person's height accurate even to the nearest millimeter, so a measured (observed) height for a person of 1.8 m means, at best, an exact height somewhere between 1.7995 m and 1.8005 m.

QUIZ ON SECTION 6.1

1. How does a *standardized* histogram differ from a relative-frequency histogram? What graphic feature conveys the proportion of the data falling into a class interval for a standardized histogram? for a relative-frequency histogram? (Section 6.1.1)
2. What are the two fundamental ways in which random observations arise?
3. How does a density curve describe probabilities?
4. What is the total area under both a standardized histogram and a density curve?
5. When can histograms of data from a random process be relied on to closely resemble the density curve for that process? (Section 6.1.2)
6. To what characteristic of the density curve does the mean correspond?
7. Does it matter whether interval endpoints are included or excluded when we calculate probabilities for a continuous random variable? Why? Are discrete variables the same or different in this regard? (Section 6.1.3)

6.2 THE NORMAL DISTRIBUTION

6.2.1 Introducing the Normal Distribution

Figure 6.2.1 presents two standardized histograms with approximating density curves superimposed. Both curves belong to a family of symmetric bell-shaped density curves referred to as the ***Normal*** (or Gaussian) distribution. Each density curve in the family can be represented by an equation of the form

$$f(x) = \frac{1}{\sqrt{2\pi}\sigma}\exp\left\{-\frac{(x-\mu)^2}{2\sigma^2}\right\}$$

for some values of μ and σ respectively. (Note that you will not have to use this formula.) It turns out that $\mu = \mu_X$ [$=$ E(X)], which is the position at which the bell is centered and $\sigma = \sigma_X$ [$=$ sd(X)], which determines the spread of the bell shape. If a random variable has a distribution of this form, we shall write $X \sim \text{Normal}(\mu_X = \mu, \sigma_X = \sigma)$.

Example 6.2.1 *Quetelet's Scottish Soldiers*

The standardized histogram in Fig. 6.2.1(a) shows the pattern of variation in the chest measurements of 5738 Scottish soldiers collected by a Belgian scholar with deep

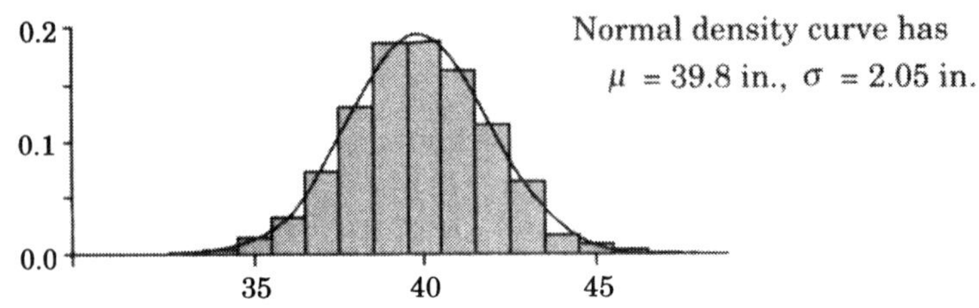

(a) Chest measurements of Quetelet's Scottish soldiers (in.)

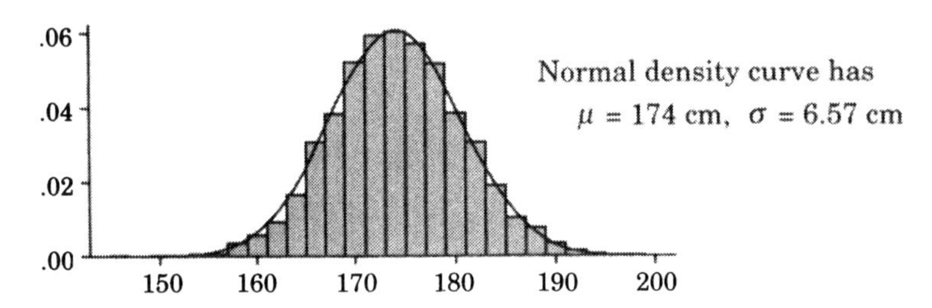

(b) Heights of the 4294 men in the workforce database (cm)

FIGURE 6.2.1 Two standard histograms with approximating Normal density curves.

statistical interests, Lambert Adolphe Jacques Quetelet (1796–1874). This data is interesting because Quetelet was the first person to apply the Normal distribution to human data, and this was his first major example.[2] The frequency table from which the histogram was drawn first appeared in an 1846 book of letters to the Duke of Saxe-Coburg and Gotha and is reproduced in Stigler [1986, p. 207]. From the histogram, there was clearly considerable variability in the soldiers' chest measurements, with the large majority of soldiers having a chest of between 35 in. and 45 in. The sample mean and standard deviation of the 5738 chest measurements were $\bar{x} = 39.8$ and $s = 2.05$, respectively. The density curve superimposed is that of a Normal($\mu = 39.8, \sigma = 2.05$) distribution.

✖ Example 6.2.2 *Male Heights from the Workforce Database*

Figure 6.2.1(b) shows a standardized histogram of the heights of the 4294 men in the workforce database. Variability in heights is one of the most obvious examples of the person-to-person differences of which we are all aware. We see from the histogram that the large majority of men have a height between 160 cm and 190 cm. The sample mean and standard deviation of the heights of the men in the database were $\bar{x} = 174.4$ cm and $s = 6.57$ cm, respectively. Superimposed is a Normal($\mu = 174.4, \sigma = 6.57$) density curve. The approximation is clearly quite good. A point of caution, however: Height was the only continuous variable in the database to produce a histogram that was not noticeably skewed, though the degree of skewness in some other physical dimensions (e.g., waist and hip measurements) was slight.

---

[2] The Normal distribution had its origin in large-sample limits to Binomial probabilities and in the theory of measurement errors. Stigler [1986] gives a fascinating account of the history.

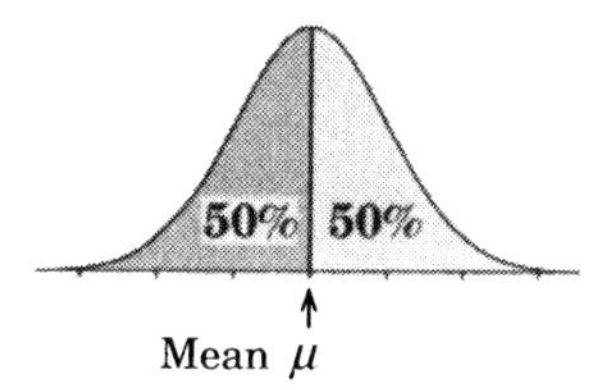

FIGURE 6.2.2

As Fig. 6.2.2 shows, the mean μ is the center of the bell. Because the distribution is symmetrical about μ, there is a 50% chance that a random observation will fall below the mean and a 50% chance that it will be greater than the mean. In this sense, the mean is also the *median* of the distribution. Figure 6.2.3 shows some other features of Normal distributions. Changing the mean shifts the curve along the axis as shown in Fig. 6.2.3(a). The standard deviation σ determines the spread, as shown in Fig. 6.2.3(b), with larger values corresponding to greater spread. As Fig. 6.2.3(c) shows, for any Normal distribution there is approximately a 68% chance of a random observation falling within one standard deviation to either side of the mean, approximately a 95% chance of falling within two standard deviations of the mean, and a 99.7% chance of falling within three standard deviations of the mean. (This is the origin of the 68%–95% rule for data used in Section 2.4.3.)

According to Stigler [1986, Chapter 5], Quetelet believed that all naturally occurring distributions of properly collected and sorted data would follow a Normal curve and that a failure to exhibit such a shape was evidence of a nonhomogeneous group. For example, a single histogram of heights for all people in the workforce database (4294 men and 1635 women) is skewed. Quetelet's proposition always had its critics and is no longer taken seriously. In fact, it is unusual for a large set of raw data on a continuous variable to pass a statistical test of the proposition that it was generated by a Normal distribution. The importance of the Normal distribution in statistics is largely because sample averages and some other quantities derived from the raw data tend to be approximately Normally distributed (see Chapter 7 on the central limit theorem). The theory underlying many statistical methods is most easily worked out under the assumption of Normal distributions, and as it turns out, many methods for analyzing data derived from this theory still work well when the Normal distribution assumption does not hold. Thus it is not so much that we believe that anything is *exactly* Normally distributed, but rather that if the approximation by a Normal distribution is not too bad, many calculations based on an assumption of Normality give good approximate answers. We revisit these issues many times in later chapters.

6.2.2 Obtaining Probabilities

Normal distribution probabilities can be obtained from all statistics packages, from some spreadsheet programs like Excel, and from graphics and some other calculators. You tell the program what Normal distribution to use by giving it the mean and standard deviation. Almost invariably, software is designed to calculate the value of pr($X \leq x$) for a given value of x. You supply an x-value, and the program returns the probability that a random observation would be smaller than that. In Fig. 6.2.3(d) we have given the value 180 to both Minitab and to Excel's function `NORMDIST`, and they

(a) Changing μ

shifts the curve along the axis

$\sigma_1 = \sigma_2 = 6$

140 160 180 200

$\mu_1 = 160$ $\mu_2 = 174$

(b) Increasing σ

increases the spread and flattens the curve

$\sigma_1 = 6$

$\sigma_2 = 12$

140 160 180 200

$\mu_1 = \mu_2 = 170$

(c) Probabilities and numbers of standard deviations

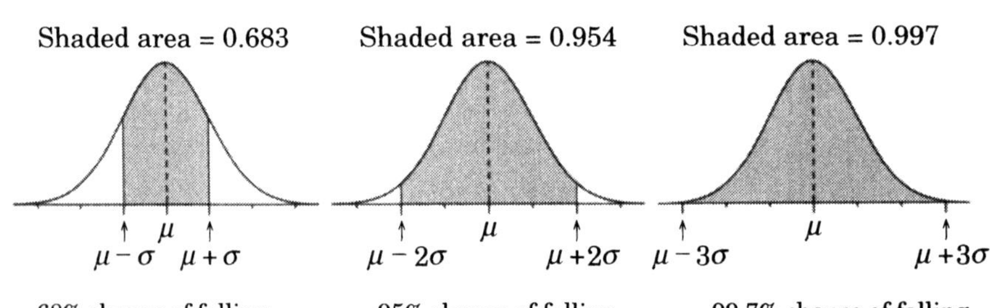

68% chance of falling between $\mu - \sigma$ and $\mu + \sigma$

95% chance of falling between $\mu - 2\sigma$ and $\mu + 2\sigma$

99.7% chance of falling between $\mu - 3\sigma$ and $\mu + 3\sigma$

(d) Probabilities supplied by computers

[Cumulative (or lower-tail) probabilities]

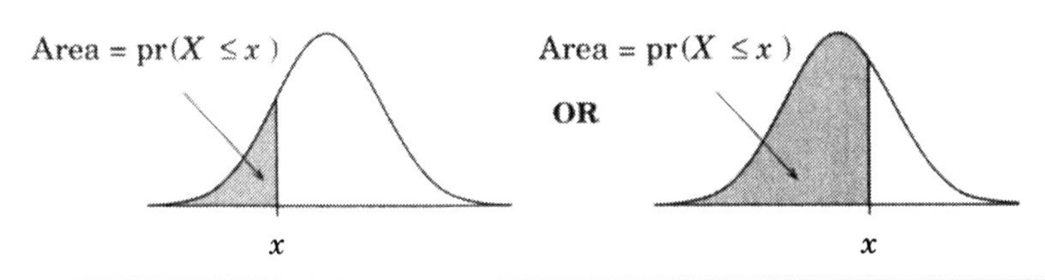

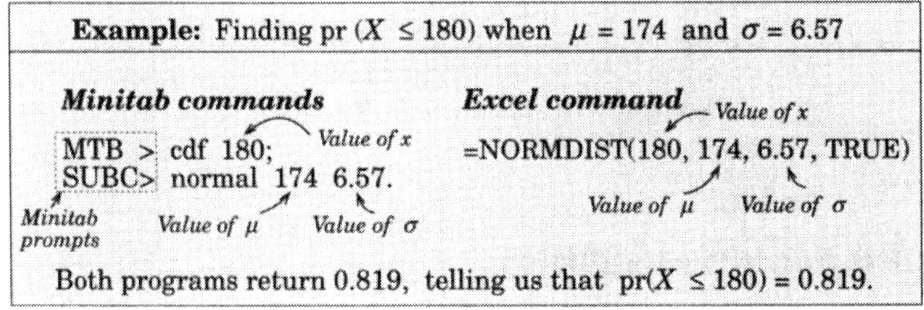

Example: Finding pr $(X \le 180)$ when $\mu = 174$ and $\sigma = 6.57$

Minitab commands

```
MTB > cdf 180;
SUBC> normal 174 6.57.
```

Minitab prompts · *Value of x* · *Value of μ* · *Value of σ*

Excel command

```
=NORMDIST(180, 174, 6.57, TRUE)
```

Value of x · *Value of μ* · *Value of σ*

Both programs return 0.819, telling us that pr$(X \le 180) = 0.819$.

FIGURE 6.2.3 Properties of the Normal distribution.

have told us that the probability that a random observation would fall below 180 is 0.819.[3]

Probabilities of the form pr($X \leq x$) are called the ***cumulative*** or ***lower-tail*** probabilities. We discussed such probabilities for discrete random variables in Section 5.2.

✖ Example 6.2.2 (cont.) *Calculating Probabilities for Intervals*

Let X be the height of a man randomly selected from a population in which heights have a Normal(μ = 174 cm, σ = 6.57 cm) distribution. (We have used the mean and standard deviation from male heights in the workforce database so that we are working with realistic values.) Earlier we saw that pr($X \leq 180$) = 0.819. This tells us that there is an 81.9% chance that a randomly selected man is no taller than 180 cm, or that almost 82% of the male population has a height below 180 cm.

We now move on to calculating the probability that X lies in an interval. First we evaluate the probability that a random man has a height between 160 cm and 180 cm.

Basic Method

1. Sketch a Normal curve, marking the mean and other values of interest.
2. Shade the area under the curve that gives the desired probability.
3. Devise a way of getting the desired area from lower-tail areas.

The right-hand plot in Fig. 6.2.4(a) incorporates the preceding steps 1 and 2. The computer program will give us the tail area below 180 as shaded in the left-hand plot and below 160 as shaded in the middle plot. The desired area is clearly the difference (step 3). We can write this as a formula:

$$\text{pr}(160 < X \leq 180) = \text{pr}(X \leq 180) - \text{pr}(X \leq 160)$$

The intermediate steps and answer are given in the table in Fig. 6.2.4(b). We see that there is a very small probability of being shorter than 160 cm (0.017 = 1.7%), and the probability of lying in the interval from 160 cm to 180 cm is 0.802, or approximately 80%.

In the table in Fig. 6.2.4(c), we present the probabilities for a set of interval boundaries involving 152.4 cm, 167.6 cm, 177.8 cm, and 182.9 cm, corresponding to 5 ft, 5 ft 6 in., 5 ft 10 in., and 6 ft respectively. Taking the final row, we see that pr($X \leq 182.9$) = 0.912, or 91.2% of men being shorter than 182.9 cm, pr($X \leq 167.6$) = 0.165, and the probability of being in the interval 167.6 to 182.9 cm is 0.747.

---

[3] pr($X \leq 180$) = pr($X < 180$) = 0.819 because endpoints don't matter. Other programs give the cumulative (or lower-tail) probabilities pr($X \leq x$) similarly, for example, the Splus and R command is `pnorm(x-val,mean,sd)`. In Minitab we generally prefer to use the menus rather than the command language. A dialogue box is obtained under `Calc` → `Probability Distributions` → `Normal` (click cumulative).

(a) Computing pr(160 < X ≤ 180)

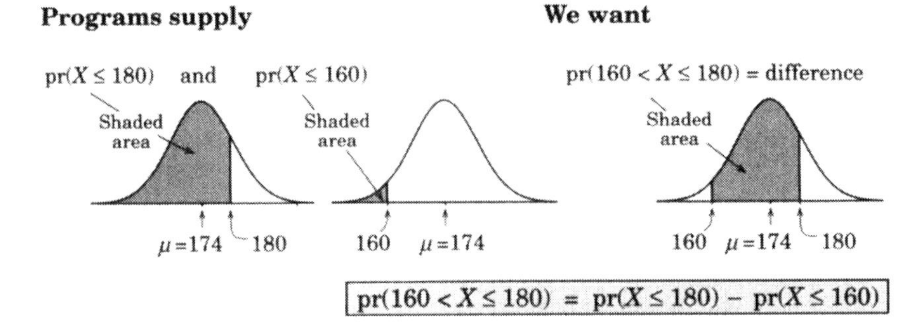

(b) Example : Implementing in Excel

	A	B	C	D
1		pr(X ≤ 180)	pr(X ≤ 160)	pr(160 < X ≤ 180)
2	Normal Probs	0.819	0.017	0.802
3				

Excel commands: =NORMDIST(180,174,6.57,TRUE) =NORMDIST(160,174,6.57,TRUE) =B2-C2

(c) More Normal probabilities *(values obtained from Minitab)*

b	pr($X \le b$)	a	pr($X \le a$)	pr($a < X \le b$) = difference
167.6	0.165	152.4	0.001	0.164
177.8	0.718	167.6	0.165	0.553
177.8	0.718	152.4	0.001	0.717
182.9	0.912	167.6	0.165	0.747

Note: 152.4cm = 5ft, 167.6cm = 5ft 6in., 177.8cm = 5ft 10in., 182.9cm = 6ft

FIGURE 6.2.4 Normal distribution probability calculations.

✖ Example 6.2.3 *Obtaining Probabilities for Intervals and Upper Tails*

Being overweight is an important risk factor for many health disorders such as heart disease, diabetes, and certain cancers. There are elaborate measures that attempt to capture the idea of overweight, such as percentage body fat. These are relatively expensive to obtain in large-scale studies, however. A "quick-and-dirty" measure employed in many large studies is *body-mass index* (BMI), defined as weight in kilograms divided by the square of the height in meters (kg/m^2). In 1997, 24 studies were initiated internationally of the weight-loss drug Xenical (Orlistat). In these studies, *obesity* was defined as a BMI of over 30, and entry to the studies was confined to people in this situation. We shall use the less emotive expression *very overweight* for people with BMI > 30. Several diabetes studies have classified men with BMI > 25 and women with BMI > 27 as being overweight. At the time of writing, Chris Wild is 88 kg and 1.82 m, a BMI of $88/1.82^2 = 26.6$, putting him in the overweight category. Chris has never thought of himself as particularly overweight. What proportions of men are overweight and very overweight?

Obtaining an upper-tail probability

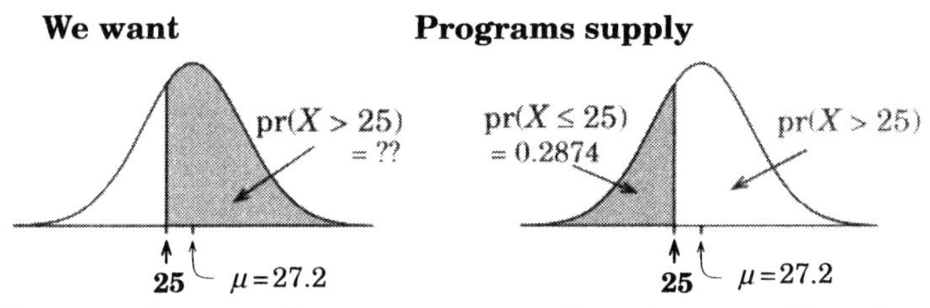

Since total area under curve = 1, $\text{pr}(X > 25) = 1 - \text{pr}(X \leq 25)$

Generally, $\mathbf{pr(X > x) = 1 - pr(X \leq x)}$

FIGURE 6.2.5 Finding an upper-tail probability.

For the men in the workforce database, BMI values have a sample mean and a sample standard deviation of $\bar{x}_M = 27.3$ and $s_M = 4.1$, respectively. Let X_M be the BMI value of a randomly chosen man. We will assume that the distribution of BMI values for the adult male population is approximately Normal($\mu_M = 27.3, \sigma_M = 4.1$) and find $\text{pr}(X_M > 25)$. The information routinely supplied by computer programs is the lower-tail probability $\text{pr}(X_M \leq 25) = 0.2874$. Since $X_M \leq 25$ is the complement of $X_M > 25$,

$$\text{pr}(X_M > 25) = 1 - \text{pr}(X_M \leq 25) = 1 - 0.2874 = 0.7126$$

This reasoning is perhaps more obvious in pictorial form, as shown in Fig. 6.2.5. We would expect about 71% of the men to have a BMI value above an overweight cutoff of 25.

In the same vein, programs tell us that $\text{pr}(X_M \leq 30) = 0.7449$. Since $\text{pr}(X_M > 30) = 1 - \text{pr}(X_M \leq 30) = 0.2551$, we see that just over 25% would fall into the very overweight group. What proportion of men are overweight (above 25) but not very overweight (below 30), that is, what is $\text{pr}(25 < X_M \leq 30)$? Using the reasoning depicted in Fig. 6.2.4(a),

$$\text{pr}(25 < X_M \leq 30) = \text{pr}(X_M \leq 30) - \text{pr}(X_M \leq 25) = 0.7449 - 0.2874 = 0.4575$$

Thus about 46% are overweight (above 25) but not very overweight.

We now turn our attention to BMI values for women, denoted X_W. Recall that the overweight cutoff for women is 27. We shall use the mean and standard deviation from the women in the workforce database to provide our μ and σ and assume that $X_W \sim \text{Normal}(\mu_W = 27.2, \sigma_W = 5.8)$. Computer programs tell us $\text{pr}(X_W \leq 27) = 0.4862$ and $\text{pr}(X_W \leq 30) = 0.6854$. Check that you can obtain these values. Then, reasoning as before, verify that $\text{pr}(X_W > 27) = 0.5138$, $\text{pr}(X_W > 30) = 0.3146$, and $\text{pr}(27 < X_W \leq 30) = 0.1991$.

QUIZ ON SECTIONS 6.2.1 AND 6.2.2

1. What features of the Normal curve do μ and σ visually correspond to?
2. What is the probability that a random observation from a Normal distribution is smaller than the mean? larger than the mean? exactly equal to the mean? Why?

3. Approximately, what is the probability that a random observation from a Normal distribution falls within one standard deviation (sd) of the mean? two sd's? three sd's?

4. Computer programs provide cumulative probabilities for the Normal distribution. What is a cumulative probability?

EXERCISES FOR SECTION 6.2.2

1. Using your statistical package or calculator, check some of the answers that we have given in Section 6.2.2 corresponding to each type of probability: lower-tail (or cumulative), interval, and upper-tail.

2. The natural gestation period for human births, X, has a mean of about 266 days with a standard deviation of about 16 days. Assume that it is also Normally distributed, that is, $X \sim \text{Normal}(\mu_X = 266, \sigma_X = 16)$. What is the proportion of women who
   (a) deliver at least two weeks early, that is, carry their babies for less than 252 days?
   (b) carry their babies for between 260 days and 280 days?
   (c) are more than two weeks overdue, that is, carry their babies for longer than 280 days?

3. IQ scores for school children are standardized so that they are approximately Normally distributed with a mean of 100 and a standard deviation of 15. What is the probability that a randomly selected child has an IQ
   (a) less than 80? (b) between 85 and 110? (c) greater than 120?

6.2.3 The Inverse Problem: Percentiles and Quantiles

We met the idea of ***percentile*** in the context of a data set in Section 2.4.2. The same idea applies to populations and distributions. For example, speaking of a population,

(a) p-Quantile

Programs supply x_p

x-value for which $\text{pr}(X \leq x_p) = p$

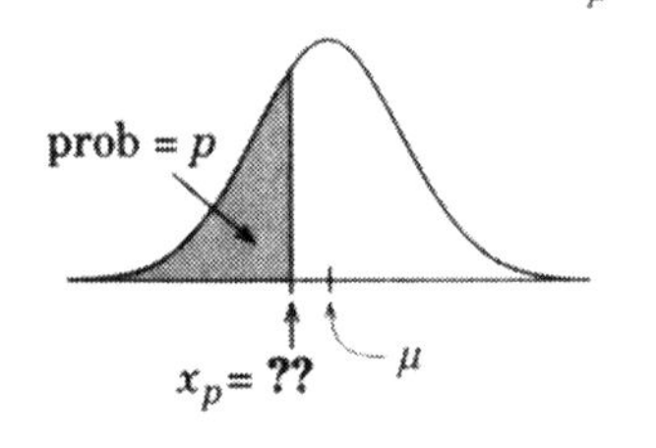

(b) 80th percentile (0.8-quantile) of women's heights

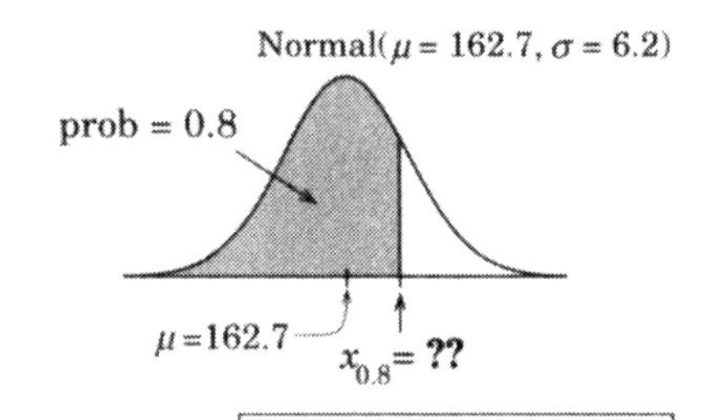

Program returns 167.9.
Thus 80% lie below 167.9.

(c) Further percentiles of women's heights

Percent	1%	5%	10%	20%	30%	70%	80%	90%	95%	99%
Propn	0.01	0.05	0.1	0.2	0.3	0.7	0.8	0.9	0.95	0.99
Percentile	*(or quantile)*									
(cm)	148.3	152.5	154.8	157.5	159.4	166.0	167.9	170.6	172.9	177.1
(ft'in")	4'10"	5'0"	5'0"	5'2"	5'2"	5'5"	5'6"	5'7"	5'8"	5'9"
(+ frac)	3/8"		7/8"		3/4"	3/8"	1/8"	1/8"	1/8"	3/4"

FIGURE 6.2.6 The inverse problem: quantiles and percentiles.

80% of women have a height that falls below the 80th percentile. Speaking of a distribution, there is an 80% chance that a random observation from the distribution will fall below the 80th percentile. The lower quartile, median,[4] and upper quartile of a distribution correspond respectively to the 25th, 50th, and 75th percentiles. *Percentile* and ***quantile*** are different words used for the same idea. When speaking in terms of percentages, the word *percentile* is used. When expressing ourselves in terms of proportions or probabilities (between 0 and 1), the word *quantile* is used. In general, the p-quantile is the value x_p for which the lower-tail probability is p. This is illustrated in Fig. 6.2.6(a). These values are provided by statistical computer packages, which invariably require that probabilities between 0 and 1 be supplied, not percentages. This is often called *the inverse problem,* and the resulting probabilities are sometimes called (as in Minitab) *inverse cumulative probabilities.* Why? Previously we gave the program an x-value, and it gave us a probability. Now we give the program a probability and it gives us an x-value.

✖ Example 6.2.4 *Percentiles (Quantiles) of Women's Heights*

Women's heights in the workforce database had a mean of 162.7 cm and a standard deviation of 6.2 cm. We shall work with the Normal(μ = 162.7, σ = 6.2) distribution depicted in Fig. 6.2.6(b) and find the 80th percentile; that is, we shall find the height $x_{0.8}$ so that pr($X \le x_{0.8}$) = 0.8. Given this information, the computer programs returned 167.9. Thus, according to this distribution, the 80th percentile of women's heights is 167.9 cm. Figure 6.2.6(c) gives the values for many other percentiles of the distribution of women's heights. In the final two rows of the table we have translated them into feet and inches to the nearest eighth of an inch. For example, the 30th percentile is $5'2\frac{3}{4}''$. To obtain such probabilities in Minitab, click `Inverse Cumulative Probability` in the Normal distribution dialogue box. Alternatively, you can use the command `InvCDF prob;` followed by the subcommand `Normal mu sigma.` (you must include the "`.`"). In Excel, use `NORMINV(prob,mean,sd)`. For example, `NORMINV(0.8,162.7,6.2)` returns 167.9. Splus and R use `qnorm(prob,mean,sd)`.

Your original interest might be in the upper extremes. What height is exceeded only by the tallest 25% of women? We have to translate such problems into what they say about lower-tail probabilities (as shown in Fig. 6.2.7) because the programs require

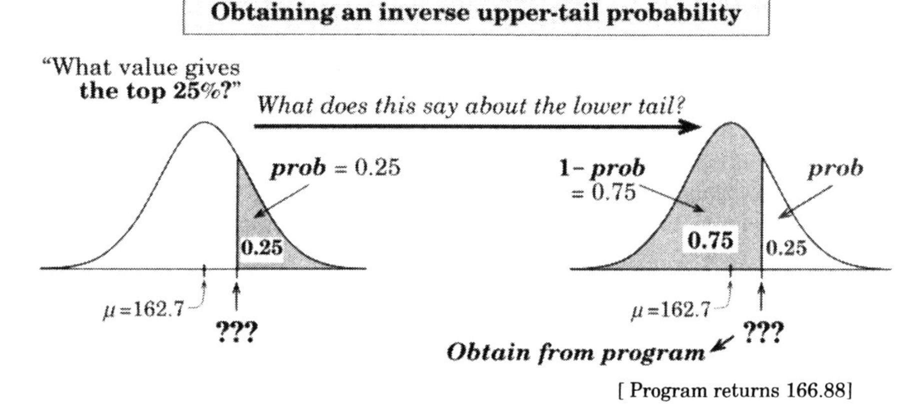

FIGURE 6.2.7 Obtaining upper-tail inverse probabilities.

[4]For the Normal distribution we have seen that the median and mean coincide.

lower-tail probabilities. An upper tail of 25% corresponds to a lower tail of 75% or 0.75. Programs tell us that when $\mu = 162.7$ and $\sigma = 6.2$, the 0.75 quantile (75th percentile or upper quartile) is 166.88. In general, if we want an upper-tail probability p, we get the program to give us the value corresponding to lower-tail probability $1 - p$. What is the cutoff value for the upper 2% of heights? The corresponding lower-tail probability is 98% or 0.98. Programs tell us that the 0.98 quantile (98th percentile) is 175.43. [For example, `NORMINV(0.98,162.7,6.2)` returns 175.43.]

QUIZ ON SECTION 6.2.3

1. What is meant by the 60th percentile of heights?
2. What is the difference between a percentile and a quantile?
3. The lower quartile, median, and upper quartile of a distribution correspond to special percentiles. What are they? Express also in terms of quantiles.
4. Quantiles are sometimes called inverse cumulative probabilities. Why?

EXERCISES FOR SECTION 6.2.3

1. Check that you can reproduce the values given in the section.
2. Let us return to the gestation period for human births (in days), which we are assuming to have an approximate Normal($\mu_X = 266$, $\sigma_X = 16$) distribution.
   (a) What is the 15th percentile of gestation periods?
   (b) What gestation period is exceeded by 2% of births?
   (c) What gestation period is exceeded by 90% of births?
   (d) Below what gestational period do 70% of births fall?
3. We return to IQ scores that are approximately Normally distributed with a mean of 100 and a standard deviation of 15.
   (a) What is the 80% percentile of IQ scores?
   (b) What IQ score is exceeded by only the top 1% of children?
   (c) Below what score do only the bottom 30% of children fall?

6.2.4 Central Ranges

We shall continue to use the women's heights and ask what range of values accounts for the central 50% of women's heights? We ask ourselves, what lower-tail areas do we need to consider to describe the central 50% of the distribution? This is done in Fig. 6.2.8. We have labeled the left- and right-hand endpoints of the desired interval a and b respectively. To have probability 0.5 in the middle, we need half of the remaining 0.5, that is, 0.25, in each tail. Thus, a is the 0.25 quantile (25th percentile) and b is the 0.75 quantile (75th percentile). Using a computer program that produces inverse cumulative probabilities, we find that $a = 158.52$ cm. We have already determined in Example 6.2.4 that the 75th percentile is 166.88. Thus, the central 50% of women's heights extends from approximately 158.5 cm to 166.9 cm. Arguing in the same way, the central 70% of the distribution extends from the 15th percentile to the 85th percentile, and the central 90% extends from the 5th percentile to the 95th percentile. The method we have used here applies to any continuous distribution. We just need to use an appropriate program to obtain our inverse probabilities.

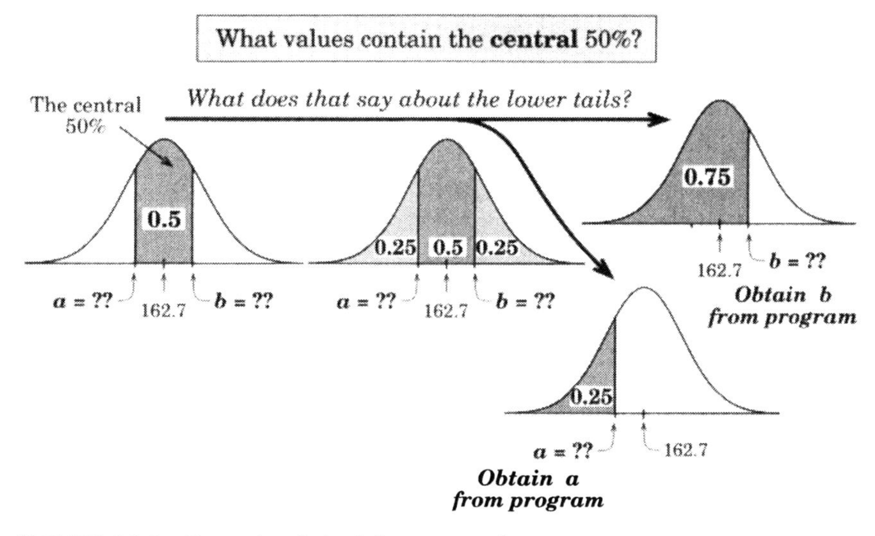

FIGURE 6.2.8 Example of obtaining a central range.

EXERCISES FOR SECTION 6.2.4

1. Find the interval encompassed by the central (a) 80% and (b) 90% of women's heights (μ = 162.7, σ = 6.2).

2. Find the interval encompassed by the central (a) 60% and (b) 80% of men's BMI values (μ = 27.3, σ = 4.1).

6.3 WORKING IN STANDARD UNITS

6.3.1 The z-Score

Recall from Fig. 6.2.3(c) that for a Normal distribution and any values of μ and σ, there is a 68.3% chance that a random observation will fall within one standard deviation of the mean, a 95.4% chance that it will fall within two standard deviations, and a 99.7% chance that it will fall within three standard deviations of the mean. More generally, the probability of falling in any interval depends *only* on how many standard deviations from the mean the interval endpoints are. Thus for Normal distributions it makes good sense to *measure distance in terms of number of standard deviations from the mean.* We use

$$z = \frac{x - \mu}{\sigma} \qquad \text{(standard units or } z\text{-scores)}$$

The ***z-score*** of x is the number of standard deviations x is from the mean.

The z-score is positive if x is larger than the mean or negative if it is smaller than the mean. Table 6.3.1 illustrates the z-score idea.

TABLE 6.3.1 Examples of z-Scores

x	z-score = $(x - \mu)/\sigma$	Interpretation
Male BMI values[a] (kg/m²)		
25	$(25 - 27.3)/4.1 = -0.56$	25 kg/m² is 0.56 sd's below the mean
35	$(35 - 27.3)/4.1 = 1.88$	35 kg/m² is 1.88 sd's above the mean
Female heights[b] (cm)		
155	$(155 - 162.7)/6.2 = -1.24$	155 cm is 1.24 sd's below the mean
180	$(180 - 162.7)/6.2 = 2.79$	180 cm is 2.79 sd's above the mean

[a]Male BMI values: $\mu = 27.3, \sigma = 4.1$.
[b]Female heights: $\mu = 162.7, \sigma = 6.2$.

It turns out that the distribution of the random variable corresponding to the z-score, namely

$$Z = \frac{X - \mu}{\sigma}$$

is Normal with mean 0 and standard deviation 1. This is called the *standard Normal* distribution.

Standard Normal distribution: mean = 0, sd = 1

[Aside: We can rewrite our expression for Z as $Z = \frac{1}{\sigma}X - \frac{\mu}{\sigma}$. The results $E(Z) = 0$ and $sd(Z) = 1$ then follow from $E(aX+b) = a\,E(X)+b$ and $sd(aX+b) = |a|sd(X)$, using $a = 1/\sigma$ and $b = -\mu/\sigma$. You might like to try this.]

Before probabilities were routinely obtained from computers, working in terms of z-scores was necessary for obtaining Normal probabilities from tables (see Section 6.3.3). Thinking in terms of numbers of standard deviations is still useful for two reasons, however. First, it lets us do quick mental calculations that give us an idea of the usual range of a variable, as we shall illustrate later. Second, this way of thinking will be generalized in forming so-called confidence intervals in future chapters.

Means and standard deviations are still the most commonly published summaries for continuous variables. For any distribution that is approximately Normal, we know that values usually (i.e., 68.3% of the time) fall within one standard deviation of the mean, almost always (i.e., 95.4% of the time) fall within two standard deviations of the mean, and hardly ever (i.e., less than 1% of the time) fall outside three standard deviations. These ideas correspond to rough calculations that can be done mentally whenever we encounter such summaries. We do not have to boot up a computer. Thus for female heights ($\mu = 162.7, \sigma = 6.2$) we can determine almost instantly that a little over 68% of women have heights in 162.7 ± 6.2 or between 156.5 and 168.9 cm, a little over 95% have heights in $162.7 \pm 2 \times 6.2$ or between 150.3 and 175.1 cm, and heights outside $162.7 \pm 3 \times 6.2$, or below 144.1 or above 181.3 cm are exceedingly rare—statements which are much more intuitively meaningful than the fact that the standard deviation is 6.2 cm. As we work with them over time, other commonly used z-scores embed themselves in our memories to similar advantage.

✖ Example 6.3.1 *Using z-Scores to Compare Performances*

Consider the results from a national or statewide examination in two successive years. You achieved a raw score of 80% this year. Your friend got 85% last year and is claiming to have beaten you. Who really did better? It is virtually impossible to set two examinations to exactly the same standard. We may be prepared to believe, however, that because of the very large numbers of students involved, the distribution of ability levels of students sitting the examination is fairly constant from year to year. We shall assume the scores were approximately Normally distributed. Suppose that the mean and standard deviation of scores this year were respectively 60% and 20%. Last year, the mean and standard deviation were 65% and 25%. Thus last year's exam was easier. Your z-score is $(80-60)/20 = 1$. Your friend's z-score is $(85-65)/25 = 0.8$. This means that you did better in the sense that your ranking among those taking the examination was higher. How do we know? In both years, the z-scores follow a Normal($\mu = 0, \sigma = 1$) distribution. The proportion of people with a z-score less than 1 is greater than the proportion with a z-score less than 0.8. [Using a Normal($\mu = 0, \sigma = 1$) distribution, the probabilities are 0.841 and 0.788 respectively.]

EXERCISES FOR SECTION 6.3.1

For the following exercises, find the z-scores of each of the values given and then interpret them as in Table 6.3.1.

1. Gestation period for human births (days) for which $\mu_X = 266, \quad \sigma_X = 16$:
(a) 280 days (b) 250 days (c) 270 days

2. Children's IQ scores, which have a mean of 100 and a standard deviation of 15:
(a) an IQ of 80 (b) an IQ of 110 (c) an IQ of 90

6.3.2 Central Ranges and Extremes Using z-Scores

Calculations toward the end of the previous subsection gave us rather approximately the central 68%, 95%, and 99.7% ranges of any Normal distribution. We can address central-range questions such as, "What range of BMI values gives us the central 90% of the population?" exactly using z-scores. First, we answer the question for the *standard Normal distribution*. Because the standard Normal is symmetrical about 0, the central 90% of the standard Normal distribution will fall between $-z$ and $+z$, where the value of z is something we have to determine. Using the method depicted in Fig. 6.3.1, we find that $z = 1.6449$. This result can then be applied to any other Normal distribution by interpreting z in terms of standard deviations from the mean. For any Normal distribution there is a 90% chance that a random observation will fall between 1.6449 standard deviations below the mean and 1.6449 standard deviations above the mean. When we apply this to male BMI values for which $\mu = 27.3$ and $\sigma = 4.1$, we find that $\mu - 1.6449\sigma = 20.56$ and $\mu + 1.6449\sigma = 34.04$. The central 90% of the distribution of male BMI values thus extends from 20.56 kg/m^2 to 34.04 kg/m^2. Alternatively, there is a 90% chance that the BMI of a randomly selected male will fall between these limits.

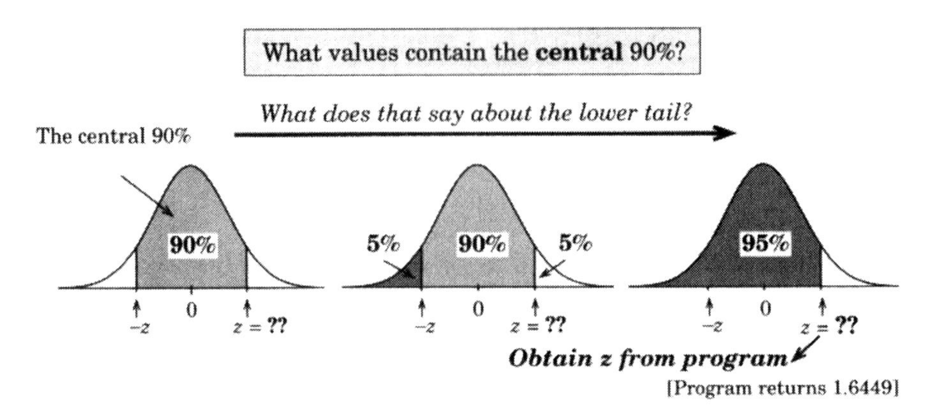

FIGURE 6.3.1 Obtaining a central range.

TABLE 6.3.2 Central Ranges

		Male BMI values[a]		Female heights[b]	
Percentage	z	$\mu - z\sigma$	$\mu + z\sigma$	$\mu - z\sigma$	$\mu + z\sigma$
80%	1.2816	22.05	32.55	154.8	170.6
90%	1.6449	20.56	34.04	152.5	172.9
95%	1.9600	19.26	35.34	150.5	174.9
99%	2.5758	16.74	37.86	146.7	178.7
99.9%	3.2905	13.81	40.79	142.3	183.1

[a]Male BMI values: $\mu = 27.3, \sigma = 4.1$.
[b]Female heights: $\mu = 162.7, \sigma = 6.2$.

We use this approach to calculating central ranges when working from tables of z-values. In Table 6.3.2 we have found the z-values corresponding to other central percentages of the standard Normal distribution using the method depicted in Fig. 6.3.1. We have then translated these in subsequent columns into the endpoints of the intervals making up that central percentage of the distribution of male BMI values, and similarly for female heights. We see, for example, that the central 80% of women's heights extends from 154.8 cm to 170.6 cm. A more extensive table of z-values for central ranges is given in Appendix A5.1.

Appendix A5.2 contains a table of z-values corresponding to various tail probabilities. For example, the upper 10% tail of the standard Normal corresponds to z-values above 1.2816. This tells us that for an arbitrary Normal distribution, the upper 10% tail corresponds to values more than 1.2816 standard deviations above the mean (i.e., above $\mu + 1.2816\sigma$). For male BMI values, $27.3 + 1.2816 \times 4.1 = 32.55$, so the upper 10% corresponds to BMI values larger than approximately 33 kg/m^2. Because the standard Normal is symmetrical about 0, the tail area above z is the same as the tail area below $-z$. The lower 10% tail of the standard Normal consists of values below -1.2816. For an arbitrary Normal distribution, the lower 10% tail corresponds to values smaller than $\mu - 1.2816\sigma$. For male BMI values, $27.3 - 1.2816 \times 4.1 = 22.05$, telling us that the lower 10% of male BMI values are those smaller than approximately 22 kg/m^2.

EXERCISES FOR SECTION 6.3.2

Use z-scores to find the following.

1. Let us return to the gestation period for human births (in days), which we are assuming has an approximate Normal($\mu_X = 266$, $\sigma_X = 16$) distribution.
   (a) What range of gestation periods corresponds to the central 95% of births? the central 80%?
   (b) What gestation period is exceeded by only 5% of births? by only 1 birth in 100?
   (c) Below what gestation period do the lowest 5% of births fall? the lowest 1%?
2. We return to IQ scores, which are approximately Normally distributed with a mean of 100 and a standard deviation of 15.
   (a) What range corresponds to the central 95% of IQ scores? the central 80%?
   (b) What is the cutoff IQ score for the top 5% of children? the top 1%?
   (c) What is the lower quartile of IQ scores? What is the cutoff value for the lowest 10%?

*6.3.3 Obtaining Normal-Distribution Probabilities from Tables

This subsection gives brief instruction in the use of tables to obtain Normal probabilities. The skill is useful whenever one does not have easy access to a computer. This subsection can, however, be omitted.

We note that

$$\text{pr}(X \le x) = \text{pr}\left(\frac{X-\mu}{\sigma} \le \frac{x-\mu}{\sigma}\right) = \text{pr}(Z \le z)$$

where $z = (x - \mu)/\sigma$. Recall from Section 6.3.1 that $Z = (X - \mu)/\sigma$ has a standard Normal($\mu = 0, \sigma = 1$) distribution. Thus no matter what Normal distribution we are working with, we can evaluate probabilities by converting interval endpoints to z-scores and using the standard Normal distribution. This makes it possible to obtain probabilities for any Normal distribution using a single set of standard Normal tables. We have supplied such tables in Appendix A4. We show how the tables are to be used in Table 6.3.3. Using the tables, we find the following (check that you can obtain these):

$\text{pr}(Z \le -1.00) = 0.159$ $\quad$ $\text{pr}(Z \le -1.02) = 0.154$ $\quad$ $\text{pr}(Z \le 1.00) = 0.841$
$\text{pr}(Z \le 1.02) = 0.846$ $\quad$ $\text{pr}(Z \le -2.10) = 0.018$ $\quad$ $\text{pr}(Z \le -2.15) = 0.016$
$\text{pr}(Z \le 2.10) = 0.982$ $\quad$ $\text{pr}(Z \le 2.15) = 0.984$ $\quad$ $\text{pr}(Z \le -0.34) = 0.367$
$\text{pr}(Z \le 0.34) = 0.633$

We deal with intervals and upper-tail probabilities in the usual way, converting interval endpoints to z-scores before looking up the tables. For example, if $\mu_X = 15$, $\sigma_X = 10$, and we want to find $\text{pr}(12 < X \le 32)$, then since 32 has z-score $\frac{(32-15)}{10} = 1.7$ and 12 has z-score $\frac{12-15}{10} = -0.3$,

$$\begin{aligned}\text{pr}(12 < X \le 32) &= \text{pr}(X \le 32) - \text{pr}(X \le 12)\\ &= \text{pr}(Z \le 1.7) - \text{pr}(Z \le -0.3)\\ &= 0.955 - 0.382 = 0.573\end{aligned}$$

TABLE 6.3.3 Using z-Score Tables

As an example, we shall find pr($Z \leq$ **1.1357**) using part of the table given in Appendix A4 (reproduced below).

STEP 1: Correct the z-value to two decimal places, that is, use $z = 1.14$.

STEP 2: Look down the z column until you find 1.1. This tells you which row to look in.

STEP 3: The second decimal place, here 4, tells you which column to look in.

STEP 4: The entry in the table corresponding to that row and column is pr($Z \leq 1.14$) = 0.873.

	z	0	1	2	3	4 ↓	5	6	7	8	9
	1.0	.841	.844	.846	.848	.851	.853	.855	.858	.860	.862
→	1.1	.864	.867	.869	.871	**.873**	.875	.877	.879	.881	.883
	1.2	.885	.887	.889	.891	.893	.894	.896	.898	.900	.901
	1.3	.903	.905	.907	.908	.910	.911	.913	.915	.916	.918
	1.4	.919	.921	.922	.924	.925	.926	.928	.929	.931	.932

Since the z-score of 4 is $\frac{4-15}{10} = -1.1$,

$$\text{pr}(X > 4) = 1 - \text{pr}(X \leq 4) = 1 - \text{pr}\left(Z \leq -1.1\right) = 1 - 0.136 = 0.864$$

EXERCISES FOR SECTION 6.3.3

1. Suppose that $X \sim \text{Normal}(\mu_X = 3, \sigma_X = 4)$.

(a) Find the z-scores of the following numbers:

(i) -5 (ii) 11 (iii) 5 (iv) 1.4

(b) How many standard deviations is each of these numbers from μ_X?

2. If $X \sim \text{Normal}(\mu_X = 7, \sigma_X = 6)$, find the following:

(a) pr($X \geq 5$) (b) pr($X \leq 9$) (c) pr($5 \leq X \leq 11$)

(d) pr($3 \leq X \leq 6$) (e) pr($X \leq 3$)

3. If $X \sim \text{Normal}(\mu_X = -3, \sigma_X = 2)$, find the following:

[Note: $7 - (-3) = 10$]

(a) pr($X \leq -4$) (b) pr($X \geq 0$) (c) pr($-3 \leq X \leq -1$)

6.4 SUMS AND DIFFERENCES OF RANDOM QUANTITIES

6.4.1 Variation and Its Propagation

As we stressed in Chapter 1, there is *always* variation from object to object and sample to sample. When we study objects produced by any manufacturing process, no two objects sampled are ever exactly identical, not even for very simple objects such as ball bearings. The question of great practical importance is whether the variation is big enough to have a significant impact on the performance of the product. The answer is that the impact is often much greater than people realize. The way forward is to try to find factors that are causing the variation and then use this information to improve the manufacturing process so that its output becomes less variable.

The following example shows how variability in components of a product leads to variability in the whole product (propagation). This is used to motivate the discussion of the random behavior of sums and differences of random variables in Section 6.4.2. The subsection ends with a discussion of independence.

EXAMPLE 6.4.1 *Quality Improvement of a Gear Assembly*

A friend working as a quality improvement consultant told us the following story. It concerns problems a North American car-manufacturing plant was having with one of its gear assemblies. Figure 6.4.1 gives an exploded view of the assembly. Our concern is with the so-called stack height, which is simply the combined length of the assembled parts or, equivalently, the sum of the thicknesses of each of the six components involved. Looking at just the first component, we find that final-drive internal gears are not all the same but unfortunately have different thicknesses! These thicknesses have a distribution, and the thickness of each new component coming off the line is like a random observation from this distribution. The same is true for each of the other five components making up the gear assembly. Their thicknesses all vary. (Their means and standard deviations are given in Table 6.4.1.) Consequently, when

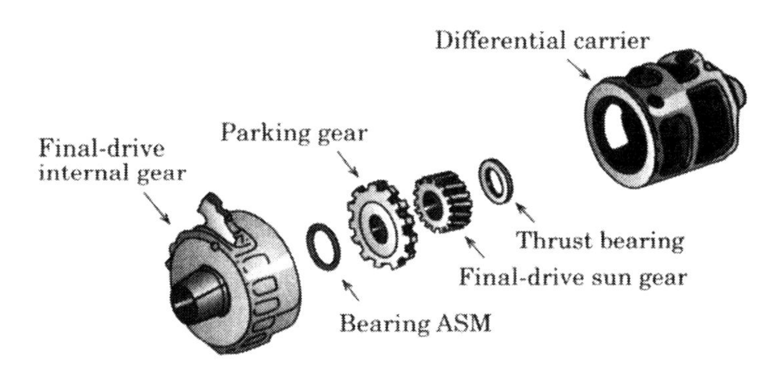

FIGURE 6.4.1 Layout of the gear assembly.

TABLE 6.4.1 The Gear Assembly Statistics

Component	Notation	Mean	Standard deviation
Final-drive internal gear	X_1	12.03	0.01
Bearing ASM	X_2	3.56	0.00
Parking gear	X_3	20.88	0.02
Final-drive sun gear	X_4	30.42	0.03
Thrust bearing	X_5	3.88	0.00
Differential carrier	X_6	81.75	0.04

the components are fitted together, the stack height (the sum of the thicknesses of the six components) also varies from assembly to assembly. Now these assemblies have to fit into a metal casing. If we make the casings large enough to accommodate the larger assemblies, the smaller ones will be too loose in the casing, leading to axle assemblies that don't work as well as they should and perhaps have a shorter lifetime. This costs the manufacturer money in higher warranty costs and results in a reputation for low-quality products, leading in turn to lost sales. If we make the casing smaller, some of the assemblies will not fit into the casing and will have to be scrapped. This costs money too. If assemblies have to be scrapped, the manufacturer has spent considerable time and money manufacturing garbage! Even if the parts can be salvaged, there are costs associated with reworking machine parts to make them usable.

Even the dimensions of the casings are variable! The company's first approach to the mismatch problem was to classify assemblies and casings coming off the line into bins by size, say, A = small, ..., F = large. Then when you had an assembly of size class D, you could get a class D casing to put it in. Even so, they still had severe mismatching problems! (We will find out why later.) Even if sorting by size works adequately, it is a clumsy procedure and has associated costs. There are space costs in storing parts until a match comes along, time costs in measuring and sorting, and organizational costs in having to use complex procedures. Instead of simply looking for a quick fix, the quality consultant prefers to look for ways to reduce variability in the system. To do this, the quality consultant needs to know how the variability in the parts affects the variability of the whole. If X_i is the (random) thickness of the ith component going into the assembly and *Sum* is the (random) stack height, this involves finding the distribution of

$$Sum = X_1 + X_2 + \cdots + X_n$$

from the distributions of the individual X_i's.

The quality consultant looks for those components that have the biggest effects on the variability of the whole and focuses the attention of the production team on ways by which some of the variability in these important components can be reduced. If successful, this can cut the costs drastically. We return to this example once we have found out how variability of the parts is transmitted into the variability of the whole.

Independence of Random Quantities

The results that we are about to see concerning the standard deviation of a sum, $\text{sd}(X_1 + X_2)$, apply only to ***statistically independent*** random variables. ***Independence*** can be defined mathematically, and you can use such a definition to prove that variables are independent. However, this works only for rather trivial models. In most statistical modeling of real-world situations, independence of random variables is an assumption that is made when it seems reasonable to do so on the basis of subject-matter knowledge or intuition and when there is no data that contradicts such an assumption. In Chapter 4 we introduced the notion of ***physical independence***. Events are ***physically independent*** if there is no physical way in which the outcome of one event can influence the outcome of the other. Variables relating to physically independent things are modeled as statistically independent. In Example 6.4.1 if the processes producing different types of components have no physical connection, that is, if there are no common factors affecting the manufacture of the different types of components, then parking gears, for example, will vary independently of other types of components. If there are common factors in the manufacture (e.g., a common source of steel), then there may be relationships between the ways in which different components vary. We can look for relationships in data using tools such as scatter plots (Section 3.1).

6.4.2 Investigating the Behavior of a Sum and a Difference

The Behavior of a Sum

For simplicity we will stack only two gears. We shall suppose that the thicknesses of type *A* gears, X_1, are Normally distributed with a mean of $E(X_1) = 25$ mm and a standard deviation of $sd(X_1) = 2$ mm, whereas the thicknesses of the type *B* gears, X_2, are Normally distributed with a mean of $E(X_2) = 50$ mm and a standard deviation of $sd(X_2) = 2$ mm independently of the value of X_1. We have exaggerated the variability of gear thicknesses (i.e., made the standard deviations larger than they would be in reality) so that the desired effects are big enough to show up in pictures. What does the distribution of the total thickness, $Sum = X_1 + X_2$, look like? Ideally, we would look at a production run to see. We have used a computer, however, to simulate the production process by generating Normally distributed random numbers with the given means and standard deviations. We simulated the manufacture of 5000 gear assemblies in this way. The results are summarized in Table 6.4.2 and plotted as histograms in Fig. 6.4.2.

We know that the X_1 measurements and X_2 measurements are Normally distributed. The histogram of total thicknesses ($Sum = X_1 + X_2$) has a similar shape but with increased variability (reflected also in the larger standard deviation). Note that because the samples are very large, the sample means agree closely with the population means, for example, for gear *A* $\overline{x}_1 = 25.01$, compared with $E(X_1) = 25$. The sample mean of the 5000 values of *Sum* was 75.02, which is very close to $E(X_1)+E(X_2)$ [$= 75$]. In fact, it can be shown mathematically that for any pair of random variables, $E(X_1 + X_2) = E(X_1) + E(X_2)$.

Additionally, because we have taken such large samples, the sample standard deviations from the data are quite close to the population standard deviations. (The sample values are $s_1 = 2.05$ and $s_2 = 1.98$ mm, compared with the population values $sd(X_1) = sd(X_2) = 2.00$ mm.) The sample standard deviation of the 5000 *Sum* measurements was 2.81. Note that this is larger than either of $sd(X_1)$ or $sd(X_2)$, indicating that assemblies are more variable than individual parts, but there is no obvious

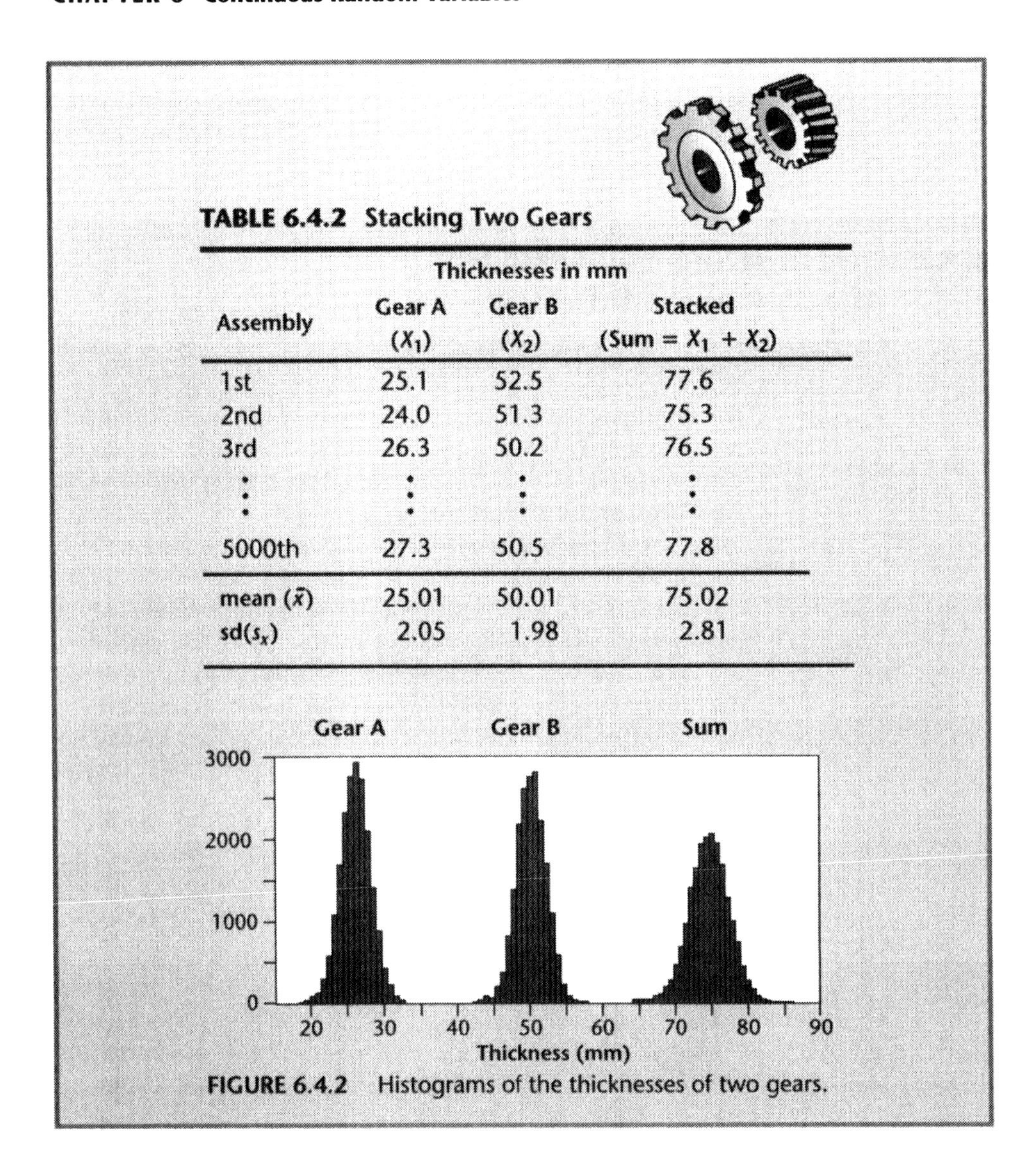

TABLE 6.4.2 Stacking Two Gears

Assembly	Gear A (X_1)	Gear B (X_2)	Stacked (Sum = $X_1 + X_2$)
	Thicknesses in mm		
1st	25.1	52.5	77.6
2nd	24.0	51.3	75.3
3rd	26.3	50.2	76.5
⋮	⋮	⋮	⋮
5000th	27.3	50.5	77.8
mean ($\bar{x}$)	25.01	50.01	75.02
sd(s_x)	2.05	1.98	2.81

FIGURE 6.4.2 Histograms of the thicknesses of two gears.

relationship. If we look at the *squares* of the standard deviations, however, we see that the square of the sample standard deviation of the 5000 measurements on the combined thickness (i.e., on *Sum* $= X_1 + X_2$) is 7.90, which is quite close to $[\text{sd}(X_1)]^2 + [\text{sd}(X_2)]^2$ [= 8.0]. It can in fact be shown mathematically that when X_1 and X_2 are independent, we have the following relationship between the squares of the standard deviations:[5]

$$[\text{sd}(X_1 + X_2)]^2 = [\text{sd}(X_1)]^2 + [\text{sd}(X_2)]^2$$

or

$$\text{sd}(X_1 + X_2) = \sqrt{\text{sd}(X_1)^2 + \text{sd}(X_2)^2}$$

[5]Recall from Section 5.4.2 that the square of a standard deviation is known as the *variance*. We have $\text{Var}(X_1 + X_2) = \text{Var}(X_1) + \text{Var}(X_2)$.

The Behavior of a Difference

Now let us look at the effect of fitting one object inside another. We stay with the automotive theme but use engine pistons inside cylinders this time. We shall measure diameters. Let X_1 be the diameter of the cylinder and X_2 be the diameter of the piston. Then the difference, $D = X_1 - X_2$, is the clearance between the cylinder and the piston. If D is negative, the piston is too big and will not fit in. Again we use Normally distributed random numbers to simulate the part-to-part variability in the dimensions of pistons and cylinders coming off different production lines. We shall use a mean diameter of 100 mm for cylinders with a standard deviation of 4 mm and a mean of 70 mm for pistons with a standard deviation of 4 mm. Variability has again been grossly exaggerated (and the average value of X_2 has been reduced) to show the effects clearly. A production run of 5000 of each component has been simulated. Histograms of the 5000 X_1 measurements (cylinder diameters), X_2 measurements (piston diameters), and D measurements (differences) are given in Fig. 6.4.3, and summary statistics from this data are given in Table 6.4.3.

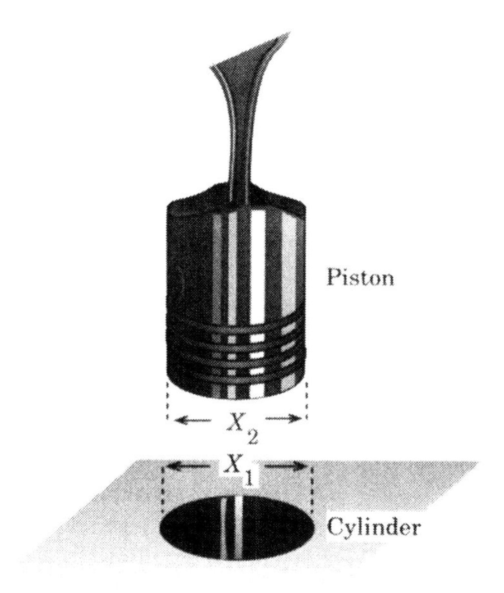

We note that the sample mean value of the differences, $\bar{x}_D[= 29.5]$, is very close to the difference in the population means of the components [= 30], thus illustrating a theoretical relationship that holds for all pairs of random variables, namely, $E(X_1 - X_2) = E(X_1) - E(X_2)$.

We see also that the variability of the differences (D measurements) has turned out to be ***greater*** than the variability of either component taken separately. This is not really surprising since we have two sources of variability, each conspiring to make the situation less predictable. It can be shown mathematically that, for population standard deviations,

$$\text{sd}(X_1 - X_2) = \sqrt{\text{sd}(X_1)^2 + \text{sd}(X_2)^2}$$

We leave it to the reader to verify that the sample standard deviations from our data come very close to obeying this relationship.

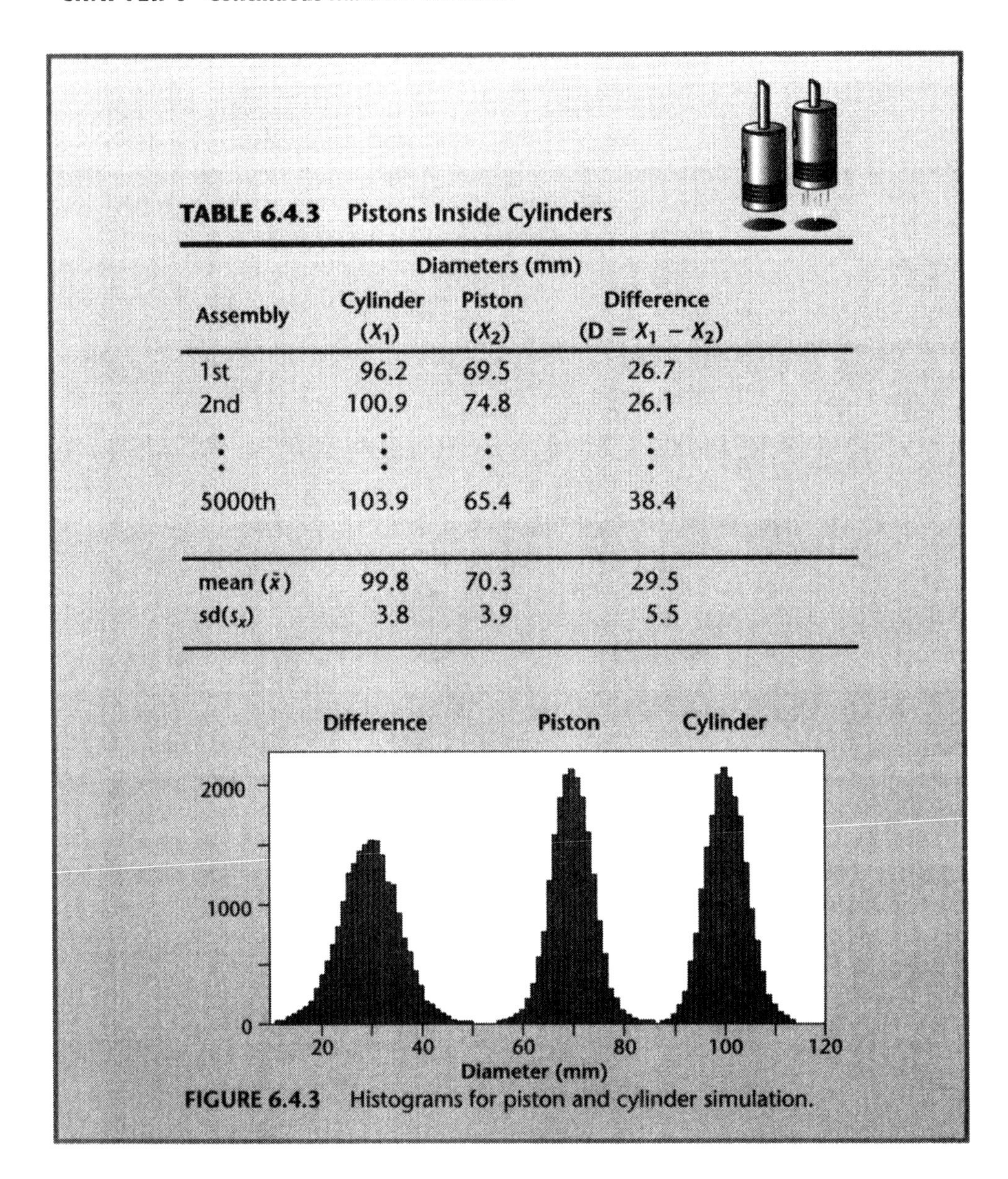

TABLE 6.4.3 Pistons Inside Cylinders

Assembly	Diameters (mm) Cylinder (X_1)	Piston (X_2)	Difference ($D = X_1 - X_2$)
1st	96.2	69.5	26.7
2nd	100.9	74.8	26.1
⋮	⋮	⋮	⋮
5000th	103.9	65.4	38.4
mean ($\bar{x}$)	99.8	70.3	29.5
sd(s_x)	3.8	3.9	5.5

FIGURE 6.4.3 Histograms for piston and cylinder simulation.

6.4.3 Working with Sums and Differences

The results that we have seen about population (or distribution) means and standard deviations of sums and differences in the context of our gear and piston examples are very general. They apply to all distributions (not just the Normal distribution), regardless of whether the distributions are discrete or continuous. We now set them out in an organized form for future use.

Means

$$E(X_1 + X_2) = E(X_1) + E(X_2)$$

The ***mean of the sum*** is the ***sum of the means.***

This result always applies. The idea extends to more than two random variables:

$$E(X_1 + X_2 + \cdots + X_n) = E(X_1) + E(X_2) + \cdots + E(X_n)$$

Moreover,

$$E(X_1 - X_2) = E(X_1) - E(X_2)$$

> The ***mean of the difference*** is the ***difference in the means.***

For example, if $E(X_1) = 3$, $E(X_2) = 1$, $E(X_3) = 8$, and $E(X_4) = 5$, then

(i) $X_1 + X_2$ has mean $3 + 1 = 4$,
(ii) $X_1 + X_2 + X_3 + X_4$ has mean $3 + 1 + 8 + 5 = 17$,
(iii) $X_1 - X_2$ has mean $3 - 1 = 2$, and
(iv) $X_1 + X_2 - X_3 + X_4$ has mean $3 + 1 - 8 + 5 = 1$.

Standard Deviations

The following results (which we saw at work in the gear and piston stories in the previous subsection) apply only if the random variables are ***independent.*** In that case we have both

$$\text{sd}(X_1 + X_2) = \sqrt{\text{sd}(X_1)^2 + \text{sd}(X_2)^2}$$

and

$$\text{sd}(X_1 - X_2) = \sqrt{\text{sd}(X_1)^2 + \text{sd}(X_2)^2}$$

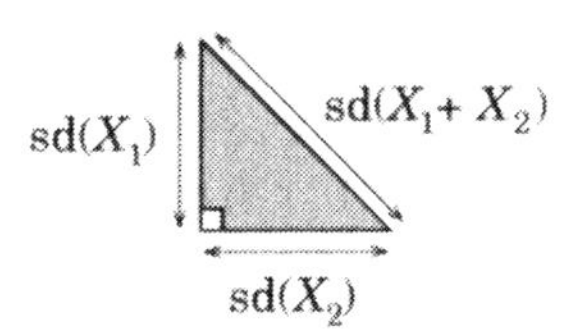

This relationship is the same as that between the lengths of the sides of a right triangle (the Pythagorean theorem). [Note: $\text{sd}(X)^2 = \text{sd}(X) \times \text{sd}(X)$.]

> For ***independent*** random variables,
> ***standard deviations add by the Pythagorean theorem.***

The same idea applies to three or more random variables,[6] for example,

$$\text{sd}(X_1 + X_2 + X_3) = \sqrt{\text{sd}(X_1)^2 + \text{sd}(X_2)^2 + \text{sd}(X_3)^2}$$

[6]Generally, $\text{sd}(X_1 + X_2 + \cdots + X_n) = \sqrt{\text{sd}(X_1)^2 + \text{sd}(X_2)^2 + \cdots + \text{sd}(X_n)^2}$.

To illustrate, suppose that $\text{sd}(X_1) = 3$, $\text{sd}(X_2) = 4$, and $\text{sd}(X_4) = 12$. Then

(i) $X_1 + X_2$ and $X_1 - X_2$ both have standard deviation $\sqrt{3^2 + 4^2} = 5$,

(ii) $X_1 + X_2 + X_3$ has standard deviation $\sqrt{3^2 + 4^2 + 12^2} = 13$, and

(iii) $X_1 + X_2 - X_3$ also has standard deviation $\sqrt{3^2 + 4^2 + 12^2} = 13$.

Constant Multipliers[7]

$$\text{E}(aX) = a\,\text{E}(X) \quad \text{and} \quad \text{sd}(aX) = |a|\,\text{sd}(X)$$

In particular, if X has mean 3 and standard deviation 4, then $2X$ has mean $2 \times 3 = 6$, whereas $-2X$ has mean $-2 \times 3 = -6$. In contrast, both $2X$ and $-2X$ have standard deviation $2 \times 4 = 8$. Sometimes we shall combine what we know about means and standard deviations of sums and differences of random variables, and what we know about the effects of constants.

Normally Distributed Random Variables

The results we saw for means and standard deviations apply regardless of the distribution of the variables. In this chapter, however, we work with Normal distributions. It turns out that

> Sums and differences of independent Normally distributed random variables also have a Normal distribution.

To be able to work with the distribution of a sum or difference, we just have to work out what values to use for the mean and standard deviation. We remind you of the short notations $\mu_X \equiv \text{E}(X)$ and $\sigma_X \equiv \text{sd}(X)$.

EXAMPLE 6.4.2 *Find the Probability that the Piston Fits into the Cylinder*

Suppose we are manufacturing parts so that engine piston diameters are approximately Normally distributed about a mean of 30.00 mm with a standard deviation of 0.05 mm, and cylinder-diameters are approximately Normally distributed about a mean of 30.10 mm with a standard deviation of 0.02 mm. What is the probability that a random piston will fit inside a random cylinder?

Let P refer to the diameter of the piston and C to the diameter of the cylinder. The probability we want to calculate is $\text{pr}(C > P)$. This is impossible for us to handle until we realize that

$$\text{pr}(C > P) = \text{pr}(C - P > 0) = \text{pr}(Y > 0), \text{ where } Y = C - P$$

[7]The boxed formulas are a special case of our earlier results: $\text{E}(aX + b) = a\,\text{E}(X) + b$ and $\text{sd}(aX + b) = |a|\,\text{sd}(X)$.

Now Y is Normally distributed because C and P are (we shall assume independence), and we can work out the mean and standard deviation of Y using the formulas at the beginning of this subsection. We do this in the following table, starting from the information given about C and P.

	Distribution	Mean	Standard deviation
Cylinder (C)	Normal	30.10	0.02
Piston (P)	Normal	30.00	0.05
Difference ($Y = C - P$)	Normal	$30.10 - 30.00 = 0.10$	$\sqrt{0.02^2 + 0.05^2} = 0.0539$

Our problem is thus to find pr($Y > 0$) where $Y \sim \text{Normal}(\mu_Y = 0.10, \sigma_Y = 0.0539)$. The answer is 0.969 (from Excel). Roughly three combinations in every hundred would fail to fit.

EXAMPLE 6.4.3 *Pricing a Job Using a Sum of Variables*

Tradespeople often break down cost estimates for doing a job into time and materials. Suppose that you estimated the time T a particular job would take as roughly 40 hours but represented your uncertainty about this estimate by imagining that the time taken is approximately Normally distributed about a mean of 40 hours with a standard deviation of 8 hours. Similarly, your estimate of the cost of materials M is approximately Normally distributed with a mean of \$1000 and a standard deviation of \$100. There are several other people competing for this job, so the price you quote cannot be too high. You want \$30 per hour, however. To be 80% sure of covering costs, what price should you quote for the job?

Let Y be the total cost of the job. Let the quoted price be c. We need to set c at the value that makes pr($Y \leq c$) = 0.8. To be able to do this, we need to know the distribution of Y.

First, we convert the time taken T to a time cost (say, C in dollars), giving us $C = 30T$. Our total cost is $Y = C + M$. All our random variables are Normally distributed. It seems reasonable to assume independence.[8] The mean and standard deviation of Y are calculated in the following table.

	Mean		Standard deviation
Time, T (hours)	(given)	40	8
Time costs, $C = 30T$	30×40 =	1200	$30 \times 8 = 240$
Materials costs, M	(given)	1000	100
Total cost, $Y = C + M$	$1200 + 1000$ =	2200	$\sqrt{240^2 + 100^2} = 260$

Thus total cost $Y \sim \text{Normal}(\mu_Y = 2200, \sigma_Y = 260)$.

[8]That is, feelings about the time the job will take are independent of the feelings about how much the materials will cost.

Our problem has now been reduced to an inverse Normal distribution problem, as in Section 6.2.3, namely, determine c so that $\text{pr}(Y \leq c) = 0.8$, where $Y \sim \text{Normal}(\mu_Y = 2200, \sigma_Y = 260)$. The answer is $c = \$2419$ (from Excel, to the nearest dollar).

Suppose that we now want to convert these figures into a different currency, say from Australian to U.S. dollars. On the day of writing, Aus \$1 = US \$0.65 so that Australian dollars are converted to U.S. dollars by multiplying by 0.65. We convert the mean and standard deviation to U.S. dollars by multiplying both of them by 0.65.

Example 6.4.4 *Mean and Standard Deviation of the Stack Height*

Let us return to the gear assembly of Example 6.4.1. The means and standard deviations of the thicknesses of the various gears were given in Table 6.4.1. The stack height, or total height Y of the assembly, is the sum of the thicknesses, that is, $Y = X_1 + X_2 + \cdots + X_6$. The mean stack height, $\mu_Y = \text{E}(X_1 + X_2 + \cdots + X_6)$, can be obtained using the fact that the mean of the sum is the sum of the means to give us

$$\mu_Y = 12.03 + 3.56 + 20.88 + 30.42 + 3.88 + 81.75 = 152.52$$

The standard deviation of assembly stack heights (assuming independence) is

$$\begin{aligned} \text{sd}(Y) &= \text{sd}(X_1 + X_2 + \cdots + X_6) = \sqrt{\text{sd}(X_1)^2 + \cdots + \text{sd}(X_6)^2} \\ &= \sqrt{0.01^2 + 0.00^2 + 0.02^2 + 0.03^2 + 0.00^2 + 0.04^2} \\ &= 0.055 \end{aligned}$$

Recall that the gear assembly story was about a quality improvement consultant trying to reduce the variability of the length of the assembly. Suppose we found a way to improve the manufacturing process to the point where the standard deviations of the final-drive internal gear (X_1) and the parking gear (X_3) were reduced to 0.00 (to two decimal places). What effect would this have on the variability of the total length of the assembly? Arguing as before, the standard deviation is reduced only to 0.05. It has hardly changed! Even if we could reduce all the standard deviations except the largest to zero, the standard deviation of the assembly length would still be 0.04. The lesson of this example is that if you want to make a substantial reduction in the variability of a combination of components, you have to be able to reduce the variability of the worst (most variable) component. Looking for ways of improving the most variable component should be your top priority.

Remember the company's solution to the problem of variability in assembly heights and casing dimensions. They sorted assemblies and casings by size, which sounds sensible enough. Why didn't sorting work? The assemblies were swung into place onto a plate, where the stack height was automatically measured. The measuring device itself was extremely accurate. The variability in the orientation of assemblies when they were swung into place on the plate, however, made the resulting measurements unreliable!

EXERCISES FOR SECTION 6.4.3

1. Suppose X_1, X_2, X_3, and X_4 are independent random variables with $\text{E}(X_1) = 2.5$, $\text{sd}(X_1) = 3$, $\text{E}(X_2) = 1.5$, $\text{sd}(X_2) = 2$, $\text{E}(X_3) = 5$, $\text{sd}(X_3) = 5$, $\text{E}(X_4) = -4$, and $\text{sd}(X_4) = 3$. For each definition of the random variable Y below, calculate $\text{E}(Y)$ and $\text{sd}(Y)$.

(a) $Y = X_1 + X_2$ (b) $Y = X_1 - X_2$ (c) $Y = X_1 + X_3 + X_4$
(d) $Y = X_2 + X_3$ (e) $Y = X_2 - X_3$ (f) $Y = X_1 + X_2 - X_3$
(g) $Y = X_3 + X_4$ (h) $Y = X_3 - X_4$ (i) $Y = X_1 + X_3 - X_4$

Suppose we want to find the mean and standard deviation of $2X_1 + 3X_2$. Note that $2X_1 + 3X_2 = W_1 + W_2$, where $W_1 = 2X_1$ and $W_2 = 3X_2$. This suggests doing the calculations in two steps (as in the table in Example 6.4.3).

Step 1: Find the mean and standard deviation of $W_1 = 2X_1$ and also of $W_2 = 3X_2$.

Step 2: Find the mean and standard deviation of $W_1 + W_2$.

Find the mean and standard deviation of

(j) $Y = 2X_1 + 3X_2$ (k) $Y = 5X_1 - 4X_2$ (l) $Y = 2X_1 + 3X_2 + 4X_3$

2. CDI measures depression in children. It is measured by the score a child obtains on a questionnaire about symptoms and feelings. (Psychologists call something like this an *instrument.*) CDI has been used for many years in research on the effects of treatments for depression, with a reduction in average CDI scores indicating a beneficial effect. But how would it rate as a tool for diagnosing depression? In one study assessing CDI, scores among clinic referrals for depression had a mean of 11.2 and a standard deviation of 6.8, whereas scores for a large group of public school children had a mean of 8.5 and a standard deviation of 7.8. Assume that these figures relate to the depressed and not depressed populations respectively and assume Normal distributions.
   (a) What is the probability that a randomly chosen depressed child actually produces a lower CDI score than a randomly chosen not depressed child? (Hint: See Example 6.4.2.)
   (b) Can you see a practical difficulty in taking the public school children as representing the not depressed?

6.4.4 Application to Random Samples

The idea of random variables forming a random sample from a distribution is fundamental to the theory underlying the statistical methods of subsequent chapters because these methods all assume that our data comes from a random sample. Technically, random variables $X_1, X_2, \ldots, X_n$ form a ***random sample*** from a distribution if

(i) they all have the same distribution and
(ii) they are independent of one another.

The ideas of this definition are easiest to grasp in the context of sampling real objects or people from a finite population. Suppose we sample n people from a real human population at random *with replacement* and measure the amount of tax they paid. Let X_i be the tax paid by the ith person who is chosen in this way. Each X_i has the same distribution, namely, the distribution of taxes paid in the population from which we are sampling. Because we are sampling at random with replacement, the results of previous draws can have no influence on any subsequent draw, so the X_i's are also independent. Thus the terms of the definition are satisfied, and we have a random sample from a distribution. In practice, we sample from finite populations *without replacement.* A sample taken at random from a finite population without replacement is called a *simple random sample* (see Section 1.1.1). For this type of sample, successive drawings of a new person are not independent because the people remaining from which to sample depend on what people have already been removed from the population. In most real sampling, however, the population size is much larger than the sample size. Removing a small number of randomly chosen individuals from the population makes very little difference to the proportions of people in the population with various characteristics (for example, paying a certain amount of tax). Choosing the next person from the remainder is thus much the same as choosing a person from the whole population (that is, sampling with replacement). When the population size is much larger than the sample size, a simple random sample thus behaves like a random sample from a distribution.

Random processes are the other major source of random data. An idealized random-number generator produces an independent stream of numbers $(X_1, X_2, \ldots)$, all from the same distribution. Thus n successive numbers from the generator provide a random sample. Where possible, we try to carry these ideas over to real processes. Consider again the diameters of pistons produced by an assembly line. A very simple model for the process imagines pistons whose diameters vary randomly and independently according to some distribution. A sequence of observations from such a process forms a random sample from a distribution, which we can use to investigate the behavior of the process. In what follows we shall usually drop the words "from a distribution" and simply talk about a random sample. Extreme caution is required, however, when treating the output of any real process as a random sample. Let X_i represent the diameter of the ith piston produced by the process since we started monitoring it. A model by which the X_i's are independent and all come from the same distribution will often be entertained because it is easy to understand and the statistical analysis of data is comparatively easy. A deeper understanding of the process or the data collected from the process, however, may force us to abandon this model in favor of a more complicated description. For real processes monitored over time we often observe patterns of dependence (for example, runs of observations that are all bigger or all smaller than average) and changes in features of the distribution, such as a pattern of change in mean level or variability over time. This kind of problem is central to the concerns of statistical process control and improvement discussed in a supplementary chapter available from our web site.

Mean and Standard Deviation of the Sum of Observations from a Random Sample

Suppose that $X_1, X_2, \ldots, X_n$ represent a random sample from a distribution with mean μ, standard deviation σ, and $Sum = X_1 + X_2 + \cdots + X_n$. Any X_i has mean μ and standard deviation σ since it is sampled from a distribution with these characteristics. The mean of the sum is the sum of the means, namely, $\mu + \cdots + \mu = n\mu$, and

$$\begin{aligned} \text{sd}(X_1 + X_2 + \cdots + X_n) &= \sqrt{\text{sd}(X_1)^2 + \cdots + \text{sd}(X_n)^2} \\ &= \sqrt{\sigma^2 + \sigma^2 + \cdots + \sigma^2} \qquad (n \text{ of them}) \\ &= \sqrt{n}\,\sigma \end{aligned}$$

> For a random sample from a distribution with mean μ and standard deviation σ,
>
> $$\mu_{Sum} = E(Sum) = n\mu \quad \textbf{and} \quad \sigma_{Sum} = \text{sd}(Sum) = \sqrt{n}\sigma$$

✱ Example 6.4.5 *Counting Coins by Weighing Them*

Suppose you have a large number of coins, all of the same denomination, and you wish to determine the number of coins by weight rather than by counting. In other words, you will obtain the number of coins by dividing the total weight of coins by the average weight of a single coin. As with all manufactured items, the weights of coins vary from coin to coin. The variability will be very small, but provided there are enough coins, the accumulated variability in total weight will be sufficiently large that you may sometimes be mistaken about the exact number of coins.

Suppose the weights of the coins are Normally distributed about a mean of 10.0 g with a standard deviation of 0.1 g. We get our estimate of the number of coins by dividing the total weight of coins by the mean weight of a single coin and rounding the answer to the nearest whole number. What would the chances of making a mistake be if we had 3000 coins? Our estimate would be correct (3000 coins) only if the total weight was between 29,995 g and 30,005 g. What is the probability that the total weight of 3000 coins falls outside this range?

We are given that the weights of coins are Normally distributed with a mean of $\mu = 10$ g and a standard deviation of $\sigma = 0.1$ g. Thus we are concerned with the distribution of the total weight of 3000 different coins, that is, of

$$Y = X_1 + X_2 + \cdots + X_n$$

where $n = 3000$. From our theory, $\mu_Y = 3000 \times 10$, and $\sigma_Y = \sqrt{3000} \times 0.1 = 5.477$, so that $Y \sim \text{Normal}(\mu_Y = 30{,}000, \sigma_Y = 5.477)$. The probability of making a mistake is thus[9]

$$\text{pr}(Y < 29{,}995) + \text{pr}(Y > 30{,}005) = 0.36$$

Counting this way, we would make a mistake approximately 36% of the time. If the weighing method is to work reliably, we need to weigh the coins in smaller batches.

Independent Individuals versus Clones

Weights of women in the workforce database have a mean of 68.1 kg with a standard deviation of 12.2 kg. When asked to find the mean and standard deviation of the total weight of a sample of, say, 100 women, many students will write

$$\text{Total weight} = 100 \times X$$

where $\text{E}(X) = 68.1$ and $\text{sd}(X) = 12.2$ instead of using

$$\text{Total weight} = X_1 + X_2 + \cdots + X_{100}$$

What is the difference, and why does it matter? It matters because $\text{sd}(100X) = 100\text{sd}(X) = 1220$, whereas $\text{sd}(X_1 + X_2 + \cdots + X_{100}) = \sqrt{100}\,\text{sd}(X) = 10\,\text{sd}\,(X) = 122$. Both expressions give the same mean (check this), but there is a huge difference in the standard deviations. If you conceptualize the problem wrongly, your measure of variability will be grossly incorrect.

There is a big difference in the meaning of the expressions. Using $Y = 100X$ is essentially saying that we have 100 *identical* copies of a single randomly sampled female. In contrast, $Y = X_1 + X_2 + \cdots + X_{100}$ is talking about 100 different individually

[9]We use $\text{pr}(Y > 30{,}005) = 1 - \text{pr}(Y \leq 30{,}005)$.

sampled women. Some will be a bit larger than average, some a little smaller. There will be some tendency for the individual deviations to cancel out. Thus it should not be surprising that the total weight of 100 different women is less variable than 100 times the weight of a single randomly chosen woman.

QUIZ ON SECTION 6.4

1. Does the rule that the mean of the sum equals the sum of the means apply to random variables that are not independent?
2. What is the rule for the standard deviation of the sum of two independent random variables? (Remember Pythagoras.)
3. Express the mean and standard deviation of $5X$ in terms of the mean and standard deviation of X. Repeat for $-5X$.
4. Why is the difference between two independent random variables more variable than either quantity alone? (This also applies to the sum.)
5. What difference, if any, is there between the standard deviation of the difference of two independent random variables and the standard deviation of the sum?
6. We treat a sequence of observations on the output of a real process as a random sample when we can because that is the simplest model we can formulate. Why is this not always possible?
7. What is the conceptual difference between $X_1 + X_2 + \cdots + X_n$ and nX? Which would you expect to be more variable? Why?

6.5 SUMMARY

6.5.1 Continuous Variables and Density Curves

1. There are no gaps between the values a continuous random variable can take.
2. Random observations arise in two main ways: (i) by sampling populations and (ii) by observing processes.
3. The probability distribution of a continuous variable is represented by a ***density curve***.
   - Probabilities are represented by areas under the curve, with the ***probability*** that a random observation falls between a and b equal to the ***area*** under the density curve between a and b.
   - The total area under the curve equals 1.
   - The population (or distribution) mean, $\mu_X = \text{E}(X)$, is where the density curve balances.
   - When we calculate probabilities for a continuous random variable, it does not matter whether interval endpoints are included or excluded.
4. For any random variable X,

$$\text{E}(aX + b) = a\,\text{E}(X) + b \qquad \text{and} \qquad \text{sd}(aX + b) = |a|\,\text{sd}(X)$$

6.5.2 The Normal Distribution

$$X \sim \text{Normal}(\mu_X = \mu, \sigma_X = \sigma)$$

1. Features of the Normal density curve:
   - The curve is a symmetric bell shape centered at μ.
   - The standard deviation σ governs the spread.
   - 68.3% of the probability lies within one standard deviation (sd) of the mean, 95.4% within two standard deviations of the mean, and 99.7% within three standard deviations.
2. Probabilities:
   - Computer programs provide *lower-tail* (or *cumulative*) probabilities of the form $\text{pr}(X \le x)$. We give the program the x-value; it gives us the probability.
   - Computer programs also provide *inverse* lower-tail probabilities (or *quantiles*). We give the program the probability; it gives us the x-value.
   - When calculating probabilities, we shade the desired area under the curve and then devise a way of obtaining it via lower-tail probabilities.
3. Standard units
   - The ***z-score*** of a value a is the number of standard deviations a is away from the mean.
   - The z-score of a is positive if a is above the mean and negative if it is below the mean.
   - The ***standard Normal*** distribution has $\mu = 0$ and $\sigma = 1$.
   - We usually use Z to represent a random variable with a standard Normal distribution.
   - Central ranges: $\text{pr}(-z \le Z \le z)$ is the same as the probability that a random observation from an arbitrary Normal distribution falls within z standard deviations on either side of the mean.
   - Extremes: $\text{pr}(Z \ge z)$ is the same as the probability that a random observation from an arbitrary Normal distribution falls more than z standard deviations above the mean. $\text{pr}(Z \le -z)$ is the same as the probability that a random observation from an arbitrary Normal distribution falls more than z standard deviations below the mean.

6.5.3 Combining Random Quantities

1. Variation and independence:
   - No two animals, organisms, or natural or human-made objects are ever identical.
   - There is always variation. The only question is whether it is large enough to have a practical impact on what you are trying to achieve.
   - Variation in component parts leads to even greater variation in the whole.
   - We model variables as being independent if we think they relate to physically independent processes and if we have no data that suggests they are related.
   - Both sums and differences of independent random variables are *more* variable than any of the component random variables.
2. Formulas:
   - For a constant number a, $E(aX) = a\,E(X)$ and $\text{sd}(aX) = |a|\,\text{sd}(X)$.
   - Means of sums and differences of random variables act in an obvious way: The *mean of the sum* is the *sum of the means;* the *mean of the difference* is the *difference in the means.*
   - For ***independent*** random variables (cf. Pythagorean theorem),

$$\text{sd}(X_1 + X_2) = \text{sd}(X_1 - X_2) = \sqrt{\text{sd}(X_1)^2 + \text{sd}(X_2)^2}$$

- Sums and differences of independent Normally distributed random variables are also Normally distributed.

REVIEW EXERCISES 6

1. A particular test for testing various aspects of verbal memory is known as the Selective Reminding Task (SRT) test. It is based on hearing, recalling, and learning 12 words presented to the client. Scores for various aspects of verbal memory such as total recall, storage into long-term memory, and so on are combined to give an overall score. Let X be the score of a female student in a particular age group, and suppose that $X \sim \text{Normal}(\mu = 126, \sigma = 10)$.

(a) Find the following probabilities:

(i) $\text{pr}(X > 141)$ (ii) $\text{pr}(120 < X < 132)$ (iii) $\text{pr}(X < 118.5)$

(b) What is the z-score of someone who scored 120 on the SRT test? Express verbally what the z-score means.

(c) Repeat (b) for someone who scored 140 on the SRT test.

(d) A firm wishes to use the test to employ a student in a job requiring similar memory skills. They wish to employ someone in the top 15% for the test. What score should a student get to be eligible?

(e) Repeat (d) for the top 10%.

(f) A student who is tested after an accident is suspected of having internal head injuries if his or her score is too low. Below what value do 1% of students' SRT test scores fall?

(g) What range of values covers the central 90% of students' test results?

(h) Repeat (g) for the central 60% of test results.

(i) Twenty students took the SRT test. Assuming that they are a random sample from the same population as given, what is the distribution of the total of these students' SRT scores? Deduce the distribution of the average of the 20 scores.

2. A medical trial was conducted to investigate whether a new drug extended the life of a patient who had lung cancer. The survival times (in months) for 38 cancer patients who were treated with the drug are as follows:

1	1	5	9	10	13	14	17	18	18	19	21	22	25	25	25	26	27	29
36	38	39	39	40	41	41	43	44	44	45	46	46	49	50	50	54	54	59

The sample mean is approximately 31.1 months and the standard deviation is approximately 16.0 months.

(a) Assume that the survival time (in months) for patients on this drug is Normally distributed with a mean of 31.1 months and a standard deviation of 16.0 months. Calculate

(i) the probability that a patient survives for no more than one year,

(ii) the proportion of patients who survive for between one year and two years,

(iii) the highest number of months that 80% of the patients survive, and

(iv) the central 80% of survival times.

(b) Draw a stem-and-leaf plot of the number of months of survival of these 38 patients.

(c) Does the plot gives us any reason to question the validity of the assumption that the survival time is Normally distributed?

3. The designer of the cockpit of a new aircraft wants to position a switch so that most pilots can reach it without having to change position. Suppose that the distribution among airline pilots of the maximum distance (measured from the back of the seat) that can be reached without moving the seat is approximately Normally distributed with mean $\mu = 125$ cm and standard deviation of 10 cm.

(a) If the switch is placed 120 cm from the back of the seat, what proportion of pilots will be able to reach it without moving the seat?

(b) What is the maximum distance from the back of the seat that the switch could be placed if it is required that 95% of pilots be able to reach it without moving the seat?

(c) If a pilot has a z-score of 1.5, what does that mean in this context? To what maximum reach does a z-score of 1.5 correspond?

4. A machine used to regulate the amount of dye dispensed for mixing shades of paint can be set so that it discharges an average of μ mL of dye per can of paint. The actual amount of dye discharged is variable. Suppose it has a Normal distribution with standard deviation equal to 0.4. If more than 6 mL of dye are discharged when making a particular shade of blue paint, the shade is unacceptable. Determine the setting for μ so that no more than 1% of the cans of paint will be unacceptable.

5. Altman [1980] discussed a divorce case heard in 1949 in which the sole evidence of adultery was that a baby was born almost 50 weeks after the husband had gone abroad on military service. The husband was appealing against the failure of his earlier petition for divorce on the grounds of his wife's adultery. The judges agreed however, that although 349 days was improbable on medical grounds, it was scientifically possible and the child could have been his. He lost the appeal. The gestation period in days for human births can be taken as Normal($\mu_X = 266, \sigma_X = 16$).

(a) Calculate the probability of a gestation period lasting at least 349 days.

(b) With the hindsight of statistical theory, what do you think of the judges' decision?

(c) Out of 10 million births, what is the probability that a gestation period of 349 or more days happens at least once? What do you think now? (Hint: What distribution is relevant?)

6. A standard test for diabetes is based on glucose levels in the blood after fasting for a prescribed period. For healthy people the mean fasting glucose level is found to be 5.31 mmol/L with a standard deviation of 0.58 mmol/L. For untreated diabetics the mean is 11.74, and the standard deviation is 3.50. In both groups the levels appear to be approximately Normally distributed.

To operate a simple diagnostic test based on fasting glucose levels we need to set a cutoff point, C, so that if the patient's fasting glucose level is at least C we say they have diabetes. If it is lower, we say they do not have diabetes. Suppose we use $C = 6.5$.

(a) What is the probability that a diabetic is correctly diagnosed as having diabetes?

(b) What is the probability that a nondiabetic is correctly diagnosed as not having diabetes?

In medicine the probability of correctly diagnosing an individual as having the condition of interest (here diabetes) is called the *sensitivity* of a test, while the probability of correctly diagnosing that a healthy individual does not have the condition of interest is called the *specificity* of the test. Ideally we want tests that are both very sensitive (very good at picking up the condition if it is there) and very specific (hardly ever signal the presence of the condition if it is not there). With blood glucose levels, as with many other tests, however, there is some overlap in levels between the two groups. Some healthy individuals have higher levels than some diabetics, so our test cannot be both 100% sensitive and 100% specific.

(c) (i) What is the sensitivity of the test if we use $C = 5.7$?

(ii) What is the specificity?

By looking at your answers to (a), (b), and (c), you will see that by making C smaller, we get a more sensitive test that is less specific, whereas by making C large, we get a more specific test that is less sensitive. In deciding what value of C to use, we have to trade off sensitivity for specificity. To do so in a reasonable way, some assessment is required of the relative "costs" of misdiagnosing a diabetic and misdiagnosing a nondiabetic. Suppose we required 98% sensitivity.

(d) What value of C gives a sensitivity of 0.98? How specific is the test when C has this value?

(e) What is the probability that a randomly chosen nondiabetic has a higher fasting glucose level than a random untreated diabetic?

7. A standard test for gout is based on the serum uric acid level. The level is approximately Normally distributed with mean 5.0 mg/100 mL and standard deviation 1.0 mg/100 mL among healthy individuals and with mean 8.5 mg/100 mL and standard deviation 1.0 mg/100 mL among individuals with gout.

Suppose we diagnose people as being healthy if their serum uric acid level is less than 6.75 mg/100 mL and as having gout if the level is more than 6.75 mg/100 mL.

(a) What is the probability that a person with gout will be diagnosed as healthy?

(b) What is the probability that a healthy person will be diagnosed as having gout?

(c) If 50 *healthy* people take the test in a given week, what is the distribution of X, the number of healthy people diagnosed as having gout in that week?

Suppose we now wish to change the serum uric acid level used to diagnose gout.

(d) What should the level be if 90% of people with gout are to be correctly diagnosed?

(e) At this new level, what proportion of healthy people will be diagnosed as having gout?

8. A very large number of fish are swimming in a lake. Their weights are Normally distributed with mean 1.30 kg and standard deviation 0.40 kg. Fish caught with weights less than 0.5 kg have to be thrown back.

(a) If an angler catches a randomly selected fish, what is the probability that the fish has to be thrown back?

(b) Find the probability that it is not thrown back and, at the same time, does not exceed the record of 2.06 kg.

(c) The local angling club decides at the beginning of the season to give prizes for any fish caught in the heaviest 5% of the fish population. What should be the minimum weight of a fish to warrant a prize?

On a certain holiday some anglers hold a competition and among them catch 900 fish. Let Y be the number of fish thrown back.

(d) Name the distribution of Y, and give values for its parameters.

(e) Find the probability that between 25 and 35 (inclusive) fish are thrown back.

9. A company sells naturally grown bean sprouts to supermarkets in bags labeled 100 g each. The price is 70 cents per bag. Data from an automatic weigher show that the weights of bags are Normally distributed with standard deviation of 5 g. In the weighing process, bags weighing less than 100 g are separated out and sold at a discount price of 65 cents each. The production cost for a bag weighing 100 g or less is 25 cents, while that for a bag weighing more than 100 g is 26.5 cents.

(a) If the target mean weight is set at 106 g (i.e., the filling machine is set at 6% overfill), find the percentage of bags sold at a discount.

(b) In a 100 randomly chosen bags, let Y be the number sold at a discount. What is the distribution of Y?

(c) Obtain an expression for P, the net profit from 100 randomly selected bags, in terms of Y. Find $E(P)$, the expected net profit.

*(d) To comply with labeling regulations, the discounted bags need to have a weight stamped on them other than 100 g. If 95 g is used, what percentage of the discounted bags would be underweight? (Hint: You will need to find a conditional probability for the condition that X is less than 100.)

10. A criminologist in the United States developed a questionnaire for predicting whether a teenager will become delinquent or not. Scores on the questionnaire can range from 0 to 100, with higher values supposedly reflecting a greater criminal tendency. As a rule of thumb the criminologist decides to classify a teenager as potentially delinquent if the teenager's score exceeds 75. She has already tested the questionnaire on a large sample of teenagers, both delinquent and nondelinquent. Among those considered nondelinquent, scores were approximately Normally distributed with a mean of 60 and a standard deviation of 10. Among those considered delinquent, scores were approximately Normally distributed with a mean of 80 and a standard deviation of 5. Suppose that 5% of the teenage population is delinquent. A teenager is chosen at random and tested.

(a) Given that the teenager is not delinquent, what is the probability that she will misclassify the teenager as delinquent?

(b) Given that the teenager is delinquent, what is the probability that she will misclassify the teenager as not delinquent?

*(c) What is the probability that the teenager is misclassified?

11. From data given in Table 2.3.2, Nova Scotian male coyotes have a mean length[10] of 92.0 cm with a standard deviation of 6.7 cm, and females have a mean length of 89.2 cm with a standard deviation of 6.6 cm. We shall assume Normal distributions.

(a) The lengths of the coyotes are important in designing a trap. What length is exceeded by only 10% of female coyotes? What proportion of male coyotes exceed this length?

(b) The first coyote caught during a radio-transmitter tagging program was a male, and the second was a female. The female was longer than the male. What is the probability that a randomly chosen female coyote is longer than a randomly chosen male coyote?

(c) Under the same assumptions, what is the probability that a random male is more than 10 cm longer than a random female?

(d) Consider the independence assumption. A male and female coyote caught in close proximity may be mates. In human couples, are the heights of males and females independent, or do they tend to be related in some way?

12. A university professor keeps records of his travel time while he is driving between his home and the university. Over a long period of time he has found that his morning travel times are approximately Normally distributed with a mean of 31 minutes and a standard deviation of 3.0 minutes; his return journey in the evening is similarly distributed but with a mean of 35.5 minutes and a standard deviation of 3.5 minutes.

(a) Find the probability that on a typical day he spends more than one hour traveling to and from work.

(b) Find the probability that on a given day his morning journey is longer than his evening journey.

(c) On what proportion of days is the evening journey more than five minutes longer than the morning journey?

(d) Over a five-day work week, what is the distribution of the total travel time for morning journeys? for evening journeys? altogether?

(e) A new route was recently opened. He has discovered that if he leaves home at 8:30 a.m. and uses this new route, on 88% of days he arrives at the university by 9:00 a.m. If the times of travel are still approximately Normally distributed with a standard deviation of 3.0 minutes, find the mean time for his morning trip using the new route.

13. The certificate on one of the elevators in our university states, "Maximum load 800 kg, capacity 11 persons."

(a) Suppose that the adult population has a mean weight of 73 kg with a standard deviation of 13 kg. What is the probability that the elevator will be overloaded when filled to

[10]Although these figures came from data from a sample, we shall assume that they are population values.

capacity? What assumptions did you make to obtain your answer? Which, if any, of these assumptions are you doubtful about?

Using values from a large survey, weights in the adult female population have a mean of 68 kg with a standard deviation of 12 kg, whereas in the adult male population the mean is 78 kg with a standard deviation of 12 kg.

(b) Eleven adult men enter the elevator for the upward journey. Assuming random selection, what is the distribution of their total weight? What is the probability that the elevator is overloaded?

(c) For the downward journey, seven men and four women enter the elevator. What is the probability that it is overloaded?

(d) What is the probability that the total weight of the group in (c) is greater than that of the group in (b)?

*(e) Can you suggest a maximum number of people that is more realistic than 11? Support your suggestions with probability calculations.

(f) The preceding weight figures are taken from working populations in which the average age is older than it is in a university population. What effect would you expect this to have?

14. For an assignment the number of correct answers obtained by students in a large class had a mean of 58 with a standard deviation of 12. Let X_i be the number of correct answers obtained by the ith student, where students appear on the list in ID-number order. Find mathematical expressions for Y_1, Y_2, and Y_3 given in the following questions in terms of the appropriate X_i's. Calculate the mean and standard deviation for each Y_i.

(a) Y_1 is the total number of correct answers obtained by the first eight students on the class list.

(b) Y_2 is the total number of correct answers obtained by the first eight students, assuming that they all copied their answers from the third student and thus got the same number correct.

(c) Y_3 is the average number of correct answers obtained by the next eight students.

(d) What have you assumed in doing the preceding problems? What worries, if any, do you have about these assumptions?

15. A smelter produces copper ingots with a mean weight of 500 g and a standard deviation of 10 g. Assume further that the weights are Normally distributed. Batches of 100 ingots are weighed together.

(a) What is the probability that the total weight of a batch lies between 49.9 kg and 50.1 kg?

*(b) A batch is accepted if the total weight is at least 49.9 kg; otherwise each of the 100 ingots is weighed separately. Let X be the total number of weighing operations required (including the initial weighing of the batch). Find the probability function of X and the expected value of X.

16. Consider an idealized stock market. Only four companies are traded on this market: Company X, Company Y, Company V, and Company W. Your broker says that the current share price for each of these is \$10 but that there have been major differences in the volatility (variability) of the share prices over time. We can consider the price that the shares will have in one month from now to be well approximated by the following random variables: $X \sim \text{Normal}(10, 1)$, $Y \sim \text{Normal}(10, 2)$, $V \sim \text{Normal}(10, 3)$, and $W \sim \text{Normal}(10, 0.5)$.

She also indicates that the stocks behave independently. You have \$10,000 to invest and are presented with three possible portfolios.

- Portfolio 1 consists of 1000 shares of W.
- Portfolio 2 consists of 500 shares of W and 500 shares of V.
- Portfolio 3 consists of 250 shares of each of the four.

Let P_i be the value of portfolio i one month from now.

(a) For each of the three portfolios, calculate the following probabilities:

(i) $\text{pr}(P_i > 12{,}000)$

(ii) $\text{pr}(P_i < 5000)$

(b) Which portfolio is the safest, and which is the riskiest?

(c) You enter a wager with a friend that requires you to make $2500 profit on your investment. Without calculating any probabilities, which portfolio should you choose and why?

17. The core of a transformer consists of 50 layers of sheet metal and 49 insulating paper layers. The thickness of a layer of a single metal sheet is Normally distributed with mean 0.5 mm and standard deviation 0.05 mm. The thickness of an insulating paper layer is Normally distributed with mean 0.05 mm and standard deviation 0.02 mm. The core of the transformer, once produced, should fit neatly into a box that has a height of 28.3 mm.

A random sample of 50 sheets was taken from the manufacturing process. The company feels that as long as the combined thickness of the 50 metal sheets is less than 26 mm, the process is working correctly.

(a) Let X be the combined thickness of the 50 metal sheets.

(i) What is the distribution of X?

(ii) What is the probability that the combined thickness of the 50 metal sheets exceeds 26 mm?

(b) Verify that Y, the total thickness of the 49 randomly chosen layers of insulating paper, is Normal($\mu = 2.45, \sigma = 0.14$).

(c) Let W be the total thickness of the combined layers of sheet metal and insulating papers that make up the core of the transformer. What proportion of the transformers will be rejected by the company because they don't fit in their boxes?

(d) The machine used for producing the insulating paper produces a large sheet of insulating paper of approximately uniform thickness. This large sheet is then cut up into 49 smaller sheets that are expected to be used in the same transformer. Let M be the total thickness of 49 sheets of insulating paper produced by this machine.

(i) What is the distribution of M?

(ii) What proportion of transformer cores will now be rejected?

(e) Identify the difference between the formulation of the random variables M and Y.

18. Some market researchers were investigating the consumer profile of two airlines, called A and B to maintain confidentiality. They asked 440 people to rate the airlines' performances from 1 to 10 on various issues. The ratings were assumed to be approximately Normally distributed. Table 1 gives the results of the survey. We shall assume that the means and standard deviations in Table 1 correspond to the true population means and standard deviations.

TABLE 1 Airline Performance

	A		B	
Issue	Mean	Standard deviation	Mean	Standard deviation
1. Value for money	6.8	0.2	7.0	0.3
2. Frequent-flyer program	5.5	0.2	7.3	0.3
3. Safety	8.8	0.4	4.1	0.8
4. Personal comfort	7.4	0.8	6.2	1.2
5. Care for environment	6.2	0.5	5.1	0.4

(a) Above what rating did 80% of those surveyed give A for safety? Above what rating did 80% of those surveyed give B for safety?

Consider the following scenario. The researchers wanted to be able to evaluate each airline using just a single number, that is, one overall index of performance. They decided to use a weighted combination of the individual issue ratings using the following formula:

$$\text{Index} = 10 \times \text{Issue } 1 + 5 \times \text{Issue } 2 + 2 \times \text{Issue } 3 + \text{Issue } 4 + \text{Issue } 5$$

The multipliers were chosen using another survey of passengers. Respondents were asked, on a 10-point scale, how important each issue was to them in determining their choice of airline. The multipliers are average importance ratings rounded to the nearest integer.

(b) Assume that the ratings given for different issues are independent.
  (i) What is the distribution of the overall index for A?
  (ii) What is the distribution of the overall index for B?
  (iii) Suppose we choose a person at random to interview. What is the probability that they give A an overall index that is higher than they give to B?

(c) Comment on the assumptions of Normality and independence. If you had all the data, how might you go about checking these two assumptions?

19. A manufacturer distributes his products in cartons, which are stacked on wooden pallets. Each carton when packed has a weight that is Normally distributed with a mean of 100 kg and standard deviation of 1.25 kg. The pallets have weights that are Normally distributed with a mean of 150 kg and a standard deviation of 3.0 kg. A loaded pallet is stacked with 64 fully packed cartons.

(a) Packed cartons with a weight exceeding 101 kg are strapped to prevent the carton from bursting when stacked on the pallet. What is the probability that a randomly chosen carton will have to be strapped? What is the distribution of the number of strapped cartons on the loaded pallet?

(b) Any loaded pallet with total weight (i.e., pallet plus the weight of the cartons) less than a prescribed value will be rejected by the purchaser of the products. What is this value if, in the long run, the purchaser rejects 5% of the pallets? (Express your answer to the nearest kilogram.)

***20.** Several years ago, an official from a New Zealand government ministry consulted one of the authors about the following problem. To reduce costs, the ministry wanted to cut down on the numbers of motor vehicles maintained in its vehicle fleet. As a first step, the official estimated the number of days in which vehicles were idle. He sampled 60 days at random from the approximately 240 working days of the previous year. In one of the services for a particular district there were 17 vehicles in the category "medium car." He took the total number of days that each car was idle, scaled it up by 240/60 to get an estimate of number of days idle in a year, and thus estimated that there were 824 idle vehicle-days in the "medium car" category. His question was, "How accurate is this estimate?"

Let's do some calculations. First, we will write down the estimate a little more formally. Let X_i be the number of days out of the 60 days observed in which the ith car was idle. Let N be the number of cars (here $N = 17$).

The estimate, then, is

$$Y = \frac{240}{60}(X_1 + X_2 + \cdots + X_N)$$

One problem in practice is to decide what we mean by "accurate." Let us first suppose that there are no trends in car usage over the years so that the year's usage is a good indication of future usage for the cars. Let us assume that the chances of any given car being idle on any given day is p, and that days and cars act independently.

(a) What is the distribution of X_i?

(b) Write down $E(X_i)$ and $sd(X_i)$ in terms of p.

(c) Next find $E(Y)$. Is this reasonable measure of the underlying idleness of the cars?

(d) Find a formula for $sd(Y)$.

(e) There were 17×240 car-days in the year, of which 824 were idle. Write down the natural estimate of p.

(f) Use your answer to (e) to give an estimate of $sd(Y)$, the standard deviation of Y.

(g) Suggest how many cars you would sell. Give reasons.

21. (Computer simulation exercise) Women's heights in the workforce database are approximately Normally distributed with mean 162.7 cm and standard deviation 6.2 cm. We can simulate the process of sampling women and measuring their heights by generating random numbers from a Normal($\mu = 162.7, \sigma = 6.2$) distribution.

(a) Generate 10 such random numbers and construct a dot plot. Think of each point on the plot as representing the height of a woman in a random sample of 10 women.

(b) Generate 15 sets of 10 observations, which you should think of as corresponding to 15 samples each of 10 women. Create a dot plot like the left-hand side of Fig. 6.1.2(a), but with 15 "lines" of dots—one for each sample. (Our graph plots against a horizontal scale. In some programs it is easier to plot against a vertical scale.) Repeat the process and construct three such sets of plots. Closely inspect each sample. Look for samples that appear left-skewed or right-skewed, or to have clusters or outliers in them. Look at differences in center and spread from sample to sample.

Note: The apparent "outliers" you see are not mistakes. Any skewness is not due to skewness in a parent distribution. What you are seeing is the variety of apparent patterns that turn up even in "perfect data" from a Normal distribution.

(c) Repeat (b), but now use sets of 15 samples each of size 40. Do you see more or fewer examples of skewness, outliers, clusters, and so on?

(d) Generate 6 samples of size 100. Obtain a histogram from each sample and compare them. What do you see?

(e) Generate 6 samples of size 1000. Obtain a histogram from each sample and compare them. What do you see? What has changed from (d)?

CHAPTER 7

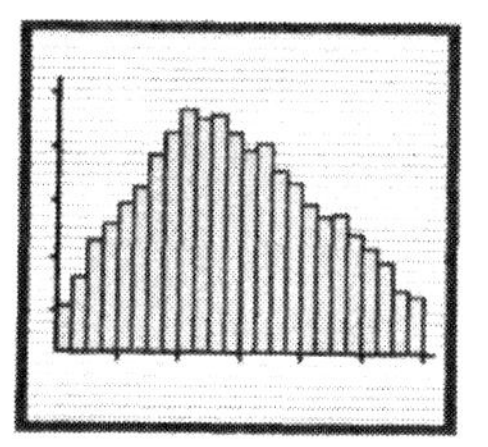

Sampling Distributions of Estimates

Chapter Overview

This is a pivotal chapter. Whereas the last few chapters concentrated on the behavior of random quantities and their distributions, the focus of the remainder of the book is on how we can use sampled data to inform us about the population from which the data are drawn. In particular, in Section 7.1 we find out about *parameters* (characteristics of a population) and *estimates* (calculated from data to estimate unknown parameters). This chapter lays the foundations for all of the discussion about estimation that follows. Large parts of all of the remaining chapters are concerned with refining and applying basic ideas introduced here.

A question we always want to ask about an estimate is, "How close is the estimate to the quantity (parameter) being estimated?" Because the sampled data come from a random experiment, the estimator that we use, being based on sample data, will also be random. In this chapter we see that the random behavior leading to the value of the estimator is best described by the so-called sampling distribution of the estimator. In Section 7.2 we talk about the sampling distribution of the sample mean. We then encounter the *central limit theorem,* which tells us that sample means from large samples are approximately Normally distributed, even when the population being sampled has a very non-Normal distribution. This lays the basis for some ways of looking at estimation that are widely applicable to estimates calculated from large samples. In particular, we extend these ideas to sample proportions in Section 7.3. Associated with estimation are the ideas of *bias* and *precision* in estimates, which we explore in Section 7.4. Along the way we introduce the *standard error* as a measure of the precision of an estimate. Preliminary versions of *interval estimates* (developed in Chapter 8) and *significance tests* (developed in Chapter 9) are introduced.

Having considered the problem of estimating a single mean and a proportion, we extend our ideas in Section 7.5 to estimating the difference between two means or two proportions from independent samples. The chapter closes with an introduction to *Student's t-distribution* in Section 7.6, which lays the groundwork for refining our methods for making inferences about means so that they work better for data from small samples.

7.1 PARAMETERS AND ESTIMATES

It is useful to think of random or unpredictable data as arising in two ways. The first is produced by sampling from a clearly defined finite population and measuring characteristics of the individuals chosen. This is the case with sample surveys and polls and with sampling from naturally occurring populations such as plants and animals. We can also include taking samples of air, soil, or water, though the underlying population is a little harder to define. A second source of random data is from a random process producing observations, for example, measured characteristics of manufactured items such as computer chips, lightbulbs, car tires, or bottles of mineral water. Here the underlying population is conceptual rather than real. This population is one that would be produced if the process was repeated a large number of times. By looking at a sample from the process, we are really looking at part of the population the process is capable of producing.

By either route we get a random observation possessing a distribution that provides a description of how the observation will vary. Numerical characteristics of that distribution are called *parameters.* For example, the mean $E(X)$ of the distribution of a random observation X is a parameter called the population mean, which we usually denote by μ. We call $E(X)$ the population mean because it turns out to be just that! We saw in Section 5.4.1 that in the case of a real, finite population, $E(X)$ is the mean of the X-values for the whole population. In the conceptual case it is again the mean, but for a very large conceptual population. Other parameters are the population standard deviation $sd(X)$, usually denoted by σ, and population proportions. If we are sampling a human population, for example, we may be interested in μ, the mean expenditure on recorded music for the whole population during 1999, or in p, the proportion of people in the population who have had an influenza vaccination. We shall meet other parameters in subsequent chapters.

A ***parameter*** is a numerical ***characteristic of a population*** or distribution.

In practice, we usually know very little about the population we are sampling or the random process generating our data. The reason for collecting data is to obtain information about these things. Therefore the parameters of interest are unknown quantities that we want to estimate. If we are interested in a population proportion, p, for example, of those who have been vaccinated, the obvious approach is to take a sample from that population and use $\hat{p}$, the proportion vaccinated in the sample (e.g., 26%) as an *estimate* of p. To estimate the mean diameter μ of bearings coming from a production line, we can take some of the bearings produced (a sample) and use the average diameter $\bar{x}$ of those bearings as an estimate of the underlying mean diameter μ for the production process. Similarly, we could estimate a population standard deviation by a sample standard deviation.

> An ***estimate*** is a quantity ***calculated from the data*** to estimate an unknown parameter.

In Section 1.1.1 we took a series of samples of people from the workforce database and recorded for each individual whether or not they thought they were overweight. Every new sample had a different proportion of people who thought they were overweight, and none of the sample values was identical with the population value of 69%. (The behavior we observed is called *sampling variation.*) So how good are the estimates produced by our intuitive ideas about estimation? How well do they work in practice? That is the subject of this chapter. In this book, however, we restrict attention to the properties of estimates calculated from *random samples* and simple random samples (see Section 6.4.4).

7.1.1 Capital Letters and Small Letters: Estimates and Estimators

> ***Capital letters*** refer to ***random variables.***
> ***Small letters*** refer to ***observed values.***

We use capital letters to represent random variables when we want to think about the way the sampling process behaves, for example, when we want to ask a question such as, "If I take a sample in this particular way, what are the chances that ... will happen?" We use the corresponding small-letter notation when we want to discuss what we actually saw when we collected the data.

Let us apply these ideas to using the proportion of people in a sample who have some characteristic, namely, having been vaccinated, to estimate the corresponding population proportion. We use two notations to refer to the sample proportion: $\widehat{p}$ (small letter) and $\widehat{P}$ (capital letter). The small letter, $\widehat{p}$, refers to the sample proportion we actually saw when we collected the data (e.g., $\widehat{p}$ would be 0.26 if we had found that 26% of our sample had been vaccinated). The capital-letter notation, $\widehat{P}$, is used when we want to think about the effects of sampling variation, that is, about the random variability involved in the process of taking a sample and obtaining a sample proportion. We might, for example, want to answer a question such as, "If I take a random sample of 1000 people, what are the chances that my sample proportion will be within 2% of the true population proportion?"

When applying these distinctions to the context of estimation, statisticians talk about an ***estimate*** (e.g., $\widehat{p}$), the number we observe, and the ***estimator*** (e.g., $\widehat{P}$), the abstraction that lets us think about the properties of the random process that produces the estimate.

QUIZ ON SECTION 7.1

1. Describe two ways in which random observations arise and give examples.
2. What is a parameter? Give two examples of parameters.
3. What is an estimate? How would you estimate the parameters you described in question 2?

4. What is the distinction between an estimate and an estimator? Why is it made? What notational device is used to communicate the distinction?

7.2 SAMPLING DISTRIBUTION OF THE SAMPLE MEAN

7.2.1 Means of Random Samples

The following numbers are the results of seven independently performed measurements of the pH level (a measure of acidity or alkalinity) of a chemical buffer solution:

5.12 5.20 5.15 5.17 5.16 5.19 5.15

Each number represents an attempt to measure the same quantity, the true pH level, but each attempt has resulted in a different answer. If the discrepancies are produced by random experimental and measurement errors and there is no systematic tendency for them to be too high or low, it would be reasonable to assume that the experimental results are randomly distributed according to a distribution whose mean, μ, is the true pH level of the buffer solution. Most people, if asked to estimate the true pH level from this data, would probably give the average of the given numbers, namely, $\bar{x} = 5.16$. In technical language, they would have used the sample mean, $\bar{x}$, to estimate the population mean, μ.

Estimating a population mean μ by drawing a random sample from that population (or distribution) and quoting the sample mean $\bar{x}$ as the estimate of μ is the obvious, naive approach. It gives us an estimate that feels intuitively plausible, but is it really any good? How accurate is our estimate? We shall use a computer simulation to investigate the behavior of the process of taking random samples and using the resulting sample mean $\bar{x}$ as an estimate of the population mean μ. In statistical jargon, what we shall be observing is the behavior of the *sampling variation* in values of $\bar{x}$.

For our simulation we shall use the chest measurements of Scottish soldiers again. We would have liked to sample directly from our population of 5738 Scottish soldiers, but we don't have the 5738 original measurements. We saw in Chapter 6, however, that the distribution of chest measurements in this group of soldiers was approximately Normal($\mu = 39.8, \sigma = 2.05$), where the measurement units are inches. Thus we shall simulate the process of sampling from the Scottish army and taking chest measurements by using a computer to generate random numbers from the preceding Normal distribution.

The chest measurements X of the first 6 "soldiers" sampled in this way were

40.5 41.5 39.5 39.6 44.3 39.3

This sample has a mean of $\bar{x} = 40.78$ in. We then took a second sample of 6 soldiers. Their chest measurements were

43.0 39.7 37.8 41.3 40.1 39.8

This second sample of 6 soldiers has a mean of $\bar{x} = 40.28$ in. In Fig. 7.2.1 we use dot plots to plot the chest measurements of samples of 6 soldiers, together with the

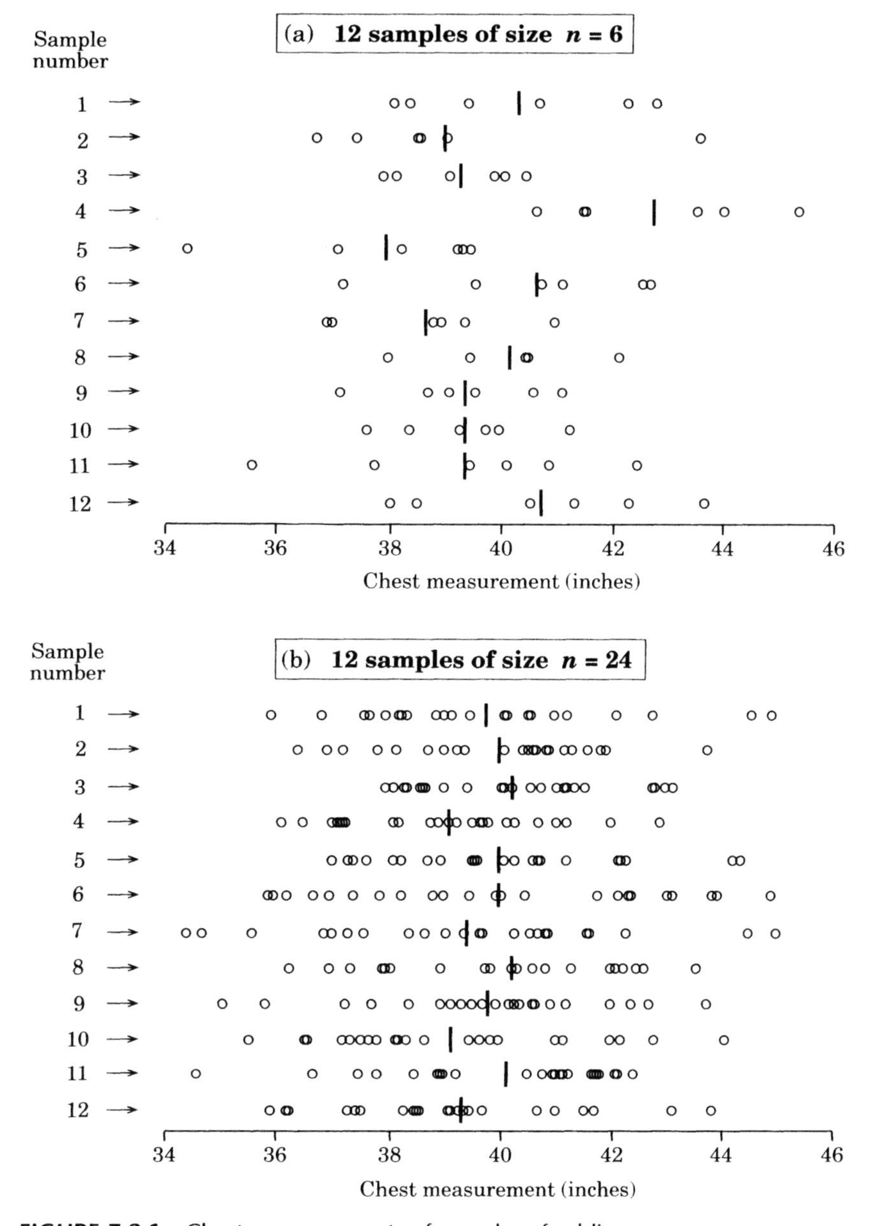

FIGURE 7.2.1 Chest measurements of samples of soldiers.

sample mean $\bar{x}$ for each sample. The position of the sample mean is given by a vertical bar "|", and we did this for 12 additional random samples of 6 soldiers. We then increased the sample size to 24 and took 12 samples each of 24 soldiers. This set of samples together with their means is also plotted in Fig. 7.2.1.

There are two important lessons to be drawn from Fig. 7.2.1. First, each sample, and therefore each sample mean, is different. The sample means, $\bar{x}$, vary in an unpredictable way, illustrating the random nature of means calculated from randomly chosen samples. Second, as we increase the sample size from 6 to 24, there appears to be a decrease in the variability of the sample means (compare the vertical bars in both sets of dot plots).

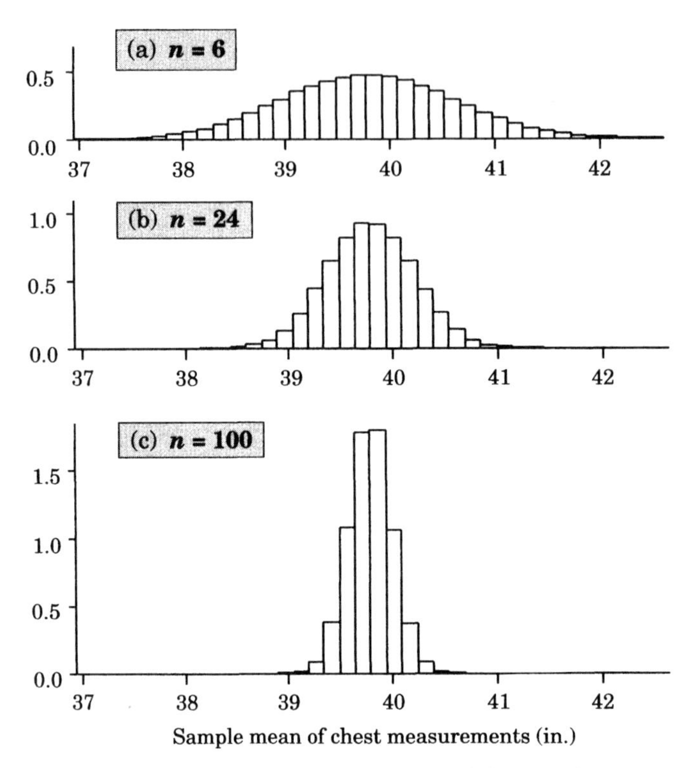

FIGURE 7.2.2 Standardized histograms of the sample means from 100,000 samples of soldiers (n soldiers per sample).

We denote the sample mean of X-measurements, thought of as a random variable, by $\overline{X}$. To further investigate the distribution of $\overline{X}$, we calculated the sample means of 100,000 samples of size 6 (giving us 100,000 sample means). A standardized histogram of these sample means is given in Fig. 7.2.2(a). The average of the sample means was 39.80, and their standard deviation was 0.837 (cf. $\mu = 39.8$ and $\sigma = 2.05$). We did the same thing using 100,000 samples each of size 24 [Fig. 7.2.2(b)]. Although the average of these sample means was again 39.80, this time their standard deviation was smaller than before, namely, 0.418 (cf. $\sigma = 2.05$) reflecting the smaller spread in the histogram for samples of size 24. Finally, we repeated the procedure with 100,000 samples of size 100. The resulting standardized histogram is given in Fig. 7.2.2(c). The average of the sample means was again 39.80, but the standard deviation was smaller again, namely, 0.205.

Regardless of the sizes of the samples we were drawing (6, 24, or 100), the average of all of the sample means had the same value as μ, the mean of the distribution we were sampling from (here 39.8). However, the variability of the sample means decreased as the sample size increased. There is an interesting pattern in the decrease, which you can verify. In each case the standard deviation of the set of sample means was (to two decimal places) $2.05/\sqrt{n} = \sigma/\sqrt{n}$, where n is the size of the sample being averaged each time (6, 24, and 100). What we have just seen illustrates two general results, which we shall now establish theoretically.

Theory

Suppose $X_1, X_2, \ldots, X_n$ is a random sample from a distribution (not necessarily Normal) with population mean μ and population standard deviation σ. As in Section 6.4.4, we use the notation $Sum = X_1 + X_2 + \cdots + X_n$. The sample mean $\overline{X}$ is then given by $\overline{X} = \frac{1}{n} Sum$. We can therefore obtain the expectation and standard deviation of $\overline{X}$ by multiplying the expectation and standard deviation of Sum by $\frac{1}{n}$ (see Section 6.4.3). Section 6.4.4 showed that $\text{E}(Sum) = n\mu$ and $\text{sd}(Sum) = \sqrt{n}\,\sigma$. Thus we have $\text{E}(\overline{X}) = \frac{1}{n}(n\mu) = \mu$ and $\text{sd}(\overline{X}) = \frac{1}{n}(\sqrt{n}\,\sigma) = \sigma/\sqrt{n}$.

To summarize:

$$\mu_{\overline{X}} = \text{E}(\overline{X}) = \mu \qquad \text{and} \qquad \sigma_{\overline{X}} = \text{sd}(\overline{X}) = \frac{\sigma}{\sqrt{n}}$$

E(sample mean)	=	Population mean
sd(sample mean)	=	$\dfrac{\text{Population standard deviation}}{\sqrt{\text{Sample size}}}$

Because the expected value (or mean value) of the sampling distribution of $\overline{X}$ is the population mean μ, it is often said that the sample mean is an *unbiased* estimate of the population mean. A useful interpretation of the expression for $\text{sd}(\overline{X})$ is that the variability in a sample mean is the variability in individual observations divided by $\sqrt{n}$. This variability becomes smaller as the sample size n increases. Since $\sqrt{4} = 2$, we halve $\text{sd}(\overline{X})$ if we take a sample 4 times as large. Similarly, we reduce $\text{sd}(\overline{X})$ to a third of its original size by taking a sample 9 times as large.

If X measurements in the population we are sampling from actually have a Normal distribution, it follows from Section 6.4 that the sampling distribution of $\overline{X}$ is also Normal, so that $\overline{X} \sim \text{Normal}(\mu_{\overline{X}} = \mu,\ \sigma_{\overline{X}} = \sigma/\sqrt{n})$. We know that a Normally distributed random variable falls within two standard deviations of its mean approximately 95% of the time (see Section 6.2.1). Thus, in the long run, the observed sample mean falls within $\pm 2\,\text{sd}(\overline{X})$ of the population mean μ for approximately 95% of samples taken. This all suggests that $\text{sd}(\overline{X})$ might be a useful measure of the "precision" of the sample mean as an estimate. Smaller values of $\text{sd}(\overline{X})$ correspond to less variable, or more "precise," estimates. The fact that, as the sample size increases, $\text{sd}(\overline{X})$ becomes smaller and the sample mean becomes a more precise estimate is in accord with the intuitive feeling that obtaining more data leads to better estimates.

We have seen that when sampling from a Normal distribution, $\overline{X}$ is also Normally distributed. It follows that $Z = (\overline{X} - \mu)/\text{sd}(\overline{X}) \sim \text{Normal}(0, 1)$. We shall need this expression several times later in this chapter.

QUIZ ON SECTION 7.2.1

1. We use both $\bar{x}$ and $\overline{X}$ to refer to a sample mean. For what purposes do we use the former and for what purposes do we use the latter?
2. What is meant by "the sampling distribution of $\overline{X}$"?
3. How is the population mean of the sample average $\overline{X}$ related to the population mean of individual observations?
4. How is the population standard deviation of $\overline{X}$ related to the population standard deviation of individual observations?
5. What happens to the sampling distribution of $\overline{X}$ if the sample size is increased?

6. What does it mean when $\bar{x}$ is said to be an "unbiased estimate" of μ?
7. If you sample from a Normal distribution, what can you say about the distribution of $\overline{X}$?
8. Increasing the precision of $\overline{X}$ as an estimator of μ is equivalent to doing what to sd($\overline{X}$)?

EXERCISES FOR SECTION 7.2.1

1. Suppose that for a given sample size of 10, sd($\overline{X}$) = 9. How many more observations would we need to take to reduce sd($\overline{X}$) (a) to 4.5? (b) to 3? (c) to 1?
2. The annual dividend paid by a large public company is calculated on the basis of the average monthly profit earned over the most recent six-month period. In fact, the company will pay this dividend only if this average monthly profit exceeds \$8.5 million. On the basis of past data, it can be assumed that the monthly profits earned by the company are approximately Normally distributed, with a mean of \$10 million and a standard deviation of \$3.5 million. What is the distribution of the average of six-month profit figures? Calculate the probability that the company will pay an annual dividend.

7.2.2 The Central Limit Effect

At the end of the previous section, we found that for Normally distributed data, the distance between $\bar{x}$ and μ is hardly ever any bigger than 2 sd($\overline{X}$), and from that it looked as if sd($\overline{X}$) might be a useful measure of the precision of our estimate $\bar{x}$. At this stage these results have no practical value, however,

(i) because nobody ever *knows* that their data are coming from a Normal distribution, and

(ii) the value of sd($\overline{X}$) = $\sigma/\sqrt{n}$ depends on the population standard deviation σ, and in practice nobody ever knows that either!

We shall consider both of these problems in turn.

We know that, no matter what distribution we sample from, E($\overline{X}$) = μ and sd($\overline{X}$) = $\sigma/\sqrt{n}$. But what about the shape of the distribution of the sample mean $\overline{X}$? If the shape of the distribution we are sampling from is unknown, we surely can't know the shape of the distribution of $\overline{X}$. Once again, computer simulation will be used to explore this question. This time we look at means of samples from some distinctly non-Normal distributions. The results will be surprising.

We shall sample from four continuous distributions whose density curves are graphed in Figs. 7.2.3 and 7.2.4. None of them look even remotely like Normal curves. The distributions used are

(a) a Triangular distribution on $0 \le x \le 1$ for which observations near 1 are much more likely than observations near 0 [Fig. 7.2.3(a), top left],

(b) a Uniform distribution for which all values between 0 and 1 are equally likely [Fig. 7.2.3(b), top right],

(c) an Exponential distribution that is more likely to give values near 0 [Fig. 7.2.4(a), top left], and

(d) a Quadratic U-shaped distribution that is very likely to give observations near to 0 or 1 and highly unlikely to give observations in the middle near 0.5 [Fig. 7.2.4(b), top right].

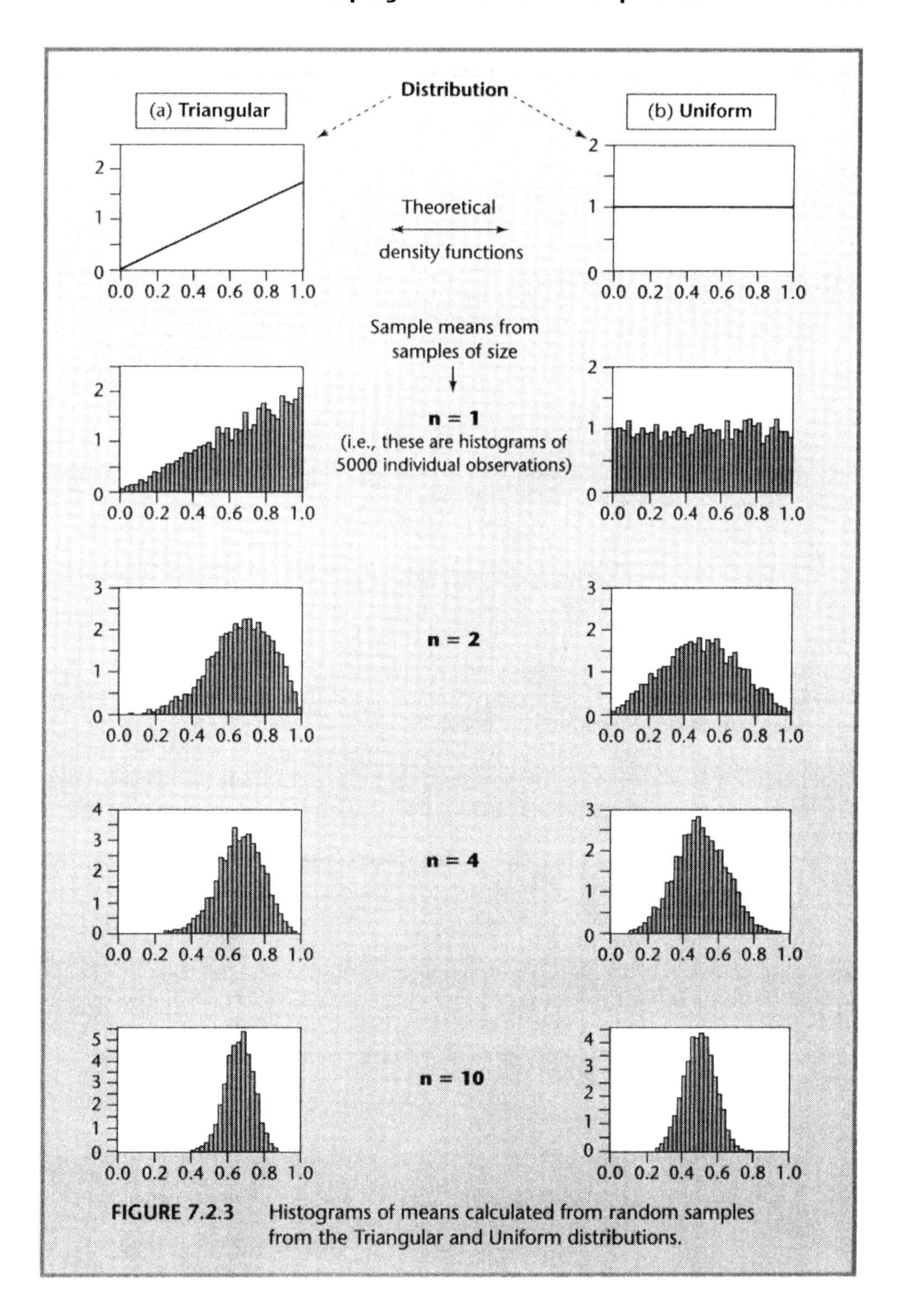

FIGURE 7.2.3 Histograms of means calculated from random samples from the Triangular and Uniform distributions.

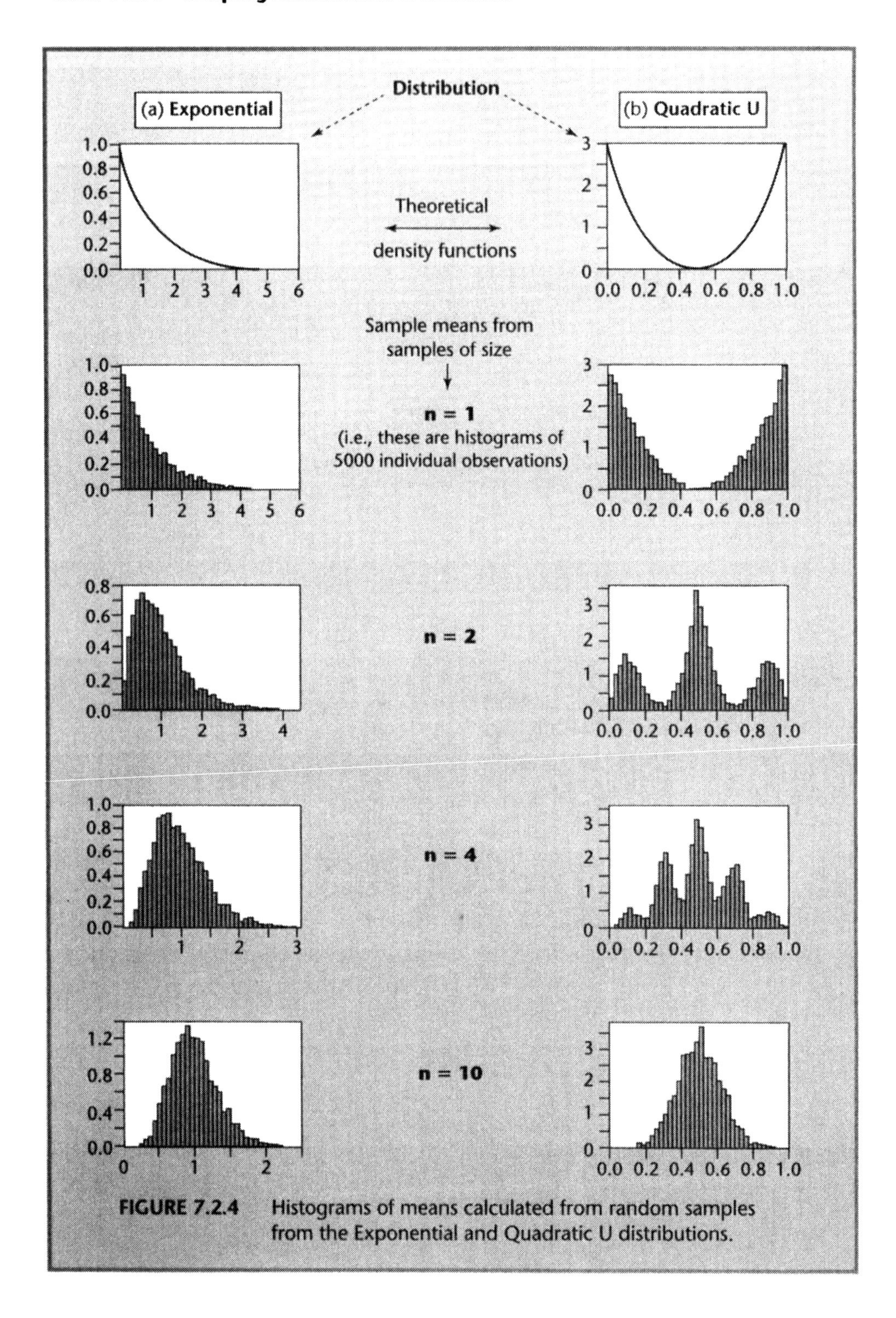

FIGURE 7.2.4 Histograms of means calculated from random samples from the Exponential and Quadratic U distributions.

What do the sampling distributions of sample means from distributions like these look like? We'll find out by doing some sampling! For each distribution we took 5000 samples of size n and averaged the n observations in each sample to get 5000 sample means. We then drew histograms of the resulting 5000 numbers. We did this for samples of size $n = 1$, $n = 2$, $n = 4$, $n = 10$, and $n = 25$ (not shown). The results for the Triangular and Uniform distributions are given in Fig. 7.2.3, and for the Exponential and Quadratic in Fig. 7.2.4.

Samples of size $n = 1$ just give single random observations from the distribution. They are included so that you can see the relationship between the theoretical density curve and a histogram of data sampled from that distribution.

The patterns we most want you to see emerge when you scan down each column of graphs in Figs. 7.2.3 and 7.2.4. What you will notice is that as the sample size n increases, the histograms begin to have more and more of the characteristic bell shape of the Normal distribution. They haven't all reached it by $n = 10$, but the rapidness of the change of shape is remarkable. Notice how the gaps are gradually filled in with the U distribution. For the Exponential distribution there is still some skewness (lack of symmetry) for $n = 10$. By $n = 25$ (not shown), however, they all look Normal.

This effect is not confined to continuous distributions. Consider rolling a single die. This gives an observation X, which is equally likely to take any of the values 1, 2, 3, 4, 5, or 6. Now what does the distribution of sample means of the results of several rolls of the die look like? Bar graphs of some of these are given in Fig. 7.2.5. For n as small as 3 the distribution is starting to look bell shaped!

In fact, no matter what distribution we sample from, the sampling distribution of $\overline{X}$ is closely approximated by the Normal distribution in large samples. This result, which can be established mathematically, is called the *central limit theorem*. For this reason we call the tendency of sample means to be approximately Normally distributed the *central limit effect*.

Central limit theorem:

$\overline{X}$ is approximately Normally distributed in large samples.

The central limit theorem is important for practical applications of statistics. Why? Because in practice we never know the exact form of the distribution we are sampling from, but the theorem tells us we can apply Normal-distribution theory for means from large samples even when the original distribution is not Normal. We note in passing

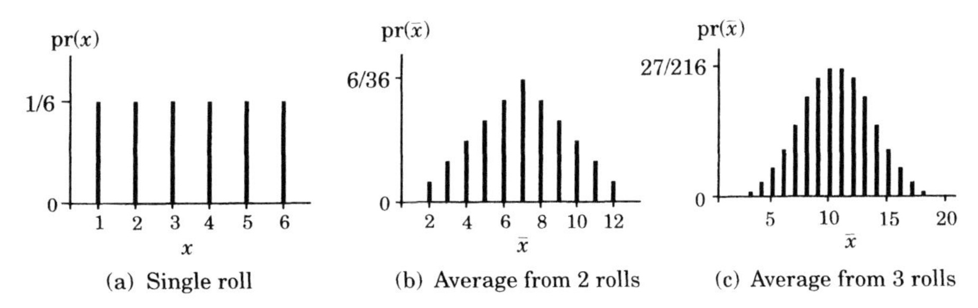

FIGURE 7.2.5 Probability function of averages of several rolls of a die.

that since $Sum = X_1 + X_2 + \cdots + X_n$ is simply $n\overline{X}$, it follows that the sum is also approximately Normally distributed.

For both sums and sample means, the bigger the sample size is, the better the Normal approximation works. But how big does the sample size n have to be before the Normal distribution provides an adequate approximation to the true distribution? Many elementary books say the approximation is good if $n \geq 30$. We can see from Figs. 7.2.3 to 7.2.5, however, that there is no simple answer. If the original distribution of X is reasonably close in shape to the Normal, the central limit effect works very fast. Even for a non-bell-shaped (but symmetric) distribution like the Uniform [see Fig. 7.2.3(b)], the distribution of $\overline{X}$ looks quite bell-shaped for n as low as 4. Indeed, many early computer programs for generating Normal random numbers worked by simply averaging 12 Uniform random numbers. The central limit effect is working much more slowly for the Exponential distribution in Fig. 7.2.4(a), however. In such a situation we would like n to be greater than 30 (e.g., 50), though how much greater would depend on the degree of skewness in the distribution of X. Heaviness of the tails of the distribution and lack of symmetry are important factors in slowing down the central limit effect. One should always plot the sample data. If the data look as though they may have come from a distribution that isn't extremely non-Normal, we can feel much more confident about calculations based on a Normal approximation with moderate-sized samples. We shall address these issues in more detail later.

EXAMPLE 7.2.1 *The Effect of Promoting Business Ethics on Decision Making*

In the late 1980s an unusual number of major business and financial figures were in serious legal trouble all over the western world for unethical business dealings such as insider trading, failure to disclose information pertinent to the sale of companies, and corporate fraud. Researchers at Marquette University tried to study whether a company's stated policy promoting ethics affected the behavior of decision makers.

MBA students were exposed to decision-making situations in a business setting, with each student being given a rating for their set of decisions on a scale of 1 ("definitely unethical") to 5 ("definitely ethical"). When no concern about ethics had been expressed, the students' ethics scores had a mean of 3.00 and a standard deviation of 1.03. A group of 30 students were read a statement from the "company president" about the company's code of business ethics before being exposed to the decision-making situations. For this group the average ethics score was 3.55 (30 scores, one for each student). This average score appears to be higher than we might have expected from previous experience, but could such a score simply be the result of sampling variation?

We shall suppose that when no reference to ethics policies have been made, ethics scores X are distributed with $\mu = 3.0$ and $\sigma = 1.03$. We shall calculate how likely an average ethics score of $\overline{x} = 3.55$ or more would have come from 30 people under these circumstances.

Here $n = 30$. Now $E(\overline{X}) = \mu = 3.0$, and $sd(\overline{X}) = \sigma/\sqrt{n} = 1.03/\sqrt{30} = 0.18805$. Thus, assuming $n = 30$ is large enough here for the central limit effect to work well, the distribution of $\overline{X}$ is approximately Normal$(\mu_{\overline{X}} = 3.0, \sigma_{\overline{X}} = 0.18805)$. We require[1]

$$\text{pr}(\overline{X} \geq 3.55) = 1 - \text{pr}(\overline{X} < 3.55) \approx 0.002$$

[1]The Minitab command "CDF 3.55;" followed by subcommand "Normal 3.0 0.18805." returns 0.99828, as does Excel's "NORMDIST(3.55,3.0,0.18805,TRUE)".

The point of this calculation is that it shows that we would have been highly unlikely to get a sample mean of 3.55 or more under the circumstances operating before reading the ethics policy. Thus the data provide some evidence that clearly stating an ethics policy is effective.

QUIZ ON SECTION 7.2.2

1. What does the central limit theorem say? Why is it useful?
2. In what way might you expect the central limit effect to differ between samples from a symmetric distribution and samples from a very skewed distribution?
3. What other important factor, apart from skewness, slows down the action of the central limit effect?
4. When you have data from a moderate to small sample and want to use a Normal approximation to the distribution of $\overline{X}$ in a calculation, what would you want to do before having any faith in the results?

EXERCISES FOR SECTION 7.2.2

1. A Japanese importer has just placed an order for coastal mussels (shellfish), the weights of which have a mean of 100 g and a standard deviation of 15 g. The mussels are packed in boxes of 50 to be freighted to Japan. Each box will be weighed on arrival and rejected if the average weights of mussels in the box is less than 97 g. What percentage of boxes would be rejected?
2. The service times for customers coming through a checkout counter in a supermarket are independent random variables with an expected value of 3.1 minutes, a standard deviation of 1.2 minutes, and a distribution that is not Normal. A random sample of 50 customers is selected.
   (a) What is the approximate probability that the average service time for these 50 customers is less than 3.3 minutes?
   (b) What is the approximate probability that the total service time for these 50 customers is less than $2\frac{1}{2}$ hours?
3. It has been reported that women smokers tend to have lower serum (blood) estrogen concentrations and earlier menopause than nonsmokers and that among pregnant women, smokers have lower serum estriol concentrations. Such associations might explain the associations between smoking and several diseases in women (e.g., osteoporosis). Dai et al. [1988] investigated associations between smoking and sex hormones in men. Several exercises over the next few sections will be drawn from their data on testosterone levels.

   Dai et al. [1988] reported that serum testosterone levels in a group of nonsmoking men aged between 35 and 57 had a mean of 620.6 ng/dL and a standard deviation of 241.5 ng/dL. We shall take these as population figures for nonsmokers. A sample of 28 heavy smokers from the same population had a sample mean of 795.1 ng/dL. Suppose that the distribution of testosterone levels was the same for smoking men as for nonsmoking men.
   (a) What is the probability that the sample mean for a sample of size 28 taken from this distribution is 795.1 ng/dL or greater? Do you believe that the distribution of testosterone levels is the same for smoking men as for nonsmoking men? Why or why not?
   (b) What proportion of nonsmokers have a serum testosterone level above 795.1 ng/dL? (Assume Normality.)
   (c) Is there any conflict between your answers to (a) and (b)? Why or why not?

7.2.3 Introducing the Standard Error

We saw that for a sample mean calculated from a random sample, $\text{sd}(\overline{X}) = \sigma/\sqrt{n}$, that is, the variability from sample to sample in sample means is given by the variability in individual observations divided by the square root of the sample size. We cannot use this expression in most practical applications because it depends on the population standard deviation, σ, which is unknown. Instead, we substitute an estimate of the variability of individual observations taken from the sample; that is, we replace the unknown true σ by the sample standard deviation, s_X. This gives us an estimate of $\text{sd}(\overline{X})$, which we call the *standard error* of $\overline{x}$, denoted $\text{se}(\overline{x})$.[2]

The **standard error** of the sample mean = $\dfrac{\text{Sample standard deviation}}{\sqrt{\text{Sample size}}}$

$$\textbf{se}(\overline{x}) = \frac{s_X}{\sqrt{n}}$$

More comprehensive versions of the central limit theorem say that for large samples, not only is $(\overline{X} - \mu)/\text{sd}(\overline{X})$ approximately Normal(0,1), but $(\overline{X} - \mu)/\text{se}(\overline{X})$ is approximately Normal(0,1) as well. Since $\text{pr}(-2 \le Z \le 2) = 0.954$ for $Z \sim \text{Normal}(0, 1)$, we see that for large samples it is quite rare to get a sample in which the observed value of $\overline{x}$ and μ are more than two multiples of $\text{se}(\overline{x})$ apart. A sample mean falling further than two standard errors from the population mean μ occurs only about 5% of the time, or for one sample in 20 on average. Since $\text{pr}(-3 \le Z \le 3) = 0.997$ (by computer[3]), samples in which the sample mean is further than three standard errors from the population mean μ are even more rare—they occur only 0.3% of the time, or for about one sample in 300 on average.[4] We identify three consequences of this for situations where we have taken a random sample and obtained a value $\overline{x}$ from our data.

(i) $\text{se}(\overline{x})$ is a measure of the precision of $\overline{x}$ as an estimate of μ, which is useful in practice.

(ii) If we want to estimate the true value of μ, it is a reasonably safe bet that it lies somewhere in the range $\overline{x} - 2\,\text{se}(\overline{x})$ to $\overline{x} + 2\,\text{se}(\overline{x})$. (We write this range of values as $\overline{x} \pm 2\,\text{se}(\overline{x})$ and describe it verbally as a two-standard-error interval for μ.)

(iii) We shall be suspicious of any theoretically specified value for μ that is more than $2\,\text{se}(\overline{x})$ from the value $\overline{x}$ we got from our data. We shall be even more suspicious of a postulated value of μ that is more than $3\,\text{se}(\overline{x})$ from our data estimate $\overline{x}$.

Example 7.2.2 *Two-Standard-Error Interval for Mean Density of the Earth*

Newton's law of gravitation states that the force of attraction (F) between two particles of matter is given by the formula $F = Gm_1m_2/r^2$, where m_1 and m_2 are the masses of

[2]Note that $\text{se}(\overline{x})$ is a fixed value for a particular sample. The corresponding random variable will be denoted by $\text{se}(\overline{X})$.

[3]The tables in Appendix A4 give .998 because of rounding errors.

[4]Use is made of this property in the optional chapter on our web site where we discuss control charts and extreme observations.

TABLE 7.2.1 Cavendish's Determinations of the Mean Density of the Earth (g/cm^3)

5.50	5.61	4.88	5.07	5.26	5.55	5.36	5.29	5.58	5.65
5.57	5.53	5.62	5.29	5.44	5.34	5.79	5.10	5.27	5.39
5.42	5.47	5.63	5.34	5.46	5.30	5.75	5.68	5.85	

Source: Cavendish [1798].

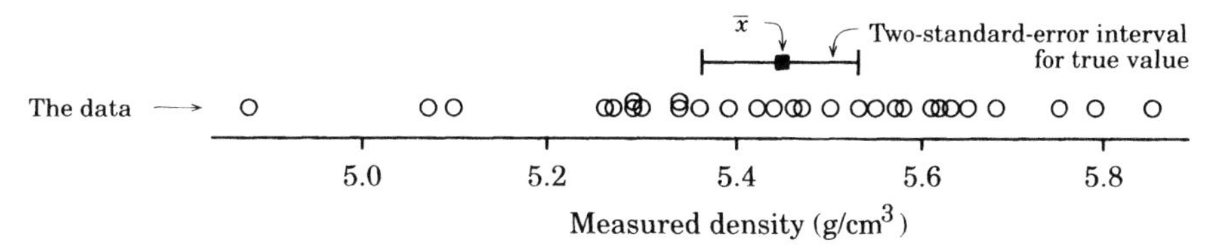

FIGURE 7.2.6 Cavendish's determinations of the mean density of the earth.

the two objects, r is the distance between their centers, and G is a number known as the gravitational constant. From the late eighteenth through the nineteenth centuries a large number of experiments were performed to determine G. These experiments were usually designed to determine the earth's attraction to masses. As a consequence, finding G could be viewed as equivalent to finding the mean density of the earth. Of all the early experiments, those performed by Cavendish in 1798 are considered the best (see Stigler [1977]). Cavendish's 29 determinations of the mean density of the earth (relative to that of water) are given in Table 7.2.1 and plotted in Fig. 7.2.6.

The 29 experimental determinations of the mean density of the earth had a sample mean of $\overline{x} = 5.447931$ g/cm^3 and a sample standard deviation of $s_X = 0.2209457$ g/cm^3. As with the pH example that started this section, we shall assume that there are no systematic errors so that μ, the underlying mean of the distribution of measurements, is the true mean density.

For these data,

$$\text{se}(\overline{x}) = \frac{s_X}{\sqrt{n}} = \frac{0.2209457}{\sqrt{29}} = 0.04102858$$

Using the preceding theory, it is a fairly safe bet that the true mean density of the earth μ lies within two standard errors of the sample mean, that is, within

$$\overline{x} \pm 2\,\text{se}(\overline{x}) = 5.447931 \pm 2 \times 0.04102858 = 5.447931 \pm 0.08205717$$

or between 5.365874 and 5.529988. Because of the length of the interval, we would not quote the endpoints so precisely, but would just say that the true value is probably between 5.37 g/cm^3 and 5.53 g/cm^3. In fact, the currently accepted value, 5.517, does lie within this interval.

To illustrate the preceding consequence (iii), suppose that some theorist had proposed 5.7 as the mean density of the earth. Now 5.7 is more than six standard errors from the data estimate $\overline{x}$ [since $(5.7 - 5.447931)/0.04102858 = 6.14$]. If we believed in the reliability of Cavendish's experimental procedures, this would make us believe that the theorist was wrong.

Item (ii) is a very simple form of *interval estimation.* A range of values is specified that we think will contain the true value of the parameter we are trying to estimate (here μ). A more sophisticated formulation of interval estimation (via *confidence intervals*) will be introduced in Chapter 8. Similarly, item (iii) is a very simple formulation of the idea of *hypothesis testing,* which will receive a more sophisticated discussion in Chapter 9. The modes of reasoning embodied in items (ii) and (iii) are thus transitional. They form a gentle introduction to the methods of Chapters 8 and 9.

QUIZ ON SECTION 7.2.3

1. Why is the standard deviation of $\overline{X}$, sd($\overline{X}$), not a useful measure of the precision of $\overline{X}$ as an estimator in practical applications? What measure of precision do we use in practice?
2. How is se($\overline{x}$) related to sd($\overline{X}$)? When we use the formula se($\overline{x}$) $= s_X/\sqrt{n}$, what is s_X, and how do you obtain it?
3. What can we say about the true value of μ and the interval $\overline{x} \pm 2\,\text{se}(\overline{x})$?
4. Increasing the precision of $\overline{x}$ as an estimate of μ is equivalent to doing what to se($\overline{x}$)?

EXERCISES FOR SECTION 7.2.3

1. After the sixth observation in Example 7.2.2, Cavendish changed his apparatus. The first six observations were 5.50, 5.61, 4.88, 5.07, 5.26, and 5.55. On the basis of these six observations:
   (a) Plot the data.
   (b) Calculate a two-standard-error interval for the true mean density of the earth and add it to your plot.
   (c) Determine how many standard errors the value 5.8 is from the mean of these six observations.
2. The sample of 28 heavy smokers in Exercises 7.2.2, problem 3, had a sample mean testosterone concentration of 795.1 ng/dL and standard deviation of 305.3 ng/dL.
   (a) Compute a two-standard-error interval for the true mean testosterone concentration in the population from which this sample was drawn.
   (b) Is 620.6 ng/dL, the mean for nonsmokers, a plausible value for the true mean for heavy smokers?
3. Suppose the daily rates charged by a random sample of 40 Seattle builders had a sample mean of $250 and a sample standard deviation of $50. In what range of values is the average daily rate of all Seattle builders likely to lie?

7.3 SAMPLING DISTRIBUTION OF THE SAMPLE PROPORTION

7.3.1 Sample Proportions and Their Standard Errors

In Section 1.1.1 we simulated the process of repeatedly choosing samples and we looked at the resulting sample proportions. It might be helpful to read Section 1.1.1 again before proceeding.

EXAMPLE 7.3.1 *Proportion of Teens Witnessing Drugs Being Sold at School*

As part of the National Center on Addiction and Substance Abuse at Columbia University's nationwide 1998 "Back-to-School Teen Survey," 1000 American teenagers aged between 12 and 17 were interviewed by telephone on many lifestyle issues.[5] Twenty-seven percent of the teens reported having seen illegal drugs being sold at their schools.

We should not take the 27% in Example 7.3.1 too literally. This is not the true percentage of all American teenagers who had seen drugs being sold at school at the time that the poll was taken. It is just an estimate based on data from a sample and is thus subject to sampling variation. How good an estimate is it? If several other researchers had each independently taken surveys of 1000 American teenagers, they would all have ended up with different people in their samples and different estimates.

Taking a random sample and observing a sample proportion is a random process. We denote the sample proportion from the one sample that we actually take by $\widehat{p}$ (e.g., $\widehat{p} = 0.27$ or 27% in Example 7.3.1) and the corresponding random variable by $\widehat{P}$. The distribution of $\widehat{P}$ tells us about the random variability in the process of taking a sample and observing a sample proportion. This gives us indirect information about the quality of the estimate we obtained from our data. The theory that we will develop applies both to random samples from a process and simple random samples from large populations.

We noted in Section 5.3 that when we take a random sample from a much larger population, the distribution of Y, the number of individuals (or objects) in the sample with a given characteristic of interest, is approximately Binomial(n, p). (The approximation is reasonable when the sample makes up less than about 10% of the population.) Here n is the sample size, and p is the true proportion of the whole population who have the characteristic of interest (e.g., the proportion of all adult Americans who believe that large corporations have too much influence on government). The Binomial(n, p) model also applies directly to any random process that behaves like tossing a biased coin (Section 5.3). Here n is the number of "tosses," and p is the probability of getting a "head" on any particular "toss." For example, if a manufacturing line is behaving in this way with respect to whether or not items coming off the line are defective, then n is the number of items in the sequence of items that we inspect, and p is the probability that any particular item produced will be defective. Even in the coin-tossing context, we shall refer to p as the (true) population proportion. On the basis of the data from our sample, the sample proportion $\widehat{p}$ is the natural estimate of the population proportion p.

The sample proportion $\widehat{\boldsymbol{p}}$ estimates the population proportion $\boldsymbol{p}$.

Figure 7.3.1 shows a histogram of pr$(Y = y)$ values for $Y \sim$ Binomial$(n = 200, p = 0.4)$. Recall that the Binomial has mean np and standard deviation $\sqrt{np(1-p)}$. Superimposed over the histogram in Fig. 7.3.1 is the curve of a Normal

[5] Full reports for this and earlier studies are available at the CASA web site, http://www.casacolumbia.org

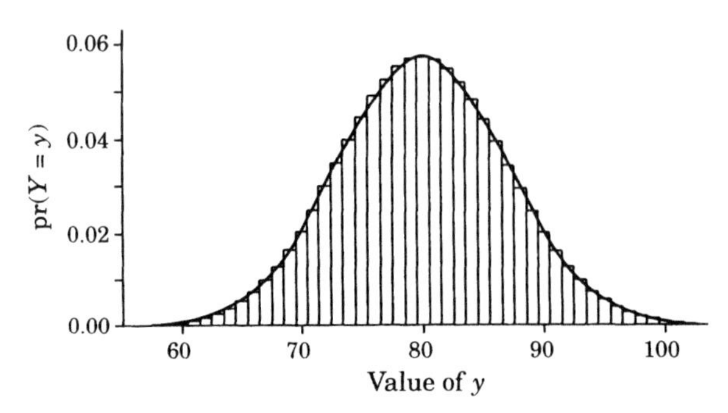

FIGURE 7.3.1 Histogram of Binomial($n = 200, p = 0.4$) probabilities with superimposed Normal curve.

distribution with the same mean $\mu = 200 \times 0.4$ and the same standard deviation $\sigma = \sqrt{200 \times 0.4 \times 0.6}$. We think you will concur that the agreement is pretty good. It turns out that, for large values of n, the Binomial distribution can be approximated by the Normal, and the larger the value of n, the better the approximation.

Now $\widehat{P}$, the sample proportion, is simply Y/n and so is also approximately Normally distributed under the same conditions. It is shown on our web site that this result is a consequence of the central limit theorem. Moreover, we showed in Example 5.4.6 that $E(\widehat{P}) = p$ and $sd(\widehat{P}) = \sqrt{p(1-p)/n}$.

> For large samples, the distribution of $\widehat{P}$ is approximately Normal with
>
> $$\text{mean} = p \text{ and standard deviation} = \sqrt{\frac{p(1-p)}{n}}$$

When we want to use a sample proportion $\widehat{p}$ to estimate an unknown population proportion p, the standard deviation $sd(\widehat{P})$ is useless as a measure of precision of the estimate because it depends on p, which is unknown. The usual estimate of $sd(\widehat{P})$ is obtained by replacing the true proportion p by the sample proportion $\widehat{p}$, giving $se(\widehat{p}) = \sqrt{\widehat{p}(1-\widehat{p})/n}$.

> Standard error of the sample proportion:
>
> $$\mathbf{se}(\widehat{p}) = \sqrt{\frac{\widehat{p}(1-\widehat{p})}{n}}$$

We can use $\widehat{p}$ and its standard error $se(\widehat{p})$ to tell us about the true value of p, in exactly the same way as we used $\overline{x}$ and $se(\overline{x})$ to tell us about the true value of the population mean μ.

EXAMPLE 7.3.1 (cont.) *Two-Standard-Error Interval*

In the poll that we used to open this section, 27% of a random sample of 1000 American teenagers had seen drugs sold at school. If the sample value is 27%, how big is the true population proportion, p, likely to be? Our estimate $\widehat{p} = 0.27$ has a standard error of[6]

$$\text{se}(\widehat{p}) = \sqrt{\frac{\widehat{p}(1-\widehat{p})}{n}} = \sqrt{\frac{0.27 \times 0.73}{1000}} = 0.014039$$

It is a fairly safe bet that the true value of p is within two standard errors of $\widehat{p}$, namely within the range $0.27 \pm 2 \times 0.014039$, or between 24% and 30% (to the nearest whole percent).

QUIZ ON SECTION 7.3.1

1. We use both $\widehat{p}$ and $\widehat{P}$ to describe a sample proportion. For what purposes do we use the former and for what purposes do we use the latter?
2. What two models were discussed in connection with investigating the distribution of $\widehat{P}$? What assumptions are made by each model?
3. What is the standard deviation of a sample proportion obtained from a Binomial experiment?
4. Why is the standard deviation of $\widehat{P}$ not useful in practice as a measure of the precision of the estimate? How did we obtain a useful measure of precision, and what is it called?
5. What can we say about the true value of p and the interval $\widehat{p} \pm 2\,\text{se}(\widehat{p})$?
6. Under what conditions is the formula $\text{se}(\widehat{p}) = \sqrt{\widehat{p}(1-\widehat{p})/n}$ applicable?

EXERCISES FOR SECTION 7.3.1

1. In a survey of undergraduate students enrolling at the University of Auckland in 1990, 39% of the 90 music students had fathers classified in the highest level of socioeconomic status (SES). We shall consider these students as a random sample from the process that produces music students in Auckland.
   (a) Calculate a two-standard-error interval for the true proportion of music students with fathers in the highest SES.
   (b) Is it plausible that the true percentage of music students with fathers in the highest SES is 23% (the value for science students)? How many standard errors is the data estimate from 23%?
2. An April 1998 survey of 2280 users of the Internet media company Yahoo! Australia & NZ found that 32% identified themselves as "professionals," and 40% had been to university. Assuming that this is a random sample of users, calculate a two-standard-error interval for the true proportion of Yahoo! Australia & NZ users who are professionals. Repeat for the proportion who have been to university.
3. In the 1998 "Back-to-School" survey (see Example 7.3.1) 825 American teachers were sampled. Of these, 48% believed they had received adequate training about how to spot substance abuse in their students. This is a sample value and thus subject to sampling variation. If someone was claiming that the majority of teachers believe they have had sufficient training, would the data prove them wrong?
4. The following presents some shocking figures about the extent of the problem of rape in the United States, and some readers may prefer to skip over it. The "National Violence Against Women" survey, jointly funded by the U.S. Departments of Justice and Health and Human

---

[6]Recall from Section 4.6.2 that a percentage is just a way of presenting a proportion; for example, 34% is equivalent to $34/100 = 0.34$.

Services, was reported in November 1998 (see Tjaden and Thoennes [1998]). This study interviewed 8000 men and 8000 women. The sheer numbers of women affected were much higher than many realized. Overall, 17.6% of the women reported having been raped (cf. 3% of men).

(a) How badly is this figure affected by sampling variation? Calculate a two-standard-error interval for the true proportion of American women who have been raped.

Another aspect of the study results that was quite frightening was just how young many of the women were. The following are percentages of the 1323 women in the study who had been raped according to the age at which they were first raped: under 12, 21.6%; 12–17, 34.2%; 18–24, 29.4%; and over 25, 16.6%. We shall regard these 1323 women as a random sample from the population of American female rape victims.

(b) The interval in (a) was fairly narrow, indicating a low level of uncertainty about the true value. This is because the sample size is very large. With the smaller sample size for rape victims we can expect more uncertainty due to sampling variation. Calculate a two-standard-error interval for the population proportion who were under 12 when first raped.

(c) What proportion were under 18? Calculate a two-standard-error interval for the population value.

7.3.2 Some Asides about Polling

You need a much larger sample when sampling from a huge population such as the whole of the United States (260 million people) than when sampling from a town of 10,000 people, right? Wrong! This is perhaps the most commonly held misconception about sampling. The Binomial-based theory of the previous subsection is an adequate approximation for sampling without replacement from a real population when the sample size n is smaller than about 10% of the population size N. This condition is satisfied for almost all real sampling. One of the consequences of the theory is that, as far as the precision of sample proportions is concerned, it is the sample size n that is important. The size of the population, N, is essentially irrelevant.[7] What is important, however, is that the sample is random, that everyone has the same chance of being selected into the sample. In practice, this is more difficult to achieve with large populations. Unfortunately, taking larger samples does nothing to alleviate nonsampling errors such as selection bias (see Section 1.1.2).

Public opinion polls are often not simple random samples, particularly if they involve face-to-face interviews rather than telephone interviews. Frequently, they are some form of two-stage cluster sample. A two-stage cluster sample is essentially sampling carried out at two levels. The population is divided up into "clusters" (e.g., towns, counties, or streets) from which a random sample of clusters is taken. Individuals are then randomly selected from within these clusters. Often there are elements of quota sampling as well (see Section 1.1.1). Unfortunately, the errors typically quoted in the media from poll results (approximately 2 se($\widehat{p}$), where se($\widehat{p}$) $= \sqrt{\widehat{p}(1-\widehat{p})/n}$ with $\widehat{p} = \frac{1}{2}$) are only really appropriate for random sampling. The idea that these errors apply to the two-stage cluster samples taken is as much an act of faith as anything else. In fact, this formula for se($\widehat{p}$) underestimates the errors in cluster samples, and

[7] As a technical aside, the formula we have given for se($\widehat{p}$) can be corrected for sampling without replacement from a finite population by multiplying it by $\sqrt{1 - n/N}$. (See our web site for details.) The correction factor is close to 1 (i.e., no change) whenever $n/N < 0.1$.

the underestimate can be severe in some situations. For example, in a 1997 survey of 2182 school children sampled from northwestern Ontario in Canada, 25.5% reported using cannabis in the past year. The survey used a regionally stratified cluster sample. Use of $\widehat{p} = 0.255$ and $\text{se}(\widehat{p}) = \sqrt{\widehat{p}(1-\widehat{p})/n}$ gives a standard error of 0.0093, or 0.9%. The researchers performed an analyis that was valid for their complex sampling scheme and obtained a standard error of 2.8%—approximately three times larger.

Another source of confusion in opinion polls is the tendency for newspapers and magazines to quote a single measure of error as though it applies universally to all estimates calculated from the poll. For example, one may see the "statistical error" quoted as $\pm 2\%$. In fact, one cannot apply the same measure of error to all estimates because $\text{se}(\widehat{p})$ varies with $\widehat{p}$. These issues are discussed in Section 8.5.3.

Finally, we remind you that in this section we have only been considering the effect of *sampling errors* (or sampling variation). The same applies to the "margins of error" and "statistical errors" encountered in news reports of polls. The *nonsampling* errors discussed in Section 1.1.2 are often larger and more important. We suggest that you reread Section 1.1.2 and use the ideas there to critically evaluate any surveys that you meet in real life.

QUIZ ON SECTION 7.3.2

1. In the American television show *Annual People's Choice Awards,* awards are given in many categories (including favorite TV comedy show and favorite TV drama) and are chosen using a Gallup poll of 5000 Americans (U.S. population approx. 260 million). At the time the 1988 Awards were screened in New Zealand, a *New Zealand Listener* journalist did "a bit of a survey" and came up with a list of awards for New Zealand (population 3.2 million). Naturally, her list differed somewhat from the U.S. list. She went on to say, "It may be worth noting that in both cases approximately 0.001 percent of each country's populations were surveyed." The inference seemed to be that because of this fact, her survey was just as reliable as the Gallup poll. Do you agree? Justify your answer.
2. Are public opinion polls involving face-to-face interviews typically simple random samples?
3. What approximate measure of error is commonly quoted with poll results in the media? What poll percentages does this level of error apply to?
4. A 1997 questionnaire intended to investigate the opinions of computer hackers was available on the Internet for two months and attracted 101 responses. For example, 82% of respondents said that stricter criminal laws would have no effect on their activities. Why would you have no faith that a two-standard-error interval would cover the true proportion in this case?

7.4 ESTIMATES THAT ARE APPROXIMATELY NORMAL

Let us reinforce the distinction between a *parameter,* which is a characteristic of a population or distribution whose value is typically unknown, and an *estimate,* which is a quantity we calculate from sampled data to estimate an unknown parameter. So far we have discussed using a sample mean $\overline{x}$ to estimate an unknown population mean

μ (a parameter) and a sample proportion $\widehat{p}$ to estimate the corresponding population proportion p (also a parameter). Single-number estimates such as sample means and sample proportions are more properly referred to as ***point estimates*** to distinguish them from ***interval estimates*** (e.g., two-standard-error intervals). Other estimates will arise later. We have uniform ways of operating with almost all of the estimates we meet in this book. To describe these in a way that exposes the underlying sameness of the formulas used in different situations, we talk in terms of some arbitrary *estimate* (denoted by $\widehat{\theta}$) of some arbitrary *parameter* (denoted by θ). By identifying the estimate with $\bar{x}$ and the parameter with μ, for example, or the estimate with $\widehat{p}$ and the parameter with p, we can apply general results to specific situations.

First, let θ be some parameter of a distribution or population that we want to estimate (e.g., μ or p). The process of using sample data to try to make useful statements about an unknown parameter θ is called ***statistical inference.*** We use $\widehat{\theta}$ (pronounced "theta hat") to denote the point estimate of θ calculated using the data from the sample that we actually obtained. Following the convention of using small letters for observed values and capital letters for random variables, the random variable corresponding to $\widehat{\theta}$ is $\widehat{\Theta}$. We use this when thinking about the effects of sampling variation (i.e., from sample to sample) in $\widehat{\theta}$ values. The degree to which we trust our data estimate $\widehat{\theta}$ (e.g., the "27% of American teenagers" in Example 7.3.1) depends on what we know about the properties of the estimation process, that is, on the properties of the estimator $\widehat{\Theta}$.

7.4.1 Bias, Precision, and Accuracy

What properties should we look for in a good estimator? It is important that the sampling distribution of the estimator be concentrated as closely as possible to the true value of the parameter θ. If that is the case, then almost every time a sample is taken, the resulting estimate $\widehat{\theta}$ is close to the true value of θ.

The ***bias*** in an estimator is the distance between the "center" of the sampling distribution of the estimator and the true value of the parameter. Stated technically, bias $= E(\widehat{\Theta}) - \theta$. If the bias $= 0$, we say that the estimator is ***unbiased.*** In other words, the expected value of the estimator is the parameter we are trying to estimate. We saw in Section 7.2.1 that $E(\overline{X}) = \mu$ and in Section 7.3.1 that $E(\widehat{P}) = p$. Thus the sample mean is an unbiased estimate of the population mean, and the sample proportion is an unbiased estimate of the population proportion. Typically, the estimators we discuss in this book are unbiased.

The ***precision*** of the estimate is a measure of how variable the estimator is in repeated sampling. Figure 7.4.1 demonstrates the interplay between the notions of bias and precision. Although both are strictly properties of the sampling distribution of the estimator, we have depicted them in terms of the behavior of individual estimates. Each point in each figure is to be thought of as an estimate resulting from an independently conducted research study (e.g., different points represent the results of different studies).

The behavior of the estimation process depicted in Fig. 7.4.1(a) is closest to our ideal. Although the estimator in Fig. 7.4.1(b) is also unbiased because the distribution of estimates is centered at the true value of the parameter, it is highly variable. Thus we are likely to get an estimate that is far from the true value. The estimator in Fig. 7.4.1(c) is reasonably precise (low variability) but is biased: There is a tendency

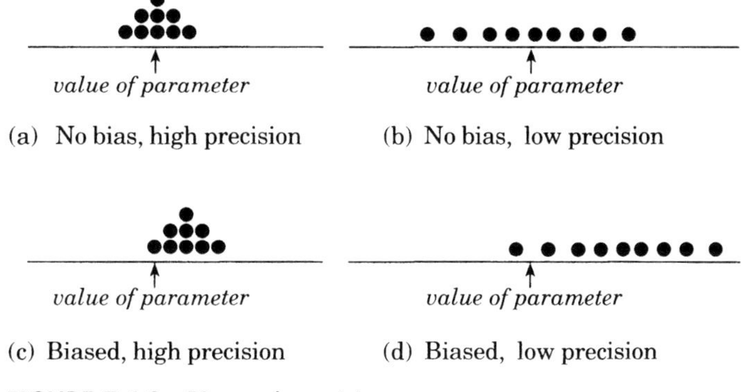

FIGURE 7.4.1 Bias and precision.

to give too many large values. Finally, Fig. 7.4.1(d) shows the worst behavior. It shows an estimation process that is both biased and highly variable (low precision). The intuitive notion of *accuracy* draws on both these ideas of bias and precision. An accurate estimator is one that generally gives estimates that are close to the parameter estimated so that it will have low bias and high precision. Clearly, we could allow a little bias in an estimator if it is compensated for by greater precision.

We have noted that a "precise" estimate is one that is subject to very little sampling variation. We would like to use the standard deviation of the sampling distribution of the estimator [i.e., $\text{sd}(\widehat{\Theta})$] as our measure of precision. In almost all practical situations, however, the actual standard deviation of the sampling distribution is unknown. We thus use an estimate of this quantity, called the standard error and denoted by $\text{se}(\widehat{\theta})$, as a measure of the precision of our estimate. Smaller standard errors correspond to more precise estimates.

> The ***standard error*** of any estimate $\widehat{\theta}$ [denoted $\text{se}(\widehat{\theta})$]
>
> - estimates the variability of $\widehat{\theta}$ values in repeated sampling and
> - is a measure of the ***precision*** of $\widehat{\theta}$.

7.4.2 Interval Estimates

For all of the estimators with which we work in detail in this book, and in fact for most estimators widely used in statistics, we find that $(\widehat{\Theta} - \theta)/\text{se}(\widehat{\Theta})$, the number of standard errors separating the data estimate and the true parameter value, is approximately Normal(0,1) in large samples. Thus the arguments we used in Section 7.2.3 apply very generally. They will now be restated.

In large samples, since $\text{pr}(-2 \le Z \le 2) = 0.954$, it is quite rare to get a sample for which the data estimate $\widehat{\theta}$ and the true value of the parameter θ are more than two standard errors apart. It happens only about 5% of the time, or for one sample in 20 on average. Since $\text{pr}(-3 \le Z \le 3) = 0.997$, it is considerably rarer to get a sample in which the data estimate is more than three standard errors from the true value of the parameter. It happens only about 0.3% of the time, or for about one sample in 300 on average. Two consequences of this are as follow:

(i) If we want to estimate the true value of a parameter θ, saying that it is somewhere in the range *estimate* $\pm$ 2 *standard errors* [or $\widehat{\theta} \pm 2\,\text{se}(\widehat{\theta})$] is a reasonably safe bet.

(ii) The data provide evidence against any theory that specifies a value for the parameter that is more than two standard errors from our data estimate. The evidence against the theory is even stronger if the postulated value of θ is more than three standard errors from our data estimate.

As stated in Section 7.2.3, idea (i) is a gentle introduction to the idea of a *confidence interval* (Chapter 8) and idea (ii) is a gentle introduction to the idea of a *test of significance* (Chapter 9). We shall use these more sophisticated tools rather than (i) and (ii) once they have been developed.

7.4.3 How Many Significant Figures?

If the true value of a parameter is imprecisely known, as signified by having an estimate with a large standard error, it is counterproductive to quote large numbers of significant figures when presenting answers. Unnecessary detail distracts the reader's attention from the parts of the numbers that matter. We therefore suggest the following rules for presentation.[8] An estimate should be presented only to the first significant figure of its standard error, and the standard error should have one more significant figure than the estimate. For example, suppose our calculations gave an estimate of 1.5436 with a standard error of 0.06923. We would present the estimate as 1.54 (as 6 is the first significant figure of the standard error), and we would present the standard error as 0.069.

Additionally, it is important not to quote too many significant figures when presenting two- or three-standard-error intervals (or confidence intervals from Chapter 8). For example, if the parameter is a proportion, then [.128654, .224103] could be reduced to [.13, .22] (or [13%, 22%]) with little loss of information. As a rough rule (and common sense needs to prevail here!), the precision of the reported interval should be *about* one more significant figure than the order of magnitude of the width of the interval. For example, we could present a calculated interval of [23.76, 45.12] as [24, 45], [0.01238, 0.01912] as [0.012, 0.019], and [5.36175, 5.42314] as [5.36, 5.42].

QUIZ ON SECTION 7.4

1. Describe what is meant by the terms *parameter* and *estimate.*
2. Why was a notation for an arbitrary estimate introduced?
3. What is statistical inference?
4. Try to describe the ideas of bias and precision in plain English.
5. What is meant when an estimate of an unknown parameter is described as unbiased? Give an example of an unbiased estimate.

---

[8]Note that *calculations* should always be performed preserving as much accuracy as possible. Thus intermediate answers that are used in further calculations should not be rounded. The rounding discussed here refers to rounding of final answers for presentation purposes.

6. What is the standard error of an estimate, and what do we use it for?
7. Given that an estimator of a parameter is approximately Normally distributed, where can we expect the true value of the parameter to lie?
8. If each of 1000 researchers independently conducted a study to estimate a parameter θ, how many researchers would you expect to catch the true value of θ in their two-standard-error interval?
9. Why can presenting too many significant figures be inappropriate when presenting estimates and their standard errors or two-standard-error intervals in the main body of a report? Do these considerations apply when they are given for reference purposes so that they can be used in calculations? Why or why not?

7.5 STANDARD ERRORS OF DIFFERENCES

Example 7.5.1 *Perceptions About Police Targeting Minorities*

The May 1998 *Star-Ledger*/Eagleton Poll asked samples of adult New Jersey residents a number of questions relating to ethnicity and treatment by state police. One question asked was the following: "Many cars on N.J. highways travel over the speed limit and could be stopped by the state police. There have been a number of stories in the news lately about something called 'profiling.' This is where the police target certain cars to stop based on the race or age of people in the car. Do you think the state police use these characteristics when deciding who to pull over, or does every person who commits a traffic violation pretty much have the same chance of being pulled over?" Of the 139 people sampled from the "black or Hispanic" group, 52% thought the police used profiling, whereas of the 378 people sampled from the "white" group, 29% thought the police used profiling. It looks as though more black or Hispanic people believe the police use profiling than is the case for whites. The figures from our survey are merely estimates from samples, however, and as such are subject to sampling variation. What can we say about the true difference? The situation is depicted in Fig. 7.5.1. The intuitive feeling of uncertainty about exactly where the true proportions for blacks or Hispanics and whites lie is depicted by the smudged horizontal lines under each estimate. We can easily see how to estimate the difference between black or Hispanic and white proportions. We need to find out how uncertainty about the true values of the individual proportions is translated into uncertainty about the difference.

In this example the parameter θ that we want to estimate is $p_1 - p_2$, the difference between the true population proportions of blacks or Hispanics and whites who believe the police use profiling. The natural estimate ($\widehat{\theta}$) of this true difference is the difference between the sample proportions for blacks or Hispanics and whites given in the data, namely,

$$\widehat{p}_1 - \widehat{p}_2 = 0.52 - 0.29 = 0.23$$

or 23%. Before we can calculate an interval of possible values for the true difference, we need the standard error of our estimate. In other words we need se($\widehat{p}_1 - \widehat{p}_2$), the standard error of a difference between two sample proportions. We can obtain standard errors for each of the individual sample proportions using the formula for se($\widehat{p}$) given Section 7.3.1, but how should the individual standard errors be combined?

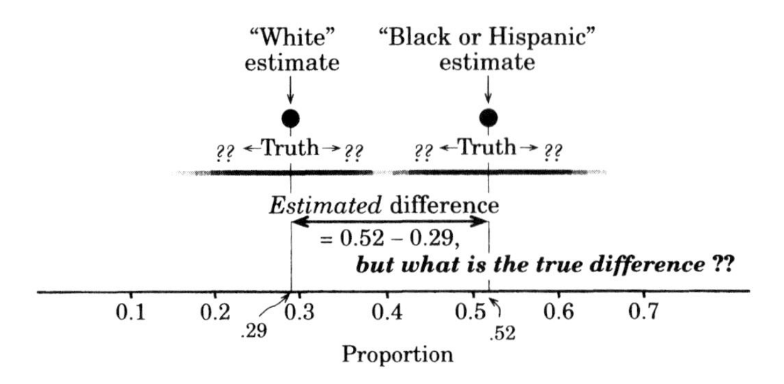

FIGURE 7.5.1 Proportions who believe that police use profiling.

7.5.1 Standard Error for a Difference (Independent Samples)

It is obvious how standard errors should be combined if you remember that standard errors are just estimates of standard deviations. They should, therefore, be combined in the same way. Recall that for independent random variables X_1 and X_2, $\text{sd}(X_1 - X_2) = \sqrt{\text{sd}(X_1)^2 + \text{sd}(X_2)^2}$. Thus,

> ***Standard error for a difference*** between independent estimates:
>
> $$\text{se}(\text{Estimate}_1 - \text{Estimate}_2) = \sqrt{\text{se}(\text{Estimate}_1)^2 + \text{se}(\text{Estimate}_2)^2}$$

or
$$\text{se}(\widehat{\theta}_1 - \widehat{\theta}_2) = \sqrt{\text{se}(\widehat{\theta}_1)^2 + \text{se}(\widehat{\theta}_2)^2}$$

"Independent" in the preceding box means *statistically* independent. Estimates from unrelated samples from different populations are statistically independent.

Standard Error for a Difference between Proportions (Independent Samples)

Let $\widehat{p}_1$ be the sample proportion from the first sample of size n_1, and let $\widehat{p}_2$ and n_2 be the sample proportion and sample size for the second sample. Recall that for a single proportion $\text{se}(\widehat{p}) = \sqrt{\widehat{p}(1-\widehat{p})/n}$. Therefore, if $\widehat{p}_1$ and $\widehat{p}_2$ are sample proportions *from independent samples,*

$$\text{se}(\widehat{p}_1 - \widehat{p}_2) = \sqrt{\text{se}(\widehat{p}_1)^2 + \text{se}(\widehat{p}_2)^2} = \sqrt{\frac{\widehat{p}_1(1-\widehat{p}_1)}{n_1} + \frac{\widehat{p}_2(1-\widehat{p}_2)}{n_2}}$$

(We shall discuss calculating standard errors for situations where the sample proportions are not independent in Section 8.5.2.)

✱ Example 7.5.1 (cont.) *Comparing Two Independent Proportions*

In the survey we used to motivate this discussion, we had for the black or Hispanic sample $n_1 = 139$, $\widehat{p}_1 = 0.52$, and for the white sample $n_2 = 378$, $\widehat{p}_2 = 0.29$.

The estimates are independent since they are drawn from different samples from different populations. The standard error associated with the estimated difference $\widehat{p}_1 - \widehat{p}_2 = 0.23$ is thus

$$\mathrm{se}(\widehat{p}_1 - \widehat{p}_2) = \sqrt{\frac{0.52 \times 0.48}{139} + \frac{0.29 \times 0.71}{378}} = 0.0483776$$

The value of 0 for $\theta = p_1 - p_2$ is approximately 4.75 standard errors away from our data estimate [because $(0.23 - 0)/0.0483776 = 4.75$]. It seems fairly clear that the true proportions are different. A two-standard-error interval of values for the true difference $p_1 - p_2$ is (*estimated difference* $\pm$ *2 standard errors*)

$$\widehat{p}_1 - \widehat{p}_2 \pm 2\,\mathrm{se}(\widehat{p}_1 - \widehat{p}_2) = 0.23 \pm 2 \times 0.0483776$$

that is, from 0.13 to 0.33. It is a fairly safe bet that the true percentage of black or Hispanic people who believe that the police use profiling is larger than the corresponding percentage for whites by between 13% and 33%.

When asked whether they actually had been pulled over by the police in the past five years, 36% of the black and Hispanic group said yes, compared with 27% of the white group. Going through the same type of calculation, the two-standard-error interval for the true difference extends from −0.3% to 18%, that is, from essentially no racial difference to an appreciable excess of black and Hispanic people being pulled over. But where in this range the truth lies, we do not know.

EXAMPLE 7.5.2 *Does Body Position for Epidural Injections Affect Labor Pain?*

A study was carried out at National Women's Hospital in Auckland to see whether the epidural injections given to deaden labor pains in women giving birth were more effective when administered when the woman was lying down or when she was sitting up. Patients were randomly allocated to the "lying" or "sitting" groups. The variable measured was the number of spinal segments that were blocked by the anesthetic. A segment is said to be blocked if all the nerves originating from it are deadened. Summary statistics for the data are (sample size, mean, standard deviation)

Lying group:	$n_1 = 48$	$\bar{x}_1 = 8.8$	$s_1 = 4.4$
Sitting group:	$n_2 = 35$	$\bar{x}_2 = 7.1$	$s_2 = 4.5$

It looks from these figures as if the anesthetic works better when the patient is lying down. Is this necessarily so? We shall consider these women to be random samples from all women who have epidural injections under these two sets of conditions. The parameter we shall estimate is $\mu_1 - \mu_2$, the difference between the true mean number of segments blocked when patients are lying down (μ_1) and the true mean when they are sitting up (μ_2). Our estimate is the difference between the corresponding sample means (i.e., the difference we see in our data), namely, $\bar{x}_1 - \bar{x}_2 = 8.8 - 7.1 = 1.7$. We now need to know how to obtain the standard error of this estimate.

Standard Error for a Difference between Means (Independent Samples)

Let n_1 be the size of the first sample and $\overline{x}_1$ and s_1 be the sample mean and standard deviation, respectively. Let n_2, $\overline{x}_2$, and s_2 denote the corresponding quantities for the second sample. Recall that for a single mean $\text{se}(\overline{x}) = s_X/\sqrt{n}$. Therefore, from the preceding theory, if $\overline{x}_1$ and $\overline{x}_2$ are sample means *from independent samples,*[9] then

$$\text{se}(\overline{x}_1 - \overline{x}_2) = \sqrt{\text{se}(\overline{x}_1)^2 + \text{se}(\overline{x}_2)^2} = \sqrt{\frac{s_1{}^2}{n_1} + \frac{s_2{}^2}{n_2}}$$

Example 7.5.2 (cont.) ***Two-Standard-Error Interval***

In the anesthetic example,

$$\text{se}(\overline{x}_1 - \overline{x}_2) = \sqrt{\frac{4.4^2}{48} + \frac{4.5^2}{35}} = 0.9909$$

Our two-standard-error interval for the true difference, $\mu_1 - \mu_2$, is thus (estimated difference $\pm$ 2 standard errors) $1.7 \pm 2 \times 0.9909$; that is, the true difference in average number of segments blocked is somewhere between -0.3 and 3.7. We do not really have enough data to tell which method is better (or to determine that there is any difference at all in effectiveness). The lying position might result in anything from *slightly* fewer blocked segments ($\mu_1 - \mu_2$ could be small and negative) to 3.7 more segments blocked on average. Because the lying position is unlikely to be much worse but could be substantially better than the sitting position, if we had to decide which to use on the basis of these data, we would choose the lying position.

7.5.2 Interpreting an Interval for a Difference

Interpretation of a two-standard-error interval, $[a, b]$, for a true difference $\theta_1 - \theta_2$ depends on the signs of the endpoints a and b.

- If *both are positive,* for example, $2 \le \theta_1 - \theta_2 \le 6$, it is a fairly safe bet that θ_1 is larger than θ_2 by between 2 and 6 units.
- If they have *opposite signs,* for example, $-2 \le \theta_1 - \theta_2 \le 6$, it is likely that θ_1 is somewhere between being smaller than θ_2 by 2 units ($\theta_1 - \theta_2 = -2$) and being larger by 6 units ($\theta_1 - \theta_2 = 6$). This includes the possibility that they are equal ($\theta_1 - \theta_2 = 0$).
- If *both are negative,* for example, $-6 \le \theta_1 - \theta_2 \le -2$, it is a fairly safe bet that θ_1 is smaller than θ_2 by between 2 and 6 units.

In the *special case of opposite signs,* for example, $-0.1 \le \theta_1 - \theta_2 \le 6$, here the negative endpoint is very much closer to 0 than the positive endpoint, and we can say that θ_1 may be larger than θ_2 by up to 6 units but is unlikely to be appreciably smaller.

[9] We learn about standard errors for a difference in one situation where the samples are not independent, namely, paired (or matched) data, in Section 10.1.2.

We have just seen an example of this, and of how the idea can be useful when making decisions in the presence of uncertainty, in Example 7.5.2.

7.5.3 Individual Two-Standard-Error Intervals and Differences

Suppose that $\widehat{\theta}_1$ is an unbiased estimate of θ_1 and $\widehat{\theta}_2$ is an unbiased estimate of θ_2.

When the Separate Intervals Do *Not* Overlap

When the intervals do not overlap (e.g., the two-standard-error intervals are for θ_1 [2, 5] and for θ_2 [7, 10]), without further calculation we *can* conclude that $\widehat{\theta}_1$ and $\widehat{\theta}_2$ are at least $2\text{se}(\widehat{\theta}_1 - \widehat{\theta}_2)$ apart,[10] and thus that it is a fairly safe bet that the true values of θ_1 and θ_2 are different.

When the Separate Intervals Do Overlap

Calculations are required before conclusions can be drawn. It is still possible that the data provide evidence that true values of θ_1 and θ_2 are different, for example, $\widehat{\theta}_1 = 14$, $\text{se}(\widehat{\theta}_1) = 1.2$, $\widehat{\theta}_2 = 10$, and $\text{se}(\widehat{\theta}_2) = 1.1$. Then our two-standard-error intervals are

for θ_1, $14 \pm 2 \times 1.2 = [11.6, 16.4]$

for θ_2, $10 \pm 2 \times 1.1 = [7.8, 12.2]$

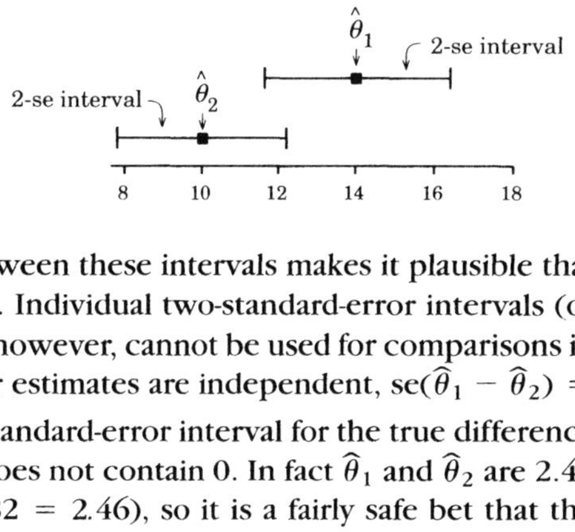

One might think that the overlap between these intervals makes it plausible that the true values of θ_1 and θ_2 are the same. Individual two-standard-error intervals (or confidence intervals from Chapter 8 on), however, cannot be used for comparisons in this way. Here $\widehat{\theta}_1 - \widehat{\theta}_2 = 4$. Provided our estimates are independent, $\text{se}(\widehat{\theta}_1 - \widehat{\theta}_2) = \sqrt{1.2^2 + 1.1^2} = 1.627882$, and the two-standard-error interval for the true difference is $4 \pm 2 \times 1.627882 = [0.7, 7.3]$, which does not contain 0. In fact $\widehat{\theta}_1$ and $\widehat{\theta}_2$ are 2.46 standard errors apart (because $4/1.627882 = 2.46$), so it is a fairly safe bet that the true values of θ_1 and θ_2 are different.

Why the apparent anomaly? For reasons explained in Section 6.4, the variability of a difference between (or sum of) two independent random quantities is smaller than the sum of the variabilities of its components [i.e., $\text{sd}(X_1 - X_2) = \text{sd}(X_1 + X_2) < \text{sd}(X_1) + \text{sd}(X_2)$].

EXERCISES FOR SECTION 7.5

1. The press release of the "Northwestern Ontario Student Drug Use Survey" gave the following figures for 1997 on the proportions of students in northwestern Ontario, Canada, who had smoked cigarettes at least once during the past year. We give the figures in the form *estimated proportion (standard error)* by grade at school—grade 7: 0.141 (.02); grade 9: 0.378 (.043); grade 11: 0.474 (.031); grade 13: 0.303 (.030). The feature that surprised us is

[10] See Review Exercises 7, problem 18.

the apparent drop in usage between grade 11 and grade 13. Do the data provide evidence of a real change, or could the drop-off be explained simply by sampling variation?

(a) Calculate the standard error of the difference between the grade 11 and grade 13 proportions.

(b) How many standard errors apart are the two proportions? Do you think that there really is a decrease?

(c) Calculate a two-standard-error interval for the true change.

(d) What, if anything, do these results suggest to you about teenage smoking?

2. After the sixth observation in Example 7.2.2 (reading across the first row of Table 7.2.1), Cavendish changed his experimental apparatus by replacing a suspension wire by a stiffer wire. Even very small changes in experimental apparatus have been known to change the results. The following plot separates out the first 6 observations from the remaining 23.

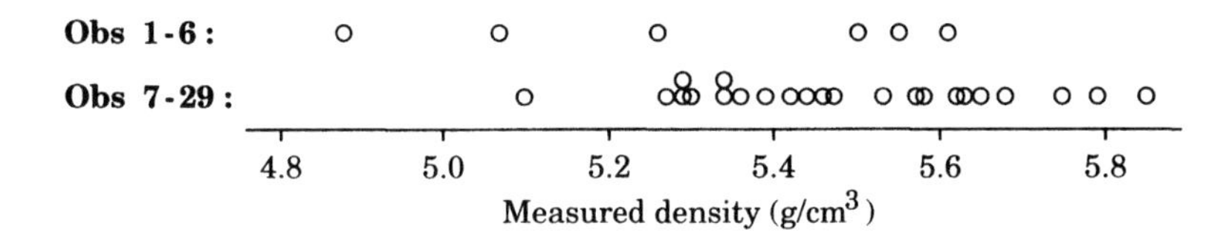

The first 6 look smaller on average. Is this real, or could it just be random?

(a) How many standard errors, se($\overline{x}_2 - \overline{x}_1$), separate the sample means for observations taken using the second wire from those taken using the first wire? Is there evidence to demonstrate that the true means are different?

(b) We may not be able to demonstrate a true difference just because of the size of the random errors. Calculate a two-standard-error interval for the difference in true means. How big could the true difference be?

3. The Centers for Disease Control and Prevention (CDCP) researchers were conducting a study at a sexual diseases clinic in Montgomery County, Maryland, when they noticed something that surprised them (*New York Times*, 29 January 1993). They were asking clients (among other things) how often in the last three months they had a one-time sexual encounter, a risky behavior. Of 186 people questioned in one period, 31% reported at least one such encounter compared with only 20% of the 97 people questioned in the next reporting period. The researchers ascribed the change to a drop in risky sexual behavior following the well-publicized announcement from basketball superstar Magic Johnson that he was HIV positive. The announcement occurred between the reporting periods.[11]

(a) Could the change be just random fluctuation? How many standard errors, se($\hat{p}_1 - \hat{p}_2$), are the sample proportions apart? Do you believe that the true proportions could be the same?

(b) Calculate a two-standard-error interval for the true difference.

(c) What populations do these results apply to?

(d) What would justify the idea that it was Magic Johnson's announcement that caused the change? What further investigation would you like to see done?

4. The motivation for the testosterone-smoking study by Dai et al. [1988] was discussed in Exercises for Section 7.2.2, problem 3. Summary statistics (in ng/dL) for three groups defined by smoking habits are given in Table 7.5.1. These statistics were obtained from men sampled from the participants in a large study investigating methods for preventing heart disease.

[11]There was a similar drop in the percentages of people with three or more partners.

TABLE 7.5.1 Smoking and Testosterone Levels

Cigarettes/day	Sample size	Mean	Standard deviation
None	62	620.6	241.5
1–30	31	715.6	248.0
31–70	28	795.1	305.3

Source: Dai et al. [1988].

(a) Calculate (separate) two-standard-error intervals for the true mean testosterone level for each of the three groups in the populations being sampled. These intervals are plotted below.

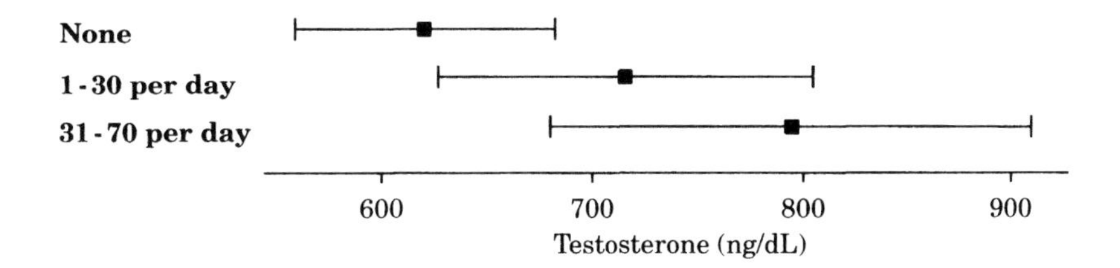

(b) Let us now investigate the differences between the true means. Is it plausible that there is no difference in mean testosterone levels between (i) nonsmokers and those smoking 31–70 cigarettes a day? (ii) nonsmokers and those smoking 1–30 cigarettes per day?

(c) The two-standard-error interval for the mean testosterone level for nonsmokers just overlaps with that for people smoking 31–70 cigarettes a day. How can you reconcile this with your answer to (b)(i)?

(d) Why can we not conclude that smoking causes testosterone levels to increase? (See Section 1.3.)

(e) Can you conclude from (b) that almost all 31- to 70-per-day smokers have higher testosterone levels than almost all nonsmokers? What is it that the two-standard-error intervals in (b) are telling you about?

To help you think more deeply about the issues involved in (e), we plotted some artificial data generated to have the means and standard deviations in the table (Dai et al. published only the summary statistics) in Fig. 7.5.2.

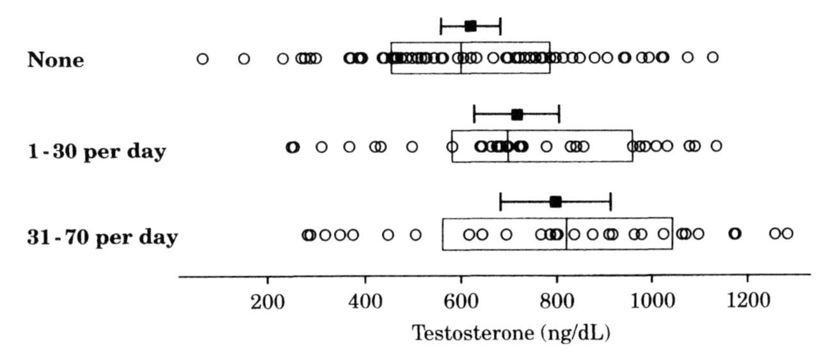

FIGURE 7.5.2 Dot–box plots of testosterone levels by smoking habits (two-standard-error intervals for true means added).

7.6 STUDENT'S *t*-DISTRIBUTION

7.6.1 Introduction

When we talked about the Normal distribution in Chapter 6, we were really talking about a whole family of distributions indexed by the parameters μ and σ. Different distributions in the family were obtained by choosing different values for μ and σ. ***Student's t-distribution***[12] similarly describes another family of distributions, indexed this time by a parameter called the ***degrees of freedom*** and abbreviated ***df.*** If the distribution of a random variable T is a member of the family, we write[13]

$$T \sim \textbf{Student}(df)$$

Density curves for several members of this family are shown in Fig. 7.6.1. These distributions are all symmetrical about the origin (0) and have a similar shape to the Normal distribution. Figure 7.6.1 also shows how the parameter df affects the curve. As df becomes large, the Student's t-distribution curve becomes indistinguishable from the standard Normal distribution [i.e., Normal($\mu_Z = 0, \sigma_Z = 1$)]. In fact, Student($df = \infty$) and the Normal(0,1) are two different ways of describing the same distribution. As df gets smaller, the tails of the distribution get fatter and the top is pushed down. Tables of the Student(df) distribution are given in Appendix A6, and their use is described in Section 7.6.3 to follow.

We might ask, "What has any of this to do with the present discussion?" It follows from Section 7.2.1 that, when we are sampling from a Normal(μ, σ) distribution, the distribution of

$$Z = \frac{\overline{X} - \mu}{\text{sd}(\overline{X})} = \frac{\overline{X} - \mu}{\sigma/\sqrt{n}} \quad \text{is exactly Normal(0, 1)}$$

If we standardize $\overline{X}$ using its standard error se($\overline{X}$) instead of its standard deviation, however, we form

$$T = \frac{\overline{X} - \mu}{\text{se}(\overline{X})} = \frac{\overline{X} - \mu}{S_X/\sqrt{n}}$$

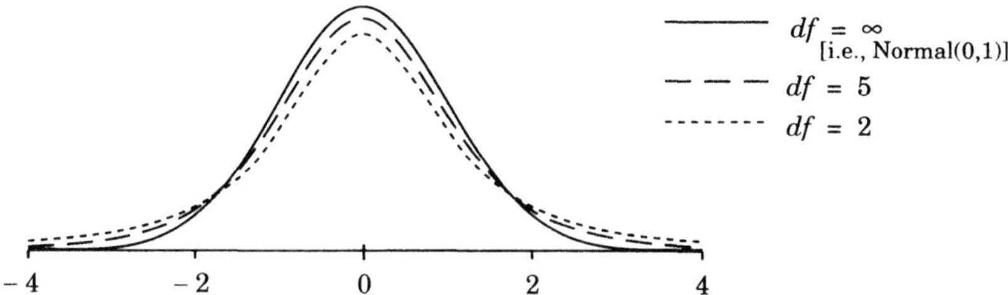

FIGURE 7.6.1 Student(df) density curves for various df.

---

[12] The t-distribution was first derived by W. S. Gosset in 1908, writing under the pseudonym "Student," for the case of a single sample mean.

[13] Most books use the notation $T \sim t_{df}$.

The difference between Z and T is that σ is replaced by the sample standard deviation S_X, thus introducing a new source of variability into T that is absent in Z. The effect of this is to make the distribution of T more variable than that of Z (which is standard Normal). It can be shown mathematically that

$$T = \frac{\overline{X} - \mu}{\text{se}(\overline{X})} \sim \text{Student}(df = n - 1)$$

where n is the sample size. The additional variability of T is reflected in its distribution being flatter and having longer tails than the standard Normal. For large sample sizes n the distribution of T is indistinguishable from standard Normal, but for smaller sample sizes the distribution of T can be a great deal more variable than Normal(0,1).

Since $T = (\overline{X} - \mu)/\text{se}(\overline{X})$, T measures how many standard errors $\overline{X}$ and μ are apart. Our previous theory based on the standard Normal distribution says that for large samples the sample mean falls within two standard errors of the population mean for approximately 95% of samples taken. Let us now consider samples of size $n = 8$. The distribution of T is then Student($df = 7$). Now,

$$\text{pr}(-2 \leq T \leq 2) = 0.9144 \qquad \text{(by computer)}$$

so that the sample mean actually falls within two standard errors of the population mean only 91% of the time, even under ideal conditions (since we are sampling from a distribution that is known to be exactly Normal). Moreover, for $T \sim$ Student($df = 7$),

$$\text{pr}(-2.365 \leq T \leq 2.365) = 0.95$$

so that to obtain limits that include μ for 95% of samples taken when $n = 8$, we need to take $\overline{x} \pm 2.365\,\text{se}(\overline{x})$, not $\overline{x} \pm 2\,\text{se}(\overline{x})$. Such modifications of our large sample theory will be discussed in detail in the next subsection and Chapters 8 and 9.

7.6.2 The Usefulness of Student's *t*-Distribution

The lesson from the theory of the previous subsection is as follows. If we want to preserve the same probability that the interval captures the population mean, we need to take a slightly larger number of standard errors on either side of $\overline{x}$ for smaller samples than we do for very large samples.

The mathematical theory we have sketched and the strict application of Student's *t*-distribution for $T = (\overline{X} - \mu)/\text{se}(\overline{X})$ apply only to data sampled from a Normal distribution. If the methods derived from the mathematical theory stopped working as soon as we got away from the Normal distribution, the theoretical result that T has a Student($df = n - 1$) distribution would have no practical relevance because in practice we never know that we have Normally distributed data. How useful is the theory in practice?

In large samples the central limit theorem tells us that, even for data sampled from non-Normal distributions, the distribution of T is approximately Normal(0,1), which for large df is virtually identical to Student(df).

How applicable is the *t*-distribution for small samples from non-Normal distributions? For inferences about means, computer simulations and other investigations have shown that some methods derived from using a Student's *t*-distribution for $T =$

$(\overline{X} - \mu)/\text{se}(\overline{X})$ work well in small samples, even when we are sampling from some quite severely non-Normal distributions. We shall explore these issues further in Chapter 10.

Most of the methods described in this book for making formal statistical inferences from data can be unified (both conceptually and notationally) by thinking in terms of taking an estimate $\widehat{\theta}$ and treating the distribution of

$$T = \frac{\widehat{\Theta} - \theta}{\text{se}(\widehat{\Theta})} = \frac{\text{Estimate} - \text{True value}}{\text{Standard error}}$$

as being Student(df). We have seen the basis for doing this in the context of using a sample mean $\overline{x}$ to estimate a population mean μ. Similar considerations apply when estimating the difference between two means or to certain estimates in regression (see Chapter 12).

The large-sample methods we have described for making inferences about proportions cannot be made to apply to small samples by a device as simple as replacing the standard Normal distribution for T by Student's t-distribution. Very different ideas are required for handling small-sample data about proportions. Nevertheless, the *large-sample* methods for proportions can be brought under the same notational umbrella as the methods for means by making the connection that Student($df = \infty$) and Normal(0,1) are just two ways of describing the same distribution. We know that $(\widehat{P} - p)/\text{se}(\widehat{P})$ is approximately Normal(0,1) in large samples.

We shall make heavy use of Student's t-distribution in subsequent chapters. How well the resulting techniques work in practice depends on the type of estimator being used, how the data have been collected, and how much data has been collected. Such details will be considered in subsequent chapters where we shall be considering particular estimators and areas of application. It is important that you make yourself very familiar with the notations and table usage discussed in the following subsection before reading Chapters 8 and 9.

7.6.3 Upper-Tail Probabilities and Percentage Points

Most statistical computer packages work consistently with lower-tail (cumulative) probabilities for all distributions, just as we did with the Normal distribution in Chapter 6. For Student's t-distribution, Splus uses a command of the form `pt(t,df)`, which returns $\text{pr}(T \leq t)$ for $T \sim$ Student(df). For example, `pt(1.5,7)` returns 0.9113508, telling us that $\text{pr}(T \leq 1.5) = 0.9113508$ when $df = 7$. Minitab operates similarly, using a dialog box obtained from the menus `Calc` → `Probability Distributions` → `t` (click cumulative). In much of our work in subsequent chapters, we need upper-tail probabilities of Student's t-distribution, which we can get immediately from the lower-tail probability using $\text{pr}(T \geq t) = 1 - \text{pr}(T \leq t)$, as in Fig. 6.2.5. Excel is an exception. The Excel function `TDIST(t,df,1)` returns the upper-tail probability directly. For example, `TDIST(1.5,7,1)` returns 0.088649, telling us that $\text{pr}(T \geq 1.5) = 0.088649$ when $df = 7$.

In Section 6.3.2 we discussed obtaining the z-values corresponding to, say, the upper 5% or 1% of the standard Normal distribution. More generally, we determined the value $z(prob)$ corresponding to an upper-tail probability of $prob$ as in the left-hand side of Fig. 7.6.2. Commonly used values are tabulated in Appendix A5.2. For example,

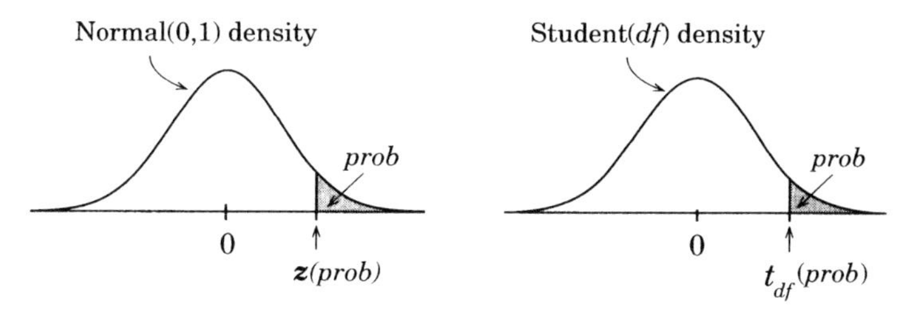

FIGURE 7.6.2 The $z(prob)$ and $t_{df}(prob)$ notations.

TABLE 7.6.1 Extracts from the Student's t-Distribution Table

prob ↓

	df	.20	.15	.10	.05	.025	.01	.005	.001	.0005	.0001
	6	0.906	1.134	1.440	1.943	2.447	3.143	3.707	5.208	5.959	8.025
	7	0.896	1.119	1.415	1.895	2.365	2.998	3.499	4.785	5.408	7.063
	8	0.889	1.108	1.397	1.860	2.306	2.896	3.355	4.501	5.041	6.442
	...	...	...	...	...	...	...	...	...	...	...
→	10	0.879	1.093	1.372	1.812	[2.228]	2.764	3.169	4.144	4.587	5.694
	...	...	...	...	...	...	...	...	...	...	...
	15	0.866	1.074	1.341	1.753	2.131	2.602	2.947	3.733	4.073	4.880
	...	...	...	...	...	...	...	...	...	...	...
	∞	0.842	1.036	1.282	1.645	1.960	2.326	2.576	3.090	3.291	3.719

the z-value corresponding to the upper 5% of the Normal(0,1) distribution is $z(.05) = 1.6449$. We also learned how to get these values from computer programs.

In subsequent chapters we shall also need the t-values corresponding to, say, the upper 5% or $2\frac{1}{2}$% of Student's t-distribution. More specifically, if we are working with a particular Student(df) distribution, we want to take a value of *prob* such as 0.05 and obtain the number t for which

$$\text{pr}(T \geq t) = prob$$

Because this number depends on both *prob* and the *df* we are using, we use the notation $t_{df}(prob)$. This is simply the number with upper-tail area *prob* above it. We shall find $t_8(0.025) = 2.306$. This means that when $df = 8$, $\text{pr}(T \geq 2.306) = 0.025$. A table for Student's t-distribution is given in Appendix A6. A segment of this table is given in Table 7.6.1 for demonstration purposes. For a given *df* (row), you enter the table with the required upper-tail probability *prob* (column), and the table supplies the $t_{df}(prob)$. For example, if $df = 10$ and $prob = 0.025$, Table 7.6.1 gives $t_{10}(0.025) = 2.228$, or equivalently, $\text{pr}(T \geq 2.228) = 0.025$. Recall that the Normal(0,1) and Student($df = \infty$) distributions are identical so that $z(prob)$ and $t_\infty(prob)$ values are identical.

From the table we find the following:

- If $df = 7$ and $prob = 0.05$, then $t_7(0.05) = 1.895$, i.e., $\text{pr}(T \geq 1.895) = 0.05$.
- If $df = 15$ and $prob = 0.1$, then $t_{15}(0.1) = 1.341$, i.e., $\text{pr}(T \geq 1.341) = 0.1$.
- If $df = 15$ and $prob = 0.05$, then $t_{15}(0.05) = 1.753$, i.e., $\text{pr}(T \geq 1.753) = 0.05$.

Recall that most specialized statistical packages work consistently with lower-tail (cumulative) probabilities. An upper-tail probability of *prob* corresponds to a lower-tail probability of 1 − *prob*. Thus in Splus we obtain $t_{df}(prob)$ as the (1 − *prob*) quantile using `qt(1-prob,df)`. In Minitab we do the same thing using the menus. In Excel, which deviates from this pattern, $t_{df}(prob)$ can be obtained using `TINV(twoprob,df)`, where `twoprob` = $2 \times prob$.

Exercise: Check that you can obtain the three preceding values of $t_{df}(prob)$ from the program you are using.

Finding Central Regions of the Student Distribution

We shall often need to find the values within which the central 90% (or 95%, etc.) of the Student(*df*) distribution lies. Like the standard Normal distribution, the Student(*df*) distribution is symmetrical about 0. Thus, if we want a probability of 0.90 in the center (see Fig. 7.6.3), we need a probability of 0.05 in the upper tail. The *t*-value that gives us this is $t_{df}(0.05)$. Therefore the central 90% of the distribution falls between $-t_{df}(0.05)$ and $t_{df}(0.05)$. For example, if $df = 7$, the central 90% of the distribution lies between −1.895 and +1.895.

More generally, if we want a probability of $(1 - \alpha)$ in the center, we need a probability of $prob = \alpha/2$ in the upper tail. Therefore the limits we require are $\pm t_{df}(\alpha/2)$ [or $\pm z(\alpha/2)$ for the Normal(0,1)].

Exercise using tables or computer: In each of the following cases, give the limits between which the central *x*% (*x*% to be given) of the Student(*df*) lies.

(a) 80%, $df = 8$ (b) 99%, $df = 15$ (c) 70%, $df = 20$

(d) 95%, $df = 40$ (e) 98%, $df = 6$

[**Answers: (a)** ±1.397 **(b)** ±2.947 **(c)** ±1.064 **(d)** ±2.021 **(e)** ±3.143]

Using the Tables to Bracket Tail Probabilities

The methods in Chapter 9 require the use of upper-tail probabilities of Student's *t*-distribution. The following applies when you do not have access to a computer package that supplies the desired probabilities. In that case you will be forced to use tables to bracket tail probabilities. Suppose $df = 10$ and we want to find pr($T \geq 1.9$). By scanning the $df = 10$ row, you will see that pr($T \geq 1.812$) = 0.05 and pr($T \geq 2.228$) = 0.025. Thus pr($T \geq 1.9$) is between 0.025 and 0.05.

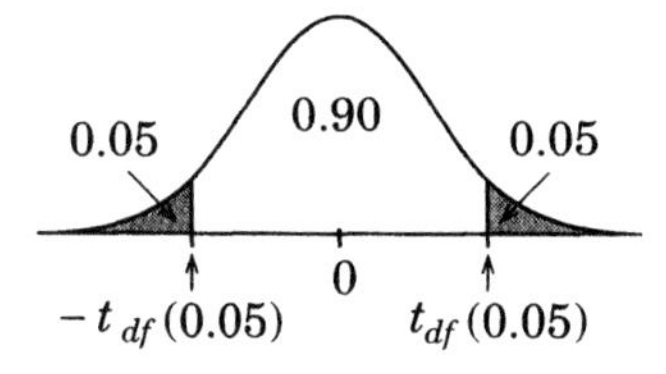

FIGURE 7.6.3 The central 90% of the Student(*df*) distribution.

Exercise: Using the same method, bracket the sizes of the following probabilities:

(a) pr($T \geq 1.6$) when $df = 8$ (b) pr($T \geq 0.9$) when $df = 15$
(c) pr($T \geq 2.7$) when $df = 20$ (d) pr($T \geq 5.4$) when $df = 40$
(e) pr($T \geq 4.0$) when $df = 6$ (f) pr($T \geq 0.6$) when $df = 13$

[**Answers:** (a) $0.05 < prob < 0.1$ (b) $0.15 < prob < 0.2$ (c) $0.005 < prob < 0.01$ (d) $prob < 0.0001$ (e) $0.001 < prob < 0.005$ (f) $prob > 0.2$]

QUIZ ON SECTION 7.6

1. Qualitatively, how does the Student(*df*) distribution differ from the standard Normal(0,1) distribution? What effect does increasing the value of *df* have on the shape of the distribution?
2. What is the relationship between the Student($df = \infty$) distribution and the Normal(0,1) distribution?
3. Why is T, the number of standard errors separating $\overline{X}$ and μ, a more variable quantity than Z, the number of standard deviations separating $\overline{X}$ and μ?
4. For large samples the true value of μ lies inside the interval $\overline{x} \pm 2\,\text{se}(\overline{x})$ for a little more than 95% of all samples taken. For small samples from a Normal distribution, is the proportion of samples for which the true value of μ lies within the two-standard-error interval smaller or bigger than 95%? Why?
5. For a small Normal sample, if you want an interval to contain the true value of μ for 95% of samples taken, should you take more or fewer than two-standard errors on either side of $\overline{x}$?
6. Under what circumstances does mathematical theory show that the distribution of $T = (\overline{X} - \mu)/\text{se}(\overline{X})$ is exactly Student($df = n - 1$)?
7. Why would methods derived from the theory be of little practical use if they stopped working whenever the data was not Normally distributed?

7.7 SUMMARY

7.7.1 Sampling Distributions

1. For random quantities we use a capital letter for the random variable and a small letter for an observed value, for example, X and x, $\overline{X}$ and $\overline{x}$, $\widehat{P}$ and $\widehat{p}$, $\widehat{\Theta}$ and $\widehat{\theta}$.
2. In estimation the random variables (capital letters) are used when we want to think about the effects of *sampling variation,* that is, about how the random process of taking a sample and calculating an estimate behaves.
3. ***Sample mean, $\overline{X}$:*** For a random sample of size n from a distribution for which $\text{E}(X) = \mu$ and $\text{sd}(X) = \sigma$, we have the following results:
   - $\text{E}(\overline{X}) = \text{E}(X) = \mu, \quad \text{sd}(\overline{X}) = \dfrac{\text{sd}(X)}{\sqrt{n}} = \dfrac{\sigma}{\sqrt{n}}$
   - If we are sampling from a Normal distribution, then $\overline{X} \sim$ Normal.
   - ***Central limit theorem:*** For almost any distribution, $\overline{X}$ is approximately Normally distributed in large samples.

4. ***Sample proportion, $\widehat{P}$:*** For a random sample of size n from a population in which a proportion p have a characteristic of interest, we have the following results about the sample proportion $\widehat{P}$ with that characteristic:
   - $\mu_{\widehat{P}} = \text{E}(\widehat{P}) = p \qquad \sigma_{\widehat{P}} = \text{sd}(\widehat{P}) = \sqrt{\dfrac{p(1-p)}{n}}$
   - $\widehat{P}$ is approximately Normally distributed for large n (e.g., $np(1-p) \geq 10$, though a more accurate rule is given in the next chapter).

7.7.2 Estimation

1. A ***parameter*** is a numerical characteristic of a population or distribution.
2. An ***estimate*** is a known quantity calculated from the data to estimate an unknown parameter.
   - For general discussions about parameters and estimates, we talk in terms of $\widehat{\theta}$ being an estimate of a parameter θ.
   - The bias in an estimator is the difference between $\text{E}(\widehat{\Theta})$ and θ.
   - $\widehat{\theta}$ is an ***unbiased estimate*** of θ if $\text{E}(\widehat{\Theta}) = \theta$.
   - The ***precision*** of an estimate refers to its variability. One estimate is *less precise* than another if it has *more variability*.
3. The ***standard error***, $\text{se}(\widehat{\theta})$, for an estimate $\widehat{\theta}$ is
   - an estimate of $\text{sd}(\widehat{\Theta})$ and
   - a measure of the precision of the estimate.
4. ***Means:***
   - The sample mean $\overline{x}$ is an unbiased estimate of the population mean μ.
   - $\text{se}(\overline{x}) = \dfrac{s_X}{\sqrt{n}}$
5. ***Proportions:***
   - The sample proportion $\widehat{p}$ is an unbiased estimate of the population proportion p.
   - $\text{se}(\widehat{p}) = \sqrt{\dfrac{\widehat{p}(1-\widehat{p})}{n}}$
6. ***Standard error of a difference:*** For ***independent estimates,***

$$\text{se}(\widehat{\theta}_1 - \widehat{\theta}_2) = \sqrt{\text{se}(\widehat{\theta}_1)^2 + \text{se}(\widehat{\theta}_2)^2}$$

TABLE 7.7.1 Some Parameters and Their Estimates[a]

	Population(s) or distribution(s) ↓ Parameters	Sampled data ↓ Estimates →	Measure of precision
Mean	μ	$\overline{x}$	$\text{se}(\overline{x})$
Proportion	p	$\widehat{p}$	$\text{se}(\widehat{p})$
Difference in means	$\mu_1 - \mu_2$	$\overline{x}_1 - \overline{x}_2$	$\text{se}(\overline{x}_1 - \overline{x}_2)$
Difference in proportions	$p_1 - p_2$	$\widehat{p}_1 - \widehat{p}_2$	$\text{se}(\widehat{p}_1 - \widehat{p}_2)$
General case	θ	$\widehat{\theta}$	$\text{se}(\widehat{\theta})$

[a]Two-standard-error intervals $\widehat{\theta} \pm 2\,\text{se}(\widehat{\theta})$ are replaced by confidence intervals from the next chapter.

7.7.3 Student's *t*-distribution

1. It is bell shaped and centered at 0 like the Normal(0,1) but is more variable (larger spread and fatter tails). As *df* becomes larger, the Student(*df*) distribution becomes more and more like the Normal(0,1) distribution.

2. Student($df = \infty$) and Normal(0,1) are two ways of describing the same distribution.

3. For random samples from a Normal distribution,

$$T = (\overline{X} - \mu)/\text{se}(\overline{X})$$

is exactly distributed as Student($df = n - 1$), but methods we shall base on this distribution for T work well even for small samples from distributions that are quite non-Normal.

4. By $t_{df}(prob)$, we mean the number t such that when $T \sim$ Student(*df*), $\text{pr}(T \geq t) = prob$; that is, the tail area above t (to the right of t on the graph) is *prob*.

REVIEW EXERCISES 7

1. The time a laser printer takes to print a picture is found to be variable, depending on a number of different aspects of the picture such as size, precision, type of computer file, and so on. Over a few weeks the time taken to print each picture on a particular laser printer is recorded and found to have a mean time of 87 seconds and a standard deviation of 23 seconds.
   (a) Let Y be the total time taken to print seven different pictures selected at random. What are the mean and standard deviation of Y?
   (b) Let W be the total time taken to print seven copies of the same randomly selected picture. Assuming the times taken to print each copy of the same picture are identical, what are the mean and standard deviation of W?
   (c) Let $\overline{X}$ be the average time taken to print 36 different pictures selected at random. What are the mean and standard deviation of $\overline{X}$?
   (d) What is the approximate distribution of $\overline{X}$? What name is given to the theory on which this is based?
   (e) What difference would we see in the distribution of $\overline{X}$ and its parameters if we had considered the average time taken to print 144 different pictures selected at random instead of 36?

2. Consider the experiment of throwing an ideal die three times (i.e., all outcomes are equally likely). Let $\overline{X}$ and M denote the mean and median, respectively, of the three scores obtained.
   (a) Find $E(\overline{X})$.
   (b) The results of 10 repetitions of the preceding experiment are given in Table 1.

TABLE 1 Tossing a Die

Experiment number									
1	2	3	4	5	6	7	8	9	10
4	2	4	2	6	2	3	5	3	2
3	6	6	3	1	5	1	5	4	1
5	5	2	5	5	2	4	2	3	6

Calculate the 10 medians and the 10 means. Find the sample means and sample standard deviations of these two sets of 10 numbers.

(c) Using the four histogram intervals [1.5, 2.5], [2.5, 3.5], [3.5, 4.5], and [4.5, 5.5], draw histograms of the two frequency distributions obtained in (b).

(d) Compare the empirical sampling distributions of $\overline{X}$ and M.

3. (Based on information in Meier [1986]) For good or ill, statistical evidence has become very important in the United States in civil rights and discrimination cases. The Supreme Court of the United States has made some important rulings that affect all subsequent trials in lower courts. In *Casteneda v. Partida* (decided March 1977) the court noted that 79% of the population of Hidalgo County had Spanish-sounding surnames. Of the 870 jurors selected to jury panels by the county, only 339 (approximately 39%) had Spanish-sounding surnames.

(a) If the 870 jurors were selected at random, what is the distribution of X, the number with Spanish-sounding surnames?

(b) Assuming random sampling, how many standard deviations is the observed proportion from 0.79? What do you conclude?

The court commented, "As a general rule for such large samples, if the difference between the expected value and the observed number is greater than 2 or 3 standard deviations, then the hypothesis that the jury drawing was at random would be suspect to a social scientist."

A subsequent case was *Hazelwood School District v. United States,* decided three months later. In this case an estimated 6% of qualified teachers in St. Louis County were black, but during the two-year period in question, only 15 of the 405 employed (3.7%) were black.

(c) How many standard deviations is 0.037 from 0.06?

The court's recommendation at this point was slightly different, namely, "If the difference exceeds 2 or 3 standard deviations, then the hypothesis that teachers were hired without regard to race would be suspect."

(d) Note some important differences between the two recommendations.

Later in the opinion, the court said, "Where gross statistical disparities can be shown, they alone may in a proper case constitute prima facie proof of a pattern or practice of discrimination."

(e) Teachers are not hired at random. The statistical argument is based on a random selection of teachers. Do you agree with the court? Can you think of some reasons for and against the position the court seems to have adopted?

4. In this problem we study the sampling behavior of a coin-tossing experiment in which a fair coin [pr(*Head*) = 0.5] is tossed an increasing number of times. We saw in Chapter 4 (Fig. 4.1.1) that if the coin is tossed n times, the proportion of heads starts to settle down and get closer to $\frac{1}{2}$ as n gets bigger and bigger. Let us now examine this experiment theoretically.

(a) What is the standard deviation of the sample proportion of heads after a sequence of (i) $n = 100$ tosses? (ii) $n = 1000$ tosses? (iii) $n = 10{,}000$ tosses?

(b) What is the probability that the sample proportion of heads is between 0.49 and 0.51 after a sequence of (i) $n = 100$ tosses? (ii) $n = 1000$ tosses? (iii) $n = 10{,}000$ tosses? (Use the approximate Normal distribution for $\widehat{P}$ given in Section 7.3.1.)

*(c) How big does n have to be for there to be a 95% chance that $\widehat{P}$ is between 0.499 and 0.501? (Use the approximate Normal distribution for $\widehat{P}$ given in Section 7.3.1.)

Although the *proportion* of heads becomes less variable as n increases, what can we say about the *difference* between the number of heads and number of tails?

(d) If D is the number of heads minus the number of tails, what are the mean and standard deviation of D?

(e) What can we say about the variability of D as n increases?

5. A politician received 60% of the vote in the last election. She now wants to gauge public opinion on a number of issues and wants to take a random sample of n voters to do this.

For all that follows, assume that the numbers are large enough to permit the use of Normal approximations.
One question she will ask is whether they voted for her in the last election.

(a) What is the approximate sampling distribution of the proportion of voters in the sample who voted for the politician?

(b) If $n = 200$, what is the (approximate) probability of getting a sample in which at least 50% voted for her?

6. Consider an acceptance-sampling scheme in which a factory takes delivery of a batch of components if a random sample of 400 components contains fewer than M defectives. Otherwise the batch is returned to the supplier. Suppose the factory manager wants to set M so that there is no more than a 10% risk of accepting a batch that has 5% or more defectives. Let $\widehat{P}$ be the sample proportion of defectives. (In the following, use the approximate Normal distribution for $\widehat{P}$.)

(a) Suppose that the batch has 5% defectives. What is the value of c for which $\text{pr}(\widehat{P} > c) = 0.10$? What value should M have?

(b) Suppose the supplier makes a batch in which the true proportion of defectives is only 2%. If the cutoff in (a) is adopted, what is the probability that the batch will be sent back?

*(c) Find the sample size n so that there is a cutoff that restricts both the factory's risk of accepting a 5% defective batch to one chance in 10, and the supplier's risk of having a 2% defective batch sent back to one chance in 10.

7. The nature of casino gambling is such that a great deal of money flows in both directions between casino and gamblers. Also, gamblers win often enough to remain optimistic and continue playing. Slowly but inexorably, however, the casino makes money! Games are structured so that the odds on a given bet favor the casino rather than the gambler, but only by a small amount. Given that this is the case, many people wonder how it is that casinos manage to make enormous amounts of money over the long run. It is all to do with the sampling distributions of averages, as this exercise will demonstrate.

We shall use a very simple roulette bet to illustrate. American roulette wheels have 18 black numbers, 18 red numbers, and 2 green numbers (0 and 00). Suppose the gambler bets \$1 on black (similarly red). The casino wins \$1 if a red or green number comes up (20 chances out of 38) and loses \$1 if a black number comes up (18 chances out of 38). Thus the casino's winnings from the ith \$1 bet is a random variable X_i with probability function

$$\text{pr}(X = 1) = \frac{20}{38} \qquad \text{pr}(X = -1) = \frac{18}{38}$$

Results from different bets are independent.

(a) Find the mean and standard deviation of X.

(b) (i) Find the mean and standard deviation of the casino's total winnings from 50 \$1 bets. (ii) Repeat for the casino's average winnings.

(c) Write the answers corresponding to those in (b) for an arbitrary number of bets, n.

For the calculations to follow, use Normal approximations to the distributions of means and totals.

(d) What proportion of players who make 50 \$1 bets over the course of an evening will make money? (Equivalently, what is the probability the casino will lose money on these 50 bets?)

(e) What is the probability that the casino loses money in servicing 1000 such bets? Repeat for 100,000 \$1 bets and also 1 million \$1 bets.

(f) Make up a table of three-standard-deviation intervals with four rows and two columns. Intervals in the four rows should refer to sets of 50 \$1 bets, 1000 \$1 bets, 100,000 \$1 bets, and 1 million \$1 bets, respectively. Place intervals for the *average* casino winnings in the first column and those for the *total* casino winnings in the second column.

Can you now see why the casino makes money, while at the same time individual gamblers win often enough to keep them playing?

8. A poll conducted on 10 July 1995 in which a random sample from the population is asked, "Should there be a total ban on smoking in restaurants?" finds that 54% of those polled agreed with a total ban.

(a) Why is 54% an unbiased estimate of the percentage in the population who favored a ban in restaurants on 10 July 1995?

(b) Why is 54% not an unbiased estimate of those in the population who favored a ban in restaurants on 10 September 1995?

(c) Why is 54% not an unbiased estimate of those in the population who favored a total ban on smoking in public eating establishments on 10 July 1995?

9. One hundred married couples are chosen at random. Each person is asked whether the state should fund abortions. Suppose 40% said that the state should never fund abortions. What standard error should be associated with this proportion

(a) if husband and wife always have the same opinion?

(b) if the opinions of husbands and wives are unconnected?

(c) What is the true situation likely to be? Which standard error should you use? Why?

10. A letter to *Time* (7 November 1994) concerning the well-publicized 1994 "Sex in America" survey said that a random sample of 3500 Americans is "by no means an accurate representation of 261 million very diverse individuals." The writer compared the report unfavorably with the Kinsey Report conducted in the late 1940s, saying that Kinsey "had the good sense to survey 11,000 people." Kinsey's was an opportunity sample in that he took his people where he could get them. What misconceptions are being perpetuated here?

11. A 1994 poll conducted in New Zealand asked a long list of questions about what was important in a successful marriage.

(a) Of the survey's 943 respondents, 66% rated "having the same moral standards" as "very important." What standard error should be associated with 0.66 as an estimate of the true population proportion?

(b) Among the 172 people surveyed who were in the 15–24 age group, 50% rated having the same moral standards as very important. What standard error should be associated with 0.50 as an estimate of the true proportion for 15–24 year olds?

(c) How many times bigger is the standard error applying to the subgroup consisting of those aged 15–24 than that applying to the whole sample?

(d) Among the 221 people in the 40–54 age group, 67% rated having the same moral standards as very important. What standard error should be associated with $0.67 - 0.50$ as an estimate of the difference between the true proportion of 40- to 54-year-olds and the true proportion of 15- to 24-year-olds who rate this as very important?

(e) How much bigger is the standard error for the difference than the standard errors of each of the individual sample proportions?

(f) In many surveys reported in the media, the same "margin of error" is used for all estimates and comparisons. What do you think of this?

12. Successful television advertising depends on being able to get viewers to recall the specific brand being advertised. Two groups of university students were shown advertisements for 10 products. One group saw advertisements with sexual content, while the other group saw advertisements with no sexual content. Two days after viewing the advertisements each student was asked to name the specific brands that were advertised. The mean and standard deviation of the number of correct answers for each group are given in Table 2.

(a) Calculate $\text{se}(\bar{x}_1 - \bar{x}_2)$.

TABLE 2 Television Advertisements

Sexual content	$n_1 = 53$	$\overline{x}_1 = 7.90$	$s_1 = 1.82$
No sexual content	$n_2 = 60$	$\overline{x}_2 = 4.30$	$s_2 = 1.53$

(b) Calculate the two-standard-error interval for $\mu_1 - \mu_2$.

(c) What conclusions can you draw from this research?

TABLE 3 Wearing Baseball Caps

	Downtown	Private university	Business school
No. observed	407	236	319
Cap backwards	174	29	107

Source: Trinkaus [1994].

13. From the late 1980s it began to become popular for young people to wear their baseball caps backwards. Are there cultural differences in this practice? Trinkaus [1994] gathered data at three locations—the downtown area of a large U.S. city, the campus of a private university, and the business school of a public university (see Table 3)—and observed how people were wearing their caps.

(a) Calculate a two-standard-error interval for the true proportion of baseball cap wearers downtown who wear their caps backwards.

(b) Is there evidence that downtown cap wearers are more likely to wear their caps backwards than business-school cap wearers? (How many standard errors are the estimates apart?)

(c) Separate two-standard-error intervals for the true proportions in each of the three locations are plotted below. The intervals for downtown wearers and business-school wearers overlap. Does this conflict with your conclusions in (b)? Why or why not?

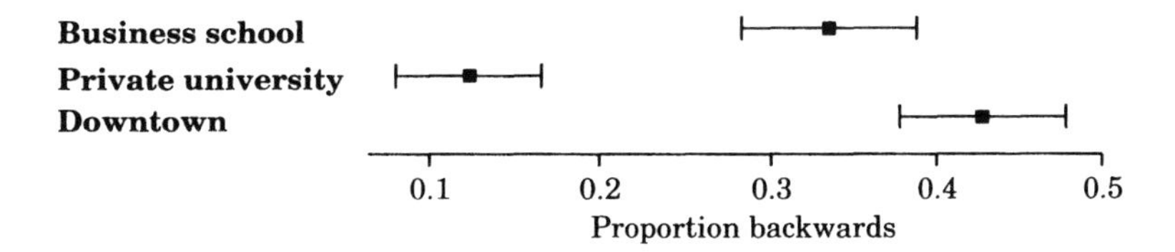

(d) Can we conclude from the plotted two-standard-error intervals that private-university wearers are less likely to wear their caps backwards than either downtown wearers or business-school wearers? Why or why not?

(e) The theory on which all the preceding conclusions rest assumes random samples from populations. This implies independence between subjects observed. Think about the process of observing individuals as they move around at any of these locations. In what ways are these assumptions likely to be violated?

14. How realistic are students in their predictions of grades in a course? Do they tend to be overly optimistic? pessimistic? Is there any relationship between this and the ability of the student? Proshaska [1994] investigated these questions using students enrolled in advanced psychology courses. Students were grouped by "ability" using their grade point averages (GPA) from the previous semester. Predicted grades were compared with actual grades at the end of the study semester, producing the data in Table 4.

TABLE 4 Student Predictions

	Prediction is			
GPA	Underestimate	Correct	Overestimate	Total
Low	4	4	33	41
Med.	6	10	24	40
High	11	12	21	44
Total	21	26	78	125

Source: Proshaska [1994].

(a) What interesting features do you see in the table?

(b) Is it clear that low-GPA students are more likely to overestimate their grades than high-GPA students?

(c) Must the reason for (b) be that low-GPA students are less realistic than high-GPA students, or can you think of another explanation? (Think about letter grades and the possibilities for overestimating when you usually get low grades versus when you usually get high grades.)

TABLE 5 Breast Feeding and IQ

	Sample size	Mean	Standard deviation
Bottle fed	90	92.8	15.18
Breast fed	210	103.0	17.39

Source: Lucas et al. [1992].

15. Does breast feeding increase the IQ of babies? Lucas et al. [1992] analyzed data collected from five special-care baby units in England (see Table 5). The babies were all preterm or had low birth weights. Mothers chose whether to provide breast milk for their infant within 72 hours of delivery. Out of 300 babies, 210 were breast fed, and 90 were not. The children were given an intelligence test after eight years of follow-up.

(a) How many standard errors separate the sample means? Are you fairly certain that there is a difference between the true means?

(b) Calculate a two-standard-error interval for the difference in true means.

(c) What populations do these results apply to?

(d) Do the results prove that breast feeding increases IQ? Why or why not?

(e) The researchers found that the children of mothers who wanted to breast feed but could not had similar IQs to the bottle-fed group. Why was it a good idea to check this?

16. Thornton and Moore [1993] investigated the effects of recently seen photographs on a person's self-rating of physical attractiveness. Sixty-seven female and 61 male undergraduate university students were each randomly divided into three groups, and each group was subject to a different set of experimental conditions. "Control" subjects were not shown photographs of other people, "negative contrast" subjects were shown photographs of highly attractive same-sex people, and "positive contrast" subjects were shown photographs of unattractive same-sex people before rating their own physical attractiveness. Ratings could range from 5 to 25; the higher the score, the greater the perceived physical attractiveness.

(a) Before inspecting the data in Table 6, how would you expect positive and negative contrasts to affect self-ratings of physical attractiveness?

TABLE 6 Physical Attractiveness

	Females			Males		
	Negative contrast	Control	Positive contrast	Negative contrast	Control	Positive contrast
n	21	20	20	22	23	22
$\bar{x}$	15.15	16.98	18.28	14.16	17.18	21.16
s_X	2.96	2.85	3.26	4.64	4.72	4.31

Source: Thornton and Moore [1993].

(b) Two-standard-error intervals for the true mean are plotted below for each of the six treatment groups. Do the data appear to bear out your expectations? What other features do you notice?

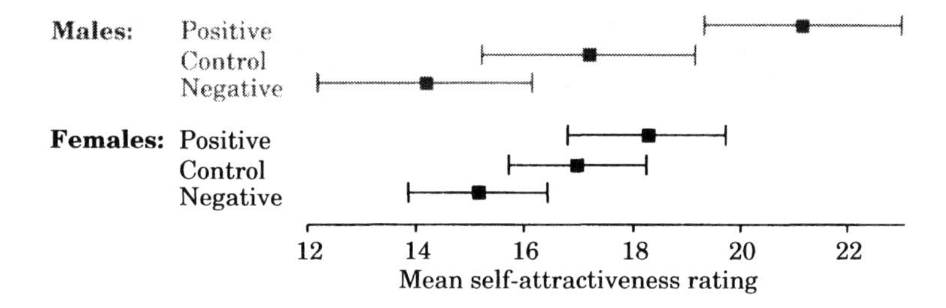

(c) Calculate a two-standard-error interval for the true difference in mean self-ratings between males and females under positive-contrast conditions. Could the true means for males and females be the same? Why does this not contradict what you saw in the plot?

(d) What other differences can you be fairly sure are real?

Males seem to be more strongly affected by the contrasts than do females. We will lead you through a partial investigation of this idea.

(e) Estimate the difference in true mean self-ratings between positive-contrast and negative-contrast conditions for females. We shall call this the estimated contrast effect for females. Calculate the standard error of this estimate.

(f) Repeat (e) for males.

*(g) Estimate the difference between the contrast effect for males and that for females. Find the standard error of this difference. Now find a two-standard-error interval for the true difference. Does this interval support the idea that the male effect is larger than the female effect? How much bigger is it?

[Hint: Your estimate of the difference between male and female contrast effects should be the difference between your male estimate from (f) and your female estimate from (e). Since this new estimate is the difference of independent estimates, you can find the standard error of this difference using the boxed formula in Section 7.5.]

(h) How would you have performed this experiment?

(i) What follow-up investigations would you like to see done?

17. The following summary statistics come from a study performed by Hamill et al. [1980] investigating how strongly people generalize from a very limited range of experience. Scores were taken of people's attitudes to prison guards as a group; the higher the score, the more favorable the attitude. A control group of 39 was tested, and the mean and standard deviation of these people's scores are given in the "control" column of Table 7. The remaining columns relate to groups subjected to experimental conditions. Those in the three "humane" columns were shown a videotaped interview of a prison guard in which the guard was portrayed as a decent, humane person. Those in the "inhumane" columns were shown a videotaped interview in which the guard was portrayed as a bitter and contemptuous person. Other experimental conditions related to what the subjects were told. "No information" is self-explanatory. Those in columns headed "typical" were told that the case was highly typical, and those in columns headed "atypical" that the case was highly atypical of the population.

(a) Before inspecting the data, what would you expect the effects of viewing the two different types of interview to be? the effects of being told the case was typical or atypical or being told nothing?

TABLE 7 Attitudes to Prison Guards

		Humane			Inhumane		
	Control	Typical	No info.	Atypical	Typical	No info.	Atypical
n	39	18	18	18	18	18	18
$\overline{x}$	10.97	12.56	13.28	11.94	9.44	10.44	10.11
s_X	2.67	2.09	1.90	2.62	1.85	2.43	2.03

Source: Hamill et al. [1980]. Copyright © (1980) by the American Psychological Association. Adapted with permission.

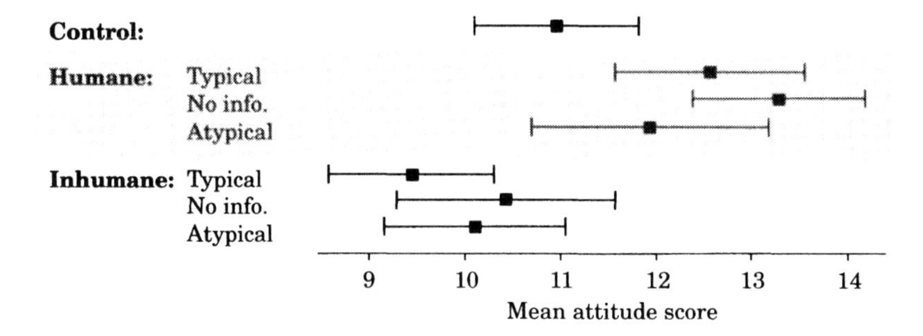

Numbers of standard errors separating sample means: There are seven columns containing means in the table. Let us number them 1 to 7. (1-2) refers to the number of standard errors between the 1st and 2nd mean, (2-6) to the number of standard errors between the 2nd and 6th, and so on (the sign of the difference is ignored). (1-2) 2.4, (1-3) 3.7, (1-4) 1.3, (1-5) 2.5, (1-6) 0.7, (1-7) 1.3, (2-3) 1.1, (2-4) 0.8, (2-5) 4.7, (3-4) 1.8, (3-6) 3.9, (4-7) 2.3, (5-6) 1.4, (5-7) 1.0, (6-7) 0.4.

(b) Using the information given and any further calculations you deem necessary, what do you conclude from these data?

(c) How would you have performed this experiment?

(d) What follow-up investigations would you like to see done?

[This problem has involved making large numbers of comparisons. (To a lesser extent, problem 16 did as well.) There are dangers in making multiple comparisons in the simple-minded way we have been doing here, which will be discussed in Section 10.3.1.]

***18.** Mathematical investigation reveals that[14]

$$\sqrt{\text{se}(\widehat{\theta}_1)^2 + \text{se}(\widehat{\theta}_2)^2} \leq \text{se}(\widehat{\theta}_1) + \text{se}(\widehat{\theta}_2)$$

Why does this enable one to say that if the separate two-standard-error intervals for $\widehat{\theta}_1$ and $\widehat{\theta}_2$ do not overlap, then $\widehat{\theta}_1$ and $\widehat{\theta}_2$ are at least $2\,\text{se}(\widehat{\theta}_1 - \widehat{\theta}_2)$ apart?

***19.** Suppose that separate health care surveys each of $n = 1000$ people are conducted in Canada, the United States, and Mexico. The percentages of people in the surveys who never visited a doctor in the past year were 10% (Canada), 20% (U.S.), and 60% (Mexico). The problem is to estimate the proportion of people in the entire region who do not visit a doctor and to calculate a standard error to accompany that estimate. Take as the populations of the three countries: Canada, 27 million; United States, 250 million; and Mexico, 90 million.

(a) Estimate the total number of people in Canada who did not visit a doctor. Repeat for the United States and Mexico. Estimate the proportion of people in the whole region who never visited a doctor.

(b) Show that your estimate is of the form

$$\widehat{p} = a_{Can}\widehat{p}_{Can} + a_{US}\widehat{p}_{US} + a_{Mex}\widehat{p}_{Mex}$$

where $a_{Can} = 27{,}000{,}000/(27{,}000{,}000 + 250{,}000{,}000 + 90{,}000{,}000)$ and similarly for the United States and Mexico.

(c) Derive a formula for $\text{sd}(\widehat{p})$ and use it to calculate a standard error for your estimate. (Hint: Since a standard error is an estimated standard deviation, you can use the rules for finding standard deviations of linear sums given in Section 6.4.3. A recent example of the situation in this problem occurred with a 1998 study about family planning, in which samples of about 1000 were taken from each of 13 European countries. The results reported in the press release were for the whole of Europe.)

***20.** A fairly typical business confidence index is based on a survey of business people. It is calculated as the difference between the percentage of those surveyed who expect improvement in general business conditions over the next 12 months and the percentage who expect a deterioration. Let n be the sample size and $\widehat{P}$ be the proportion expecting improvement. Let I be the value of the index.

(a) Write I in terms of $\widehat{P}$. (Assume that respondents must opt for either "improvement" or "deterioration.")

(b) Derive a formula for the standard deviation of I from that for $\widehat{P}$.

(c) Suppose that $n = 1000$ and that the summer quarter survey gave a value of $I = 40\%$. Find **(i)** $\widehat{P}$ and **(ii)** $\text{se}(I)$, the standard error to be associated with I.

(d) Why might someone prefer to report an index constructed in this way rather than just the percentage expecting improvement?

21. (Computer-simulation exercise) We will investigate the central limit effect when sampling people from the workforce database and measuring their dietary intake of carbohydrates (cf. Fig. 6.1.1). Random numbers from the Chi-square distribution are available in statistical packages. (This distribution is described in Section 11.1.) If we take a random number from the Chi-square distribution with 17 degrees of freedom, multiply it by 15, and add 4.5, we get sampling behavior that mimics very well the carbohydrate levels of people sampled from the workforce database.

[14]Equality holds only when at least one standard error is 0.

(a) Generate 1000 "carbohydrate levels" as described above and construct a histogram of your "data." Compare it with Fig. 6.1.1.

(b) Generate 100 samples each of size 25. Obtain the sample mean of each sample and construct a histogram of the 100 sample means using the same horizontal scale as in (a) to facilitate comparisons. Compare your new histogram with the histogram in (a). What do you see?

(c) Repeat (a) and (b) using random numbers from a Chi-square distribution with 4 degrees of freedom.

22. (Computer simulation exercise) Men's heights in the workforce database are approximately Normally distributed with mean 174 cm and standard deviation 6.57 cm [cf. Fig. 6.2.1(b)]. We can simulate the process of sampling men and measuring their heights by generating random numbers from a Normal(μ = 174, σ = 6.57) distribution.

(a) Obtain "heights" for 10 samples each of nine males, as described above. Find the sample mean for each sample and construct a dot plot of the 10 sample means. Reflect on what each point on your plot represents.

(b) Now obtain 100 samples, each of size 9. Find the sample mean for each sample and obtain a histogram of the 100 sample means.

(c) Repeat (b), but using samples of size 25. Use the same horizontal scale for the histogram as in (b). Compare the histograms. What do you see?

(d) Find the (sample) standard deviation of the set of 100 sample means in (b). What value would you expect it to have? How close is the observed value to the theoretical value? Repeat for the samples in (c).

(e) For each sample in (b), find the sample standard deviation of the observations. Obtain a histogram of the 100 standard deviations. Repeat for the samples in (c) and compare the two histograms. What do you see?

(f) For each sample (of size 9) in (b), obtain a two-standard-error interval for the true mean. The true mean is 174. What proportion of the intervals from your samples contain 174? Find the average width of the 100 intervals.

(g) Repeat (f) for the samples of size 25 taken in (c). What do you conclude?

23. (Computer simulation exercise) This question involves guessing the answers to a true-false question when you do not know the answer (which is just like tossing a coin).

(a) Intuitively, are you more likely to obtain at least 7 correct answers on a 10-question test or at least 70 correct answers on a 100-question test, or is there no difference? Why do you think this?

(b) Simulate the 10-question test by tossing a coin 10 times with heads corresponding to a correct answer. How many questions did you get right? We can get the computer to do the same thing by generating 10 Binomial(n = 1, p = 0.5) random numbers. Each 1 corresponds to a correct answer and each 0 to an incorrect answer. When you used the computer to simulate the test, how many did you get right?

(c) Using the computer method in (b), simulate 15 people guessing answers to the 10-question test. Count the number of answers each person got right.

(d) In (c), the number correct in a 10-question test for each person corresponds to a random observation from a Binomial(n = 10, p = 0.5) distribution. Why? Generate the number of correct answers for 15 people by generating 15 Binomial(n = 10, p = 0.5) random numbers.

(e) Simulate 100 people taking the 10-question test. Obtain a bar graph of the results and calculate the proportion of people who got at least 7 of the 10 questions correct. Repeat for 100 people and the proportion who got at least 70 of the 100 questions correct.

(f) In both situations, we would expect people to get about half of the questions right by guessing. What theoretical result in this chapter about variability should have led us to expect the type of results we saw in (e)?

(g) For the people doing the 100-question test in (e), convert the 100 values of "number correct" to values of "proportion correct." Find the standard deviation of these sample proportions. What should you expect the standard deviation to be? Why?

(h) Obtain a histogram of the proportions in (g) and comment on its shape.

(i) Obtain a bar graph (chart) of the Binomial($n = 10, p = 0.5$) probability distribution. Repeat for $n = 40$ and $n = 100$. What do you notice?

CHAPTER 8

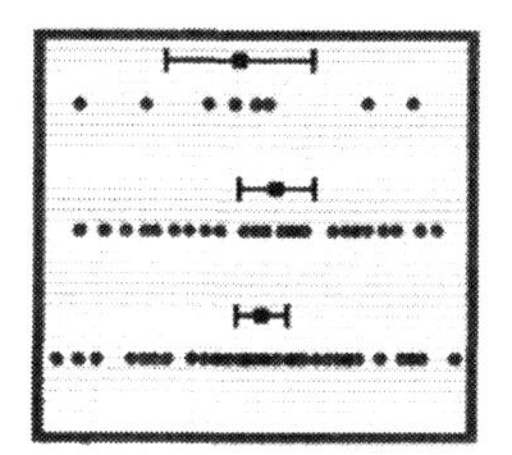

Confidence Intervals

Chapter Overview

We extend the simple idea, introduced in Chapter 7, of using a two-standard-error interval to provide a range of likely values for an unknown parameter to the more sophisticated concept of a confidence interval. The behavior of such intervals is discussed, and a widely applicable formula for confidence intervals based on either a Student's t-distribution or Normal(0,1) distribution is developed in Section 8.1 and applied to the estimation of means (Section 8.2), proportions (Section 8.3), differences between means (Section 8.4), and differences between proportions (Section 8.5). The range of situations in which one can investigate differences between proportions is broadened, with particular attention being given to distinguishing between different types of sampling situations so that appropriate standard errors can be used. The chapter concludes with a discussion of estimating the sample sizes required to give sufficiently precise interval estimates of means and proportions in Section 8.6.

An understanding of the central ideas presented in Chapter 7 and the detailed material on Student's t-distribution in Section 7.6 will be necessary for an understanding of this chapter. The primary emphasis of the chapter is on the nature of confidence intervals and interpreting confidence intervals in the context of a practical investigation.[1]

8.1 INTRODUCTION

In Chapter 7 we talked informally about using a two-standard-error interval as an interval estimate of an unknown parameter. We now take a close look at this idea

[1]Note for teachers: Considerations that are specific to inferences about means, such as Welch versus pooled two-sample t intervals, the paired-data problem, and robustness are postponed until Chapter 10. Robustness is touched on very lightly in this chapter.

and consider intervals of varying widths. Here the width will depend on how much "confidence" we want to have in the interval as a means of capturing the true value of the parameter. The following example introduces the ideas.

✱ EXAMPLE 8.1.1 *Estimating the Speed of Light*

Below are 20 replicated measurements from an 1882 experiment performed by Simon Newcomb to measure the speed of light. He used a variant of Foucault's method, in which light is passed from its source onto a rapidly rotating mirror and picked up by a distant fixed mirror (see Stigler [1977]). The observations are passage times for the light to travel from Fort Myer on the west bank of the Potomac river to a fixed mirror at the foot of the Washington Monument 3721 m away and back (in millionths of a second). There are 20 observations corresponding to 20 repetitions of the measurement procedure.

24.824	24.828	24.837	24.832	24.820	24.825	24.825	24.836	24.836	24.821
24.828	24.826	24.832	24.828	24.826	24.830	24.836	24.829	24.830	24.822

So how long does it take light to travel across the river to the Washington Monument and back? We see that every time Newcomb tried to measure it, he got a slightly different answer. If there is no systematic error (bias) in the experimental procedure, the true passage time will be μ, the population mean of the distribution producing the measurements. A natural estimate of this quantity is the sample mean of the data, $\bar{x}$. In Chapter 7 we learned to place a two-standard-error interval around the estimate to allow for the uncertainty due to sampling variation.

Figure 8.1.1 gives two plots from this data. For Fig. 8.1.1(a), imagine that Newcomb was less persistent and performed only the first 8 replications. Figure 8.1.1(b) uses the full set of 20 measurements. The two-standard-error interval for the true pas-

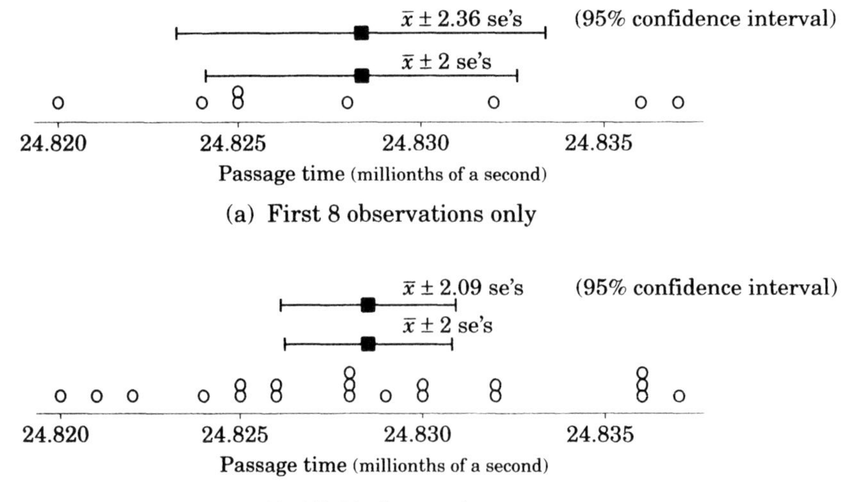

FIGURE 8.1.1 Passage times for light across the Potomac River.

sage time μ from the 8 observations in (a) extends from 24.824 to 24.833 millionths of a second. By using all 20 observations, the uncertainty about where the true value lies has been reduced. We have a narrower two-standard-error interval extending from 24.826 to 24.831 millionths of a second. (We might expect more precision from more data.) The ideas that this example is intended to introduce, however, relate to the intervals labeled *confidence intervals.*

From Sections 7.2.3 and 7.6 we know that a two-standard-error interval captures the true mean for approximately 19 out of every 20 samples taken (actually 95.4% of the time) when we have *large samples.* Section 7.6 demonstrated that if we want to preserve this property with *small samples,* we need to take more than two standard errors either side of our estimate. The "95% confidence interval" in Fig. 8.1.1(a) uses 2.36 standard errors (with $n = 8$) and is noticeably wider than the two-standard-error interval. With the larger sample size of $n = 20$ in Fig. 8.1.1(b), however, the 95% confidence interval uses 2.09 standard errors and is almost the same as the two-standard-error interval.

In this chapter, instead of always blindly using *two* standard errors to either side of our estimate, we begin to use a slightly more sophisticated way of choosing the number of standard errors. The way we do it takes the sample size into account, with smaller sample sizes giving bigger values, such as 2.36 when $n = 8$ versus 2.09 when $n = 20$, as shown for the confidence intervals in Fig. 8.1.1(a and b). Nothing else changes. We still interpret the resulting intervals in the same way. The key to determining the number of standard errors that we should use is Student's t-distribution, discussed in Section 7.6.

We shall think generally about estimating a *population parameter* θ using a *data estimate* $\widehat{\theta}$. The defining property of a 95% confidence interval for a parameter θ is as follows.

> A type of interval that contains the true value of a parameter for 95% of samples taken is called a ***95% confidence interval*** for that parameter.

Recall that $(\widehat{\theta} - \theta)/\text{se}(\widehat{\theta})$ measures how many standard errors our estimate $\widehat{\theta}$ is away from the parameter θ. The random variable $T = (\widehat{\Theta} - \theta)/\text{se}(\widehat{\Theta})$ enables us to explore the behavior of this quantity in the repeated taking of samples. We saw in Section 7.6.2 that for many important situations the distribution of T is well approximated by a Student t-distribution. From Section 7.6.3 we know that 95% of the time a random variable that has a Student(df) distribution falls in the range $\pm t_{df}(0.025)$. Thus $\widehat{\theta}$ is within $t_{df}(0.025)$ standard errors of the true θ for 95% of samples taken. Equivalently, the true θ falls within the interval $\widehat{\theta} \pm t_{df}(0.025)\,\text{se}(\widehat{\theta})$ for 95% of samples taken.[2] This motivates the use of

(95% CI for θ) $$\widehat{\theta} \pm t_{df}(0.025)\,\text{se}(\widehat{\theta})$$

as our interval estimate of the unknown θ when we have data from a sample. In other words, we obtain the number of standard errors to use from the Student(df)

---

[2]If we solve the two inequalities in $-t\,\text{se} \le \widehat{\theta} - \theta \le t\,\text{se}$ separately for θ, we get $\theta \le \widehat{\theta} + t\,\text{se}$ and $\theta \ge \widehat{\theta} - t\,\text{se}$, giving $\widehat{\theta} - t\,\text{se} \le \theta \le \widehat{\theta} + t\,\text{se}$.

TABLE 8.1.1 Value of the Multiplier, t, for a 95% CI

df:	7	8	9	10	11	12	13	14	15	16	17
t:	2.365	2.306	2.262	2.228	2.201	2.179	2.160	2.145	2.131	2.120	2.110
df:	18	19	20	25	30	35	40	45	50	60	∞
t:	2.101	2.093	2.086	2.060	2.042	2.030	2.021	2.014	2.009	2.000	1.960

distribution. For a single mean the sample size enters the picture via $df = n - 1$. We shall refer to $t_{df}(0.025)$ as the multiplier and generally abbreviate it to t.

> A confidence interval (CI) for the true value of a parameter is given by
>
> ***estimate*** $\pm$ ***t standard errors***

This formula can be used in any situation in which the distribution of $(\widehat{\Theta} - \theta)/\text{se}(\widehat{\Theta})$ is well approximated by the Student(df) distribution [or the Normal(0,1) distribution, which is the same as the Student($df = \infty$)]. The endpoints of a confidence interval are called ***confidence limits.*** The piece being added and subtracted is often called the ***margin of error*** so that the form of the interval is *estimate* $\pm$ *margin of error.* Values of the multiplier, t, are given for a range of values of df in Table 8.1.1. Recall that Student($df = \infty$) is simply a way of referring to the standard Normal(0,1) distribution. In this case, it is conventional to represent the multiplier by z rather than t. For a 95% confidence interval, $z = 1.96$.

EXAMPLE 8.1.1 (cont.) *Confidence Interval for a Mean*

The preceding theory applies under the assumption that the data are Normally distributed. There is nothing in Fig. 8.1.1 to rule out Normality as a credible possibility. (These issues will be addressed very briefly in Section 8.2 and with more detail in Section 10.1.1.) Here the parameter (θ) is the true mean μ. For the full set of $n = 20$ observations our estimate ($\widehat{\theta}$) is the sample mean $\overline{x} = 24.82855$. Moreover, the standard deviation of the data is $s_X = 0.0051245$ so that $\text{se}(\overline{x}) = 0.0051245/\sqrt{20} = 0.0011458$. The degrees of freedom are $df = n - 1 = 19$ so that the t-multiplier is $t = 2.093$, giving an interval of (*estimate* $\pm$ 2.093 *standard errors*)

$$\overline{x} \pm 2.093\,\text{se}(\overline{x}) = 24.82855 \pm 2.093 \times 0.0011459 = [24.8262, 24.8310]$$

This is the interval labeled "95% confidence interval" in Fig. 8.1.1(b).

From the previous example we saw that the calculated confidence interval is determined by the sample estimates $\overline{x}$ and s. Clearly, a different sample will lead to different values of these estimates and consequently a different confidence interval. We shall now use a computer simulation to illustrate how the process of taking samples and calculating intervals behaves. This is done by repeatedly taking samples in a situation where we know how the data are generated and know the true value of the parameter. We shall calculate the corresponding 95% confidence intervals and then observe what happens. Here we have used samples of size $n = 10$ ($df = n - 1 = 9$) for which the t multiplier is 2.262, giving the formula $\bar{x} \pm 2.262\,\text{se}(\bar{x})$ for the interval. For the purposes of the simulation we shall use Newcomb's measurements on the passage time of light to provide a context[3] and will suppose that the true passage time for light is 24.83 millionths of a second and the nature of the error in our experimental determinations is such that each measurement we make is Normally distributed about a mean of 24.83 with a standard deviation of 0.005. Thus for the simulation the parameter, θ, that we are trying to estimate is the true value of the mean, namely 24.83. (We also assume that replicated measurements are independent.)

Under these conditions we took 1000 samples each of size 10. Figure 8.1.2 displays the first 10 samples obtained; the 100th, 500th, 501st, 502nd samples; and the last 10 samples. The vertical dashed line marks the position of the true "concentration" in the simulation, namely 24.83 millionths of a second (this would be unknown in practice). For each sample the resulting 95% confidence interval for the mean is displayed immediately above the dot plot of the sample data.

We note the randomness of the intervals and their calculated endpoints, a direct result of the random sampling. We note also that almost every time we take a sample and calculate an interval, we obtain an interval that captures (or covers) the true value of the mean. For two of the depicted samples (the 9th and 991st), the calculated 95% confidence intervals failed to cover the true value. But over the course of 1000 samples, very close to 95% of samples led to intervals that captured the true value. In practice, it is this property that gives us a measure of confidence in the particular interval, such as the interval [24.826, 24.831] we calculated from the passage time data in Example 8.1.1. *Our confidence in the interval* [24.826, 24.831] *comes from the fact that it was produced by a method that works 95% of the time.* We shall use phrases such as, "With 95% confidence the true passage time is between 24.826 and 24.831 millionths of a second" to express such intervals verbally.

We conclude this part of the discussion by outlining how confidence intervals operate under three scenarios.

(i) *Repeated sampling.* In the long run, if we repeatedly take samples and calculate an interval from each sample, 95% of the samples will give intervals that contain the true value (as illustrated in Fig. 8.1.2.)

(ii) *Many studies.* If a large number of researchers independently perform studies, 95% of the researchers will catch the true value of the parameter in their intervals, while 5% of them will fail to do so.

(iii) *Planning a study.* If I plan to do a study in the future in which I will take a random sample and calculate an interval, there is a 95% chance that I will catch the true value of the parameter in my interval and a 5% chance that I will miss it.

8.1.1 Adjusting the Confidence Level

There is nothing hard and fast about using methods that work 95% of the time. Round-number success rates such as 9 times out of 10 (90%), 19 times out of 20 (95%), or 99 times out of every 100 (99%) are usually chosen. For interval estimates the success rate is variously called the ***confidence level***, the *confidence coefficient*, and the *coverage*

[3] When planning the simulation, we purposely chose values for our population mean and standard deviation that were similar to the mean and standard deviation observed in the data of Example 8.1.1.

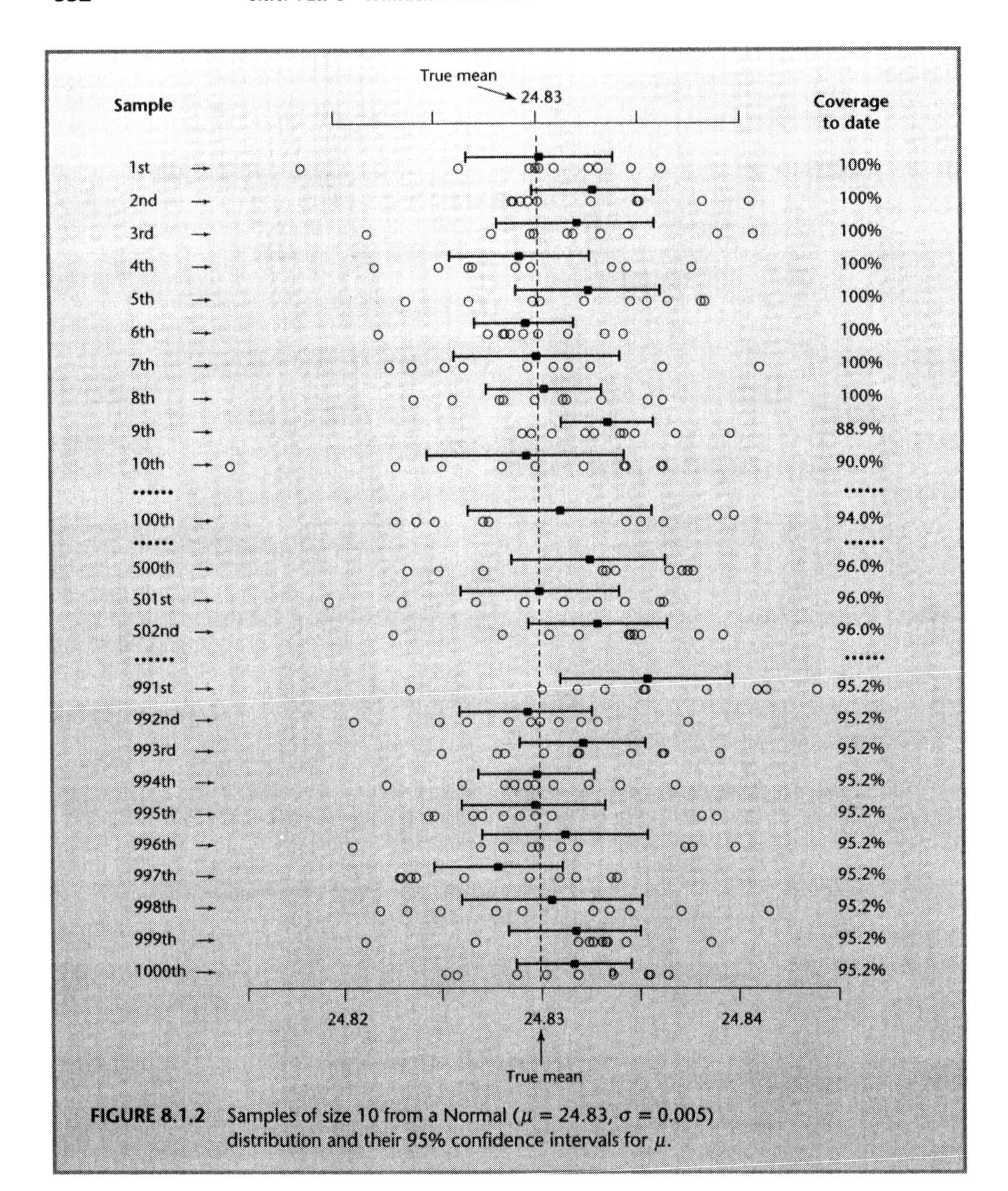

FIGURE 8.1.2 Samples of size 10 from a Normal ($\mu = 24.83$, $\sigma = 0.005$) distribution and their 95% confidence intervals for μ.

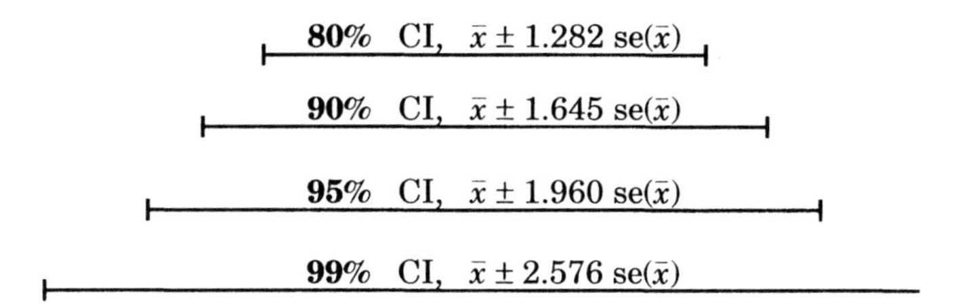

FIGURE 8.1.3 The greater the confidence level, the wider the interval.

frequency. We can obtain the value of t for other confidence levels using the methods described in Section 7.6.3. A 90% confidence interval corresponds to a central probability of 0.9 or an upper-tail probability of 0.05, so we use $t_{df}(0.05)$ (see Fig. 7.6.3) as our multiplier. A 99% confidence interval corresponds to a central probability of 0.99 or an upper-tail probability of 0.005, so we use $t_{df}(0.005)$ as our multiplier.

If we want to increase our level of confidence in an interval for a particular sample, we do so by using a method with a higher success rate. In practice this just means that we use a greater number of standard errors, giving a wider interval. We illustrate this in the $df = \infty$ case. (Any large value of df will give very similar multipliers.) The numbers of standard errors one should take to obtain an interval with the required confidence level are as follows:

Confidence level	80%	90%	95%	99%
Multiplier ($df = \infty$ case)	1.282	1.645	1.960	2.576

The relative widths of the resulting intervals are shown pictorially in Fig. 8.1.3 (in the context of estimating a mean).

To be able to make a statement in which we have more confidence or to run a smaller risk of being wrong, we have to make a less precise statement by quoting a wider interval. Unfortunately, very imprecise statements are of little practical use. For example, 99.9% confidence intervals have an error rate of 1 in 1000. For the 8 observations in Fig. 8.1.1(a) the lower 99.9% confidence limit for the true passage time is 24.817, which is smaller than any of the actual measurements we have seen, and the upper limit is 24.840, which is larger than any of the 20 measured values we have seen.

8.1.2 How Can We Make a Precise Statement with Confidence?

As you might expect, to make very precise interval estimates with a given level of confidence, say, 95%, we need a lot of data. This shows up in the mathematics as follows. Recall from Chapter 7 that

$$\text{se}(\bar{x}) = \frac{s_X}{\sqrt{n}} \qquad \text{and} \qquad \text{se}(\hat{p}) = \sqrt{\frac{\hat{p}(1-\hat{p})}{n}}$$

The standard errors of estimates are almost always at least approximately proportional to $1/\sqrt{n}$, so the standard error gets smaller as the sample size increases. The width of the interval is a constant multiple of the standard error. If we replace n by $4n$

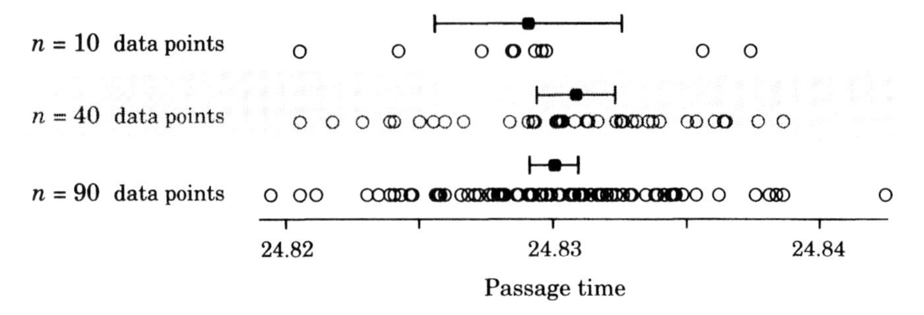

FIGURE 8.1.4 Three random samples from a Normal(μ = 24.83, σ = 0.005) distribution and their 95% confidence intervals for μ.

in either preceding standard error formula, you will see that you get a standard error that is half as big. Therefore, a consequence of the inverse square root relationship is that to double the precision of an estimate (or, equivalently, to halve the width of the confidence interval), you need to take four times as many observations. This type of behavior is illustrated in Fig. 8.1.4.

> To ***double the precision*** we need ***four times*** as many observations.

8.1.3 Conservative Confidence Intervals

There are many instances in the statistical literature when a method for calculating confidence intervals is described as being *conservative*—usually in the context of describing the performance of a confidence interval formula in a situation where the assumptions used in its mathematical derivation do not apply. A method that nominally produces 95% confidence intervals is said to be conservative if the intervals produced by the method cover the true parameter value for *at least* 95% of samples taken. More generally, intervals are conservative if their actual coverage frequency is at least as big as the nominal confidence level. In practical terms this usually means that the intervals produced are somewhat wider, or less precise, than they need to be.

QUIZ ON SECTION 8.1

1. Why do we calculate interval estimates and not just point (single-number) estimates?
2. When we calculate an interval from a set of data and say that it is a 95% confidence interval, what does that statement mean? (What property is possessed by the method used to obtain the interval?)
3. Reinterpret question 2 in the context of many research studies and in the context of planning a study.
4. If we collect data and obtain [6.3, 7.5] as our 95% confidence interval for μ, what gives us confidence in this particular interval as an interval estimate of the true value of μ?
5. In calculating 95% confidence intervals using *estimate* $\pm$ *t standard errors,* does the multiplier t get larger or smaller as the sample size increases? Why?
6. What is meant by the terms (i) *confidence limits,* and (ii) *margin of error?*

7. If we calculate both a 99% confidence interval and a 95% confidence interval for a given parameter from the same data set, which is wider? Why?
8. Suppose you are using the 95% confidence level. If we wanted to double the precision of an estimate how many more observations would be needed? How many more would we need to take to treble the precision? What do we mean by "double the precision" anyway?
9. What is meant when a method for calculating confidence intervals is said to be conservative?

8.2 MEANS

As part of our discussion of Example 8.1.1 in the previous section, we developed and used the following formula for a confidence interval for a population mean μ. The interval follows from using the sample mean $\bar{x}$ as a point estimate of the population mean, μ, and calculating *estimate* $\pm$ *t standard errors.*

> Confidence interval for the true (population) mean μ:
>
> ***sample mean*** $\pm$ ***t standard errors***
>
> or $\quad \bar{x} \pm t\,\text{se}(\bar{x})$, where $\text{se}(\bar{x}) = \dfrac{s_X}{\sqrt{n}}$ and $df = n - 1$

Recall that the multiplier used, t, depends on the value of $df = n - 1$. Almost all confidence intervals used in this book are 95% confidence intervals for which values of t are given in Table 8.1.1. Use of the boxed formula is justified when the distribution of $(\overline{X} - \mu)/\text{se}(\overline{X})$ is well approximated by a Student($df = n - 1$) distribution. As we saw in Section 7.6.2, mathematical theory tells us that the distribution of T is exactly Student($df = n - 1$) when

(i) we have a random sample from a distribution with mean μ and
(ii) the distribution being sampled is a Normal distribution.

It turns out that assumption (i) is critically important in practice. One implication of (i) is that the observations have to be independent. Assumption (ii) is much less important. In fact, even for small samples, the t-intervals work well in practice except when the distribution being sampled from is severely non-Normal. In practical situations the types of data behavior that would make us doubt the usefulness of confidence intervals based on the preceding formula are the presence of either outliers or clusters in the data (see Section 10.1.1 for further discussion).

Example 8.2.1 *Computer-Produced Confidence Interval*

This is a continuation of Example 8.1.1, which concerned Newcomb's measurements of the passage time for light across the Potomac River. In Example 8.1.1 (cont.), we computed a 95% confidence interval for the true passage time by hand using the formula in the box before the example. We found that, with 95% confidence, the true passage time μ was between 24.826 and 24.831 millionths of a second. Figure 8.2.1

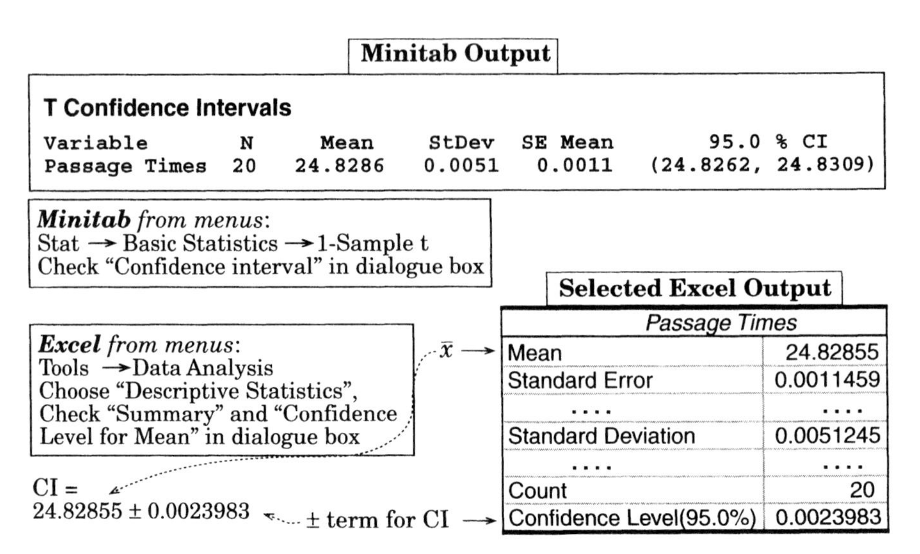

FIGURE 8.2.1 Computer output for Newcomb's passage time data.

gives the output produced automatically by Minitab and Excel when we input the data in Example 8.1.1. As is typical with statistical packages, both programs allow the user to specify the desired confidence level. Excel is unusual in that, rather than give the whole confidence interval, it gives the margin of error (the $\pm$ term) from which the confidence interval can be constructed. This is shown in the lower left-hand corner of Fig. 8.2.1. We note that there is a tiny discrepancy between the upper limit for μ in the Minitab output and our previously calculated value due to rounding errors in the hand calculation.

QUIZ ON SECTION 8.2

1. Under what conditions are confidence intervals of the form *sample mean* $\pm$ *t standard errors* justified by mathematical theory?
2. When do they work in practice? When should we be suspicious of them?

EXERCISES FOR SECTION 8.2

1. Calculate a 95% confidence interval for the true mean density of the earth from Cavendish's first 6 observations, namely: 5.50, 5.61, 4.88, 5.07, 5.26, and 5.55.
2. The following are the results of 10 attempts to measure the concentration of nitrate ions (in μg/mL) in a specimen of water:

   0.513 0.524 0.529 0.481 0.492 0.499 0.518 0.490 0.494 0.501

   (a) Calculate a 95% confidence interval for the true concentration.
   (b) Construct a dot plot of the data (by hand or computer). Add your confidence interval to the plot by hand.
3. Consider the data in Exercises for Section 7.5, problem 4.
   (a) Calculate a 95% confidence interval for the true mean testosterone level for the population of men (i) who were nonsmokers and (ii) who smoked 31–70 cigarettes per day.

(b) Give (i) 90% and (ii) 99% confidence limits for the true mean testosterone level for 31- to 70-per-day smokers.

8.3 PROPORTIONS

Having developed the idea of a confidence interval for a mean, we now turn our attention to doing the same for a population proportion or probability. We set the scene by considering an example.

EXAMPLE 8.3.1 *Do CEOs Disapprove of Office Romances?*

We have heard a number stories over the years of companies discouraging relationships between co-workers, even to the extent of prohibiting both members of a married couple from working for the company. If a couple married, one had to leave. In 1994 *Fortune* reported a survey of 200 chief executives of American companies that explored the issues surrounding relationships between co-workers. When asked, "Do you approve or disapprove of office romances between unmarried employees—or would you say it is none of the company's business?" 70% of the chief executives said that it was none of the company's business. We shall assume that the 200 chief executives surveyed were the CEOs of a random sample of major American companies. We want to construct a confidence interval for p, the true proportion of all such CEOs who believe that office romances are none of the company's business, based on the survey's estimate $\widehat{p} = 0.7$.

Here, our parameter is p, the proportion of all individuals in the population with the characteristic of interest. Our estimate of p is the sample proportion $\widehat{p}$, the proportion of the individuals surveyed with that characteristic. The theory applies when n is large because then, as we saw in Section 7.3, the distribution of $(\widehat{P} - p)/\text{se}(\widehat{P})$ is approximately Normal(0,1) [equivalently, Student($df = \infty$)]. The specialized version of *estimate* $\pm$ *t standard errors* is as given in the box (using z instead of t, which we have noted is conventional when $df = \infty$).

> Confidence interval for the true (population) proportion p:
>
> ***sample proportion*** $\pm$ ***z standard errors***
>
> or $\widehat{p} \pm z\,\text{se}(\widehat{p})$, where $\text{se}(\widehat{p}) = \sqrt{\dfrac{\widehat{p}(1-\widehat{p})}{n}}$

For 95% confidence intervals the multiplier is $z = 1.96$. This, of course, gives virtually identical intervals to the two-standard-error intervals calculated in Chapter 7. Other levels of confidence require other multipliers, such as 1.645 for 90% confidence and 2.576 for 99% confidence. The standard-error formula was given in Section 7.3.1. The interval is based on large-sample theory. How large the sample has to be before the intervals perform adequately depends on the size of the proportion being worked with. The following small table of values gives the minimum sample size n for various values of $\widehat{p}$.[4] An extensive table is given in Appendix A3, where you should use the column headed "10% rule." Note that $\widehat{p}$-values 0.05 and 0.95 have the same minimum n (namely, 960) as do $\widehat{p}$-values 0.1 and 0.9, and so on.

Value of $\widehat{p}$	0.05	0.1	0.15	0.2	0.25	0.3	0.35	0.4	0.45	0.5
Minimum n	960	400	220	125	76	47	23	13	11	10
Value of $\widehat{p}$	0.95	0.9	0.85	0.8	0.75	0.7	0.65	0.6	0.55	0.5

✖ EXAMPLE 8.3.1 (cont.) *Confidence Interval for a Proportion*

Here the sample proportion is $\widehat{p} = 0.7$, but this value is subject to sampling error. The sample size is $n = 200$, which is considerably larger than the minimum allowable sample size of 47 given by the preceding table. We can therefore use our formula. For a 95% confidence interval we use $z = 1.96$ standard errors. The resulting confidence interval for the true proportion is (*estimate* $\pm$ *z standard errors*)

$$\widehat{p} \pm 1.96\,\mathrm{se}(\widehat{p}) = 0.7 \pm 1.96 \times \sqrt{\frac{0.7 \times 0.3}{200}} = [0.636, 0.764]$$

Thus, with 95% confidence, the true proportion of all American CEOs who believe that office romances are none of the company's business is somewhere between 64% and 77%.

When the value of n is smaller than the minimum prescribed by our tables, the "exact" interval (sometimes known as the Clopper-Pearson interval) or the more complex approximate methods described by Bohning [1994] and Newcombe [1998b] can be used. These topics are beyond the scope of this book.

EXERCISES FOR SECTION 8.3

1. Consider Example 8.3.1 where p is the true proportion of CEOs who believe that office romances are none of the company's business. Within what limits could you say that the true p lies
   (a) with 90% confidence? (b) with 99% confidence?
2. Of the 139 black or Hispanic people sampled in Example 7.5.1, 36% had been pulled over on the roads by the police. Calculate a 95% confidence interval for the population proportion.

8.4 COMPARING TWO MEANS

We now go on to apply the idea of a confidence interval calculated using *estimate* $\pm$ *t standard errors* to a difference between two means. Again, the calculations differ

[4] The rules commonly quoted in elementary books, such as $n\widehat{p}(1-\widehat{p}) \geq 10$ or both $n\widehat{p} \geq 5$ and $n(1-\widehat{p}) \geq 5$, are inadequate. Our table is based on Samuels and Lu [1992], who compared the simple approximate interval above with that based on exact Binomial theory. The value of n quoted is the minimum required to ensure that the sum of the absolute errors in the endpoints constitutes less than 10% of the length of the interval. For further discussion, see our web site.

from those in Chapter 7 only in terms of how many standard errors we take on either side of the estimate. The paragraphs for interpreting two-standard-error intervals for a difference described toward the end of Section 7.5 also apply here to 95% confidence intervals. If the separate 95% confidence intervals for θ_1 and θ_2 do *not* overlap, then the 95% confidence interval for $\theta_1 - \theta_2$ will *not* contain 0. Overlap of the separate intervals, however, does not necessarily imply that the 95% confidence interval for the difference will contain 0.

Example 8.4.1 *Higher Blood Thiol Concentrations with Rheumatoid Arthritis*

Banford et al. [1982] noted that thiol concentrations within human blood cells are seldom determined in clinical studies, in spite of the fact that they are believed to play a key role in many vital processes. They reported a new reliable method for measuring thiol concentration and demonstrated that, in one disease at least (rheumatoid arthritis), the change in thiol status in the lysate from packed blood cells is substantial. There were two groups of volunteers, the first group being "normal" and the second suffering from rheumatoid arthritis. The data (and some summary statistics) are given in Table 8.4.1 and plotted in Fig. 8.4.1.

We see that the rheumatoid group has thiol concentrations that are both larger and more variable than those in the normal group. Interestingly, there is complete separation between the two groups. Every person in the rheumatoid group had a substantially higher thiol concentration than *any* person in the normal group. We shall estimate how much higher thiol levels are on average in rheumatoid people. We shall treat the two groups as random samples from the normal and rheumatoid populations respectively (for the area in which the study was undertaken) and will concentrate on

TABLE 8.4.1 Thiol Concentration (mmol)

	Normal	Rheumatoid
	1.84	2.81
	1.92	4.06
	1.94	3.62
	1.92	3.27
	1.85	3.27
	1.91	3.76
	2.07	
Sample size	7	6
Sample mean	1.92143	3.46500
Sample standard deviation	0.07559	0.44049

Data from Banford et al. [1982].

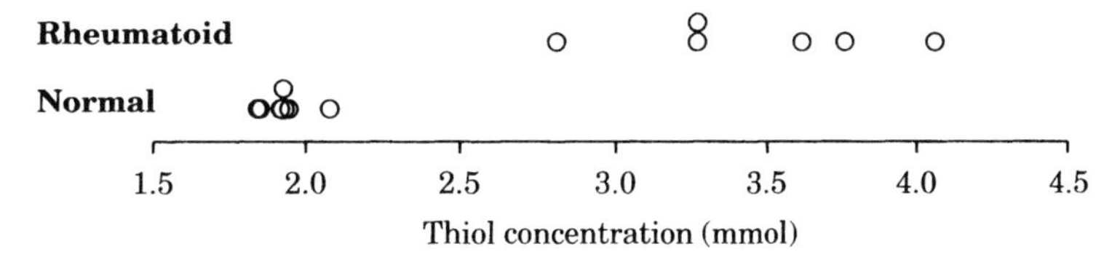

FIGURE 8.4.1 Dot plot of thiol concentration data.

estimating $\mu_R - \mu_N$, the difference in true mean thiol levels between the rheumatoid and normal populations. The natural point estimate is the difference in mean thiol concentrations observed in the data, namely, $\bar{x}_R - \bar{x}_N = 3.46500 - 1.92143 = 1.54357$ mmol. However, like any data estimate, this one is subject to sampling error. We want to compute a confidence interval for $\mu_R - \mu_N$ as a means of representing the uncertainty about the true value.

We wish to estimate the difference of two population means using the difference between the corresponding sample means. Here our parameter is $\mu_1 - \mu_2$, the difference between the population means, and our estimate is $\bar{x}_1 - \bar{x}_2$, the difference between the sample means. The formula *estimate* $\pm$ *t standard errors* gives the following.

> Confidence interval for a difference between population means ($\mu_1 - \mu_2$):
> ***difference between sample means*** $\pm$ ***t standard errors of the difference***
>
> or
>
> $$\bar{x}_1 - \bar{x}_2 \pm t\,\text{se}(\bar{x}_1 - \bar{x}_2)$$

Where the two samples are *independent* (see Section 7.5),

$$\text{se}(\bar{x}_1 - \bar{x}_2) = \sqrt{\frac{s_1^2}{n_1} + \frac{s_2^2}{n_2}}$$

For hand calculation only, we shall use $df = \text{Min}(n_1 - 1, n_2 - 1)$, the smaller of $n_1 - 1$ and $n_2 - 1$. This is a conservative approximation (see end of Section 8.1) to the Welch procedure discussed later in Section 10.2, which gives intervals that are somewhat wider than those produced by computer packages. The Welch procedure has a complicated *df* formula.

✖ EXAMPLE 8.4.1 (cont.) *Computing the Confidence Interval for a Difference*

The sample sizes are $n_R = 6$ and $n_N = 7$. Our estimate of the true difference is $\bar{x}_R - \bar{x}_N = 1.54357$, which is subject to sampling variation. Our estimate has standard error

$$\text{se}(\bar{x}_R - \bar{x}_N) = \sqrt{\frac{s_R^2}{n_R} + \frac{s_N^2}{n_N}} = 0.18208$$

Using $df = \text{Min}(n_R - 1, n_N - 1) = 5$, the multiplier, t, for a 95% confidence interval is 2.571. The resulting 95% confidence interval for $\mu_R - \mu_N$ is given by (*estimate* $\pm$ *t standard errors*)

$$1.54357 \pm 2.571 \times 0.18208 = [1.08, 2.01]$$

Thus, with 95% confidence, the true mean thiol concentration for the rheumatoid population exceeds that of the normal population by somewhere between 1.08 mmol and 2.01 mmol.

EXERCISES FOR SECTION 8.4

1. Problem 2 of Exercises for Section 7.5 concerned comparing the underlying true mean of the 6 determinations of the mean density of the earth that Cavendish took before replacing the wire with that of the 23 determinations made after changing the wire.
   (a) Calculate a 95% confidence interval for the difference between the two means.
   (b) Compare it with the two-standard-error interval from Exercises for Section 7.5, problem 2.
   (c) Are the conclusions you would draw from the data affected in any real way?
2. Calculate a 95% confidence interval for the difference for the true mean testosterone levels for 1- to 30-per-day smokers and 31- to 70-per-day smokers from the data given in the Exercises for Section 7.5, problem 4.

8.5 COMPARING TWO PROPORTIONS

Having seen how to construct a confidence interval for the difference of two means, we now direct our attention to finding a confidence interval for the difference of two proportions. We shall find that we have to contend with three separate sampling situations. We begin with the simplest case of comparing proportions from independent populations.

8.5.1 Independence Case

EXAMPLE 8.5.1 *Relationship between Smoking and Grades at School*

In the 1998 "Back-to-School Teen Survey" (see Example 7.3.1), 68% of the 870 teenagers who did not smoke got good grades (mostly A's and B's) compared with 41% for the 130 smokers. We can regard these teenagers as a random sample of size 870 from the population of teenagers who do not smoke and a random sample of 130 from the population of teenagers who do smoke. It looks as if a considerably higher percentage of nonsmokers get good grades than do smokers, with the observed difference being 68% − 41% = 27%. The percentages quoted for nonsmokers and smokers are sample estimates, however, and are thus subject to sampling error. We shall calculate a 95% confidence interval for the true difference between the proportion of all nonsmokers who get good grades and the proportion of all smokers who get good grades.

In this subsection we consider situations in which we have independent samples of size n_1 and n_2, respectively, from two different populations. Unknown proportions, p_1 of the first population and p_2 of the second population, have some characteristic of interest. We wish to form a confidence interval for the difference in population proportions, $p_1 - p_2$, based on the difference seen in the data, namely, $\widehat{p}_1 - \widehat{p}_2$, the difference between the corresponding sample proportions.

Here our parameter is $p_1 - p_2$, and our estimate is $\widehat{p}_1 - \widehat{p}_2$, so the formula *estimate* ± *t standard errors* gives the following.

> Confidence interval for a difference between population proportions ($p_1 - p_2$):
>
> ***difference between sample proportions*** $\pm$ ***z standard errors of the difference***
>
> or
>
> $$\widehat{p}_1 - \widehat{p}_2 \pm z \operatorname{se}(\widehat{p}_1 - \widehat{p}_2)$$

Our intervals are based on large-sample Normal approximations to the distribution of each sample proportion (i.e., $df = \infty$). Thus, as with a single proportion, it is conventional to use z rather than t to represent the multiplier. All examples in this section are 95% confidence intervals and hence use $z = 1.96$ standard errors.

When comparing proportions using *two independent samples* (see Section 7.5)

$$\operatorname{se}(\widehat{p}_1 - \widehat{p}_2) = \sqrt{\frac{\widehat{p}_1(1-\widehat{p}_1)}{n_1} + \frac{\widehat{p}_2(1-\widehat{p}_2)}{n_2}}$$

Such intervals can be used when each n_i is greater[5] than the minimum given for the corresponding $\widehat{p}_i$ in the column headed "10% rule" in Appendix A3.

Example 8.5.1 (cont.) *Computing the Confidence Interval*

We shall now calculate a 95% confidence interval for $p_{NS} - p_{Sm}$, the difference between the true proportions of nonsmokers (NS) and smokers (Sm) who get good grades. The sample sizes are $n_{NS} = 870$ and $n_{Sm} = 130$, and sample proportions are $\widehat{p}_{NS} = 0.68$ and $\widehat{p}_{Sm} = 0.41$. Applying the 10% rule, Appendix A3 gives us a minimum value of $n = 38$ when $\widehat{p} = 0.68$ (cf. $n_{NS} = 870$) and a minimum value of $n = 12$ when $\widehat{p} = 0.41$ (cf. $n_{Sm} = 130$). Since our sample sizes are greater than these minimum values,[6] we can use our confidence interval formula. Our point estimate of the true difference is $\widehat{p}_{NS} - \widehat{p}_{Sm} = 0.68 - 0.41 = 0.27$, which has standard error

$$\operatorname{se}(\widehat{p}_{NS} - \widehat{p}_{Sm}) = \sqrt{\frac{0.68 \times 0.32}{870} + \frac{0.41 \times 0.59}{130}} = 0.04594436$$

Our 95% confidence interval for $p_{NS} - p_{Sm}$ is given by (*estimate* $\pm$ *z standard errors*)

$$0.27 \pm 1.96 \times 0.04594436 = [0.18, 0.36]$$

We can say with 95% confidence that the true percentage of nonsmokers who get good grades is greater than the corresponding percentage for smokers by between 18 and 36 percentage points.

It is quite clear that nonsmokers are considerably more likely to get good grades than smokers. This is an observational study rather than a designed experiment (see

---

[5]For smaller values of n_1 and n_2 more complex methods are needed. See Hauck and Anderson [1986], and Newcombe [1998a]; for exact methods see Chapter 13 of Mehta and Patel [1995] and recent releases of the SAS package.

[6]We shall not explicitly provide such checks in future examples.

Section 1.3), however, so we cannot conclude that the act of smoking causes them to get lower grades. It could just be that the type of people who are more likely to smoke are also more likely to get bad grades. (We shall see later that smokers are also more likely to drink to excess and to use drugs. Both of these could conceivably be causal influences for poor grades.)

A case study involving differences in proportions is given on the web site. This case study is based on research showing how fairly minor differences in the way survey questions are worded can affect the results obtained.

8.5.2 Other Common Comparisons

In the previous subsection we presented the confidence interval formula for estimating a difference between population proportions. The formula given there for se($\widehat{p}_1 - \widehat{p}_2$) applies only when the two sample proportions come from *two independent samples,* however. We shall distinguish between some commonly occurring situations where the independence-case standard-error formula is applicable and where it is not applicable. When confronted with reports about surveys in newspapers, in magazines, or on television, we have found that we want to compare proportions relating to the same sample of people at least as often as we wish to compare proportions from independent samples. The standard-error formula for the independence case is often wrongly applied for such comparisons. Tables 8.5.1 to 8.5.4 provide some examples of comparing proportions that are not from independent populations. Two cases are considered, and simple standard-error formulas are provided.

The independence case is like situation (a), depicted in Fig. 8.5.1(a). We have two separate samples of individuals or objects, for example, nonsmokers and smokers in Example 8.5.1. We look at each nonsmoker and ask whether she or he gets good

TABLE 8.5.1 1996 U.S. Election

		Pre-election polls				Election results		
State	*n*	Clinton	Dole	Perot	Other/undecided	Clinton	Dole	Perot
New Jersey	1000	51	33	8	8	53	36	9
New York	1000	59	25	7	9	59	31	8
Connecticut	1000	51	29	11	9	52	35	10

Source: Quinnipiac College Polling Institute press release, 8 Nov. 1996.

TABLE 8.5.2 Why Radio Stations Play Public Service Ads

Reason	%
Message is relevant to station or viewer needs	52
Nonprofit organization is well known or represents a worthwhile cause	25
Spot length fits into available slots	15
Celebrity spokesperson offers credibility	5
Production quality, chance, right format	3

Source: West Glen's 1997 Annual Survey of Public Affairs Directors on their use of PSAs.

TABLE 8.5.3 People's Reactions to Their Health Care System

(Table entry is % agreeing)	Australia	Canada	N.Z.	UK	U.S.
Difficulties getting needed care	15	20	18	15	28
Recent changes will harm quality	28	46	38	12	18
System should be rebuilt	30	23	32	14	33
No bills not covered by insurance	7	27	12	44	8
Sample size	1000	1000	1000	1000	1000
Health care expenditure (US$/person)[a]	1805	2095	1352	1347	4090

[a]Adjusted for cost-of-living differences.

Source: Commonwealth Fund International Health Policy Survey press release, 22 Oct. 1998.

TABLE 8.5.4 Characteristics by Smoking and Drinking Status

(Table entry is % of group saying yes)	Smoker[a]	Nonsmoker	Drinker[b]	Nondrinker
Get mostly A's or B's?	41	68		
Read one or more hours/day?	54	72	56	75
Get drunk at least once a month?	63	10		
Have smoked marijuana?	79	14	52	12
Likely to try illegal drug in future?	42	14	35	11
n	130	870	260	740

[a]Smoked cigarettes in last 30 days.

[b]Consumed more than a few sips of alcohol in the last 30 days.

Source: 1998 "Back-to-School Teen Survey" (see Example 7.3.1).

grades. Nonsmokers who do get good grades fall into the "yes" tank, and the rest fall into the "no" tank. The sample of smokers is classified in the same way. We then compare the proportions of "yes" people from each group.

Table 8.5.1 gives results from separate polls conducted in three states just before the 1996 U.S. presidential election. If we want to compare the proportion of New Jersey voters who supported Bill Clinton (51%) with the proportion of New York voters who supported Clinton (59%), we are again making a situation (a) comparison. There are two separate samples of people being classified, and we focus on whether or not each individual supports Clinton (yes for supporting Clinton, no for any other response). If we wanted to estimate how far Clinton was ahead of Bob Dole in New Jersey (51% compared with 33%), however, we are in situation (b), depicted in Fig. 8.5.1(b). We have a single sample of individuals, and each individual is slotted into one category (tank) chosen from a set of several categories (here determined by the candidate they want to vote for). Because the proportions relate to the same set of people, they are not independent, and it is incorrrect to use the standard-error formula for the independence case. (The answers it gives are too small.)

A good clue that you have a situation (b) comparison is that you are comparing two categories from a set for which the percentages add to 100%. Table 8.5.2 gives the percentages of a sample of 162 radio stations falling into a set of categories describing "the most critical factor" used in deciding whether to air public service

SITUATION (a) Two independent samples

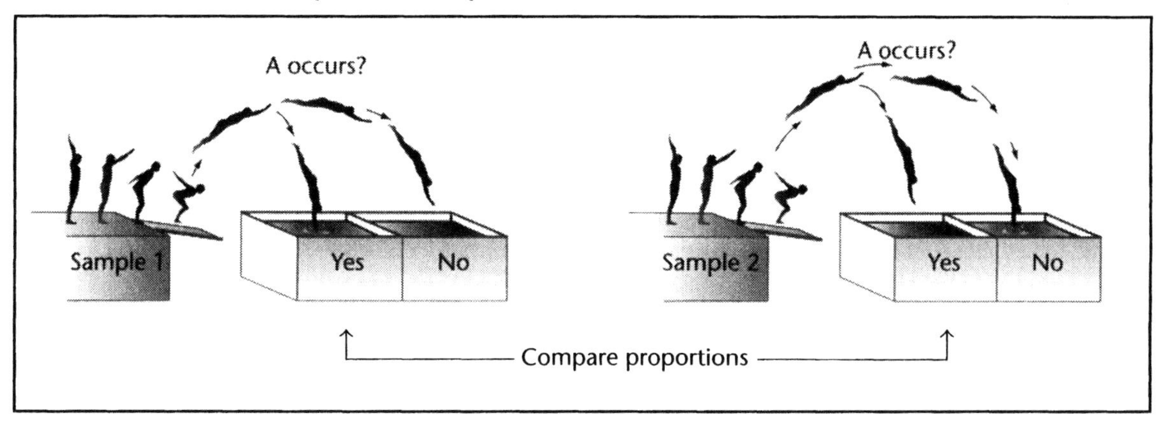

SITUATION (b) Single sample, several response categories

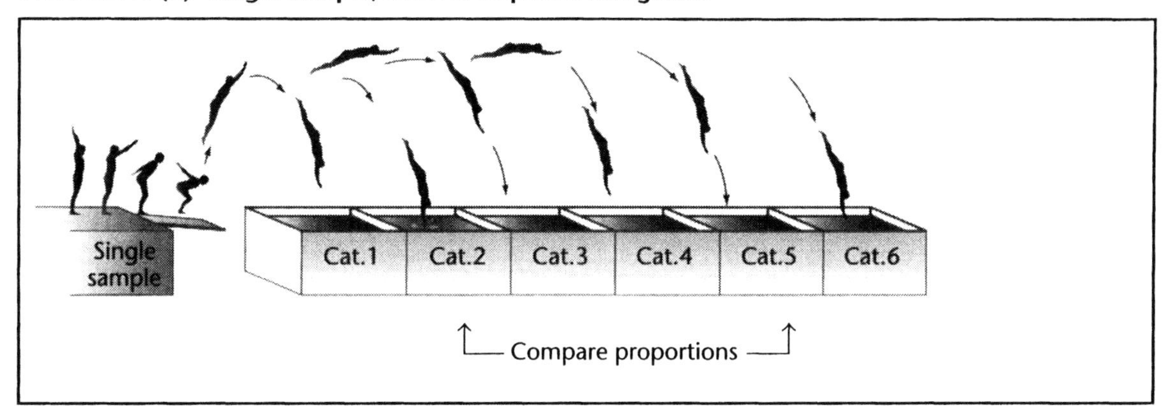

SITUATION (c) Single sample, two or more Yes/No items

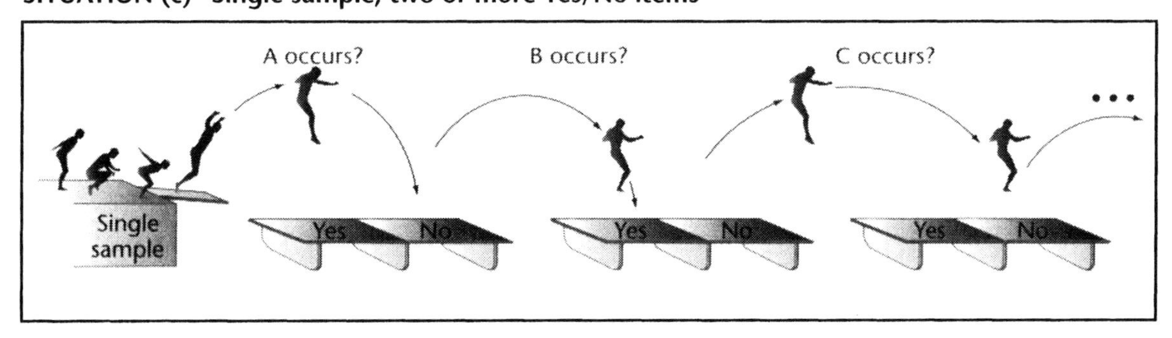

FIGURE 8.5.1 Three sampling situations for comparing proportions. (For corresponding standard error formulas, see Table 8.5.5)

advertisements. A comparison between any pair of percentages listed there is a situation (b) comparison.

Table 8.5.3 summarizes data obtained from an international study of health care systems in which a random sample of 1000 people was taken from each of five countries. If we want to compare the proportions of people from different countries agreeing to a particular statement, for example, the 30% of Australians agreeing to "the system should be rebuilt" with the 23% of Canadians agreeing to the same statement, then we again have a situation (a) comparison [Fig. 8.5.1(a)]. If we look at the data on a single country and want to compare, say, the proportion of Americans reporting "difficulties with getting needed care" (28%) with the proportion of Americans believing that "the system should be rebuilt" (33%), however, we are again comparing proportions that come from the same sample (but for different questions) and thus are not independent. This latter comparison is a situation (c) comparison, depicted in Fig. 8.5.1(c). The Americans being surveyed either agree (yes) or disagree (no) with each of a set of statements. They are not being forced to choose one category out of a set of categories as in situation (b).

Turning our attention to Table 8.5.4 from the "1998 Back-to-School" survey of Example 8.5.1, we get a more compelling example of a situation (c) [Fig. 8.5.1(c)] comparison when we compare the percentage of smokers who get drunk at least once a month (63%) with the percentage of smokers who have smoked marijuana (79%). Comparisons between smokers and nonsmokers on the same item are situation (a) comparisons [Fig. 8.5.1(a)].

Confidence Intervals

A confidence interval for a same-sample difference in proportions is calculated using the same formula as before, namely,

$$(\text{CI for } p_1 - p_2) \qquad \widehat{p}_1 - \widehat{p}_2 \pm z\,\text{se}(\widehat{p}_1 - \widehat{p}_2)$$

All that changes is the way we calculate $\text{se}(\widehat{p}_1 - \widehat{p}_2)$, and formulas are given in Table 8.5.5.

TABLE 8.5.5 Standard Errors for Differences in Proportions
(These correspond to the situations depicted in Fig. 8.5.1.)

(a) ***Proportions from two independent samples*** of sizes n_1, n_2, respectively

$$\text{se}(\widehat{p}_1 - \widehat{p}_2) = \sqrt{\frac{\widehat{p}_1(1-\widehat{p}_1)}{n_1} + \frac{\widehat{p}_2(1-\widehat{p}_2)}{n_2}}$$

(b) ***One sample of size n, several response categories***

$$\text{se}(\widehat{p}_1 - \widehat{p}_2) = \sqrt{\frac{\widehat{p}_1 + \widehat{p}_2 - (\widehat{p}_1 - \widehat{p}_2)^2}{n}}$$

(c) ***One sample of size n, many yes/no items***

$$\text{se}(\widehat{p}_1 - \widehat{p}_2) = \sqrt{\frac{\text{Min}(\widehat{p}_1 + \widehat{p}_2, \widehat{q}_1 + \widehat{q}_2) - (\widehat{p}_1 - \widehat{p}_2)^2}{n}}$$

where $\widehat{q}_1 = 1 - \widehat{p}_1$ and $\widehat{q}_2 = 1 - \widehat{p}_2$.

Use the 10% rule for each proportion.

The use of the standard-error formula for two independent samples [Situation (a)] was illustrated in Example 8.5.1. We now demonstrate the use of the formulae for the other two situations.[7]

EXAMPLE 8.5.2 *Differences in Levels of Support for Clinton and Dole*

From Table 8.5.1, 51% of the $n = 1000$ voters polled in New Jersey supported Clinton, and 33% supported Dole. On the face of it Clinton was ahead by 18%. These results are subject to sampling error, however. We shall estimate the true difference in support levels by calculating a 95% confidence interval for $p_C - p_D$, the difference between the proportions of all New Jersey voters supporting Clinton and Dole at the time the poll was taken. Our point estimate is the difference seen in the poll, that is, $\widehat{p}_C - \widehat{p}_D = 0.51 - 0.33 = 0.18$. As this is a situation (b) comparison (see Fig. 8.5.1), the standard error of this difference is

$$\text{se}(\widehat{p}_C - \widehat{p}_D) = \sqrt{\frac{0.51 + 0.33 - (0.51 - 0.33)^2}{1000}} = 0.0284183$$

Our 95% confidence interval for the true difference is *estimate* $\pm$ *z standard errors* with $z = 1.96$, or $0.18 \pm 1.96 \times 0.0284183 = [0.12, 0.24]$. Thus with 95% confidence, Clinton was leading Dole in percentage support among New Jersey voters by somewhere between 12 and 24 percentage points. The actual difference at the election several days later was $53 - 36 = 17$ percentage points.

EXAMPLE 8.5.3 *Comparing Proportions Who Use Marijuana and Get Drunk*

Returning to Table 8.5.4, we see that 79% of the 130 smokers say they have tried marijuana, while 63% of them say that they get drunk at least once a month. It looks as if a higher percentage have tried marijuana than get drunk regularly, with the difference being 16%. These are sample results, however, and as such are subject to sampling variation. We shall calculate a 95% confidence interval for $p_{Mar} - p_{Dr}$, the true difference between the proportions of all American teenage smokers who would report having engaged in these activities. Our point estimate is the difference seen in the data, $\widehat{p}_{Mar} - \widehat{p}_{Dr} = 0.79 - 0.63 = 0.16$. We have $n = 130$, which is sufficiently large. This is a situation (c) comparison (see Fig. 8.5.1). Since $\widehat{p}_{Mar} = 0.79$, $\widehat{q}_{Mar} = 1 - 0.79 = 0.21$. Similarly, $\widehat{q}_{Dr} = 1 - 0.63 = 0.37$. We note that $\widehat{q}_{Mar} + \widehat{q}_{Dr} < \widehat{p}_{Mar} + \widehat{p}_{Dr}$, so that

$$\text{se}(\widehat{p}_{Mar} - \widehat{p}_{Dr}) = \sqrt{\frac{0.21 + 0.37 - (0.79 - 0.63)^2}{130}} = 0.06530402$$

[7]The formula for situation (b) was derived by Scott and Seber [1983], and that for situation (c) was derived by Wild and Seber [1993]. The latter formula is a conservative approximation that should be used when the only information available is just the individual proportions answering *yes* to the two items. This is almost always the case when we try to compare proportions from news reports. The investigators who collect the original data should retain information on the proportions who answer *yes* to one item and *no* to the other. With this additional information, exact standard errors and narrower confidence intervals can be obtained.

Our 95% confidence interval for the true difference is *estimate* $\pm$ *z standard errors* with $z = 1.96$, or $0.16 \pm 1.96 \times 0.06530402 = [0.03, 0.29]$. The true difference could be as small as 3 percentage points or as large as 29.

EXERCISES FOR SECTIONS 8.5.1 AND 8.5.2

This set of exercises has two main purposes: to give practice in distinguishing between the sampling situations in Fig. 8.5.1 and to give some experience with using the corresponding formulas. The data in Table 8.5.6 all come from the 1998 "Back-to-School Teen Survey" (see Example 7.3.1), a telephone survey of 1000 Americans aged between 12 and 17. They have been spilt into two age groups, 12–14 and 15–17, each of size approximately 500. This enables us to see how some patterns change with age. Data are given for three questions. The study also included surveys of 825 teachers and 822 principals. Their responses have been included for the second question.

1. For each of the following differences (comparisons), say whether the sampling situation falls into situation (a), situation (b), or situation (c) depicted in Fig. 8.5.1.
   (a) First question: The difference between the proportion of 15- to 17-year-olds who know a student who sells illegal drugs and the proportion who know a teacher who uses illegal drugs.
   (b) First question: The difference between the proportions of 15- to 17-year-olds and 12- to 14-year-olds who know a student who sells illegal drugs.
   (c) Second question: The difference between the proportion of principals who think students can use marijuana every weekend and still do well at school and the corresponding proportion for 15- to 17-year-olds.

TABLE 8.5.6 More Results from the 1998 "Back-to-School Teen Survey"

Do you know a . . . ? (entry is % saying yes)

	12–14	15–17
friend or classmate who uses illegal drugs	39%	67%
student who sells illegal drugs	22%	51%
teacher who uses illegal drugs	6%	21%

Is it possible for students to use marijuana every weekend and still do well in school?

	Teachers	Principals	12–14	15–17
Yes	43%	35%	10%	23%
No	46%	55%	80%	62%
Don't know/refused	11%	10%	10%	15%

What are you most likely to do in the afternoon after school?

	12–14	15–17
Hang out with friends	21%	19%
Go home and do homework	22%	16%
Go home and watch TV	13%	12%
Go home and do something else	16%	13%
Play on sports team	18%	18%
Go to a job	2%	13%
Other organized activity	7%	8%
Don't know/refused	1%	1%

(d) Third question: The difference between the proportion of 12- to 14-year-olds who are most likely to hang out with friends after school and the proportion who go home and watch TV.

(e) Third question: The difference between the proportion of 12- to 14-year-olds who go home and do homework and the proportion of 15- to 17-year-olds who do this.

2. Calculate 95% confidence intervals for the differences specified in each part of problem 1. Express your answers verbally in terms of percentages.

3. Calculate 95% confidence intervals for the following differences, expressing your answers verbally in terms of percentages.
   (a) (From Table 8.5.1) The difference in support between Clinton and Dole in New York.
   (b) (From Table 8.5.1) The difference in support for Dole between New Jersey and Connecticut.
   (c) (From Table 8.5.3) The difference between the proportion of Americans and Canadians worried about difficulties in getting needed care.
   (d) (From Table 8.5.3) The difference between the proportion of New Zealanders who think that recent changes will harm the quality of health care and the proportion believing the system should be rebuilt.
   (e) Of the countries in Table 8.5.3, per-person expenditure on health care is lowest in the UK. What do you notice about health care satisfaction levels in the UK in comparison with the other countries?

8.5.3 "Margins of Error" in Media Reports

This nontechnical subsection discusses some shortcomings of the ways polls are often reported in the media. It discusses the many situations where the reported "margin of error" is misleading and gives some simple rules of thumb that you can use to mentally adjust the reported margin in your recreational reading when you have neither the time, energy, nor inclination to reach for a calculator.

The press release for the 1998 "Back-to-School Teen Survey" included the statement, "margin of error ± 3.1%." Most polls are accompanied by some similar statement. The language used varies. In addition to "margin of error," we have seen the phrases "sampling error," "maximum poll error," and "statistical error" used for the same purpose. The reporting of a margin of error for a poll is a good thing in that it reminds readers that the percentages produced by the poll are not the true percentages and that there is uncertainty about the true percentages. It is common practice, however, to report a single "margin of error" for a whole survey (e.g., ± 3.1%). A single margin of error would seem to indicate that this is the likely error for any percentage quoted from that survey. Unfortunately, this is not true, as the error will change with the percentage. This problem can be particularly acute for percentages relating only to subgroups of those sampled and for handling differences between percentages.

What Is the "Margin of Error" Typically Reported?

We know from Section 8.1 that the margin of error technically is the term added to and subtracted from the estimate to form the confidence interval. Using the 95% level of confidence, the margin of error for a proportion $\hat{p}$ is thus $margin = 1.96\sqrt{\hat{p}(1-\hat{p})/n}$. The size of the actual margin of error therefore depends on the size of the proportion. The single "margin of error" accompanying a poll in most media reports, however, is typically the value of *margin* when $\hat{p} = 0.5$, with the result expressed as a percentage. In other words, it is the margin of error appropriate for a sample percentage of 50%. It turns out that the value of $\sqrt{\hat{p}(1-\hat{p})}$ is fairly constant over the range $0.3 \leq \hat{p} \leq 0.7$, so applying the reported margin of error gives a reasonable approximation to the 95% confidence interval for any sample percentage within the range 30% to 70%.

When Is the Reported "Margin of Error" Inadequate or Misleading?

For any single percentage that is much outside the range 30% to 70%, the reported margin of error is too big. It grossly *overstates* the likely error in a percentage that is very close to either 0 or 100%.

With data like that in Tables 8.5.1 to 8.5.4, it is entirely natural and commendable to look for interesting features. Journalists often do this when they write stories based on poll results. Frequently, the interesting features relate to a *difference* between two percentages or to a *subgroup* of the whole sample. In these cases the quoted margin of error can appreciably *understate* the true level of error for a difference. This leads to news headlines making claims such as, "Candidate A moves ahead of candidate B in new poll" in situations where there is simply not enough information in the data to reliably determine who has more support in the population as a whole (because the confidence interval for the true difference contains both negative and positive values and includes 0).

The press release for the 1998 teen survey reported percentages for subgroups, for example, stating that 54% of 17-year-olds say that alcohol is available at most parties that they attended in the last six months. Those aged 17 constitute one-sixth of those surveyed. It also compares proportions from subgroups, for example, "56% of pot smokers get drunk at least once a month compared with 15% of non-pot smokers." Situations in which we often see differences reported in the media occur when the level of political support for a candidate or party from a current poll is compared with the level in some previous poll or when the support for different candidates in the same poll are compared.

Mental Adjustments to the Margin of Error for Differences

(i) From *two independent samples:* Multiply the average of the margins of error from the two polls by $1\frac{1}{2}$.

(ii) From *one sample,* either situation (b) or (c): *Double* the quoted margin of error.

These rules are obtained by inspecting the relative sizes of standard errors for proportions that are close to 0.5 and, in (i), assuming $n_1 \approx n_2$. See Review Exercises 8, problem 13.

Mental Adjustments to the Margin of Error for Subgroups

Multiply the margin of error by a "guesstimate" of $\sqrt{\text{Whole-group size/Subgroup size}}$. Thus, if you think that the subgroup constitutes about $\frac{1}{4}$ of the whole, then you should double the quoted margin of error ($\sqrt{4} = 2$). If you think that the subgroup is likely to constitute about $\frac{1}{10}$ of the sample, then triple the quoted margin of error ($10 \approx 9$ and $\sqrt{9} = 3$).

QUIZ ON SECTION 8.5.3

All of the following questions refer to a poll in which the quoted margin of error is ±4%. You should perform at most only the quick mental calculations previously described to obtain approximate margins of error that are more appropriate in the situation outlined.

1. What is the margin of error typically quoted with a poll? What are some other names for it?
2. The percentage unemployed in the survey was 10%. Is applying the quoted margin of error to this a reasonable approximation, or is it too big or too small?
3. The sample was 65% female. Is applying the quoted margin of error to this a reasonable approximation, or is it too big or too small?
4. Approximately a quarter of those sampled came from rural areas. Of rural respondents, 35% had a car that was less than two years old. What margin of error should be associated with this?
5. For 30% of those sampled the main source of international news was the local newspaper, and for 35% the main source was television news broadcasts. What margin of error should be associated with the difference?
6. Approximately one-tenth of those sampled were in the age range 18–24, and another tenth were in the range 25–31.
   (a) Of 18- to 24-year-olds, 30% owned a car under five years old. What margin of error should be associated with this?
   (b) Of 25- to 31-year-olds, 65% owned a car less than five years old. What margin of error should be associated with the difference between the percentage of 25- to 31-year-olds and 18- to 24-year-olds who own a car less than five years old?

*8.6 HOW BIG SHOULD MY STUDY BE?

*8.6.1 Proportions

"How big should my sample be?" Statisticians are often asked this question. When studies involve data in the form of counts or proportions, the best answer is probably, "As big as you can afford." The reason for this is that there is surprisingly little information in such data, even from quite big studies.

Suppose that we want to estimate a population proportion, for example, the proportion of the population carrying hepatitis B, using a random sample from that population. We want the size of our sample to be large enough so that the margin of error for our estimate will be acceptably small, say no larger than 0.03 (or 3%). Recall that the margin of error is the term added and subtracted from the estimate to form the confidence interval. Let us use m to denote the maximum size we want the margin of error to have, in this case, $m = 0.03$ (or 3%). For a single proportion the margin of error is $margin = z\sqrt{\hat{p}(1-\hat{p})/n}$. We substitute a guess, p^*, for the unknown proportion. We then need to solve the equation $margin \leq m$, or more explicitly $z\sqrt{p^*(1-p^*)/n} \leq m$, for n. A little algebra shows that the sample size, n, must be at least $(z/m)^2 \times p^*(1-p^*)$. We have used this formula to produce the values given in Table 8.6.1. For example, when $m = 0.03$ and $p^* = 0.5$, using $z = 1.96$ (for a 95% confidence interval), the formula gives a minimum sample size

TABLE 8.6.1 Minimal Sample Sizes—95% CI for a Proportion

Margin of error (m)	*(Percent)*	10%	5%	3%	1%
	(Proportion)	0.10	0.05	0.03	0.01
Minimum sample size	using $p^* = 0.5$	96	384	1067	9604
(p^* is guessed value	using $p^* = 0.3$ or 0.7	81	323	896	8067
of the proportion)	using $p^* = 0.1$ or 0.9	35	138	384	3457

of $(1.96/.03)^2 \times 0.5 \times 0.5 \approx 1067$. The treatment here is very brief. A more detailed exposition is given on the web site.

Sample size for a desired margin of error: Proportions

For a margin of error no greater than m, use a sample size of approximately

$$\left(\frac{z}{m}\right)^2 \times p^*(1 - p^*)$$

- p^* is a guess at the value of the proportion—err on the side of being too close to 0.5.
- z is the multiplier appropriate for the confidence level.
- m is expressed as a proportion (between 0 and 1), not a percentage.

Many will find these numbers surprisingly large. As Table 8.6.1 shows, when p^* moves away from 0.5 toward the extremes of 0 or 1, the required sample sizes become smaller. Use of $p^* = 0.5$ is the "safe guess" that ensures that the margin of error will be no larger than m no matter what the value of the true proportion. If you knew that the true proportion was close to either 0 or 1, using $p^* = 0.5$ would lead to your taking a sample that was larger, and thus more expensive, than strictly necessary. To be safe, we should substitute a value as close to 0.5 as we can imagine the true proportion having. In other words, we use the biggest conceivable value of a small proportion and the smallest conceivable value for a large proportion. For example, if we are estimating the proportion of teenagers who used cocaine and believed the proportion could not be any bigger than 0.05 (or 5%) we would use $p^* = 0.05$. If we were estimating the proportion who were adequately fed and believed it could not possibly be smaller than 0.9 (90%) we would use $p^* = 0.9$.

*8.6.2 Means

We now turn our attention to the problem of determining the sample size for estimating a mean. A simple example points the way.

EXAMPLE 8.6.1 *Sample Size Calculations for Passage Times*

In Example 8.1.1 if we use just the first 8 observations to estimate the true passage time of light, the margin of error for the estimate is 0.0051. If we use all 20 observations, the margin of error is reduced to 0.0024. How many observations do we need to obtain a margin of error of only 0.001?

The formula for a confidence interval for a mean is $\bar{x} \pm t\,\text{se}(\bar{x}) = \bar{x} \pm t\,s_X/\sqrt{n}$. The margin of error is thus $ts_X/\sqrt{n}$. Here s_X is a sample estimate of the population standard deviation, σ, which measures the variability of individual measurements. When trying to calculate sample sizes, we need some estimate of variability, σ^*. We shall discuss how we might obtain σ^* shortly. When n is large, $t \approx z$. If we want to ensure that the margin of error is no greater than m, arguments similar to those of the previous subsection establish that the sample size n should be at least $(z\sigma^*/m)^2$.

Sample size for a desired margin of error: Means

For a margin of error no greater than m, use a sample size of approximately

$$\left(\frac{z\sigma^*}{m}\right)^2$$

- σ^* is an estimate of the variability of individual observations.
- z is the multiplier appropriate for the confidence level.

There are three commonly used ways of obtaining a value for σ^*: (1) as the standard deviation of "similar" data collected in the past, (2) by doing a small pilot study to help plan the main study and using the standard deviation of the pilot-study observations, and (3) by using $\sigma^* =$ (maximum imaginable measurement − minimum imaginable measurement)/6. The rationale for this is that virtually all observations typically fall within 3σ on either side of the mean.

EXAMPLE 8.6.1 (cont.) *Determining the Sample Size*

In our opening example, we can treat the 20 measurements we already have as constituting a pilot study that we can use to determine an appropriate size for a larger study. For σ^* we use the standard deviation from those 20 measurements, namely 0.00512. If we want the margin of error from a 95% confidence interval to be no larger than $m = 0.001$, we need a sample size of

$$n \geq \left(\frac{1.96 \times 0.00512}{0.001}\right)^2 = 100.7$$

We would need to take about 100 observations.

The sample sizes obtained from these calculations are not very reliable. The weak link is σ^*, the estimate of variability. Sample standard deviations from small pilot studies are notoriously variable. The assumption of method (1), that the variability in the future data set will be the same as in past data sets deemed "similar" is also often unreliable. Method (3) is even worse. The preceding calculations just give us rough ballpark sample sizes for planning purposes, although our colleague Brian McArdle jokes that the sample size estimates these procedures give us are not so much in the right "ballpark" as in the right "national park."

In situations like Example 8.6.1, taking large samples (e.g., many repeat measurements of the passage time for light) helps us deal only with random measurement errors leading to observations distributed about the true value. If there is any systematic bias in the measurement process itself, taking a lot of measurements will obviously not give us more accurate answers.

EXERCISES FOR SECTION 8.6

In all of the following, use the 95% level of confidence.

1. Calculate the sample size required so that the margin of error for a true proportion p is no larger than 0.02 (2%) under the following circumstances:
   (a) You want to make no assumptions about the value of p.
   (b) You are sure that $p < 0.15$.
   (c) You are sure that $p > 0.85$.
2. Suppose that the data in Example 8.4.1 (see Table 8.4.1) come from a pilot survey being used to plan a definitive study. How large a sample would you need to obtain a margin of error for the population thiol concentration that is no larger then 0.025 mmol (a) for the normal population and (b) for the rheumatoid population. Why is the difference between these two sample sizes so big?

8.7 SUMMARY

1. We construct an interval estimate of a parameter to summarize our level of uncertainty about its true value. The uncertainty is a consequence of the sampling variation in point estimates.
2. If we use a method that produces intervals that contain the true value of a parameter for 95% of samples taken, the interval we have calculated from our data is called a 95% confidence interval for the parameter.
3. Our confidence in the particular interval comes from the fact that the method works 95% of the time (for 95% CIs).

TABLE 8.7.1 Standard Errors and Degrees of Freedom[a]

Parameter		Estimate	Standard error of estimate	*df*
Mean,	μ	$\overline{x}$	$\frac{s_X}{\sqrt{n}}$	$n-1$
Proportion,[b]	p	$\widehat{p}$	$\sqrt{\frac{\widehat{p}(1-\widehat{p})}{n}}$	∞
Difference in means,[c]	$\mu_1 - \mu_2$	$\overline{x}_1 - \overline{x}_2$	$\sqrt{\frac{s_1^2}{n_1} + \frac{s_2^2}{n_2}}$	$\text{Min}(n_1 - 1,\ n_2 - 1)$
Difference in proportions,	$p_1 - p_2$	$\widehat{p}_1 - \widehat{p}_2$	(see Table 8.5.5)	∞

[a] $df = \infty$ means we use a multiplier obtained from the Normal(0,1) distribution.

[b] CI's work well when sample sizes are big enough to satisfy the 10% rule in Appendix A3.

[c] Applies to means from independent samples.

df given is a conservative approximation for hand calculation (see Section 10.2).

4. For a great many situations, an (approximate) confidence interval is given by ***estimate*** $\pm$ ***t standard errors.*** The size of the multiplier, t, depends both on the desired confidence level and the degrees of freedom (df).
[With proportions, we use the Normal distribution (i.e., $df = \infty$), and it is conventional to use z rather than t to denote the multiplier.]
Examples are given in Table 8.7.1.

5. The *margin of error* is the quantity added to and subtracted from the estimate to construct the interval (i.e., t standard errors).

6. If we want greater confidence in an interval calculated from our data, we have to use a wider interval.

7. To double the precision of a 95% confidence interval (i.e., halve the width of the confidence interval), we need to take four times as many observations.

REVIEW EXERCISES 8

1. A new batch of ore is to be tested for its nickel content to determine whether it is consistent with the usual mean content of 3.25% that has been found for previous batches. Ten samples are taken with the following results:

   3.27 3.23 3.31 3.34 3.26 3.24 3.25 3.37 3.29 3.33

   (a) Obtain a dot plot for these data. Is there anything to suggest that the data may not be Normally distributed?
   (b) Find the sample mean and standard deviation for the 10 observations.
   (c) Obtain a 95% confidence for the mean content of this batch. Is there any evidence that this batch differs from previous batches?
   (d) Draw your confidence interval on your dot plot. If you had 40 observations with the same sample mean and standard deviation as the preceding 10 observations, what would your confidence interval look like?
   *(e) If the original 10 observations represent a pilot sample, what sample size would you need to obtain a 95% confidence interval with margin of error no greater than 0.015?

2. Much behavioral research on humans relies on volunteers. It has been long suspected that volunteers are often not representative of people in general, tending to be more intelligent, better educated, and more achievement oriented than nonvolunteers. In research conducted by Wright and Bonett [1991], 79 supervisory employees in a municipal department were asked to return a questionnaire examining such things as mood state, job satisfaction, and coping behavior. The time taken to return the questionnaire was used as a surrogate for tendency to volunteer. The 38 who returned it at the first request were labeled "enthusiastic volunteers," the 28 who returned it after the second request were termed "reluctant volunteers," and the 13 who did not return it were termed "nonvolunteers." The most recent job-performance ratings for each subject were then examined. Employees had been rated on three factors: "goal emphasis," "support," and "team building," each on a 5-point scale (from 1 = never to 5 = always) according to the extent to which each employee's work showed an emphasis on the factor. Summary statistics are given in Table 1.
   (a) On simple inspection, what trends or features do you see in these data?
   (b) Calculate 95% confidence intervals for the difference in true mean goal emphasis between

TABLE 1 Behavioral Research on Volunteers

Group	n	Goal emphasis		Support		Team building	
		$\bar{x}$	s_X	$\bar{x}$	s_X	$\bar{x}$	s_X
Nonvolunteers	13	3.15	0.689	2.92	0.493	2.85	0.689
Reluctant	28	3.32	0.723	3.57	0.879	3.39	0.831
Enthusiastic	38	3.82	0.729	3.61	0.718	3.52	0.951

Source: Wright and Bonett [1991].

(i) enthusiastic volunteers and reluctant volunteers?
(ii) enthusiastic volunteers and nonvolunteers?
(iii) reluctant volunteers and nonvolunteers?

Using the same breakdown as in (b), confidence intervals for the other two factors are:

SUPPORT: (i) [−0.38, 0.46], (ii) [0.30, 1.08], (iii) [0.18, 1.12]

TEAM BUILDING: (i) [−0.32, 0.58], (ii) [0.13, 1.21], (iii) [0.00, 1.08]

(c) Very briefly, how would you summarize the relationship between volunteerism and job performance?

(d) In what ways does this research imperfectly address the issue of nonrepresentativeness of volunteers in behavioral research? Can you think of any other ways of investigating this issue?

3. As part of an experiment undertaken to investigate the effects of survey format on the response rate for a mailed questionnaire, Childers and Ferrell [1979] compared a survey printed on both sides of a single piece of paper with the same survey printed on two separate sheets.

(a) Why did they think that this might make a difference? Which format would you imagine would yield the higher response rate?

Childers and Ferrell used a sample of 440 members of the American Marketing Association. Half were randomly selected to get the single-sheet version, resulting in a 36% response rate. The rest got the two-sheet version, resulting in a 30% response rate.

(b) Calculate both a 90% and a 95% confidence interval for the difference in response rates. Write down in plain English what your intervals tell you.

(c) If the information in this question was all that was available to you, and you had to conduct a similar survey, would you use a single-sheet or two-sheet version? Why?

4. For more than 50 years, Gallup polls in the United States have been asking the question: "Do you approve of the way that . . . is handling his job as president?" Nowadays it is the first question asked. Prior to 1956 and for several polls during the Johnson administration, it came toward the end of the interview schedule. Some concern has been expressed about a possible question-order effect on this item, so that current approval ratings may not be comparable with earlier approval ratings. If previous questions remind participants of unpopular actions of the president, the approval rating may drop. For example, the Johnson polls in the 1960s followed questions about the Vietnam war. Prior questions could also influence whether people are prepared to express an opinion. Sigelman [1981] conducted an experiment in Kentucky in which half the sample in a random telephone poll got the approval question as the first question, while the rest got it as the 33rd question. None of the intervening questions related directly to the president, but many related to politically charged issues such as home heating costs, energy problems, pollution, drugs, and the bribing of politicians. In all, 665 interviewees provided usable responses, giving the results in Table 2.

TABLE 2 Question-Order Effect

	First question	33rd question
	Percentage (*n*)	Percentage (*n*)
% expressing opinion		
All respondents	80.4% (337)	88.7% (328)
Less educated	79.5% (224)	91.6% (214)
More educated	84.2% (101)	83.2% (113)
% approving (of those with opinion)		
All respondents	52.8% (271)	51.5% (291)

Source: Sigelman [1981].

Using appropriate 95% confidence intervals, answer the following:

(a) How big is the true difference between the proportions of people in Kentucky who would express an opinion when the question is asked first and those who would when it is asked late

(i) among all respondents?

(ii) among the less educated?

(iii) among the more educated?

(b) Sigelman [1981, p. 205] states that the contention of some earlier authors that the results from Gallup presidential polls conducted the two different ways are not directly comparable were justified "but for the wrong reasons." He went on to argue that the difference comes in the percentage expressing an opinion, not in the approval rating among those who do express an opinion. Is this last statement justified? How big could the difference in true approval ratings be in the Sigelman study?

(c) Suppose that in this study there was no difference in approval ratings among those expressing an opinion. Suggest a reason why we could still not apply such a finding to the Johnson polls.

*5. A survey is planned to determine what proportion of high school students in your city have parents who are separated or divorced. Let p be this proportion.

(a) What is the smallest sample size needed to guarantee that the margin of error for a 99% confidence interval for p is no greater than 0.04?

(b) If we believed from prior research that p is between 0.3 and 0.4, how would this change your answer for (a)?

(c) If we believed from prior research that p is between 0.3 and 0.6, how would this change your answer for (a)?

(d) What questions would you need to resolve before planning the survey?

6. Breast cancer is the most commonly occurring of the so-called women's cancers. In the United States it kills more women than any cancer except lung cancer. Of course, the earlier it is detected, the greater the chances of long-term survival. Until the mid-1980s the standard treatment for breast cancer had been total (or radical) mastectomy, in which the entire breast is surgically removed together with some surrounding tissue and muscle. It gradually became suspected that such extensive surgery might not be necessary. Increasingly, a much more limited operation called segmented mastectomy was performed, in which just enough tissue surrounding the tumor is removed so that the specimen margins are tumor free. There are clear advantages in the more limited operation. There can be psychological problems associated with radical mastectomy (feelings of disfigurement and loss of sexual

TABLE 3 Breast Cancer and Surgery

	TM	SM	SM + R
Five-year disease-free proportion	0.719	0.681	0.814
Correct standard error	0.035	0.035	0.029

Source: Fisher et al. [1985].

attractiveness) and also physical problems caused by the removal of associated muscle. After a limited operation the breast can often be "rebuilt" using plastic surgery.

The first large-scale randomized clinical trial comparing the effectiveness of the two operations reported its findings in Fisher et al. [1985]. Over a five-year period, 1843 patients with breast tumors under 4 cm in diameter were randomly allocated to several treatments. Of these patients 1148 had negative nodes (no apparent cancer in the lymph nodes). Of the patients with negative nodes 362 were allocated to total mastectomy (TM), 390 to segmented mastectomy (SM), and 396 to a combination of segmented mastectomy and radiation therapy (SM + R). The proportions still free of the disease after five years were 0.719 (TM), 0.681 (SM), 0.814 (SM + R).

(a) Give a 90% confidence interval for the five-year disease-free rate for total mastectomy patients.

(b) Calculate a 95% confidence interval for the differences in disease-free rates between the TM and SM treatments.

In fact, the five-year disease-free rates quoted were really not simple proportions. They were estimates "actuarially adjusted" to allow for the fact that not all women were at risk for the whole five-year period. The effective sample sizes were thus smaller and the correct standard errors associated with the quoted rates somewhat larger than those obtained from ordinary sampling theory [as in (a)]. The rates and the correct standard errors are given in Table 3.

(c) Using the correct standard errors, calculate 95% confidence intervals for the true differences between the five-year disease-free proportions for each treatment.

(d) What are your conclusions about the relative effectiveness of the treatments?

7. Does breast feeding increase the IQs of babies? These questions relate to the data and story in Review Exercises 7, problem 15.

(a) If you had the raw data, what would be the first thing you would do?

(b) Compute a 95% confidence interval for the difference in mean IQ between breast-fed and non-breast-fed children. What does the result suggest?

*(c) How large a sample size would be required if you wanted a 95% confidence interval for the mean IQ of breast-fed babies to be no wider than 2 (i.e., a margin of error no greater than 1)?

(d) Do you think that the results are applicable to the general population? Why or why not?

(e) Is this an observational study or an experiment? What are the consequences of this?

(f) Repeat the calculation in (b), but assuming a sample size of only 15 in the bottle-fed group. Compare the two intervals.

(g) The results of (b) make it look as though a breast-fed baby will be smarter than one who is not breast fed. Take the preceding means and standard deviations as if they were the true population values, assume Normal distributions, and calculate the probability that a random non-breast-fed baby has a higher IQ than a random breast-fed baby. How can you reconcile the message given by the results of this calculation and the preceding confidence interval for the difference?

8. Tuberculosis (TB) is known to be a highly contagious disease. A study in 1995 (Sanches et al. [1995]) was carried out on a random sample of 1074 Spanish prisoners. The study

TABLE 4 Possible Factors Associated with Tuberculosis

Variable		Prisoners with tuberculosis	Total number of prisoners
Sex	Male	556	984
	Female	36	90
Race	White	496	886
	Gypsy	74	152
	Other	22	36
Intravenous drug users	Yes	361	629
	No	231	445
HIV positive	Yes	186	294
	No	406	780
Re-imprisonment	Yes	272	456
	No	320	618

Source: Sanches et al. [1995].

investigated what factors might be associated with the tuberculosis infection. The results are given in Table 4.

(a) Use the information given in Table 4 to find the following:

(i) What proportion of prisoners had TB?

(ii) What proportion of prisoners who were intravenous drug users had TB?

(iii) For these prisoners, which race (white, Gypsy, or other) had the highest rate of TB?

(b) Calculate a 95% confidence interval for the difference between the proportion of males infected with TB and the proportion of females infected with TB.

(c) Calculate a 95% confidence interval for the difference between the proportion of HIV-positive prisoners who have TB and the proportion of HIV-negative prisoners who have TB.

(d) Previous studies have shown a link between race and TB infection. Calculate a 95% confidence interval and a 90% confidence interval for the difference between the proportion of white prisoners who were infected with TB and the proportion of Gypsy prisoners who were infected. Does this study confirm previous findings?

Breaking down a sample of people who contract a disease like TB into various groups (factors) as in Table 4 is an early step in a search for causes of the disease. Statisticians tend to look for those factors that are most strongly associated with disease contraction as a way of narrowing down the search. The following situation exemplifies a common type of problem with any sort of causal reasoning from observational data. There is evidence of sex differences, with males more likely to contract TB than females. There is also evidence of HIV differences, with HIV-positive prisoners more likely to contract TB than HIV-negative prisoners. The question arises as to whether the sex differences might just be an artifact of more males being HIV positive than females.

(e) How would you attempt to explore this question and what additional data would you need?

(f) What considerations, if any, would make you nervous about transferring the findings from this study to the general population?

(g) Identify the sampling situation as (a) two independent samples; (b) single sample, several response categories; or (c) single sample, two or more Yes/No items, and give a 95% confidence interval in the following cases:

(i) Of those prisoners who had TB, we want to compare the proportion of intravenous drug users with the proportion of HIV-positive prisoners.

(ii) Of those prisoners who had TB, we want to compare the proportion of Gypsy prisoners with the proportion of white prisoners.

(iii) We want to compare the proportion of Gypsy prisoners who were infected with TB with the proportion of white prisoners who were infected with TB.

9. In Section 1.3 we stated that some of the differences in health problems such as lung cancer could possibly be caused by differences in diet between smokers and nonsmokers. Do such differences exist? Whichelow et al. [1988] performed a study in which they compared the dietary habits of smokers, ex-smokers, and nonsmokers using a nationwide survey of 9003 randomly chosen British adults. The data in Table 5 relate to male smokers and ex-smokers[8] who had ceased smoking at least one year previously. Manual workers and nonmanual workers are treated separately.

(a) What patterns do you see in these data?

(b) Using only the data on nonmanual workers, obtain a 95% confidence interval for the true difference in proportions between ex-smokers and smokers consuming breakfast. Repeat for the other dietary habits. Interpret your intervals.

TABLE 5 Dietary Habits of Smokers and Ex-smokers

	Nonmanual		Manual	
	Smokers	Ex-smokers	Smokers	Ex-smokers
Number in group	404	517	931	629
(%) *Consuming*				
Breakfast	62.9	82.4	55.3	81.1
Brown bread not white	35.4	53.6	19.2	32.4
Fresh fruit "frequently"	59.4	77.6	51.1	65.3
Fried food "frequently"	28.2	16.2	35.3	22.1

Source: Whichelow et al. [1988].

10. The data in Table 6 came from a large survey of car owners conducted by the N.Z. Consumers Institute published in *Consumer* in October 1996. We have assembled the results from owners of Japanese cars with respect to a question asking car owners to report on any problems they had experienced with their cars in the previous 12 months that were not due to accidents or part of routine maintenance. As an example of how the table is to be read, there were 152 returns relating to 1991–1993 Honda cars, of which 82 reported trouble-free running over the past year and 70 reported having had problems. We shall assume that the responses constitute a random sample of Japanese cars.

(a) Find the percentages of cars of each type experiencing trouble-free running. Which makes of car appear to be the most reliable for 1991–1993 models? for 1994–1996 models? Which appear to be least reliable?

(b) There is a problem of comparability between information on 1996 models and earlier models. Why?

[Note: Ignore this problem in answering the remaining parts of the question.]

The comparisons made in (a) ignore sampling variation. The percentages you used to make those comparisons are sample percentages, which will have sampling errors.

Answer each of (c) through (e) by calculating the relevant 95% confidence interval and then stating in words what that confidence interval tells you.

---

[8]Corresponding data on nonsmokers are not given in this paper, but from the graphs given in the paper, patterns for nonsmokers and ex-smokers are very similar.

TABLE 6 Japanese Car Data

Make	1991–1993 models			1994–1996 models		
	Trouble free	Had problems	Total	Trouble free	Had problems	Total
Honda	82	70	152	80	68	148
Mazda	44	41	85	46	33	79
Mitsubishi	110	134	244	89	84	173
Nissan	88	120	208	80	74	154
Subaru	37	36	73	22	13	35
Toyota	212	196	408	123	87	210
Total	573	597	1170	441	358	799

Source: Calculated using information digitized from graphs in *Consumer,* October 1996.

(c) What is the true proportion of 1991-1993 Toyotas that are trouble free? (Calculate also for any other type of car that interests you.)

(d) What is the difference between the true proportions of 1991-1993 Toyotas and 1991-1993 Nissans that are trouble free? Make any other comparisons of this type that interest you.

(e) What change has there been in true proportions of Nissans that are trouble free between the 1994-1996 batch and the 1991-1993 batch? Make any other comparisons of this type that interest you.

(f) One factor in any changes seen between 1994-1996 cars and 1991-1993 cars of the same make is the effect of the aging of cars. What other factors might be operating?

(g) Suppose that 1991-1993 Toyotas were fairly evenly spread across years but that Subarus in this class were mainly 1993s. What effect would this have on a comparison between Toyotas and Subarus?

(h) What questions would you want to ask before applying the results of this survey to Japanese cars in another country? To what extent do the results inform you if you want to buy a new Japanese car?

If respondents are representative of car owners in general, the numbers in each total column can be used to tell us about the market share of the various brands in the market for Japanese cars.

Answer each of (i) through (k) by calculating the relevant 95% confidence interval and then stating in words what that confidence interval tells you.

(i) What was Honda's market share (i.e., the true proportion of Japanese cars that are Hondas) for 1994-1996? Do the same computation for any other combination of make of car and year that interests you.

(j) What was the difference between Toyota's market share and Honda's market share for 1994-1996 models? Make any other comparisons of this type that interest you.

(k) How did Honda's market share change between 1991-1993 and 1994-1996? Make any other comparisons of this type that interest you.

(l) There are considerably fewer 1994-1996 cars than there are 1991-1993. Does this demonstrate a slowdown in sales for Japanese cars?

11. In the 1998 U.S. midterm elections, support for the Republican Party was down from what it had been in the previous election. This was interpreted as something of a slap in the face for the Republicans and their leadership, who were in the process of advancing impeachment proceedings against President Bill Clinton (a Democrat). As might be expected, almost all of the voter swing occurred among voters who considered themselves Independents, rather than among Democrats or Republicans. Among the approximately 2700 Independent voters in the 1998 exit polls, 48% voted Republican, compared with 55% in the 1994 exit polls.

(a) Calculate a 95% confidence interval for the true proportion of Independents who voted Republican in 1998.

(b) Calculate a 95% confidence interval for the difference between the true proportions of Independents who voted Republican in 1994 and 1998 (assume 2700 Independents in both sets of exit polls).

(c) Of the approximately 2700 four-year college graduates in the 1998 exit polls, 53% voted Republican versus 45% of the 1800 who had done postgraduate study. Calculate a 95% confidence interval for the true difference.

(d) The ethnic breakdown was (number, percentage Republican): white (8200, 55%), black (1000, 11%), Hispanic (500, 35%), Asian (100, 42%). In each case, calculate a 95% confidence interval for the true proportion. Calculate a 95% confidence interval for the true difference between the proportions of Asian and Hispanic voters who voted Republican. Repeat for any other differences that interest you.

12. The data in Table 7 were collected in 1996 by a large ongoing study, the Australian Longitudinal Study on Women's Health (courtesy of Prof. Annette Dobson, and Drs. Wendy Brown and Gita Mishra). They are summaries from responses to questions by a random sample of Australian women aged 18–23. The women have been divided into groups according to their

TABLE 7 Percentages of Women Aged 18–23

	Living status					
Issue	**With parents**	**Share**	**Alone**	**Partner/ spouse**	**Partner and child**	**Single parent**
Percentage very or extremely stressed by issue						
Money	20.9	30.5	29.8	28.1	31.6	41.1
Living arrangements	10.2	17.3	16.2	11.4	10.5	21.1
Relationship with boyfriends	10.0	11.6	14.5	4.8	2.6	14.1
Relationship with partner	6.4	6.9	10.1	7.6	10.2	18.6
Relationship with parents	9.6	8.8	10.3	8.2	8.2	12.9
Percentage reporting unhealthy lifestyle factors						
Smoking[a]	28.0	34.3	32.5	34.5	42.4	54.1
Binge drinking[b]	18.4	28.0	20.8	12.7	6.5	18.4
Unhealthy eating practice[c]	29.7	32.6	36.5	32.2	29.0	43.2
Weight category (from body mass index)						
Underweight (bmi < 20)	30.7	27.1	28.8	27.4	25.3	28.3
Healthy weight (20–25)	51.7	53.6	46.9	50.3	41.2	45.5
Overweight (>25–30)	12.3	15.1	16.2	16.0	21.6	16.9
Obese (>30)	5.3	4.3	8.2	6.3	12.0	9.3
Number in group	6678	3125	875	2313	915	575

[a]Includes occasional smoking.

[b]Five or more drinks on one occasion each week.

[c]Any unhealthy eating practice, including vomiting after meals, laxatives, diuretics, and fasting.

Source: The Australian Longitudinal Study on Women's Health, University of Newcastle, New South Wales.

living arrangements. We have reported data from some of the questions asked about causes of pronounced stress (e.g., 20.9% of young women living with parents reported being very stressed by money matters), on unhealthy lifestyle factors (e.g., 28% of those living with their parents smoked), and on weight categories (e.g., 30.7% of those living with parents were underweight).

(a) What interesting and/or surprising features can you see in the data in Table 7?

(b) Are there some living-status groups that you might expect to contain women who are older (or younger) on average than other groups? What problems, if any, might this create when we interpret differences between groups?

(c) What are some other sources of stress affecting many people that should be considered? (The issues reported here are only a subset of those investigated in the study.)

(d) Calculate a 95% confidence interval for the true proportion of young women living in shared accommodations who are stressed by their living arrangements.

(e) Calculate a 95% confidence interval for the true difference between the proportion of those living alone who are stressed by relationships with boyfriends and the proportion of those living in shared accommodations who are stressed by relationships with boyfriends.

(f) For those living alone, calculate a 95% confidence interval for the true difference between the proportion stressed by money problems and the proportion stressed by their living arrangements.

(g) For those living with a partner and child, calculate a 95% confidence interval for the true difference between the proportion in the underweight category and the proportion in the overweight category.

13. A standard question in New Zealand political polls allows respondents to pick their "preferred Prime Minister." Party leaders whose ratings slide to 3% or below are colorfully said "to have slid below the margin of error." (They sometimes reemerge!) If a poll samples 1000 people so that the reported margin of error is approximately 3%, what is the actual margin of error associated with a support rating of 3%?

14. These questions relate to the data and story in Review Exercises 7, problem 16.

(a) Calculate a 95% confidence interval for the true mean rating for females under "control" conditions.

(b) Calculate a 95% confidence interval for the difference in true mean ratings for males between "control" conditions and "negative contrast" conditions.

(c) Repeat (b) for the difference between "positive contrast" conditions and "control" conditions.

15. These questions relate to the data and story in Review Exercises 7, problem 17.

(a) Calculate a 95% confidence interval for the true mean score under "control" conditions.

(b) Calculate a 95% confidence interval for the difference in true mean scores between "humane/no info." conditions and "control" conditions.

(c) Calculate a 95% confidence interval for the difference in true mean scores between "control" conditions and "inhumane/no info." conditions. What do you conclude?

***16.** In Example 3.4.2 we described the capture-recapture methodology for estimating animal numbers. Suppose that we have an unknown number N animals in the population. We catch, tag, and release M of them. We allow time for the tagged and untagged animals to become well mixed. We then capture n animals and observe that x of them are tagged. By equating the sample proportion of tagged animals x/n with the population proportion M/N, we obtain nM/x as an estimate of N. But how can we associate a confidence interval with such an estimate? The following problem will lead you through one way of doing it.

Seber [1982, p. 105] describes an experiment in which the numbers of *Lasius flavus* ants in several colonies were estimated this way. The ants were tagged with radioactive

P^{32}, and mixing took place over 5 to 10 days. Seber explains how ants were sampled and caught and how the underlying assumptions were approximately met. In the sixth colony 600 tagged ants were released, and after the mixing period 321 were captured. Of the captured ants, 89 were tagged. Thus $M = 600$, $n = 321$, and $x = 89$.

(a) What is the estimated population size N?

(b) Calculate a 95% confidence limits for p, the population proportion of tagged animals.

(c) Noting that with 95% confidence $p = M/N = 600/N$ lies between the limits calculated in (b), obtain the corresponding confidence limits for N.

***17.** Suppose that we are planning a study in which sample proportions will be taken from two independent samples, each of size n. Suppose, moreover, that we want the margin of error for the difference between the two true proportions $p_1 - p_2$ to be no greater than w. (As an example, consider estimating the true difference between the proportions of people carrying hepatitis B in each of two different cities.)

(a) Arguing as in Section 8.6.1, show that n must satisfy

$$n \geq \left(\frac{z}{w}\right)^2 \times \{\widehat{p}_1(1-\widehat{p}_1) + \widehat{p}_2(1-\widehat{p}_2)\}$$

(b) What values of $\widehat{p}_1$ and $\widehat{p}_2$ lead to the largest n (each sample has size n)?

(c) If nothing is known about p_1 and p_2, what minimum value of n is required before you can be sure that the margin of error for $p_1 - p_2$ is no greater than w?

(d) Repeat this derivation for sampling situation (b) of Fig. 8.5.1 (one sample of size n, several response categories) where you want to limit the margin of error for the difference between the true proportions falling into two of the categories to being no greater than w.

***18.** The purpose of this exercise is to derive the rules of thumb given for obtaining mentally the margins of error for differences between poll proportions, as discussed in Section 8.5.3.

(a) Write a formula for the margin of error associated with a sample proportion for a single poll.

(b) Suppose that we have two independent samples, both of size n, and that $\widehat{p}_1 = \widehat{p}_2 = 0.5$. What margin of error should be associated with the difference?

(c) How many times larger is the margin of error for the difference than that for a single poll?

(d) Repeat this line of reasoning for sampling situation (b) of Fig. 8.5.1 (one sample of size n, several response categories).

19. (Computer simulation exercise) This is a continuation of a problem in Review Exercises 7 where we generated random numbers from a Normal($\mu = 174$, $\sigma = 6.57$) distribution to simulate the behavior of the heights of males randomly sampled from the workforce database.

(a) Take 100 samples each of size 9. For each sample, obtain a 95% confidence interval for the true mean, which we know to be 174. What proportion of the intervals from your samples contain the true value? Find the average width of the 100 intervals.

(b) Repeat (a), but using samples of size 25. How much smaller is the average width of your intervals this time?

20. These questions relate to the thinking used in planning a simulation.

(a) For 95% confidence intervals obtained from M independent samples, the number of intervals that cover the true value has a Binomial distribution. Why is this? What are the values of the parameters?

(b) With 95% confidence intervals from 100 independent samples, what is the probability that between 91 and 98 intervals (inclusive) contain the true parameter value.

(c) With 95% confidence intervals from 1000 independent samples, what is the probability that between 935 (93.5%) and 965 (96.5%) of intervals (inclusive) contain the true parameter value.

*(d) Often, as in the next problem, we use computer simulation to investigate the behavior of confidence intervals when an assumption used in deriving the confidence interval formula is violated. Assume that, despite the violated assumption, the coverage frequency will still be roughly 95%. How many samples do we have to take in our simulation for the margin of error associated with our experimental coverage frequency to be no greater than 1%?

21. (Computer simulation exercise) The confidence interval formula given in Section 8.2 was obtained assuming that we are sampling from a Normal distribution. Will it still work if we apply it to observations sampled from a skewed distribution? We will try it out and see. The Chi-square distribution with 4 degrees of freedom is a skewed distribution with population mean $\mu = 4$.

(a) Generate 500 observations from a Chi-square distribution with 4 degrees of freedom and plot a histogram. (This will show you that the distribution is indeed very skewed.)

(b) Take 100 samples each of size 9. For each sample, obtain a 95% confidence interval for the true mean. What proportion of your intervals contain the true mean (4 here)?

(c) Repeat (b) using samples of size 25.

(d) If it is easy to do so with your computer software, repeat both (a) and (b) using 1000 samples rather than just 100 samples.

(e) How well does the confidence interval formula work for data such as these?

CHAPTER 9

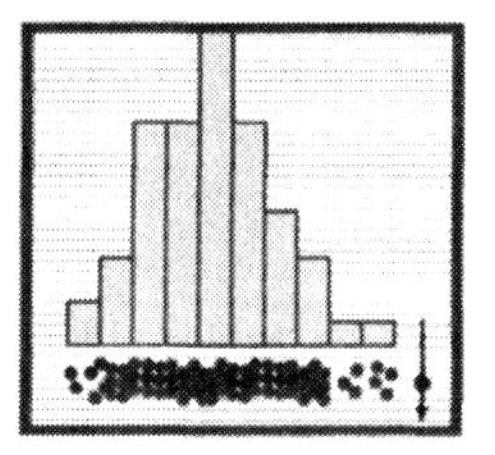

Significance Testing: Using Data to Test Hypotheses

Chapter Overview

This chapter introduces the significance test, an important new way of making inferences about an unknown parameter. Significance testing is a very visible area of statistics. The expression "doing statistics" is almost regarded as synonymous with performing significance tests by many researchers in other disciplines. The development here will build on the Chapter 7 idea that our data provides evidence against any theoretically specified value for a parameter that is more than two or three standard errors away from the data estimate.

We begin in Section 9.1 by using simulation to motivate the idea of testing a hypothesis. The difference between hypothesis testing and confidence intervals is then explained. Section 9.2 describes various types of hypotheses, their formulation, and how they fit into the research process. We introduce the t-test and its close relative the z-test as a unified procedure in Section 9.3, where we develop the idea of a P-value and the interpretation of a P-value as a measure of evidence against the null hypothesis. We then apply the methodology to means, proportions, and differences between means and proportions and use examples to introduce applications. Our focus here is on the interpretation of test results in the context of a real problem. Section 9.4 discusses testing viewed as decision making (the Neyman-Pearson approach) in relation to the use of P-values. The relationship between tests and intervals is further developed in Section 9.5, and there is a discussion about misconceptions and communication problems that have arisen from the use of the word *significant.* This forms part of a larger discussion that includes practical versus statistical significance, the interpretation of nonsignificant results, and the desirability of supplementing tests with intervals. We conclude with two very short sections: a brief introduction to design issues in Section 9.6 and in Section 9.7 a generalization of the notion of the test statistic to prepare for the F- and Chi-square tests in later chapters.

9.1 GETTING STARTED

9.1.1 Examples

We know that the estimate of a population parameter is never exact because of sampling variation and possible nonsampling errors. We shall show that significance testing, like confidence intervals, however, provides a way of dealing with uncertainty about the true parameter value due to sampling variation. We consider two examples that informally introduce some fundamental ideas underlying the significance testing approach. In both examples we begin by hypothesizing that a population parameter has a particular value. To test the hypothesis, we see whether our data estimate of the parameter is consistent with the pattern of variation we would expect to get if the hypothesis was true.

EXAMPLE 9.1.1 *ESP or Just Guessing?*

In a famous experiment conducted by Pratt and Woodruff in 1938 to investigate the existence of extrasensory perception (ESP), an experimenter and a subject sat at opposite ends of a table. The experimenter used a deck of so-called Zener (or Rhine) cards in which there were equal numbers of cards depicting each of five very different shapes, shown below. One by one the experimenter turned up a random card from the shuffled deck and looked at it, whereupon the subject (who could not see the card) indicated what shape he or she "thought" it depicted.

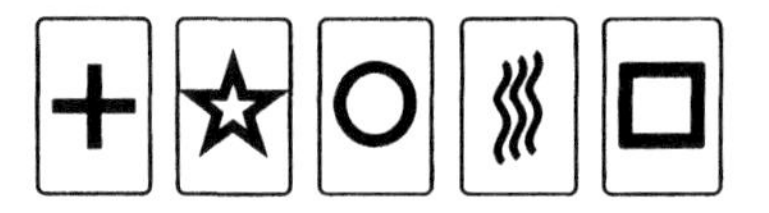

Using a number of students as experimental subjects, the experiment was repeated 60,000 times, resulting in 12,489 correct guesses. If all subjects were just guessing, then since each of the five shapes is equally likely to turn up at any given draw, there is 1 chance in 5 of correctly identifying a card, and we would expect to get about one fifth, or 20%, of our guesses correct. So out of 60,000 guesses, we'd expect to get 12,000 correct. Pratt and Woodruff's experiment did a little better than this. Their success rate was 12,489/60,000, or just over 20.8%. This has been held up as evidence of ESP. But couldn't the subjects all just have been guessing? The 20.8% success rate is only slightly better than the 20% we would expect by chance, and we would expect some sampling variation, even with such a huge sample size. So can sampling variation alone account for Pratt and Woodruff's success rate, or should we look for some other explanation?

How big *is* the sampling variation in an experiment like this? We will use a computer to simulate making 60,000 guesses with a 1 in 5 chance of guessing correctly each time and calculate the proportion of correct guesses. We will call this a "just-guessing" experiment. We shall conduct several just-guessing experiments and see what happens. The success proportions for the first seven just-guessing experiments are plotted below.

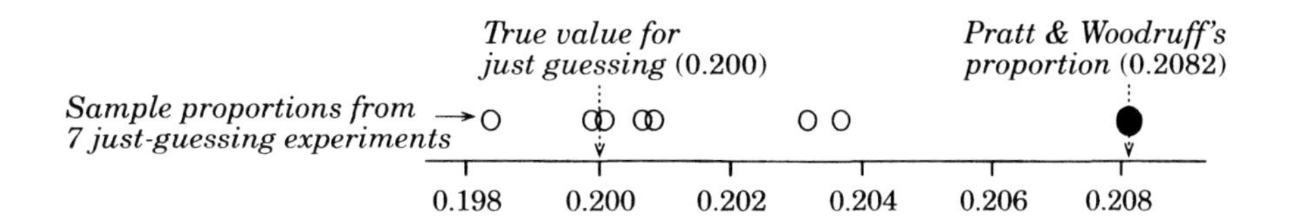

There is variation between the results of the different experiments, but Pratt and Woodruff's result is bigger than any we saw. However, seven experiments isn't very many, so we performed 400 just-guessing experiments. The resulting sample proportions are plotted in Fig. 9.1.1. They are plotted in two ways, first as a dot plot with a little vertical jitter to give you an impression of the 400 different sample proportions and second as a histogram. None of the proportions from the 400 just-guessing experiments was as big as Pratt and Woodruff's. Being persistent, we then tried 4,000 experiments. Again, none of them gave a result as big as Pratt and Woodruff's! Guessing coupled with sampling variation just doesn't seem to be able to account for Pratt and Woodruff's success rate. Their estimate is not consistent with the pattern of variation that we would expect to get if the students were guessing.

As one might expect, the Pratt and Woodruff study aroused a great deal of controversy and was often quoted in support of the existence of ESP. A great deal of effort was expended by others to find non-ESP mechanisms that seemed to explain the results (see Hansel [1966, Chapter 8]).

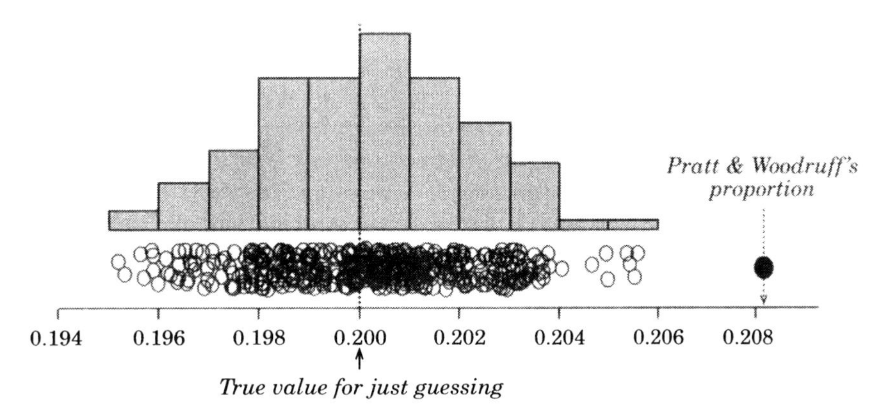

FIGURE 9.1.1 Sample proportions from 400 just-guessing experiments.

EXAMPLE 9.1.2 *Was Cavendish's Experiment Biased?*

The experimental procedures used in a number of the famous early experiments that measured physical constants have later been shown to have had systematic biases. In other words, the distribution generating the measurements was centered at a value that is detectably different from the true value of the physical constant. Let us return to Cavendish's determinations of the mean density of the earth (Example 7.2.2, data in Table 7.2.1) and see whether there is any evidence that his method was biased. We will use the last 23 observations (after the wire was changed), namely:

5.36 5.29 5.58 5.65 5.57 5.53 5.62 5.29 5.44 5.34 5.79 5.10

5.27 5.39 5.42 5.47 5.63 5.34 5.46 5.30 5.75 5.68 5.85

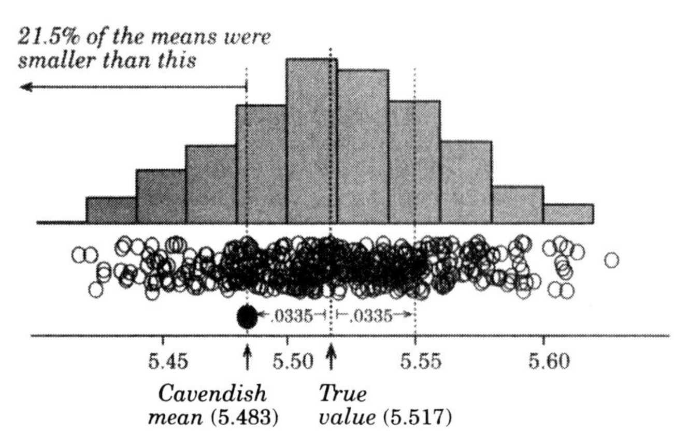

FIGURE 9.1.2 Sample means from 400 sets of observations from an unbiased experiment.

These observations have a sample mean and standard deviation of 5.4835 and 0.1904, respectively. The currently accepted figure for the mean density of the earth is 5.517 g/cm^3. We want to find out whether the data mean is far enough from the true value to show that the experiment was biased or whether the difference can be explained simply in terms of sampling error.

Again, we will use a computer simulation to answer some questions. We want to compare the sample mean from Cavendish's 23 observations with the sample means of many sets of 23 observations produced by an *unbiased* experimental procedure. We will decide that Cavendish's experiment was biased if the mean his experiment produced is much further from the true value than the means produced by the unbiased experiment. The population mean of measurements from an unbiased experiment will be the true mean density of the earth, namely 5.517. The standard deviation of the above data is 0.1904. This is an estimate of the variability of individual measurements. We will therefore generate Normally distributed "measurements" with a population mean of 5.517 and standard deviation of 0.1904.

We simulated taking 400 sets of 23 measurements. The resulting sample means are plotted in Fig. 9.1.2 using, once again, both a dot plot and a histogram. The position of the mean from Cavendish's data has been added. What happened? The value from Cavendish's data is clearly not unusual for sample means from an *unbiased* experiment with the observed level of measurement-to-measurement variability. We see that the Cavendish mean lies at the edge of the interval that stretches 0.0335 units on either side of the true value. This interval encompasses about 60% of the histogram so that the Cavendish mean lies within the central 60% of the distribution. This mean is therefore not inconsistent with the pattern of variation we would expect to get, so we see no evidence that Cavendish's procedure was biased.

We have just seen two examples of the use of sampling variation. In Example 9.1.1, the success rate observed in the data was further from the "just-guessing" value than could be explained simply in terms of sampling variation. This leads us to doubt just guessing as a credible explanation and to consider other explanations, one of which is ESP. In Example 9.1.2, the estimate of the mean density of the earth from Cavendish's experiment was well within the sampling variation that one would get about the true value using an unbiased experimental procedure. The data cannot establish that

Cavendish's experiment was biased, so the possibility that it was unbiased remains credible.

9.1.2 Measuring Distance in Numbers of Standard Errors

We saw in the previous section that if a hypothesized value for a population parameter is correct, then a data estimate of the parameter will not be too far from its hypothesized value. We can measure the distance between the estimate and parameter value in terms of the number of standard errors of the estimate. If this number is unacceptably large, we would reject the hypothesized value. To determine what is an unacceptable distance, we compute t_0 = *(estimate − parameter value)/standard error*. We use this number because the sampling distribution of

$$T = \frac{\text{estimator} - \textbf{true} \text{ parameter value}}{\text{standard error}}$$

for many examples in this book is found to conform (at least approximately) to a Student t-distribution or a Normal(0, 1) distribution. If our hypothesized value of the parameter happened to be the true value, then we would know the true distribution of T, and the observed value t_0 would be a "typical" value from this distribution. If the value of t_0 was not typical, but was well out in the tails of the distribution, we would conclude that we don't have the true value of the parameter. We now clarify this idea by returning to the Cavendish data.

✖ Example 9.1.2 (cont.) ***Using the t-Distribution***

The mean of our 23 observations is 5.4835 and the standard error (sd/$\sqrt{n}$) is 0.0397. The mean from the Cavendish data is thus 0.844 standard errors below the true value of 5.517 [since $(5.4835 - 5.517)/0.0397 = -0.844$]. Can this discrepancy be explained simply in terms of sampling error? If 5.517 was the true mean for the process producing Cavendish's observations, then assuming Normality, t_0-values of the form

(sample mean − 5.517)/standard error

will have a Student($df = n - 1 = 22$) distribution. Figure 9.1.3 plots the t_0-values from 400 sets of 23 observations from unbiased experiments.

Recall that the Cavendish mean was 0.844 standard deviations below the true value. We see from Fig. 9.1.3 that approximately 20% of unbiased experiments gave us sample means that were even more than 0.844 standard errors below the true value. Approximately another 20% were more than 0.844 standard errors above the true value. In all, approximately 40% of the means were further from the true value (*in the sense of more standard errors*) than was Cavendish's. We reach the same conclusions as before.

In fact, there is no need to simulate. We can just use probabilities from the Student($df = 22$) distribution directly. The values of $\text{pr}(T < -0.844)$ and $\text{pr}(T > 0.844)$ are given in Fig. 9.1.4. We see that when $n = 23$ and therefore $df = 22$, the probability of falling outside ± 0.844 is approximately 40%. Thus, in long-run repeated sampling, 40% of sample t_0-values from samples of size 23 fall outside ± 0.844. We can

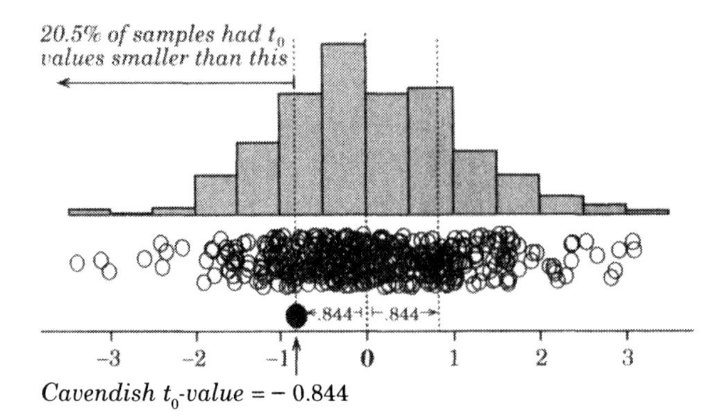

FIGURE 9.1.3 Sample t_0-values from 400 unbiased experiments (each t_0-value is distance between sample mean and 5.517 in standard errors).

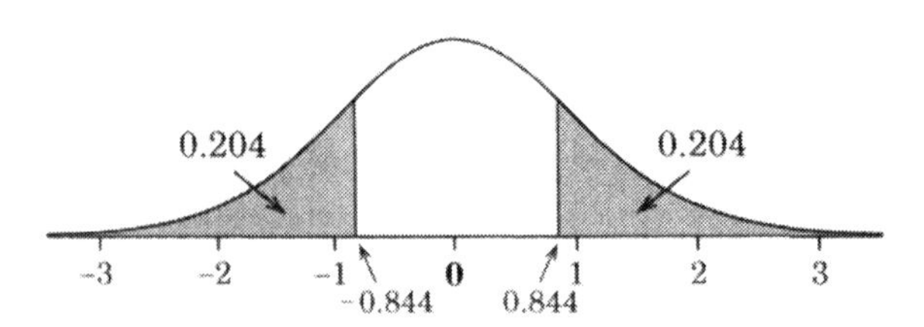

FIGURE 9.1.4 Student(df = 22) density.

interpret this latter statement as saying that, in the long run, 40% of sample means fall more than 0.844 standard errors away from the true mean. Consequently, when a sample mean is only 0.844 standard errors from a postulated value for the population mean (here 5.517), it is entirely plausible that the postulated value is true and the difference we see between sample mean (5.4835) and postulate (5.517) is simply due to sampling variation. We are not saying that the explanation is a proven truth, only that it is credible or that the data do not rule it out as a possibility.

The two examples that we have just seen have introduced fundamental issues underlying significance testing. We will now go on to refine the ideas introduced here. Throughout the chapter we will measure distance in terms of numbers of standard errors. We will use Student's t-distribution to calculate tail probabilities such as those depicted in Fig. 9.1.4 as a way of determining whether a data estimate is within the usual range of sampling variation that we would expect to see about a *hypothesized* value if it was the true value of the parameter.

9.1.3 The Difference between Tests and Intervals

A confidence interval gives a *range of possible values* for a parameter that are consistent with the data at a given level of confidence. Recall Cavendish's 23 measurements on the mean density of the earth in Example 9.1.2. If we calculate the 95% confidence interval for the unknown population mean μ from this data, we obtain [5.40, 5.57].

In this procedure the level of confidence is fixed at 95%, and we determine that the true mean, μ, is somewhere between 5.40 and 5.57. Every value in the interval is a credible value for the true mean μ at the 95% level of confidence.

In Example 9.1.2 we asked a quite different question of the data. We wished to know whether there was any evidence of bias, that is, whether there was any evidence to show that the population mean, μ, for the process producing Cavendish's measurements was not 5.517. Here, we were not interested in a range of possible values for μ. We were concerned only with *evaluating a single possibility:* whether or not μ could be 5.517. In statistical terminology we speak of "testing the hypothesis that the true value of μ is 5.517."

The parameter of interest in the ESP example, Example 9.1.1, was the probability of correctly identifying a card. We will denote this probability by p. If people were just guessing, then p would be 0.2, so we looked at the data to see whether it provided any evidence against the proposition that the probability of a correct identification was 0.2. We informally tested the hypothesis that the true value of p was 0.2. Again, our interest is not in estimating a range of values for p but in evaluating whether the single specified value of 0.2 was credible in the light of the data. *Significance testing* provides a way of measuring the strength of the evidence that the data provides against a specified or *hypothesized* value of the parameter.

In Example 7.5.2 we compared the mean number of spinal segments blocked in patients who were anesthetized with an epidural injection while sitting with the mean number blocked in patients who were anesthetized while lying down. We calculated a confidence interval for the true difference $\mu_{lying} - \mu_{sitting}$. This true difference is the unknown parameter that we are interested in. We found that at the 95% level of confidence the true difference was somewhere between -0.3 and 3.7 segments. Note again that with this interval the confidence level is fixed at 95%, and we have determined the range of values of the parameter that are consistent with the data (at the 95% level of confidence). Often, however, we want to focus on the question, "*How strong is the evidence* that body position makes a difference to the number of segments blocked?" More simply, "To what extent does the data rule out the possibility that body position makes no difference at all?" In this case the specified, or *hypothesized,* value for the true-difference parameter is 0. We want to test the hypothesis that the true difference is 0. Significance testing will give us a measure of the "strength of the evidence" against this proposition.

Comparing confidence intervals with significance tests

These are different methods for coping with the uncertainty about the true value of a parameter caused by the sampling variation in estimates.

Confidence interval: A fixed level of confidence is chosen. We determine *a range of possible values* for the parameter that are consistent with the data (at the chosen confidence level).

Significance test: *Only one possible value* for the parameter, called the hypothesized value, *is tested.* We determine the *strength of the evidence* provided by the data against the proposition that the hypothesized value is the true value.

The way that we measure the strength of the evidence will be developed from the ideas we saw in Examples 9.1.1 and 9.1.2. But first we will learn some general principles about how to choose an appropriate hypothesis to test.

QUIZ ON SECTION 9.1

1. What intuitive criterion did we use to determine whether the hypothesized parameter value ($p = 0.2$ in the ESP Example 9.1.1 and $\mu = 5.517$ in Example 9.1.2) was credible in the light of the data? (Section 9.1.1)
2. Why was it that $\mu = 5.517$ was credible in Example 9.1.2, whereas $p = 0.2$ was not credible in Example 9.1.1?
3. Why in Section 9.1.2 did we switch to measuring distance between estimate and hypothesized parameter value in terms of numbers of standard errors?
4. What do t-values tell us? (Section 9.1.2)
5. What is the essential difference between the information provided by a confidence interval and by a significance test? (Section 9.1.3)

9.2 WHAT DO WE TEST? TYPES OF HYPOTHESES

9.2.1 We Cannot Prove That a Hypothesized Value Is True

We repeat the following dot plot from Fig. 9.1.1. Each point in the dot plot is a sample proportion of correct answers from the 60,000 guesses of a simulated ESP experiment. The pattern was generated using $p = 0.2$. We have obtained estimates ranging all the way from 0.195 to almost 0.206.

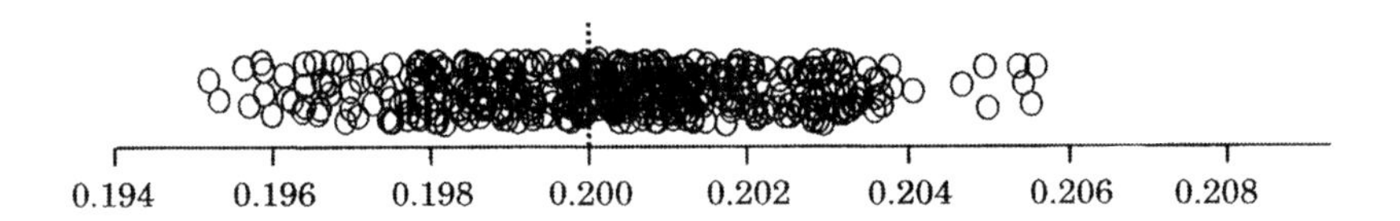

If we had used $p = 0.2001$ the pattern would have been shifted very slightly to the right but would basically have been visually indistinguishable. An estimate from 60,000 guesses is obviously far too variable (or unreliable) for us to be able to tell that the true value of p was 0.2 and could not have been 0.2001. There is far too much uncertainty in the information we can produce to enable us to make such fine distinctions. If we increased the sample size tenfold, the pattern of variation would narrow to about one third as wide as it is here, but we would not be able to tell the difference between 0.2 and 0.20001. In fact, we can never prove that a hypothesized value for a parameter (e.g., $p = 0.2$) is exactly true. Since we can never demonstrate that a hypothesized value for a parameter is the true value, we approach the problem from the opposite vantage point. We determine whether there is compelling evidence *against* a hypothesized value for the parameter. This is the approach we took in both Example 9.1.1 and Example 9.1.2. In the first, there was clear evidence against the value $p = 0.2$ (we could rule it out as a credible possibility). In the second, we could not rule out the possibility that $\mu = 5.517$.

> **Guiding principle**
>
> We cannot rule in a hypothesized value for a parameter, we *can only* determine whether there is evidence *to rule out* a hypothesized value.

Because we cannot rule in and can only rule out such hypotheses, the hypotheses that we test in statistics (called *null hypotheses*) are like straw people. They are proposed in order to see whether we can knock them down.

9.2.2 The Research Hypothesis and the Null Hypothesis

The hypothesis we test using a statistical test is called the *null hypothesis* and is commonly denoted H_0. What null hypotheses can profitably be tested?

Researchers conduct a study because there is something they think or suspect is true about the world that has not previously been established. They are checking out a hunch. This hunch will often be that some experimental intervention will change the way experimental subjects respond or that different groups of subjects differ in important ways. The researchers in an example we will meet shortly (Example 9.3.3) suspected that, on average, the sorts of companies targeted for takeover tend to be those with a worse-than-average return on investors' capital. In many subject areas a hypothesis like this will be stated as a formal *research hypothesis* before the research begins. ***Research hypotheses*** lay out the conjectures that the research is designed to investigate and, if the researchers' hunch proves correct, establish as being true.

The doctors in Example 7.5.2 had a research hypothesis that the patient's body position when the epidural anesthetic was being injected would affect how effectively it worked. When they performed their study, they observed that more spinal segments' nerves were blocked, on average, in the lying group than in the sitting group. There is great person-to-person variability in reactions to anesthetic. If the observed difference in sample means (8.8 − 7.1) often arose when sampling from a population in which body position made no difference at all, the researchers' results would not provide evidence that body position is important. So we test a null hypothesis that says "body position makes no difference," or H_0: $\mu_{lying} - \mu_{sitting} = 0$, and check out whether the difference observed in the data can credibly be explained in terms of sampling variation.

The most commonly tested hypotheses are of the "it makes no difference" variety. We test hypotheses such as there is no difference between the effects of a set of treatments, there is no difference in worker morale between companies who do operate a new management system and those who do not, there are no differences in promotion rates between males and females, or an advertising campaign made no difference to sales. In each case we want to know whether observed differences or effects seen in the data can be explained simply in terms of sampling variation. If they can, nobody is going to be impressed. Thus an important step in showing that differences or effects seen in data are real is to show that they cannot be explained simply in terms of sampling variation. To summarize:

- The chief use of significance testing is to check whether apparent differences or effects seen in data can be explained simply in terms of sampling variation.

- The most commonly tested null hypotheses are of the "it makes no difference" variety.
- Differences seen in data that are easily explainable in terms of sampling variation do not provide convincing evidence that real differences exist.

Generally, the null hypothesis tested encapsulates a skeptical reaction to the hunch or research hypothesis that the study was set up to investigate or confirm. The preceding "it makes no difference" hypotheses are all of this form. The research hypothesis underlying the ESP example is the idea that some people have an ability to receive mentally transmitted information about the shape of an image on a card and thus will be able to do better than chance in identifying cards. The skeptical reaction is, "They were all just guessing." In Example 9.1.1 we informally tested the null hypothesis H_0: $p = 0.2$ (the just-guessing value) by looking to see whether there was information in the data that would rule this value out as a credible possibility. The skeptical reaction to a suggestion that Cavendish's measurement process for the mean density of the earth was biased is, "No, it wasn't. That's just sampling error." So we tested the null hypothesis H_0: $\mu = 5.517$, where 5.517 is the true value.

> The ***null hypothesis*** tested is typically a skeptical reaction to a ***research hypothesis***

9.2.3 The Alternative Hypothesis (H_1)

The *alternative hypothesis,* conventionally denoted by H_1, specifies the type of departure from the null hypothesis we are expecting to detect. Whereas the null hypothesis typically encapsulates the skeptical reaction to the hunch or research hypothesis the study was set up to investigate, the alternative corresponds to the research hypothesis itself. We have to think clearly about the type of data behavior that would constitute evidence against the null hypothesis and confirm the alternative.

In the ESP experiment the research hypothesis was the idea that some subjects could receive information about the identity of the cards through ESP and would thus have a success rate that was better than just guessing. The null hypothesis (skeptical reaction) is H_0: $p = 0.2$. The alternative hypothesis (research hypothesis) is that the true success rate is higher than that, or H_1: $p > 0.2$. The data behavior that would provide evidence against the null and confirm the alternative is a sample success rate sufficiently bigger than 0.2 that it could not be explained away simply in terms of sampling variation.

In Cavendish's measurements on the mean density of the earth, our null hypothesis was H_0: $\mu = 5.517$. We suspected bias, but not bias in any particular direction, so the alternative is simply that the true mean is not 5.517, that is, H_1: $\mu \neq 5.517$. The data behavior that would provide evidence against the null and confirm the alternative is a sample mean sufficiently far from 5.517 (either above or below 5.517) that it could not be explained away simply in terms of sampling variation.

With the experiment comparing the effectiveness of epidural anesthetic administered lying and sitting, the research hypothesis is that body position makes a difference. The null hypothesis comes from the skeptical "it makes no difference" reaction, producing H_0: $\mu_{lying} - \mu_{sitting} = 0$. If the investigators believed from the outset that the lying position was better, the alternative would be that more spinal segments have their nerves blocked on average when the person is lying rather than sitting, that is, H_1: $\mu_{lying} - \mu_{sitting} > 0$. Evidence against the null confirming the alternative would be a sample mean for the lying group that was sufficiently *larger* than that for the sitting group that the difference could not be explained away simply in terms of sampling error. In fact, they had no strong prior belief and used the alternative H_1: $\mu_{lying} - \mu_{sitting} \neq 0$. In the latter case, a difference between sample means in either direction that was sufficiently large would provide evidence against the null hypothesis and in favor of the alternative.

QUIZ ON SECTION 9.2

1. Why can't we prove that a hypothesized value of a parameter is exactly true? (Section 9.2.1)
2. Section 9.2.1 discussed a principle about ruling in and ruling out values, which guides the formulation of a null hypothesis. What was it?
3. Why is significance testing typically used? (Section 9.2.2)
4. What is a null hypothesis?
5. What is the most common form of null hypothesis? Give examples.
6. How does a null hypothesis typically relate to a belief, hunch, or research hypothesis that initiates a study?
7. How can researchers try to demonstrate that effects or differences seen in their data are real?
8. How does the alternative hypothesis typically relate to a belief, hunch, or research hypothesis that initiates a study? (Section 9.2.3)

EXERCISES FOR SECTION 9.2

Each of the following briefly describes a situation. The three problems that go along with each situation are as follows:

(a) What null hypothesis, H_0, would you test?

(b) What should the alternative hypothesis, H_1, be?

(c) What behavior would you be looking for in the data to provide evidence against the null hypothesis and confirm the alternative?

You will need to define your parameters and estimates in each case. All hypotheses can be specified in terms of the value of a mean μ, a proportion p, a difference between means $\mu_1 - \mu_2$, or a difference between proportions $p_1 - p_2$.

1. A consumer advocate suspects that a bottling plant is systematically underfilling its 750-mL wine bottles. A sample of 40 bottles is taken and the volume of wine in each bottle measured.
2. A public health researcher believes that males in white-collar jobs tend to have more stressful lives than males in blue-collar jobs. This should cause white collar workers to have higher blood pressures on average. Samples of 100 men in each category are taken, and their blood pressures measured.
3. A researcher thinks that learning a closely related foreign language should improve the comprehension of English. Sixty students were randomly split into two classes of 30 students. The educational program of both classes was the same except that one class studied French and the other took mathematics. At the end of the year all students took a reading comprehension test.
4. Is a manufacturing process on target? The target value for the diameters of ball bearings is 10 mm. A sample of 50 bearings is taken, and the diameter of each is measured.

5. As for (4), but the proportion of the 50 bearings whose diameter lies above the target value is measured.

6. Is a coin biased? One thousand tosses are made, and the proportion of heads obtained.

7. Are bank customers attracted by a free-giveaway promotion less loyal than those who became customers without inducements? Samples of 1000 of each type of customer are taken and the proportion of each group who are still doing business with the bank five years later is obtained.

8. (Continuation of 7) Do the inducements cost more than they are worth, or do they make money? For each member of the sample of 1000 customers attracted by inducements, the net amount of money earned by the bank from that customer is obtained.

9.3 MEASURING THE EVIDENCE AGAINST A NULL HYPOTHESIS

9.3.1 *t*-Tests and *P*-Values

Significance tests work by comparing a data result with the pattern of sampling variation that would be generated if the hypothesis was true. Evidence against the hypothesis is provided by a data value that is outside the usual range of that variation. We saw examples of reasoning in this way in Examples 9.1.1 and 9.1.2. In both examples the conclusions we should reach were readily apparent. We will often be in situations that are much less clear-cut.

We now discuss a formal procedure for measuring the evidence against H_0 in situations where the sampling distribution of

$$T = \frac{\widehat{\theta} - \theta}{\text{se}(\widehat{\theta})} = \frac{\text{estimator} - \textbf{true} \text{ parameter value}}{\text{standard error}}$$

is approximately Student(df) for some appropriate df [or Normal(0,1), which corresponds to $df = \infty$].The type of test used is a *t-test.* A step-by-step procedure for such a test is given in Table 9.3.1. Figure 9.3.1 is a pictorial version of this table. In fact, in the continuation of Example 9.1.2 (see Section 9.1.2) we essentially followed these steps. We now elaborate on the steps in Table 9.3.1.

We first calculate the *t-test statistic,*

$$t_0 = (\text{estimate} - \text{hypothesized value})/\text{standard error}$$

which tells us how many standard errors the data estimate is from the hypothesized value. Conventionally, Z and z_0 are used in place of T and t_0 for tests involving proportions because the distribution used is the Normal(0,1). We are using the t-notation for both means and proportions, however, to emphasize the sameness of the procedure.

Our measure of evidence against the null hypothesis is called a *P-value.* For the t-test, the P-value is the probability that, if the hypothesis was true, sampling variation would produce an estimate that is further away from the hypothesized value than the estimate we got from our data.[1]

[1] Note that, since the Student(df) distribution is continuous, pr($T \geq t$) and pr($T > t$) are equal.

TABLE 9.3.1 The t-Test[a]

Using $\widehat{\theta}$ to test H_0: $\theta = \theta_0$ versus some alternative H_1.

STEP 1 Calculate the ***test statistic,***

$$t_0 = \frac{\widehat{\theta} - \theta_0}{\text{se}(\widehat{\theta})} = \frac{\text{estimate} - \text{hypothesized value}}{\text{standard error}}$$

[This tells us how many standard errors the estimate is above the hypothesized value (t_0 positive) or below the hypothesized value (t_0 negative).]

STEP 2 Calculate the P-value using the following table.

STEP 3 Interpret the P-value in the context of the data.

Alternative hypothesis	Evidence against H_0: $\theta = \theta_0$ provided by	P-value
H_1: $\theta > \theta_0$	$\widehat{\theta}$ too much bigger than θ_0 (i.e., $\widehat{\theta} - \theta_0$ too large)	$P = \text{pr}(T \geq t_0)$
H_1: $\theta < \theta_0$	$\widehat{\theta}$ too much smaller than θ_0 (i.e., $\widehat{\theta} - \theta_0$ too negative)	$P = \text{pr}(T \leq t_0)$[b]
H_1: $\theta \neq \theta_0$	$\widehat{\theta}$ too far from θ_0 (i.e., $\lvert\widehat{\theta} - \theta_0\rvert$ too large)	$P = 2 \times \text{pr}(T \geq \lvert t_0\rvert)$

where $T \sim \text{Student}(df)$.[a]

[a]For proportions we use Normal(0, 1), which is the same as Student($df = \infty$).

[b]For H_1: $\theta > \theta_0$, if t_0 is positive, the P-value is greater than 50%, and there is no need to do a calculation. If t_0 is negative, $P = \text{pr}(T \leq t_0) = \text{pr}(T \geq \lvert t_0\rvert)$.

TABLE 9.3.2 Interpreting the Size of a P-Value[a]

Approximate size of P-value	Translation
$>$ 0.12 (12%)	**No** evidence against H_0
0.10 (10%)	**Weak** evidence against H_0
0.05 (5%)	**Some** evidence against H_0
0.01 (1%)	**Strong** evidence against H_0
$\leq$ 0.001 (0.1%)	**Very strong** evidence against H_0

[a]These translations are the authors' and are not universally accepted. For further discussion, see the surrounding text.

One-Tailed versus Two-Tailed Tests

The meaning of "further away" in the definition of the P-value depends on the *alternative hypothesis.* We decided in Section 9.2.3 that for the ESP example (Example 9.1.1) the alternative hypothesis should be $H_1: p > 0.2$ and that we would need a data estimate sufficiently *bigger* than 0.2 to provide evidence against the null hypothesis and confirm the alternative. From Table 9.3.1 and its pictorial equivalent, Fig. 9.3.1, the P-value is then $\text{pr}(T \geq t_0)$. A hypothesis of the form H_1: $\theta > \theta_0$ is called a *one-sided hypothesis* and the resulting test is called a *one-tailed test* (because the P-value is the area of a single tail of the distribution).

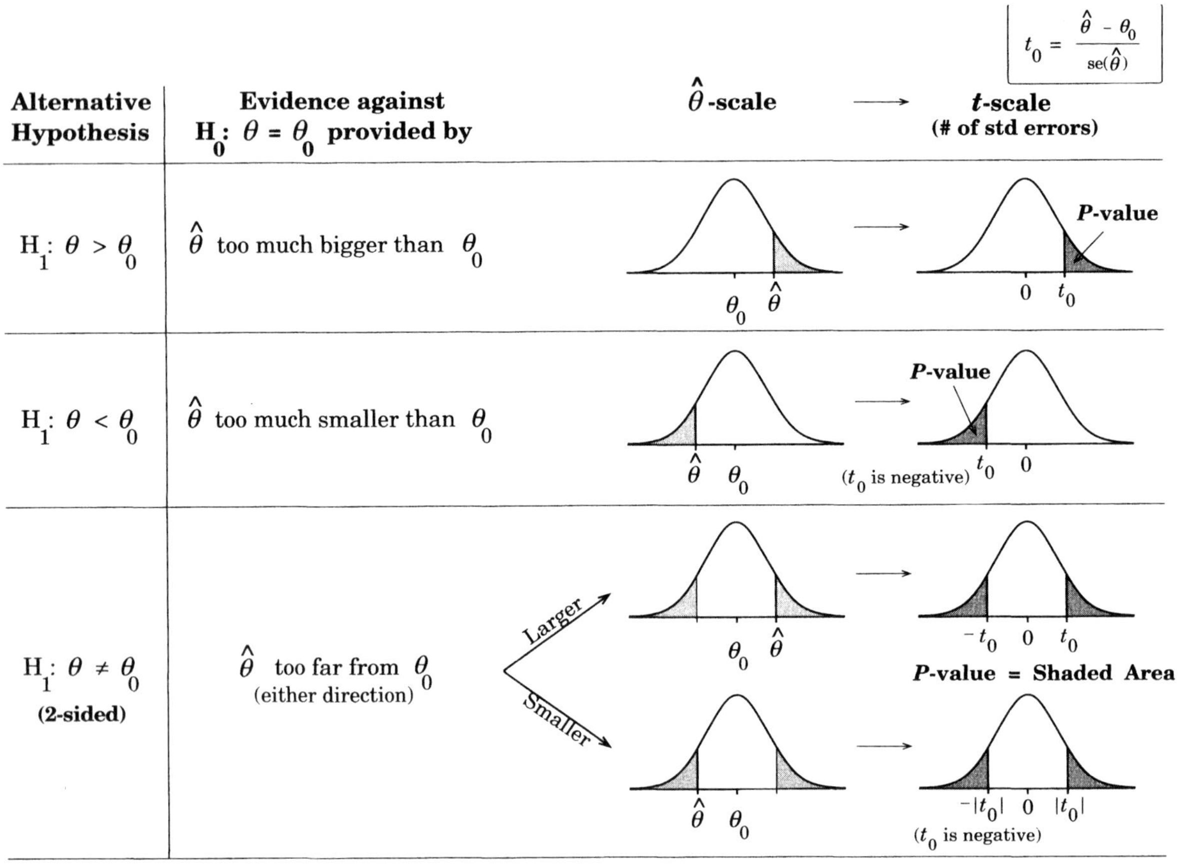

$t_0 = \frac{\hat{\theta} - \theta_0}{se(\hat{\theta})}$

Alternative Hypothesis	Evidence against H_0: $\theta = \theta_0$ provided by	$\hat{\theta}$-scale	→	t-scale (# of std errors)
H_1: $\theta > \theta_0$	$\hat{\theta}$ too much bigger than θ_0	θ_0 $\hat{\theta}$	→	**P-value**; 0 t_0
H_1: $\theta < \theta_0$	$\hat{\theta}$ too much smaller than θ_0	$\hat{\theta}$ θ_0	→	**P-value**; (t_0 is negative) t_0 0
H_1: $\theta \neq \theta_0$ **(2-sided)**	$\hat{\theta}$ too far from θ_0 (either direction)	Larger: θ_0 $\hat{\theta}$	→	$-t_0$ 0 t_0; **P-value = Shaded Area**
		Smaller: $\hat{\theta}$ θ_0	→	$-\lvert t_0 \rvert$ 0 $\lvert t_0 \rvert$ (t_0 is negative)

FIGURE 9.3.1 Testing H_0: $\theta = \theta_0$ (Pictorial version of Table 9.1.1)

We also decided in Section 9.2.3 that for Cavendish's measurements in Example 9.1.2 the alternative hypothesis should be H_1: $\mu \neq 5.517$ and that a data estimate that is sufficiently far from 5.517 *in either direction* would suffice to provide evidence against H_0 and confirm the alternative. The *P*-value is then pr($T \geq |t_0|$), which corresponds to the sum of the tail areas at both ends of the distribution (see Fig. 9.3.1). Here, H_1 is a *two-sided alternative* and the resulting test is called a *two-tailed test.* We use two-tailed tests unless there was a prior (i.e., before the data was collected) research hypothesis or strong prior reason for believing the result should go in a particular direction.

Interpreting *P*-Values

Recall that the *P*-value is the probability, under the assumption that the null hypothesis is true, that sampling variation would produce an estimate that is further away from the hypothesized value than the estimate we obtained from our data. Recall also that we are measuring distance in numbers of standard errors.

Interpreting a large P-value (e.g., > 0.12). Suppose that the *P*-value is large. Then, if the hypothesis was true, sampling variation about the hypothesized value *would often give values that were even further away* from the hypothesized value than our data estimate is (cf. Figs. 9.1.3 and 9.1.4). Thus the data provide no evidence against the hypothesized value being the true value. This was the situation we found ourselves in with Cavendish's measurements in Example 9.1.2. It does not prove that the hypothesized value is correct, only that the data does not rule it out as a credible possibility.

Interpreting a small P-value. Suppose that the *P*-value is small. Then, if the hypothesis was true, sampling variation about the hypothesized value *would almost always give values that were closer to the hypothesized value* than our data estimate. (Figure 9.1.1 is an extreme example.) The smaller the *P*-value, the less plausible the idea becomes that the hypothesized value is the true value of the parameter. Put another way, the smaller the *P*-value, the *stronger the evidence* becomes that the hypothesized value is not the true value of the parameter.

P-values from t-tests

- The ***P-value*** is the probability that, if the hypothesis was true, sampling variation would produce an estimate that is further away from the hypothesized value than our data estimate.
- The ***P-value*** measures the strength of the evidence against H_0.
- The ***smaller*** the *P*-value, the ***stronger*** the evidence against H_0.

(The second and third points are true for significance tests generally, and not just for *t*-tests.)

Table 9.3.2 summarizes the language that this book will use to interpret the size of the *P*-value in the context of the question that originally motivated the study. This form of language emphasizes that (1) a *P*-value is a measure of evidence against the null hypothesis, and (2) the smaller the *P*-value, the stronger the evidence against H_0. Translation of the *P*-values in precisely this way is by no means universally accepted. Other approaches to the interpretation of *P*-values are discussed in Section 9.4. We conclude this subsection by using Examples 9.1.1 and 9.1.2 to illustrate working through the procedure in Table 9.3.1 and the language in Table 9.3.2.[2]

Example 9.3.1 *Obtaining the P-Value for the ESP Example*

This is a continuation of the ESP example, Example 9.1.1. From Section 9.2.3, the null and alternative hypotheses in this example are respectively $H_0\colon p = 0.2$ and $H_1\colon p > 0.2$. Pratt and Woodruffe's sample proportion from $n = 60{,}000$ trials was $\hat{p} = 12{,}489/60{,}000 = 0.20815$, which has standard error $\text{se}(\hat{p}) = \sqrt{\hat{p}(1-\hat{p})/n} = 0.001657$. The test statistic is

$$t_0 = \frac{\text{estimate} - \text{hypothesized value}}{\text{standard error}} = \frac{0.20815 - 0.2}{0.001657} = 4.92$$

The estimate is thus 4.92 standard errors above the hypothesized value. For a proportion from a large sample, we use the Normal(0,1) distribution [same as Student(df =

[2]The *P*-values in our examples are obtained by using a computer program to compute $\text{pr}(T \geq t_0)$, and so on, as in Section 7.6.3. Readers using hand calculators and the Student *t*-tables in this book can use the bracketing methods of Section 7.6.3 to get a rough idea of the size of the *P*-value that will be good enough for interpretive purposes.

$\infty)]$. The one-sided P-value is, from Table 9.3.1,

$$P\text{-value} = \text{pr}(T \geq 4.92) \quad \text{where } T \sim \text{Normal}(0, 1)$$
$$< 10^{-6}$$

With a P-value of less than 1 in a million, we have very strong evidence against H_0, that is, very strong evidence that something other than just guessing was operating.

Example 9.3.2 *Obtaining the P-Value for the Cavendish Data*

This is a continuation of Example 9.1.2. From Section 9.2.3, the null and alternative hypotheses in this example are respectively H_0: $\mu = 5.517$ and H_1: $\mu \neq 5.517$. From Example 9.1.2, the sample mean and the standard error for the $n = 23$ observations were 5.4835 and 0.0397, respectively. The t-test statistic is

$$t_0 = \frac{\text{estimate} - \text{hypothesized value}}{\text{standard error}} = \frac{5.4835 - 5.517}{0.0397} = -0.84426$$

The estimate is thus approximately 0.844 standard errors below the hypothesized value. The P-value should be two-tailed since we have a "$\neq$" alternative (see Table 9.3.1). We obtain, using the Student($df = n - 1 = 22$) distribution,

$$P\text{-value} = 2 \times \text{pr}(T \geq |-0.84426|) = 2 \times \text{pr}(T \geq 0.84426) = 0.408$$

Since the P-value is very large, the data provides no evidence against the hypothesized value of 5.517. In fact, there is not enough information in the data to estimate the true mean at all precisely. A 95% confidence interval for the true mean for the process producing the measurements is given by *sample mean* $\pm$ *t standard errors* with $t = 2.073873$, namely [5.40, 5.57].

QUIZ ON SECTION 9.3.1

1. What does the t-statistic tell us?
2. What is the definition of a P-value?
3. When do we use a two-tailed rather than a one-tailed test?
4. Write down the three types of alternative hypothesis involving the parameter θ and the hypothesized value θ_0. We now lead you through constructing your own Table 9.3.1. For each alternative, think through what would constitute evidence against the hypothesis and in favor of the alternative. Then write down the corresponding P-values in terms of t_0 and represent these P-values on hand-drawn curves (cf. Fig. 9.3.1).
5. What does the P-value measure?
6. What do very small P-values tell us? What do large P-values tell us?
7. Complete the phrase: "The ______ the P-value, the ______ the evidence ______ the null hypothesis."
8. Do large values of t_0 correspond to large or small P-values? Why?
9. What is the relationship between the Student(df) distribution and the Normal(0,1) distribution?

9.3.2 Further Examples

This subsection is composed entirely of examples. In analyzing real data you will almost always be obtaining the quantities here using the automatic capabilities of a computer package. In each example to follow we will be determining the null and alternative hypotheses, working through the t-test procedure given in Table 9.3.1, and then interpreting the resulting P-value using the language of Table 9.3.2. The simple calculations done in these examples are easily performed on a hand calculator. We used our computers as glorified calculators and also to produce the relevant tail areas of Student's t-distribution (as in Section 7.6.3). In Chapter 10 we do virtually all of our analyses using the automatic features commonly found in packages such as Minitab and elaborate on details glossed over here, including the checking of assumptions like Normality and the robustness of methods of analysis when the assumptions are not obeyed.

EXAMPLE 9.3.3 *Do Takeover Targets Give a Lower Return on Investor Capital?*

It is often thought that companies that have been achieving relatively poor returns on shareholders' capital are those most likely to attract takeover bids. The accounting measure *abnormal returns* standardizes the rate of return so that it averages to 0 over all companies. The value of the variable *abnormal returns* for a particular company is the company's return on capital minus the average return for all companies. Thus a company with poorer-than-average return on capital has a negative value of *abnormal returns,* while a company with a good return has a positive value. Researchers Krummer and Hoffmeister hypothesized that *abnormal returns* for companies targeted for takeover by other companies would be negative on average, corresponding to a worse-than-average performance. They collected data on a sample of 88 such takeover targets and reported their results in Krummer and Hoffmeister [1978]. For returns reported prior to the takeover bids, the sample mean value of *abnormal returns* for the 88 takeover targets was -0.0029, and the standard deviation was 0.0169. Does this support the research hypothesis?

Let μ be the unknown true mean for the variable *abnormal returns* for takeover targets. The skeptical reaction to the research hypothesis is, "Those companies are no different from any others." If takeover targets were no different from companies in general, the true mean value of *abnormal returns* for takeover targets would also be 0. Thus we test the null hypothesis H_0: $\mu = 0$ versus the alternative hypothesis (the research hypothesis) H_1: $\mu < 0$, using $\bar{x} = -0.0029$ and $s_X = 0.0169$.

We follow the steps in Table 9.3.1. The standard error of $\bar{x}$ is $\text{se}(\bar{x}) = s_X/\sqrt{n} = 0.0018015$. The t-test statistic is

$$t_0 = \frac{\text{estimate} - \text{hypothesized value}}{\text{standard error}} = \frac{-0.0029 - 0}{0.0018015} = -1.610$$

This tells us that our estimate is 1.61 standard errors below the hypothesized value $\mu_0 = 0$. Because the alternative hypothesis is H_1: $\mu < 0$, we are looking for large negative sample means as evidence against the null hypothesis. The P-value is thus

$$\begin{aligned} P\text{-value} &= \text{pr}(T \leq -1.61) \quad \text{where } T \sim \text{Student}(df = n - 1 = 87) \\ &\approx 0.056 \end{aligned}$$

As the P-value is approximately 5%, this does give us some evidence against H_0 in favor of the research hypothesis that takeover targets have lower values of abnormal returns on average than do companies in general.

EXAMPLE 9.3.4 *Can a Faster Playback Speed for an Ad Increase Recall?*

It is possible to play back a television commercial faster than the speed at which it was filmed without the voice pitches changing perceptibly (which is important unless you want everyone to sound like Donald Duck!). This shortens the time occupied by the advertisement and thus reduces the cost of screening it. It sometimes makes the ad appear more "dynamic." It may have an even more important effect. MacLachlan and Siegel [1980] made a study of the effects of time-compressed commercials. As part of the study a random sample of 57 people was shown an ordinary (30-second) version of an Allstate Insurance commercial, and 15 were able to recall the name two days later (approximately 27%). An independent sample of 74 people was shown the sped-up version (24 seconds), and 32 were able to recall the name two days later (approximately 43%). Is the recall rate for the sped-up version significantly higher?

Let p_S be the population proportion of people who would be able to remember the sped-up version, and p_N be the proportion who would remember the normal version. The parameter of interest is the unknown true difference, $p_S - p_N$. The research hypothesis is that speeding up the ad might change recall levels, but there was no prior belief that it would improve recall. Thus we will test a null hypothesis of no difference in true proportions (H_0: $p_S - p_N = 0$) versus the alternative that a real difference exists (H_1: $p_S - p_N \neq 0$).

Our data is as follows. The sample sizes are $n_S = 74$ and $n_N = 57$. The sample proportions recalling the ad are $\hat{p}_S = 32/74 \approx 0.4324$ and $\hat{p}_N = 15/57 \approx 0.2632$. Our estimate of the true difference in proportions is the difference in sample proportions, $\hat{p}_S - \hat{p}_N \approx 0.16927$. This estimate is subject to sampling error. We will use the large-sample Normal theory even though n_N is a little small (see Appendix A3). Since these two proportions come from independent samples [cf. Fig. 8.5.1(a)], we use the formula in Table 8.5.5(a) to obtain $\text{se}(\hat{p}_S - \hat{p}_N) = 0.0819667$. The test statistic is

$$t_0 = \frac{\text{estimate} - \text{hypothesized value}}{\text{standard error}} = \frac{\hat{p}_S - \hat{p}_N - 0}{\text{se}(\hat{p}_S - \hat{p}_N)} = 2.0652$$

Using the Normal(0,1) distribution, we obtain a two-tailed P-value of

$$P\text{-value} = 2 \times \text{pr}(T \geq 2.0652) = 0.039$$

The small P-value tells us that we have evidence against H_0: $p_S - p_N = 0$. Thus we have evidence that a real difference exists. Inspection of the estimates shows that it is the sped-up version of this particular advertisement that people are more likely to remember. But how much more likely? A 95% confidence interval for the true difference in recall proportions ($p_S - p_N$) is given by [0.01, 0.33], which is not very precise. The true difference may be very small. One percent more of the viewing public being able to remember your ad would probably not make a real impact on the effectiveness of your ad campaign. On the other hand, the true difference may be very large. Having 33% more of the public being able to remember your ad, while lowering the screening cost, would be a very nice bonus!

✖ EXAMPLE 9.3.5 *Are Those Answering Yes More Certain than the No's?*

If people have a strong opinion about a survey question, they will tend to answer quickly. If they do not have a strong prior opinion about the question, they will tend to take longer to think about their response. Researchers sometimes analyze the reciprocals of the judgment or answering times, which they call *certainty values.* La Barbera and MacLachlan [1979] report on a study where, in a telephone poll, a random sample of 143 people were asked: "Assuming it is convenient for you, do you expect to get a swine flu shot?" (There was a swine flu epidemic in parts of the United States at about this time.) We want to see whether the people answering yes were any more or less certain about their responses than those answering no. The summary statistics in Table 9.3.3 were obtained from the certainty values of the responses (units = sec^{-1}).

Using the order that makes the estimated difference positive, the parameter of interest is $\mu_{No} - \mu_{Yes}$, the unknown difference between the population mean certainty values, and we estimate this by the difference between the sample means, $\bar{x}_{No} - \bar{x}_{Yes} = 1.41 - 0.76 = 0.65$. This estimate is subject to sampling error. We want to test H_0: $\mu_{No} - \mu_{Yes} = 0$. Having no particular prior reason to believe that, as a group, the no's should tend to be any more or less certain than the yes's, we perform a two-sided test (i.e., our alternative is H_1: $\mu_{No} - \mu_{Yes} \neq 0$). Since they come from two entirely separate groups of people, $\bar{x}_{No}$ and $\bar{x}_{Yes}$ are independent and (see Section 8.4)

$$\text{se}(\bar{x}_{No} - \bar{x}_{Yes}) = \sqrt{\frac{s_{No}^2}{n_{No}} + \frac{s_{Yes}^2}{n_{Yes}}} = 0.1206475$$

The t-test statistic is

$$t_0 = \frac{\text{estimate} - \text{hypothesized value}}{\text{standard error}} = \frac{0.65 - 0}{0.1206475} = 5.387597$$

Using $df = \text{Min}(n_{No} - 1, n_{Yes} - 1) = 42$, the approximation introduced in Section 8.4, the corresponding two-tailed P-value is

$$P\text{-value} = 2 \times \text{pr}(T > 5.387597) = 0.000003$$

Thus we have very strong evidence against H_0: $\mu_{No} - \mu_{Yes} = 0$, that is, very strong evidence of a real difference. It is clear from the sample means that the "no" group has higher certainty values on average, or in other words, they are answering more quickly. How much bigger is μ_{No} than μ_{Yes}? A 95% confidence interval for $\mu_{No} - \mu_{Yes}$ is given by [0.41, 0.89]. Therefore, with 95% confidence, the mean certainty value in the "no" group is between 0.41 and 0.89 s^{-1} higher than the mean value in the

TABLE 9.3.3 Certainty Values Data

	Answering yes	Answering no
Sample size	100	43
Sample mean	0.76	1.41
Standard deviation	0.50	0.72

Source: La Barbera and MacLachlan [1979].

"yes" group. Translating the certainty values back to times ($certainty = 1/time$), this interval roughly corresponds to yes's answering more slowly on average by between 1.1 and 2.4 seconds.

Question: Can you think of any possible explanations for these results?

EXAMPLE 9.3.6 *Is a Daughter More Likely if the Previous Child Was a Girl than if It Was a Boy?*

We stated in Chapter 4 that there is some evidence that subsequent births tend to be of the same sex. We decided to use local data to confirm a version of this idea. Our research hypothesis concerned the sex of a second child. Our hypothesis was that a mother whose first child was a girl is more likely to have a girl than is a mother whose first child was a boy. To test this research hypothesis, we used the data in Table 9.3.4, drawn from 20 years of records from National Women's Hospital in Auckland. The data is a cross-tabulation from all records in which the mother's first two children were born at the hospital.

TABLE 9.3.4 First and Second Births by Sex

		Second child		
		Male	Female	Total
First child	Male	3,202	2,776	5,978
	Female	2,620	2,792	5,412
	Total	5,822	5,568	11,390

We now reformat the information in Table 9.3.4 so that it better addresses the research hypothesis.

	Second child	
Group	Number of births	Number of girls
1 (previous child was girl)	5412	2792 (approximately 51.6%)
2 (previous child was boy)	5978	2776 (approximately 46.4%)

Recall that we are looking at the sexes of *second children.* Let p_1 be the true proportion of girls in births from mothers whose first child was a girl and p_2 be the true proportion of girls in births from mothers whose first child was a boy. The parameter of interest is the difference between these true proportions, namely $p_1 - p_2$. The null hypothesis we should test is the skeptical reaction to our research hypothesis, namely, that there is no difference in true proportions between the two groups, or H_0: $p_1 - p_2 = 0$. The alternative, taken from the research hypothesis, is that births from the first group are more likely to be girls so that the true difference is positive, or H_1: $p_1 - p_2 > 0$.

Our sample sizes are $n_1 = 5412$ and $n_2 = 5978$. The sample proportions of girls are $\widehat{p}_1 = 2792/5412 \approx 0.5159$ and $\widehat{p}_2 = 2776/5978 \approx 0.4644$. We estimate the difference in true proportions using the difference in sample proportions, namely, $\widehat{p}_1 - \widehat{p}_2 \approx 0.0515$. This estimate is subject to sampling error. The samples are clearly large enough to use large-sample Normal theory. Since the two sample proportions come from separate groups of people [cf. Fig. 8.5.1(a)] they are independent, and we use the formula in Table 8.5.5(a) to obtain $\text{se}(\widehat{p}_1 - \widehat{p}_2) = 0.0093677$. The test statistic is

$$t_0 = \frac{\text{estimate} - \text{hypothesized value}}{\text{standard error}} = \frac{\widehat{p}_1 - \widehat{p}_2 - 0}{\text{se}(\widehat{p}_1 - \widehat{p}_2)} = 5.49986$$

Using the Normal(0,1) distribution, the one-tailed *P*-value is

$$P\text{-value} = \text{pr}(T \geq 5.49986) = 1.9 \times 10^{-8}$$

Thus we have very strong evidence against the null hypothesis and confirming the alternative. Stated positively, we have extremely strong evidence that mothers whose first child was a girl *are* more likely to have a girl the second time than mothers whose first child was a boy. How much more likely? A 95% confidence interval for $p_1 - p_2$ is given by [0.033, 0.070] which tells us that the population percentage of girls born is larger in the former group by between 3 and 7 percentage points than it is in the latter group.

Let us now ask another question.

Is having a boy followed by a girl any more or less common than having a girl followed by a boy? If you look at the off-diagonal counts in Table 9.3.4, you will see that there are more occasions in which a boy is followed by a girl than the other way around. This time we need to think in terms of mothers and the birth orders of their first two children. There are four possible birth orders, namely, {BB, BG, GB, GG}, where "BG" means a boy followed by a girl. We have data from 11,390 mothers represented in Table 9.3.4, and each can fall into any one of the four categories defined by the birth order of her children.[3] Thus our sampling situation corresponds to Fig. 8.5.1(b) (a single sample, several response categories).

We will perform a test for no difference. The test will be two-tailed since we had no idea about the direction in which the difference should lie before looking at the data, that is, we test H_0: $p_{BG} - p_{GB} = 0$ versus H_1: $p_{BG} - p_{GB} \neq 0$. Our sample size is $n = 11{,}390$,

$$\widehat{p}_{BG} = \frac{2776}{11{,}390} = 0.24372 \quad \text{and} \quad \widehat{p}_{GB} = \frac{2620}{11{,}390} = 0.23003$$

Since we have a single sample and several response categories, we use the formula in Table 8.5.5(b) to obtain $\text{se}(\widehat{p}_{BG} - \widehat{p}_{GB}) = 0.006448$. The test statistic is

$$t_0 = \frac{\text{estimate} - \text{hypothesized value}}{\text{standard error}} = \frac{\widehat{p}_{BG} - \widehat{p}_{GB} - 0}{\text{se}(\widehat{p}_{BG} - \widehat{p}_{GB})} = 2.124098$$

[3]When considering birth orders, a single "individual" can be thought of either as a mother who has had (at least) two children or as a unit defined as the first two children from the same mother. We have used the former.

Using the Normal(0,1) distribution, the two-tailed P-value is

$$P\text{-value} = 2 \times \text{pr}(T \geq 2.124098) = 0.034$$

Thus we do have evidence against H_0: $p_{BG} - p_{GB} = 0$ and therefore evidence that a boy followed by a girl is more common than a girl followed by a boy. How much more common? A 95% confidence interval for $p_{BG} - p_{GB}$ is given by $[0.001, 0.026]$. With 95% confidence, the excess percentage of BG over GB is between 0.1 (or essentially 0) and 2.6 percentage points.

A case study applying significance tests to data from research into the effects of differences in survey-question wording is given on our web site.

EXERCISES FOR SECTION 9.3

1. It has been estimated that in western countries between 10% and 15% of people are left-handed. There have been a number of theories about what causes left-handedness. One theory is that "birth stress," including birth conditions such as breech births, prolonged labor, premature and cesarean births, tends to push infants toward left-handedness. According to a 1991 *New Zealand Herald* article, researchers at Queen's University, Belfast, have investigated this theory by looking at babies in the womb using ultrasound. Of 224 fetuses, 12 sucked their left thumbs while the rest favored their right thumbs.
   (a) Does this show that less than 10% of fetuses are left-handed?
   (b) What assumptions are being made here?

2. There is a lot of anecdotal evidence that people often seem to postpone their deaths until after some event of importance to them has passed, be it an 80th birthday celebration, a wedding, the birth of a grandchild, or whatever. To test this theory, Phillips [1972] used the birth dates and death dates of 348 people found in the pages of a book called *400 Notable Americans*. They found that 16 of them died in the month before their birthdays.
   (a) Does this provide evidence that people are less likely to die in the month before their birthday?
   (b) What population is being sampled here? What assumptions are being made in extrapolating the results to people in general?

3. An advertisement for a certain brand of cigarettes claims that there is no more than 18 mg of nicotine per cigarette on average. A test of 12 cigarettes gave a sample mean of 19.1 mg of nicotine with a sample standard deviation of 1.9 mg. Do you think that the claim is true?

4. In 1991, *First for Women* magazine described some Norwegian research that investigated 885 women and their newborn children. Looking at breast-fed infants, the researchers found that among approximately 200 smoking mothers, 40% had infants who suffered from colic, compared with 26% for 400 nonsmoking mothers. (Colic was defined as crying for more than two hours a day at least four days a week.) Are smoking mothers more likely to have colicky babies?

5. Although Easter is the most important festival of the Christian calendar, a 1991 nationwide poll of 1101 adult Britons found that many Britons were ignorant about why Easter is celebrated. Although 85% of those sampled named a Christian denomination as their religion, only 61% were able to say that Christ was crucified on Good Friday. Slightly more, 66%, said that he was resurrected on Easter Sunday. Does this provide evidence that more Britons know that Christ was resurrected on Easter Sunday than know he was crucified on Good Friday?

6. The *Times Higher Education Supplement* (7 October 1988) told the story of Dr. William Epstein, a Washington-based consultant on social policy, who submitted a fictitious article

to 140 social work journals to investigate editorial bias.[4] The article supposedly analyzed the value of social work intervention in the case of a child temporarily separated from its parents in an effort to relieve the symptoms of an illness, which is often psychosomatic. In the version sent to half the journals, he claimed the action of the social workers was beneficial, and in the version sent to the other half he claimed they brought the child no benefit. Of the journals sent the positive version, 53% accepted the paper for publication, compared with only 14% of those sent the version casting doubt on the effectiveness of the social work. Does this provide evidence that editors of social work journals tend to be biased in favor of articles that show that the intervention of social workers is beneficial? What assumption are you making about the way Epstein decided what version of the paper he should send to which journal? What other assumptions are being made?

7. Cuckoos like to lay their eggs in other birds' nests. It has been suggested that the length of an egg laid may be related to the type of nest. In research to investigate this, the lengths (in mm) of 58 eggs laid in hedge-sparrow nests were found to have a mean of 22.6 and a standard deviation of 0.8759, whereas 91 eggs laid in garden-warbler nests had a mean of 21.9 and a standard deviation of 0.7860. What do you conclude? Would a significant difference mean that the suggested theory is true? What other explanations are possible?

9.4 HYPOTHESIS TESTING AS DECISION MAKING

Suppose that whenever we tested a hypothesis and obtained a *P*-value that was less than or equal to 5%, we made a decision to reject the hypothesis (i.e., say that it is false). This procedure is called ***testing at the 5%*** (or 0.05) ***level of significance.*** When H_0 is true, *P*-values less than or equal to 0.05 occur only 5% of the time, so if we operated this policy consistently, in the long run we would reject true hypotheses only 5% of the time, or one time in 20. Similarly, if we consistently tested at the 1% significance level, which means reject H_0 if *P*-value ≤ 0.01, then we would reject only one true hypothesis in 100 on average. Almost all of the time, our decision to reject would be correct. On the other hand, if we never rejected a hypothesis, we would never make an error of this sort! It is an unfortunate fact that if we test at a significance level set very low to prevent rejecting true hypotheses, we tend to end up being unlikely to reject important false hypotheses as well. Recall that the null hypothesis tends to represent a skeptical point of view. Because we do not wish to abandon such a viewpoint rashly, a reasonably low significance level is used. The significance level, be it 0.01 or 0.05, is conventionally denoted by α. We will call the approach to testing introduced in this paragraph *fixed-level hypothesis testing*.

The correspondence between *P*-values and the significance level as a long-run error rate helps improve the understanding of *P*-values and helps calibrate our "measure of evidence." In this book we are not interested in viewing testing as a decision process ("reject" or "do not reject" H_0). We prefer to use testing in the context of simply assessing the strength of the evidence that the data provide against the hypothesis. If decisions really have to be made, they will draw on more than just the evidence in the data set.[5]

[4]For his actions Epstein was charged with unethical conduct, charges that may have seen him expelled from the National Association of Social Workers! Unfortunately, we don't know how this saga ended.

[5]Bayesian decision theory enables the subjective opinions or knowledge of the decision maker to be combined with the data in order to try to make a rational decision. That subject is beyond the scope of this book, however.

In the literatures of many research areas, effects or differences for which the hypothesis test is significant at the 5% level (i.e., P-value ≤ 0.05) are said to be ***significant.*** When the test is not significant at the 5% level, the effect or difference is said to be ***nonsignificant.*** (There are important misconceptions caused by this language, which we discuss in Section 9.5.2.) Widespread use of the 5% significance level originated because R. A. Fisher, the most influential statistician this century, felt intuitively that probabilities less than about one chance in 20 were of about the right order for demonstrating that something unusual was going on. The 5% significance level, together with 10% and 1% significance levels, were then enshrined in tables because most sets of tables have severe space restrictions on the amount of detail about a distribution that can be included. Nowadays data analysts are not constrained by having to use tables. Most practical data analysis is done using computer packages, which automatically provide exact P-values.

Unfortunately, use of the 5% significance level as a litmus test for whether or not results are "significant" has become too rigidly applied in many fields. This can lead to ludicrous anomalies such as treating a P-value of 0.052 differently from a P-value of 0.048. It can be hard to get a study published in some journals without being able to demonstrate significance at the 5% level.[6]

When we view testing as a decision-making process, we can construct Table 9.4.1, contrasting the state of the world with the decision that is made. It is used in almost every discussion of fixed-level significance testing. The *significance level* is defined to be the probability of making a Type I error. If we make the decision to reject a hypothesis whenever the P-value is less than α, the significance level is α. We believe that the accept/reject language reinforces some common misconceptions about testing, and we will not use it. The expression "accept H_0" is particularly dangerous. There are many scientific situations in which one will continue to accept a theory or hypothesis unless evidence has been found against it. In fact, the whole of science tends to proceed in this manner. We want to concentrate just on what the data tell us, however. As we have mentioned before, the data never tell us that the null hypothesis is true. The idea of a Type II error, very slightly modified, is very useful in the planning stages of a study, as we shall see in Section 9.6.

The approach to testing that uses P-values to measure the strength of evidence against H_0 is close to the thinking of R. A. Fisher, the originator of significance testing. Fixed-level hypothesis testing was introduced by J. Neyman and E. S. Pearson in the 1930s. Another approach to statistical inference that should be mentioned in passing

TABLE 9.4.1 Decision Making

	Actual situation	
Decision made	**H_0 is true**	**H_0 is false**
Accept H_0 as true	OK	Type II error
Reject H_0 as false	Type I error	OK

[6]This practice can be dangerous. Significant results can arise simply from sampling variation. For example, there is one chance in 20 that H_0 is rejected given that it is true. In every 20 experiments investigating a *true* H_0, we would expect on average to find one where H_0 was rejected. By only publishing studies that show significant results, journals can give their readers a very biased view of the evidence on questions of interest. It would be much better to publish all well-designed studies of important questions, irrespective of whether or not they gave significant results.

is *subjective Bayesian inference.* Here the analyst's subjective ideas about the size of a parameter are expressed in the form of a probability distribution representing degree of certainty. Bayesian statistical theory tells you how to update this "certainty" distribution to take account of the new information contained in the data. Parameter values can be tested, and interval estimates formed from the resulting distribution. A discussion of the Bayesian approach would require the development of too many new ideas for us to make more of it here.

How one should make inferences from data in an uncertain world is a very deep question. There has been a history of vociferous, and sometimes even bitter, philosophical arguments in the statistical literature about the ways in which knowledge should be abstracted from data. We believe, however, that these philosophical differences have little impact on the practical conclusions that most statisticians would draw from data structures of the form considered in this book. The methods we advocate here are comparatively simple and are those most widely used in practice.

QUIZ ON SECTION 9.4

1. What is the difference between fixed-level hypothesis testing and the *P*-value approach?
2. For fixed-level hypothesis tests, why are low significance levels chosen?
3. If you wanted to perform a fixed-level hypothesis test, for what values of the *P*-value would you "reject the null hypothesis at the 1% level?"
4. Give another name for the significance level.
5. If you test at the 5% significance level, the 5% can be thought of as an error rate. For what type of error is it the error rate?
6. If 100 researchers each independently investigated a *true* hypothesis, how many researchers would you expect to obtain a result that was significant at the 5% level (just by chance)?
7. What was the other type of error described? What was it called? When is the idea useful?
8. Why is the expression "accept the null hypothesis" dangerous?
9. What is meant by the word *nonsignificant* in many research literatures?
10. In fixed-level testing, what is a Type I error? What is a Type II error?

9.5 WHY TESTS SHOULD BE SUPPLEMENTED BY INTERVALS

9.5.1 A Relationship between Tests and Intervals

We have noted that the big difference between confidence intervals and significance tests is that an interval gives *a range of values* that are credible possibilities for the true value of the parameter (θ) *at a given* confidence level, whereas a significance test gives a measure of *the strength of the evidence against a single hypothesized value* (θ_0) being the true value. However, there is an important similarity between the results of a test and the values in a confidence interval, which we now state and will justify shortly.

> A ***two-sided*** test of H_0: $\theta = \theta_0$ is ***significant*** at the 5% level if and only if θ_0 lies ***outside*** a 95% confidence interval for θ.

Thus the data provide evidence at the 5% level of significance against any value θ_0 that lies outside the 95% confidence interval for θ and provides no evidence at the 5% level against any value θ_0 that lies inside the confidence interval. Using fixed-level testing we would formally "reject" H_0: $\theta = \theta_0$ for any θ_0 that lies outside the confidence interval. These ideas are not confined to 95% confidence intervals and 5%-level tests. The same relationship applies to 90% intervals and tests at the 10% level, or 99% intervals and tests at the 1% level.

In complex situations where confidence intervals are hard to obtain but tests of H_0: $\theta = \theta_0$ are manageable, statisticians actually construct a 95% confidence interval for θ by collecting together all values θ_0 for which a test of H_0: $\theta = \theta_0$ does not result in significance at the 5% level. This process is called *inverting* the test.

EXAMPLE 9.5.1 *Relating Test and Interval Results*

This is a continuation of Example 9.3.2. A 95% confidence interval for the true mean, μ, from Cavendish's 23 measurements of the mean density of the earth is [5.40, 5.57]. A formal hypothesis test of H_0: $\mu = \mu_0$ conducted at the 5% level of significance would reject any hypothesized value μ_0 that was bigger than 5.57 or smaller than 5.40. It would not reject any hypothesized value μ_0 that lay in the interval.

We conclude by justifying the result in the preceding box. All probabilities here come from the Student(*df*) distribution. The 95% confidence interval for the true value of θ is given by $\widehat{\theta} \pm t\ \text{se}(\widehat{\theta})$, where t is chosen so that $\text{pr}(T \geq t) = 0.025$. If a value of θ, say, $\theta = 0.3$, lies outside the confidence interval, then we know that 0.3 is greater than t standard errors away from $\widehat{\theta}$. This means that the magnitude of $t_0 = (\widehat{\theta} - 0.3)/\text{se}(\widehat{\theta})$ (i.e., its absolute value $|t_0|$) is greater than t. Therefore a two-sided test of H_0: $\theta = 0.3$ will have $|t_0| > t$, that is, H_0 is rejected at the 5% level of significance (since we multiply $\text{pr}(T \geq t) = 0.025$ by 2). The argument can be reversed so that the two statements in the box are equivalent.

A two-sided test of H_0: $\theta = \theta_0$ gives a result that is significant at the 5% level if $P\text{-value} = 2\,\text{pr}(T > |t_0|)$ is less than 0.05. This occurs if and only if $|t_0| > t$. Now $|t_0|$ tells us how many standard errors $\widehat{\theta}$ and θ_0 are apart (without specifying the direction of the difference). If $|t_0| > t$, then θ_0 is more than t standard errors from $\widehat{\theta}$ and thus lies outside the 95% confidence interval.

9.5.2 Interpreting the Word *Significant*

Women with scleroderma have significantly higher levels of non-self fetal cells circulating in their blood decades after pregnancy than healthy women who have previously given birth.

(From a 1998 press release from the U.S. National Institutes of Health)

Beta blockers have no effect on the risk of death.

(From the draft of a medical paper about heart attack patients)

What do these statements say to you? It is entirely natural to take them literally and imagine that, in the first case, women with scleroderma have substantially more fetal cells circulating in their blood than do healthy women, and that in the second case, a patient will have exactly the same risk of death whether or not he or she takes beta blockers. In fact, we can conclude nothing of the sort from such statements.

Statements like the first are commonly used in writing about research findings, both in technical reports and in the popular media, when a significance test has been applied and a *P*-value smaller than 0.05 has been obtained. Statements like the second are often used when a significance test has been applied and a *P*-value greater than 0.05 has been obtained.

Practical versus Statistical Significance

Using the same pattern as the first statement, we can truthfully say, "Data on all births over a 20-year period at Auckland's National Women's Hospital shows a very significant excess of boys." This sounds serious, almost as if young Auckland males must experience real difficulties in finding girlfriends! The evidence that the population process produces more boys than girls is extremely strong (P-value $< 10^{-6}$). But what about the size of the effect? Among the 92,208 births at National Women's Hospital between 1968 and 1987, 47,824 of the babies were boys. The sample percentage of boys was 51.87%, and a 95% confidence interval for the true percentage of boys produced by the process was [51.5%, 52.2%]. Although the test result was very highly significant, the actual percentage excess of boys is really quite small.

In using the word significant or significance to talk about test results, the statistical and research literatures have taken the ordinary English word *significant* and given it a technical meaning, which is similar but not the same as its usual meaning. This causes confusion. Because most people associate significant with *important,* the reader of the newspaper thinks that "significantly better survival" means that lots of people will live longer, that something of proven *practical significance* has been found. But we have seen that significant in the statistical jargon sense need not mean that at all. It is better to use the term *statistically significant* for technical usage, although even this does not entirely solve the problem.

The statistical significance of an effect relates to the strength of evidence that a real effect *exists,* for example, the evidence that a true difference is not exactly 0. Now, if we have found a treatment that "significantly increases survival rates," it is not enough to know that the increase is not 0—which is all that statistical significance tells you. To know how important the discovery will be in practice, we have to know how big the increase in survival rates is. Speaking generally, the *practical significance* of an effect depends on how large it is.[7]

- ***Statistical significance*** relates to the ***existence*** of an effect.
- The ***practical significance*** of an effect depends on its ***size.***
- A small *P*-value provides evidence that the effect ***exists*** but says ***nothing*** at all about the ***size*** of the effect.
- To estimate the ***size*** of an effect, ***compute a confidence interval.***

[7]One of the authors was an expert witness in a legal dispute concerned with a comparison of fertilizers. A new but more costly fertilizer for increasing grass production was under investigation. Even if the new fertilizer provided a growth rate that was significantly higher in a statistical sense, it would not replace the standard treatment unless the profit from the increased growth exceeded the additional cost of using the new product. Practical significance in this context also involves notions of "cost effectiveness."

Demonstrating statistical significance is like getting to first base. At the state of knowledge at which we are still interested in statistical significance, there is often insufficient information in the data to determine whether the effect is practically significant or not. For example, if the *P*-value from a test for a true *difference* is close to 5%, it follows from the last subsection that one end of the 95% confidence interval will be very close to 0. We now give a slightly more complex example.

EXAMPLE 9.5.2 *Determining Whether an Effect Is Practically Significant*

In the heart study that gave rise to the beta-blocker quotation, it was found that whether or not the patient smoked was significantly related to risk of death (*P*-value = 0.048) among patients who did not undergo surgery. A 95% confidence interval gave the risk for smokers as being between 1.004 times and 2.5 times of the risk for nonsmokers. (Note that when a risk is multiplied by 1, it does not change.) While we do seem to have detected that there *is* an increase in risk, the actual increase could be tiny and thus of little practical importance, or it could be quite large. In fact, the risk may have more than doubled.

One more word of warning. When a press report quotes an increase in risk, it is merely quoting an estimate. It could well be that the margin of error about that estimate is large. We cannot recall ever having seen a newspaper story on health include confidence limits!

Interpreting Nonsignificant Results

We have noted that when people write about research where a test gives a *P*-value greater than 5%, they might say that "the effect/difference is nonsignificant" or "doesn't reach significance." They then might add, very misguidedly, that "there is no effect/difference." In the jargon of fixed-level testing, people often then say that they "*accept* H_0." This has helped to feed the dangerous misconception that a nonsignificant result is evidence that the null hypothesis H_0 is true, a misconception that is harbored even by many researchers who use statistical methods regularly. What can we reasonably conclude from nonsignificant results?[8] All we can say is that the data do not rule out the possibility that H_0 is true. To check what other parameter values, other than that specified by H_0, are consistent with the data, we use *confidence intervals.*

One of the authors was involved in a study that looked at risk factors for death in some period of time following a heart attack. After adjusting for the effects of various measures of the internal viability of the heart muscle, the taking of beta blockers was nonsignificant as a risk factor for death following coronary heart disease. When we tested H_0: "beta blockers make no difference," the *P*-value was 0.35. Most doctors whom we have met would take this as evidence that beta blockers make no difference to a patient's chances of survival. While our data provided no evidence against the possibility that beta blockers make no difference, a 95% confidence interval said that the true risk of death for someone on beta blockers was somewhere between 0.29

[8]We prefer to use the phrase a *nonsignificant result* in the loose sense of a test result in which the *P*-value is sufficiently large that we believe it doesn't provide evidence against H_0. The comments made in this subsection apply whether you prefer to apply the phrase in our loose sense or rigidly as *P*-value > 0.05.

times and 1.56 times the risk for someone who was not taking them. This is very different from saying that beta blockers have no effect. They may be protective and cut the risk by up to a third of what it was before. They might not make any difference. They might also be harmful and increase the risk of death by anywhere up to 56%.

A nonsignificant test does not imply that the null hypothesis is true.

How Can We Prevent People from Making These Mistakes?

To stop the people who read your reports from mistaking the nonsignificance of an effect for its nonexistence, or the statistical significance for practical significance, adopt the simple rule which follows.

Never quote a P-value about the existence of an effect ***without*** also ***providing a confidence interval*** estimating its size.

9.5.3 How Do You Show That a Hypothesis Is True?

Suppose that you really do wish to show that the true value of a parameter θ is θ_0. How do you go about it? We saw in the Section 9.2.1 that the significance testing methodology is quite useless for this purpose. In fact, in the presence of randomness we can ***never*** demonstrate that the true value of θ ***is exactly*** θ_0. The closest we can come to showing that $\theta = \theta_0$ is when all values contained in an appropriate *confidence interval* for θ are *essentially the same as* θ_0 *for all practical purposes.* Here the meaning of "for all practical purposes" is a subject-matter decision, not a statistical one. But as laypeople, we might imagine that if a 95% confidence interval for the headache cure rate of an aspirin-type drug was [39%, 41%], all of the values contained in the interval are 40% for all practical purposes.

A field where this methodology is very useful is called *bio-equivalence.* Name-brand and different generic drugs for the same purpose often have somewhat different "recipes." Regulatory agencies need to be convinced they have the same effect on patients. Similar ideas are explored in Fleming [1992, section titled "Active Control Designs"].

QUIZ ON SECTION 9.5

1. What is the relationship between a 95% confidence interval for a parameter θ and the results of a two-sided test of H_0: $\theta = \theta_0$? (Section 9.5.1)
2. If you read, "research shows that ... is significantly ... than ...," what is a likely explanation? (Section 9.5.2)
3. If you read, "research says that ... makes no difference to ...," what is a likely explanation?
4. Is a significant difference necessarily large or practically important? Why?
5. What is the difference between statistical significance and practical significance?

6. What does a P-value tell us about the size of an effect?
7. What tool do we use to gauge the size of an effect?
8. If we read that a difference between two proportions is nonsignificant, what does this tell us? What does it not tell us?
9. What general strategy can we use to help prevent misconceptions about the meanings of significance and nonsignificance?
10. What is the closest you can get to showing that a hypothesized value is true, and how could you go about it? (Section 9.5.3)

*9.6 DETECTING DIFFERENCES WHEN THEY MATTER

The ideas that follow are useful at the planning stage of an investigation for purposes such as determining an appropriate sample size. We will need to think in terms of fixed-level significance testing, which means saying that the data provide evidence against H_0: $\theta = \theta_0$ if the P-value is smaller than some threshold. We will use the conventional threshold, 5%, and say that we have obtained a significant result if we get $P\text{-value} \leq 0.05$.

Let θ_{true} be the unknown true value of the parameter of interest. Ideally, when $\theta_{true} \neq \theta_0$, we would hope to get a significant result whenever we collected a sample of data. Knowing that this is impossible, we would want the probability of getting a significant result to be as high as possible whenever $\theta_{true} \neq \theta_0$. In terms of Table 9.4.1, we would want the probability of making a Type II error to be as small as possible.[9]

Two ideas that we have already met are important enough to expand upon here.

(i) ***When the sample size is comparatively small,*** the sampling variability of an estimate is typically large. Even when the difference between θ_{true} and θ_0 is reasonably large, we may not be able to reliably detect that the two are different because estimates from small samples are subject to so much sampling error. The probability of getting a significant result can be quite small in such circumstances.

(ii) ***When the sample size is very large,*** the sampling variability of an estimate is typically small. We can reliably detect that $\theta_{true} \neq \theta_0$ in situations where the difference between θ_{true} and θ_0 is moderate to large. We may even be able to detect fairly reliably that $\theta_{true} \neq \theta_0$ in situations where the difference between θ_{true} and θ_0 is reasonably small. The probability of getting a significant result can be quite high in such circumstances.

Some influential funding agencies require sample-size calculations in all applications for funding for research studies. The process involves the following thinking. If θ_{true} is sufficiently close to θ_0, it is not practically important that we know that the two values are different, so first we determine δ, the smallest value of $\theta_{true} - \theta_0$ for which it *is* practically important that a difference be detected. We then assume that

[9]The probability of getting a significant result is called the *power* of the test. The power of the test is = 1 − pr(Type II error).

the true difference is δ and calculate the minimum sample size required so that the probability of getting a significant result is high (e.g., 80%). One rationale for these requirements is the idea that if a study cannot even be relied on to demonstrate *the existence* of a difference between θ_{true} and θ_0 when the true difference is important, the study is probably not worth doing.

Sample-size calculations for significance tests are similar to but more complicated than those for confidence intervals. They suffer from the same difficulties in that they have to use available information about the variability of measurements, and this information tends to be unreliable. New difficulties are introduced; for example, it is often very hard to come up with a sensible and generally acceptable δ. Further information on this topic is given on our web site.

9.7 THE TEST STATISTIC AND THE *P*-VALUE AS GENERAL IDEAS

Until now we have developed the computation of *P*-values to use as a measure of evidence against a null hypothesis only for a very restricted class of problems. We have dealt just with hypotheses just about a single parameter in situations where the sampling distribution of *(estimate − true parameter value)/standard error* is well approximated by a Student(*df*) or Normal(0,1) distribution. We will meet two important statistical tests that are much more complicated than this. The first is the so-called one-way analysis of variance *F*-test in Chapter 10, which tests a hypothesis that says that all members of a set of three or more group means are identical. The second is the Chi-square test in Chapter 11, which tests hypotheses about the set of probabilities that govern how individuals fall into a table of counts. We need to generalize our conception of how tests are developed to accommodate these more complicated tests.

We start with a *test statistic,* which is some sort of *measure of discrepancy* between what we see in data and what we would expect to see if the null hypothesis H_0 was true. With the *t*-tests in this chapter, that measure of discrepancy was the distance between the estimate and hypothesized value expressed in numbers of standard errors. In the hypothesis in Chapter 10, which specifies equality of the true means for each of a set of groups, the discrepancy measure will involve a combined measure of the distances between all the sample means. In the Chi-square test we add terms of the form *(observed count − expected count)*^{2}*/expected count* for all cells in a table.

> A ***test statistic*** is a measure of discrepancy between what we see in data and what we would expect to see if H_0 was true.

As with the *t*-tests, we can determine what the sampling distribution of the test statistic would be if the null hypothesis was true. (This is sometimes called the ***null distribution*** of the test statistic.) We then obtain the *P*-value as the probability that sampling variation alone would produce data that is more discrepant, or further from the expected, than the data set we actually got. Here, "further from the expected" means producing a bigger test statistic.

> The ***P*-value** is the probability, calculated assuming that the null hypothesis is true, that sampling variation alone would produce data that is more discrepant than our data set.

Everything we have learned about interpreting P-values carries over to these more complicated tests. The basic idea is still that if sampling variation alone would often produce data that is even further than the actual data is from what we would expect under the null hypothesis, then the data provides no evidence against the null hypothesis. If sampling variation hardly ever produces such discrepant results, then we do have evidence against the null hypothesis.

9.8 SUMMARY

The chief use of significance testing is to check whether apparent differences or effects seen in data can be explained away simply in terms of sampling variation. The essential difference between confidence intervals and significance tests is as follows:

- *Confidence interval:* A *range* of possible values for the parameter is determined that are consistent with the data at a specified confidence level.
- *Significance test:* Only one possible value for the parameter, called the hypothesized value, is tested. We determine the strength of the evidence provided by the data against the proposition that the hypothesized value is the true value.

9.8.1 Hypotheses

1. The ***null hypothesis,*** denoted by H_0, is the hypothesis tested by the statistical test.
2. *Principle guiding the formulation of null hypotheses:* We cannot rule a hypothesized value in; we can only determine whether there is enough evidence to rule it out.
3. ***Research hypotheses*** lay out the conjectures that the research is designed to investigate and, if the researchers' hunches prove correct, establish as being true.
4. The *null hypothesis* tested is typically a skeptical reaction to the research hypothesis.
5. The most commonly tested null hypotheses are of the "it makes no difference" variety.
6. Researchers try to demonstrate the existence of real treatment or group differences by showing that the idea that there are no real differences is implausible.
7. The ***alternative hypothesis,*** denoted by H_1, specifies the type of departure from the null hypothesis, H_0, that we expect to detect.
8. The *alternative hypothesis* typically corresponds to the research hypothesis.
9. We use one-sided alternatives (using either H_1: $\theta > \theta_0$ or H_1: $\theta < \theta_0$) when the research hypothesis specifies the direction of the effect, or more generally, when the investigators had good grounds for believing the true value of θ was on one particular side of θ_0 before the study began. Otherwise a two-sided alternative, H_1: $\theta \neq \theta_0$, is used.

9.8.2 *P*-Values

1. Differences or effects seen in data that are easily explainable in terms of sampling variation do not provide convincing evidence that real differences or effects exist.
2. The ***P-value*** is the probability that, if the hypothesis was true, sampling variation would produce an estimate that is further away from the hypothesized value than the estimate we got from our data.
3. The *P*-value measures the strength of the evidence against H_0.
4. The *smaller* the *P*-value, the *stronger* the evidence against H_0.
5. A large *P*-value provides no evidence against the null hypothesis.
6. A large *P*-value does *not* imply that the null hypothesis is true.
7. A small *P*-value provides evidence that the effect *exists* but says *nothing* at all about the *size* of the effect.
8. To estimate the ***size*** of an effect, *compute a confidence interval.*
9. Never quote a *P*-value about the existence of an effect without also providing a confidence interval estimating its size.
10. Suggestions for ***verbal translation of P-values*** are given in Table 9.3.2.
11. ***Computation of P-values:*** Computation of *P*-values for situations in which the sampling distribution of $(\widehat{\theta} - \theta_0)/\text{se}(\widehat{\theta})$ is well approximated by a Student(df) distribution or a Normal(0,1) distribution is laid out in Table 9.3.1.

    The *t-test statistic* tells us how many standard errors the estimate is from the hypothesized value.

    Examples given in this chapter concerned means, differences between means, proportions, and differences between proportions. For the corresponding estimates, standard errors, and degrees of freedom, see Table 8.7.1.

    In general, a *test statistic* is a measure of discrepancy between what we see in the data and what we would have expected to see if H_0 was true.

9.8.3 Significance

1. If, whenever we obtain a *P*-value less than or equal to 5%, we make a decision to reject the null hypothesis, this procedure is called *testing at the 5% level of significance.* The *significance level* of such a test is 5%.
2. If the *P*-value $\leq \alpha$, the effect is said to be *significant at the* α *level.*
3. If you always test at the 5% level, you will reject one true null hypothesis in 20 over the long run.
4. A *two-sided* test of H_0: $\theta = \theta_0$ is significant at the 5% level if and only if θ_0 lies *outside* a 95% confidence interval for θ.
5. In reports on research the word *significant* used alone often means "significant at the 5% level" (i.e., *P*-value ≤ 0.05). "Nonsignificant," "does not differ significantly," and even "is no different" often mean *P*-value > 0.05.
6. A nonsignificant result does *not* imply that H_0 is true.
7. A ***Type I error*** is made when one concludes that a true null hypothesis is false. The *significance level* is the probability of making a Type I error.
8. ***Statistical significance*** relates to having evidence of the ***existence*** of an effect.
9. The ***practical significance*** of an effect depends on its ***size.***

REVIEW EXERCISES 9

Recall that *P*-values from significance tests tell us about the strength of evidence for the *existence* of an effect but say nothing at all about its size. To find out what the data say about the *size* of an effect, we compute a confidence interval. These review exercises do not explicitly ask you to perform a hypothesis test or compute a confidence interval. You are expected to work that out for yourself by applying these two simple ideas. The confidence intervals given in our answers to these exercises are all 95% confidence intervals.

Additionally, one or two exercises involving proportions have sample sizes that are a little small for applying the (approximate) large-sample theory of this chapter. Work them out anyway, and just note your concerns about sample size. A starred exercise near the end of the set gives you some idea about how exact theories can be developed to cope with small samples.

1. Allen et al. [1980] investigated the effect of prior phone contact on the response rate to mailed surveys. A control group consisting of a random sample of 836 residents of Malmö, Sweden, were mailed a marketing research questionnaire. The control group produced 186 completed questionnaires in response. The treatment group was made up of another random sample of 239. Trained interviewers contacted these people by telephone, explained the purposes of the survey, and invited them to participate. Of the treatment group, 196 agreed to participate, and 134 actually returned completed questionnaires.

(a) Is there any evidence that the telephone calls make a difference to response rates?

(b) How big a difference, if any, do they make?

(c) Would you trust the results of the market research questions asked in the preceding survey from the treatment (i.e., phone-contact) group to be representative? Justify your answer.

2. Some people in the United States are concerned about what they perceive as a steady erosion of the demands made on the language skills of students by the reading materials used in schools, a process often referred to as "dumbing down." Lendvoy [1996] compared new editions of two of the popular Nancy Drew series of mystery stories with the older editions that they were replacing. We will use only one of the stories, "The Secret at Shadow Ranch." The publisher cited several reasons for the changes, including racial and social stereotypes that may be offensive to today's readers. But were they also "dumbed down?" Two important factors in easier reading are shorter sentences and fewer words with many syllables. Lendvoy sampled 100 sentences from each book and counted *numbers of words per sentence* and also *number of syllables per sentence*. These numbers are given in Table 1.

(a) Is there any evidence that the average number of words per sentence has been reduced in the book as a whole? To what extent, if any, has the number of words per sentence been reduced?

TABLE 1 Nancy Drew Data

	Words per sentence		Syllables per sentence	
	Old version	New version	Old version	New version
Number of sentences	100	100	100	100
Sample mean	15.31	11.88	21.02	16.34
Standard error[a]	0.71	0.65	0.97	0.95

[a] $se(\bar{x}) = s_X/\sqrt{n}$.

Source: Lendvoy [1996].

(b) Repeat (a) using syllables per sentence.

(c) We have said very little about how sentences were sampled. What strategies can you think of for sampling sentences from a book?

3. A common strategy used by manufacturers to encourage consumers to switch to their brand is to distribute coupons offering savings on the purchase price. How cost effective such promotions are depends on whether enough people who have bought the brand because of the coupon will buy it again. A study by Dodson et al. [1978] looked at brand switching and incentives including coupons. For margarine they found that among 23,794 purchases in which no deal was offered, the same brand was bought the next time in 87% of cases. Among 671 purchases in which a coupon was redeemed, the same brand was bought the next time in 49% of cases.

   (a) Is there any evidence that brand loyalty is lower on purchases made after redeeming a coupon?

   (b) Use a 95% confidence interval to estimate any change in brand loyalty (as measured by the difference in probabilities of a repeat purchase) between coupon transactions and no-deal transactions. Write down in words what your interval tells you.

   (c) Among the 671 observations on purchases involving coupons, a reasonable number probably came from people who regularly buy the brand regardless of coupons. The marketing people's real concern was probably about the brand loyalty of new customers attracted by the promotion. Among new customers, will the reduction in brand loyalty be bigger or smaller than that seen for all 671 purchases? Why?

   (d) Is the question of whether new customers attracted by the promotion are more or less loyal than the old customers really the important issue here? What do you see as the central issue, and how would you go about addressing it?

Note: In this exercise we have treated all transactions as independent. A closer reading of the paper revealed that all the data had come from following the spending patterns of 459 families, so the data will include observations on several transactions from each person. Repeat observations on the same person are likely to be positively associated (similar). Thus we do not really have 27,394 independent observations on no-deal transactions, for example. The effective sample sizes could be much smaller. The authors of the original paper appear to have made no correction for this fact.

4. It is generally believed that red-cockaded woodpeckers require or preferentially select old-age pine trees and stands for constructing nest and roost cavities. This hypothesis is important since the need for suitable habitat for nesting and roosting is vital to the continued existence of the species. The information for this exercise came from part of a study conducted by DeLotelle and Epting [1988]. Trees were sampled from stands of trees occupied by colonies of woodpeckers. Trees with current or abandoned woodpecker holes were termed *cavity trees*. Untouched trees in a defined neighborhood of cavity trees were called *colony trees*. The ages (in years) of cavity trees and colony trees were compared, giving the following summary statistics:

Cavity trees: $n = 54$ $\bar{x} = 104.1$ $s_X = 24.1$

Colony trees: $n = 143$ $\bar{x} = 83.6$ $s_X = 38.3$

   (a) Is there evidence that cavity trees are older than colony trees on average?

   (b) Why can we not immediately conclude that cavity trees being older implies that woodpeckers prefer older trees? Can you think of some other way of investigating this question?

   (c) The technique used in (a) assumed independent samples. What problems, if any, can you see with the independence assumption?

5. Sudman and Ferber [1974] investigated the effects of a $5 payment on the proportion of people who were prepared to cooperate with marketing surveys. The treatment group of

111 were offered the payment for cooperation, and 79.3% cooperated in full. A control group of 116 were not offered the payment, and 67.2% cooperated in full.

(a) Is there any evidence that the payment makes any difference to cooperation levels?

(b) How big is any difference likely to be?

(c) What issues do you see as being important in deciding whether the payments are cost effective?

TABLE 2 Marijuana Usage and Schizophrenia

Consumption (no. of occasions)	Number of soldiers	Number of cases of schizophrenia
0	41,280	197
1–10	2,836	18
11–50	702	10
>50	752	21

Source: Andréasson et al. [1987].

6. Andréasson et al. [1987] investigated links between marijuana use and the subsequent development of schizophrenia. A total of 45,570 Swedish soldiers were questioned about the number of occasions on which they had used marijuana at the time of their induction into military service in 1969–70. Subsequent cases of schizophrenia over a 15-year period were traced through Sweden's national psychiatric register. Table 2 gives a break down of schizophrenia cases by reported marijuana use.

(a) Is there evidence of a difference in the proportions of soldiers developing schizophrenia between the no-marijuana-use group and the group who had used it 1–10 times? How big is any such difference?

(b) Is there evidence of a difference in the proportions of soldiers developing schizophrenia between the no-marijuana-use group and the group who had used it 11 or more times? How big is any such difference? [Note: We have pooled the 11–50 and >50 groups because of the small numbers of cases.]

(c) Does the data give us any proof that marijuana use causes schizophrenia? Justify your answer. Can you think of any alternative explanations?

(d) Can you see any difficulties with the type of study conducted here as a way of investigating any relationship between marijuana use and subsequent development of schizophrenia?

Note: The significant increase in schizophrenia risk with marijuana use was still present under sophisticated analyses that allowed for imbalances between the usage groups on such factors as previous psychiatric diagnosis and social factors such as divorced parents.

7. For quite a long time now, North American banks have used free gift promotions (or heavily discounted items) to attract new account holders. An important factor in working out the cost effectiveness of such promotions is how long people retain accounts attracted in this way. One would expect such customers to be less loyal. Preston et al. [1978] conducted a study on two Midwestern banks in which heavily discounted calculators or cookware were used as an inducement to open accounts. Two hundred accounts were randomly sampled from those that were opened during the first two weeks of promotions and were compared with a random sample of 200 accounts from those opened in the two weeks prior to the promotion. Of the accounts opened prior to the promotion, 178 were still open six months later compared with 158 of those opened during a promotion.

(a) Do these data confirm our suspicion that customers attracted by promotions will be less loyal?

(b) If you wanted to evaluate the cost effectiveness of the promotional campaign, what other types of information (other than information on proportion of accounts retained) would you want to obtain?

TABLE 3 Attractiveness and Similarity

	Attraction ratings			Perceived similarity ratings		
	Dissimilar	No info.	Similar	Dissimilar	No info.	Similar
n	39	43	40	39	43	40
$\bar{x}$	4.64	6.20	6.36	2.28	5.98	6.60
s_X	1.33	1.38	1.28	1.15	1.52	1.37

Source: Hoyle [1993].

8. Hoyle [1993] investigated the effect of prior information on a person's reaction to a stranger after meeting him or her. A sample of 122 university students were divided into three groups and either provided with no explicit information about a stranger, or given information indicating that the stranger was attitudinally similar or dissimilar to them. After contact, subjects rated their attitudinal similarity to the stranger and their attraction toward them, both on a 1-to-9 scale. The results are summarized in Table 3.

(a) Is it plausible that the true mean attraction rating is the same for

(i) those given no information and those told the stranger had similar attitudes?

(ii) those given no information and those told the stranger was attitudinally dissimilar?

(b) Repeat (a) for the perceived similarity ratings.

(c) How large are the differences between the true means in (a)(i), (a)(ii), (b)(i), and (b)(ii)?

(d) These data were interpreted as supporting the theory that the relationship between attitudinal similarity and attraction can be attributed to repulsion invoked by *dissimilarity* rather than attraction invoked by similarity. Is this reasonable?

(e) How would you have wanted to conduct this experiment?

(f) What follow-up investigations would you like to see done?

9. Abbott et al. [1986] looked at the relationship between cigarette smoking and stroke. Investigators at the Honolulu Heart Program followed up a cohort of 8006 men of Japanese ancestry enrolled during 1965 to 1968. Of subjects who had not had a stroke at study entry, 3435 were cigarette smokers and 4437 were nonsmokers. Over 12 years of follow-up, 171 smokers and 117 nonsmokers had strokes.

(a) Does this provide any real evidence of an association between smoking and having a stroke?

(b) How big is the increase (or decrease) in risk of having a stroke over the 12-year period for smokers?

10. A market researcher interested in the relative effectiveness of various advertising media conducted a survey in Auckland, New Zealand, in 1985. Five hundred randomly chosen washing machine purchasers were contacted and asked what advertising medium had the greatest influence on their choice of machine. Their responses were as follows:

Television	Word of mouth	Magazines (incl. newspapers)	Brochures	Other
105	171	90	85	49

(a) Does this provide evidence that television advertising was more influential than magazine advertising with respect to choice of washing machine in Auckland in 1985?

(b) What factors are involved in deciding the mode of advertising in which you should put most of your resources?

(c) From the raw figures, word of mouth seems to be the most influential source. How could you improve your word-of-mouth advertising?

TABLE 4 pH Measurements

Morning patients	$n_1 = 50$	$\bar{x}_1 = 3.94$	$s_1 = 2.51$
Afternoon patients	$n_2 = 49$	$\bar{x}_2 = 2.93$	$s_2 = 2.39$

11. Surgery patients are fasted, that is, not allowed food for some period of time, before they are anesthetized. This is done to reduce the risk of patients inhaling their stomach contents. If this happens, they may develop a severe pneumonia-like condition called Mendelson's syndrome, which often results in death. The risk of getting Mendelson's syndrome is increased if the inhaled stomach contents are particularly acidic (low pH).

A study at Auckland's Green Lane Hospital led by Dr. Alan Merry looked at the effect of length of fast on stomach pH levels. Patients operated on in the *morning* have shorter fasts than those operated on in the *afternoon*. An interesting question is, "Does the additional fasting change average stomach pH levels?" The data from the study are given in Table 4.

(a) Is there any evidence of a difference between mean pH levels for morning and afternoon patients? How big is any such difference?

It is widely believed that the risk of Mendelson's syndrome is increased if the inhaled stomach contents have a pH lower than 2.5. In this study 21 of the 50 morning patients and 31 of the 49 afternoon patients had a stomach pH below 2.5.

(b) Is the proportion of afternoon patients with a pH below 2.5 significantly larger than the proportion of morning patients with a pH below 2.5? How big is any difference in the underlying proportions?

(c) Entry to the morning or afternoon operating lists was not randomized but was haphazard. Discuss the implications of this.

12. Vasil [1987] described a 1968 study, carried out by the California Department of Corrections, in which 7712 parolees were classified as being "potentially aggressive" or "less aggressive" on the basis of offender histories and psychiatric reports. Of the 20% classified as being potentially aggressive, 3.1 per thousand were reconvicted for violent offenses committed within one year after release from prison. For the 80% classified less aggressive, the rate was 2.8 per thousand. The difference between the rates for the two groups looks very small. Is the difference statistically significant? Do you think it is practically significant? What are the practical implications of this?

13. This exercise concerns practical statistical issues involved in studying the prediction of reoffending. Vasil [1987] surveyed studies in which psychologists tried to classify prisoners as being at high or low risk of reoffending. In one such study 31% of low-risk offenders reoffended, compared with 41% of high-risk offenders, over a three-year period. Comment on any implications of these figures.

(a) Comment on any implications of these figures.

Only some of the prisoners classified high risk were released by the parole board (for obvious reasons), however, so the preceding comparison necessarily involved comparing reconviction rates for low-risk prisoners with reconviction rates for that subset of high-risk prisoners who were released by the parole board.

(b) What sort of bias would you predict would be introduced by the mechanism we have described?

In many of the studies at least two thirds of those predicted to be at high risk have not reoffended. There are worries about civil-liberty implications of not releasing prisoners on the strengths of predictions that "vastly overpredict dangernousness." Another criticism is that follow-up times have been too short.

(c) What would be the effect on the figures of longer follow-up times?

14. (a) In Western Australia in the early 1990s, approximately 28% of the adult population smoked. A statewide survey studying participation in sport contained 130 members of football clubs. Of these, 32% smoked. Does this provide evidence that football club members are more likely to smoke than the general population?

(b) The survey questioned 2600 people. Nonparticipation in sporting or outdoor recreational activities was more prevalent among males (11.6%) than females (8.9%). Could this just be due to sampling variation? Assuming 1300 males and 1300 females, test for a difference between female and male nonparticipation rates. Estimate the size of any such difference.

(c) Among the 1200 people sampled who lived in rural areas, 47% belonged to a sports or outdoor recreation club, versus 31% for the 1400 people sampled who lived in urban areas. Is there any evidence of a difference in membership rates between rural and urban areas? How big is any difference?

15. In the 1998 exit polls following the 1998 U.S. midterm elections, 42% of the approximately 100 voters who classified themselves as Asian voted Republican versus 35% of the approximately 500 who classified themselves as Hispanic. Is there any evidence that Asian Americans are more likely to vote Republican than Hispanic Americans?

16. This question refers to the data and story in Review Exercises 8, problem 2. Use two-tailed significance tests to investigate any differences in mean team-building scores between the following groups.

(a) Enthusiastic volunteers and reluctant volunteers

(b) Enthusiastic volunteers and nonvolunteers

(c) Reluctant volunteers and nonvolunteers

Before you begin, try to guess the approximate size of the *P*-values you will get from the confidence intervals given in Review Exercises 8, problem 2.

17. The questions to follow refer to the data and story in Review Exercises 8, problem 12.

(a) Is there any evidence of a difference between the proportion of young women living as single parents who are stressed by their relationships with their parents and the corresponding proportion for young women living alone?

(b) Among young women living as single parents, is there any significant difference between the proportion who smoke and the proportion who report unhealthy eating practices?

(c) Is there any significant difference between the proportion of young women living with a partner and child(ren) falling into the underweight category and the proportion falling into the overweight category?

18. This is a question about interpreting statements similar to those you might meet in news reports and the scientific literature. In the following, a series of claims are made. If you think a claim is valid, write "true." If you think that it is false or not supported by the data, write "false" and very briefly give a reason.

(a) CLAIM: The phrase, "a test for a difference (in means) between two groups" refers to a test of H_0: $\mu_1 = \mu_2$ versus some alternative.

(b) ***Scenario:*** The average examination score for the 1200 male students in a large university course is 62.1 with a standard deviation of 13. The average mark for the 1200 females is 63.8 with a standard deviation of 15. A researcher checking for a sex difference obtained a P-value of about 0.01.

(i) CLAIM: Female exam scores are significantly higher than male exam scores.

(ii) CLAIM: In general, females do much better than males. The course is discriminatory.

(c) ***Scenario:*** In an experiment conducted by a large company to test whether salespeople sell more when on commission than when employed on fixed salaries, a group of 16 newly hired salespeople was randomly split into two groups of size 8. One group was put on commission, and the other on fixed salary for a six-month probationary period. A t-test comparing dollar sales figures gave a P-value of greater than 20%.

(i) CLAIM: The data showed no significant difference in sales between salespeople on commission and those on fixed salaries.

(ii) CLAIM: Commissions have no effect on sales performance.

(d) (i) CLAIM: If there is no such thing as ESP and 100 experimenters independently did experiments to test a "just-guessing" hypothesis, we would expect about 5 of them to get significant results (P-values below 5%).

(ii) CLAIM: The P-value tells us nothing about the size of an effect.

(e) CLAIM: If the P-value from testing for a difference in sales figures between commissioned and salaried salespeople in (c) had been 8%, we could interpret this as an 8% probability that chance caused the observed difference in sales and a 92% probability that the difference was real.

(f) CLAIM: A test of H_0: $\mu_1 = \mu_2$ is significant at the 5% level if the P-value for the test is less than 5%.

(g) CLAIM: A two-sided test of H_0: $\mu_1 = \mu_2$ is significant at the 5% level if and only if 0 does not lie within a 95% confidence interval for $\mu_1 - \mu_2$.

(h) The following claims relate to the scenario of simple random sampling from a human population to estimate a population proportion.

(i) CLAIM: For a random sample the expected value of the sample proportion is the population proportion, p, and the standard deviation is $\sqrt{p(1-p)/n}$.

(ii) CLAIM: By taking a large enough sample we can make the standard deviation of the sample proportion as small as we like.

(iii) CLAIM: In real life a survey sampler with a huge budget can estimate a population proportion almost perfectly accurately by taking a large enough sample.

19. In a light-hearted piece, staff at *Consumer* magazine conducted a test of people's ability to discriminate between butter and margarine (August 1993). A trial consisted of three pieces of bread being presented to a volunteer, two covered with the same spread and one with the other. The subject had to try to identify by taste which slice of bread was different. With random order of presentation and random choice of the spread (butter or margarine) to be used twice, 9 volunteers attempted the identification on 3 different occasions for a total of 27 trials. "Of the 27 results, 16 were right and 11 were wrong. This is not statistically significant."

(a) What would the probability of a correct identification be, for any trial, if all subjects had no ability to discriminate between butter and margarine and thus were just guessing?

(b) Obtain a one-tailed P-value for the just-guessing hypothesis using the standard large-sample methods presented in this chapter. (We will do it another way in a subsequent problem.)

(c) Is there any evidence that the subjects did better than just guessing? Are the magazine's conclusions correct?

(d) What practical problems do you see in carrying out this experiment and how would you overcome them?

(e) Why did *Consumer* get it wrong? What other value for p_0 could someone easily but mistakenly use to encapsulate "just guessing" for this experiment?

*20. ***Exact tests for a proportion.*** Suppose that we wish to test $H_0: p = p_0$ versus $H_1: p > p_0$ and obtain an observed proportion from our data of $\widehat{p}_{obs}$ that is bigger than the hypothesized value p_0. The *P*-value for the exact test is the probability that sampling variation would produce a value of $\widehat{p}$ at least as big as $\widehat{p}_{obs}$ if H_0 was true. Consider the Binomial (biased-coin-tossing) model. Let $\widehat{P}$ be the proportion of heads in n tosses and Y be the number of heads in the n tosses. Then $\widehat{P} = Y/n$, and

$$\begin{aligned} P\text{-value} &= \text{pr}(\widehat{P} \geq \widehat{p}_{obs}) \qquad \text{calculated assuming } H_0 \text{ true} \\ &= \text{pr}(Y \geq y_{obs}) \qquad \text{where } Y \sim \text{Binomial}(n, p_0) \end{aligned}$$

and $y_{obs} = n\widehat{p}_{obs}$ is the number of heads observed in the data. Thus we can use the Binomial distribution to calculate the *P*-value exactly, and this is what we do in small samples.

(a) We shall apply this idea to the ESP experiment (Example 9.1.1). Suppose one subject obtained 8 correct identifications out of 20 attempts. What is the exact *P*-value for a test of the just-guessing hypothesis $H_0: p = 0.2$ versus $H_1: p > 0.2$ provided by these data?

(b) Obtain the exact *P*-value for problem 19(b). Compare your answer with the *P*-value obtained from large-sample theory in problem 16(b). Do your conclusions change?

21. (Computer simulation exercise) These simulations are intended are intended to show something of the way *t*-tests operate in repeated sampling. We will use the context of Cavendish's experiment for estimating the mean density of the earth. The true mean density is 5.517, but our experiment is not necessarily measuring the true value because there may be a systematic bias. The true mean underlying our experiment has some value μ_{expt}. The variability of our experiment is such that observations have standard deviation 0.19. We will simulate performing experiments of n observations and testing H_0: $\mu = 5.517$.

For parts (a) and (b) we assume that the experiment is unbiased so that $\mu_{expt} = 5.517$.

(a) Generate 100 samples of size $n = 10$. For each sample, obtain the *t*-statistic and the *P*-value for testing H_0: $\mu = 5.517$. Obtain a histogram of the *t*-test statistics. What proportion of *P*-values would you expect to be less than 0.05? Why? Check using your data. Plot a histogram of the *P*-values. What do you see and what does this tell you?

(b) Repeat (a) using samples of size $n = 40$. What has changed from (a)? What do the results tell you?

Suppose now that the experiment is slightly biased and that $\mu_{expt} = 5.45$. Again, we will test H_0: $\mu = 5.517$.

(c) Repeat both (a) and (b) using $\mu_{expt} = 5.45$. What has changed? What have you learned from the results?

(d) Repeat both (a) and (b), but using $\mu_{expt} = 5.3$. What has changed? What have you learned from the results?

CHAPTER 10

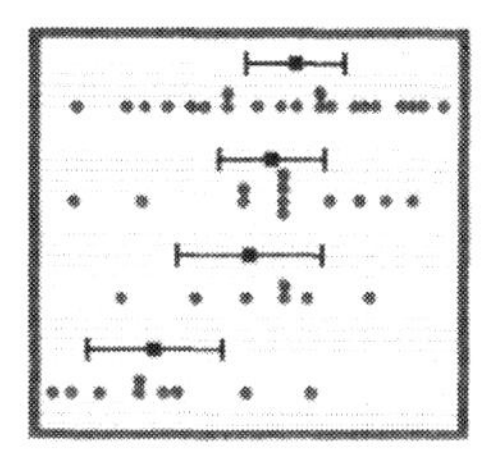

Data on a Continuous Variable

Chapter Overview

This chapter discusses methods for analyzing data on a continuous variable. It brings together material from previous chapters on plotting methods (Chapters 2 and 3), the tests, and making inferences (Chapters 8 and 9). This is done to encourage you to take an integrated approach to data analysis. In addition, underlying models and their assumptions are examined together with the consequences of violating those assumptions.

In Section 10.1 we concentrate on one-sample problems and discuss when t-procedures can be safely used. Included in this topic is the method of paired comparisons as it reduces to a one-sample problem. One way of coping with any violation of the underlying assumptions is to use a so-called *nonparametric* (or distribution-free) approach. Several nonparametric methods are touched on in the chapter, but we explore only the *sign test* in any detail. Section 10.2 discusses methods of comparing two independent samples and begins by raising the issue of an observational study versus an experiment. Some alternative t-methods are introduced for comparing two means. More than two independent samples from different groups or populations are the subject of Section 10.3, and the term *one-way analysis of variance* is introduced. The F-test for comparing the means is developed, and the underlying assumptions on which the test is based are discussed in detail. In Section 10.4 we discuss some aspects of study design, such as using pairing, blocking, and stratification as a means of getting more precise answers to research questions.

10.1
ONE-SAMPLE ISSUES

10.1.1 Tests and Confidence Intervals for a Mean

One message you had from Chapter 9 is that a t-test for a hypothesis about a mean will show any statistically significant departure from the null hypothesis (i.e., where we have evidence that the hypothesized value is not the true value), whereas a confidence interval will tell us about the *magnitude* of the departure. These two methods should always go hand in hand. However, the burning question is, "When can we use the t-tests and intervals?" We have seen that these are justified by theory when the underlying sample comes from a Normal distribution, but that if the sample size n is large, we generally do not need the assumption of Normality because the central limit theorem ensures that $T = (\overline{X} - \mu)/\text{se}(\overline{X})$ is Normal(0,1), which in turn is approximately Student($df = n - 1$). If n is small, however, how critical is the Normality assumption? We shall look at t-based methods in the example to follow, which we will then use to investigate this question.

✖ Example 10.1.1 *A t-Test and Confidence Interval for a Small Data Set*

The following data set consists of 10 attempts to measure the nitrate ion concentration (in μg/mL) in a specimen of water.

0.513 0.524 0.529 0.481 0.492 0.499 0.518 0.490 0.494 0.501

Dotplot of Concentration

(with Ho and 95% t-confidence interval for the mean)

Ho

$\overline{X}$

Hypothesized value

0.48 0.49 0.50 0.51 0.52 0.53

Concentration

T-Test of the Mean

se($\bar{x}$) t_0 *P-value*

```
Test of mu = 0.49200 vs mu not = 0.49200
Variable      N      Mean     StDev   SE Mean      T        P
Concentr     10   0.50410   0.01600   0.00506   2.39    0.040
```

T Confidence Intervals

```
Variable      N      Mean     StDev   SE Mean       95.0 % CI
concentr     10   0.50410   0.01600   0.00506 (0.49265, 0.51555)
```

FIGURE 10.1.1 Minitab output for the nitrate ion concentration data.

Each data point arises from taking a small sample of the water and subjecting it to chemical analysis. There was concern that the concentration had changed from the desired concentration of 0.492. If there are no biases in the measurement procedure, the population mean, μ, is the true nitrate ion concentration of the water.

Figure 10.1.1 gives annotated output from the Minitab package. We filled out a dialog box located under `Stat` → `Basic Statistics` → `1-Sample t` and asked for a dot plot, a test for the null hypothesis H_0: $\mu = 0.492$ versus the alternative H_1: $\mu \neq 0.492$, and a confidence interval. Other packages give similar output. (The relevant command in R and Splus is `t.test`.) The packages allow the user to choose the form of the alternative hypothesis (in Minitab `not equal`, `less than`, or `greater than`).

There is nothing in the dot plot in Fig. 10.1.1 to suggest that the data *couldn't* have come from a Normal distribution. Since the *P*-value is small (0.04), we have evidence that the concentration has changed from 0.492. With 95% confidence, the true concentration lies between 0.493 and 0.516. Note that the hypothesized value is very close to the left-hand endpoint of the confidence interval, so the change may have been small.

Can we trust this analysis from such a small sample? We do not believe that data is ever exactly Normally distributed, and even if it is, we shall never *know* whether it is exactly Normal or not. Fortunately, *t*-tests and *t*-intervals can be used for small samples because they are surprisingly *robust* against the Normality assumption. What does the word *robust* mean here? It means that if we apply the methods when sampling from some distribution other than the Normal, they still "work." For a 95% confidence interval formula, "working" means that the resulting intervals cover the true μ for approximately 95% of samples taken. The robustness of tests is usually investigated for testing at the conventional 5% and 1% levels. "Working" there means that, when H_0: $\mu = \mu_0$ is true, we get a result that is "significant at the 5% (or 1%) level" for approximately 5% (or 1%) of samples taken.

The behavior of *t*-tests and *t*-intervals has been investigated using computer simulation and also mathematically. For example, we conducted simulations like those depicted in Fig. 8.1.2 for the distribution shown below [Chi-square(df = 4); see Chapter 11]. We took 10,000 samples for each of 4 sample sizes and obtained the following frequencies for the percentage of times the *t*-interval covered the population mean.

Sample Size	6	8	10	15
Coverage (%)	92	92	93	94

Even though the distribution is very badly skewed, the intervals are performing well for very small sample sizes. Studies such as this have shown that *t*-tests and intervals are robust against all but severe non-Normality.[1] One type of non-Normal behavior that *t*-tests and intervals are generally not robust against is the presence of outliers. This should not be surprising, as we saw in Chapter 2 that both $\bar{x}$ and s_X were sensitive to outliers.

---

[1] See, for example, Posten [1979].

✱ Example 10.1.1 (cont.) *The Effect of an Outlier on a t-Test and Interval*

We consider the effect of shifting the largest data point (0.529) in Example 10.1.1 to the right (0.581) so that it becomes an obvious outlier.

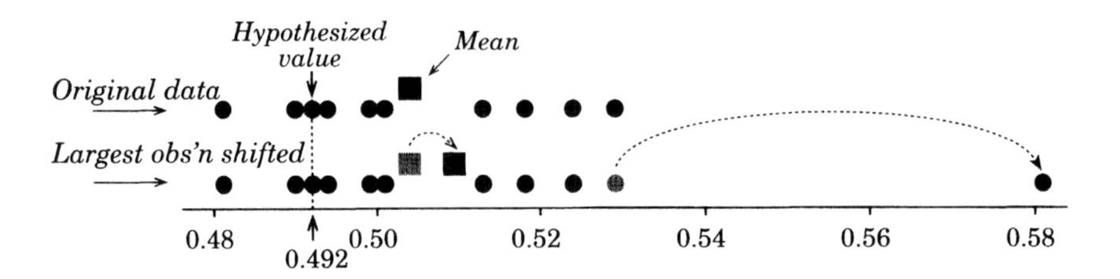

Note how the sample mean has shifted to the right, away from the hypothesized value. We would then perhaps expect the null hypothesis to be even more firmly rejected, but the opposite is true! This is shown in Fig. 10.1.2, which also includes edited computer output for the data set with the outlier. The *P*-value is actually increased. This is caused by the increase in the sample standard deviation (from 0.0160 to 0.0285) and the consequent widening of the confidence interval.

We have seen how an outlier can have a considerable effect on our inferences. In fact, if an outlier is present, we can no longer trust *t*-tests and intervals. We shall see shortly that there are tests for Normality. Such tests detected non-Normality in the Example 10.1.1 data that contained the outlier. What about the effect of the shape of the underlying distribution? Fortunately, two-sided *t*-tests and confidence intervals can tolerate a great deal of skewness in the distribution, even in quite small samples, but one-sided *t*-tests can be badly affected by pronounced skewness in samples smaller than about $n = 40$; the smaller the sample, the smaller the degree of skewness that can be tolerated. The procedures are also badly affected by distributions with very heavy tails, that is, distributions that, like the Normal, still have tails going off to infinity in both directions but have more area out in the distant tails. Such distributions tend to occasionally produce outlying observations that are not mistakes but instead are

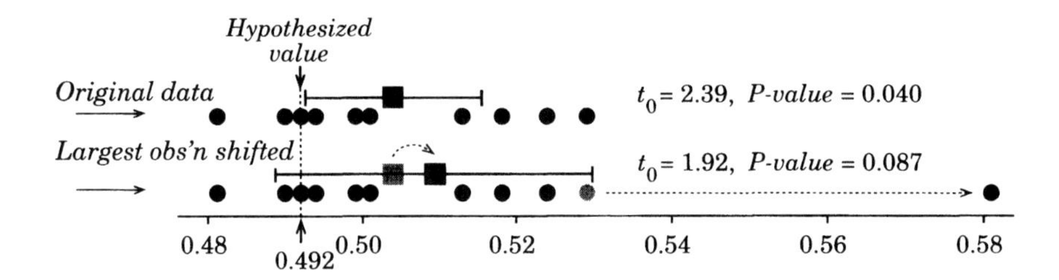

Data with largest obs'n shifted

```
Test of mu = 0.49200 vs mu not = 0.49200
Variable      N      Mean     StDev    SE Mean   t-stat.   P-val
Outlier      10   0.50930   0.02853    0.00902    1.92     0.087
                                 95.0 % CI   (0.48889,0.52971)
```

FIGURE 10.1.2 Effect of an outlier on the confidence interval and test.

an inherent feature of the random process. Unfortunately, heavy tails in a distribution cannot be reliably detected with small samples.

To investigate the appropriateness or otherwise of the t-procedures, we have seen that the first step, as with any data analysis, is to look at a plot of the data. This should *precede* the use of formal tools like t-tests and intervals.

Always plot your data before using formal tools of analysis.

To obtain a general impression of the shape of the data set, we can use the tools of Chapter 2, namely, dot plots and stem-and-leaf plots. The ***Normal probability plot*** (also known as the *Normal quantile plot*)[2] is a specially designed graphic tool provided by most statistical packages for detecting lack of Normality in data. If the data come from a Normal distribution, the Normal quantile plot should approximate a straight line. For small data sets (e.g., $n < 15$) and even moderate data sets ($15 \leq n \leq 40$), however, plots of data that really are Normal can be far from straight with lots of "wobbles" and the odd "staggered" value, particularly at each end of the plot, as shown in Fig. 10.1.3. It is good practice to compare your Normal probability plot with plots from a number of "data sets" of the same size consisting of computer-generated Normal data.

In addition to plots, statistical tests for Normality are available in many computer packages.[3] (Minitab gives a choice of three under `Stat` → `Basic Statistics` → `Normality Test`, which also provides Normal probability plots). A *small P-value* from a test for Normality signals evidence against the null hypothesis that the data come from a Normal distribution.

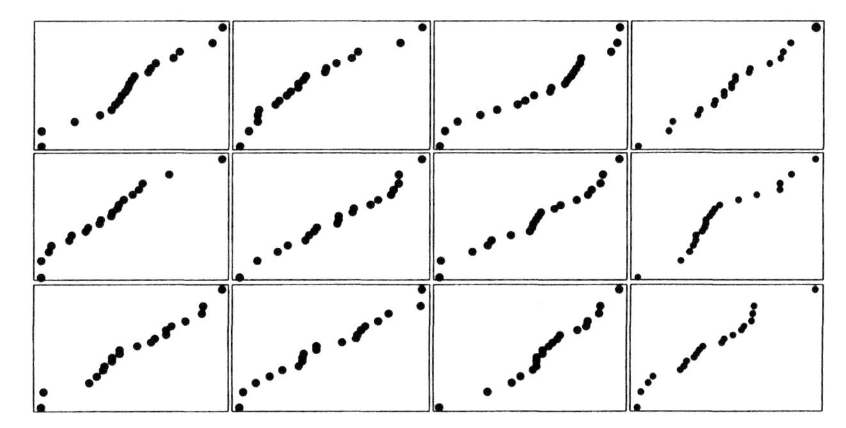

FIGURE 10.1.3 Normal probability plots for samples of size 20 generated from a Normal distribution.

EXAMPLE 10.1.1 (cont.) *Normal Plot and Test*

In Fig. 10.1.4, obtained from Minitab as previously described, we have Normal plots for the original data and the "modified" data (with the outlier introduced by shifting

[2]See, for example, D'Agostino [1986a] or Chambers et al. [1983, pp. 194–199].

[3]Tests for Normality are described in D'Agostino et al. [1990] and D'Agostino [1986b].

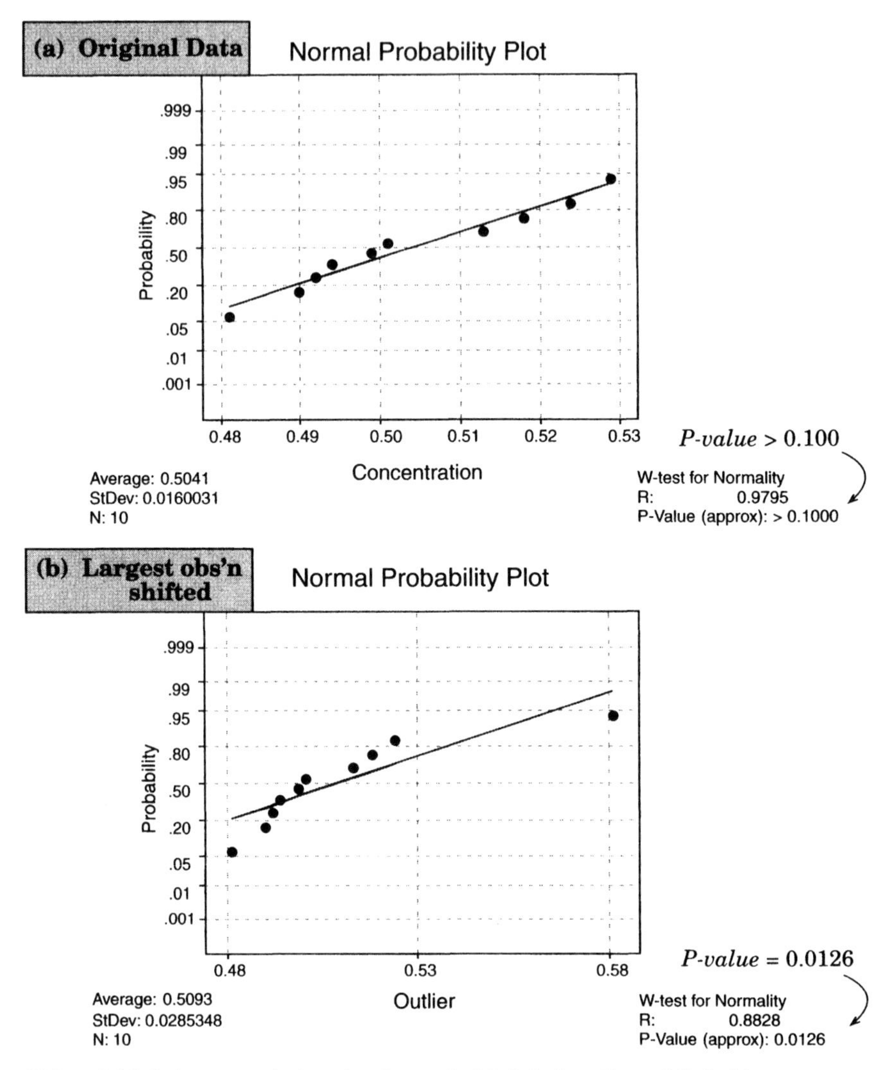

FIGURE 10.1.4 Normal plots for Example 10.1.1 data (from Minitab).

the largest observation). Of the three tests of Normality, we chose Minitab's version of the Wilk-Shapiro W-test. For the original data, Fig. 10.1.4(a) suggests that the Normality assumption is not unreasonable, though with only 10 points we cannot expect too much from such a plot. The test of Normality gave a P-value greater than 10%, providing no evidence against the Normality hypothesis. In Fig. 10.1.4(b), however, the outlier stands out as being off the trend of the other points. We occasionally get behavior somewhat similar to this in Normal data, even with 20 data points (see the lower right-hand plot of Fig. 10.1.3). The test of Normality, however, gives a P-value of approximately 0.01, signaling strong evidence against the Normality assumption.

A Commonsense Approach to Using t-Procedures

The t-procedures seem to be widely used throughout the research literature. This is not surprising, given how robust they are. Because of their universal use, however, it is perhaps appropriate to highlight when they are *not* applicable. Clearly they shouldn't be used if the data show separation into clusters or the stem-and-leaf plot has more than one mode. As we stated in Section 2.3.4, such behavior indicates the presence of two or more distinct groups, and a single measure of center virtually never makes sense in such circumstances. In addition:

- *Small samples:* Do not use t-procedures if there are outliers present. Do not use one-sided t-procedures if the data is more than slightly skewed.
- *Moderate samples* (e.g., $15 \leq n \leq 40$): Do not use t-procedures if there are outliers present. Do not use one-sided t-procedures if the data is strongly skewed.
- *Large samples* (e.g., $n \geq 40$): t-procedures work well even for clearly skewed data, although they can still be adversely affected by gross outliers.
- *Dealing with outliers:* If there appear to be outliers in data, the first thing you should try is to check back to original sources. These data points may be mistakes. Observations that we know are mistakes cause no difficulties. If possible, they should be corrected. Otherwise, they should be removed from the data set. If there is doubt about whether outliers are real observations or mistakes, check whether their presence makes any substantial difference to the conclusions reached from the data (delete them, reanalyze, and check).

*Coping with Non-Normality

There are four main ways of coping with non-Normality when it appears severe enough to affect the t-procedures:

(i) You can use so-called *nonparametric* methods,[4] which make no assumptions about the distribution from which the data are sampled and which are insensitive to the presence of outliers. One such method, called the sign test, is discussed later in this chapter as an illustration of the idea.

(ii) You can use so-called *robust methods,* which have been designed to be insensitive to the presence of outliers. Such methods are outside the scope of this book.[5]

(iii) The t-procedures are based on a Normal distribution model. You can change to using *another model* that provides a better description of the data, for example, by replacing the Normal distribution by some other distribution. We shall not discuss this advanced topic.

(iv) You can *transform* the data, for example, work with $x_i^* = \log_e(x_i)$ or $x_i^* = \sqrt{x_i}$ instead of the original data points. Here the transformation is often chosen so that the transformed data points conform better to a Normal distribution.

The following example introduces an experiment that will have a prominent place in the remainder of this section.

✖ Example 10.1.2 *Looking for a Moon Illusion Using a t-Test and Interval*

Perhaps you have noticed that at times the moon seems to be incredibly large. What did those times have in common? A series of idyllic romantic evenings? For most of us, the actual explanation is something a lot more mundane: The moon appears bigger

[4]For general references see Conover [1980] and Sprent [1993].

[5]See, for example, Tiku et al. [1986].

when it is low in the sky and close to the horizon than it does when it is high in the sky. This is the so-called moon illusion. The Greek astronomer Ptolemy discussed the moon illusion and suggested a possible explanation in the second century A.D. In a now classic paper, Kaufman and Rock [1962] described a series of experiments investigating the moon illusion and some possible contributing factors.

They devised an optical device to be used outdoors that projected two disks of adjustable size so that one appeared like a moon near the actual horizon and the other like a moon near the zenith (directly overhead). One disk was kept at a standard size and the other adjusted until both disks *appeared* to the experimental subject *to be of the same size.* Each measurement given is the ratio of the actual diameter of a zenith disk to the actual diameter of a horizon disk. Where a moon illusion exists, the ratio should be greater than 1, as the actual zenith disk has to be increased in size to compensate for the effect of the illusion. The following data give the ratios for each of 10 experimental subjects. To avoid confusion as to which is greater than which, these ratios can also be regarded as the ratio of the apparent diameter of the horizon disk to the apparent diameter of the zenith disk. The ratios are plotted in Fig. 10.1.5.

2.03 1.65 1.00 1.25 1.05 1.02 1.67 1.86 1.56 1.73

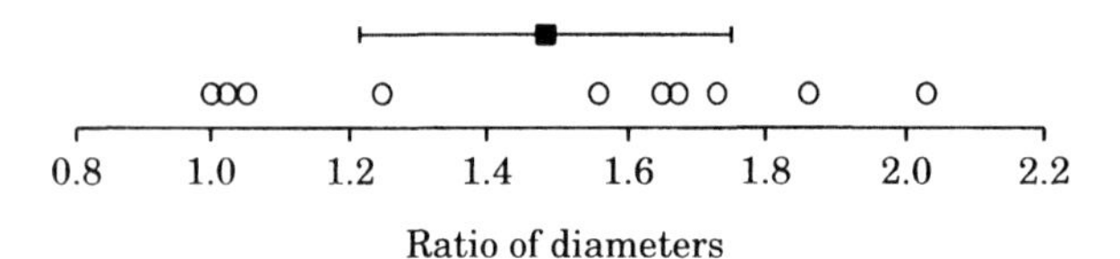

FIGURE 10.1.5 Dot plot of moon illusion data with a 95% confidence interval for the mean.

In the plot there are no outliers and no pronounced skewness in the data. If there was no illusion, the ratios should all be within random error of unity. The main reason that we cannot expect them to be exactly 1 is as follows. Subjects could not see the moon in both places at the same time and therefore had to match the size of the moon in one place to the remembered size in the other. We cannot expect such comparisons to be exact. The observed ratios range from 1.00 (no apparent illusion) to 2.03. Although we would not expect all of the observed ratios to be exactly 1.00 if there was no moon illusion, we *would* expect the *underlying mean* of the observed ratios to be 1.00.

For formal inference we assume that these experimental subjects constitute a random sample from the larger population. This is usually the strongest and most critical assumption. Most experiments like this are performed on volunteers. The 10 subjects here were neighbors of one of the researchers. We should always ask the questions, "Are these subjects special in some way? How far can we reasonably extrapolate from the results of this experiment?" The fact that these questions cannot be answered entirely satisfactorily for most real experiments provides just one more reason for science's insistence on results being repeatable. We should not set too much store on the results of any one isolated study.

Do these data provide any real evidence that people on average do experience a moon illusion? The answer seems fairly clear from the plot, but we shall still formally test the skeptical hypothesis H_0: $\mu = 1$ (no illusion) versus the alternative H_1: $\mu > 1$. Since the dot plot shows no overtly non-Normal features, we shall apply our t-procedures. (Also, formal tests show no evidence against the Normality assumption.)

```
Test of mu = 1.000 vs mu > 1.000

Variable    N     Mean     StDev     SE Mean     t-stat  P-value
Elevated   10    1.482     0.374       0.118     4.07     0.0014
                                   95.0 % CI:   (1.214, 1.750)
```

The one-sided *P*-value is 0.0014. We therefore have very strong evidence that the population mean of the ratios is greater than unity, that is, people do on average experience a moon illusion. A 95% confidence interval for μ is given by [1.21, 1.75]. Thus we can say with 95% confidence that the population mean of the ratios is somewhere between 1.2 and 1.8. This interval has been plotted on Fig. 10.1.5. Because the sample is small, we see that the interval is quite wide.

Other issues about this experiment will be raised, and more data will be given as this section proceeds. We shall see, for example, that different people experience illusions of very different magnitude. Under such circumstances formal inferences such as confidence intervals about μ, which is the average of everybody's illusions, are not of compelling interest. We are more interested in individual behavior and comparisons within rather than between people. The next topic to be discussed is *paired comparisons*, which are included in this section because the technical methods for making the comparison use one-sample techniques.

QUIZ ON SECTION 10.1.1

1. What assumptions about the data are made by the theory underlying *t*-tests and confidence intervals for a population mean μ?
2. When we say that a *t*-confidence interval for μ is robust against some particular form of non-Normality, what do we mean by *robust?* What do we mean when we say that a *t*-test is *robust* against some departure from the assumptions?
3. What should you always do with data on a continuous variable *before* performing formal significance tests or intervals?
4. Under what circumstances should you *not* use *t*-tests and intervals?
5. If there are outliers in a data set, what should you do?

*6. Four approaches to dealing with severe non-Normality (including the presence of outliers) were outlined. What were they?

10.1.2 Paired Comparisons

Sometimes two sets of data are not independent of each other but rather match up in pairs. Usually the members of the pair are related in some way, such as two similar measurements on the same person or measurements on twins. The following example illustrates this.

EXAMPLE 10.1.2 (cont.) ***An Example of Paired Data***

We continue with Kaufman and Rock's study of the moon illusion. An earlier paper by other researchers showed that, under some experimental conditions, objects appeared bigger with eyes level than they do with eyes raised. This was suggested as a factor in explaining the moon illusion. In their moon illusion experiment, Kaufman and Rock investigated whether eye position played a part. Each subject was tested under two conditions: eyes raised ("elevated") with respect to the head and eyes level with respect to the head. "Eyes elevated" refers to normal behavior in which a person first looks at the horizon "moon" (disk) and then tilts their head and raises their eyes to see the zenith disk overhead. The ratios of the apparent size of the horizon disk to the apparent size of the zenith disk for the eyes-elevated case are given in Table 10.1.1, and this data set was investigated in the previous section.

Under the eyes-level, or control, condition, a physical device forced the subjects to tilt their heads backward to view the zenith disk with eyes looking straight ahead so that they looked at both the horizontal and zenith "moons" with the same orientation of the eyes with respect to the head. The corresponding ratios are also given in Table 10.1.1. If the "moon" really does look smaller with eyes elevated, what effect will this have in addition to any possible moon illusion? In this situation the effect of the moon illusion, as measured by the ratio of the apparent size of the horizon disk to that of the zenith disk, will be greater for the eyes elevated than for eyes level because we are dividing by a smaller number in the former case. The observations in Table 10.1.1 suggest that this might be the case for some subjects but not for others, however.

TABLE 10.1.1 The Moon Illusion

Subject	Eyes elevated	Eyes level	Difference (elevated − level)
1	2.03	2.03	0.00
2	1.65	1.73	−0.08
3	1.00	1.06	−0.06
4	1.25	1.40	−0.15
5	1.05	0.95	0.10
6	1.02	1.13	−0.11
7	1.67	1.41	0.26
8	1.86	1.73	0.13
9	1.56	1.63	−0.07
10	1.73	1.56	0.17

Source: Kaufman and Rock [1962].

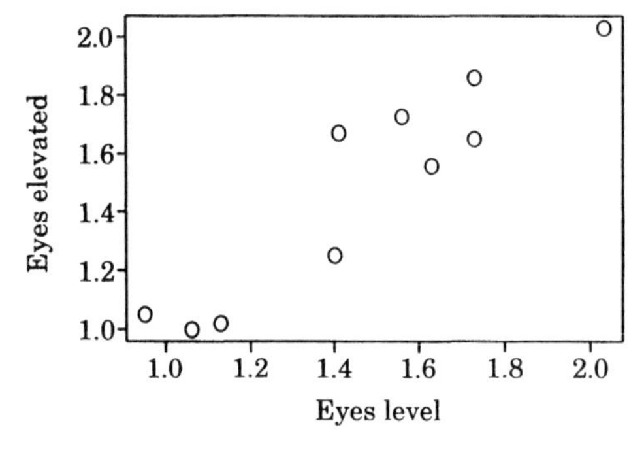

FIGURE 10.1.6 Eyes-elevated ratios versus eyes-level ratios.

Data sets such as this are often wrongly analyzed by treating the two sets of measurements as if they came from two independent samples. They are not independent here, however. The eyes-elevated ratio and the corresponding eyes-level ratio in the same row come from the same person, and therefore we might expect them to be related. Figure 10.1.6 shows that in this example there is quite a strong relationship between a person's eyes-elevated ratio and eyes-level ratio.

This example is typical of a large class of studies in which observations are made on the same person or object (often before and after some experimental treatment has been applied) or on matched pairs (e.g., twins, brothers, or just people matched for the purposes of the study on factors such as age, occupation, and sex). Such an experiment is called a ***paired-comparison*** experiment. Pairing is also used in observational studies. In Section 1.3 we described studies on pairs of twins in which one twin smoked and the other did not. We shall explain the advantages of pairing as a strategy in Section 10.4.1 (which can be read at this point if you wish). Here we concentrate on analysis.

To investigate changes under two conditions with data such as that in Example 10.1.2, we ***analyze the differences;*** that is, we take the set of differences between measurements within each pair as our basic data to be analyzed (see the column to the right of Table 10.1.1). We thus have a one-sample problem. If there is no systematic tendency for the first observation (e.g., eyes elevated) in a pair to be greater or smaller than the second, then μ_{diff}, the underlying mean of the differences, should be 0. Thus we can test for no difference between the effects of the two experimental conditions by testing H_0: $\mu_{diff} = 0$ using the ordinary one-sample t-test statistic applied to the differences. Similarly, we use the single sample of differences to calculate a confidence interval for μ_{diff} in the usual way. There is no new theory here. The only thing you have to learn is to look closely at the way data was produced to see whether you have two independent samples or paired data. If your data is paired, it is the differences you should be plotting and analyzing.

For ***paired*** data, ***analyze the differences.***

✱ Example 10.1.2 (cont.) *Inferences about Differences*

The differences (*eyes elevated − eyes level*) are given in Table 10.1.1. A dot plot of these differences is given in Fig. 10.1.7. The points seem to be equally scattered either side of 0. There is definitely no strong tendency for the differences to be either positive (eyes elevated gives the bigger illusion on average) or negative (eyes level gives the bigger illusion on average), so there is probably no evidence here that eye position

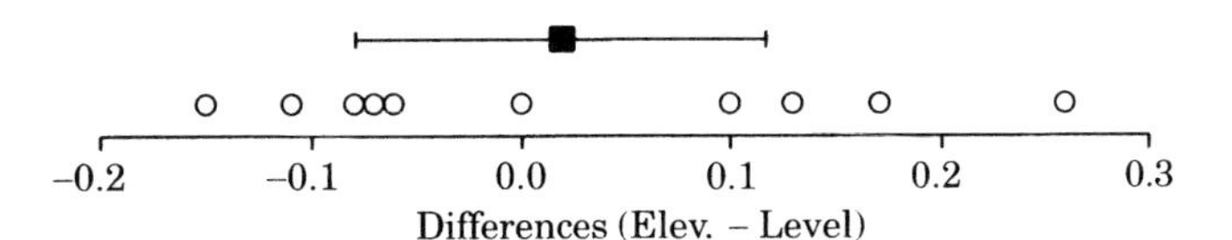

FIGURE 10.1.7 Dot plot of differences for the moon illusion data (with a 95% confidence interval for the mean difference).

makes a difference. Since there are no overt signs of non-Normality, we shall use our t-procedures. (Also, formal tests show no evidence against the Normality assumption.)

The skeptical null hypothesis is H_0: $\mu_{diff} = 0$. Since this experiment was conducted to try to confirm the implications of earlier research that suggested that the moon illusion should be stronger under the eyes-elevated experimental conditions, the alternative hypothesis is H_1: $\mu_{diff} > 0$.

```
Test of mu = 0.0000 vs mu > 0.0000

Variable      N      Mean      StDev     SE Mean     t-stat    P-value
Difference   10    0.0190     0.1371      0.0434       0.44       0.34
                                       95% CI ( -0.0791,      0.1171)
```

Since the sample mean of the differences is positive, we have to see whether it is sufficiently positive to reject H_0 in favor of H_1. The one-sided P-value for t_0 is 0.34, which provides no evidence against H_0. The data provide no evidence that the average size of the moon illusion that people experience under the eyes-elevated condition differs from that under the eyes level condition. This is not to say that there is no difference. The 95% confidence interval for the true mean difference between the sizes of illusion that people experience under the two experimental conditions extends from -0.08 to 0.12, so there could be a difference in perceived size of up to about 0.1 (10%) in either direction between looking with eyes elevated and looking with eyes level. Note that this is small compared with the variation we have seen in ratio measurements between people (see Fig. 10.1.5).

Although the theory of paired testing says that we should analyze the differences, most of the statistical packages do it for you if you give them two variables and tell the program that they are paired. In Minitab, fill out a dialog box located under `Stat` → `Basic Statistics` → `Paired t`. Minitab gives slightly more output for this option (see Fig. 10.1.9), though the test results and confidence interval are identical to what you get by forming the differences yourself as in the previous example. In Splus and R, the command is of the form `t.test(variable1,variable2,paired=T)`. Paired t-tests are also available in Excel under `Tools` → `Data Analysis`; choose `t-test: Paired Two Sample for Means`. This performs the test only; it does not supply a confidence interval.

EXAMPLE 10.1.3 *Looking at Pairs of Head Measurements*

After buying a batch of flying helmets that did not fit the heads of many pilots, the N.Z. Air Force decided to measure the head sizes of all recruits. Initially, information was collected to determine the feasibility of using cheap cardboard calipers to make the measurements instead of metal ones, which were expensive and uncomfortable. Table 10.1.2 gives the head diameters of 18 recruits measured once using cardboard calipers and again using metal calipers. The question we shall try to answer here is whether there is any systematic difference in the measurements produced by the two sets of calipers. The answer is not immediately obvious because of the variability in measurements taken on the head of the same person. An important cause of this variability is variability in the positioning of the calipers on the head.

TABLE 10.1.2 Air Force Head Size Data

Recruit	Cardboard (mm)	Metal (mm)	Difference (Card − Metal)	Sign of difference
1	146	145	1	+
2	151	153	−2	−
3	163	161	2	+
4	152	151	1	+
5	151	145	6	+
6	151	150	1	+
7	149	150	−1	−
8	166	163	3	+
9	149	147	2	+
10	155	154	1	+
11	155	150	5	+
12	156	156	0	0
13	162	161	1	+
14	150	152	−2	−
15	156	154	2	+
16	158	154	4	+
17	149	147	2	+
18	163	160	3	+

Data courtesy of Dr. Stephen Legg.

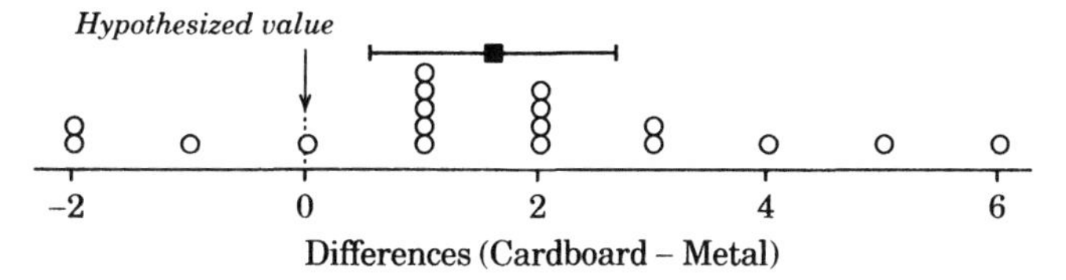

FIGURE 10.1.8 Dot plot of differences in head size (with a 95% confidence interval).

These are again paired-comparison data, with the pairs being measurements of two different types on the same recruit. Does one type of caliper tend to give measurements that are above or below those given by the other? We should work with the differences. A dot plot of the differences is given in Fig. 10.1.8. The most noticeable features of the plot are (1) the fact that the data have been rounded to the nearest millimeter, which has caused some stacking in the dot plot, and (2) the center of the data is to the right of 0, showing a tendency for the cardboard calipers to give larger measurements. The question remains as to whether there really is a shift or whether what we are seeing could just be due to sampling variation.

The plot gives us no reason to doubt the applicability of *t*-tests and intervals. In fact, the Normal probability plot (not shown) is very close to linear, and formal tests reveal no evidence against the Normality assumption. Minitab output for the paired *t*-test of H_0: $\mu_{diff} = 0$ versus the two-sided alternative H_1: $\mu_{diff} \neq 0$ is given in Fig. 10.1.9. The *P*-value of 0.005 tells us that we have very strong evidence that caliper type makes a difference. The confidence interval tells that with 95% confidence the

```
Paired T-Test and Confidence Interval
paired T for cardboard - metal
                  N      Mean     StDev   SE Mean
cardboard        18    154.56      5.82      1.37
metal            18    152.94      5.54      1.30
Difference       18     1.611     2.146     0.506
95% CI for mean difference: (0.544, 2.678)
T-Test of mean difference=0 (vs not=0): T-Value=3.19  P-Value=0.005
```

FIGURE 10.1.9 Minitab paired-t output for the head size data.

cardboard calipers are giving measurements that are bigger on average than those provided by the metal calipers by between 0.5 mm and 2.7 mm.

In the next subsection we introduce an alternative method of analysis for paired data, a nonparametric method called the sign test.

QUIZ ON SECTION 10.1.2

1. What is a paired-comparison experiment?
2. In a paired-comparison experiment, why is it wrong to treat the two sets of measurements as independent data sets?
3. How do you analyze the data from a paired-comparison experiment?
4. Can you think of some situations in which the paired-comparison method is the appropriate way to analyze the data?

EXERCISES FOR SECTION 10.1.2

1. While we are thinking about head measurements, Table 10.1.3 gives a famous set of data obtained by carrying out measurements on the heads of the first two sons of 25 families.

TABLE 10.1.3 Head Lengths of the First and Second Adult Sons

Family	First son	Second son	Difference	Family	First son	Second son	Difference
1	191	179	12	14	190	195	−5
2	195	201	−6	15	188	187	1
3	181	185	−4	16	163	161	2
4	183	188	−5	17	195	183	11
5	176	171	5	18	186	173	13
6	208	192	16	19	181	182	−1
7	189	190	−1	20	175	165	10
8	197	189	8	21	192	185	7
9	188	197	−9	22	174	178	−4
10	192	187	5	23	176	176	0
11	179	186	−7	24	197	200	−3
12	183	174	9	25	190	187	3
13	174	185	−11				

Source: Frets [1921].

(a) Plot the data using some appropriate method. Is the Normality assumption tenable here?
(b) It has been conjectured that there is a difference in the average head lengths of the first two sons in families with at least two sons. Test this conjecture. How big is the difference?
(c) What else would you want to know about this experiment?

2. In Example 10.1.3 measurements were taken by one person using one set of cardboard calipers and one set of metal calipers. When the air force makes such measurements routinely, what additional sources of variation would be introduced? The cardboard calipers are cheap, so they can be thrown away without cost being an important factor. But what might you expect to happen to measurements obtained from cardboard calipers as they started to wear out?

10.1.3 A Nonparametric Test

In the previous sections we used methods that have been derived on the assumption that the original data have come from a Normal distribution. Although the t-methods are robust with regard to this assumption, there will be occasions when the data are so non-Normal-looking that we shall have serious misgivings about using such methods. What we need, then, is a method that does not depend on the underlying distribution, that is, one that is *distribution free*. Such a method is said to be *nonparametric* in contrast to the t-methods, which are *parametric*. Nonparametric methods are generally based only on the order in which observations fall and not on how far apart the observations are. For this reason the median (rather than the mean) is used as the standard measure of center or location.

We shall briefly encounter several nonparametric tests in this chapter: the sign test, the Wilcoxon signed-rank test, the Wilcoxon-Mann-Whitney test, and the Kruskal-Wallis test. Programs also invert all but the last of these to give nonparametric confidence intervals. Nonparametric methods have the great advantage of not making any assumptions about the distribution producing the data. They also tend to be insensitive to the presence of outliers. Why don't we use them all the time? It turns out that when the assumptions of parametric procedures such as the t-test hold, the parametric tests are superior in that they are more likely to detect departures from the null hypothesis (in statistical jargon, they are *more powerful*) and the parametric confidence intervals tend to be shorter. We shall now discuss the *sign test,* which is the only nonparametric test that we shall develop in any detail. In doing so, we shall also expand on the ideas introduced here.

The Sign Test

In the same way that the *sample* median for a random variable X is the middle value for the data (half above, half below), the population median ($\tilde{\mu}$, say) locates the middle of the distribution of X. You are just as likely to obtain a value of X above $\tilde{\mu}$ as below it (as depicted in Fig. 10.1.10).

If we were to test that the median is 0, for example, that is, test H_0: $\tilde{\mu} = 0$, under the null hypothesis we would expect an observation to be just as likely to be positive as negative. If we are interested only in whether an observation has a + or − sign, it would be like tossing a coin each time we recorded the sign of an observation (assuming that 0s are ruled out). We would record a head for a + and a tail for a −. Testing H_0 would then be like testing whether a coin is unbiased, so that if p is the probability

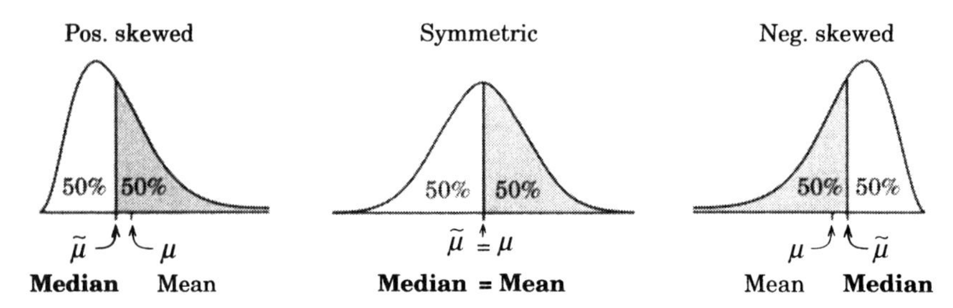

FIGURE 10.1.10 The population median.

of getting a head, we would be testing H_0: $p = \frac{1}{2}$. Evidence against H_0 would be provided by a gross imbalance of $+$'s and $-$'s. The testing can be done using exact Binomial theory and a computer. Notice that there is no mention of the distribution of the original data; in this sense the test is distribution free. This does not mean, however, that we don't use a distribution to do the test; we use the Binomial distribution to do it. Clearly, we can apply this approach to both one-sided and two-sided alternatives. For example, if we test H_0: $\tilde{\mu} = 0$ versus H_1: $\tilde{\mu} > 0$, this is equivalent to testing H_0: $p = \frac{1}{2}$ versus H_1: $p > \frac{1}{2}$. We can also use the method for testing whether the median is equal to some value other than 0. Can you think how you might do that? For example, if the data points are 3, 6, 2, 7, ... and we want to test H_0: $\tilde{\mu} = 5$, we can apply the test to the signs of $3 - 5, 6 - 5, 2 - 5, 7 - 5, \ldots$, namely, to $-, +, -, +, \ldots$, to test H_0: $\tilde{\mu} = 0$.

The method we have just described is, for obvious reasons, called the *sign test*, and we have discussed it in detail to give you the flavor of nonparametric methods. Such methods are readily available from most software packages, and we shall freely use computer output from them in future analyses. In addition to the test, it is also possible to obtain a nonparametric confidence interval[6] for the median. You will recall from Section 9.5.1 that there is a close link between a confidence interval and a hypothesis test. A 95% confidence interval can be constructed by collecting together all the hypothesized values of the parameter that are not rejected at the 5% level of significance. This method of obtaining a confidence interval is often called *inverting the test*, and it can be used for all nonparametric tests. We are now going to apply the sign test to the more common situation of paired data.

✖ Example 10.1.3 (cont.) *Using the Signs of Paired Differences*

We previously analyzed the air force head size data, which is paired data, by using a t-test and confidence interval applied to the differences. A key assumption made by the underlying theory was that the differences were Normally distributed. We now avoid this assumption by using the sign test to test that the median of the differences is 0. Because of the inaccuracy of the measuring devices, there is one individual for whom the recorded measurements are the same so that the difference is 0. It transpires that we are able to ignore that individual and just focus on the remaining 17 individuals. Of these, the cardboard caliper gave the bigger measurement in 14 cases, whereas the metal caliper gave the bigger measurement in only 3 cases. From the last column of Table 10.1.2 we have 14 plus signs and 3 minus signs. This fact alone looks as if it

[6] More properly called *confidence sets*, as the resulting set may not be a single interval. It may have holes or gaps in it.

should constitute evidence that the cardboard caliper tends to give bigger measurements. It would be like tossing a coin 17 times and getting 14 heads. Without even doing any testing, common sense and our experiences of tossing coins would indicate to us that the coin is biased. The probability of getting 14 or more heads from 17 tosses of a fair coin is $\text{pr}(Y \geq 14)$ where $Y \sim \text{Binomial}(n = 17, p = 0.5)$. This probability is 0.00636. Doubling this (thinking in terms of a two-tailed test) gives 0.0127.

In fact, if we form the differences as a column and then apply the Minitab commands `Stat` → `Nonparametric` → `1-Sample sign`, we get a two-sided *P*-value of 0.0127. This confirms our suspicions. A similar instruction gives us a 95% confidence interval of [1.0, 2.5] for the median of the differences, which the computer obtains by inverting the test. There is another nonparametric one-sample test that is a little more complicated but better at detecting departures from the null hypothesis called the *Wilcoxon signed-rank test.*[7] This test is also available from Minitab, and it gave *P*-value = 0.011 and a 95% confidence interval of [0.5, 2.4]. Thus the two nonparametric tests and the previous *t*-test have all given us strong evidence against the null hypothesis that caliper type makes no difference.

When Do We Use the Sign Test?

The sign test can always be used, provided that the observations are independent (one-sample case) or that different pairs are independent (paired-data case). The test is very sensitive to departures from this assumption. The main advantages of the test are that it is still valid for data from *any* distribution, whether Normal or not, and it is insensitive to outliers. We noted that any ties are ignored, whether they are due to observations being equal to a hypothesized median or to some paired differences being 0.[8] When the data are entered into the computer, the ties are automatically taken care of when the test is computed. You might then ask, "Why don't we always use the sign test (or the Wilcoxon signed-rank test)?" It transpires that when the assumptions of the *t*-test hold, the *t*-test is superior to these tests in that it is more likely to detect departures from the null hypothesis. Clearly, when we have very small samples, such as $n \leq 10$, we are less able to check whether the assumptions of the *t*-test are valid or to detect outliers. We might also want to perform a one-tailed test in a situation where we suspect a skewed distribution. We may have a suspiciously large or small data point that we think may be real rather than just a mistake. A nonparametric test is clearly useful in these situations.

QUIZ ON SECTION 10.1.3

1. How is the population median defined? In what sense is it similar to the sample median? How does it differ?
2. Under what circumstances is the population median the same as the population mean?
3. Why do we use the population median rather than the population mean in the sign test?
4. Why is the model for the sign test like tossing a fair coin?
5. What independence assumption must hold before the sign test is applicable? How important is it that this assumption is true?

---

[7]See Bhattacharyya and Johnson [1977, p. 519].

[8]Ignoring ties is an approximate procedure (see Emerson and Simon [1979]).

6. What advantages and disadvantages does the sign test have in comparison with the t-test?
7. Why is the sign test called a *distribution-free* test? Does this mean that distributions are not used in performing the test?
8. In applying the sign test to paired data, how do you handle situations where both observations are tied (indistinguishable)?

EXERCISES FOR SECTION 10.1.3

1. Referring to Exercises for Section 10.1.2, problem 1, use the sign test to see whether there is a difference in average head length between first and second sons.
2. Cyclozocine was an alternative to methadone for treating heroin addiction. The following data came from 14 males who were chronic heroin addicts. After cyclozocine had removed the addicts' physical dependence on heroin, they were asked a battery of questions designed to assess their psychological dependence. The test scores are called Q-scores, and high values represent less psychological dependence. The Q-scores were as follows (Resnick et al. [1970]):

   51 53 43 36 55 55 39 43 45 27 21 26 22 43

   From past experience the median score for addicts who have not had a cyclozocine treatment is 28. Is cyclozocine an effective treatment for psychological dependence?

10.1.4 Inferences about Spread

✖ EXAMPLE 10.1.2 (cont.) *Assessing the Variability in Moon Illusion Ratios*

If you look back at the moon illusion data in Table 10.1.1, you will see that the two measurements on the same person are quite similar (even though made under different experimental conditions). On the other hand, there is more variability between ratios measured for different people. This indicates that a good deal of the variability we see in the ratios (under either experimental condition) represents actual variability in what people see. Figure 10.1.11 shows the extent of moon illusion represented by three ratios within the range of the data. The visual differences are more pronounced than one might think by looking at the ratios. If the moon is seen as a two-dimensional disk (as represented here), visual impression of size is much more closely related to area than it is to diameter. Area is proportional to the square of diameter, so it is the squares of the ratios that best tell us about perceived size. If the moon is seen as

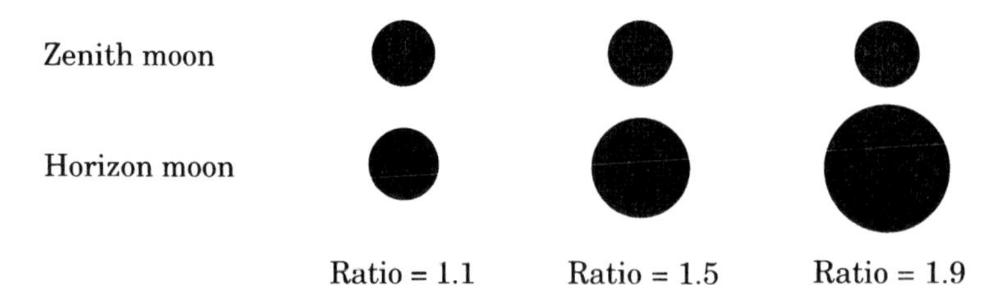

FIGURE 10.1.11 Different moon illusions.

a three-dimensional sphere, the cubes of the ratios best tell us about perceived size. (These considerations are the root cause of the problems in many examples of misleading statistical graphics in the media, as addressed by a problem in Review Exercises 12.)

As far as we are concerned, the most interesting thing that we have seen in these data is the enormous variability in the extent of the illusion that different people experience. Such a feature in the data then raises the question, "*Why* are there such big differences in what people see?" which is a starting point for further research, a new iteration of the investigative cycle (Section 1.4). If we wanted to do formal statistical inference about the moon illusion data, the features that we would most want to estimate would be related to the variability in illusions seen and not just to μ, the average of all these disparate illusions. One could, for example, wish to calculate a confidence interval for σ, the population standard deviation of the ratios.

In the past the standard method used for inferences about σ was based on the sample standard deviation and used a distribution called the Chi-square distribution. (We use this distribution for another purpose in Chapter 11.) In contrast to inferences about the mean μ using the t-distribution, these inferences are extremely sensitive to any non-normality in the data. Unfortunately, the situation does not improve as the sample size n gets bigger: There is no large-sample theory for the sample standard deviation that corresponds to the central limit theorem for sample means. For these reasons this particular approach is no longer recommended. What then can be done? Not very much at this level.[9] The main problem is that σ is not a very good measure of spread, and it is virtually impossible to visualize; it is not an intuitive concept like interquartile range. It is also sensitive to outliers.

Another way of approaching the variability in what people see would be to try to estimate the range of ratios experienced by the central 90%, say, of the population. We would like to make a statement with some level of confidence, such as 90% of people experience a ratio of between 1.10 and 1.6. Such an interval is called a *tolerance interval.* The most commonly used methods for calculating tolerance intervals either are based on Normal distribution assumptions or are nonparametric (distribution free). Tolerance intervals based on Normality assumptions are sensitive to non-Normality.[10] Some people mistakenly interpret the confidence interval for μ as providing tolerance-interval-type information, but of course the confidence interval for μ tells us only about the *position* of the population *mean.* It says nothing at all about the rest of the distribution.

QUIZ ON SECTION 10.1.4

1. Why are inferences for the mean μ straightforward, whereas inferences for the standard deviation σ difficult?
2. What large sample theory can we invoke when making inferences about a mean?
3. Is there a large-sample theory for sample standard deviations?
4. What information does a 90% tolerance interval convey?

[9]See Solomon and Stephens [1988] for a brief discussion. The "bootstrap" method is one possibility (Efron and Tibshirani [1993]).

[10]For further discussion of tolerance intervals, see Hahn and Meeker [1991] and Vardeman [1992].

10.2 TWO INDEPENDENT SAMPLES

10.2.1 Questions of Study Design and Analysis

In this section we look closely at the sorts of questions raised in Sections 8.4 and 9.3.2 when comparing independent samples from two different populations. For example, we would typically like to compare the two population means by applying t-methods to the difference in the means. Another question, generally of lesser interest, is how we compare the two population spreads. We are also interested in the role that the experimental design has on our conclusions from the data relating to cause and effect. To help us focus on the design issue, we introduce two examples.

EXAMPLE 10.2.1 *Two Independent Samples from an Observational Study*

The possible factors determining sexual preference have been widely debated over the years. Factors such as environment, genetics, and chance have been put forward as possibilities. In a study during the late 1960s, Margolese [1970] looked at the relationship between hormone levels and homosexuality. The data given in Table 10.2.1 present measured levels of urinary androsterone (mg/24 hr) for each of 11 male heterosexuals and 15 male homosexuals in good physical health. The data are plotted in Fig. 10.2.1.

The samples are unrelated and thus independent. What do we see? First, the androsterone levels of the homosexual sample are lower on average than the heterosexual levels. We shall see later that a t-test for the difference in the means has P-value 0.004, with a 95% confidence interval for $\mu_{Het} - \mu_{Hom}$ given by [0.4, 1.7]. There is a good deal of overlap between the two samples. Thus, even though androsterone

TABLE 10.2.1 Urinary Androsterone Levels (mg/24 hr)

Homosexual	2.5	1.6	3.9	3.4	2.3	1.6	2.5	3.4	1.6	4.3	2.0
	1.8	2.2	3.1	1.3							
Heterosexual	3.9	4.0	3.8	3.9	2.9	3.2	4.6	4.3	3.1	2.7	2.3

Source Margolese [1970].

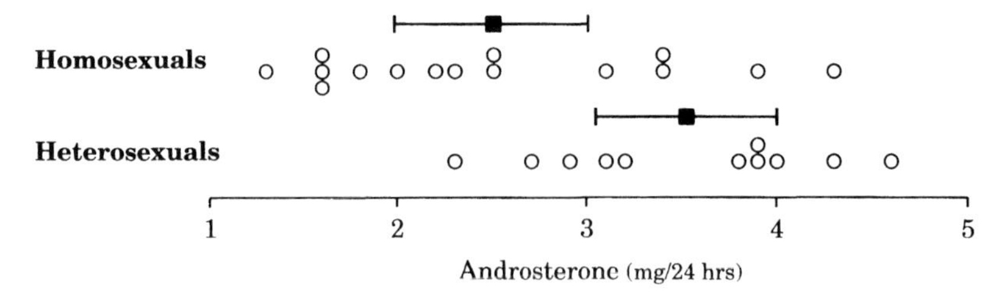

FIGURE 10.2.1 Dot plots of the androsterone data (with 95% confidence intervals).

levels are lower in homosexuals on average, some homosexual males have higher levels than most heterosexuals. There are some weak suggestions of non-Normal behavior, for example, a slight positive skewness in the homosexual data. Would this be sufficient to make us doubt the applicability of t-tests and intervals? There is an appearance of greater spread in the homosexual group. Does this demonstrate a real difference, or is this within expected levels of variation?

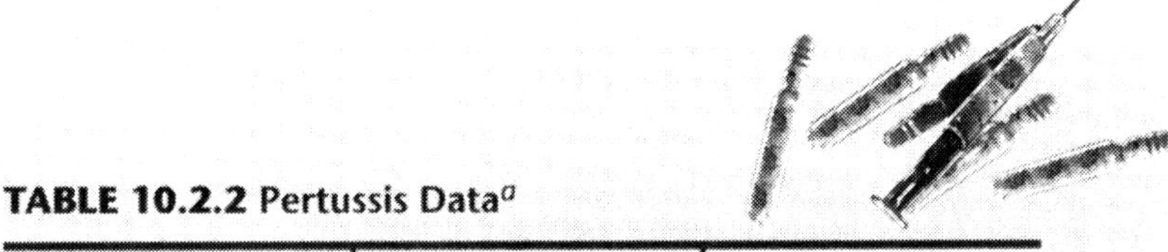

TABLE 10.2.2 Pertussis Data[a]

APV (Single)			DAPV (Double)			WCV (Control)		
3.91	5.21	3.96	5.76	5.14	3.97	6.17	5.73	1.21
4.86	5.16	4.08	3.31	4.92	4.62	5.78	5.32	1.68
3.98	5.17	-2.30	5.04	4.07	4.63	2.87	4.66	3.76
5.11	4.92	4.09	3.62	4.50	4.63	4.49	4.82	2.17
3.27	4.60	2.58	3.73	4.20	4.23	3.07	3.06	1.36
2.26	4.26	3.49	5.28	4.38	3.64	2.90	1.29	1.47
3.81	3.71	5.17	4.10	3.63	3.45	2.41	1.92	2.47
3.78	3.50	4.83	3.46	3.45	4.87	2.20	2.01	1.80
3.96	4.30		4.92	3.59	4.09	4.68	4.07	
4.88	4.27		3.41	2.34		5.45	2.56	
4.74	2.34		5.66	4.05		5.70	1.74	

[a] Concentrations of pertussis antibodies in ln(IU/mL).
Data courtesy of Clinical Trials Research Centre, Dalhousie University; the IWK-Grace Health Centre; and Connaught Laboratories Limited.

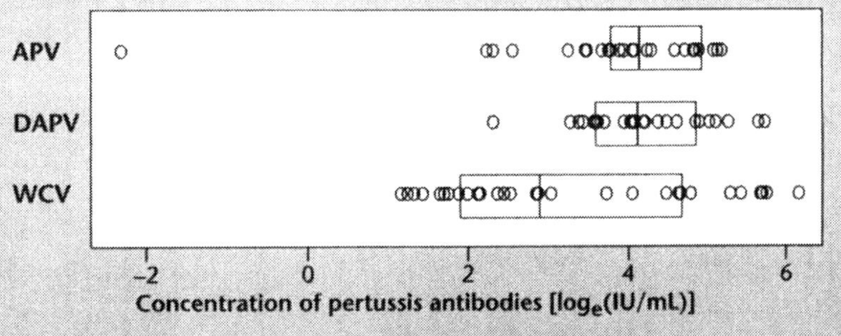

FIGURE 10.2.2 Dot-box plots of pertussis data.

✖ EXAMPLE 10.2.2 *Comparing Two Independent Samples from an Experimental Study*

Pertussis, commonly called whooping cough because of the characteristic sound victims make, is a childhood disease that earlier in this century was a feared killer of infants. One hears little about it today because effective vaccines have greatly reduced the incidence of pertussis. In fact, the proportions contracting pertussis today are about about one tenth of what they were 50 years ago. Nevertheless, whooping cough still kills about 50,000 infants each year in North America, and researchers continue to search for better vaccines. The data in Table 10.2.2 and plotted in Fig. 10.2.2 were collected as part of a study by Halperin et al. [1994], which compared a standard vaccine (WCV) with a new vaccine called APV and with DAPV (the APV vaccine administered at double the strength). This was a randomized study. Each of 91 infants in the age range from 17 to 19 months was assigned a vaccine at random, and the study was conducted as a double-blind experiment (see Section 1.2). The variable measured here is the concentration of pertussis antibodies in the blood of the infant one month after immunization. Such concentrations are usually measured on a log scale. The units of measurement here are log (IU/mL).

There are a number of issues that arise out of Examples 10.2.1 and 10.2.2 that we need to discuss in some detail.

Issues of Study Design

- *Experiments versus observational studies*
  As noted in the titles, Example 10.2.1 was an observational study while Example 10.2.2 was a randomized experiment. Does this make a difference to the way we analyze the data? The answer is no. We analyze data from experiments and observational studies in the same way. The distinction becomes important when we come *to interpret* the results of the analysis. We know from Section 1.3 that we should be very wary of making causal inferences from observational data. We cannot conclude from Example 10.2.1 that androsterone levels are a reason why some people are homosexual and others are heterosexual. Although it is one possible explanation, there may be other differences between these groups besides sexual orientation, and one of these may be the real cause of the differences we are seeing. On the other hand, in Example 10.2.2 the researchers have used randomization to prevent imbalances between their treatment groups with respect to other factors that might affect antibody levels. We are on much safer ground concluding that the differences we are seeing are caused by the vaccines administered.
- *Investigative strategy*
  In Example 10.2.2 treatments were compared using several samples of people, with different people getting different treatments. Why did the researchers do it this way? There are other ways of comparing treatments, such as trying all treatments on every individual. What investigative strategy gives the best-quality information at lowest cost?

Issues of Analysis

- *The importance of plotting data*
  Our first reaction to data is to plot it and look at it. There is no quicker way of seeing what is happening in data than by looking at an appropriate plot. Plots can

prevent dangerous misconceptions from arising. If we had not plotted the data in Example 10.2.1 but had only done a *t*-test and found that homosexuals have significantly lower androsterone levels than homosexuals, we would not have noticed that there is still a substantial overlap in androsterone levels between the two groups, a fact with important practical implications.

- *The applicability of formal tools of analysis*
  All formal methods of analysis make assumptions. Are the assumptions of the method we wish to use satisfied, and if not, is this fact likely to hurt us?
- *Features other than means*
  Are there features of the data other than the means that it would be interesting to compare? How do we go about this?

Study design issues are discussed in Section 10.4. This section deals with statistical analysis, focusing on differences between means. The method used for making inferences about the difference between two means in Chapters 8 and 9 was a conservative approximation. We now wish to discuss this two-sample problem in more detail and then move on to discussing problems involving more than two samples.

QUIZ ON SECTION 10.2.1

1. At what stage of an analysis is the distinction between a randomized experiment and an observational study important?
2. The importance of plotting data was stressed. Plots have two important roles. What are they?

10.2.2 Comparing Two Means

Suppose that we have two random samples, the first from a Normal(μ_1, σ_1) distribution and the second from a Normal(μ_2, σ_2) distribution. In Chapters 8 and 9 we used the following formulas for making inferences about $\mu_1 - \mu_2$:

$$\text{Confidence interval for } \mu_1 - \mu_2\text{:} \quad \overline{x}_1 - \overline{x}_2 \pm t\,\text{se}(\overline{x}_1 - \overline{x}_2)$$

$$t\text{-test for } H_0\text{: } \mu_1 - \mu_2 = 0: \quad t_0 = \frac{(\overline{x}_1 - \overline{x}_2) - 0}{\text{se}(\overline{x}_1 - \overline{x}_2)}$$

In the examples, we used the standard error formula $\sqrt{s_1^2/n_1 + s_2^2/n_2}$ together with degrees of freedom, $df = \text{Min}(n_1 - 1, n_2 - 1)$. As we noted in Section 8.4, this choice of standard error and *df* gives a conservative approximation (for hand calculation only) to the so-called *Welch* procedure. The Welch procedure has a very complicated *df* formula.[11]

In this chapter the focus has shifted from ease of hand calculation to use of computer packages. We now give the background to one of the major choices the programs offer you. You are asked to choose between assuming *equal variances* (equivalently, equal standard deviations) and *unequal variances*. Here the term variance refers to the square of the standard deviation. A variant of the Welch procedure is used if you choose unequal variances. The so-called pooled method described shortly is

[11] See Roth [1988] and Best and Rayner [1987] for further technical details.

used for equal variances. In Minitab and R "unequal variances" is the default (you have to act to change it). In Splus "equal variances" is the default. We recommend routine choice of unequal variances for reasons we now describe.

There is no exact theory supporting the Welch method. The sampling distribution of *(estimate − true value)/standard error* using the preceding formulas does not have an exact Student's t-distribution for any value of df, even when both samples come from Normal distributions. The df value used by the Welch method depends on the sample sizes and also on the sample standard deviations, so that in repeated sampling the value it gives changes from sample to sample. You will often see noninteger values in computer output. What has been shown[12] is that confidence intervals and tests obtained using the Welch df values have properties that are very close to the desired properties; for example, the nominally 95% confidence intervals obtained cover the true difference in means for very close to 95% of samples taken.

If, however, we assume Normal distributions and also equal population standard deviations ($\sigma_1 = \sigma_2$), then the sampling distribution of $(\overline{x}_1 - \overline{x}_2 - \textit{true difference})/\textit{standard error}$ has an exact Student($df = n_1 + n_2 - 2$) distribution if you use the following formula for the standard error:

$$\text{se}(\overline{x}_1 - \overline{x}_2) = s_p\sqrt{\frac{1}{n_1} + \frac{1}{n_2}} \quad \text{where } s_p^2 = \frac{(n_1 - 1)s_1^2 + (n_2 - 1)s_2^2}{n_1 + n_2 - 2}$$

Here s_p^2 is called the *pooled estimate of variance* because it pools information from both samples to form a combined estimate of the single variance $\sigma_1^2 = \sigma_2^2 = \sigma^2$, say. The equal-variances method is the one described in most introductory books. The critical factor in its use is the need to make one additional assumption, namely, $\sigma_1 = \sigma_2$.

The equal-variances assumption will seldom, if ever, be true in practice. With experimental data, when we compare a treatment group with a control group and we are interested in the null hypothesis that the treatment makes no difference, calculating a P-value under the twin assumptions that $\mu_1 = \mu_2$ and $\sigma_1 = \sigma_2$ makes a good deal of sense. Unfortunately, this idea does not carry over to confidence intervals. If the mean changes, there is no reason to think that the standard deviations will remain the same. Although the pooled method is reasonably robust against departures from this assumption when the two sample sizes are the same, little is gained in practical terms by making the assumption. A good deal can be lost, however, when the assumption is false and the sample sizes are unequal. Thus our recommendation is for routine use of the Welch (unequal-variances) procedure.

Robustness of Two-Sample t-Tests

The qualitative statements made about the robustness of the one-sample t-procedures in Section 10.1.1 still hold. In fact, two-sample t-tests and intervals are even more robust against non-Normality than one-sample t-tests, particularly when the shapes of the two distributions are similar and the sample sizes are the same, that is, $n_1 = n_2$. In that case the two-sample t-procedures have been shown to work well for a broad class of distributions with samples as small as $n_1 = n_2 = 5$. The sample-size recommendations given in the one-sample case can be taken as applying to $n_1 + n_2$.

[12]For technical details see Algina et al. [1994].

Nonparametric Two-Sample Tests

The best-known nonparametric alternative to the two-sample *t*-test is variously called the ***Wilcoxon (rank sum) test*** and the ***Mann-Whitney test.*** The two tests are equivalent and produce the same *P*-values. Computer packages invert these tests to produce nonparametric confidence intervals. Recall from Section 10.1.3 that nonparametric methods are based only on the orderings of the data. The notion of average used is the median. Thus the Wilcoxon or Mann-Whitney test is a test for the equality of two population medians,[13] and the intervals provided are intervals for the difference in two medians. For a discussion of the way the Wilcoxon test works, see our web site.

Using Packages

The two-sample *t*-procedures (both equal- and unequal-variance versions) are available in Minitab under `Stat` → `Basic Statistics` → `2-Sample t`, in Excel under `Tools` → `Data Analysis` (no confidence intervals are produced by Excel), from the Splus and R function `t.test` (see the on-line help files for syntax), and in all other general-purpose statistical packages. The output in Fig. 10.2.3 was produced by Minitab. The Wilcoxon/Mann-Whitney method is available in Minitab under `Stat` → `Nonparametrics` → `Mann-Whitney`, using the Splus function `wilcox.test`, and in many other packages.

EXAMPLE 10.2.1 (cont.) ***Computer Output for a Difference of Means***

We suggested the possibility of non-Normal behavior with these data. In fact, it transpired that the Normal probability plots looked fine, and formal tests showed no evidence against the Normality assumption. We are also protected by the robustness of the *t*-tests and intervals. Minitab output for a test of no difference in true mean androsterone levels between heterosexuals and homosexuals (using the unequal-variances option) versus a not-equal-to alternative is given in Fig. 10.2.3. The very small *P*-value (0.004) tells us that we have very strong evidence of a difference between the true means. A 95% confidence interval for the true difference, $\mu_{Het} - \mu_{Hom}$ is given by [0.4, 1.7] (the units are mg/24 hr). We note that the pooled test gives a very similar *P*-value and interval. The Mann-Whitney test gave a *P*-value of approximately 0.01 and a 95% confidence interval for the difference in true medians of [0.4, 1.8], which is again very similar to the values given in Fig. 10.2.3 for means.

Two Sample T-Test and Confidence Interval

```
Two sample T for androsterone
             N       Mean      StDev    SE Mean
hetero      11      3.518      0.721       0.22
homose      15      2.500      0.923       0.24
95% CI for mu (hetero) - mu (homose): ( 0.35,   1.69)
T-Test mu (hetero) = mu (homose) (vs not=): T=3.16  P=0.0044  DF=23
```

Confidence interval — *t-test statistic* — *P-value*

FIGURE 10.2.3 Minitab two-sample *t*-output for the androsterone data.

[13]Technically it is a test that the two underlying distributions are the same (which implies that their medians are equal).

QUIZ ON SECTION 10.2.2

1. The pooled form of the two-sample t-test makes one additional assumption that the Welch form of the test does not. What is it?
2. How sensitive is the two-sample t-test to non-Normality in the data?
3. What are the names of the nonparametric alternatives to the two-sample t-test?
4. What difference is there between the quantities tested and estimated by the two-sample t-procedures and the nonparametric equivalents?

10.3 MORE THAN TWO SAMPLES

10.3.1 One-Way Analysis of Variance and the *F*-Test

Having looked at the problem of comparing the means of two populations using a random sample from each population, we now consider the more general problem of comparing the means of several populations or groups using independent samples. This topic comes under the general heading of *analysis of variance,* commonly abbreviated *ANOVA. One-way* analysis of variance refers to situations where there is only *one factor* (or categorical variable) defining group membership, as was the case with the pertussis data in Example 10.2.2 and is also the case in the following Example 10.3.1. In this section we discuss a new test called the ***F*-test for one-way ANOVA.** The null hypothesis tested by the F-test is that all of the underlying true means are identical. The alternative hypothesis is that differences exist between some of the true means.

Hypotheses for the one-way analysis-of-variance *F*-test
Null hypothesis: All of the underlying true means are identical.
Alternative: Differences exist between some of the true means.

Written mathematically, the null hypothesis is H_0: $\mu_1 = \mu_2 = \cdots = \mu_k$.

EXAMPLE 10.3.1 *Comparing Four Reading Methods*

An investigation was performed at Carmel College, Auckland, New Zealand, of the effects of different reading methods on reading comprehension. As part of a study-skills program, fifty 13- and 14-year-old students learned about mapping (using diagrams to note and relate main points) and scanning (reading introductions and skimming for an overview before reading in detail). The teacher wanted to know how effective these methods were. The Gapadol test gives a measure of reading age. There are two similar but different forms of the test. Reading ages were obtained before and after the course of instruction (a period of about six weeks). One version of the Gapadol test was used for "before" testing and the other for "after" testing. Students were classified into four groups, depending on which of the reading methods they said they had used when taking the second test. Table 10.3.1 gives the increases in test score (years of reading age).

TABLE 10.3.1 Increase in Reading Age

Both	0.1	3.2	4.3	−0.5	1.9	3.3	2.5	3.6	0.4	2.3	−1.4	−0.7
	−0.1	0.2	0.4	0.9	1.2	1.4	1.8	1.8	2.4	3.1		
Map only	1.0	−0.5	1.0	0.6	0.6	1.0	1.0	−1.4	2.2	3.6	3.1	2.6
Scan only	1.0	3.3	1.4	−0.9	1.0	0.0	0.6					
Neither	−0.3	−1.3	1.6	−0.4	−0.7	0.6	−1.8	−2.0	−0.7			

Kindly provided by Mary Matthews, Carmel College.

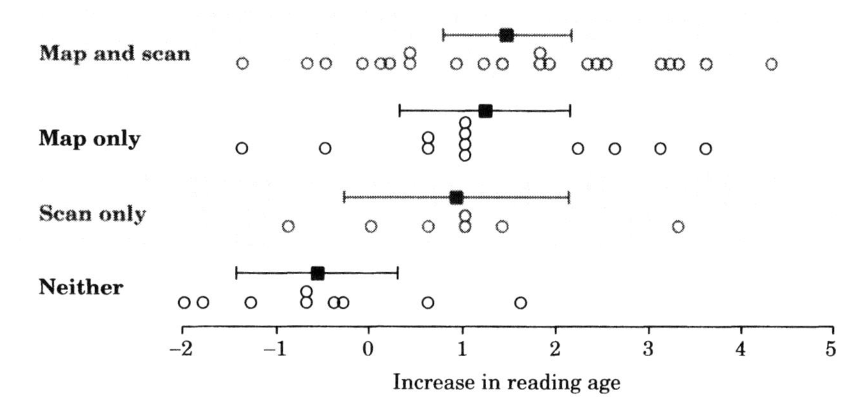

FIGURE 10.3.1 Increases in reading ages with individual 95% confidence intervals.

One-way Analysis of Variance

```
Analysis of Variance for Increase          F-statistic  P-value

Source     DF        SS        MS        F        P
Grp         3     27.06      9.02     4.45    0.008      } Anova Table
Error      46     93.35      2.03
Total      49    120.41
                                 Individual 95% CIs For Mean
                                 Based on Pooled StDev
Level       N      Mean     StDev  ------+---------+---------+---------+
MapScan    22     1.459     1.544                            (------*-----)
MapOnly    12     1.233     1.441                        (-------*--------)
ScanOnly    7     0.914     1.302                  (----------*----------)
Neither     9    -0.556     1.135   (--------*---------)
                                   ------+---------+---------+---------+
Pooled StDev =    1.425               -1.0       0.0       1.0       2.0
```

FIGURE 10.3.2 Minitab analysis-of-variance output for reading ages.

Are the increases better for students who used mapping, scanning, or both? Is it better to use just one of these methods or to use them in combination? This is an observational study. If "reading age after the course of instruction" was used as the variable to be analyzed, any differences might just be due to differences between the groups in reading ability. We get around this problem to some extent by looking at the *increases* in reading ages to see how much the reading age had changed.

The data from Table 10.3.1 are plotted, together with 95% confidence intervals for the true means, in Fig. 10.3.1. One-way analysis-of-variance output from Minitab is given in Fig. 10.3.2. From the dot plots in Fig. 10.3.1, the scores for each of the three groups that had used mapping, scanning, or both appear to be shifted to the right by a similar amount from those for the group that used neither. The visual impression is that the treatments all seem to have some effect and that their effects appear to be similar. The confidence intervals plotted are individual intervals for means. We noted in Chapter 8 that when individual 95% confidence intervals do not overlap, the 95% confidence interval for a difference will not contain 0. Equivalently, the difference will be significant at the 5% level. We therefore seem to see at least one statistically significant difference in the plot.

Figure 10.3.2 gives output automatically generated by Minitab's one-way analysis-of-variance procedure. The portion of the output inside the dotted box is called an analysis-of-variance table. We shall explain the contents of the table in the next subsection. However, the *P*-value indicated at the top right-hand corner is the *P*-value for the *F*-test for a one-way analysis of variance. Recall that the null hypothesis tested states that all of the underlying true means are identical. The *P*-value is very small (0.008), showing very strong evidence against this hypothesis. We have clear evidence that differences exist between at least some of the true means.

Although in Example 10.3.1 the *F*-test has told us that we have evidence that true differences exist, the test gives no indication of where the differences are or how big they are. This information comes from confidence intervals for differences between the true means.

Interpreting the *P*-value from the *F*-test

(The null hypothesis is that all underlying true means are identical.)

- A ***large*** *P*-value indicates that the differences seen between the sample means could be explained simply in terms of sampling variation.
- A ***small*** *P*-value indicates evidence that real differences *exist* between at least *some* of the true means but gives no indication of where the differences are or how big they are.
- To find out how big any differences are, we need confidence intervals.

We note in passing that, as with the equal-variances approach to the two-sample problem, the confidence intervals printed by one-way ANOVA programs (e.g., those in Figs. 10.3.2 and 10.3.3) assume that all groups have the same true variance or, equivalently, the same standard deviation. A pooled estimate of this standard deviation (i.e., one that uses information from all of the groups) is employed in the calculation of standard errors for single means and for differences between means.

Figure 10.3.3 gives further Minitab output for the reading age data. The additional output is available when requesting a one-way analysis of variance in Minitab by checking requests in dialogue boxes that offer additional information. The output consists of 95% confidence intervals for all differences between the underlying true means of pairs of groups. Similar information is available from other packages. The interval given in the dotted rectangle, $[-1.255, 0.803]$, estimates the true difference in mean scores between "mapping and scanning" and "mapping only." Note that there are two entire sets of confidence intervals. The first is headed "Fisher's pairwise comparisons." These intervals are 95% confidence intervals. You will see "Individual error rate = 0.0500" near the top of the printout. We already know that 95% intervals have a 5% error rate. However, "Family error rate = 0.198" indicates the presence of a new problem.

Intervals for differences between means (from Minitab)

```
→ Fisher's pairwise comparisons
      Family error rate = 0.198
  Individual error rate = 0.0500
  Critical value = 2.013
  Intervals for (column level mean) - (row level mean)
                 MapOnly       MapScan      Neither
        MapScan  -1.255
                  0.803
        Neither   0.524         0.880
                  3.053         3.149
        ScanOnly -1.047        -0.700       -2.915
                  1.683         1.789       -0.025

→ Tukey's pairwise comparisons
      Family error rate = 0.0500
  Individual error rate = 0.0106
  Critical value = 3.77
  Intervals for (column level mean) - (row level mean)
                 MapOnly       MapScan      Neither
        MapScan  -1.589
                  1.137
        Neither   0.114         0.512
                  3.463         3.517
        ScanOnly -1.487        -1.103       -3.384
                  2.125         2.193        0.444
```

FIGURE 10.3.3 Confidence intervals for differences between means.

Each 95% confidence interval has a 95% chance of capturing the true value of the parameter it is estimating. When we calculate a set of intervals, the chance that they all simultaneously catch their true values can be a good deal less than 95%. The more simultaneous comparisons, the worse the problem gets. Minitab has calculated that the particular set of six comparisons for the reading age data has an error rate of almost 20%. Put another way, there is 1 chance in 5 that at least one interval in the set will fail to catch its true value compared with 1 chance in 20 for a single interval.

A number of methods for *multiple comparisons* have been worked out to ensure that the error rate for a whole set of intervals remains low at 5%. We chose "Tukey's pairwise comparisons." This is the second half of the printout. You will note that the "Family error rate" is now 0.05, that all of the intervals have become wider, and that the "Individual error rate" is now 0.0106. In essence, to get an overall error rate of 5% for the whole set of intervals, the program has had to use 99% confidence intervals for each comparison.

✱ Example 10.3.1 (cont.) *Confidence Intervals for Differences between Means*

Looking at the set of intervals in Fig. 10.3.3 headed "Tukey's pairwise comparisons," we see that all but two intervals contain 0. Thus, at the 95% confidence level, these differences can plausibly be explained simply in terms of sampling errors. The exceptions are the comparison between "map only" and "neither" and between "map and scan" and "neither." In the first case our interval estimates that the true increase in test score is higher on average for the "scan only" group than for the "neither" group by between 0.1 and 3.5. Other intervals can be read similarly. Note that we cannot conclude that any treatment is better than any other treatment. While the confidence interval comparing "scan only" to "neither" contains 0, one confidence limit is very close to 0 so that the P-value for an equivalent test comparing the "scan only" and "neither" groups would probably give a P-value very little larger than 5%. Overall, we stand by our initial visual impression. There is evidence that the methods are effective but insufficient information in the data for us to tell which of them is best.

10.3.2 The *F*-Test and the Analysis-of-Variance Table

The F-test that we have used earlier applies to the situation where we have independent samples from each of k populations. We assume that the data from the ith group is a random sample from a Normal(μ_i, σ) distribution. We note the important *assumption* implicit here that the underlying population standard deviation is the same for each group.

When you ask for a one-way analysis of variance,[14] most statistical computer packages will supply you with the information in Table 10.3.2. The labeling of the rows varies between programs, but we shall see that the first row relates to between-groups variability and the second row (often labeled "error") relates to within-groups variability.

The following notation is used. There are n_i observations in the ith group and $n_{tot} = \sum n_i$ is the total number of observations. The jth observation in group i is x_{ij}. Group i has sample mean and standard deviation $\bar{x}_{i\cdot}$ and s_i, respectively. The overall sample mean of all of the observations, regardless of which group they come from, is $\bar{x}_{\cdot\cdot}$. The information in the "Mean sum of squares" column is obtained by dividing the entry in the "Sum of squares" column by the information in the df column. Thus

$$s_B^2 = \frac{\sum n_i(\bar{x}_{i\cdot} - \bar{x}_{\cdot\cdot})^2}{k-1} \quad \text{and} \quad s_W^2 = \frac{\sum(n_i - 1)s_i^2}{n_{tot} - k}$$

The test statistic for the F-test, which we denote f_0, is given in the "F-statistic" column, and the P-value for the test is printed in the right-hand column.

To understand the F-test, it helps to approach the issue of obtaining evidence against H_0 intuitively. We might expect having sample means that are well separated to be a factor in demonstrating the existence of real differences, but this is not the whole story. Figure 10.3.4 shows three examples of situations in which three groups of data are being compared. All samples are of size $n = 8$. The position of the sample

TABLE 10.3.2 Typical Analysis-of-Variance Table for One-Way ANOVA

Source	Sum of squares	df	Mean sum of squares[a]	F-statistic	P-value
Between	$\sum n_i(\bar{x}_{i\cdot} - \bar{x}_{\cdot\cdot})^2$	$k - 1$	s_B^2	$f_0 = s_B^2/s_W^2$	pr($F \geq f_0$)
Within	$\sum(n_i - 1)s_i^2$	$n_{tot} - k$	s_W^2		
Total	$\sum\sum(x_{ij} - \bar{x}_{\cdot\cdot})^2$	$n_{tot} - 1$			

[a]Mean sum of squares = (sum of squares)/df

[14]For example, in Minitab under `Stat` → `ANOVA` → `One-way`; in Excel under `Tools` → `Data Analysis`, choose `Anova: Single Factor`; or in Splus using aov followed by `summary`.

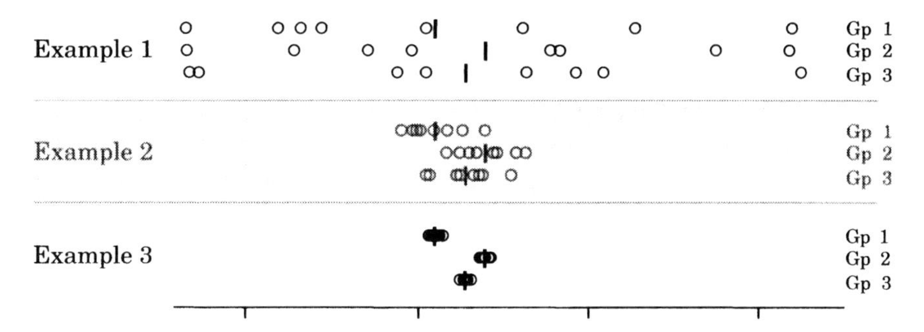

FIGURE 10.3.4 Examples of comparing groups.

mean for each group is plotted with a vertical line (|). Note that the observations in example 3 are so close together that they are overprinting one another. From our experience with students, virtually nobody looking at example 1 is convinced that it provides any evidence that the true means are different. On inspecting example 2, some believe that the true means are probably different, but others are not sure. Finally, almost everybody believes that the true means underlying example 3 are different. The examples have been constructed so that the set of sample means is identical for each example. What distinguishes example 3 from example 1 is that in example 3, the differences between the sample means are *large compared to the internal variability* within the samples, whereas in example 1 they are small compared to the internal variability.

Close inspection of the formula $s_B^2 = \sum n_i(\overline{x}_{i\cdot} - \overline{x}_{\cdot\cdot})^2/(k - 1)$ reveals that s_B^2 is a measure of the variability of the sample means, namely, how far apart they are. We know that s_i measures variability within the ith sample. Also, $s_W^2 = \sum(n_i - 1)s_i^2/(n_{tot} - k)$ is a weighted average of the s_i^2 and thus reflects average internal variability within samples. Thus the ***F-statistic,*** $f_0 = s_B^2/s_W^2$, compares the variability of the sample means with the internal variability within the groups. (In fact, s_W^2 is the natural generalization to k samples of s_p^2, the pooled estimate of the common σ^2 given in Section 10.2.2. Moreover, it can be shown that when $k = 2$, the F-statistic f_0 is the same as the square of the pooled two-sample t-statistic and that the P-value obtained from the F-test is the same as that obtained from the *two-sided* pooled t-test.)

- The F-test statistic, f_0, tests H_0 by comparing the variability of the sample means (numerator) with the variability within the samples (denominator).
- Evidence against H_0 is provided by values of f_0, which would be unusually *large* if H_0 was true.

When H_0 is true and the model assumptions hold, the formula for the sampling distribution of F_0, the random variable corresponding to f_0, is known and is called the F-distribution, with numerator degrees of freedom $df_1 = k - 1$ and denominator degrees of freedom $df_2 = n_{tot} - k$. We write this distribution as $\text{F}(df_1 = k - 1,\ df_2 = n_{tot} - k)$. The P-value for the test, which is printed out by computer programs, is

$$P\text{-value} = \text{pr}(F \geq f_0), \quad \text{where } F \sim \text{F}(df_1 = k - 1,\ df_2 = n_{tot} - k)$$

Hand-calculator computation of the F-statistic and the use of F-distribution tables is discussed on our web site.

Analysis of Variance Procedure					
Dependent Variable: PERTUSS				*F-statistic* ↓	*P-value* ↓
Source	DF	Sum of Squares	Mean Square	F Value	Pr > F
Model	2	15.81677991	7.90838996	6.17	0.0031
Error	87	111.47241564	1.28129213		
Corrected Total	89	127.28919556			
With the outlier included, the *P*-value increases to 0.023					

FIGURE 10.3.5 ANOVA output for the pertussis data from SAS.

✖ EXAMPLE 10.3.2 *An Outlier and the F-Test*

This is a continuation of Example 10.2.2. In the pertussis (whooping cough) data presented in Table 10.2.2, there were three treatment groups made up of randomly assigned subjects: One group received the standard vaccine WCV, another group received the experimental vaccine APV, and the third received DAPV (a double dose of APV). The investigators had good reasons for believing that the outlier (-2.30) in the APV group (see Fig. 10.2.2) was a mistake.

Figure 10.3.5 gives computer output from the well-known SAS package for the one-way ANOVA *F*-test, which tests the null hypothesis that the underlying means for each of the groups are identical. The output refers to analyzing the data with the outlier removed. The *P*-value for the test is 0.003, signaling clear evidence that differences do exist between the true means. (If the outlier is included, the *P*-value increases considerably to 0.023.) From Fig. 10.2.2 it appears that there is a difference between the experimental vaccine and the standard vaccine but little if any difference between the two different strengths of the experimental vaccine. It would be interesting to estimate the sizes of the differences. We should go on to obtain confidence intervals for differences (cf. Fig. 10.3.3) to further investigate this.

Assumptions Underlying the *F*-Test

We have noted that the theory that produces the sampling distribution for the *F*-statistic (f_0) given earlier assumes that the ith sample of data is a random sample from a Normal distribution with mean μ_i and standard deviation σ and that the samples are independent.

- ***Independence of the samples***

This assumption is critical. Usually, the only way of knowing that this assumption is true is when the samples consist of physically unrelated individuals, objects, or processes. In Section 10.1.2 we saw some paired data sets. Paired data, we noted, are often wrongly analyzed as coming from independent samples, when in fact the two sets of observations are often very closely related. In the same way, when we scrutinize the way that the data were obtained, data that at first glance look like a ***k***-independent-samples problem are sometimes seen to consist of sets of observations that are closely related.

- ***Normally distributed data***

The *F*-test enjoys a similar robustness against departures from the Normality assumption as does the two-sample *t*-test (particularly when the group sizes are very similar; see Section 10.2.2).

- ***Equality of standard deviations***

The theory assumes that the underlying level of variability, as measured by the population standard deviation or its square, the variance, is exactly the same for each group. A usually conservative rule of thumb is that departures from this assumption are unlikely to do much harm if the ratio of the largest to smallest sample standard deviation is no more than 2.

Although the F-test is reasonably robust against differences in spreads, the confidence intervals printed by analysis-of-variance programs are not. The problem occurs because the programs calculate standard errors assuming that all groups are equivariable and thus incorporate contributions from all groups in a pooled estimate of variability. The result is that confidence intervals for the least-variable groups are too wide and those for the most-variable groups are too narrow. One solution is to abandon the equal-variance assumption and go back to individual confidence intervals for means and Welch (unequal-variance) confidence intervals for differences. We shall discuss testing for equal variances (and standard deviations) shortly.

What to do if the assumptions are badly violated. The options discussed in Section 10.1.1, including the advice about the treatment of outliers, also apply here. One can use robust methods, change the model, transform the data, or use nonparametric methods. The best known nonparametric alternative to the F-test is called the *Kruskal-Wallis test.*[15] This test is usually available from statistics packages, for example, under `Stat` → `Nonparametrics` → `Kruskal-Wallis` in Minitab.

Welch and others have generalized the Welch procedure for comparing more than two means without making the equal-standard-deviations assumption, but the procedure is not widely available.[16] When the sample sizes are large, weighted analyses are often used to cope with severe departures from the equal-standard-deviations assumption. All of these topics are beyond the scope of this book.

****Testing for differences in spread.*** There are a number of standard tests for the equality of population standard deviations based on the sample standard deviations. Unfortunately, these tests are very sensitive to any non-Normality of the data and are therefore suspect.[17] There are several robust tests,[18] however, for testing whether the populations have different standard deviations using a different (and more robust) measure of spread. One test (available in Minitab under `Stat` → `ANOVA` → `Homogeneity of Variance`) is called Levene's test. The background of this test and similar tests follows.

Recall that we first thought of spread as being some sort of average of the distances between the data points and their center. Here we work with this idea directly through the so-called *absolute deviations.*[19] For each group we calculate

$$y_1 = |x_1 - \overline{x}|,\ y_2 = |x_2 - \overline{x}|,\ \dots,\ y_n = |x_n - \overline{x}|$$

where $\overline{x}$ is the mean for the group. If the spread is larger for one group than it is for another, this will be reflected in the former group having deviations that are larger

[15]See, for example, Bhattacharyya and Johnson [1977] or a general text book on nonparametric statistics such as Conover [1980].

[16]See Roth [1988].

[17]See Markowski and Markowski [1990].

[18]See Conover et al. [1981].

[19]With absolute distances we use the size of the difference and ignore the sign. Recall that the standard deviation is essentially the square root of the average of the squared distances (with $n - 1$ instead of n).

on average. It transpires that we can test a null hypothesis of equal spreads (cf. equal standard deviations) by testing that the true (i.e., population) mean absolute deviations for all of the groups are identical. This can be accomplished by using an F-test for the one-way analysis of variance using the y-data instead of the x-data. We prefer the robust version of Levene's test,[20] which uses the sample median as its measure of center in place of the sample mean. You do not need new software for the robust Levene's test. You need only form the absolute deviations and run them through an analysis-of-variance program.

When applied to the pertussis data of Examples 10.2.2 and 10.3.2, Levene's test gives a P-value that is 0.000 (to three decimal places, from Minitab). This signals strong evidence of a difference in spreads between the groups. Looking back at Fig. 10.2.2, we can see that the control group looks much more variable than the treatment groups. Its standard deviation is only about twice as big, so we would be prepared to trust the F-test. We would not, however, trust the confidence intervals from a one-way ANOVA program.

EXAMPLE 10.3.1 (cont.) *Checking out the Assumptions*

Let us return to the reading methods data. We have separate groups of students so that we need not worry about any lack of independence. The sample standard deviations are very similar. Inspection of the dot plots in Fig. 10.3.1 shows no evidence of grossly non-Normal behavior. The most worrying feature is some clumping of data points in the middle of the "map only" group, which we would want to discuss with the investigator. If inspected in isolation, the biggest observation in the "scan only" group might appear to be an outlier, but in view of the variation seen in other groups we do not believe that this is an outlier. The ratio of the largest to smallest standard deviation is approximately 1.4, which is considerably less than 2. Thus, to the extent that you have learned to check the model assumptions, we see no important violations of the assumptions and would be happy to present the results that we obtained previously concerning differences between the groups. The P-value from Levene's test for equal spreads is 0.46, thus confirming our previous assessment.

We complete this example with a brief sketch of further investigations of model adequacy that go beyond the scope of this book. Some readers may wish to omit this material.

Our basic model for observations in group i can be written as

$$\text{(Population model)} \qquad X_{ij} \sim \text{Normal}(\mu_i, \sigma_i), \quad j = 1, \ldots, n_i$$

Another common way of expressing this model is

$$\text{(Measurement model)} \quad X_{ij} = \mu_i + U_{ij}, \quad \text{where } U_{ij} \sim \text{Normal}(0, \sigma_i)$$

Here μ_i is the underlying true mean for the ith group, and the U_{ij}'s (the distances of the observations X_{ij} from their mean μ_i) are thought of as random perturbations (or random errors). Reverting to lowercase letters because we wish to refer to the observed data, we note that $u_{ij} = x_{ij} - \mu_i$, and we can estimate this by the so-called *residual* $\widehat{u}_{ij} = x_{ij} - \overline{x}_i$. Plotting the $\widehat{u}_{ij}$'s versus group i gives a way of looking for

[20] See Glaser [1983].

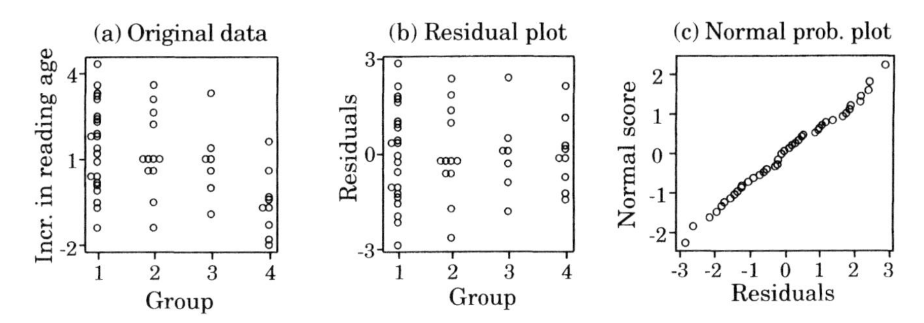

FIGURE 10.3.6 Diagnostic plots for the reading scores data.

differences in spreads between groups without the distraction of changing mean levels. Figure 10.3.6(a) is the same as Fig. 10.3.1, except the scale is running vertically rather than horizontally. Figure 10.3.6(b) plots the residuals. Visually, the difference between (a) and (b) is that the patterns of points have been shifted so that their mean values are all at the same level (namely 0). (Recall that Levene's test showed no evidence of differences in spread.)

If we make the working assumption that the standard deviations are the same and now wish to examine the Normality of the errors, we can use the complete set of residuals $\widehat{u}_{ij}$. The actual errors, u_{ij}, are a random sample from a Normal(0, σ) distribution, and we would expect their estimates, the residuals, to behave similarly.[21] It therefore makes sense to check whether the residuals look as though they could have come from a Normal distribution. A Normal probability plot of the residuals is given in Fig. 10.3.6(c) and looks well behaved. Formal tests for Normality (see Section 10.1.1) applied to the residuals showed no evidence against the Normality assumption.

QUIZ ON SECTION 10.3

1. What is a *one-way* analysis of variance?
2. When do we use the one-way ANOVA F-test?
3. What null hypothesis does it test? What is the alternative hypothesis?
4. Qualitatively, how does the F-test obtain evidence against H_0?
5. Qualitatively, what type of information is captured by the numerator of the F-statistic? What about the denominator?
6. Qualitatively, what values of f_0 provide evidence against H_0?
7. What does a large P-value from the F-test tell us about differences between means? How about a small P-value?
8. What does a small P-value tell us about which means differ from one another? about how big the differences between means are?
9. How do we obtain information about the sizes of differences between means?
10. What assumptions are made by the theory on which the F-test is based? How important is each of these assumptions in practice?
11. What new problem arises when we need to obtain and inspect a large set of confidence intervals?

[21] We form the $\widehat{u}_{ij}$'s by subtracting the sample means, which are random, and this changes the distribution somewhat.

12. Which is affected worst by departures from the equal-standard-deviations assumption, the F-test or the confidence intervals? Why?

EXERCISES FOR SECTION 10.3

Monitoring human exposure to environmental agents that might cause genetic damage, and determining the type and extent of such damage, is a subject that is commanding increasing attention. To pursue these objectives, scientists need to develop methods of measuring damage to human genetic structures. One method that has been suggested as a possible measure of the genetic damage present in an individual's DNA is derived from counting the number of SCEs (sister chromatid exchanges) observed per cell. An SCE results from a reciprocal exchange of DNA between two spiral filaments that constitute a chromosome (sister chromatids). For each individual a sample of cells is taken, and the average of the resulting SCE measurements is recorded. This gives the value of the variable we call MSCE for that individual. Other scientists have suggested using a variable we call DISPERSION, which relates to the variability of SCE measurements within an individual compared to that individual's average SCE level. Preliminary work in investigating the usefulness of these measures involved investigating whether racial differences could be detected. The data reported in Margolin [1988] is given as Table 10.3.3. Plot the data and obtain separate analysis-of-variance output for investigating racial differences in MSCE and also for investigating racial differences in DISPERSION. Then answer the questions that follow.

1. What do you conclude about racial differences in MSCE? Is there anything in the MSCE plot that makes you doubt the assumptions of the F-test?
2. Assuming the applicability of Normal theory methods, what would you conclude about racial differences in the mean DISPERSION?
3. There is an outlier in the black group. Intuitively, do you think that the evidence for a difference in mean DISPERSION between the groups would be weakened or strengthened if the outlier was omitted?
4. Reanalyze the data with the outlier omitted. Are your intuitions confirmed?
5. Are there any other features of the data on DISPERSION that would make you doubt the applicability of a one-way analysis of variance?

TABLE 10.3.3 Genetic Data

Black		Caucasian		Native American		Asian	
MSCE	DISPERSION	MSCE	DISPERSION	MSCE	DISPERSION	MSCE	DISPERSION
8.44	1.09	8.27	0.58	8.50	0.45	9.84	1.63
7.40	0.62	8.20	0.52	9.48	0.75	9.40	1.29
8.15	0.91	8.25	0.65	8.65	0.85	8.20	1.39
7.36	0.97	8.14	0.45	8.16	0.50	8.24	1.28
7.60	0.75	9.00	0.56	8.83	0.61	9.20	0.66
8.04	1.02	8.10	0.44	7.76	0.63	8.55	2.06
9.04	2.63	7.20	0.58	8.63	0.93	8.52	1.24
9.76	1.18	8.32	0.79			8.12	1.30
		7.70	0.53				

Source: Margolin [1988].

10.4 BLOCKING, STRATIFICATION, AND RELATED SAMPLES

10.4.1 Pairing

In Section 10.1.2 we referred to the situation where the data can occur in pairs. This does not usually happen accidently but is often orchestrated as part of the experimental design. The following example focuses on this design aspect and asks some questions about when the pairing can be useful.

Example 10.4.1 *Issues in Designing an Experiment*

A shoe company wanted to compare two synthetic materials (A and B) for use for constructing the soles of boys' shoes. We could design an experiment to compare the two materials in a number of ways, of which two are depicted in Fig. 10.4.1. For the *completely randomized design,* we recruit a number of boys and randomly split them into two groups. This leads to independent samples. Boys in the first group are given shoes soled with material A, and the second group would have shoes soled with material B. Then after a suitable length of time, say three months, we could measure the wear on each boy's shoes.[22]

We can also exploit the fact that each boy has two feet to construct a *paired design.* We make up pairs of shoes with one material on the sole of the left shoe and the other material on the sole of the right shoe. We would want to randomize which shoe gets material A and which gets B so that if, for example, right shoes tend to get more wear than left shoes, it does not bias the results. What is there to be gained by pairing? We shall use the data set in Table 10.4.1, taken from a trial using a paired

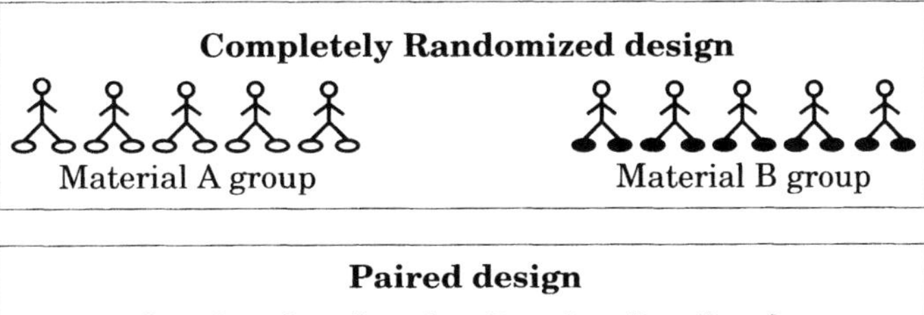

FIGURE 10.4.1 Comparing two treatments: Two experimental designs.

[22]Measuring wear on shoe soles is not a simple thing to do, as the amount of wear varies over different parts of the sole (check your own shoes). One possibility is to measure the reduction in depth of sole at a number of prespecified positions on the sole and to use the average of those measurements to provide an average measure of wear for the whole sole. To be able to use the independent-samples theory of Section 10.2 to analyze the completely randomized design in Fig. 10.4.1, we need the observations within each group to be independent. Since measurements on the same boy will be related, we shall want to use only one measurement per boy, for example, the average wear for the two feet.

TABLE 10.4.1 Boy's Shoe Wear Data

Boy	1	2	3	4	5	6	7	8	9	10
Material A	13.2	8.2	10.9	14.3	10.7	6.6	9.5	10.8	8.8	13.3
Material B	14.0	8.8	11.2	14.2	11.8	6.4	9.8	11.3	9.3	13.6

Source: Box et al. [1978, p. 100].

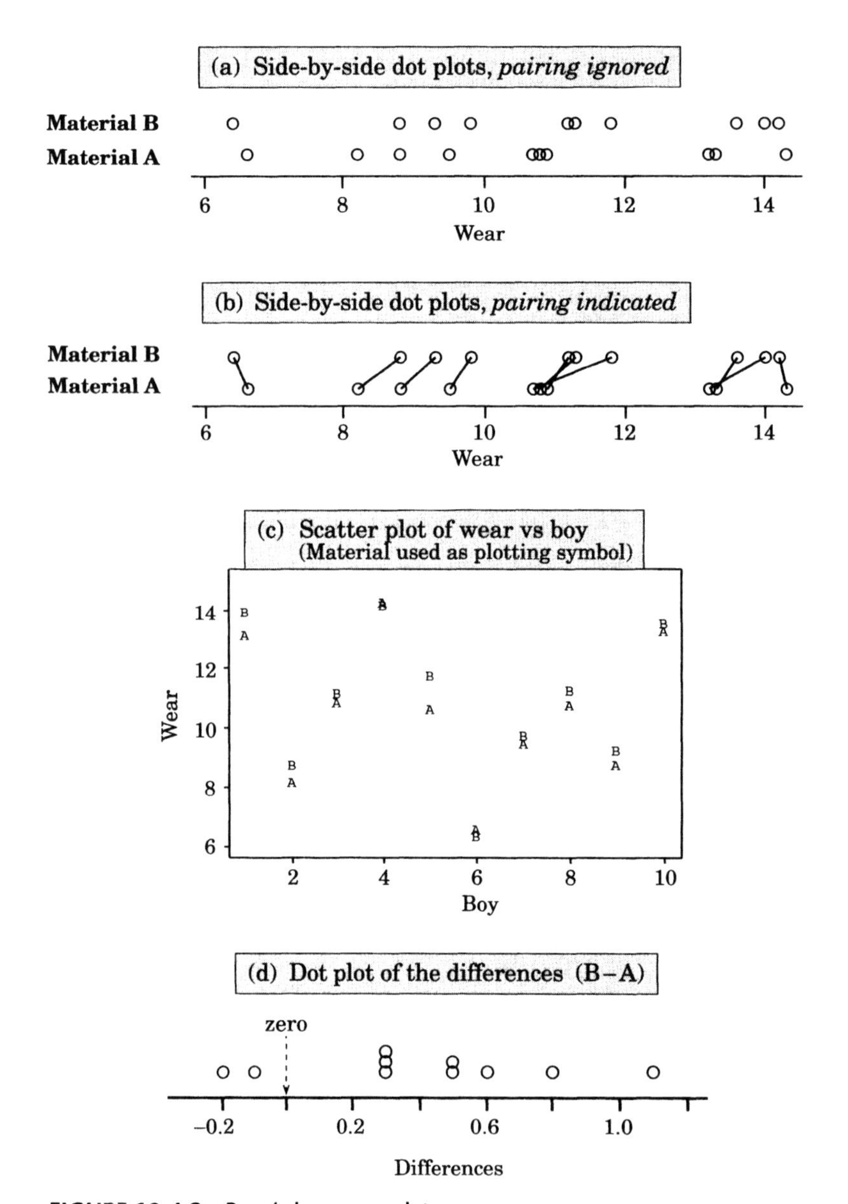

FIGURE 10.4.2 Boys' shoe wear data.

design described by Box et al. [1978, pp. 97–100], to explore some of the issues. We plot this data in various ways in Fig. 10.4.2.

Figure 10.4.2(a) presents side-by-side dot plots of the data on each material. Although this plot ignores the fact that we have relationships between data points, it gives us a reasonably good idea of what we might expect to see if we used a completely randomized design to compare the materials.[23] Does choice of material make a difference? There is nothing in Fig. 10.4.2(a) to suggest that it makes any difference at all. In Fig. 10.4.2(b) we indicate the pairing by drawing a line to join the two data points from the same boy in each case. For all but two boys there is more wear with material B than with material A. This suggests that choice of material does make a difference and that material B wears faster. For each boy Fig. 10.4.2(c) plots wear on both shoe soles against a vertical scale using an A to plot wear on the material A sole and B for wear on the material B sole. Again we see the tendency for there to be more wear on a boy's material B sole than on his material A sole. The most notable feature in both Fig. 10.3.2(b) and (c) is that there is a great deal of variability from boy to boy in the amount of wear in their shoes. Differences due to materials are very small in comparison (with material B tending to show slightly more wear).

As advocated in Section 10.1.2 for paired data, we analyze the differences. Figure 10.4.2(d) gives a dot plot of the differences. We see a pronounced shift to the right of 0. Formal testing provides strong evidence that material *does* make a difference to the wearing quality of the shoes.[24] A 95% confidence interval for the mean difference tells us that material B wears faster by between 0.13 and 0.69 units, whereas we see wear measurements ranging between 6.4 and 14.3. This quantifies the visual impression that the difference made by the material is small compared with the variability between boys.

When Is Pairing Beneficial?

What can we learn from this example? When the variability between experimental units is large, treatment effects are obscured in a completely randomized design so that we may not even be able to detect that they are there [cf. Fig. 10.4.2(a)]. If we can find a way of forming pairs so that there is much less variability within a pair than there is between pairs, a paired design will be much more effective for detecting the existence of treatment effects and estimating their size.

> Pairing is beneficial when the variability within pairs is small compared with variability between pairs.

Paired designs are ineffective when the pairing is ineffective. This occurs when members of the same pair are not more similar with respect to the variable of interest than individuals from different pairs.

---

[23] Figure 10.4.2(a) mimics what we might expect to see if we compared the wear on the right shoes of 10 boys wearing material A with the wear on the right shoe of 10 boys wearing material B.

[24] A paired t-test of the null hypothesis of no difference in mean wear has two-sided P-value = 0.009. A sign test of the null hypothesis that the population median of the differences is 0 has two-sided P-value = 0.02.

Examples of the Use of Pairing

We saw that the behavior in the measurements of air force head diameters in Example 10.1.3 was very similar to the behavior of the boys' shoe data. The pairing in the air force data related to repeat measurements on the same head. Within pairs, variability resulted largely from small changes in placement of the calipers, but this source of variation was much smaller than the variation between the head sizes of different recruits. The paired design allowed us to detect a small systematic difference between sizes measured by the cardboard calipers and sizes measured by the metal calipers. A completely randomized design would have required much larger sample sizes before we could expect to detect such an effect. In medical experiments to compare two treatments, paired designs are often used in which everyone gets one treatment for a period of time and then the other treatment. The order in which the treatments are given is randomized. (Why?) These are called crossover designs. The method can be very effective, but there are practical problems that limit its applicability. For example, if drugs are being used, some of the first drug may still be in the bloodstream when it comes time to try the second drug. This is called a *carryover* effect. (How might we reduce such an effect?)

We cannot use two treatments on the same experimental unit if treatments tend to destroy or damage the units. In such situations, rather than use the same unit twice, we can try to find pairs of units that are very similar. There are many situations in which there are naturally occurring pairings that can be exploited in an investigation. In Example 10.4.1 we exploited the fact that people have two feet. We can use human twins, animals from the same litter, plots of ground that are very close together, and so on. Medical experimenters often form so-called *matched pairs* by matching people as closely as possible on a set of variables such as age, sex, and general health status.

Pairing and Observational Studies

The preceding discussion applies to experiments in which pairing can be used to reduce the effect of natural variation on any treatment comparison. (By *natural variation* we mean variation that exists even without applying the treatments.) Pairing, including pairing by matching, is also useful in observational studies for similar, but subtly different, reasons. Section 1.3 discussed the use of identical twins in looking at the effects of smoking on health problems. With any observational study there is always the chance that in any comparison between smokers and nonsmokers the groups are different with respect to other factors and that these other factors might be the real cause of the health problems. When we compare identical twins in which one member smokes and the other does not, many other factors are automatically accounted for. For example, identical twins are the same genetically and will usually have experienced the same environment while growing up. Comparisons between a smoking twin and a nonsmoking twin are unaffected by factors that are the same, or very similar, for both twins.

*10.4.2 More General Blocking and Stratification

When comparing two treatments, whenever we could find pairs of experimental units for which within-pair variability was considerably smaller than pair-to-pair variability, we saw that we could detect and estimate treatment effects much more efficiently by

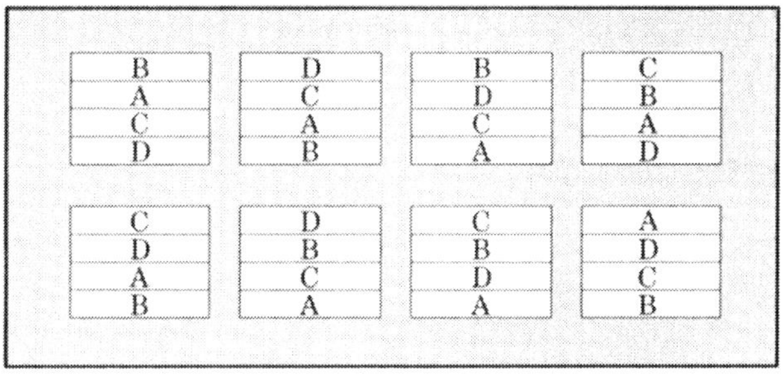

B	D	B	C
A	C	D	B
C	A	C	A
D	B	A	D

C	D	C	A
D	B	B	D
A	C	D	C
B	A	A	B

FIGURE 10.4.3 Comparing 4 treatments in 8 blocks.

using a paired design and doing our treatment comparisons within pairs. The same ideas carry over to comparisons of several treatments where the available experimental units vary among themselves. The basic idea is to find groups of experimental units (called ***blocks***) that are as homogeneous as possible, and to make comparisons of treatments within blocks. (Thus pairing is a special case of blocking in which the block size is 2.)

We introduced the idea of blocking in Section 1.2 in the context of agricultural experimentation as a means of coping with variability of land with respect to such factors as fertility and drainage. There the available land is divided up into physical blocks. The blocks are subdivided into strips called plots, and each treatment (e.g., variety) is applied to a plot in random order. Blocking permits comparisons between varieties within pieces of land that are close together and therefore more likely to be similar. Randomization is done to prevent unknown factors from biasing the treatment comparisons. Technically, this type of experimental design is called a ***randomized block design.***[25] Figure 10.4.3 shows a tract of land laid out according to a randomized block design for comparing 4 treatments (labeled A, B, C, and D) in 8 blocks.

Blocking works best when within-block variability is smaller than between-block variability and when the effects of the treatments are reasonably constant from block to block. If treatments have vastly different effects in different blocks, we cannot pool information about treatment effects across blocks (as is done in the paired t-test, for example). In such a situation blocks and treatments are said to *interact* (or *interactions* are said to be present). For an in-depth discussion of experimental strategies and the analysis of the resulting data, the reader is referred to Box et al. [1978].

The preceding ideas can be useful in observational studies as well as designed experiments. Suppose we were interested in the effect of marijuana use on human concentration spans. Here we would generally have to rely on information about current users and nonusers because it would be unethical to perform a designed experiment in which randomly chosen people were instructed to use marijuana. There will be many differences between marijuana users and nonusers besides drug use. These could explain any differences seen in concentration spans between them. We would wish to to get as close as possible in an observational setting to observing whether smoking marijuana would change a person's concentration span. Using twins (as with smoking in the previous subsection) is one approach. More generally, we can try to think of other variables that may affect concentration spans and try to form groups consisting of individuals who have similar values for these variables. Comparisons between users and nonusers can then be made within these relatively homogeneous

[25] Experimental design was originally formulated for agricultural applications, and hence a good deal of agricultural terminology (such as *blocks* and *plots*) has stuck. (See our web site.)

groups. In an observational setting this process is called ***stratification*** rather than blocking.

QUIZ ON SECTION 10.4

1. What is a completely randomized experimental design for comparing treatments? What is a randomized block design? How do they differ?
2. If you are interested in differences between the effects of treatments on mean levels of a variable, what type of analysis do you use for data from a completely randomized design?
3. In what circumstances is a blocked (including paired) experimental design more effective than a completely randomized design for detecting treatment differences?
4. When is blocking (including pairing) ineffective?
5. Using pairs in an experiment to compare two treatments corresponds to using two units in each block. Where does the randomization come in?
6. What are some naturally occurring pairs? some naturally occurring homogeneous groups with a block size larger than 2?
7. How does one form matched pairs? Could the same method be used to form triples, quadruples, and so on?
8. Stratification in observational studies plays a very a similar role to blocking in experiments. For what purposes can stratification be useful?

10.5 SUMMARY

10.5.1 Some General Ideas about Analyzing Data

1. Always ***plot*** your data ***before*** using formal tools of analysis (tests and confidence intervals).
   - It is the quickest way to see what the data say.
   - It often reveals interesting features that were not expected.
   - It helps prevent inappropriate analyses and unfounded conclusions.
   - Plots also have a central role in checking the assumptions made by formal methods.
2. All formal means of analysis make assumptions.
   - If the assumptions are false, the results of the analysis may be meaningless.
   - A formal method is ***robust*** against a specific departure from an assumption if it still behaves in the desired way despite that assumption being violated. For example it gives "95% confidence intervals" that still cover the true value of θ for close to 95% of samples taken.
   - A formal method is ***sensitive*** to departures from an assumption if even a small departure from the assumption causes it to stop behaving in the desired way.
   - Many types of assumptions are seldom, if ever, obeyed exactly, so methods that are sensitive to departures from such assumptions are of limited use in practical data analysis.
   - You must check whether the data contradict the assumptions made to an extent that the tests and intervals no longer behave properly. (Plots are a useful tool here.)

3. ***Outliers:*** If these appear to be present, try to check the original sources. Any observations that you know to be mistakes should be corrected or removed. If in doubt, do the analysis with and without the outliers to see whether you come to the same conclusions.

4. You can use ***nonparametric,*** or distribution-free, methods. They are less sensitive to outliers and do not assume any particular distribution for the original observations but do assume random samples from the populations of interest. Their measure of center is the median rather than the mean. They tend to be somewhat less effective at detecting departures from a null hypothesis and tend to give wider confidence intervals.

10.5.2 Normal Theory Techniques

One-Sample Methods

1. Two-sided *t*-tests and *t*-intervals for a single mean are quite robust against non-Normality but can be sensitive to the presence of outliers in small to moderate-sized samples. One-sided tests are reasonably sensitive to skewness. Normality can be checked graphically using Normal quantile plots or, formally, using a test such as the Wilk-Shapiro test.

2. ***Paired data:*** We have to distinguish between independent and related samples because they require different methods of analysis. Paired data (Section 10.1.2) are an example of related data. With paired data we analyze the differences, thus converting the initial problem into a one-sample problem.

3. The *sign test* and *Wilcoxon signed-rank test* are nonparametric alternatives to the one-sample or paired *t*-test.

Two-Sample Methods

1. Two-sample *t*-tests and intervals for the difference of two means $\mu_1 - \mu_2$:
   - They ***assume statistically independent*** random samples from the two populations of interest.
   - The pooled method also assumes that $\sigma_1 = \sigma_2$. The Welch method does not.
   - Two-sample *t*-methods are remarkably robust against non-Normality but can be sensitive to the presence of outliers in small to moderate-sized samples. One-sided tests are reasonably sensitive to skewness.

2. The ***Wilcoxon (Mann-Whitney) test*** is a nonparametric alternative to the two-sample *t*-test.

More than Two Samples

1. For testing whether more than two means are different, we use the ***F-test.***

2. The method of comparing several means is referred to as a ***one-way analysis of variance.***

The *F*-Test for One-Way Analysis of Variance

1. The formal null hypothesis (H_0) tested is that all k ($k \geq 2$) underlying population means μ_i are identical.

2. The alternative hypothesis (H_1) is that differences exist between at least some of the μ_i's.

3. The numerator of the *F*-statistic f_0 reflects how far apart the sample means are. The denominator reflects average variability within the samples.

4. Evidence against H_0 is equivalently provided by
   - sample means that are further apart than expected from the internal variability of the samples or
   - large values of the F-statistic.

   A small P-value demonstrates evidence that differences exist between some of the true means. To estimate the size of any differences, we use confidence intervals.
5. The test assumes
   - independent samples,
   - Normality, and
   - equal population standard deviations.
6. The test is robust to non-Normality. It is also reasonably robust to differences in the standard deviations when there are equal numbers in each sample, but not so robust if the sample sizes are unequal. The test can be used if the usual plots are satisfactory and the largest sample standard deviation is no larger than twice the smallest. The test is not robust to any dependence between the samples.
7. The ***Kruskal-Wallis test*** is a nonparametric alternative to the one-way ANOVA F-test.
8. ***Levene's test*** provides a robust method of comparing spreads.

10.5.3 Study Design and Interpretation

1. Significant differences between means *cannot* be interpreted as providing evidence of causation when the data come from an observational study, but they can if the data come from a well-executed randomized experiment.
2. In a completely randomized design for comparing treatments each experimental unit is randomly assigned to exactly one group, and all units in a group are given one of the treatments. (This results in independent samples.)
3. In a randomized block experiment experimental units are subdivided into blocks of more homogeneous units. Separate random assignment of treatments is performed within each block. (In a paired comparison experiment each pair is a block.)
4. Randomized block experiments are more effective than completely randomized experiments for detecting treatment differences if the blocks really are more homogeneous, that is, if there is appreciably less variability between units within the same block (e.g., pair) than between units from different blocks.

REVIEW EXERCISES 10

Instructions

There are features common to all of the exercises to follow that, in the interests of brevity, we list here rather than with each individual question.

- Where the actual measurements are given, rather than just summary statistics, we expect you to *create appropriate plots of the data* (Chapters 2 and 3) and comment on anything of interest that you see in the plots, in particular whether you see any grossly non-Normal behavior that would make you uneasy about using the standard tests and confidence intervals. To remind you of this standard requirement *always* to look at your data, we shall often use a phrase such as, "What interesting features do you see in these data?" This does not mean that there will always be something in particular that we want you to see, but just that we want to encourage you always to look.

- Where the exercise asks whether some effect occurs or whether there is any evidence for the existence of an effect or difference, the answer in the back of the book uses a significance test, whereas when the exercise asks how big an effect or difference is, the answer employs a 95% confidence interval. Most often the question will ask both whether an effect occurs and how big any such effect is, and you are expected to use both of these tools in your answer.

TABLE 1 Running Times (seconds)

Glooscap:	14.37	14.58	13.02	12.73	15.14	13.16	12.04	12.81	13.03
	13.13	13.85	11.89						
Coldbrook:	12.01	13.17	11.13	13.95	13.28	12.30	13.65	11.42	12.01
	11.08	11.73	12.13	11.11					

Source: Data courtesy of Dr. Vera Eastwood, University of Auckland.

1. As part of a study to compare the physical education programs at two Canadian schools, running times (in seconds) over a set distance were recorded for random samples of sixth grade students. The data are given in Table 1.

(a) Is there any real difference in running times on average between the two schools? How big is the difference? Is there anything in the data to make you doubt the applicability of Normal theory techniques?

*(b) Use a nonparametric technique to compare median running times. Does it lead you to the same conclusions?

(c) Gooscap Elementary children were taught by a specialist sprint coach. Do we have evidence that the running times of children are faster when coached by a specialist running coach than when coached by a generalist physical education teacher? To what extent does a study like this answer such a question?

2. The following are two sets of replicated measurements from an 1882 experiment performed by Simon Newcomb to measure the speed of light. We used set 2 in Example 8.1.1, and details of the experiment are described there. The numbers are times for the light to travel from Fort Myer on the west bank of the Potomac river to a fixed mirror at the foot of the Washington monument 3721 meters away and back. If the given values are multiplied by 10^{-3} and added to 24.8, they give the time in millionths of a second.

Set 1:	28	−44	29	30	26	27	22	23	33	16	24	29	24	40	21	31	34	−2	25	19
Set 2:	24	28	37	32	20	25	25	36	36	21	28	26	32	28	26	30	36	29	30	22

The "true" time, calculated using a modern estimate of the speed of light, is 33.02 in the unit used in the table. The second set of measurements looks fairly consistent with the true time. We initially chose the first set because it looked pretty bad and then added the second to be fair to Newcomb.

(a) Construct dot plots of both sets of data and mark the position of the true time on the plot.

(b) Check whether the first set of 20 measurements is consistent with the true time.

(c) The values −44 and −2 look like gross outliers. Omit these values and do (b) again.

(d) Test whether the two sets of measurements appear to be estimating the same quantity, both when the two outliers are present, and again when they have been omitted.

(e) Does omitting the two outlying values make much difference to the test results? What difference does it make to confidence intervals for the difference between the two means?

TABLE 2 Electromagnetic to Electrostatic Ratios

Group						
1	2	3	4		5	
62	65	65	62	65	66	64
64	64	64	66	63	65	65
62	63	67	64	63	65	64
62	62	62	64	63	66	
65	65	65	63	61	67	
64	63	62	62	56	66	
65	64		64	64	69	
62	63		64	64	70	
62			66	65	68	
63			64	64	69	
64			66	64	63	
			63	65	65	

Source: Rosa and Dorsey [1907].

3. The data in Table 2 came from Rosa and Dorsey [1907]. The quantity being measured was the ratio of the electromagnetic to the electrostatic units of electricity, a quantity that is equal to the velocity of light. The numbers are deviations in the third and fourth decimal places from 2.9900, so that the values in Table 2 are (measurement − 2.99) $\times 10^4$. The groups referred to in Table 2 correspond to successive dismantling and reassembling of the experimental apparatus.

   (a) What interesting features do you see in these data?

   (b) Is there any evidence of differences between the quantities being measured in each group?

   (c) What differences between groups are there (if any), and how big are any such differences?

4. The so-called fog index is intended to measure how difficult it is to read a piece of writing. It is calculated as[26]

$$\text{fog index} = 0.4 \times (\text{average number of words per sentence} + \text{the percentage of words with 3 or more syllables})$$

Shuptrine and McVicker [1981] made a study of the readability of magazine advertisements in which they obtained fog indices for a random sample of six advertisements from each of a number of magazines, obtaining the results in Table 3 for three of them. What do you conclude from the data?

[26] This is a slight simplification. See the original paper for details.

TABLE 3 Fog Indices

Scientific American	15.75	11.55	11.16	9.92	9.23	8.20
Newsweek	10.21	9.66	7.67	5.12	4.88	3.12
Sports Illustrated	9.17	8.44	6.10	5.78	5.58	5.36

Source: Shuptrine and McVicker [1981].

TABLE 4 Factor V and Blood Sterilization

Donor number	Presterilization	Poststerilization	Donor number	Presterilization	Poststerilization
1	1073	916	9	957	809
2	1064	1030	10	829	773
3	967	923	11	821	786
4	849	892	12	1257	1106
5	810	628	13	1095	832
6	855	759	14	1098	863
7	1047	828	15	932	783
8	1008	784	16	1440	869

Source: Data courtesy of Dr. C. M. Triggs, University of Auckland.

5. Factor V is a protein involved in the forming of blood clots. The higher the level of factor V, the faster the blood clots. The Auckland Blood Transfusion Service was interested in the effects of sterilization of blood plasma because factor V is known to be unstable and may break down during sterilization. Table 4 gives measured levels of factor V in blood samples from 16 blood donors. Both pre- and poststerilization measurements are given for each blood sample.

(a) How strong is any relationship between presterilization measurements and corresponding poststerilization measurements? Use an appropriate plot.

(b) Is there any evidence that sterilization reduces factor V levels in blood plasma? How large is any such effect?

(c) Is there anything in the data that makes you doubt the applicability of Normal theory methods? Remove any outlier(s) and repeat (b). Do your conclusions change?

6. In a study of the effects of diabetes on problems associated with the wearing of contact lenses, 16 diabetics and a control group of 16 nondiabetics wore contact lenses for a prescribed length of time. The swelling of their eyes was measured (as percentage swelling) at the time the contact lenses were removed and periodically thereafter. The data are given in Table 5.

(a) Would you expect any relationship between the measurements in column 2 (immediate) and column 3 (1 hour)? column 2 and column 5? Justify your answers. If you are using a computer, construct relevant plots to check whether what you expect is borne out by the data.

(b) Is there any evidence of a difference on average in eye swelling between diabetics and nondiabetics immediately after removing the lenses? How large is any such difference? Is there anything in the data that makes you doubt the applicability of standard parametric methods?

*(c) Use a nonparametric technique to gauge the extent of evidence in (b). Does this change your conclusions?

(d) Repeat (b) but for swelling 1 hour after removal.

TABLE 5 Percentage Swelling of the Eye

	Diabetic subjects			Control subjects		
	Immediate	1 hour	2 hours	Immediate	1 hour	2 hours
1	6.1	2.8	0.6	8.5	3.4	−0.3
2	7.8	4.6	2.9	10.9	5.8	2.5
3	8.4	3.2	0.4	10.4	6.9	1.7
4	9.0	4.4	2.7	8.9	6.3	2.6
5	7.6	3.6	—	10.6	5.1	0.5
6	9.1	6.0	2.9	8.6	2.5	−0.9
7	6.8	2.2	−0.1	12.0	7.6	−1.7
8	10.0	6.7	3.7	9.1	6.1	1.6
9	9.9	6.8	3.3	17.9	11.0	4.0
10	10.2	5.2	1.1	9.0	4.4	2.7
11	6.8	3.2	−0.1	10.1	4.2	−0.5
12	6.7	4.0	−0.4	8.9	5.6	2.0
13	5.3	1.8	0.5	11.5	2.3	−3.2
14	11.3	4.5	1.7	12.3	6.1	3.4
15	9.1	5.1	2.3	13.4	7.9	2.4
16	9.2	6.0	3.0	7.1	—	1.2

Source: Data courtesy of Erin Harvey, University of Waterloo. (Note two missing values.)

(e) If you wished to look for a change between column 2 conditions and column 4 conditions, what techniques would be relevant?

(f) There are negative values in columns 4 and 7. Speculate about possible reasons for this.

7. A common yardstick used in testing the effectiveness of television commercials is the 24-hour recall score. Advertisers are required to air the ad and interview by phone the next day a sample of viewers, say, 200, who were watching the ad. Interviewees are asked whether they remember seeing a commercial for the brand the night before, and if they do, they are asked a series of follow-up questions. From the answers a recall score is computed. This is a somewhat slow, cumbersome, and expensive procedure. If certain types of brain activity were strongly predictive of later recall, advertising professionals could get much more immediate feedback about what works and what doesn't by measuring the brain activity of experimental subjects while they are watching the commercial.

   Alpha-wave brain activity is characteristic of an awake but inattentive state. With increased stimulus, alpha activity tends to decrease as the alpha waves are replaced by waves with higher frequencies. The following data come from a study described by Appel et al. [1979]. The measurements in Table 6 (total-activity indices) are averages over groups of experimental subjects of measures of *reduction* in alpha-wave activity when watching the commercial compared with watching a blank screen. For each brand/product type two commercials were used. In each case one commercial had a considerably higher recall score than the other.

   One of the investigators' hypotheses before starting the study was that the indices would be higher for high-recall commercials (a higher index means a greater reduction in alpha-wave activity) than for low-recall commercials.[27]

[27]This study was the first important study of the relationship between advertising effectiveness and a type of physiological measurement. Physiological monitoring of experimental subjects watching advertisements, both commercial and political, now seems to be widespread.

TABLE 6 Total-Activity Indices from 20 Advertisements

Brand	Low-recall commercial	High-recall commercial
1	53	137
2	114	135
3	81	83
4	86	125
5	34	47
6	66	46
7	89	114
8	113	157
9	88	57
10	111	144

Source: Appel et al. [1979].

(a) Is there any evidence that high-recall commercials tend to have higher activity indices than low-recall commercials?

(b) Would a significant result in (a) necessarily mean that 24-hour recall scores can be discarded and total-activity index scores used instead?

(c) Create a scatter plot of each low-recall activity index against the corresponding high-recall value. What pattern, if any, do you see? Is there a strong relationship between the scores?

(d) The experimenters made up four videocassettes. Cassette A had brands 1-5 in the order 1H, 1L, 2L, 2H, . . . ; Cassette B had brands 5-1 in the order 5L, 5H, 4H, 4L, . . . ; Cassette C had brands 6-10 in the order 6L, 6H, 7H, 7L, . . . ; while Cassette D had brands 10-6 in the order 10H, 10L, 9L, 9H, . . . (1H means the high-recall commercial for brand 1; 1L means the low-recall commercial). They divided their 30 experimental subjects into 6 groups of 5. Each group saw one sequence labeled AC, DB, CA, BD, CB, or AD. Each subject was first exposed to 30 seconds of blank screen (X), then exposed to each of the pair of cassettes three times, and finally exposed to another 30 seconds of blank screen. For example, the AC group saw X-A-C-A-C-A-C-X. Why did the investigators go to all this trouble? What are some of the factors they were trying to guard against?

8. There has been a growing practice in sport of inhaling pure oxygen to speed recovery after intensive exercise. It has been particularly widespread in American football, where specialist squads continually come on and go off the field. While players are resting, they inhale pure oxygen so that they supposedly recover more completely before they are called on to play again. Other obvious applications are for boxers or wrestlers between rounds. But does it speed recovery? One of the causes of feelings of fatigue and soreness in the muscles is the buildup of lactic acid, or lactate, in the blood. Winter et al. [1989] conducted a study to see whether the use of pure oxygen after exercise tended to reduce blood lactate levels. They used 12 professional soccer players in peak physical condition and put them through a defined exercise program. This was done twice for each person. On one occasion they breathed pure oxygen for four minutes after exercise, and on another occasion they breathed ordinary air for four minutes after the exercise. Blood lactate levels were measured immediately after the exercise and again after the four-minute recovery time. The data are given in Table 7.

(a) Probably the most relevant question is whether the *change* in lactate levels over the recovery period when pure oxygen is used is different from the *change* in lactate levels when ordinary air is used. Is there any evidence of a difference?

TABLE 7 Blood Lactate Levels

	Ordinary air		Pure oxygen	
Subject	Immediately after exercise	4 minutes after exercise	Immediately after exercise	4 minutes after exercise
1	6.8	9.3	8.5	9.3
2	8.0	8.1	7.3	8.4
3	9.6	8.3	12.3	11.1
4	6.7	5.3	6.3	7.0
5	10.9	11.4	7.6	7.0
6	5.9	5.0	5.9	5.7
7	6.5	8.8	9.0	8.5
8	6.2	6.6	5.6	6.0
9	7.9	8.9	8.2	9.7
10	8.0	10.8	8.1	10.5
11	7.7	7.9	7.5	9.2
12	6.9	9.0	7.5	10.2

Source: Winter et al. [1989].

(b) Actually, the paper and the press reports stemming from it said that oxygen makes no difference. Are they justified in claiming this?

(c) How would you conduct this experiment?

9. University of Waterloo biologist Dr. Ian Martin wanted to undertake a series of experiments involving a snail (*Physella gyrina*) that lives in rivers and feeds on algae as it grazes the limestone bedrock at the bottom of the river. He wanted to check whether clay tiles would mimic the natural bedrock habitat of the snails. (He needed something that was suitable for experimental manipulation and laboratory work.) He placed 11 tiles in a river colonized by algae and snails. After three weeks the number of snails on each tile and the number of snails on 11 sections of bedrock, each of the same area as a tile, were counted. The results are given below:

Numbers on tiles:	65	60	56	64	54	62	64	67	59	53	77
Numbers on bedrock:	57	63	54	57	83	57	82	58	63	85	54

(a) Is there any evidence of a real difference in mean numbers of snails between tiles and bedrock? Is there anything in the data to make you doubt the applicability of Normal theory methods?

(b) Do nonparametric methods show any evidence of a real difference in median numbers of snails between tiles and bedrock?

(c) How would you carry out this experiment in terms of the placement of tiles and the choice of sections of bedrock for comparison?

10. As one of his experiments, Dr. Martin investigated moth larvae (genus *Petrophilia*). The larvae graze on algae growing on the surfaces of underwater stones, just as the snails do. Moth larvae construct a silk canopy over an area of rock and graze under the canopy. They also graze on uncovered rock at the periphery of the canopy. Dr. Martin's hypothesis was that competition for algae food resources from snails may result in moth larvae building bigger canopies (and thus protecting more food from competitors). He conducted the

TABLE 8 Area Covered by Canopy

Bottom before	Bottom after	Platform before	Platform after
184.8	177.5	200.2	423.0
153.7	244.0	292.4	188.6
283.2	233.2	171.4	218.2
46.5	117.9	292.7	251.4
208.3	196.5	251.0	190.0
63.5	238.4	313.4	208.0
164.0	265.6	166.6	470.5
126.6	292.5	171.5	126.8
110.6	111.5	219.0	68.3
115.3	267.3	192.9	209.5
200.5	164.1	144.5	253.2
247.3	466.1	146.3	146.9
412.5	318.9	197.2	156.9
156.1	215.9	161.4	147.7
245.6	160.5	214.0	106.3
339.1	413.8	396.4	412.3
237.4	131.4	57.6	104.3
239.5	130.3	98.3	159.6
366.3	724.5	110.4	274.7
269.9	855.8	233.8	120.2
72.0	714.2	260.2	382.1
412.8	499.1	268.4	341.2
113.1	547.7	294.1	241.0
119.1	397.5		
116.3	245.3		
116.4	1094.7		

Source: Dr. Ian Martin, University of Waterloo.

following experiment using ceramic tiles colonized by moth larvae. Treatment tiles were placed on the bottom of the river where snails could graze. Control tiles were placed on little platforms supported by legs 4 cm above the river bottom. Snails could very seldom find their way up onto these tiles. The area covered by the canopy over each tile was measured when the tiles were initially laid and then two weeks later. The question of interest is whether the canopies on the river-bottom tiles have grown more. The data are given in Table 8.

(a) Is there any evidence of a real change in mean canopy size for platform tiles? Is there anything in the data to make you doubt the applicability of Normal theory methods?

(b) Is the *change* in canopy size larger on average for river-bottom tiles than for platform tiles as predicted?

(c) Is there anything in the data to make you doubt the applicability of Normal theory methods in (b)? Would a nonparametric test be better here?

(d) How would you carry out this experiment in terms of the placement of tiles in the river?

(e) Is there anything that concerns you about this experiment? How could you check out whether this was a problem?

11. The following information comes from a study that suggests that homosexuality may be in part a biological phenomenon. Salk Institute neuroscientist Simon LeVay (see LeVay [1991])

TABLE 9 Volume of INAH-3 Region of Brain (0.01 mm^3)

Heterosexual men:	1.25	2.00	4.00	10.00	10.25	10.25	11.50	11.75
	12.5	13.00	13.5	15.75	15.75	17.00	17.25	20.25
Homosexual men:	0.00	0.75	1.25	1.50	1.75	2.00	2.25	2.50
	3.00	3.25	3.50	3.75	4.00	4.75	6.50	8.25
	12.50	14.00	19.00					

Source: Digitized from a graph in LeVay [1991].

stated that in homosexual men, the anatomical form of part of the anterior hypothalamus in the brain called the INAH-3 region was more typical of women than of heterosexual men. LeVay examined the brains of 41 subjects who had died in seven hospitals in New York and California. Of these, 19 were homosexual men who had died of AIDS, 16 were heterosexual men (of which 6 died of AIDS and 10 of other causes), and 6 were women. The data on the volumes of the INAH-3 region in the men (units = 0.01 mm^3) are given in Table 9.

(a) What do you notice about these two samples?

(b) Is there any evidence that INAH-3 volumes are different on average for homosexual and heterosexual men? How different are they?

(c) What weaknesses can you see in this study as a demonstration of INAH-3 differences between homosexuals and heterosexuals?

12. The Cancer Research Unit attached to the University of Auckland Medical School has been developing a technique to test chemicals for their ability to affect the division of cells in the body and so potentially to cause cancer. Human blood cells were grown in cell culture and then treated with various chemicals: chloral hydrate, hydroquinone, diazepam, econidazole, and colchicine, some of which are known to be potent carcinogens. The carcinogens act by breaking chromosomes, and this disrupts cell division. The broken fragments of chromosome are left as micronuclei, and the ratio of the size of a micronucleus to its parent cell nucleus is measured. The more carcinogenic the chemical, the higher the ratio tends to be. The data in Table 10 give measurements on 50 of these ratios when there has been no chemical treatment of the cells (control sample) and 50 ratios from cells treated with each of the five chemical treatments. We want to look at the effects of these chemicals on the ratios.

(a) Plot the data. What does the plot suggest to you about differences between cell ratios for the six different conditions?

(b) Perform the F-test. What do the results of this test tell us?

(c) Perform an F-test (i) for just samples 2–5 and (ii) for just samples 1–5. Do you now have a clearer picture of where the real (in the sense of demonstrated) differences lie?

(d) Obtain confidence intervals for any interesting differences between the true means. Can we conclude that one chemical is more carcinogenic than the others?

(e) What does your plot in (a) suggest about the shape of the distribution for each chemical?

(f) How would you investigate distributional shape? Do this for the control sample.

(g) Is there anything to make you doubt the applicability of one-way analysis of variance to this set of data?

13. In Table 11 are house prices (in thousands of dollars) taken from those advertised by Ace Realty in the city of Hamilton at three different times.

(a) Plot the data and obtain means and standard deviations for each of the three time periods. What do the plot and summary statistics suggest about the movement of house prices over time?

(b) Perform the F-test. What do the test results tell you?

TABLE 10 Cell Nucleus Ratios

Sample[a]						Sample[a]					
1	2	3	4	5	6	1	2	3	4	5	6
0.08	0.08	0.10	0.10	0.07	0.17	0.21	0.24	0.27	0.27	0.28	0.41
0.08	0.10	0.08	0.10	0.08	0.19	0.21	0.25	0.29	0.28	0.29	0.43
0.09	0.08	0.12	0.10	0.09	0.22	0.22	0.25	0.29	0.29	0.29	0.45
0.12	0.09	0.13	0.12	0.10	0.24	0.23	0.27	0.30	0.29	0.32	0.46
0.12	0.10	0.14	0.13	0.10	0.24	0.24	0.28	0.30	0.31	0.32	0.47
0.13	0.10	0.15	0.13	0.10	0.24	0.24	0.29	0.30	0.31	0.32	0.47
0.15	0.13	0.15	0.13	0.11	0.24	0.24	0.29	0.31	0.35	0.32	0.48
0.16	0.13	0.16	0.14	0.11	0.25	0.24	0.30	0.31	0.35	0.33	0.48
0.16	0.14	0.16	0.14	0.12	0.27	0.24	0.32	0.31	0.35	0.33	0.48
0.16	0.14	0.17	0.15	0.12	0.27	0.25	0.32	0.32	0.36	0.35	0.49
0.16	0.15	0.18	0.15	0.12	0.28	0.25	0.33	0.33	0.38	0.37	0.50
0.17	0.15	0.18	0.15	0.13	0.29	0.25	0.33	0.34	0.38	0.37	0.52
0.18	0.15	0.18	0.15	0.14	0.29	0.26	0.33	0.35	0.39	0.37	0.59
0.18	0.17	0.18	0.17	0.15	0.30	0.26	0.34	0.37	0.40	0.39	0.59
0.19	0.17	0.19	0.18	0.16	0.35	0.28	0.34	0.38	0.42	0.39	0.60
0.19	0.19	0.20	0.20	0.17	0.36	0.29	0.37	0.40	0.48	0.39	0.65
0.19	0.21	0.21	0.21	0.17	0.36	0.29	0.38	0.40	0.49	0.39	0.67
0.19	0.21	0.23	0.22	0.18	0.36	0.30	0.38	0.43	0.53	0.40	0.69
0.19	0.22	0.23	0.22	0.19	0.38	0.31	0.38	0.46	0.54	0.40	0.71
0.20	0.23	0.23	0.24	0.22	0.39	0.33	0.42	0.46	0.55	0.40	0.71
0.20	0.23	0.24	0.25	0.24	0.40	0.42	0.43	0.47	0.66	0.41	0.72
0.20	0.23	0.24	0.25	0.25	0.40	0.44	0.54	0.47	0.68	0.42	0.76
0.20	0.24	0.24	0.25	0.26	0.40	0.51	0.62	0.50	0.69	0.42	0.78
0.21	0.24	0.25	0.25	0.27	0.41	0.59	0.62	0.55	0.71	0.45	0.79
0.21	0.24	0.25	0.27	0.27	0.41	0.62	0.69	0.56	0.72	0.59	0.80

[a]Sample: 1 = control; 2 = chloral hydrate; 3 = hydroquinone; 4 = diazepam; 5 = econidazole; 6 = colchicine.

Source: Data courtesy of Dr. C. M. Triggs, University of Auckland.

TABLE 11 House Prices ($1000s)

Nov. '87	93	85	60	62	63	66	70	71	72	73	72	75	76
	80	86	96	97	69	135	165	177	98	115	120	120	125
	79	145	152	159	160	169	69	146	65	100	94		
Sep. '89	185	210	249	395	139	248	97	148	88	265	150	198	110
	179	65	79	74	69	82	85	93	97	82	95	81	64
	119	88	112	85									
Aug. '91	119	170	92	139	175	128	69	80	114	349	265	75	148
	123	140	137	186	125	105	66	121					

Source: Data courtesy of Dr. M. Jorgenson, Waikato University.

(c) Obtain appropriate confidence intervals for true differences in average selling prices between the three time periods.

(d) Do you see anything in the results of (a) that makes you doubt the applicability of the F-test or confidence intervals for differences obtained from an analysis-of-variance program?

(e) Instead of plotting the house prices themselves, plot the logarithms of house prices. Do the logged house prices more closely obey the Normal theory assumptions than the untransformed observations?

(f) Using the logged data, is there any real evidence that house prices sought by Ace Realty have changed on average? In what way have they changed?

For the purposes of the next part, assume that the preceding data are selling prices rather than advertised prices. Rather than being interested in movement in the average price of houses currently for sale, people tend to be more interested in what has happened to the value of their own house.

(g) List any reasons supporting the idea that an increase in the average price of houses sold by Ace Realty may not imply that any particular house in the city has increased in value. Which of these reasons still apply if the figures apply to all houses that have sold in the city (and not just those being advertised by Ace Realty)? What sort of information would you want in order to have a better idea about changes in value for particular houses?

14. The data in Table 12 are measurements on the rate at which oxygen is consumed by biological respiratory processes in two different layers of Lake Awassa in Ontario on 13 different occasions. The two layers are the epilimnion (0–10 m deep) and the hypolimnion (cold water at bottom of lake), and the units are $\mu g\ O_2$ per liter per hour.

(a) Test the hypothesis of no difference in dark oxygen consumption between the two layers.

(b) Is there anything in the dot plot of the differences (hypolimnion − epilimnion) that makes you doubt the applicability of the t-test here?

The underlying mean of the differences (hypolimnion − epilimnion) would be useful if we could get a good idea of the hypolimnion value, which would be harder to measure, by adding μ_{diff} to the epilimnion value.

(c) Plot the differences (vertical) versus the epilimnion values (horizontal). What do you see?

15. You are interested in comparing the effect of a new treatment for headaches with that of a standard treatment, and you have enough money to see just 20 patients twice. Three possible designs are as follows:

- DESIGN I: The subjects are randomly assigned to treatments in such a way that 10 subjects are assigned to each treatment. Each time a subject has a headache, the treatment is applied, and the duration time of the headache is measured. Using the same treatment the duration time for their second headache is recorded.
- DESIGN II: The 20 subjects are randomly assigned a beginning treatment. When they have their first headache, they are given that treatment, and the duration time is mea-

TABLE 12 Consumption Rate of Oxygen ($\mu g \cdot L^{-1} \cdot hr^{-1}$)

	Nov	Jan	Feb	Feb	Mar	Apr	Apr	Apr	Apr	Mar	Mar	Jan	Apr
Epilimnion	29	23	20	21	28	25	38	28	32	33	42	20	19
Hypolimnion	36	45	20	25	37	46	58	44	35	75	79	25	29

Source: Data courtesy of Dr. Jeanette O'Hara-Hines, University of Waterloo.

sured. For their next headache they are given the other treatment, and the duration time measured.

- DESIGN III: The subjects are randomly assigned to treatments in such a way that 10 subjects are assigned to each treatment. When the subjects have their first headache, the treatment is applied and the duration time of the headache is measured. When they have their second headache, the other treatment is used, and the duration time measured.

(a) How would you carry out the randomization for design I?

(b) Explain how you could analyze the data from design II?

(c) Can you use the same method for design I? Justify your answer.

(d) How would you analyze the data from design III?

(e) Explain the purpose for randomizing the initial treatment in design II.

(f) Which design do you prefer and why?

16. We have data on the energy required to penetrate the wall of an unfertilized egg from each of 300 women. We divide the women into three groups according to whether the usual diet is low fat, medium fat, or high fat. The following series of claims are made. If you think a claim is valid, just write "true." If you think it is false, write "false" and give a reason.

(a) CLAIM: We could use a one-way analysis of variance to investigate differences in mean energy levels between the groups.

(b) CLAIM: The alternative hypothesis for the F-test is that every group mean differs from every other group mean.

(c) CLAIM: A small P-value for the F-test would indicate that there was little or no difference between the groups in the mean energy level required to penetrate the egg.

(d) CLAIM: The analysis would be useful because, if significant differences emerge, we would be able to recommend to women diets that would improve their ability to become pregnant.

17. For each of the following experiments, answer these questions:

(i) State the most appropriate design for the experiment. (Pick one of one sample, two independent samples, more than two independent samples, paired data.) State which statistical method—namely, a hypothesis test or confidence interval—is more appropriate.

(ii) State the assumption(s) that must be satisfied so that you can apply your chosen method.

(iii) Briefly describe a potential problem with the experiment.

(a) A research institute is investigating new medication for the relief of migraine headaches. They have 40 available subjects who suffer from acute migraine pain. The drug being investigated is intended to start taking effect immediately and is advertised as having a sizable effect after 60 minutes. The research group decides to ask each patient to record the percentage of their migraine headaches that they feel have been relieved after one hour. They wish to find out if the drug is effective or not.

(b) A horticulturist is comparing two methods (A and B) for growing potatoes. Standard potato cuttings will be planted in small plots of ground. The variables of interest (response variables) are the number of potato tubers and the fresh weight (weight when just harvested) of vegetable growth per plant. The horticulturist has 20 plots available for the experiment and is interested in determining the size of the difference, if any, between planting methods in the resulting weight of vegetable growth.

(c) Biologists are interested in performing an experiment to determine the best color for attracting cereal leaf beetles to boards on which they will be trapped. Four colors are to be compared: yellow, white, green, and blue. The response variable is the number of beetles trapped. A board will be attached to each of 16 poles spaced evenly throughout the center of a field of oats (16 boards in total, 4 of each color).

(d) The design of controls for instruments has a large effect on how easily people can use them. A student project investigated this effect by asking 25 right-handed people to turn a knob (with their right hands) that moved an indicator by a screw action. There were two identical instruments: one with a right-hand thread and the other with a left-hand thread. Each of the 25 people used both instruments, first with the left-hand thread and then with the right-hand thread. The student wanted to discover whether the average time to move the indicator a prescribed distance depended on the thread.

TABLE 13 Fresh Weights of Paspalum Grass Tops (g)

Temperature	Control						Fungal infected					
14°	1.7	2.3	3.2	1.5	2.0	2.9	2.0	2.7	1.8	2.4	1.1	2.6
18°	23.0	15.4	22.8	21.5	20.8	20.3	14.2	14.7	21.4	9.7	12.2	8.2
22°	30.8	14.6	36.0	23.9	20.3	27.7	23.6	28.1	13.3	24.6	19.3	31.5
26°	10.2	15.6	14.7	20.5	14.3	23.2	13.3	8.5	11.8	7.8	15.1	11.8

Source: Data courtesy of Peter Buchanan.

18. The introduction of a fungal infection has been suggested as one way of reducing the growth of paspalum grass, a weed in pasture grazed by farm animals. Scientists at the Mt. Albert Research Centre in Auckland performed an experiment to compare the growth of plants infected with the fungus with a control group at each of four different temperatures (14, 18, 22, and 26°C). Six measurements were made at each of the eight treatment-temperature combinations. Results, measured as fresh weight of tops in grams, are given in Table 13.

(a) What are some of the practical problems that would have to be considered in planning this experiment?

(b) Why is a two-sample t-test comparing 24 fungal-infected and 24 control plants not appropriate here?

(c) Why is a one-way analysis of variance comparing all the plants at each of the four temperature settings not appropriate here?

We have plotted the data in Fig. 1(a), plotting weight of top versus temperature.

(d) What does the graph suggest to you about the effect of temperature? about the effect of the treatment? What other features do you see?

We shall now go on to use formal tests and intervals to check these impressions.

(e) Is a one-way analysis of variance appropriate to compare the effects of the temperatures in a situation like this if we use control plants only or fungal-infected plants only? Why?

(f) Even without statistical testing, it seems clear that there is much less growth of paspalum at 14°C than at the higher temperatures. Is there evidence of true differences in top weights at the other three temperatures? Apply a one-way ANOVA F-test to the control plants in the 18, 22, and 26°C groups to answer this question. What have you learned?

(g) How large is the true difference, if any, in top weights produced by control plants at 22°C and at 26°C? Compare 18°C and 22°C in the same way.

(h) For what technical reason would we be reluctant to use one-way analysis of variance to compare all four temperature groups of control plants?

(i) Is a two-sample t-test appropriate to compare the effects of the treatments if we compare only plants grown at the same temperature?

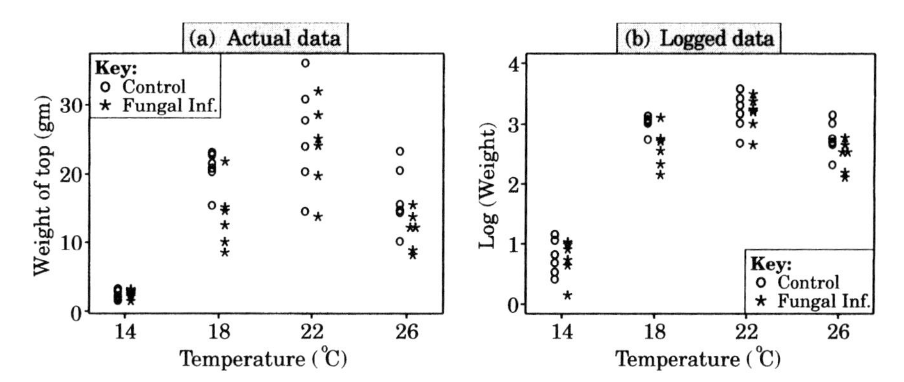

FIGURE 1 Fresh top weights of paspalum grass.
(Control plants: "o"; fungal plants: "*.")

(j) Is there evidence of a treatment effect at 18°C? How large is any such effect? Repeat for plants grown at 22°C.

[If we are prepared to entertain the idea that the true treatment effect is the same at all temperature levels and the data do not contradict such an idea, the *two-way* analysis-of-variance F-test combines evidence about treatment differences from all four temperature groups and thus provides a much more sensitive way of detecting small treatment differences.]

When large measurements are considerably more variable than small measurements, statisticians often try to work with the logarithms of the measurements rather than the measurements themselves. The logarithmic data are plotted in Fig. 1(b).

(k) Is the variability of the logarithmic data reasonably constant?

(l) Using the logarithmic data and control plants only, conduct a one-way analysis-of-variance F-test to compare the four temperature groups. What have you learned?

(m) How useful do you think the fungal treatment would be as a practical measure for combating paspalum growth? Why?

Further exercises, some involving large data sets, are available on our web site.

CHAPTER 11

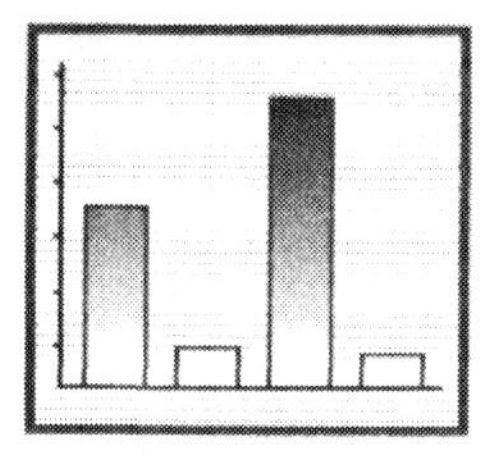

Tables of Counts

Chapter Overview

In the previous chapter we brought all the tools we had collected thus far out of our statistical toolbox in order to analyze sets of data from quantitative variables. Further tools were developed for handling data sets from more than two populations. We now want to do the same sort of thing for qualitative variables, namely, variables that define classes or group membership (see Section 2.1). Tables of measurements are now replaced by tables of counts, and we focus on proportions (probabilities) rather than means as we did in Chapter 10. We emphasize that these tables are based on samples, and the sample proportions in the tables will be estimates of the true underlying population proportions. Such tables of counts are often described under the heading of *categorical data* as the counts are organized under various categories.

In the same way that the t- and F-distributions are used for making inferences about samples of quantitative variables, the Chi-square distribution is the main distribution used for handling inferences about categorical data. It is based on comparing the observed counts in the tables with those that one might expect to get under some hypothesis about the way the underlying population proportions or probabilities are structured. If the differences between the observed and expected counts are sufficiently large, then the hypothesis will be rejected.

In Section 11.1 we consider so-called one-dimensional tables, introduced in Section 2.6, which are simply tables for a single qualitative variable consisting of a single column (or row). Our main interest here is in testing whether or not the underlying population proportions in such a table follow a hypothesized pattern, the so-called *goodness-of-fit* problem. We introduce the Chi-square distribution and show how to use it for carrying out such tests. In Section 11.2 we extend our ideas to two-dimensional tables, which have at least two rows and two columns. These are the *two-way tables* introduced in Section 3.3. We focus on testing whether certain types of patterns exist among the underlying population probabilities, and we again use the Chi-square distribution. Two situations are of particular interest. The first situation arises when we have several different populations, each represented by a one-dimensional table. Here we want to test whether the populations have the same underlying proportions, and the test we use is called the test for *homogeneity.* The second arises when there is just one population, but the categories can now be arranged in a two-dimensional array. Here we are mainly interested in testing whether the row and column categories are independent of each other. The test is then called a test for *independence.* Because of the relationship to the theory for comparing two proportions, so-called 2×2 tables get special mention in Section 11.2.4. The chapter ends in Section 11.3 on a cautionary note about the perils of collapsing complex tables to simpler ones.

11.1 ONE-DIMENSIONAL TABLES

Recall from Section 2.1 that qualitative variables define groups, in contrast to quantitative variables, which are measurements. In this chapter we usually call qualitative variables *grouping factors,* or just factors, to stress the role that they play. Frequency tables are the simplest tables of counts. We saw in Sections 2.5.1 and 2.6 how frequency tables were obtained from the raw data as a means of summarizing data on single discrete or qualitative variable. When we have sample data and we want to make inferences about the population those data were drawn from, our interest tends to center on the *proportions* of the observations that fall into each level of the grouping factor rather than on the frequencies themselves. We see this in Table 11.1.1 where we have the sample proportions of people falling into different blood types. Such a table can be variously referred to as a ***one-dimensional table*** or a *one-way classification* (or *table* or *layout*). Note that there is no particular reason to write the table in terms of a *row* of counts (except for saving space). We could equally well write it as a column.

TABLE 11.1.1 Proportions of Three Blood Types

	A	AB	B	Total
No. observed	39	70	42	151
Proportion observed	0.258	0.464	0.278	1.000

TABLE 11.1.2 210 Rolls of a Die

Outcome	1	2	3	4	5	6	Total
Count	26	40	37	26	43	38	210
Proportion	0.124	0.190	0.176	0.124	0.205	0.181	1.000

Chapters 8 and 9 showed us how to make inferences about differences between pairs of proportions. When looking at differences between proportions from a one-dimensional frequency table, we have to use the theory for comparing proportions from the *same sample with several response categories* (situation (b) in Fig. 8.5.1). In this chapter our focus is different. We now want to test a hypothesis about the whole set of underlying population proportions. For example, in Table 11.1.1 we might want to test whether the data are consistent with a set of probabilities obtained from a genetic law (see Exercises for Section 11.1, problem 2).

EXAMPLE 11.1.1 ***Is Our Die Symmetrical?***

One morning one of our children rolled a die 210 times and recorded the results. This gave us the simple table of counts in Table 11.1.2. We want to know whether our die is a fair one, that is, whether each face is equally likely.

This is a situation in which we have a single sample and 6 response categories. Assuming the die to be symmetrical and the rolling sufficiently vigorous, we would then expect that all 6 outcomes would be equally likely (with probability $\frac{1}{6}$). Therefore we would expect the observed proportions to be roughly equal to 0.1667. Instead of looking at proportions, we can also look at the counts. We would then expect one sixth of the rolls (namely $210 \times \frac{1}{6} = 35$ rolls) would fall into each outcome category. Of course, we don't expect any of the observed counts to be exactly 35 because of sampling variation. In fact, they vary from 26 to 43. We can now ask whether the differences between the observed counts and the expected counts lie within the usual limits of sampling variation or whether they are sufficiently big that we should doubt the equally likely outcomes assumption.

We can formulate this problem of testing whether the die is fair as a test of the hypothesis

$$H_0\colon p_1 = p_2 = p_3 = p_4 = p_5 = p_6 = \frac{1}{6}$$

where p_j is the probability of rolling a j.

11.1.1 The Chi-Square Test for Goodness of Fit

To test a hypothesis like H_0 in the previous example, we compare the observed counts in a table with the counts one would *expect* to get if the hypothesis was true. For this reason it is called a ***goodness-of-fit test*** as we test whether the observed counts "fit" the expected counts adequately. Since we are interested in the magnitude of each difference *observed count* − *expected count,* that is, each difference without its sign, we square the differences and add together scaled versions of the squares. We then end up with the so-called ***Chi-square test statistic*** with calculated value

$$x_0^2 = \sum_{\text{all cells in the table}} \frac{(\text{observed count} - \text{expected count})^2}{\text{expected count}}$$

Let X^2 be the random variable corresponding to x_0^2. This random variable tells us about the behavior of x_0^2 in repeated sampling. It can be shown that when H_0 is true, the statistic X^2 has a distribution that is known as the Chi-square[1] distribution on *df* degrees of freedom, which we denote by ***Chi-square(df)***. We write $X^2 \sim$ Chi-square(*df*) or, more briefly, $X^2 \sim \chi^2_{df}$. If the observed counts tend to be too far from the counts expected under H_0, then x_0^2 will be large. Thus it is large values of x_0^2 that provide evidence against H_0, so the *P*-value for the test is

$$P\text{-value} = \text{pr}(X^2 \geq x_0^2), \qquad \text{where } X^2 \sim \text{Chi-square}(df)$$

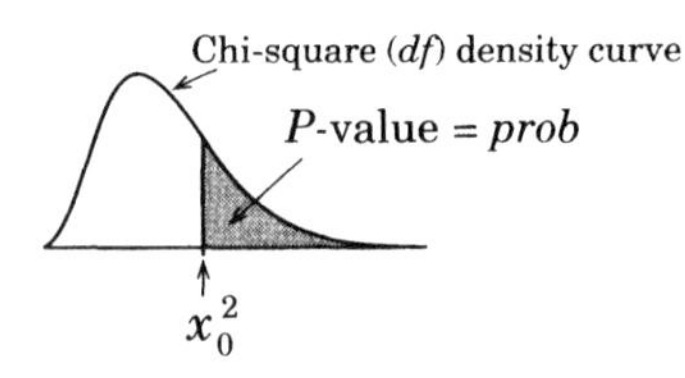

When we have obtained the test statistic x_0^2 and df, we can obtain the *P*-value from a computer package. (Table usage is considered shortly.) In Minitab, look under `Calc` → `Probability Distribution` → `Chi-Square`. Splus and R use `pchisq(x,df)`. These packages give cumulative, or *lower-tail,* probabilities. As the *P*-value is an upper-tail probability, we obtain it by subtracting the value printed by the program from one, *P*-value $= 1 - \text{pr}(X^2 \leq x_0^2)$. This is the pattern with most statistical packages. Excel's `CHIDIST(x0sq,df)` is unusual in that it gives the *P*-value directly. (Here, `x0sq` represents x_0^2.)

How do we find *df*? The value of *df* for Example 11.1.1 turns out to be $df = 5$. This is arrived at by first noting that there are 6 probabilities associated with the 6 categories, and these probabilities always add to 1. When we test H_0 we are testing whether these 6 probabilities have certain values ($\frac{1}{6}$). We have to specify only 5 of them, however, as the remaining one we get by subtracting the rest from 1. We therefore have the following rule.

df = number of categories − 1

We also note that the expected count for each cell was obtained by multiplying each specified probability by the *total* number of observations n.

expected cell count = total × specified cell probability

[1]Chi is pronounced "kie" (rhymes with "pie"). It is the name of the Greek letter χ.

This way of thinking about constructing a test statistic is used throughout this chapter. We always compare the observed and expected counts using a Chi-square statistic, and the P-value always has the same form, irrespective of whether the data table is one dimensional (as in Example 11.1.1), two dimensional (see Section 11.2), or of higher dimension. All that changes from situation to situation is the way we compute the expected counts and the way we determine df. In examples it is often convenient to include the expected counts with the observed counts in the table (especially in the more usual case when the expected counts are all different, as in the following Example 11.1.2).

Example 11.1.1 (cont.) *Calculating the Test Statistic*

For this problem, $df = 6 - 1 = 5$. We have

$$x_0^2 = \frac{(26-35)^2}{35} + \frac{(40-35)^2}{35} + \frac{(37-35)^2}{35} + \frac{(26-35)^2}{35} + \frac{(43-35)^2}{35} + \frac{(38-35)^2}{35}$$

$$= 7.54$$

Using the Chi-square($df = 5$) distribution, we obtain $P-$value $= 0.18$. Thus we do not have any evidence against the hypothesis that all outcomes for this die are equally likely, each with probability $\frac{1}{6}$.

Example 11.1.2 *Comparing the Age Structures of Voters and the Population*

A so-called exit poll is carried out by sampling voters as they leave polling booths. Results of exit polls of approximately 10,000 voters were widely reported after the 1998 U.S. midterm elections. The age distribution of sampled voters given in the first row of Fig. 11.1.1(a) came from press reports of the exit polls. The population age distribution in the second row of Fig. 11.1.1(a) was constructed from information obtained from the U.S. Bureau of the Census web site. We have plotted both age distributions in Fig. 11.1.1(b). There are clear differences. In particular, the younger age groups appear to be underrepresented among voters, and the older age groups appear to be overrepresented. Are these discrepancies real, or could they be just due to sampling variation?

To answer this question, we conducted a Chi-square test of the hypothesis that the true proportions in the four age groups in the population *of people who voted* were the same as for the adult population as a whole; that is, we tested H_0: $p_{18-29} = 0.22$, $p_{30-34} = 0.32$, $p_{45-59} = 0.24$, and $p_{60+} = 0.22$. The observed counts and the expected counts for a sample of 10,000 have been reconstructed from the percentages in Fig. 11.1.1(a) to form the table in Fig. 11.1.1(c) (they are clearly heavily rounded). We find that

$$x_0^2 = \sum \frac{(\text{observed count} - \text{expected count})^2}{\text{expected count}} = 709.9423$$

We have

$$df = \text{number of groups} - 1 = 4 - 1 = 3$$

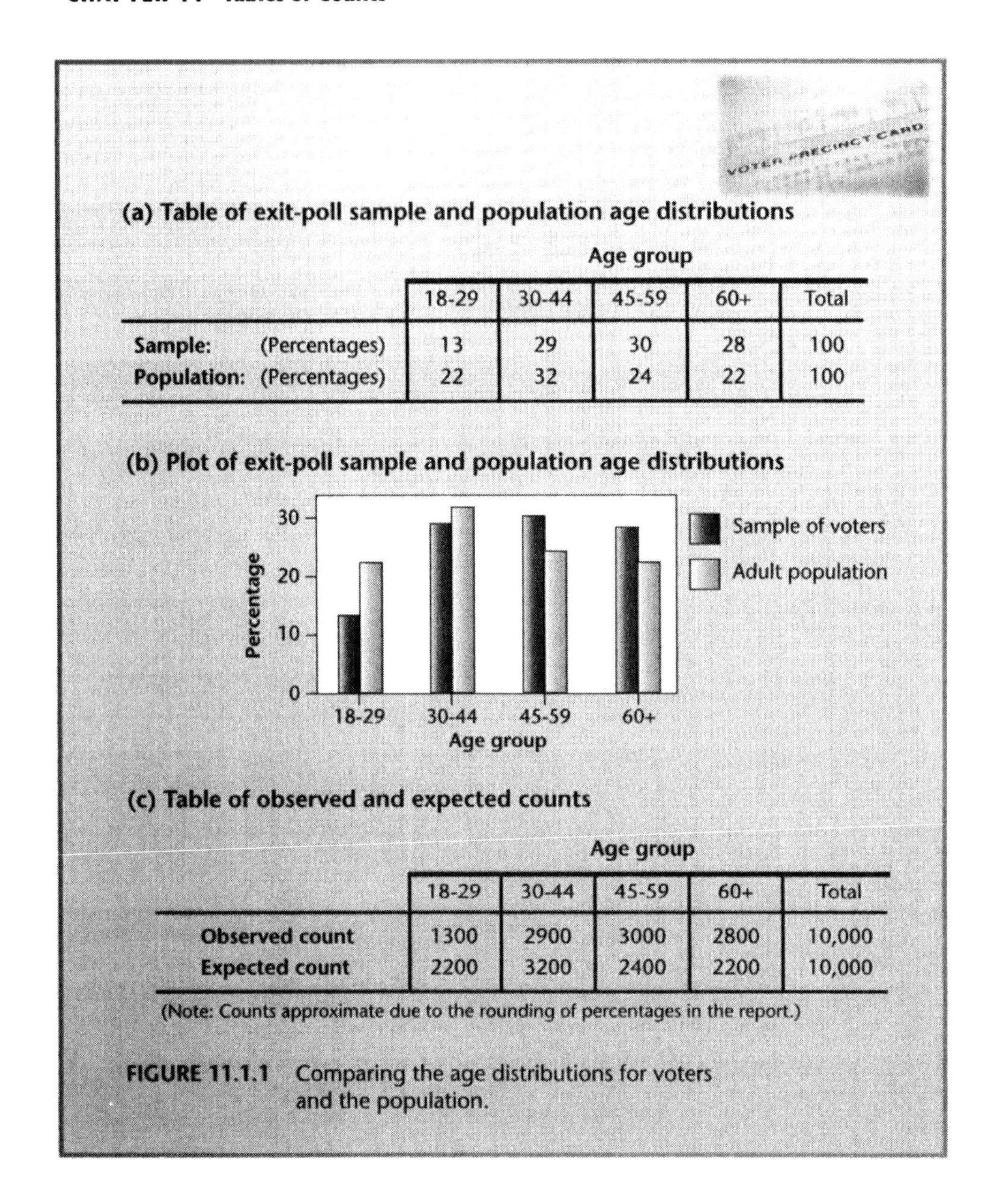

(a) Table of exit-poll sample and population age distributions

	Age group				
	18-29	30-44	45-59	60+	Total
Sample: (Percentages)	13	29	30	28	100
Population: (Percentages)	22	32	24	22	100

(b) Plot of exit-poll sample and population age distributions

(c) Table of observed and expected counts

	Age group				
	18-29	30-44	45-59	60+	Total
Observed count	1300	2900	3000	2800	10,000
Expected count	2200	3200	2400	2200	10,000

(Note: Counts approximate due to the rounding of percentages in the report.)

FIGURE 11.1.1 Comparing the age distributions for voters and the population.

giving

$$P\text{-value} = 0.0000$$

The *P*-value is very small, so we have extremely strong evidence against the null hypothesis. It is clear that the age distribution for voters is different from the age distribution for the over-18 population as a whole. We note that the Chi-square statistic from the real counts would have been somewhat different from what we have been able to reconstruct from the heavily rounded information in press reports. The statistic would still have been very large, however, and we would have reached the same conclusions.

It should be noted that the exit-poll sample size is very large, and consequently the degree of sampling variation in the estimated percentages is relatively small. We calculated 95% confidence intervals for the true percentages of voters in each of the four age groups and obtained [12.3, 13.7], [28.1, 29.9], [29.1, 30.9], and [27.1, 28.9], respectively. In particular, the percentage of voters in the 18–29 age group is clearly considerably smaller than the population percentage. In fact, it is well known that in the United States, younger people are less likely to vote.

11.1.2 Tables for the Chi-Square Distribution

As with Student (df), Chi-square(df) refers to a family of distributions. Each value of df gives us another member of the family. Examples are depicted in Fig. 11.1.2. We see from this figure that the distribution is very skewed when df is small. It becomes more and more symmetric as df increases, however, until for very large df the distribution is well approximated by a Normal distribution.

Suppose that $X^2 \sim$ Chi-square(df) (or χ^2_{df} for short). Tables of upper-tail probabilities for this distribution (see Fig. 11.1.3) are given in Appendix A7. They are very similar in format to the Student (df) tables in Appendix A6. For given *prob* (column) and df (row), the tabulated value is the number $\chi^2_{df}(prob)$ indicated in Fig. 11.1.3. Although in practice we generally use P-values obtained from computer packages, the table in Appendix A7 can be used to bracket the P-value in the same way as we used the Student's t-distribution table to bracket tail probabilities in Section 7.6.3. From the table in Appendix A7 we find the following:

$$\begin{aligned}
&\text{If } df = 7, \; prob = 0.05, \quad \text{then } \chi^2_7(0.05) = 14.07\\
&\text{If } df = 10, \; prob = 0.15, \quad \text{then } \chi^2_{10}(0.15) = 14.53\\
&\text{If } df = 15, \; prob = 0.95, \quad \text{then } \chi^2_{15}(0.95) = 7.261
\end{aligned}$$

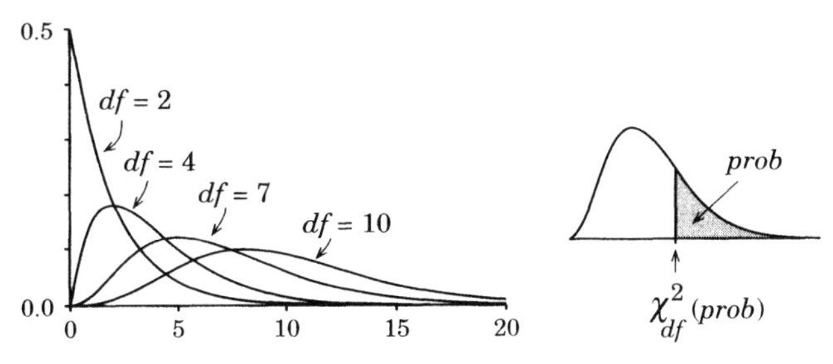

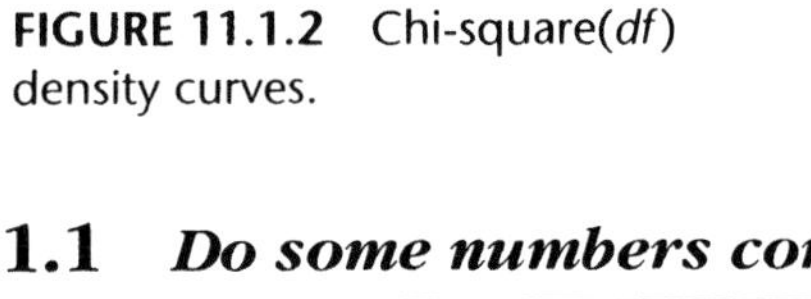

FIGURE 11.1.2 Chi-square(df) density curves.

FIGURE 11.1.3 The χ^2_{df} (*prob*) notation.

CASE STUDY 11.1.1 *Do some numbers come up more frequently in Lotto?*

Variants of the gambling game Lotto are used by governments in many countries and states to raise money for the arts, charities, and sporting and other leisure activities. Such games date back to the Han Dynasty in China over 2000 years ago (Morton [1990]). The basic form of the game is universal. Only small details change, depending on the size of the population involved. In the N.Z. version of the game, a person chooses six different numbers between

TABLE 11.1.3 Frequency of Winning Numbers in Lotto

1. (7)	**2.** (10)	**3.** (8)	**4.** (9)	**5.** (13)	**6.** (8)	**7.** (12)	**8.** (16)	**9.** (11)	**10.** (6)
11. (13)	**12.** (10)	**13.** (9)	**14.** (11)	**15.** (11)	**16.** (6)	**17.** (11)	**18.** (13)	**19.** (6)	**20.** (13)
21. (7)	**22.** (9)	**23.** (8)	**24.** (12)	**25.** (6)	**26.** (4)	**27.** (10)	**28.** (8)	**29.** (14)	**30.** (12)
31. (11)	**32.** (12)	**33.** (9)	**34.** (11)	**35.** (6)	**36.** (8)	**37.** (14)	**38.** (10)	**39.** (15)	**40.** (10)

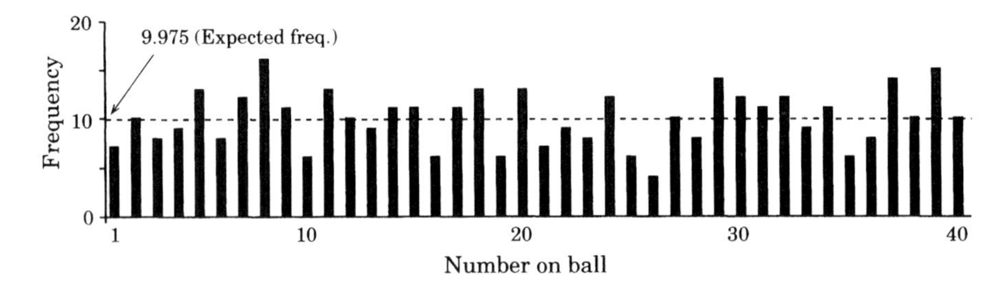

FIGURE 11.1.4 Frequency of Lotto winning numbers.

1 and 40. If the person's six chosen numbers match the six numbers drawn by the Lotto machine, that person wins a share of the first prize.

For the first year or so that Lotto was run in New Zealand, there was a lot of talk in the press about some numbers being lucky because they had appeared relatively often in Lotto draws and others being unlucky because they had seldom appeared. Now, the machine that does the sampling is supposed to be randomly sampling the numbers between 1 and 40. Let's check this out!

Table 11.1.3 gives the number of times each of the 40 numbers occurred in the first 57 Lotto draws. Each number from 1 to 40 is listed with the number of times it occurred in parentheses, for example, the number 19 turned up on 6 occasions. These frequencies are plotted in Fig. 11.1.4.

At each draw, 7 numbers were selected (including a "bonus" number), giving 399 numbers selected in all. We shall ignore the fact that at each draw numbers were selected without replacement[2] and assume that for every one of the 399 times a ball is selected, each of the numbers from 1 to 40 has an equal chance of being drawn. Note that this is a one-dimensional table with 40 response categories. It is just too big to write as a single row. The frequencies range from 6 to 16. Is there any evidence of nonrandomness in the draw? Under random sampling each number has a probability of being chosen of $\frac{1}{40}$ each time. Thus the expected count in each cell is $399 \times \frac{1}{40} = 9.975$. (Note that the expected count need not be an integer.) The observed Chi-square statistic is $x_0^2 = 30.97$ with $df = 40 - 1 = 39$ giving a P-value $= 0.817$, so there is no evidence of nonrandomness.[3]

QUIZ ON SECTION 11.1

1. The test statistic for the Chi-square test compares observed and expected frequencies. In what sense are the *expected* frequencies expected?

2. What shape does the Chi-square distribution generally have? What happens to its shape as the degrees of freedom increase?

3. What values of the Chi-square test statistic (large or small) provide evidence against the null hypothesis? Why?

[2]The sampling without replacement actually reduces slightly the chances of any particular number being drawn too often so that the counts should look slightly closer to uniform than for sampling with replacement.

[3]In this particular situation the observed Chi-square statistic can be corrected for sampling without replacement by multiplying χ_0^2 by (39/33), which brings the exact P-value down to 0.58. This type of correction factor is beyond the scope of this book.

4. For one-dimensional tables, how do you compute the degrees of freedom *df*?
5. Do the expected counts have to be whole numbers?

EXERCISES FOR SECTION 11.1

1. It has been suggested that the telephone book can be used as a source of random digits 0, 1, ..., 9 by selecting telephone numbers and noting the last digit. To test this out, the 10th number was selected from each of the first 110 pages of a local telephone directory, giving the following data.

Outcome	0	1	2	3	4	5	6	7	8	9	Total
Count	12	17	8	7	13	21	6	7	12	7	110

   (a) What do you conclude?
   (b) What do you think of the sampling procedure?

2. The frequencies of three blood types A, AB, and B among 151 children from parents whose blood types are both AB are

	A	AB	B	Total
No. observed	39	70	42	151

   A law of genetics postulates that the ratios A:AB:B are 1:2:1.
   (a) What are the values of the three probabilities postulated by the law?
   (b) Do the observations support the law?

11.2 TWO-WAY TABLES OF COUNTS

11.2.1 Introduction

Section 3.3 introduced the two-way table of counts as a first step toward exploring the relationship between two qualitative variables, that is, between two grouping factors. There the emphasis was descriptive in that we simply calculated various proportions or probabilities. In this section we look for structure in such tables and make more-detailed comparisons and inferences.

Suppose we have two qualitative variables that we are using to classify individuals into groups, giving us two ***grouping factors.*** For example, in Example 3.3.1 we classified a group of women by their ETHNICITY and BODY IMAGE. Suppose that the purpose of the investigation is to see what relationship, if any, there is between the two factors. The obvious way to investigate a question like this is to take a sample of individuals and count how many fall into each classification. This produces a two-way table of counts, as in Fig. 3.3.1(b) and also in the two examples to follow. Figure 3.3.1 illustrates the process of forming a two-way table of counts by cross-tabulating raw data on two factors. The positions inside the table are called ***cells.***

EXAMPLE 11.2.1 *Comparing Site and Type of Melanoma*

We revisit a data set we first saw in Example 4.6.2. Four hundred patients with malignant melanoma (a form of skin cancer) were cross-classified by both TYPE (the type of abnormal cells making up the malignancy) and SITE (where on the body the cancer appeared). The number of patients falling into each cell, determined by the levels of these two factors, is given in Table 11.2.1. The table is said to be a 4×3 table because it has 4 rows and 3 columns.

Before rushing in and attempting to make formal inferences about the data, we need to do what we encouraged you to do in Chapter 10 with quantitative data—have a good look at the data using any derived tables and graphs to highlight any features. The first questions that occur to us are those such as, "What is the most common of these skin cancers?" "What are the most common places to get skin cancers?" "What is the most common combination of skin cancer and site?"

In making comparisons, the numbers of people falling into each cell of Table 11.2.1 depend on the sample size (400 people); the bigger the sample, the bigger all the cell counts. If we are to think about what might be happening in the larger population, the *proportions* of people falling into each cell in the table are more useful. These proportions are given in Table 11.2.2. They are calculated by taking the cell count in Table 11.2.1 and dividing it by the grand total for the whole table. We call such proportions ***whole-table proportions.***

To give a feeling for these proportions, we have plotted them in two ways in Fig. 11.2.1. In the thermometer plot, Fig. 11.2.1(a), the length of the black bar gives the size of the proportion. The bars have been scaled so that the largest proportion occupies a whole "thermometer." We see that superficial spreading melanoma on the extremities is the most common combination (29%) with nodular cancers on the extremities coming next. A Hutchinson's freckle on the trunk is a very rare combination, and there is a lot of visual information about the relative sizes of intermediate values. The three-dimensional plot, Fig. 11.2.1(b), is more attractive. It looks like a collection of high-rise buildings with the size of the proportion being represented by the height

TABLE 11.2.1 400 Melanoma Patients by Type and Site

	Site			
Type	Head and neck	Trunk	Extremities	Row totals
Hutchinson's melanomic freckle	22	2	10	34
Superficial spreading melanoma	16	54	115	185
Nodular	19	33	73	125
Indeterminate	11	17	28	56
Column totals	68	106	226	400

Source: Roberts et al. [1981].

TABLE 11.2.2 Proportions of Melanoma Patients by Type and Site (whole-table proportions)[a]

Type	Site			Row totals
	Head and neck	Trunk	Extremities	
Hutchinson's melanomic freckle	0.055	0.005	0.025	0.085
Superficial spreading melanoma	0.040	0.135	0.288	0.463
Nodular	0.048	0.083	0.183	0.313
Indeterminate	0.028	0.043	0.070	0.140
Column totals	0.170	0.265	0.565	1.00

[a]All numbers rounded to three decimal places.

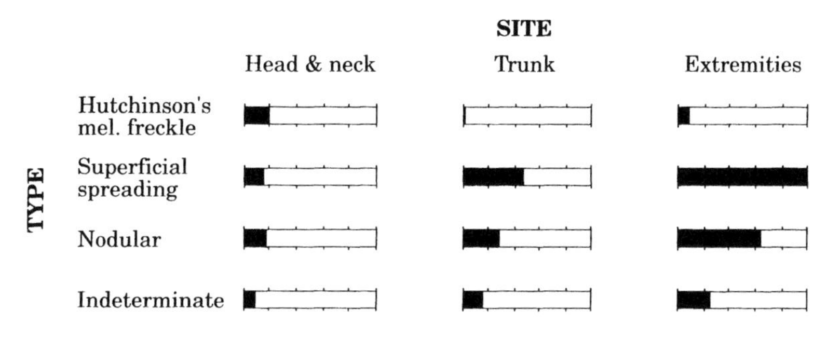

(a) Thermometer plot

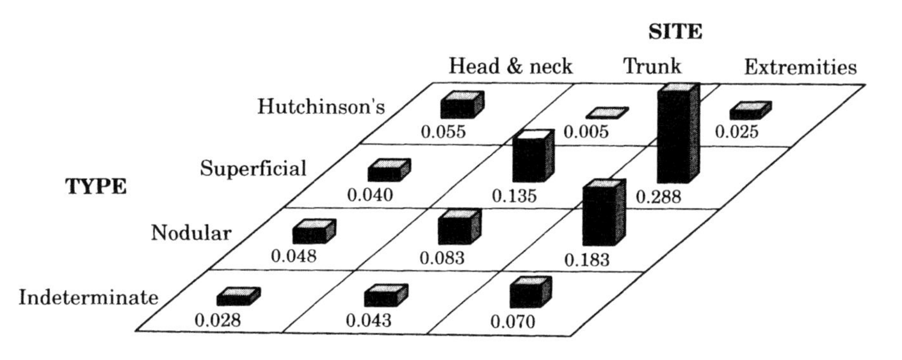

(b) 3-dimensional plot

FIGURE 11.2.1 Proportion of all patients with a skin cancer with the given type and site (whole-table proportions).

of the building. Unfortunately, the three-dimensional appearance can introduce visual distortions. Because of the appearance of distance (perspective) in looking from the "front" of the plot to the "back," blocks toward the back appear bigger than blocks of the same actual size placed toward the front. As is so often the case with statistical graphics, simpler is better.

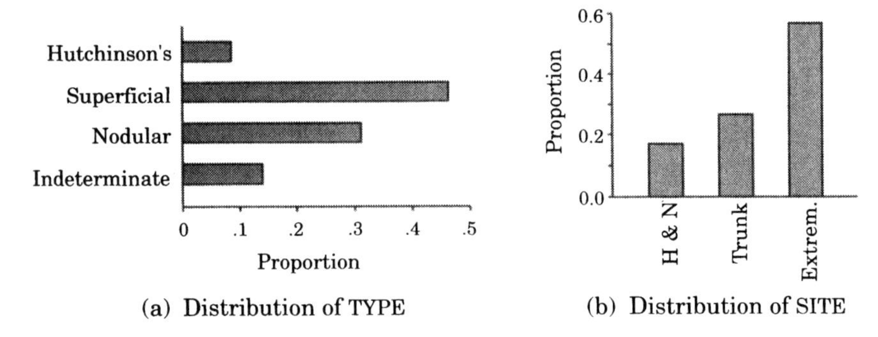

FIGURE 11.2.2 Distributions for SITE and TYPE.

Having looked at the big picture, we now consider the TYPE and SITE separately. The *row* totals of Tables 11.2.1 and 11.2.2 give us the results of a one-way classification according to TYPE of cancer (ignoring SITE). The former table gives us the counts for such a classification, and the latter gives the proportions. We see that superficial spreading melanomas are most common (46% of skin cancers) and Hutchinson's melanomic freckles are least common (less than 9%). The *column* totals give us a one-way classification according to SITE of cancer (ignoring TYPE). We see that these skin cancers usually appear on the extremities (57%) and are much more rare on the head and neck (17%). Figure 11.2.2 uses bar graphs to plot these proportions so that we can get a visual feel for their relative sizes. We have oriented the plots in such a way as to preserve their identification with the rows and columns of the data, though horizontal or vertical orientation is generally a matter of taste.

We must remember that what we are looking at here are sample proportions and not the true proportions for the population from which these 400 people were drawn. It is possible that the differences we are seeing here are in part or wholly due to sampling variation. We might test for equality of the TYPE proportions formally using a Chi-square test as in the previous section. We would compare each of the observed counts—34, 185, 125, and 56 in the row totals column of Table 11.2.1—with the expected count of 100. Such a test gives $x_0^2 = 141.42$ on $df = 3$ so that *P*-value $< 10^{-6}$. It is clear that these four cancers are not equally likely, so real differences between the true proportions exist. We could use confidence intervals or significance tests to investigate differences between these proportions, but we must remember that we are comparing proportions from the *same sample* [situation (b) in Fig. 8.5.1; see Section 8.5 for confidence intervals and Section 9.3.2 for tests]. Although useful for a thorough analysis of this data, these ideas relate to techniques that we have already used many times, and we shall not pursue them here.

Up until now we have concentrated on individual proportions. In this chapter, however, we are more concerned with taking a broad approach. For example, a question relating to the whole table is whether certain types of cancer are more likely to appear in certain sites or whether the SITE and TYPE of the cancer are independent. We shall find that we can test this hypothesis of independence using a Chi-square test. Questions like this will be our main focus for the next few sections.

✖ EXAMPLE 11.2.2 *Comparing Blood Types for Different Regions*

Blood contains a number of genetic factors that can be used to see whether the populations in some regions have different racial origins from those in others. Samples of blood are collected from all the regions and the phenotype proportions are compared using various blood-grouping systems. One such study was carried out by Mitchell et al. [1976] using blood donor data from southwest Scotland. Data from three regions, which were classified using the ABO system, are given in Table 11.2.3. This is a 3×4 table, and the two grouping factors in this example are REGION and PHENOTYPE. The question of interest is whether there are regional differences in the phenotype structure.

In this example we have three populations and a sample from each population. This is an observational study (see Section 1.3). The sample sizes are the sizes of the donor registers at the particular time in the three regions.

Let us assume that the people represented in the table constitute a random sample of people from their region as far as ABO blood types are concerned.[4] Consider the region of Nithsdale. The proportions of Nithsdale people in the table who have type-A blood, type-B blood, type-AB blood, and type-O blood give us an estimate of the distribution of blood phenotypes for all people living in Nithsdale. The total number of people in the sample from Nithsdale, 253, forms the denominator for each one of these proportions. These proportions form the first row of Table 11.2.4. Each proportion in Table 11.2.4 is calculated from Table 11.2.3 by dividing the cell count by the row total. We call proportions made up in this way ***row proportions.*** Table 11.2.4 also gives the corresponding row proportions of people with each blood phenotype for people from Cree (row 2) and for people from Rhinns (row 3).

We have plotted these estimated distributions, one for each row (region), in Fig. 11.2.3 using bar graphs. The vertical scale for each of the three graphs has been kept the same so that valid comparisons can be made. If we look at the plotted row proportions, we see similarities in the phenotype distributions between regions. Each distribution is of a roughly similar shape. In all three regions, type-O blood is most common, followed by type-A blood, whereas type-AB blood is rarest in all three regions. We also see differences, however. The type-O bar is taller in Cree than in the other two regions suggesting that type-O blood is more common in Cree than in Nithsdale or Rhinns. Similarly, the shorter type-A bar for Cree suggests that type-A blood is rarer in Cree than in Nithsdale or Rhinns.

At this point we must remember that the distributions we are looking at in Fig. 11.2.3 are not the true population distributions. They are sample estimates. The differences in proportions that we are seeing may not correspond to real differences between the populations. They may simply be due to sampling variation. We therefore need a way of looking at the data that takes sampling variation into account. If we compare the proportion of people in Nithsdale with type-A blood with the proportion of Cree people with type-A blood, we are comparing proportions from two independent samples. We know how to calculate a confidence interval for the true difference in proportions; the confidence interval for $p_1 - p_2$ is given in Section 8.5. We also know how to test for equality of the two true proportions; testing $H_0: p_1 - p_2 = 0$ is given in Section 9.3. Before snooping around looking at various pairs of proportions,

[4]There are a number of practical complications with blood donor data. We don't know whether the blood donors are representative of the population as a whole in each region with respect to phenotype because they were self-selecting rather than randomly selected by the experimenter. Unfortunately, this means that we can't be sure that regional differences shown by the samples accurately reflect population differences in blood types between the regions or just regional differences in patterns of blood donations. There is some evidence that sometimes blood donor data is not representative of population data, particularly when less-common blood genes are involved. This is not surprising, as blood centers would encourage donors with rare blood types to give blood, and some regions may do a better job of this than others. We shall ignore these complications in the analysis that follows.

TABLE 11.2.3 Regional Data for the ABO System

Region	Phenotype				Total
	A	B	O	AB	
Nithsdale	98	35	115	5	253
Cree	38	9	79	6	132
Rhinns	36	9	47	7	99
Total	172	53	241	18	484

Source: Mitchell et al. [1976].

TABLE 11.2.4 ABO Distributions from Three Areas (row proportions)[a]

Region	Phenotype				Total
	A	B	O	AB	
Nithsdale	0.39	0.14	0.45	0.02	1.00
Cree	0.29	0.07	0.60	0.05	1.00
Rhinns	0.36	0.09	0.47	0.07	1.00

[a]All entries are the accurate figures rounded to two decimal places.

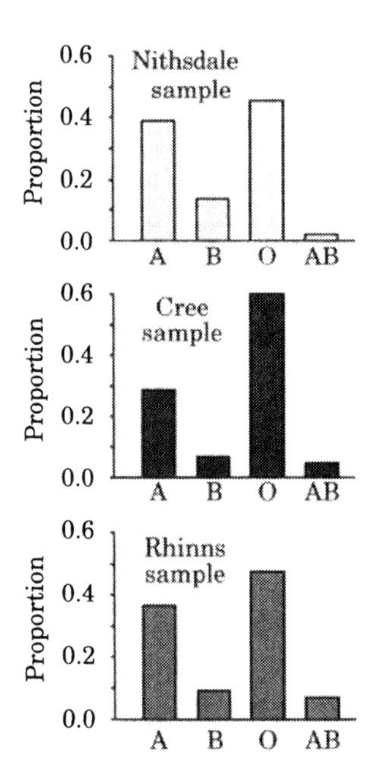

FIGURE 11.2.3 ABO distributions from three areas (row proportions).

however, we want to look at the overall picture first. This means testing an overall hypothesis that there are no differences between the three regions, that is, that the population blood phenotype distributions are identical for all three regions.

We have seen from the preceding two examples that two-way tables can arise in two ways. In the skin cancer example a *single* sample was taken from a single population, and the counts were categorized according to two grouping factors. In the second example on blood phenotypes we have three samples from three populations to form the three rows of the two-way table. In spite of these differences, however, we shall find that there is an underlying similarity about the hypotheses tested for the two models. The Chi-square test that we use in the following section turns out to be the same in *both* situations.

11.2.2 The Chi-Square Test of Homogeneity

Table 11.2.5 sets up a general notation for two-way tables. Our two grouping factors are A, which has I levels, and B, which has J levels. The observed number (frequency) of individuals in the jth column of the ith row is O_{ij}. We call the position in the ith row and jth column the *cell* (i, j). The sum of the frequencies across the ith row is R_i, and the sum of the frequencies down the jth column is C_j. The total number of individuals classified in the table is n, where $n = \sum R_i = \sum C_j = \sum\sum O_{ij}$. We call n the ***grand total*** for the table. A two-way table with I rows and J columns is called an $I \times J$ table.

We want to test whether there is any evidence that the row distributions are not identical. Somewhat more formally, the null hypothesis H_0 that we wish to test states that the underlying true distribution of B is the same for every level of A. This is the so-called homogeneity hypothesis.

H_0: the distribution of B is the same for every level of A.

In terms of Example 11.2.2 this means that the PHENOTYPE distribution (i.e., the population proportions having types A, B, O, and AB blood) is the same for each of

TABLE 11.2.5 General Notation for a Two-Way Table

	Level of B						
Level of A	**1**	**2**	...	**j**	...	**J**	**Total**
1	O_{11}	O_{12}	...	O_{1j}	...	O_{1J}	R_1
2	O_{21}	O_{22}	...	O_{2j}	...	O_{2J}	R_2
⋮	⋮	⋮	⋮	⋮	⋮	⋮	⋮
i	O_{i1}	O_{i2}	...	O_{ij}	...	O_{iJ}	R_i
⋮	⋮	⋮	⋮	⋮	⋮	⋮	⋮
I	O_{I1}	O_{I2}	...	O_{Ij}	...	O_{IJ}	R_I
Total	C_1	C_2	...	C_j	...	C_J	n

the three regions. As for one-dimensional tables, the observed value of the Chi-square test statistic is based on

$$x_0^2 = \sum_{\text{all cells}} \frac{(\text{observed count} - \text{expected count})^2}{\text{expected count}} = \sum_{ij} \frac{(O_{ij} - \widehat{E}_{ij})^2}{\widehat{E}_{ij}}$$

where the *expected count* in cell(i,j) turns out to be

$$\text{expected count} = \frac{\text{row total} \times \text{column total}}{n}$$

or mathematically[5]

$$\widehat{E}_{ij} = \frac{R_i C_j}{n}$$

We shall also find that the associated degrees of freedom df is given by

$$\begin{aligned} df &= (\text{number of rows} - 1) \times (\text{number of columns} - 1) \\ &= (I-1)(J-1) \end{aligned}$$

As before, the P-value for the test is the upper-tail probability

$$P\text{-value} = \text{pr}(X^2 \geq x_0^2) \qquad \text{where } X^2 \sim \text{Chi-square}(df)$$

We motivate these formulas on our web site.

EXAMPLE 11.2.2 (cont.) *Homogeneity of the Phenotype Distributions*

We shall now perform the Chi-square test of no difference between the PHENOTYPE distributions for each of the three regions. The data in Table 11.2.3 have been reprinted in Table 11.2.6. This time, however, the expected counts (or frequencies)

TABLE 11.2.6 Phenotype Regional Data (with expected counts in parentheses)

	Phenotype				
Region	A	B	O	AB	Totals
Nithsdale	98 (89.91)	35 (27.70)	115 (125.98)	5 (9.41)	253
Cree	38 (46.91)	9 (14.45)	79 (65.73)	6 (4.91)	132
Rhinns	36 (35.18)	9 (10.84)	47 (49.30)	7 (3.68)	99
Totals	172	53	241	18	484

[5] $\widehat{E}_{ij} = R_i\widehat{p}_j$ where $\widehat{p}_j = C_j/n$ is the sample proportion of those in the jth level of B.

have been included in the table in parentheses with the observed counts, which is standard practice with such tables. You can check that for each entry

$$\widehat{E}_{ij} = \text{expected count in cell}(i,j) = \frac{R_i C_j}{n}$$

For example, the cell corresponding to region Cree and phenotype O has an expected count of

$$\widehat{E}_{23} = \frac{132 \times 241}{484} = 65.73$$

As a check on the calculations, the expected counts $\widehat{E}_{ij}$ in parentheses should have the same row and column totals (apart from rounding error) as the observed counts. The observed value of the Chi-square test statistic is computed as follows:

$$\begin{aligned} x_0^2 &= \frac{(98-89.91)^2}{89.91} + \frac{(35-27.70)^2}{27.70} + \frac{(115-125.98)^2}{125.98} + \frac{(5-9.41)^2}{9.41} \\ &\quad + \frac{(38-46.91)^2}{46.91} + \cdots + \frac{(7-3.68)^2}{3.68} \qquad \text{(12 terms)} \\ &= 15.78 \end{aligned}$$

Now $df = (3-1)(4-1) = 6$, so we have P-value $= \text{pr}(X^2 \geq 15.78) = 0.015$ (by computer). We therefore have good evidence that the phenotype distributions are different for the different regions. This suggests that we should now start looking at individual differences between proportions using Table 11.2.4. We can do this using the methods of Section 8.5. Proportions relating to different regions are independent [cf. Fig. 8.5.1(a)], and so we use the standard error for differences between independent proportions [Table 8.5.5(a)]. For example, 95% confidence intervals for the true differences between the proportions of people having type-A blood are Nithsdale − Cree, [0.002, 0.197]; Nithsdale − Rhinns, [−0.088, 0.136]; and Rhinns − Cree, [−0.046, 0.198]. From the two latter intervals we see that we cannot demonstrate a difference between Nithsdale and Rhinns or between Rhinns and Cree. Moreover, from the first interval the difference between Nithsdale and Cree could be tiny. On the other hand, the differences could be quite large. Unfortunately, there is not enough information in the data to be more precise. Other blood groups could be compared similarly.

One of the problems with count data is that it does not contain much information. This means that we shall often find ourselves in a position where we can demonstrate that differences between the distributions exist, but we can't say much that is very useful about the nature of the differences because any associated confidence intervals are so wide.

The preceding intervals, which are ordinary 95% confidence intervals, ignore the fact that we are making multiple comparisons (cf. Section 10.3.1). Thus any set of 95% confidence intervals will incur a "family error rate" that is greater than the 5% error rate that individual intervals have. How such sets of intervals can be widened to compensate for multiple comparisons and give a family error rate of 5% is beyond the scope of this book.

Testing the Equality of Column Distributions

We arranged the blood phenotype data in Table 11.2.3 so that REGION defined the rows of the table and PHENOTYPE defined the columns. This choice was entirely arbitrary. We could equally well have arranged the data so that REGION defined the columns of the table and PHENOTYPE defined the rows. In the latter case, testing that the underlying PHENOTYPE distribution was the same for each of the three regions would correspond to testing for equality of the *column* distributions (which are made up of column proportions). We note the symmetry of the Chi-square statistic. It gives the same answer irrespective of whether factor A represents independent row distributions and B represents columns or whether A represents independent column distributions and B represents rows.

Notes:

- Provided one of the factors represents independent distributions, we can use the ***Chi-square test*** to test for the equality of the distributions, irrespective of whether the distributions are listed as rows or columns.
- Testing for the equality of row (or column) distributions is called testing for ***homogeneity.***[6]

EXAMPLE 11.2.1 (cont.) *Homogeneity Test for the Melanoma Data*

We now revisit the melanoma data. Although the data are not sampled as separate groups, we can estimate the row distributions using row proportions, and we may be interested in testing whether these distributions are equal. The data are presented again in Table 11.2.7. We first ask where the various cancers tend to be found. In other words, we want to look at the SITE distribution for each TYPE of cancer. This leads us to look at the row proportions, which are calculated using the row totals as denominators. These are given in Table 11.2.8 and graphed in Fig. 11.2.4. We see that the SITE distributions of superficial spreading, nodular, and indeterminate melanomas all look very similar, with most of these melanomas appearing on the extremities. Hutchinson's melanomic freckles appear to behave quite differently, however, with most of them appearing on the head and neck.

Are these differences real, or could we just be seeing the results of sampling variation? Let us now perform a Chi-square test of homogeneity to test the hypothesis that corresponding true population distributions are identical. Figure 11.2.5 is the output from Minitab and includes the expected counts underneath the observed counts. Recall that each expected count is given by the row sum times the column sum, divided by the grand total of individuals within the table (400 here). We see from this output that $x_0^2 = 65.813$ with $df = (3-1)\times(4-1) = 6$ and P-value $= 0.000$. We therefore have extremely strong evidence against the identical-distributions hypothesis (homogeneity); that is, we have extremely strong evidence that differences exist between the SITE distributions. However, we previously noticed that only the SITE distribution of Hutchinson's melanomic freckle stood out as different, whereas those of the other three types looked fairly similar. This suggests using a Chi-square test to test for equality of the remaining three distributions. When we did this, we obtained $x_0^2 = 6.5619$, $df = 4$, P-value $= 0.16$, so we have no evidence that these distributions differ.

[6]From the Greek words *homo* for same and *genus* for family.

TABLE 11.2.7 Melanoma Data (reproduces Table 11.2.1)

Type	Site			Row totals
	Head and neck	Trunk	Extremities	
Hutchinson's	22	2	10	34
Superficial	16	54	115	185
Nodular	19	33	73	125
Indeterminate	11	17	28	56
Col. totals	68	106	226	400

TABLE 11.2.8 Row Proportions (site distribution for each type of cancer)

Type	Site			Row totals
	Head and neck	Trunk	Extremities	
Hutchinson's	0.65	0.06	0.29	1.00
Superficial	0.09	0.29	0.62	1.00
Nodular	0.15	0.26	0.58	1.00
Indeterminate	0.20	0.30	0.50	1.00

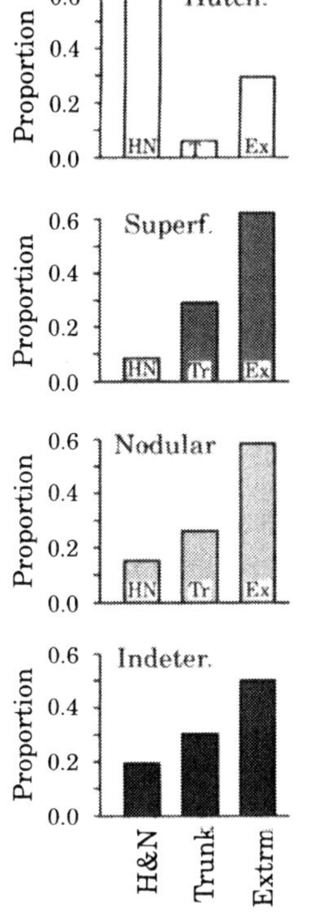

FIGURE 11.2.4 Row proportions.

Chi-Square Test

```
Expected counts are printed below observed counts
                  Head & N    Trunk   Extremit    Total
Hutchinson's  1         22        2         10       34
                      5.78     9.01      19.21
Superficial   2         16       54        115      185
                     31.45    49.03     104.53
Nodular       3         19       33         73      125
                     21.25    33.13      70.62
Indeterminate 4         11       17         28       56
                      9.52    14.84      31.64
Total                   68      106        226      400

Chi-Sq = 45.517 +  5.454 +  4.416 +
          7.590 +  0.505 +  1.050 +
          0.238 +  0.000 +  0.080 +
          0.230 +  0.314 +  0.419 = 65.813

DF = 6, P-Value = 0.000
```

FIGURE 11.2.5 Minitab output for the melanoma data.

TABLE 11.2.9 Column Proportions for Melanoma Data (distribution of types at each cancer site)

	Site		
Type	Head and neck	Trunk	Extremities
Hutchinson's	0.32	0.02	0.04
Superficial	0.24	0.51	0.51
Nodular	0.28	0.31	0.32
Indeterminate	0.16	0.16	0.12
Column totals	1.00	1.00	1.00

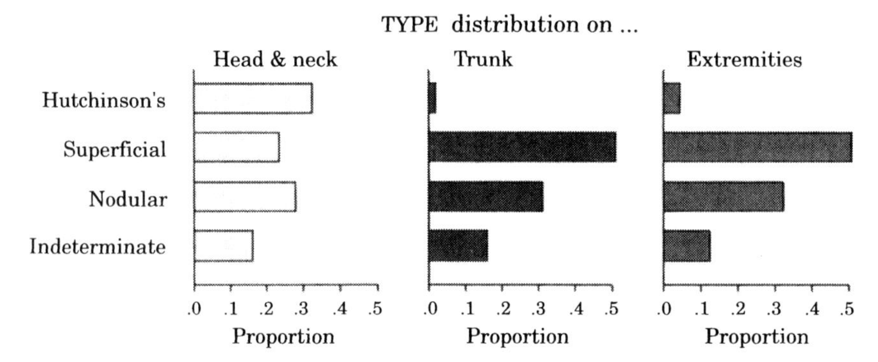

FIGURE 11.2.6 Column proportions (distribution of Types of cancer found at each of four different sites).

We may have wanted to look at the data in the converse way. What is the distribution of types of melanoma at a given site, and how does this differ from site to site? This leads to inspection of the column proportions, which are calculated using column totals as denominators. These are given in Table 11.2.9 and depicted in Fig. 11.2.6. We see similar distributions on the trunk and extremities; most melanomas are superficial, with smaller numbers of nodular and then indeterminate and with Hutchinson's freckles being very rare. The distribution on the head and neck is quite different, with the Hutchinson's freckles being the most common type of melanoma, closely followed by nodular and then superficial.

The Chi-square test performed earlier can also be regarded as a test of equality of column distributions. Recall that it gave a vanishingly small *P*-value, signaling extremely strong evidence of the existence of real differences between the underlying true distributions.

Computing

All general-purpose statistical packages will perform the Chi-square test for a two-way table of counts. When working from the raw data, `Cross-tabulation` programs will both form the two-way table of counts (cf. Section 3.3) and perform the Chi-square test. In Minitab, for example, look under `Stat` → `Tables` → `Cross Tabulate`. Splus uses the `crosstabs` function. If you already have the table, then you can use the procedure under `Stat → Tables → Chi-Square test` in Minitab, and `chisq.test` in Splus and R. The Excel function CHITEST will also produce a *P*-value for the test, but the user has to form the table of expected counts. (This is easy in a spreadsheet or any of the packages.)

11.2.3 The Chi-Square Test of Independence

So far in Section 11.2 we have looked at two examples of a two-way table of data. The first example, Example 11.2.1, discussed a situation where a sample of 400 melanoma patients from a single population were classified according to two categories: SITE and TYPE. This is an example of *situation 1: one sample cross-classified by two factors,* depicted in Fig. 11.2.7. The second example, Example 11.2.2, discussed a situation where we have three independent samples, each from a different population. This is an example of *situation 2: I samples, single response factor* shown in Fig. 11.2.7 with $I = 3$. Here we were only interested in estimating and comparing the row (REGION) distributions using a test of homogeneity. Whole-table proportions are meaningless if either rows or columns correspond to separate samples, as is the case here. In the first example, however, we can estimate the row distributions using row proportions, the column distributions using column proportions, and the joint probabilities of the form pr($A = i$ and $B = j$) using the whole-table proportions. All these quantities have meaning, and most of the time there will be one of these three different ways of thinking about the table that is most natural for the problem at hand. For example, we saw that we can carry out a test of homogeneity using row distributions, and this test is exactly the same as the test for homogeneity using column distributions. Using the whole-table proportions, however, we can test another hypothesis of interest for situations like situation 1, namely, that the row that an individual falls into is independent of the column he or she falls into. In the melanoma example this is the hypothesis that the SITE and TYPE of cancer are independent. On our web site we derive the expected counts under both the homogeneity hypothesis for situation 2 and under the hypothesis of independence (of row and column position) for situation 1 in Fig. 11.2.7. We show that you get the same expected counts in both cases and thus show that the two Chi-square test statistics are the same.

QUIZ ON SECTIONS 11.2.1 TO 11.2.3

1. What information do the row sums of a two-way table of counts give you? What about the column sums? (Section 11.2.1)
2. How do you calculate whole-table proportions? When does it make sense to calculate them? What information do such proportions give you?
3. What sort of information do the row sums of the whole-table proportions give you? What about the column sums?
4. What are the denominators of the row proportions? What information do they give you? Repeat for column proportions. (Section 11.2.2)

The following sequence of questions concern the application of the Chi-square test to a two-way table, as discussed in Sections 11.2.2 and 11.2.3. Recall that the same test can be used to test three different null hypotheses.

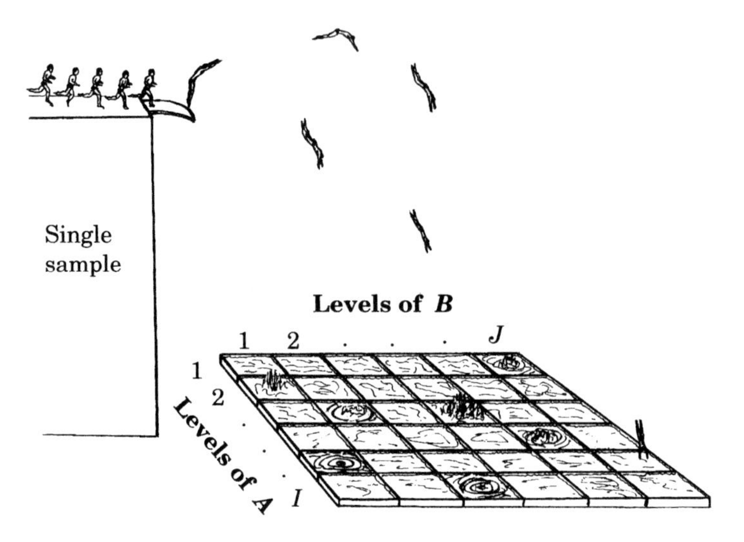

Situation (1): One sample cross-classified by 2 factors

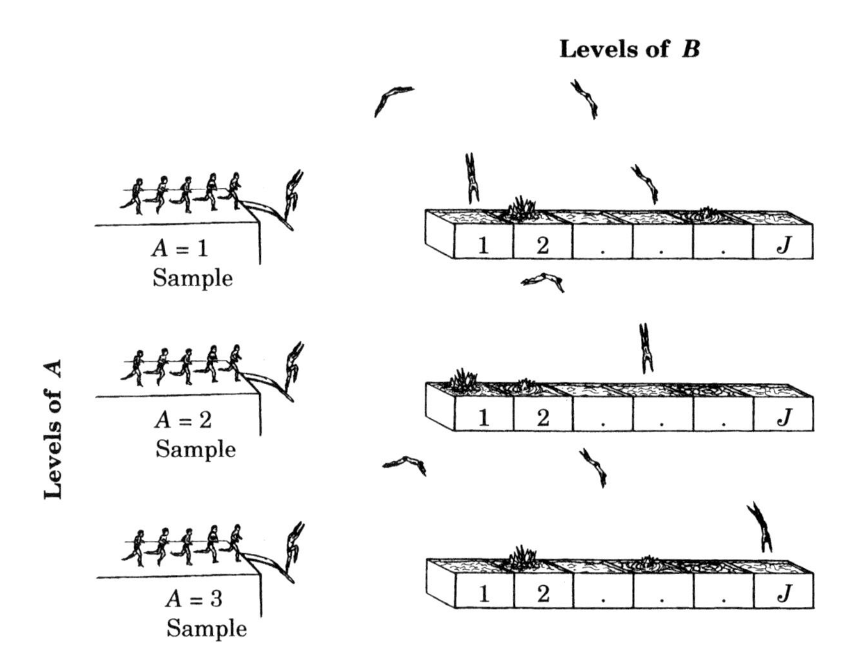

Situation (2): I samples, single response factor

FIGURE 11.2.7 Two types of two-way table.

5. Suppose that we are interested in comparing row distributions. In what way(s) can we sample to obtain our data? Express in words the null hypothesis tested by the Chi-square test. Repeat for column distributions.

6. If we do not want to think in terms of row distributions or column distributions but just want to sce whether there is any relationship between the row an individual falls into and the column he or she falls into, express in words the null hypothesis tested by the Chi-square test.

7. Express in words how one calculates the expected count for cell(i, j).

8. Qualitatively, would a large value or a small value of x_0^2 make you think that there was evidence of a relationship between row and column classifications? Why?

9. Qualitatively, would a large P-value or a small P-value make you think that there was evidence of a relationship between row and column classifications?

EXERCISES FOR SECTIONS 11.2.1 TO 11.2.3

1. Consider the data given in Fig. 3.3.1(b) where women are classified by two factors, ETHNICITY and BODY IMAGE. Test whether the two factors are independent. Repeat the test omitting the Pacific group. (This group stood out as looking very different in Fig. 3.3.1.)

2. Six hundred patients were used to test a new drug. The 600 people were randomly divided into 3 groups of 200. The first group was given a placebo, the second was given the drug at single-dose strength, and the third was given the drug at double-dose strength. The patients were classified according to their improvement levels over a period of a week. The data, which classify patients by the two factors TREATMENT and RESPONSE, are laid out as a 3×3 table in Table 11.2.10 with the levels of TREATMENT defining rows.
   (a) Which of the two situations given in Fig. 11.2.7 applies?
   (b) Carry out an appropriate test. What do you conclude?

TABLE 11.2.10 Drug Trial

	Response			
Treatment	**Improve**	**No change**	**Worse**	**Total**
Placebo	35	70	95	200
Single dose	62	76	62	200
Double dose	88	80	32	200
Total	185	226	189	600

11.2.4 2 × 2 Tables

Example 11.2.3 *Is There Sexual Discrimination at Berkeley Campus?*

Bickel and O'Connell [1975] investigated whether there was any evidence of gender bias in graduate admissions at the University of California at Berkeley. Table 11.2.11

TABLE 11.2.11 Berkeley Admissions (SEX × ADMISSION)

		Admission		
		Yes	**No**	**Totals**
Sex	Men	3,738	4,704	8,442
	Women	1,494	2,827	4,321
	Totals	5,232	7,531	12,763

Source: Bickel and O'Connell [1975].

comes from their cross-classification of all 12,763 applications to graduate programs in 1973 by SEX (male or female) and ADMISSION (whether or not the applicant was admitted to the program).

A table like this is called a ***2 × 2 table*** because each factor has only two levels. For a 2 × 2 table, the Chi-square test has $df = (2 - 1) \times (2 - 1) = 1$.

Chi-square test for a 2×2 table: $\boldsymbol{df = 1}$.

If we had asked you in Chapter 9 whether the data in Table 11.2.11 showed any evidence that men were more likely to be admitted than women, you would have looked at the proportions of men and women being admitted and performed a test for no difference in the underlying proportions using the estimated difference in proportions and its standard error.

This chapter has given you another way of testing for no relationship between SEX and ADMISSION, namely, the Chi-square test. What is the relationship between the two tests? It turns out that for 2×2 tables, the Chi-square test gives *P*-values almost identical to the *two-sided* test for a difference between two proportions.

More technically, the relationship between the two tests is as follows. The value of x_0^2 is the square of t_0 to a very close approximation. In fact, if $se(\hat{p}_1 - \hat{p}_2)$ is calculated in a slightly different way (which comes from assuming $p_1 = p_2$), then x_0^2 is exactly t_0^2 (see D'Agostino et al. [1988]). It can also be shown that if $T \sim$ Normal(0, 1), then T^2 has a Chi-square($df = 1$) distribution. Because the Chi-square test looks at *squared* differences between observed and expected counts, it loses information on the direction of the differences. That is why it is equivalent to the two-tailed test of H_0: $p_1 - p_2 = 0$ and not to a one-tailed test.

EXAMPLE 11.2.3 (cont.) *Comparing Two Proportions*

We begin by comparing the proportions of men and of women who were admitted. We have $n_{Men} = 8{,}442$, of which a proportion $\hat{p}_{Men} = 3738/8442 = 0.4428$ were admitted, and $n_{Wom} = 4{,}321$, of which $\hat{p}_{Wom} = 1494/4321 = 0.3458$ were admitted. So we have 44% of men admitted versus 35% of women. With the huge sample sizes here, you will probably guess that that difference will be highly statistically significant. These two proportions come from independent samples. The *t*-test statistic using the standard-error formula in Table 8.5.5(a) is 10.74. Using the Normal approximation, the two-sided *P*-value is essentially 0. We have exceedingly strong evidence that Men were more likely to be admitted to a graduate program at Berkeley than women.

Now let us use the Chi-square test to see whether the chances of admission are independent of sex. The Chi-square test statistic is $x_0^2 = 112.2$, which may be compared with $t_0^2 = 115.3$. The *P*-value is essentially 0 again. In situations where the hypothesis H_0: $p_1 - p_2 = 0$ is more plausible than it is here, the agreement between t_0^2 and x_0^2 is even closer.

We have seen that the admission rate for women to Berkeley graduate programs was lower than the admission rate for men and that the difference was extremely significant statistically. A 95% confidence interval for the difference in underlying admission rates is given by [0.08, 0.11], or between 8% and 11%. Does this mean that the Berkeley administration was operating an admissions policy that discriminated against women? We shall answer this question in Section 11.3.

QUIZ ON SECTION 11.2.4

1. What is a 2×2 table?

2. The Chi-square test is essentially identical to what test for a difference in two proportions?

11.2.5 The Validity of the Chi-Square Test

The Chi-square test for a one-way or two-way table is a large-sample test requiring n, the total count for the table, to be sufficiently large. A number of rules are available in the literature, and as with other large-sample recommendations, there isn't complete agreement over the rules. They all relate to the expected counts in each of the categories, however. The rule we shall use, due to Cochran [1954], is that each expected count should be greater than 1, and 80% of the expected counts should be at least 5. This rule is very conservative in that it requires n to be larger than it needs to be in many situations (Moore [1986]). For 2×2 tables we use the rule for comparing two proportions.[7]

11.3 THE PERILS OF COLLAPSING TABLES

We concluded Section 11.2.4 by asking whether the data in Table 11.2.11 provided evidence of sex discrimination in admissions to graduate study at Berkeley. Without doubt, significantly fewer women were admitted, but can we be sure that discrimination was the cause? Table 11.3.1 looks at admissions to Berkeley's six largest graduate programs. The applications are broken down not just by ADMISSION and SEX, but also by the PROGRAM applied for. (Thus there are three factors defining group membership at work here.)

If you look just at the row labeled "Total," which tells us what is happening overall, you will see the same sort of pattern as in Table 11.2.11. Here 45% of men who applied to the six biggest programs were admitted, compared with only 30% of women. If you look at the percentages admitted program by program, however, you will see that

TABLE 11.3.1 Berkeley Admissions Data by Faculty

	Men		Women	
Program	No. of applicants	Percent admitted	No. of applicants	Percent admitted
A	825	62	108	82
B	560	63	25	68
C	325	37	593	34
D	417	33	375	35
E	191	28	393	24
F	373	6	341	7
Total	2691	45	1835	30

Source: Bickel and O'Connell [1975].

[7]There is a long and continuing controversy surrounding exact methods for 2×2 tables (see Barnard [1988] and Mehta and Patel [1995]).

TABLE 11.3.2 Artificial Three-Way Table

	Nonsmokers			Smokers	
	Not irradiated	Irradiated		Not irradiated	Irradiated
No cancer	950	9000	No cancer	5000	5
Cancer	50 (5%)	1000 (10%)	Cancer	5000 (50%)	95 (95%)

TABLE 11.3.3 The Collapsed Table

	Not irradiated	Irradiated
No cancer	5950	9005
Cancer	5050 (46%)	1095 (11%)

there is usually very little difference between the admission rates for men and women except in one case (Program A), where the admission rate for women is considerably higher. The difference in overall admission rates is being caused not by overt sexism, but by the majority of men applying to programs where there is the least competition for places (Programs A and B, where the overall percentages admitted are high) and the majority of women applying to programs where competition for places is strongest (Programs C to F).

Similar behavior is shown in the artificial Table 11.3.2. This table is called a *three-way cross-classification* or three-way table because we have three factors defining groups, namely, SMOKING (yes/no), IRRADIATION (yes/no) and CANCER (yes/no).[8]

If we wanted to look at the relationship between IRRADIATION and CANCER, it would be tempting just to look at CANCER versus IRRADIATION and not distinguish between whether a person smokes or not. This is called *collapsing* the table with respect to the factor SMOKING. In the collapsed table, Table 11.3.3, it looks as though being irradiated is *good* for you, even though it is clear from the separate tables for smokers and nonsmokers that it is bad. This type of phenomenon is called ***Simpson's paradox.***

In both examples, the factor we want to collapse to look at the relationship of interest is associated with the two factors of interest. In the first example PROGRAM is associated with SEX (men are more likely to apply to A and B, whereas women are more likely to apply to C through F), and PROGRAM is also associated with ADMISSION (different programs have different admission rates). In the second example SMOKING is associated with both CANCER (more smokers in Table 11.3.2 are getting cancer than nonsmokers) and IRRADIATION (a much larger proportion of nonsmokers are irradiated than of smokers). In situations like this, collapsing the table and looking at the two factors of interest in isolation can be very misleading. With observational studies (as opposed to designed experiments; see Sections 1.2 and 1.3), there is always the possibility of there being a factor like PROGRAM in the first example and SMOKING in the second example that you never even thought of taking into account. This is just another manifestation of the impossibility of reliably concluding that a change in one factor *caused* the change in another using data from an observational study.

[8]Although Table 11.3.1 also had three factors, the data are not organized there to show the numbers corresponding to each combination of the three factors.

QUIZ ON SECTION 11.3

1. What is Simpson's paradox?
2. When can it occur, and what consequences can it have?

11.4 SUMMARY

11.4.1 General Ideas about Chi-Square Tests

The Chi-square test statistic has observed value

$$x_0^2 = \sum_{\text{all cells in the table}} \frac{(\text{observed} - \text{expected})^2}{\text{expected}}$$

The P-value for the test is

$$P\text{-value} = \text{pr}(X^2 \geq x_0^2) \qquad \text{where } X^2 \sim \text{Chi-square}(df)$$

- ***Observed*** refers to the count observed in the cell (i.e., what the data say).
- ***Expected*** refers to the count that would be expected if H_0 was true.
- ***Large values*** of x_0^2 provide evidence ***against*** H_0. Such values occur when we get observed counts far from what H_0 would lead us to expect.
- The ***degrees of freedom (df)*** depend on the dimensions of the table and the hypothesis being tested.
- The individual terms in the preceding sum (one for each cell) are called the *components of Chi-square.* When we have a statistically significant test result, inspecting the large components can lead to insight into how the hypothesis is failing to describe the data.

Warning: The use of the Chi-square distribution as the sampling distribution of X^2 when H_0 is true is a large-sample approximation. Where expected counts are small, P-values from the Chi-square distribution begin to become unreliable. Our rule is that expected counts should be greater than 1 and 80% of the expected counts should be at least 5. If this rule is not satisfied, we can often amalgamate rare categories (i.e., treat two or more similar classes as a single class) in order to increase the expected counts. For 2×2 tables we use the rule for comparing two proportions.

11.4.2 One-Dimensional Tables

A single sample of units or individuals is classified into groups by a single factor (with J levels).

- We summarize the data using a one-way frequency table and plot it using a bar graph.

- Chi-square tests are useful when we have a hypothesis defining the values of the set of probabilities (or population proportions) from which the data were sampled.
- The degrees of freedom are $df = J - 1$.
- A common hypothesis is that all the probabilities (or population proportions) are identical.
- If the preceding hypothesis is rejected, we can investigate the nature of the differences by looking at the differences between pairs of proportions.
- When constructing confidence intervals for differences between proportions, use standard errors for a single sample and several response categories.

11.4.3 Two-Way Tables

Chi-Square Test[9]

Whether H_0 specifies *equality of row distributions, equality of column distributions,* or *independence of row and column classifications,* the Chi-square test uses

$$\text{Expected count in cell}(i,j): \quad \widehat{E}_{ij} = \frac{R_i C_j}{n} = \frac{i\text{th row total} \times j\text{th col total}}{\text{grand total}}$$

and

$$df = (I-1)(J-1)$$

Two Types of Tables

We distinguished between situation 1, single sample cross-classified by two factors and situation 2, separate samples, each classified according to one response factor (see Fig. 11.2.7).

Row Distributions

- Row distributions tell us about the chances that an individual who belongs to a given row will fall into each of the column classes.
- They are estimated by the row proportions of the table (using row totals as denominators).
- They are not meaningful if columns are separate samples.
- When constructing confidence intervals for differences between proportions, proportions from different rows are statistically independent.

[9]Chi-square tests as described in this book are appropriate only when the data are collected as a single random sample or as I independent random samples. Social scientists have often used it on two-way tables constructed using data from complex surveys that employ devices such as cluster sampling. The Chi-square test is not appropriate under such circumstances.

Column Distributions

- Column distributions tell us about the chances that an individual who belongs to a given column will fall into each of the row classes.
- They are estimated by the column proportions of the table (using column totals as denominators).
- They are not meaningful if rows are separate samples.
- When constructing confidence intervals for differences between proportions, proportions from different columns are statistically independent.

Whole-Table Proportions

- Whole-table proportions are formed using the grand total of the table as the denominator.
- They tell us about the chances of an individual experiencing a given *combination* of the two factors.
- They are meaningful only when we have a single sample cross-classified by two factors. (They are not meaningful if rows are separate samples or if columns are separate samples.)
- When constructing confidence intervals for differences between proportions, use standard errors for a single sample and several response categories.

REVIEW EXERCISES 11

1. A sample of 300 voters living in a certain district is drawn at random, and the people are asked which of three mayoral candidates they would vote for. Results were

Preferred mayor	A	B	C
Number of voters	119	97	84

Do you think that the three candidates are equally preferred?

2. In a genetics experiment the types BC, Bc, bC, and bc are expected to occur with relative frequencies in the ratios 9:3:3:1. The observed frequencies were

	BC	Bc	bC	bc	Total
Observed freqs	102	16	35	7	160

(a) What are the values of the four probabilities postulated by genetics?

(b) Do the data contradict the genetic frequencies, or can the differences easily be explained in terms of sampling variation?

TABLE 1 Predicting Re-offending

	Risk group			
	Low	Medium	High	Total
Reconvicted	23	50	53	126
Not reconvicted	52	25	22	99
Total	75	75	75	225

Source: Adapted from Schumacher [1972].

3. Schumacher [1972] built a statistical model for predicting reconviction based on three years of postprison follow-up for each of 347 men who had been imprisoned for crimes against persons or property. A classification rule was formed by linking recorded inmate characteristics to what happened to them after prison using a statistical method called discriminant analysis. The rule enabled Schumacher to classify the prisoner as having either a low risk, medium risk, or high risk of re-offending. How useful is the resulting classification rule? To investigate this, Schumacher looked at the experiences of a second sample of 225 inmates. These are given in Table 1.[10]
   (a) Use the Chi-square test to determine whether there is any evidence of a difference in reconviction rates between the three risk groups.
   (b) Investigate the nature of any differences.
   (c) The outcome of real interest was re-offending, not reconviction. Is a method for predicting reconviction necessarily good for predicting re-offending?

4. Market researchers and pollsters worry that having chosen a sample, the people that they fail to contact may differ in important ways from the ones that they do contact. The data in Table 2 were taken from Ward et al. [1985]. Interviewers called at designated houses up to three times in an attempt to make contact with the residents. The table gives the numbers of people in each income level with whom contact was made on the first call, those for whom a second call was needed to make contact, and those for whom a third call was needed.
   (a) Is there any evidence of a relationship between income and the number of callbacks? Investigate the nature of any such relationship.
   (b) What are the practical implications of what you have found?
   (c) What category of people is not addressed in Table 2? Why is this category important?

TABLE 2 Income by Callback

	Respondents			
Income	1st call	2nd call	3rd call	Total
$<$\$10,000	79	15	5	99
\$10,000–14,999	41	17	4	62
\$15,000–24,999	88	27	11	126
$\geq$\$25,000	96	55	36	187
Total	304	114	56	474

Source: Ward et al. [1985]

[10]We have used the published figures of 225 prisoners and the published reconviction rates and have assumed equal numbers in each risk group to obtain our table.

TABLE 3 Undergraduate Students Enrolling at Auckland

Faculty	Father's socioeconomic status[a] 1	2	3	4	5	6	Total
Arts	895	1095	739	616	189	157	3691
Commerce	446	617	320	242	69	49	1743
Science	242	330	213	194	51	36	1066
Engineering	254	261	190	146	39	26	916
Architecture	177	144	106	96	19	15	557
Law	170	140	106	80	15	15	526
Medicine	184	120	78	39	9	9	439
Total	2368	2707	1752	1413	391	307	8938

[a]1 = highest, ..., 6 = lowest.
Data courtesy of Dr. Tony Morrison.

TABLE 4 Snoring and Nightmares

	Frequency of fantastic nightmares: Never	Seldom	Occasionally	Frequently	Always	Total
Nonsnorers	22	45	35	7	4	113
Snorers	16	31	27	10	2	86
Total	38	76	62	17	6	199

Reproduced with permission of authors and publisher from Hicks, R. A., and Bautista, J. "Snoring and nightmares." *Perceptual and Motor Skills*, 1993, **77**, 433–434. ©Perceptual and Motor Skills 1993.

5. In 1990, 8938 undergraduate students enrolled in the six largest colleges[11] at the University of Auckland. The enrollments in the six colleges are broken down according to the socioeconomic status (SES) of their fathers in Table 3. We shall consider these students as a random sample from the process that produces undergraduate students in Auckland.

(a) Is there any evidence that the SES distribution differs between colleges?

(b) If so, what are the important differences?

6. Hicks and Bautista [1993] sought to replicate the findings of an earlier study that had concluded that snoring and disturbing dreams were strongly associated. A total of 199 university students responded to a questionnaire that assessed their level of snoring and the frequency with which they experienced fantastic nightmares. Do the results of Hicks and Bautista given in Table 4 concur with those of the earlier study? What issues are raised here?

7. The data in Table 5 come from a case-control study conducted by Auckland Medical School researchers and their collaborators investigating possible causes of cot or crib death in infants.[12] These data from the first year of the study compare the circumstances of the 162 infants who died and were classified as crib deaths with those of 589 randomly selected control infants. The basic idea is that if a higher proportion of cases than controls have some

[11]Called Faculties in Auckland.

[12]More correctly known as sudden infant death syndrome (SIDS).

TABLE 5 Case-control Study of Cot (Crib) Deaths

Smoke[a]	Cases	Controls
Yes	79	168
No	46	324
Total	125	492

[a]During pregnancy.

Sleep prone[b]	Cases	Controls
Yes	93	216
No	35	287
Total	128	503

[b]Prone = on stomach.

Breast feed[c]	Cases	Controls
Yes	85	429
No	43	74
Total	128	503

[c]Solely, at discharge.

Mother's smoking[d]	Cases	Controls
Nil	51	334
1–9	18	63
10–19	25	62
20+	34	44
Total	128	503

[d]Cigarettes/day in last 2 weeks of pregnancy.

Socioecon status	Cases	Controls
I, II	30	167
III, IV	62	249
V, VI	36	87
Total	128	503

Season	Cases	Controls
Jan–Feb	11	75
Dec–Mar	15	96
Nov–Apr	20	74
Oct–May	27	88
Sep–Jun	23	83
Aug–Jul	32	87
Total	128	503

Data courtesy of Dr. Ed Mitchell.

specified characteristic, then that characteristic is a risk factor, and possible cause, of the problem (here cot or crib death).[13]

(a) The totals for one table are different from the others. Can you think of any reasons for this? What sort of biases might result?

(b) There are two types of variables here: (i) variables that future mothers[14] could control (if it would help to prevent the death of their child) and (ii) those that they could not. Which variables fall into which category? What type of variable is of most practical interest?

(c) Analyze the data in each of the tables. Investigate how, if at all, the distribution of each variable differs between cases and controls. If there is evidence of differences, investigate the nature of those differences.

(d) One of the tables, that for season, needs further explanation. In the first instance January and February are summer months in Auckland, and the pairs Dec-Mar etc. are equidistant from Jan-Feb. For deaths, season gives the season in which the infant died. Control babies

[13]Furthermore, from case-control sampled data one can estimate the relative risks of death associated with a characteristic, that is, make statements of the form: "Infants with the characteristic are 2.4 times more likely to die than those without the characteristic."

[14]Variables that cannot be controlled may still give insights into the mechanisms of the problem and lead to benefits in the longer term.

TABLE 6 Income versus Job Satisfaction

	Job satisfaction				
Income ($)	Very dissatisfied	Little dissatisfied	Moderately satisfied	Very satisfied	Total
<6000	20	24	80	82	206
6000–15,000	22	38	104	125	289
15,000–25,000	13	28	81	113	235
>25,000	7	18	54	92	171
Total	62	108	319	412	901

Source: 1984 General Social Survey; see Norušis [1988].

had a "nominated date" for comparative purposes on a whole range of variables not shown here. In analyzing the table for season you have been comparing the distribution of death dates for cases with the distribution of nominated dates for controls. Look now at the one-way classifications treating cases and controls separately.

(i) Is there any evidence that cot deaths do not occur uniformly through the year? If so, when are the high and low points?

(ii) Do the nominated dates for controls occur uniformly through the year?

(e) We have been looking at the relationship between each variable and[15] case-control status separately. In terms of Section 11.3, what have we been doing, and what dangers does this expose us to?

8. The data in Table 6 were taken from the General Social Survey of the U.S. National Data Program as quoted by Norušis [1988]. Use the Chi-square test to test for a relationship between income and job satisfaction. You will find that there is no evidence of a relationship. Now look at plots of the row proportions. Does there seem to be a pattern in the direction one might expect?[16]

9. The success or failure of new products often varies regionally, so it is important for marketing strategists to understand differences in the tastes and values of people in different regions. Kahle [1986] investigated differences in people's values between geographical segments of the U.S. market. Samples of people from each region were asked what their most important value was. The results for the larger regions are given in Table 7.

(a) Is there is any evidence of regional differences in values between these regions? Comment on any differences you see.

(b) The idea of distinguishing between market segments is so that different strategies can be used for different segments. Do you therefore want to divide up your market into segments that are as similar as possible or as different as possible? Why?

The following five exercises make use of data you have seen in previous chapters and fill gaps in previous analyses.

10. Using the data in Review Exercises 4, problem 5, test whether there is any evidence of differences in response rates between the four questionnaire formats. If so, investigate any important differences.

[15]The researchers also did *multivariate analyses* in which they considered the effects of many variables jointly. Such analyses are beyond the scope of this book.

[16]There is a lesson here. The Chi-square test makes no use of the fact that the row and column categories are ordered. More sensitive (or powerful) tests that do take account of this information can detect a significant relationship. These data have been used by Agresti [1990] to demonstrate the importance of such order information.

TABLE 7 Market Segments

Most important value	Region: New Engl.	The Foundry	Dixie	Bread-basket	Mex.-Amer.	Eco-topia	Total
Self-respect	27	154	147	55	34	34	451
Security	26	147	152	62	26	37	450
Warm relationships	17	125	90	63	27	35	357
Sense of accomplishment	17	88	65	38	17	23	248
Self-fulfillment	11	74	55	23	24	24	211
Being well respected	10	65	72	31	4	8	190
Sense of belonging	6	63	49	24	10	15	167
Fun, enjoyment	6	34	23	11	8	13	95
Total	120	750	653	307	150	189	2169

Source: Kahle [1986].

11. Using the data in Review Exercises 7, problem 13, test whether there is any evidence of differences between the three locations in the true proportions of people wearing their baseball caps facing backward. (You will have to change the format of the table before applying the Chi-square test.)

12. Using the data in Review Exercises 7, problem 14, test whether there is any evidence that the ability to predict one's grade changes with grade point average in the previous semester.

13. Use the data in Table 8.5.6 and surrounding story regarding the ability of students to smoke marijuana every weekend and still do well at school. The percentages given are column percentages.

(a) Plot the percentages for each of the four groups (teachers, principals, students aged 12–14, and students aged 15–17). What do you see?

(b) Test whether there is any evidence of differences between teachers and principals with respect to beliefs about the ability of students to smoke marijuana every weekend and still do well at school. (You will have to reconstruct the counts from the percentages given, which related to 825 teachers and 822 principals.)

(c) Repeat (b), comparing students aged 12–14 with students aged 15–17. (There were 500 in each group.)

14. Refer back to the data on 1994–1996 model cars in problem 10 of Review Exercises 8. Is there any evidence of differences between makes of car in trouble-free running?

15. Apply the Chi-square test to the data in Table 9.3.4 to see whether there is any evidence of a relationship between the sex of the first child in a family and the sex of the second child.

16. Refer back to problem 6 of Review Exercises 9, and test for a relationship between marijuana consumption and the development of schizophrenia. (Hint: You will have to be careful about how you set up the table of counts. It should *not* look like the one in problem 6 in this chapter. Why?)

17. Set up the data in problem 9 of Review Exercises 9 as a 2×2 table, and test for a relationship between smoking and having a stroke. What is the approximate relationship between the test statistic from problem 9 of Review Exercises 9 and your Chi-square test statistic?

***18.** Table 8 is from a survey of American attitudes. Here 1397 people have been cross-classified both by their attitude to the death penalty and their attitude to gun registration. We now examine if there is any difference between the underlying population values.

TABLE 8 American Attitudes

Favors gun registration	Favors death penalty: Yes	No	Total
Yes	784	236	1020
No	311	66	377
Total	1095	302	1397

1982 General Social Survey as reported by Clogg and Shockey [1988].

(a) If this survey was reported in the news media, you would typically only be told that 1020/1397 ≈ 73% favored gun registration, while 1095/1397 ≈ 78% favored the death penalty. Use the Chapter 9 method for testing for a true difference between the proportions of people answering yes to the two questions. Use the standard error formula for situation (c): single sample, two yes/no questions.
(Take particular note of your standard error, which we noted is a conservative approximation. We are now going to derive an improved method.)

The following tables set up some notation to help you think through the problem using the important new information about the numbers of people who said yes to one question and no to the other.

Counts

Qu. 2	Qu. 1: Yes	No	Total
Yes	X_{11}	X_{12}	X_{1+}
No	X_{21}	X_{22}	X_{2+}
Total	X_{+1}	X_{+2}	n

Proportions

Qu. 2	Qu. 1: Yes	No	Total
Yes	$\widehat{p}_{11}$	$\widehat{p}_{12}$	$\widehat{p}_{1+}$
No	$\widehat{p}_{21}$	$\widehat{p}_{22}$	$\widehat{p}_{2+}$
Total	$\widehat{p}_{+1}$	$\widehat{p}_{+2}$	1

(b) Note that the difference between the sample proportions answering yes to the two questions is $\widehat{p}_{+1} - \widehat{p}_{1+}$. Show that this difference is equal to $\widehat{p}_{21} - \widehat{p}_{12}$.

(c) Note that we have n individuals falling into four cells in this table so that in terms of Fig. 8.5.1 this corresponds to situation (b): single sample, several response categories. Thus the standard error we should be using for $\widehat{p}_{21} - \widehat{p}_{12}$ when we have all the information in the table is that for situation b. Use this standard error to test the hypothesis of no difference between the proportions again.

(d) How much smaller is your standard error from (c) than the one from (b)? Have your conclusions changed?

Further exercises, some involving large data sets, are available on our web site.

CHAPTER 12

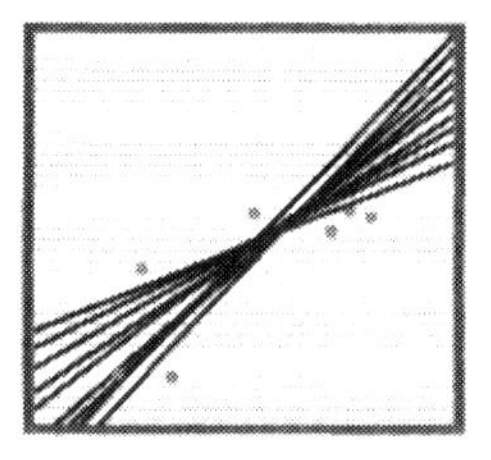

Relationships between Quantitative Variables:

Regression and Correlation

Chapter Overview

This chapter is a continuation of the first part of Chapter 3, where scatter plots were introduced as a primary tool for exploring relationships between two variables and the basic ideas of regression were introduced intuitively. Here we develop those ideas formally for the simple linear model. We also consider, to a lesser extent, the concept of correlation.

In Section 12.1 we talk quite broadly about the place of regression modeling in scientific investigation. We give special attention to regression methods in relation to questions of causality. We recall that a regression relationship consists of a trend plus scatter in Section 12.2 and describe three types of commonly used trend curve: the straight line, the quadratic curve, and the exponential. This is followed in Section 12.3 by a discussion of the least-squares method for fitting lines to data. We point out that the same ideas carry over to the fitting of other curves.

Section 12.4 is concerned with inference for the simple linear model. Section 12.4.1 defines the simple linear model and then illustrates the sampling behavior of the least-squares estimates through computer simulation. Significance tests and confidence intervals for the regression coefficients are discussed in Section 12.4.2. This is followed in Section 12.4.3 by a brief explanation of the meanings of a prediction interval and a confidence interval for the mean. In Section 12.4.4 we discuss the nature and practical importance of the various model assumptions and the checking of assumptions using residual plots. The chapter concludes with a discussion of correlation in Section 12.5.

12.1 WHY DO REGRESSION?

12.1.1 Correlation versus Regression

This chapter can be thought of as a continuation of Section 3.1, and we advise that you reread Section 3.1 before proceeding. Section 3.1 introduced some of the most important ideas in regression in a very informal and intuitive way. This chapter will take those ideas and develop them formally for the so-called simple linear model, a model that applies to data with linear trend and constant scatter [e.g., Fig. 3.1.6(a)]. As the name *simple linear model* suggests, there are many regression models, and this is the simplest of them. Before developing statistical inference for the simple linear model, we discuss the role of regression modeling in scientific investigation and place the simple linear model in context.

Chapter 3 was concerned with exploring relationships between variables. Section 3.1 discussed relationships between a continuous variable and another quantitative variable (discrete or continuous). The tool we used for exploring such relationships was the scatter plot. To permit general discussion, we will again call our variables X and Y. There are two main approaches to considering a relationship between X and Y, namely, *correlation* and *regression*.

Section 3.1.1 distinguished between random and nonrandom variables. Correlation applies only if both of our variables are random, for example, the systolic and diastolic volumes of the heart (SYSVOL and DIAVOL in Fig. 3.1.1) of a randomly selected person. In this case we may not wish to single out one variable to have a special role but may prefer to treat the two variables symmetrically, that is, on an equal footing. The focus would then be just on the strength of the relationship. One simple way of doing this is to use *correlation*. This method is discussed in Section 12.5.

Section 3.1 concentrated on *regression*, and regression is again our primary concern. In regression we are particularly interested in one variable, denoted Y and called the *response variable*, and we want to use the other variable X (called the *explanatory variable*) to help us predict or explain the behavior of the response variable.

- ***Regression*** singles out one variable as the response and uses the explanatory variable to explain or predict its behavior.
- ***Correlation*** treats both variables symmetrically.

12.1.2 Causal Relationships

As you read this, thousands of people all around the world are fitting regression models to data. Most of them are doing it because they are trying to *predict* Y behavior or as part of a strategy for uncovering the reasons (*causes*) for why different individuals produce different values of Y. Often both aspects are intertwined. We will use some of our own recent involvements in research projects to illustrate.

One project involved children who had been sexually abused. As far as can be told from psychological tests, there is variation in the amount of lasting psychological damage done to children by such abuse. In other words, some children seem to cope with life after abuse better than others. The researchers wanted to know why different children were affected to different extents, partially in a predictive sense—to be able to identify those abused children who will need the most help—and partially to get a better understanding of why some cope better than others. They measured many variables, capturing aspects of the child's environment before and after the abuse, the relationship with the abuser, and the nature and duration of the abuse. As a way of finding out what was important, they looked for variables that were most strongly related to a response variable that measured the child's current psychological state. Typically, there are a lot more ideas about what might be important than end up being supported by the data. In another project the researcher was looking at the cost of upkeep on stretches of highway and how they were related to aspects of the geology of the ground the road was running through, climate and drainage variables, and past maintenance policies.

An ongoing medical project involves finding risk factors for babies being born very small. Low birth weight is associated with increased risk of subsequent health, developmental, and educational problems. The research team measured many variables that they thought might either be causes of low birth weight or need to be taken into account. Having data on all these variables, we then need ways to identify the variables that are most important and to tell the researchers when they are "barking up the wrong tree." In the low-birth-weight study, the researchers are looking for *causes* of low birth weight, and not just any causes, but preferably causes that permit corrective action. The last research project of this same team was a major success. It was an investigation of possible causes for cot or crib death in infants. Among explanatory variables that showed up as being important were several factors that something could indeed be done about. These included maternal smoking, having babies in the bed with sleeping parents, and the then-common practice of putting babies down to sleep on their stomachs. After a nation-wide publicity campaign the death rate halved. We have seen similar applications in industry where the objective is to find causes of deficiencies in manufactured goods.

We introduced the *prediction problem* in Section 3.1.3 and showed intuitively how we could use trends in data to make predictions and use the scatter about the trend to tell us about the uncertainty in the prediction [see Figs. 3.1.7(b) and 3.1.12]. We now elaborate further on *questions of causation.*

Searching for causes is about trying to *understand why* different individuals produce different Y-values. Regression modeling, that is, using information in explanatory variables to explain the behavior of a response, is the major contribution of statistics to the detective work that goes into uncovering causes. We identify variables that are related to the response as *possible causes* that merit further investigation. To maintain the detective-story theme, these were the "suspects." In the examples we have discussed the researchers were considering many X-variables. Even with many X-variables, preliminary analyses often look at the relationship of each variable (considered alone) with Y; that is, at any instant we are considering only a single Y and a single X. These preliminary analyses are looking for X-variables that have a particularly strong relationship with Y. These are suspects that merit further investigation. A question we often need to ask early in the process is whether an apparent relationship seen in data is telling us about something real or whether it could have just been produced by sampling variation. Significance tests and confidence intervals help us answer such questions.

Sounding a Warning Bell

If Y is plotted against a variable X and a strong relationship between the two is seen, we have an indication that X *may* be a cause and is worthy of further investigation. It is definitely ***not*** *proof* of a causal relationship, however, unless the data come from a designed experiment. The following example reinforces this point.

EXAMPLE 12.1.1 *Infant Death Rates and Breast Feeding*

The data in Fig. 12.1.1(a) plot the infant death rates for each of 14 countries (1989 figures, deaths per 1000 of population) against the percentage of mothers in those countries who are still breast feeding at six months. There is quite a strong linear pattern of increasing death rates with increasing levels of breast feeding. One might naively conclude that breast feeding is dangerous. Figure 12.1.1(b), however, plots the breast-feeding rates against the percentage of people in each country who have access to safe water. Those countries with the highest breast-feeding rates are those with lowest access to safe water. Inspection of the countries themselves showed that the highest breast-feeding rates were in countries such as Ethiopia, Sudan, and Bangladesh. The lowest breast-feeding rates in the 14 countries belonged to the United States, Brazil, and Australia. It is not that higher breast-feeding rates increase infant mortality, but that these rates are closely related to poverty and lack of sanitation, which are the real causes of high infant mortality. (In fact, it is particularly important to breast feed in places where the water is not safe.)

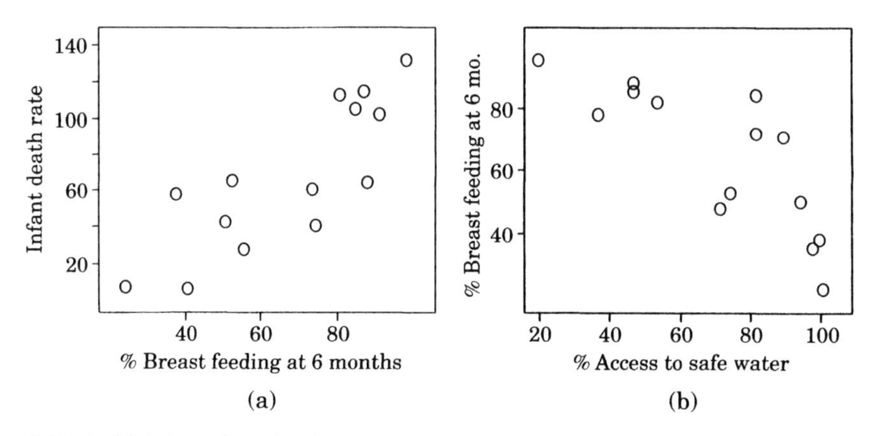

FIGURE 12.1.1 Infant death rates (14 countries).

To generalize, when one has observational data, there may be no direct causal relationship between X and Y, but there may be a so-called ***lurking variable*** (e.g., clean water) that is the real cause of changes in Y but also is associated with X, thus giving rise to the observed relationship between X and Y. Time is often a lurking variable. If two things (e.g., road deaths and chocolate consumption) just happen to be increasing over time for totally unrelated reasons, a scatter plot of one against the other will show a relationship that suggests that when one increases, so does the other. It is also possible that the relationship goes the other way around, that changes in Y (the variable we are interested in) cause changes in X.

With real observational studies there is no guarantee that you will ever find the important lurking variables, so there is always the possibility that your observed relationships are artifacts of something else. It is virtually impossible to reliably conclude a cause-and-effect relationship between variables using data from observational studies. We discussed some of the issues involved in Sections 1.3 and 10.2.1.

In observational data, strong relationships are ***not*** necessarily causal.

Causation and Controlled Randomized Experiments

The only situations in which we can *reliably* conclude that changes in X caused the changes in Y occur when the data come from a randomized experiment in which the *experimenter has control* of the level of X applied. Either the experimental conditions have to be identical, apart from the application of a prescribed level of X, or levels of X must be randomly allocated to the available experimental units. This includes time. The time order in which the various levels of X are tested should be randomized to prevent unexpected changes over time in other factors affecting the response Y from confounding the results.

A problem in Review Exercises 12 gives data from an experiment in which batches of concrete were made with varying levels of cement (X) incorporated. The response variable Y is the hardness of the resulting concrete. There other factors in producing concrete besides cement, such as the amount of water added, the ratio of stone chips to sand, and drying conditions, which are hard to control completely. To prevent uncontrolled differences between batches from confounding our impression of the effect of the cement added, we should choose which batch gets what amount of cement at random. Then if X had no effect on Y, the distribution of Y would not depend on the level of X. We could use a significance test to test this. If the pattern or trend we observed was significantly different from a random pattern, then we have some evidence of a causal relationship. We note that such evidence is strengthened if the experimental results are repeatable.

Experimentation May Not Be Possible

As noted in Section 1.3, it is often physically or ethically impossible to do the experiments that would conclusively answer the question of whether a relationship is causal or not. When we see a relationship in observational data, we have to brainstorm to think of any plausible explanations for the observed relationship apart from changes in X causing changes in Y. Then, through strategies involving mixtures of experimentation, other observational studies, and analysis, we try to confirm or eliminate these other explanations (see Section 1.3 for further ideas). But this takes us outside the scope of this book. In practical situations, such as in business, we often have to act on intuition without conclusive proof. In the crib-death study previously discussed, there was no conclusive proof that things such as not laying babies on their stomachs would cut crib-death rates. What had been established was a close association in data between death rates, implicating some very plausible causal suspects. The health authorities acted, and in this case it worked spectacularly.

Although the simple linear model is useful its own right, an understanding of the simple linear model also lays many of the important foundations needed to understand the more-complicated models used in the preceding research. We must walk before we can run. But Chapter 12 is only about "walking."

12.1.3 Inferences about Theories

Another use for regression, over and above prediction and explanation, is in testing scientific hypotheses. A scientific theory will often lead to a mathematical model that describes how one variable is related to another. The theory can then be tested by seeing whether observations on these variables are well fitted by the model. Several examples using simple models now follow.

Hooke's law states that the amount of stretch in a spring Y is related to the applied weight X by $Y = \alpha + \beta X$, where α and β are constants particular to the spring. We might be interested in estimating the values of α and β for a particular spring.

The theory of gravitation states that the force of gravity F between two objects is given by $F = \alpha/(D^{\beta})$. Here D is the distance between the objects, and α is a constant related to the masses of the two objects. The famous inverse square law says that $\beta = 2$. We might want to test whether this is consistent with experimental measurements.

Theoretical chemistry predicts that, under constant temperature, the pressure P exerted by a gas is related to the volume V in which it is contained by $P = \alpha/V$, where α is a constant for each gas. We might want to estimate the value of α for some particular gas.[1]

The allometric model used to relate two measures of size X and Y on the same animal, for example, the lengths of two bones, is given by $Y = \alpha X^{\beta}$, where α and β are constants to be estimated from data. We might want to check whether a curve of this form does in fact describe the relationship between the dimensions of a sample of bones that we are investigating.[2]

Economic theory uses a so-called production function, $Q = \alpha L^{\beta} K^{\gamma}$ to relate Q = PRODUCTION to L = QUANTITY OF LABOR and K = QUANTITY OF CAPITAL. Here α, β, and γ are constants that depend on the type of goods and the market involved. We might want to estimate these parameters for a particular market and use the relationship to predict the effects of infusions of capital on the behavior of that market.

These mathematical models are just a few of the multitude of relationships between variables postulated by scientific, economic, social, or other theories. Many of these models do not completely specify the relationship. They contain *parameters*—numbers such as α and β that are unknown and have to be estimated from data. When we collect data, the theoretical relationships are never exactly satisfied. The model may describe the trend, but we still see residual scatter caused by factors such as measurement error or sampling variation. There are three main reasons why we want to fit the relationships predicted by these theories to data:

(i) To test the model itself by checking whether the data is in reasonably close agreement with the relationship predicted by the model (as discussed with the allometric model).

---

[1] We can also test the theory by plotting P versus X $(= 1/V)$, which should give a straight line through the origin.

[2] A quick visual check is to plot $\log(Y)$ versus $\log(X)$ and look for a linear trend: If $Y = \alpha X^{\beta}$, then $\log(Y) = \log(\alpha) + \beta \log(X)$, which is of the linear form $Y' = \alpha' + \beta X'$.

(ii) Assuming the model is true, to test whether theoretically specified values of the parameters appear consistent with the data (as discussed with the gravitational model).

(iii) Assuming the model is true, to estimate the unknown constants in the model so that the relationship is completely specified (as discussed with the gas-pressure model). Once the parameters have been estimated, the model can then be used for prediction and other uses (as discussed with the production function model).

EXERCISES FOR SECTION 12.1.3

According to Ohm's law, the voltage V in volts measured across a resistance of $r = 4$ ohms when a current of I amps is passed through is given by the formula $V = rI$.

(a) How would you set up an experiment to test Ohm's law using a scatter plot? What sort of trend line or curve should fit the plot? What point should the trend pass through?

(b) Suppose that the resistance r is unknown. Assuming Ohm's law to be true, how could you use your trend to estimate r?

QUIZ ON SECTION 12.1

1. What essential difference is there between the correlation and regression approaches to a relationship between two variables? (Section 12.1.1)
2. What are the most common reasons why people fit regression models to data? (Section 12.1.2)
3. Can you conclude that changes in X caused the changes in Y seen in a scatter plot if you have data from an observational study? Justify your answer.
4. What is a "lurking variable," and why is this idea important?
5. When can you reliably conclude that changes in X cause the changes in Y? Why is this method not always used?
6. If the experimenter has control of the levels of X used, how should these levels be allocated to the available experimental units?
7. Can you think of some theories that you might want to explore using regression methods? (Section 12.1.3)
8. People fit theoretical models to data for three main purposes. What are they?

12.2 INTRODUCTION TO RELATIONSHIP MODELING

In Section 3.1.2 we identified the major components of a regression relationship as the *trend* and *scatter* about that trend. (We also warned you to look out for outliers.) In all the examples given, the trends were placed onto the scatter plot by eye. It is an underrated method. The approach encouraged you to *look* at the data to see what is going on and enabled us to introduce intuitively some important statistical ideas in a general setting without getting bogged down in technical details. Although the eye is a powerful tool for smoothing scatter plots and seeing a trend, different people looking at a plot will always draw slightly different trend curves or lines. We need automated (nonsubjective) ways to derive trend curves that always give the same answer.

There are *two main approaches* to summarizing trends in statistics. The still-dominant approach is the *fitting of mathematical functions,* and this is the approach we will pursue. An emerging approach is via so-called *smoothing techniques*. The way smoothers work is perhaps closer to what the eye does. Smoothers like `lowess` available in Minitab, Splus, and R are easy to use and produce nice trend curves. The way they work is rather complicated, however, and we will not pursue them further here.

Our main concern is with fitting lines to data. The data set in Example 12.2.1 carries much of the discussion. Following the example we discuss the features of a straight line.

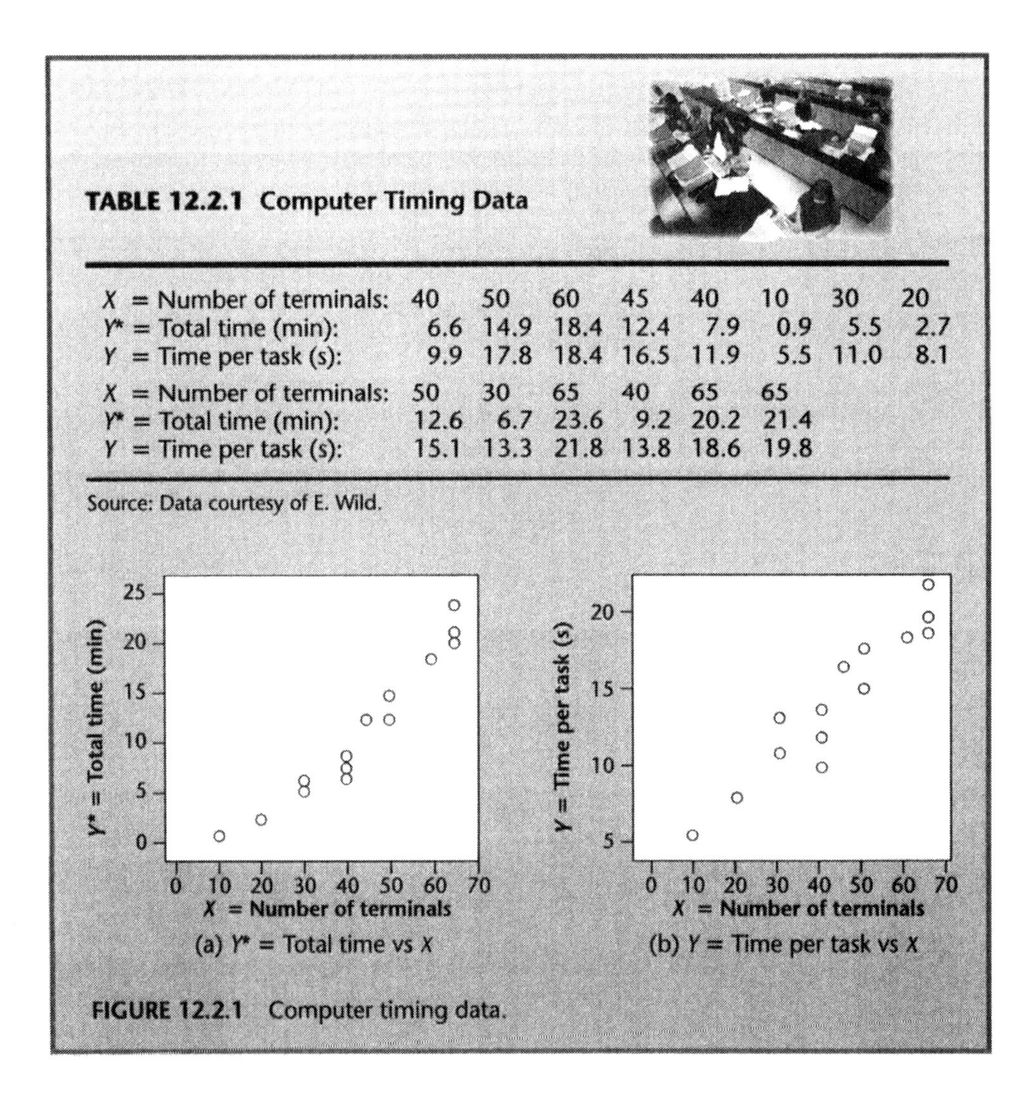

TABLE 12.2.1 Computer Timing Data

X = Number of terminals:	40	50	60	45	40	10	30	20
Y^* = Total time (min):	6.6	14.9	18.4	12.4	7.9	0.9	5.5	2.7
Y = Time per task (s):	9.9	17.8	18.4	16.5	11.9	5.5	11.0	8.1
X = Number of terminals:	50	30	65	40	65	65		
Y^* = Total time (min):	12.6	6.7	23.6	9.2	20.2	21.4		
Y = Time per task (s):	15.1	13.3	21.8	13.8	18.6	19.8		

Source: Data courtesy of E. Wild.

(a) Y^* = Total time vs X

(b) Y = Time per task vs X

FIGURE 12.2.1 Computer timing data.

EXAMPLE 12.2.1 ***Time Taken versus the Number of Computer Terminals***

This is an extension of Example 3.1.2 from Chapter 3 that we will be using often in this chapter. It refers to the practical situation where many tasks are being run by users at different terminals on a multiuser computer system. The way the system works is that it does not complete one task before going on to the next one (except for very small tasks): It does some of one task, then swaps to doing some of another, and so on until tasks are completed. The effect of this is that when there are more people using the system, the elapsed time from the initiation of any particular task until it has been completed increases. The data in Table 12.2.1 came from an investigation of the completion times experienced by users at various levels of total workload for the system. Each data point corresponds to an experiment in which x terminals all initiated the same task. The variable measured, $Y^* =$ *total time*, was the sum of the times for all of the x terminals to finish the task. From this, the variable $Y =$ *time per task* was calculated as total time divided by the number of terminals. This gives the average time taken to complete the task in that particular experimental run and is a reasonable measure of what the users experience.

Each of these variables is graphed against X in Fig. 12.2.1. We note that Y^* increases with X [Fig. 12.2.1(a)]. The pattern is not quite linear but curves very slightly upwards. In contrast, the graph of Y versus X [Fig. 12.2.1(b)] shows a linear trend, though there appears to be more scatter about the trend. Even repeat experiments using the same number of terminals give quite different values for Y. In the pages to come we will fit a model to these data and use it to make predictions.

12.2.1 The Straight Line

We have seen several examples where the trend curve looked like a straight line, for example Fig. 12.2.1(b). In elementary algebra the equation of a straight line is often written $y = mx + c$, where m is the slope and c is the intercept. Statisticians usually write an exact linear relationship between x and y as

$$y = \beta_0 + \beta_1 x$$

where β_0 is the intercept and β_1 is the slope, as depicted in Fig. 12.2.2. The definitions for the intercept and slope given to the left of Fig. 12.2.2 are our key to interpreting

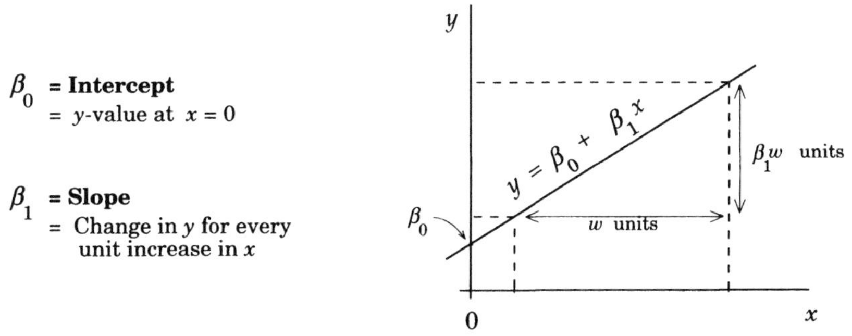

β_0 = **Intercept**
= y-value at $x = 0$

β_1 = **Slope**
= Change in y for every unit increase in x

FIGURE 12.2.2 Statistical notation for the straight line.

values for the slope and intercept in the context of the data. When we use the equation of a line to describe a trend in a scatter plot, we will put a "hat" over the y and write $\hat{y} = \beta_0 + \beta_1 x$. (This follows a common statistical usage of using "hats" to denote estimates.)

EXAMPLE 12.2.1 (cont.) *Interpreting a Slope*

Suppose that the linear trend in Fig. 12.2.1(b) was well summarized by $\hat{y} = 3 + 0.25x$, or $\widehat{\textit{time per task}} = 3 + 0.25 \times \textit{number of terminals}$. The intercept of a line (3 here) is the y-value when $x = 0$. The intercept is not a meaningful quantity for this example. It relates to the predicted time per task when no terminals are running jobs. The slope is the change in y associated with a 1-unit increase in x. For this example, a slope of 0.25 would be telling us to expect the time per task to increase by an additional 0.25 seconds for every additional terminal added to the system. Thus, if we increased the load on the system by 10 terminals, we would expect the time per task experienced by each terminal to increase by 2.5 seconds.

*12.2.2 Extensions

The preceding notation for a line is often used in statistics because it easily extends to more complex curves and the inclusion of more than one X-variable. For example, if we extend it to $y = \beta_0 + \beta_1 x + \beta_2 x^2$, we obtain the equation of a quadratic curve (see Fig. 12.2.3). Gently curved trends in data can often be summarized by some segment of a quadratic (examples are shown in Fig. 12.2.3). The trends in Figs. 12.2.1(a) and 3.1.4 (apart from an outlier there) are quite well described by quadratics.

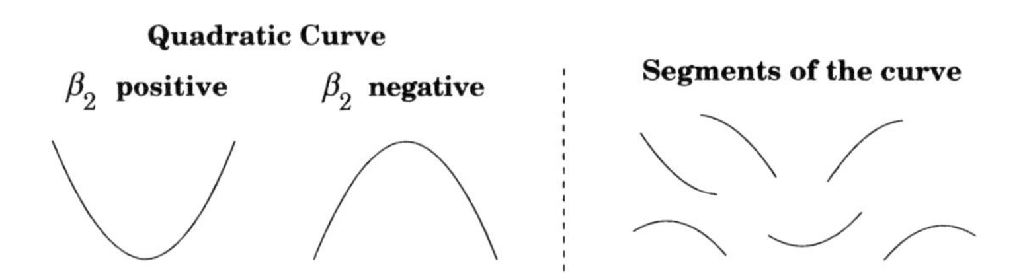

FIGURE 12.2.3 The quadratic curve ($y = \beta_0 + \beta_1 x + \beta_2 x^2$).

*12.2.3 The Exponential Curve

Another commonly used mathematical equation for describing trends is the exponential curve, $y = ae^{bx}$. Figure 12.2.4 shows the characteristic shape of the exponential function for both positive and negative b. Like the intercept of a line, a in this equation is the value y takes when $x = 0$.

The property of a straight line that gives it its characteristic shape is that for every unit increase in x, we *add* a constant amount to y (namely β_1). The characteristic property of the exponential curve is that for every unit increase in x, y changes by a constant *multiple*, namely, e^b. For example, with the exponential equation $y = 2e^{0.4x}$, $y = 2$ when $x = 0$, and for every unit increase in x the value of y is multiplied

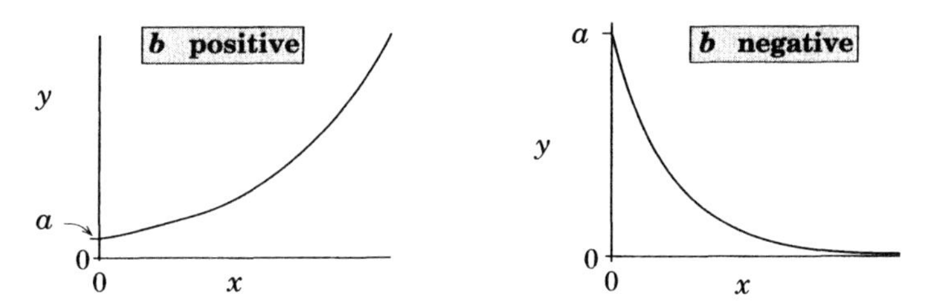

FIGURE 12.2.4 The exponential curve ($y = ae^{bx}$).

by $e^{0.4} \approx 1.5$. Thus if we increase x by 1, we multiply y by approximately 1.5 (which is a 50% increase).

> A straight ***line*** changes by a fixed ***amount*** with each unit change in x.
> An ***exponential*** changes by a fixed ***percentage*** with each unit change in x.

An exponential function is useful for describing the early stages of growth of organisms before the size of the organism starts being constrained by external factors such as space and the availability of nutrients. Here X is time, and Y is some measure of size. It also describes the growth in size of a bank deposit over time under compound interest at a fixed rate.

*How can I tell whether the trend is exponential?

It is virtually impossible to tell whether or not a trend can be well described by an exponential curve just by looking at it. If we take the equation $y = ae^{bx}$ and take natural logarithms of both sides, however, we get $\log(y) = \log(a) + bx$. Therefore if we plot $\log(Y)$ versus X instead of Y versus X, we should have a linear relationship with slope $\beta_1 = b$ and intercept $\beta_0 = \log(a)$, as in Fig. 12.2.5. Figure 3.1.6(b) looks as if it could conceivably be exponential. A plot of $\log(y)$ versus x for this example (not shown) has an obviously curved trend, however.

> To check for an exponential trend, see whether a plot of **log(y) versus x** has a linear trend.

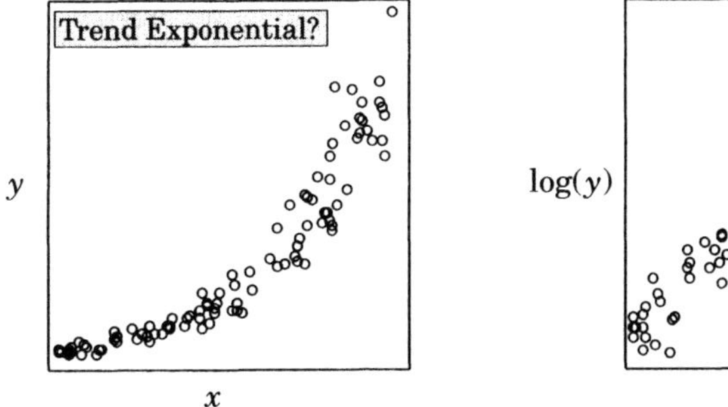

FIGURE 12.2.5 Checking for an exponential trend.

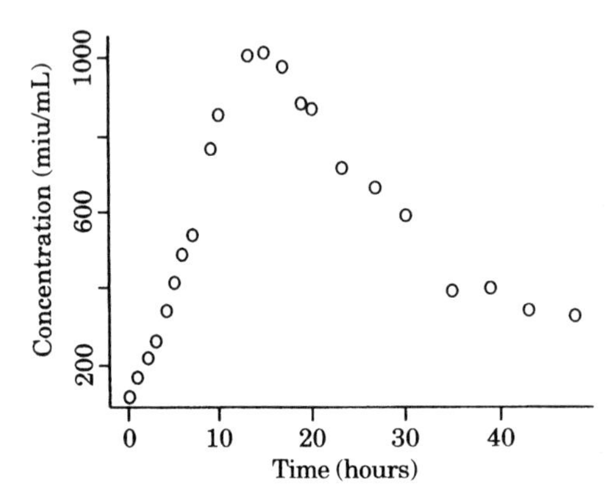

FIGURE 12.2.6 Creatine kinase concentration in a patient's blood.

12.2.4 Other Types of Trend Curves

As we have already noted, trends in scatter plots come in many different shapes, and particular shapes prompt particular questions from the data. For example, Fig. 12.2.6 plots, for a set of heart attack patients, a Y-variable that is the creatine kinase concentration in the blood versus an X-variable that is the time since the heart attack. The trend curves up to a peak and then falls away, but is not quadratic because it is not symmetric. A trend with a maximum prompts the question (among others), "At what X-value can I expect to get the maximum response Y?" Some trends oscillate, and these prompt questions about the locations of the peaks and troughs. Some appear to be approaching an asymptote; that is, there is a value that they appear to be getting closer and closer to without ever quite reaching. Can we estimate the size of this value?

In this chapter we develop some technical tools for fitting just a straight line to a scatter plot and for making inferences and predictions from the line. When you are exploring real data, however, you have to be creative and think beyond just straight lines!

> ***You should not let the questions you want to ask be dictated by the tools you know how to use.***

The most important part of the investigative process is asking the right questions, though you may need professional help to answer them. The next section continues with this theme of fitting trends.

QUIZ ON SECTION 12.2

1. In statistics what are the two main approaches to summarizing trends in data?
2. In $y = 5 + 2x$, what information do the 5 and the 2 convey?
3. In $y = 7 + 5x$, what change in y is associated with a 1-unit increase in x? with a 10-unit increase? Repeat for $y = 7 - 5x$.

TABLE 12.2.2 DEA Budget and U.S. Drug Deaths

Year	1981	1982	1983	1984	1985	1986	1987	1988	1989	1990	1991
Budget ($millions)	216	239	255	292	344	372	486	493	543	558	692
Deaths	7,106	7,310	7,492	7,892	8,663	9,976	9,796	10,917	10,710	9,463	10,388

Source: Duncan [1994].

*4. For a line, y changes by a fixed *amount* for every 1-unit increase in x. What is the corresponding idea for an exponential curve?

*5. How can we tell whether a trend in a scatter plot is exponential?

EXERCISES FOR SECTION 12.2

Duncan [1994] looked at drug law enforcement expenditures and drug-induced deaths. Table 12.2.2 gives figures from 1981 to 1991 on the U.S. DEA (Drug Enforcement Agency) budget and the numbers of drug-induced deaths in the United States.

(a) Plot deaths versus DEA budget. Do you think the budget causes deaths? Why not? Plot budget versus deaths. What do you think now?

(b) The variables deaths and budget are affected by a third variable—year.

(i) Plot budget versus year. Do you think that a straight line would adequately fit this scatter plot?

*(ii) What other trend might fit?

(c) Plot deaths versus year. What do you conclude from the three plots?

12.3 CHOOSING THE BEST LINE

We now go on to decide how to choose the line that best summarizes the trend in the computer timing data in Fig. 12.2.1(b). For the straight line we have the following equation for the trend:

$$\hat{y} = \beta_0 + \beta_1 x$$

This expression has two *parameters:* the intercept β_0 and the slope β_1 of the line. By changing the values of these parameters, we change the position of the line [cf. Fig. 12.3.1(b)]. When we say that we want to choose the best line to summarize the trend, we are also saying that we want to find the intercept β_0 and the slope β_1 of the best-fitting line.

This basic idea applies to any curve. Suppose we want to choose the function having the exponential form $\hat{y} = ae^{bx}$ that fits the data best. This curve has two parameters, a and b. By changing these values, we change the position and shape of the curve. Saying we want to find the best-fitting exponential curve is equivalent to saying we want to select the values of the parameters a and b that determine the best-fitting exponential.[3]

[3]For the quadratic function $\hat{y} = \beta_0 + \beta_1 x + \beta_2 x^2$ we need to select the values of three parameters to define the best-fitting curve.

12.3.1 What Line Fits the Data Best? Defining What We Mean by Best

The ideas we want to discuss are illustrated in Fig. 12.3.1. We have a set of points illustrated in Fig. 12.3.1(a). Which line [Fig. 12.3.1(b)] best summarizes the trend? Alternatively, which line *fits the data best?* The equation of our line is $\hat{y} = \beta_0 + \beta_1 x$. Choosing a line is equivalent to choosing values for β_0 and β_1. Let us put a line onto the scatter plot [Fig. 12.3.1(c)]. We will use the line that we have drawn to predict the y-value at the ith data point, which has $X = x_i$. Let us denote this predicted value by $\hat{y}_i$. In using $\hat{y}_i$ to predict the actual result y_i, we make a ***prediction error*** of size $y_i - \hat{y}_i$. This is the vertical distance from line to data point (x_i, y_i) depicted in Fig. 12.3.1(c). The same sort of thing is happening for all of the data points (see the dotted arrows). A line that fits the data well will be one for which these prediction errors, or vertical distances, are as small as possible in some overall average sense.

In the most commonly used criterion, called the *least-squares* criterion, the best-fitting line is the one with the smallest average squared prediction error $\frac{1}{n}\sum(y_i - \hat{y}_i)^2$, or equivalently, the smallest sum of squared prediction errors $\sum(y_i - \hat{y}_i)^2$.

> ***Least squares criterion:*** Choose the values of the parameters to *minimize the sum of squared prediction errors,*
>
> $$\sum_{i=1}^{n}(y_i - \hat{y}_i)^2$$

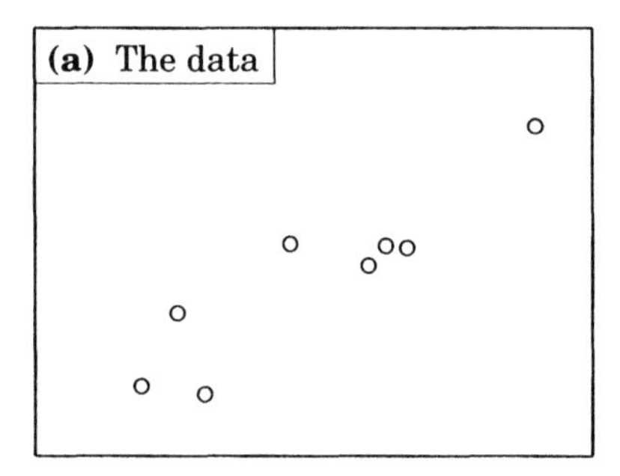

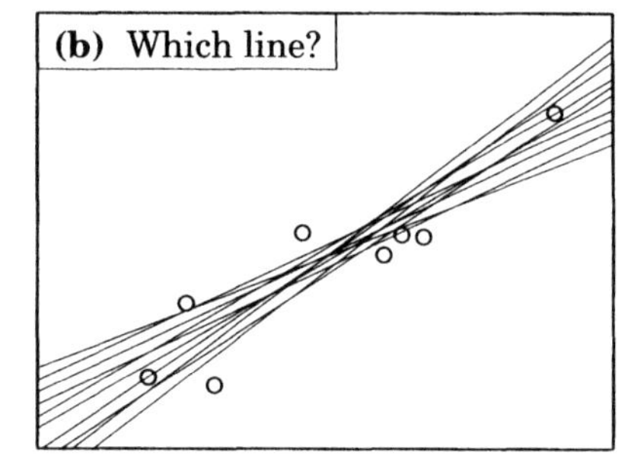

Least-squares line

Choose line with smallest sum of squared prediction errors

Min $\Sigma(y_i - \hat{y}_i)^2$

Its parameters are denoted:

Intercept: $\hat{\beta}_0$

Slope: $\hat{\beta}_1$

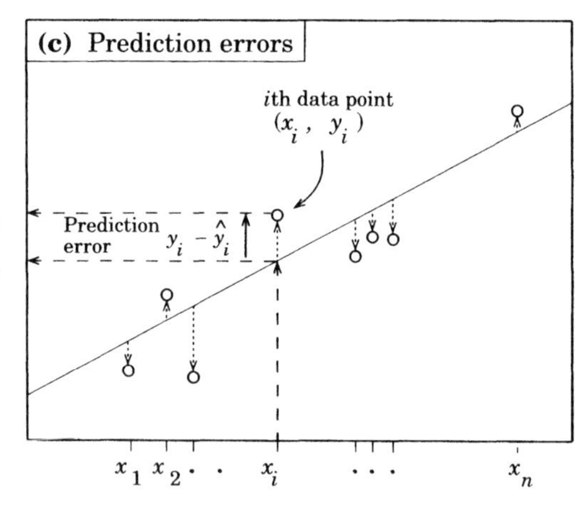

FIGURE 12.3.1 Fitting a line by least squares.

In Fig. 12.3.1 this process is depicted for the special case of fitting a line. The least-squares idea is general, however. It applies to the fitting of any type of curve. We note that in the least-squares criterion, all prediction errors $y_i - \widehat{y}_i$ are treated alike. This is appropriate when the level of scatter in the relationship is constant with x. This will be discussed in more detail in Section 12.4.4.

✖ EXAMPLE 12.3.1 *Prediction Errors for the Computer Timing Data*

In Fig. 12.3.2 we have replotted the computer timing data from Example 12.2.1 and placed two lines on the plot. One, $\widehat{y} = 3 + 0.25x$, clearly fits the data better than the other, $\widehat{y} = 7 + 0.15x$. The data are given again in Table 12.3.1, together with some

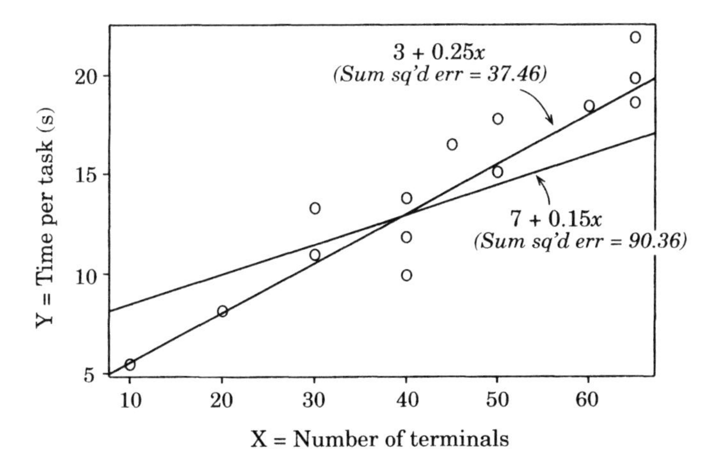

FIGURE 12.3.2 Two lines on the computer timing data.

TABLE 12.3.1 Prediction Errors

		3 + 0.25x		7 + 0.15x	
x	y	$\widehat{y}$	$y - \widehat{y}$	$\widehat{y}$	$y - \widehat{y}$
40	9.90	13.00	−3.10	13.00	−3.10
50	17.80	15.50	2.30	14.50	3.30
60	18.40	18.00	0.40	16.00	2.40
45	16.50	14.25	2.25	13.75	2.75
40	11.90	13.00	−1.10	13.00	−1.10
10	5.50	5.50	0.00	8.50	−3.00
30	11.00	10.50	0.50	11.50	−0.50
20	8.10	8.00	0.10	10.00	−1.90
50	15.10	15.50	−0.40	14.50	0.60
30	13.30	10.50	2.80	11.50	1.80
65	21.80	19.25	2.55	16.75	5.05
40	13.80	13.00	0.80	13.00	0.80
65	18.60	19.25	−0.65	16.75	1.85
65	19.80	19.25	0.55	16.75	3.05
Sum of squared errors			37.46		90.36

additional information. If we use the line $\widehat{y} = 3 + 0.25x$ to predict the y-value at $X = 20$, we obtain $\widehat{y} = 8$. The only observation taken at $X = 20$ gave a y-value of 8.1. Thus, in using the value predicted by the trend, we make a *prediction error* of $y - \widehat{y} = 8.1 - 8.0 = 0.1$. In Table 12.3.1 we have used both lines to predict all the data points and worked out all the prediction errors. For example, the second data point has $X = 50$. The prediction from the second line, $\widehat{y} = 7 + 0.15x$, is 14.5 compared with the observed y-value of 17.8, so the prediction error is $y - \widehat{y} = 17.8 - 14.5 = 3.3$. We have also added up the squared prediction errors. For the first line this gives $(-3.10)^2 + 2.30^2 + 0.40^2 + \cdots = 37.46$. The sum of squared prediction errors for the second line ($\widehat{y} = 7 + 0.15x$) is considerably larger at 90.36. We could see from the plot that the second line fits much worse, and the sum of squared prediction errors is reflecting this.

12.3.2 The Least-Squares Line

Let us denote our n pairs of data points $(x_1, y_1), (x_2, y_2), \ldots, (x_n, y_n)$. It can be shown (using calculus) that the line $\widehat{y} = \beta_0 + \beta_1 x$ that fits the data best (in the sense of having minimum sum of squared prediction errors) has intercept $\beta_0 = \widehat{\beta}_0$ and slope $\beta_1 = \widehat{\beta}_1$ given by the following formulas:

Slope: $$\widehat{\beta}_1 = \frac{\sum (x_i - \overline{x})(y_i - \overline{y})}{\sum (x_i - \overline{x})^2}$$

Intercept: $$\widehat{\beta}_0 = \overline{y} - \widehat{\beta}_1 \overline{x}$$

The values obtained are called the ***least-squares estimates*** of the intercept and slope for reasons that will become apparent in the next section. They are also often called ***regression coefficients.***

Least-squares line: $\widehat{y} = \widehat{\beta}_0 + \widehat{\beta}_1 x$

An interesting feature of the line given by the preceding formulas is that it passes through the point $(\overline{x}, \overline{y})$. We do not actually have to use these formulas. Not only do statistical packages and Excel automatically compute the intercept and slope of the least-squares line, but so do many hand calculators.

✖ EXAMPLE 12.3.1 (cont.) *Least-Squares Estimates and the Least-Squares Line*

We return to the computer timing data. The data have been replotted in Fig. 12.3.3. A small extract from Minitab regression output is also given in the lower right-hand corner. In the Minitab output the least-squares estimates are to be found in the `Coef` column. Minitab refers to an explanatory variable as a *predictor.* It uses `Constant` to label the intercept and the name of the variable (which was `nterm` in our data file) to label the slope. As we see from the Minitab output, the intercept is estimated by $\widehat{\beta}_0 = 3.050$, and the slope is estimated by $\widehat{\beta}_1 = 0.26034$. The least-squares line

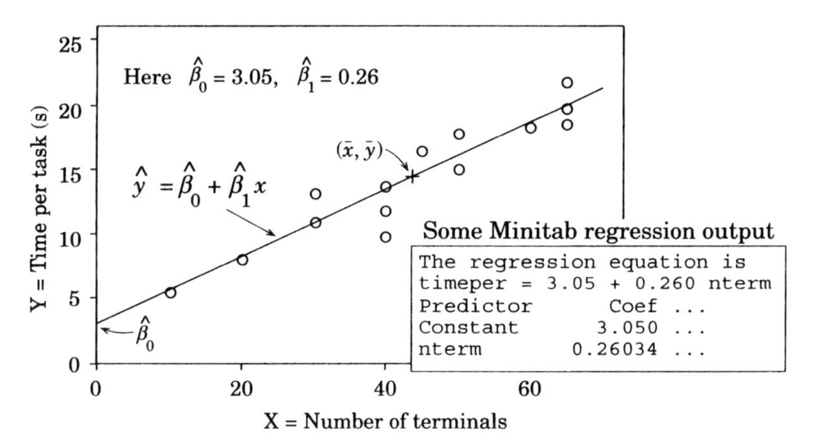

FIGURE 12.3.3 Computer timing data with least-squares line.

is therefore (with some rounding) $\hat{y} = 3.05 + 0.26x$. We have added this line to the plot. Note that the line passes through the estimated intercept and the point $(\bar{x}, \bar{y})$ $[= (43.57, 14.39)]$.

To produce a scatter plot with a fitted least-squares line on it in Minitab, look under `Stat` → `Regression` → `Fitted Line Plot`. It will alternatively put a least-squares quadratic or cubic curve on the scatter plot if you click the appropriate button. Excel will add a least-squares line to an existing scatter plot using the `trendline` capability. (Buttons also allow you use quadratic, other polynomial, and exponential trends fitted by least squares.) In Excel 97 `trendline` is located under the `Chart` menu item that appears when a chart is activated. You can achieve the same ends, though often with a little more work, in other packages.

QUIZ ON SECTION 12.3

1. What are the quantities that specify a particular line?
2. Explain the idea of a prediction error in the context of fitting a line to a scatter plot. To what visual feature on the plot does a prediction error correspond?
3. What property is satisfied by the line that fits the data best in the least-squares sense?
4. The least-squares line $\hat{y} = \hat{\beta}_0 + \hat{\beta}_1 x$ passes through the points $(x = 0, \hat{y} = ?)$ and $(x = \bar{x}, \hat{y} = ?)$. Supply the missing values.

EXERCISES FOR SECTION 12.3

The data in Table 12.3.2 come from a study of the effects of dissolved organic matter in water. It gives mean dissolved organic carbon (DOC, in mg/L) and a measure of mean *optical absorbance* in water from 13 lakes in northwest England.

(a) Plot DOC versus ABSORBANCE. Does there seem to be a relationship?

(b) Determine the least-squares line and superimpose it on your plot.

(c) Plot ABSORBANCE versus DOC. What do you think now? Determine the least-squares line and superimpose it on your plot. Superimpose the same line on your plot in (a).

(d) Which line would you use?

TABLE 12.3.2 DOC and Optical Absorbance in Lake Water

DOC	1.70	0.70	1.90	0.90	1.50	1.40	2.60	1.70	2.00	1.60	2.00	0.90	1.80
Absorbance	.089	.015	.049	.027	.053	.030	.078	.067	.036	.065	.074	.014	.048

Source: Tipping et al. [1988].

12.4 FORMAL INFERENCE FOR THE SIMPLE LINEAR MODEL

12.4.1 The Simple Linear Model

The approximate linear relationships that we have seen so far have had two features, a linear trend and fluctuations (scatter) about that trend. To make any statistical statements about the underlying linear relationship, we have to either know something or make some assumptions about the behavior of these chance fluctuations. It is convenient in what follows to think of X as *fixed,* because even if it is random, we are ultimately interested in the behavior of Y at various fixed X-values.

Because of random variations in experimental conditions, we do not expect to get the same Y value even if we keep repeating the experiment at the *same* value of $X = x$. We saw this in Example 12.2.1, where repeat experiments using the same numbers of terminals resulted in quite different values of the Y-variable time per task. It is even more evident in Fig. 12.4.1, plotted from data given in Table 12.2 of Johnson and Leone [1977]. It concerns the relationship between the lifetime of a lathe tool bit (the Y-variable) and the speed at which it is working (the X-variable). There is a downward linear trend (i.e., negative slope). Quite different cutting tool lifetimes (Y) are observed at the same cutting speed (X), however. Reasons for this probably include differences from lathe tool to lathe tool in terms of things such as the hardness of the steel and slight differences in the material they were cutting. Experimental differences always exist, no matter what steps are taken to standardize experimental conditions, and these will cause some degree of variation.

A natural way of modeling data like that in Fig. 12.4.1 is to say that observations (Y) at $X = x$ (e.g., $X = 90$) are random, having a distribution with some population mean μ_Y and standard deviation σ (this explains the scatter). We model

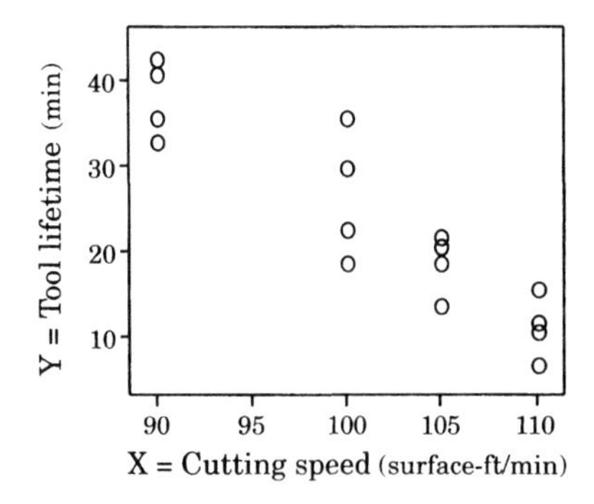

FIGURE 12.4.1 Lathe tool lifetimes.

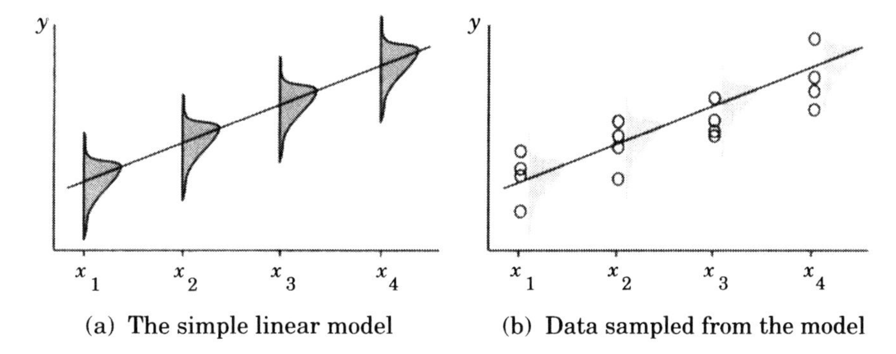

FIGURE 12.4.2 The simple linear model.

the underlying true means as falling on a line (this explains the linear pattern). We have depicted this idea in Fig. 12.4.2(a) using little Normal distribution curves rising up vertically out of the page.

Each Normal distribution has the same spread, that is, the same standard deviation σ. If we were generating data from Normal distributions with their means positioned as in the diagram, what would the resulting data look like? We generated four random observations at each x_i-value from the Normal distributions in Fig. 12.4.2(a) and plotted the resulting data in Fig. 12.4.2(b). We left the "ghosts" of the Normal curves as a reminder of how these data points arose. As you can see, the model is generating the sort of data we have been seeing.

Figure 12.4.2 is a visual image of the *simple linear model.* We can write the model down in various ways. The most direct translation of the picture is

$$\text{when } X = x, \quad Y \sim \text{Normal}(\mu_Y, \sigma) \qquad \text{where } \mu_Y = \beta_0 + \beta_1 x$$

The model is more commonly presented as follows. Consider $U = Y - \mu_Y$. Then $E(U) = 0$, and because subtracting a fixed constant does not affect variability, $sd(U) = sd(Y) = \sigma$. We can thus write $Y = \mu_Y + U$, where $U \sim \text{Normal}(0, \sigma)$. This together with the model assumption that the means lie on the line $\mu_Y = \beta_0 + \beta_1 x$ gives the most common description of the simple linear model.

$$\textbf{When } \boldsymbol{X = x, Y} = \underbrace{\boldsymbol{\beta_0 + \beta_1 x}}_{\substack{\textbf{linear trend} \\ \boldsymbol{E(Y)}}} + \underset{\substack{\uparrow \\ \textbf{random error}}}{\boldsymbol{U}}$$

where $U \sim \text{Normal}(0, \sigma)$. With this latter version of the model we can think of an observation at $X = x$ being obtained by adding a random (e.g., measurement) error or fluctuation to the value on the line $\beta_0 + \beta_1 x$. These random errors give rise to the scatter we see in our plots. Because the standard deviation σ is the same for all observations, the model produces "constant scatter." Observations have the same level of variability regardless of the value of X. For the model to be applicable we should not notice any pronounced change in the variability of Y in a scatter plot of the data as X changes (see Section 12.4.4).

We now lay out the assumptions implicit in the simple linear model and the way it is used.

(i) There is a linear relationship between x and the mean value of Y at $X = x$, namely, $\mu_Y = \beta_0 + \beta_1 x$.

(ii) This is the underlying true line, and the values of its intercept and slope parameters (β_0 and β_1 respectively) are unknown.

(iii) The least-squares estimates $\widehat{\beta}_0$ and $\widehat{\beta}_1$ estimate the unknown true values β_0 and β_1.

(iv) The random errors ($U = Y - \mu_Y$) are Normally distributed. They all have the same standard deviation, σ, and are all statistically independent.

Least-Squares Estimates Have a Sampling Distribution

Figure 7.2.1 used computer simulation to show the random nature of sample means in the repeated taking of samples. We now do something similar for least-squares estimates. In Fig. 12.4.3 we repeatedly generated data from the model $Y = 6 + 2x + U$, where $U \sim \text{Normal}(0, \sigma = 3)$. Initially, five data sets were generated, each with observations taken at $X = 1, 2, \ldots, 8$. For each data set we calculated the least-squares estimates $\widehat{\beta}_0$ of the intercept and $\widehat{\beta}_1$ of the slope. On each scatter plot (labeled "Sample 1," ..., "Sample 5") we placed the least-squares line (solid) calculated from the data and also the true line (dotted).

Although each data set was generated using the same mechanism, the unpredictability of the random errors produces scatter-plot patterns that look quite different. The different point clouds give rise to quite different least-squares estimates and thus different least-squares lines, all of which differ from the true line. Although $\beta_0 = 6$, the estimates from data, $\widehat{\beta}_0$, vary from 3.63 (Sample 1) to 9.14 (Sample 5). Similarly, although $\beta_1 = 2$, $\widehat{\beta}_1$ varies from 1.13 (Sample 5) to 2.26 (Sample 1). In other words $\widehat{\beta}_0$ and $\widehat{\beta}_1$ are random in repeated sampling and have sampling distributions. A picture of these sampling distributions is given by the histograms at the bottom of Fig. 12.4.3, which we will discuss shortly.

There is one last scatter plot in Fig. 12.4.3, labeled "Combined." Here we accumulated the data from all of Samples 1–5 to give five observations at each x-value. We see a similar pattern of scatter as in Figs. 12.4.1 and 12.4.2(b). The least-squares line (solid) for the combined data set is closer to the true line (dotted) than in any of the other scatter plots, largely because it is based on more data.

Next we used computer simulation to generate 1000 data sets like those labeled "Sample 1" to "Sample 5," producing 1000 intercept estimates and 1000 slope estimates. Histograms of these estimates are given at the bottom of Fig. 12.4.3 together with summary statistics. The histograms look roughly Normal in shape. The sample mean of the 1000 intercept estimates was 6.05, which is close to the true value of 6. Similarly, the sample mean of the least-squares estimates of slope was 1.96, which is close to the true value of 2. In fact, it can be shown mathematically that when the simple linear model is true, the least-squares estimates are unbiased (the mean of the sampling distribution is the true value) and Normally distributed.

For the simple linear model, ***least-squares estimates are unbiased and Normally distributed.***

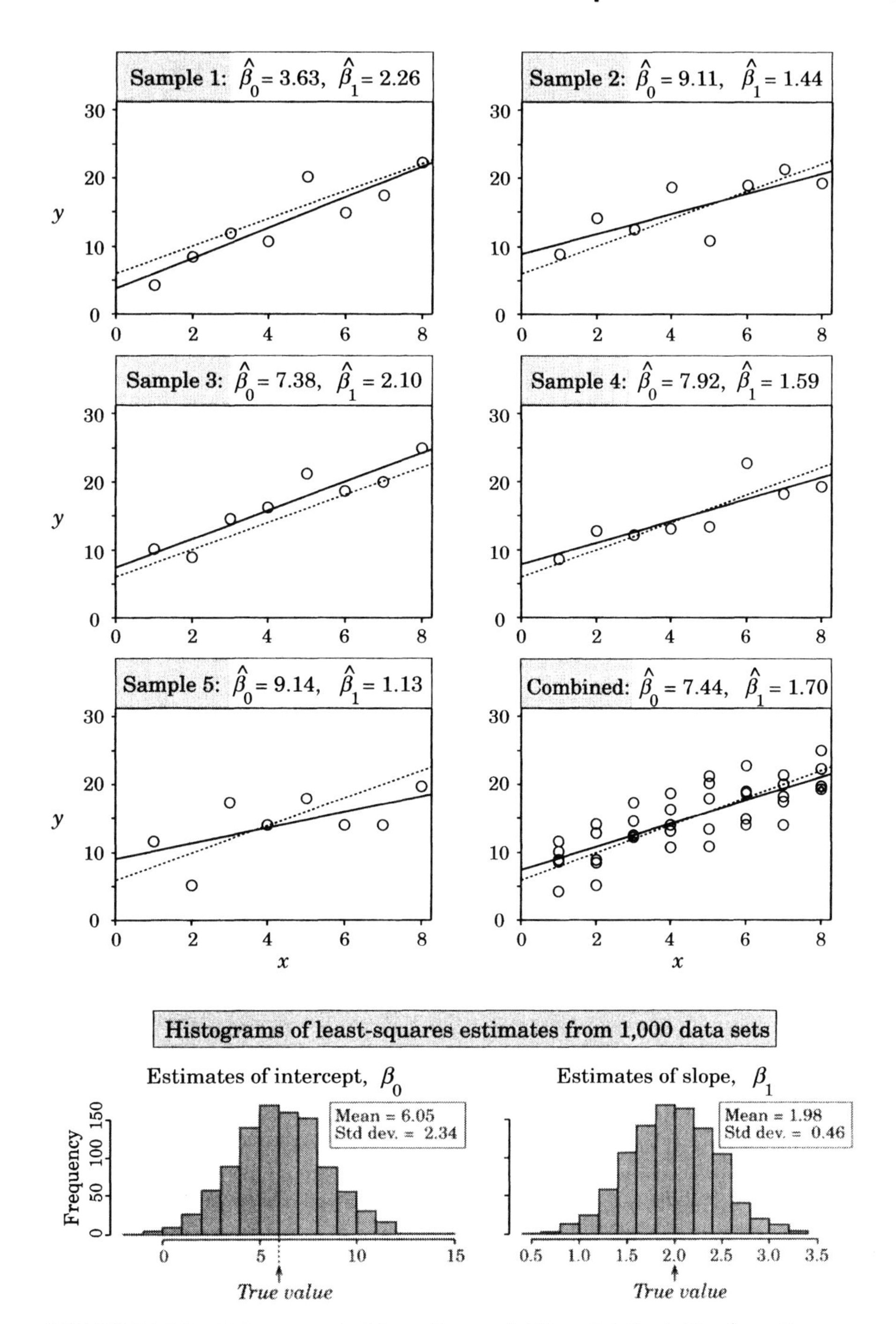

FIGURE 12.4.3 Data generated from the model $Y = 6 + 2x + U$, where $U \sim \text{Normal}(\mu = 0, \sigma = 3)$.

The formulas for the standard deviations of the sampling distributions are as follows:

$$\text{Intercepts: } \frac{\sigma}{s_X}\sqrt{\frac{\sum x_i^2}{n(n-1)}} \qquad \text{Slopes: } \frac{\sigma}{s_X}\frac{1}{\sqrt{n-1}}$$

Here s_X is the standard deviation of the x-values in the data set. Recall that σ is the population standard deviation of the random errors. Viewed another way, σ is the population standard deviation of Y-values observed at the same value of X. As σ gets larger, the scatter in the data gets larger. Note the role of σ in both of the preceding formulas. As σ gets larger, the standard deviations of the least-squares estimates get larger. In the following box we translate this into terms of the sort of behavior we observe in data. (By *noisy* we mean having a lot of scatter about the trend.)

> Noisier data produce *more-variable* least-squares estimates.

As we found in Chapter 7 when estimating the population mean μ and population proportion p, the population standard deviation of a least-squares estimate (e.g., $\widehat{\beta}_1$) does not give us a measure of its variability that we can use in practice. This is because the standard deviation formula contains a population quantity, σ, which we will never know. If we replace σ by an estimate from the data, however, we get a standard error for $\widehat{\beta}_1$ that *can* be calculated from the data. The usual estimate of σ is given by

$$s_U = \sqrt{\frac{1}{n-2}\sum(y_i - \widehat{y}_i)^2}$$

This can be thought of as the sample standard deviation of the set of prediction errors from the least-squares line. The only difference[4] is the use of a divisor of $n - 2$ instead of $n - 1$. The divisor $n - 2$, called the degrees of freedom, is introduced so that S_U^2 is an unbiased estimator of σ^2. We will not be calculating standard errors by hand. We will be using standard errors produced by computer packages.

QUIZ ON SECTION 12.4.1

1. Before considering using the simple linear model, what sort of pattern would you be looking for in the scatter plot? (There are two aspects to this.)
2. What assumptions are made by the simple linear model?
3. If the simple linear model holds, what do you know about the sampling distributions of the least-squares estimates?
4. In the simple linear model, what behavior is governed by σ?
5. Our estimate of σ can be thought of as a sample standard deviation for a certain set of quantities. What quantities?

[4] Let $u_i = y_i - \widehat{y}_i$. It can be shown that the prediction errors sum to zero (positive and negative errors cancel each other out). Hence the u_i's have sample mean $\overline{u} = 0$ and sample standard deviation $\sqrt{\sum(u_i - \overline{u})^2/(n-1)} = \sqrt{\sum u_i^2/(n-1)} = \sqrt{\sum(y_i - \widehat{y}_i)^2/(n-1)}$.

12.4.2 Inferences about the Slope and Intercept

In Chapter 8 we developed methods for constructing confidence intervals for various parameters such as μ and p, and in Chapter 9 we saw how we could test a hypothesis about an unknown parameter. We can apply exactly the same methods to the simple linear model. The one new ingredient is that for the simple linear model we use $df = n - 2$.

For the simple linear model $\boldsymbol{df = n - 2}$.

Confidence intervals for the true value of β_1 are of the form

$$\text{estimate} \pm t \text{ standard errors} = \widehat{\beta}_1 \pm t\,\text{se}(\widehat{\beta}_1)$$

The *t-test statistic* for testing a hypothesized value for the true value of β_1, that is, testing H_0: $\beta_1 = c$, is

$$t_0 = \frac{\text{estimate} - \text{hypothesized value}}{\text{standard error}} = \frac{\widehat{\beta}_1 - c}{\text{se}(\widehat{\beta}_1)}$$

Generally, there is a greater interest in the slope than in the intercept. (Sometimes what is happening at $x = 0$ does not relate to anything that is interesting or meaningful for the problem. We saw this with the computer timing data.)

Testing for No Linear Relationship between X and Y

Figure 12.4.4(a) shows a scatter plot of a computer-generated set of 1000 (x, y) pairs of data points. There is no relationship between X and Y (we used independent sets of Normal(0,1) random numbers). We then obtained 12 small data sets by taking random samples of size 20 from the 1000 data points shown in Fig. 12.4.4(a). Scatter plots for these 12 data sets are given in Fig. 12.4.4(b). The patterns you see there relate to nothing real. They simply result from "the luck of the draw." Yet a number of these data sets seem to be showing patterns. We want you to look at each and ask yourself, "What does this plot suggest to me?" Several seem to show distinct positive relationships (Y increasing as X increases). The lower right-hand plot looks like a distinct negative relationship with an outlier. But these "relationships" have all just come from sampling from Fig. 12.4.4(a). You see things that are even "weirder" with smaller samples, and much less in the way of apparent relationships with larger samples.

Whenever we see a pattern in a scatter plot, we have to ask ourselves, "Is there something real here, or could what I am seeing just have been produced by sampling variation?" In the context of the simple linear model we test this by testing whether the true value of the slope, β_1, could be 0. (Since in the model $\mu_Y = \beta_0 + \beta_1 x$, if $\beta_1 = 0$, the mean value of Y does not change with X.)

Testing for no linear relationship: Test $\boldsymbol{H_0 : \beta_1 = 0}$

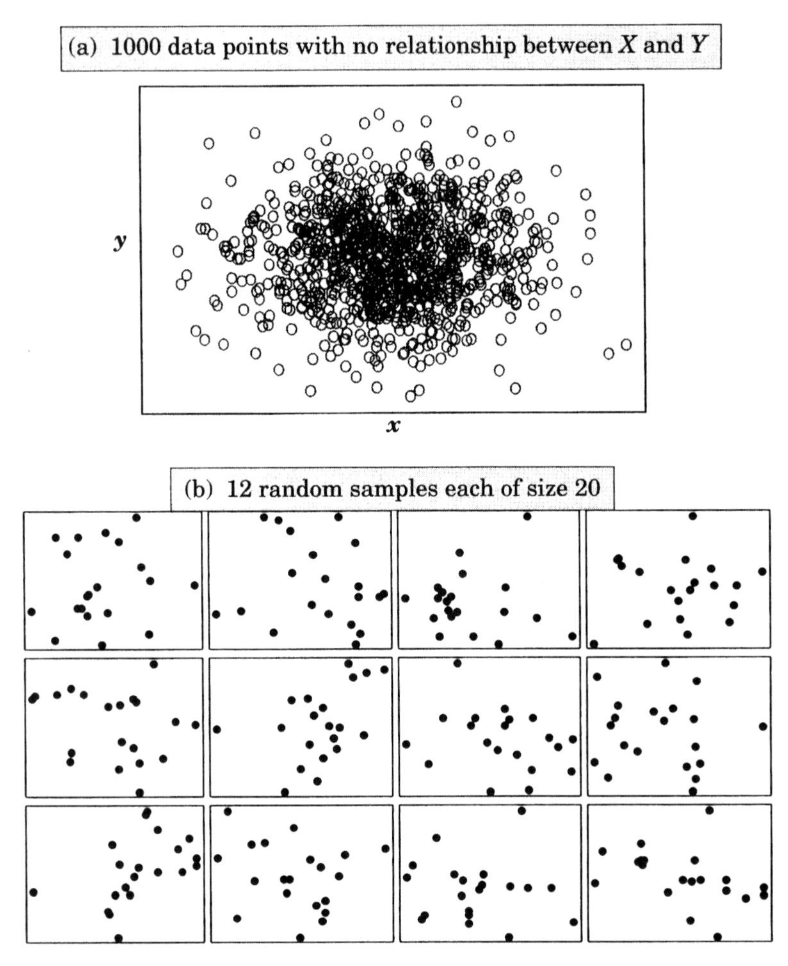

FIGURE 12.4.4 Sampling from a large data set with no X-Y relationship.

This is the most commonly tested hypothesis in regression; it is so common that the results for such tests appear on all standard computer output for regression.

EXAMPLE 12.4.1 *Is There a Relationship?*

We reproduce Fig. 3.1.10 as Fig. 12.4.5. It comes from a study of 58 sexually abused children conducted by Auckland psychiatrists Drs. Sally Merry and Leah Andrews. The children were rated on a measure of psychological disturbance by both their teacher and by the nonabusing parent.

In the scatter plot there appears to be a weak relationship between the parent's rating and the teacher's rating. Is there a real relationship, or could this just be sampling variation? We address this question by testing the no-relationship null hypothesis, H_0: $\beta_1 = 0$. Figure 12.4.6 gives some annotated regression output from Excel. From the t-test statistic we see that that $\widehat{\beta}_1$ is about 2.3 standard errors from the hypothesized value of 0. The P-value for the test is 0.0236, showing that we do indeed have evidence against the no-relationship hypothesis. We do have evidence that the parent's and teacher's ratings are related. (As we noted near Fig. 3.1.10, the surprising thing is just how weakly related they are.)

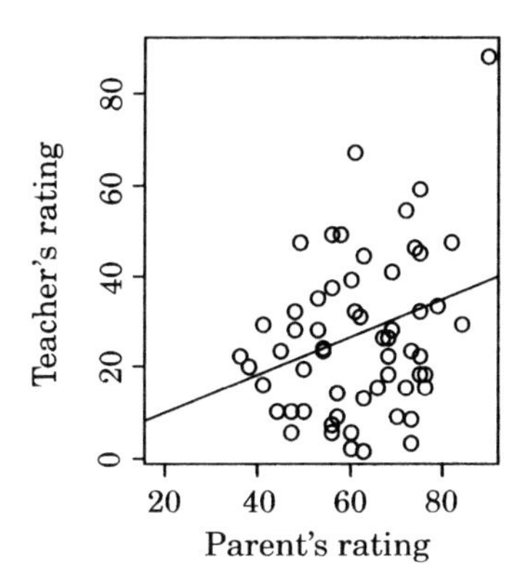

FIGURE 12.4.5 Parent's rating versus teacher's rating for abused children (with least-squares line).

testing $H_0: \beta_i = 0$ — CI's for true β_i's

$\hat{\beta}_0$ — $se(\hat{\beta}_0)$

	Coefficients	Standard Error	t Stat	P-value	Lower 95%	Upper 95%
Intercept	1.3659	11.3561	0.1203	0.9047	-21.3831	24.1149
parent	0.4188	0.1799	2.3277	0.0236	0.0584	0.7792

Name of *X*-variable — $\hat{\beta}_1$ — $se(\hat{\beta}_1)$ — *P*-value for $H_0: \beta_1 = 0$

FIGURE 12.4.6 Excel regression output for the child abuse data.

Note that the Excel output also includes 95% confidence limits for the true intercept and the true slope. With 95% confidence the true slope is between 0.06 and 0.78. For $df = n - 2 = 56$ the *t*-multiplier for a 95% confidence interval is 2.003. You will note that *estimate* $\pm$ *t standard errors* gives you the same confidence interval as in the printout.

If parent's ratings and teacher's ratings were identical, the true slope would be 1. We will test H_0: $\beta_1 = 1$ for experience, even though we know what the answer will be (because 1 is well outside the 95% confidence interval). We cannot read the result of this test off the printout and must construct the test statistic ourselves. The *P*-values on the printout are just for null hypotheses of the form H_0: $\beta_i = 0$. The *t*-test statistic for testing H_0: $\beta_1 = 1$ is

$$t_0 = \frac{\text{estimate} - \text{hypothesized value}}{\text{standard error}} = \frac{0.4188 - 1}{0.1799} = -3.2307$$

The two-tailed *P*-value for the test is $2 \times \text{pr}(T \leq -3.2307)$ where $T \sim \text{Student}(df = 56)$, giving *P*-value $= 0.002$. We have very strong evidence that the true slope is not 1.

To fit a regression model in Minitab, look under `Stat` → `Regression` → `Regression`. In Excel look under `Tools` → `Data Analysis` and choose `Regression`. In Splus and R use the function `lm` followed by `summary`. See the on-line help files for syntax. Often these programs provide more extensive output than we explain here. The P-values given in regression output are almost always two tailed.

✖ Example 12.4.2 *Calculating and Interpreting a Confidence Interval*

We recall Example 12.3.1 about the computer timing data. The regression output in Fig. 12.4.7 comes from Minitab, which refers to an explanatory variable as a *predictor* and labels the intercept with `Constant`. Recall that the Y-variable was the measured time per task. The X-variable, the number of terminals active, was called `nterm` in our data file on the computer. Reading the `nterm` row in the output, we note that the estimated slope is 0.2603, the test of no effect due to the number of terminals (H_0: $\beta_1 = 0$) has t-statistic 9.62, and the two-tailed P-value is 0 to three decimal places. There is clear evidence that the number of terminals on the system has an effect on Y. (We could have guessed this from the very strong relationship in the scatter plot.) No confidence interval was provided for the slope, so we will calculate one.

The degrees of freedom is $df = n - 2 = 12$, so that the t-multiplier of a 95% confidence interval is $t = 2.179$. A 95% confidence interval for the true slope is given by

$$\text{estimate} \pm t \text{ standard errors} = 0.26034 \pm 2.179 \times 0.02705 = [0.20, 0.32]$$

Recall that the slope tells us about the change in Y associated with a one-unit increase in X. Thus with 95% confidence there is an increase in time per task of between 0.20 and 0.32 seconds for every additional terminal in operation. For every additional 10 terminals this amounts to an increase in Y of between 2 and 3.2 seconds on average. We will not make any inferences about β_0 here, as time per task has no meaning when $X = 0$ terminals are initiating tasks.

Regression Analysis *Standard errors* *t-statistics* *P-values*

```
The regression equation is
timeper = 3.05 + 0.260 nterm
```

Predictor	Coef	StDev	T	P
Constant	3.050	1.260	2.42	0.032
nterm	0.26034	0.02705	9.62	0.000

testing H_0: $\beta_i = 0$

FIGURE 12.4.7 Minitab output for the computer timing data.

✖ Example 12.4.3 *Testing a Hypothesized Value That Is Not Zero*

The following data were collected by a friend's father who lives in the United States and has an interest in meteorology. He regularly collects precipitation data (rainfall and snowfall) at home. You cannot use an ordinary plastic rainfall gauge for snow because it can crack when melted snow freezes. For snow, the makers of the rain gauge advise using a standard-sized (2.5-inch diameter) can, allowing the snow to melt, and pouring it into the plastic rain gauge to take a reading, which you should then multiply by 0.44. Our friend's father was skeptical about the 0.44 multiplier and decided to check it out.

Over a summer he collected the rainfall data given in Table 12.4.1 using both the can and the rain gauge, which were mounted next to each other at the same height. Each day on which there was any rain, he took a reading from the rain gauge (in millimeters, which we call the "gauge reading"), then emptied the gauge and poured the water from the can into it. This gave the "can reading."

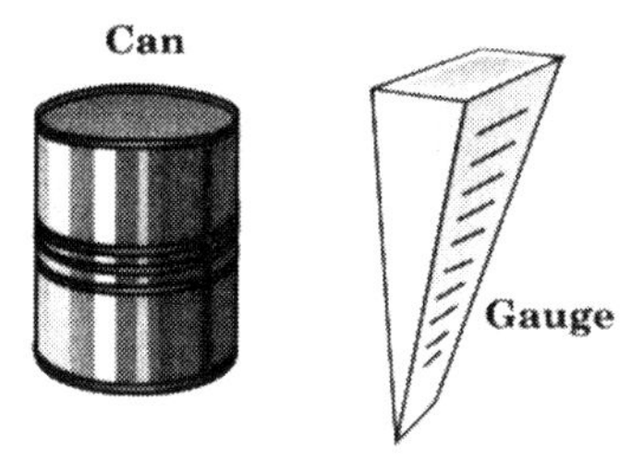

The data are plotted in Fig. 12.4.8 and show a very strong linear relationship except for the 35th observation, which is an outlier. We will suppose that some sort of mistake was made on that day and omit it from all of our analyses. Figure 12.4.8 also gives the regression output from fitting the simple linear model. We have used the gauge reading as our Y-variable and the can reading as our X-variable because we are translating can readings into gauge readings (or predicting gauge readings from can readings). This time we obtained our output from the statistical package SAS. SAS takes variable names of only up to eight characters in length, so we called our X-variable CAN_READ. Output is given at the bottom of Fig. 12.4.8.

These data call for testing for a value of β_1 that is not 0. If the gauge manufacturer is right, the intercept (β_0) should be 0 and the slope (β_1) should be 0.44. We can see that the least-squares estimates are very close to these values. We will assume that the simple linear model provides a reasonable model for these data. (We examine whether this is so in Section 12.4.4.) We see from the INTERCEP row of the output that the t-value for testing H_0: $\beta_0 = 0$ is $t_0 = 4.52$. SAS has printed a P-value of 0.0001, which is very small, so H_0 is very strongly rejected. (Actually, SAS prints 0.0001 whenever a P-value is ≤ 0.0001.) A 95% confidence interval for the true intercept β_0 is given by [0.019, 0.050], however. We conclude that the true value of the intercept is small but nonzero.

Turning our attention to the CAN_READ row of the output, the t-test statistic for testing H_0: $\beta_1 = 0$ is $t = 120.5$. The P-value is essentially 0 (SAS has printed 0.0001 again). It is clear that there is a relationship between the variables. (Again this is obvious from the plot.) The null hypothesis we are really interested in testing, however, is H_0: $\beta_1 = 0.44$. We have to construct this t-test statistic ourselves. The t-test statistic is

$$t_0 = \frac{\text{estimate} - \text{hypothesized value}}{\text{standard error}} = \frac{0.439803 - 0.44}{0.00364925} = -0.054$$

Since $\widehat{\beta}_1$ is only $\frac{1}{20}$ of a standard error from the hypothesized value, the data clearly provide no evidence against a slope of 0.44 (the two-sided P-value is 0.98). A 95% confidence interval for the true slope is given by [0.432, 0.447].

The manufacturer's 0.44 conversion factor is not quite consistent with the data, as $\beta_0 \neq 0$. Use of the can with this conversion factor systematically underestimates

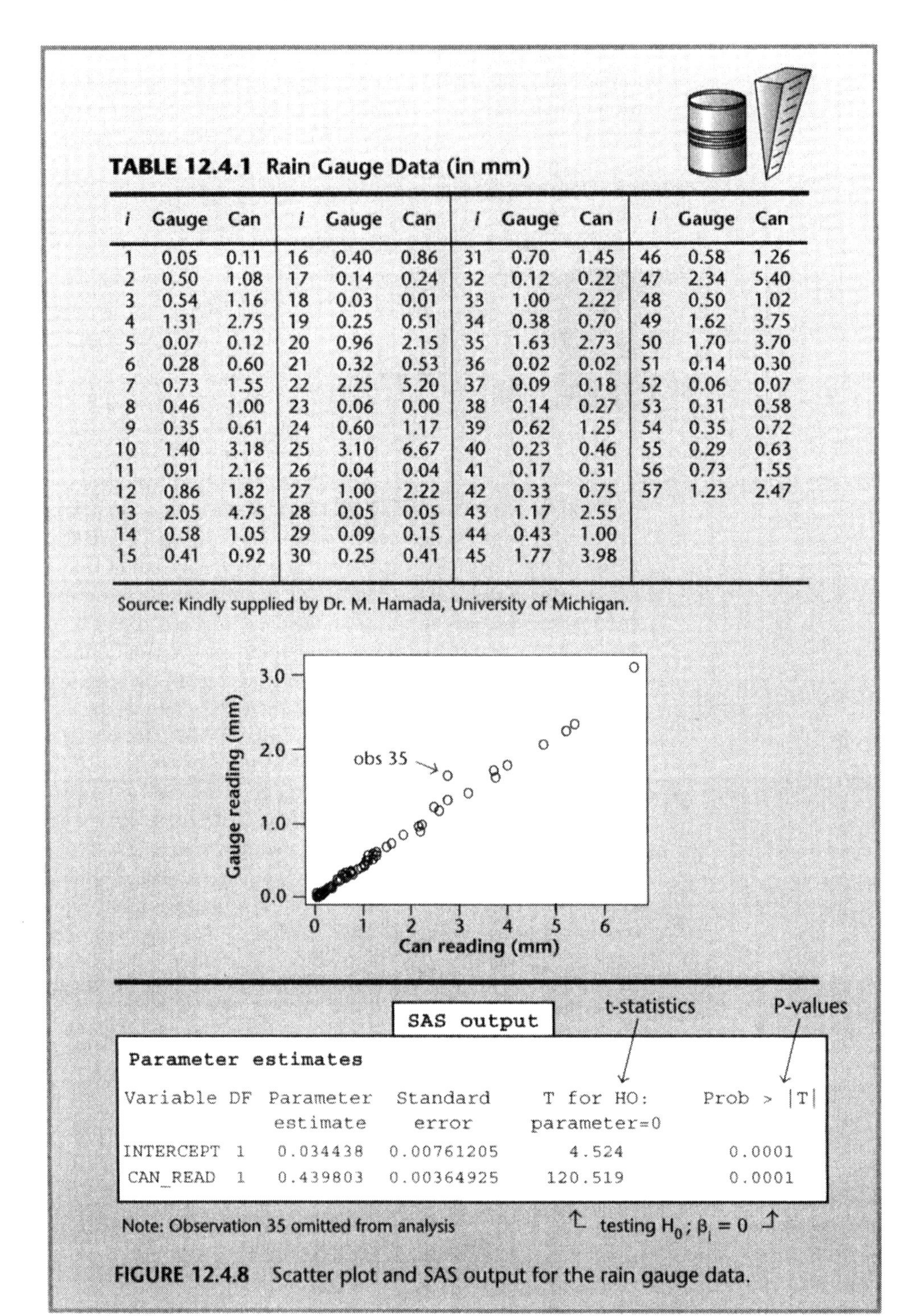

TABLE 12.4.1 Rain Gauge Data (in mm)

i	Gauge	Can	*i*	Gauge	Can	*i*	Gauge	Can	*i*	Gauge	Can
1	0.05	0.11	16	0.40	0.86	31	0.70	1.45	46	0.58	1.26
2	0.50	1.08	17	0.14	0.24	32	0.12	0.22	47	2.34	5.40
3	0.54	1.16	18	0.03	0.01	33	1.00	2.22	48	0.50	1.02
4	1.31	2.75	19	0.25	0.51	34	0.38	0.70	49	1.62	3.75
5	0.07	0.12	20	0.96	2.15	35	1.63	2.73	50	1.70	3.70
6	0.28	0.60	21	0.32	0.53	36	0.02	0.02	51	0.14	0.30
7	0.73	1.55	22	2.25	5.20	37	0.09	0.18	52	0.06	0.07
8	0.46	1.00	23	0.06	0.00	38	0.14	0.27	53	0.31	0.58
9	0.35	0.61	24	0.60	1.17	39	0.62	1.25	54	0.35	0.72
10	1.40	3.18	25	3.10	6.67	40	0.23	0.46	55	0.29	0.63
11	0.91	2.16	26	0.04	0.04	41	0.17	0.31	56	0.73	1.55
12	0.86	1.82	27	1.00	2.22	42	0.33	0.75	57	1.23	2.47
13	2.05	4.75	28	0.05	0.05	43	1.17	2.55			
14	0.58	1.05	29	0.09	0.15	44	0.43	1.00			
15	0.41	0.92	30	0.25	0.41	45	1.77	3.98			

Source: Kindly supplied by Dr. M. Hamada, University of Michigan.


```
Parameter estimates

Variable DF Parameter   Standard      T for HO:      Prob > |T|
            estimate    error         parameter=0
INTERCEPT 1  0.034438   0.00761205      4.524          0.0001
CAN_READ  1  0.439803   0.00364925    120.519          0.0001
```

Note: Observation 35 omitted from analysis testing $H_0: \beta_i = 0$

FIGURE 12.4.8 Scatter plot and SAS output for the rain gauge data.

the true level of precipitation.[5] If you plot the line with 0 intercept and slope 0.44 (not shown), all but 6 of the 56 observations lie above the line. The errors are very small, however, compared with the day-to-day variability in rainfall. The answers you get with the 0.44 multiplier are still probably good enough for almost all practical purposes.

---

*EXAMPLE 12.4.4 *Is the Relationship Curved?*

We now investigate another variable in the computer timing data in Example 12.2.1 and Table 12.2.1. In Fig. 12.2.1(a) we plotted Y^*, which was the total time all the tasks took, against the number of terminals operating and thought we saw a slight curve in the relationship. Figure 12.4.9 plots these data again together with an added least-squares quadratic curve. (As we have noted previously, Minitab and Excel will do this automatically.) Is the relationship really curved, or could the apparent curve just be a result of sampling variation? If we take a quadratic trend curve $\beta_0 + \beta_1 x + \beta_2 x^2$ and set $\beta_2 = 0$, the curve reduces to the line $\beta_0 + \beta_1 x$. So one way to test whether a relationship is curved is to fit a quadratic and test whether the coefficient of the squared term is 0.

In the data file our X variable is called `nterm`. We created a new variable `ntermsq` containing the squares of the numbers of terminals. Then we gave the regression program the two X-variables `nterm` and `ntermsq`. The result was the output in Fig. 12.4.9. The P-value for `ntermsq` is 0.003, showing strong evidence against the hypothesis that the true coefficient of `ntermsq` is 0. We have strong evidence that the real relationship is curved.

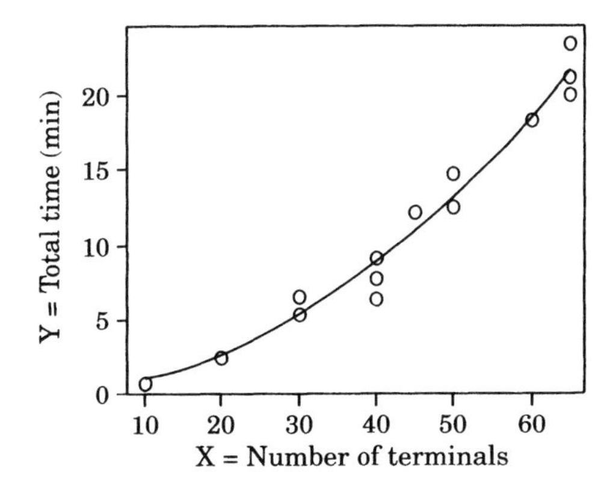

R Output

```
Coefficients:
             Estimate Std. Error t value Pr(>|t|)
(Intercept) 0.215067 1.941166    0.111   0.91378
nterm       0.036714 0.100780    0.364   0.72254
ntermsq     0.004526 0.001209    3.745   0.00324   **
```

FIGURE 12.4.9 Quadratic model for Y^* = total time.

[5] Thinking question: What physical factor(s) involved in the measurement process using the can would result in a small positive intercept?

TABLE 12.4.2 Hours of Training and Blood Lead Levels

Hours of training	8	10	10	12	15	18	18	21	25	25
Blood lead (μmol/L)	0.53	0.25	0.34	0.25	0.29	0.30	0.53	0.53	0.53	0.87

Source: Atkinson et al. [1994].

QUIZ ON SECTION 12.4.2

1. What value of df is used for inference for $\widehat{\beta}_0$ and $\widehat{\beta}_1$?
2. Within the context of the simple linear model, what formal hypothesis is tested when you want to test for no linear relationship between X and Y?
3. What hypotheses do the t-test statistics and associated P-values on regression output test?
4. What is the form of a confidence interval for the true slope?
5. What is the form of the test statistic for testing H_0: $\beta_1 = c$?

EXERCISES FOR SECTION 12.4.2

1. Atkinson et al. [1994] investigated the extent to which potentially toxic lead particulates emitted from motor vehicles are absorbed by competitive cyclists. Table 12.4.2 constructed from a graph given in their paper gives blood lead levels and hours of training for 10 cyclists.
   (a) Plot the data. What are your impressions?
   (b) Test whether there is a relationship between blood lead levels and hours of training.
   (c) Cyclist 10 has very high levels. Is our evidence for a relationship coming almost entirely from this data point? Repeat (b) omitting cyclist 10.
   (d) Does what we have done in (c) seem reasonable?
   (e) It is clear from the graph that there is variation in the data that is not explained by hours of training. (What tells us this?) Perhaps the hours-training effect does not show up as strongly as it might because we are failing to take into account other important variables. Suggest some other variables that might be important.
2. Referring to Exercises for Section 12.3, calculate a 95% confidence interval for the slope of a straight line fitted to the plot of DOC versus ABSORBANCE.

12.4.3 The Regression Model and Prediction

We now revisit the prediction problem introduced in Section 3.1.3. You may wish to reread that section, where we drew some freehand prediction curves and obtained some rough prediction intervals for predicting the liver length of a fetus using its gestational age. We now discuss the technical tools for constructing prediction intervals in cases where the simple linear model seems reasonable.

EXAMPLE 12.4.2 (cont.) *Making a Prediction*

We return to the computer timing data. When we performed the analysis for regressing time per task on the number of terminals for Example 12.4.2, one of the optional extras Minitab offered was prediction. We asked for a prediction when X, the number of terminals, was 70. We got the following output.

```
Predicted Values

    Fit   StDev Fit          95.0% CI              95.0% PI
 21.273       0.843    ( 19.436,   23.110)   ( 17.193,   25.353)
```

What do these numbers mean?

In general terms we want to make a prediction at $X = x_p$. For the simple linear model the mean value of Y at $X = x_p$ is $\mu_Y = \beta_0 + \beta_1 x_p$. We can use the fitted least-squares line to make a prediction, that is, $\widehat{y}_p = \widehat{\beta}_0 + \widehat{\beta}_1 x_p$. Using the least-squares estimates for the computer timing data from Fig. 12.4.7 and $x_p = 70$, $\widehat{y}_p = 3.050 + 0.26034 \times 70 = 21.2738$, so the number labeled `Fit` is the prediction from the line. (The slight discrepancy from 21.273 is due to rounding errors.)

Confidence Interval for the Mean

Because $\widehat{\beta}_0$ is an unbiased estimate of the true β_0 and $\widehat{\beta}_1$ is an unbiased estimate of β_1, $\widehat{y}_p = \widehat{\beta}_0 + \widehat{\beta}_1 x_p$ is an unbiased estimate of $\mu_Y = \beta_0 + \beta_1 x_p$. The number labeled `StDev Fit` is the standard error of this estimate. Taking $\widehat{y}_p \pm t\ \text{se}(\widehat{y}_p)$ gives you the interval labeled `95% CI`. This interval is called a 95% *confidence interval for the mean.* Suppose we repeated the computer timing experiment with 70 terminals an enormous number of times and averaged all the results. The confidence interval for the mean estimates this average. It estimates the *mean value* of Y at $X = x_p$.

Prediction Interval

If we knew the true values of β_0 and β_1, we could tell you the mean value of Y at $X = 70$ *exactly.* Suppose that instead we were interested in predicting the *actual* measured time per task for the next experiment run at $X = 70$. We could not tell what that value would be, even if we knew the true line, because of the randomness of the system as shown by the scatter about the line. A prediction interval also has to allow for the usual extent of the scatter. A *prediction interval* is of the form $\widehat{y}_p \pm t\ \text{sep}(\widehat{y}_p)$, where $\text{sep}(\widehat{y}_p)$ is called the *standard error of prediction.*[6] It allows for two sources of uncertainty: (1) the uncertainty about the true values of β_0 and β_1 and (2) uncertainty due to the random scatter about the line. The prediction interval is always wider than the confidence interval for the mean because the confidence interval allows only for the first source of uncertainty.

Predicting at $X = x_p$

- The ***confidence interval for the mean*** estimates the ***average*** Y-value at $X = x_p$ (averaged over many repetitions of the experiment).
- The ***prediction interval*** (PI) tries to predict the next ***actual*** Y-value.

---

[6]Formulas: $\text{se}(\widehat{y}_p) = s_U\sqrt{\frac{1}{n} + \frac{(x_p - \overline{x})^2}{(n-1)s_X^2}}$, while $\text{sep}(\widehat{y}_p) = s_U\sqrt{1 + \frac{1}{n} + \frac{(x_p - \overline{x})^2}{(n-1)s_X^2}}$.

We have found that, in practice, prediction intervals are relevant more often than are confidence intervals for the mean. (In some practical situations the idea of averaging over many repetitions does not make much sense.)

EXAMPLE 12.4.2 (cont.) *Prediction Intervals*

From the 95% prediction interval in the Minitab output previously given, we predict that an experiment with 70 terminals will result in a measured time per task that lies between 17 and 25 seconds.

In fact, the experiment was rerun using $X = 70$ terminals, and a measured time per task of 22.5 seconds was obtained. On another earlier run with 70 terminals, warning messages aroused concern that the job had not run properly and that some processes had been terminated prematurely. The resulting measured time per task was 14.6 seconds, which falls below the prediction interval. This reinforced the investigators' belief that this was not a genuine observation. Figure 12.4.10 plots the original timing data and the two new observations. The least-squares line and prediction intervals for each possible value of x_p are also shown. Such plots are automatically generated in Minitab under `Stat` → `Regression` → `Fitted Line Plot`. They can also be obtained from other packages, though in Minitab it is particularly easy, as we need only click a button to get the prediction-interval curves. Curves for both prediction intervals and confidence intervals for the mean tend to look "narrow waisted." The intervals are narrowest in the center of the data (i.e., at $X = \overline{x}$) and widen out as you move away from the center in either direction. This effect is more dramatic in larger data sets than it is in Fig. 12.4.10.

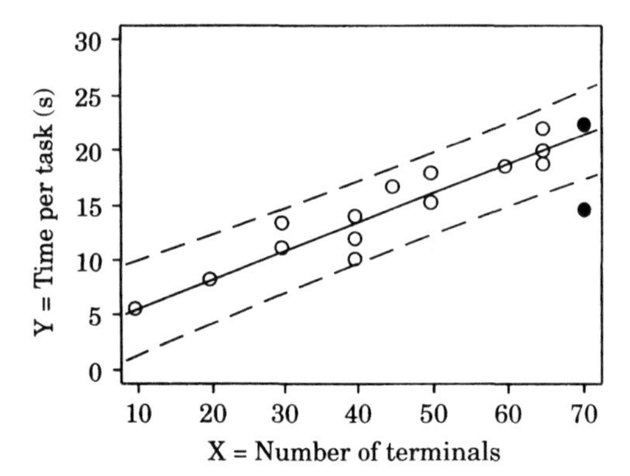

FIGURE 12.4.10 Time per task versus number of terminals (with the least-squares line and 95% prediction intervals superimposed; • = new observations).

QUIZ ON SECTION 12.4.3

1. What is the difference between a confidence interval for the mean and a prediction interval?
2. Prediction intervals make allowances for two sources of uncertainty. What are they? How does a confidence interval for the mean differ in this regard?
3. At what point along the X-axis are these intervals narrowest?

4. We gave some general warnings about prediction back in Section 3.1.3. They are relevant here. What were those warnings?

EXERCISES FOR SECTION 12.4.3

Referring to the rain gauge data of Table 12.4.1, obtain a 95% confidence interval for the mean and a 95% prediction interval for Y (a) when $x = 0.5$ and (b) when $x = 6$.

12.4.4 Model Checking

We conduct this discussion in terms of the random error description of the simple linear model. We have data $(x_1, y_1), (x_2, y_2), \ldots, (x_n, y_n)$ on n individuals. The simple linear model is

$$Y_i = \beta_0 + \beta_1 x_i + U_i \qquad \text{where } U_i \sim \text{Normal}(0, \sigma)$$

independently for $i = 1, 2, \ldots, n$.

The statistical inferences discussed in the previous subsections depend for their validity on the simple linear regression model being at least a good approximation to the true situation. Fitting any model (e.g., a straight line) is only sensible when the trend in the data is adequately summarized by that model. We should never fit a linear regression to data until we've looked at the scatter plot. This should prevent us from doing something inappropriate such as fitting a straight line to a relationship that is clearly curved. The other assumptions that affect the applicability of our formal methods of inference concern the behavior of the random errors U_i. Least-squares theory assumes the following:

(i) *Normality:* The U_i's are Normally distributed (with mean 0).

(ii) *Constant spread:* The U_i's all have the same standard deviation σ regardless of the value of x.

(iii) *Independent observations:* The U_i's are independent.

All tests and intervals are very sensitive to departures from the independence assumption and also to moderate to large departures from the constant-spread assumption. As with t-tests and intervals for means, t-tests and intervals for β_0 and β_1 are surprisingly robust against departures from the Normality assumption, particularly in moderate and large samples. Prediction intervals are sensitive to departures from Normality, and inferences about σ are very sensitive to departures from Normality.

Model Checking

Model checking involves checking the model assumptions using the residuals $\widehat{u}_i = y_i - \widehat{y}_i$ from the fitted model since these are estimates of the unobserved errors U_i. One can usefully check the preceding assumptions by plotting the residuals in various ways. The plots to follow are available from all general-purpose statistical packages and even from Excel's regression tool. In Excel and Minitab you merely need to click choice buttons to tell the program that you want the plots included in your output. The programs store the residuals for you in case you want to investigate them in other ways.

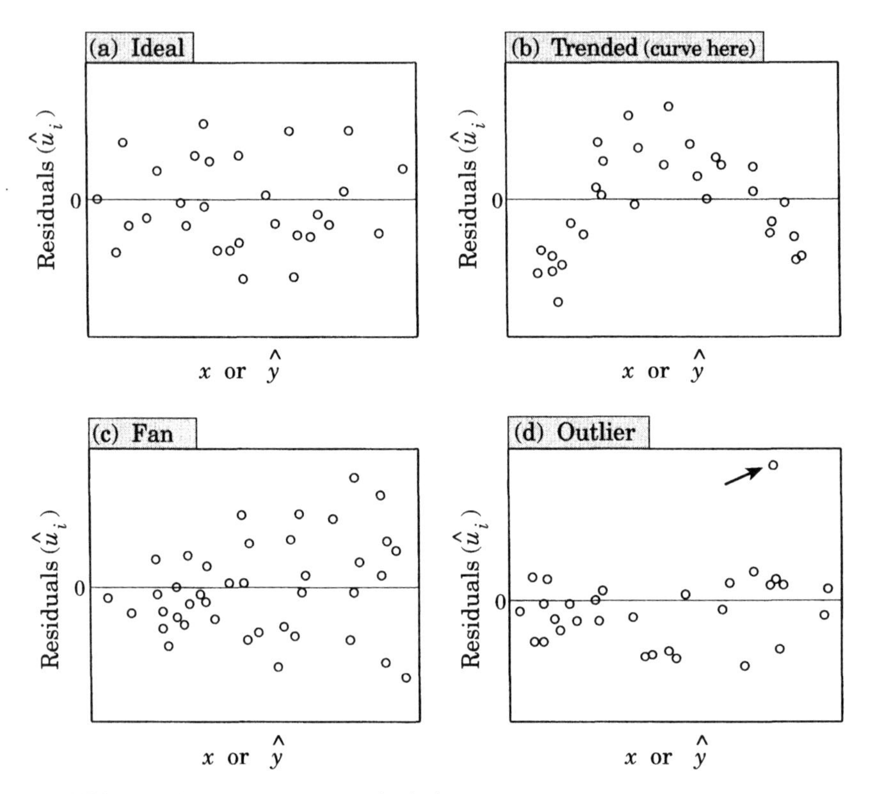

FIGURE 12.4.11 Patterns in residual plots.

Checking for Normality of the errors. We investigate approximate Normality of the errors by looking at a *Normal probability plot* of the residuals. In Minitab the Normal probability plot in the regression module does not include formal test results. To obtain these, you need to save the residuals and proceed as in Section 10.1.1.

Plots of residuals versus X and residuals versus $\hat{Y}$. A great deal of information can be gained from these plots. *You should see a patternless horizontal band* as in Fig. 12.4.11(a).

Any trend in these residual plots indicates that the model fitted has not adequately summarized the trend or pattern in the data. The fitted trend should capture all of the pattern in the data. What is left over should look like random noise. If there is still pattern left in the residuals, it means that the fitted trend has not done this pattern-capturing job properly. If you fit a line and see a curved trend in the residuals, as in Fig. 12.4.11(b), it means that the actual trend in the data is curved. A quadratic may provide a better fit.

A fan (or *funnel*) pointing in either direction, as in Fig. 12.4.11(c), indicates that the variability of the errors is not constant. There is considerably more variability at high values of x (or $\hat{y}$) than at low values. Sometimes judging the constancy of scatter can be difficult when there are large numbers of data points and many more observations in the vicinity of some X-values than in other regions. "Football shapes," as in Fig. 12.4.4, need not indicate any problems. When the model is true, we get the appearance of greater spread in residuals in regions of X where there are more observations.

Outliers show up clearly on such a residual plot, as in Fig. 12.4.11(d).

Plots against time. Plots of residuals against the times at which the observations were made can bring to light breakdowns in an experiment such as unplanned changes in experimental conditions that are occurring over time. If the actual times are not known but data are recorded in the order that they were collected, a plot of residuals versus time order (that is, first, second, etc.) is almost as good.

****Looking for lack of independence.*** When our data have been collected sequentially over time, we should look for any relationship between successive residuals (i.e., so-called *serial correlation*) by plotting $\widehat{u}_i$ (y-coordinate) versus $\widehat{u}_{i-1}$ (x-coordinate) [i.e., plot the points $(\widehat{u}_1, \widehat{u}_2)$, $(\widehat{u}_2, \widehat{u}_3)$, ..., $(\widehat{u}_{n-1}, \widehat{u}_n)$]. Again, we hope to see a patternless horizontal band. The packages tend not to perform this simple plot. Many packages, however, print the results for a formal test for serial correlation called the Durbin-Watson test.[7]

EXAMPLE 12.4.5 *Residual Plots*

Figure 12.4.12 shows residual plots for several of the examples we have been working with. We look first at the residuals from fitting the simple linear model to the computer timing data using time per task as the Y-variable (Example 12.4.2). Figure 12.4.12(a) is a plot of residuals versus X (the number of terminals, called `nterm` in the computer file). We see nothing that particularly worries us. A Normal probability plot

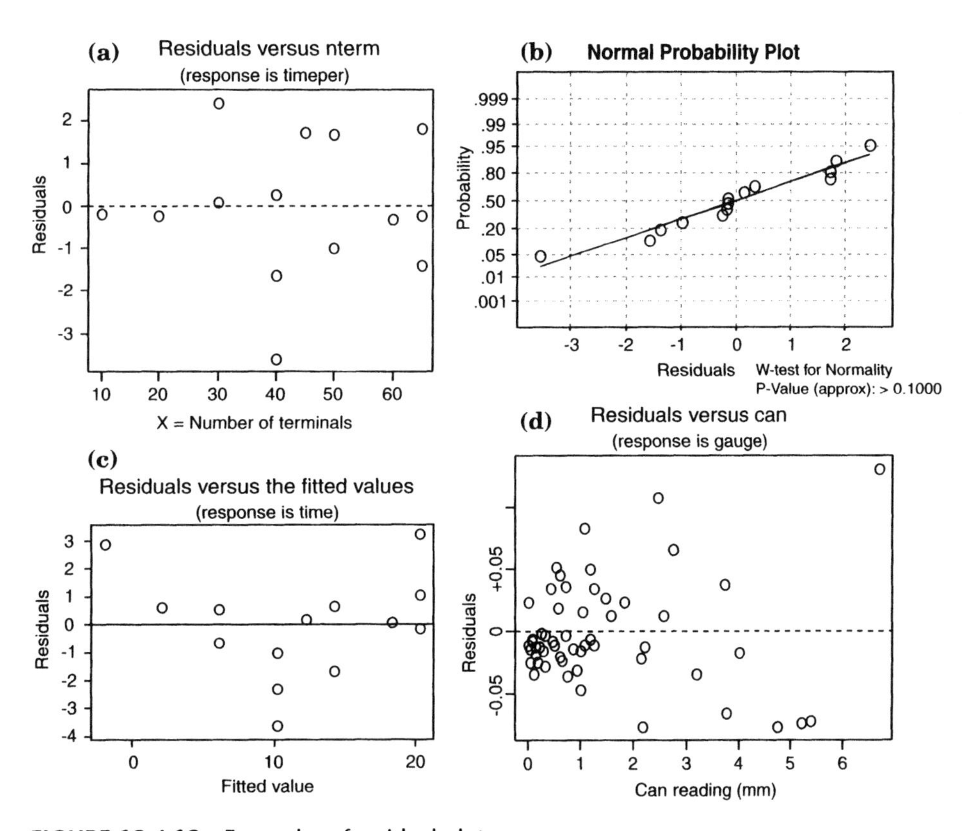

FIGURE 12.4.12 Examples of residual plots.

[7]For a brief summary of the method see Pettit [1982].

from Minitab is given as Fig. 12.4.12(b). Minitab's W-test gives a P-value greater than 0.1, so there is no evidence against the Normality assumption. In Fig. 12.4.12(c) we stay with the computer timings data but change our Y-variable. Figure 12.4.12(c) plots the residuals from a simple linear model for the *total time* to complete the tasks (see Examples 12.2.1 and 12.4.4). There was a slight curve evident in Fig. 12.2.1(a). The residual plot [Fig. 12.4.12(c)] has magnified this and made the curve more obvious. We clearly should not be fitting a linear model to this relationship.

We now turn our attention to the rainfall data (Example 12.4.3). Figure 12.4.12(d) is another residuals versus X plot, where X is the reading taken from the can. It shows a distinct fan in the residuals with more variability at large values of X. Other plots such as residuals versus time and residuals versus lagged residuals ($\widehat{u}_i$ versus $\widehat{u}_{i-1}$) reveal nothing worrying. Since the residuals in Fig. 12.4.12(d) reveal different spreads at different x-values, it is not meaningful to lump them all together and use the resulting single sample to investigate Normality. We would not be prepared to trust formal statistical inferences obtained from fitting the simple linear model to the rainfall data. For example, the P-values and confidence intervals produced using this model would be suspect. Also, changes in spread are particularly damaging to the prediction intervals produced by the simple linear model. They give intervals that are too wide in regions of X-values where the spread is smallest and too narrow in regions where the spread is greatest. To make adequate statistical inferences, we would need to come up with a new, more complex model that made allowance for the pattern of increasing spread that we see in the residuals.

Outliers in X

By an outlier in X we mean an observation with an X-value that shows up as an outlier in a plot of the X's alone. Observations like this can have a big influence on the position of the least-squares line. This aspect is shown in Fig. 12.4.13. Here the same cluster of points is used each time except for one point out to right of the others, which we move. Note the dramatic changes made as this outlying x observation moves and drags the fitted least-squares regression line after it. (Such observations are said to be ***influential.***) In the left-hand plot, the isolated point falls on the same general trend as the remainder of the observations, so it does not adversely affect the least-squares line. In the remaining two plots the isolated point is far from the general trend, and the least-squares line becomes a bad summary of the trend in the main body of observations. In such circumstances it is usually better to analyze the relationship only over the range of values covered by the bulk of the data and not include outlying X-values. The range of X-values considered should be reported together with the results of the analysis.

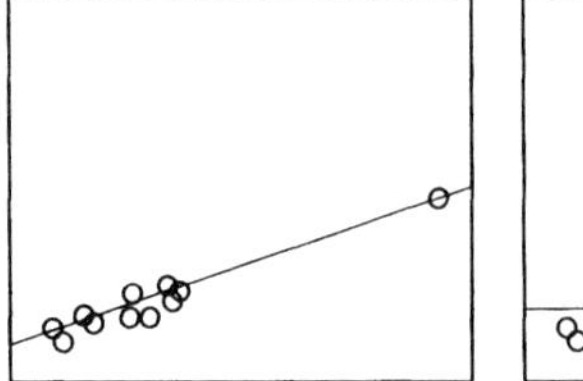
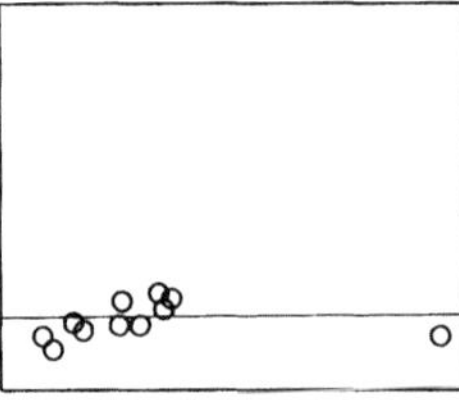
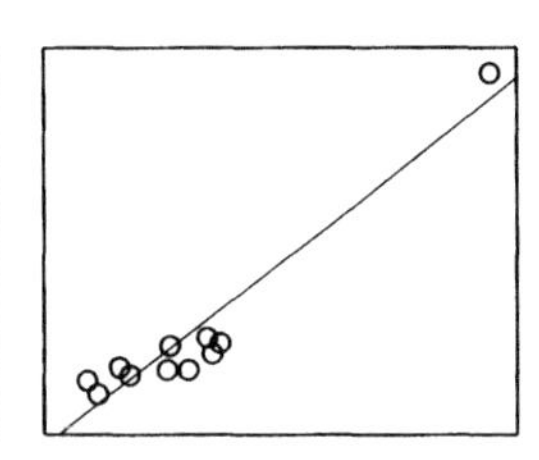

FIGURE 12.4.13 The effect of an X-outlier on the least-squares line.

QUIZ ON SECTION 12.4.4

1. What assumptions are made by the simple linear model?
2. Which assumptions are critical for all types of inference?
3. What types of inference are relatively robust against departures from the Normality assumption?
4. Four types of residual plot were described. What were they, and what can we learn from each?
5. What is an outlier in X, and why do we have to be on the lookout for such observations?

EXERCISES FOR SECTION 12.4.4

1. This exercise, designed to be done by hand, is aimed at establishing the idea of a residual and a residual plot. The least-squares estimates from fitting a line to the following data points are $\widehat{\beta}_0 = 6$ and $\widehat{\beta}_1 = 3$.

x	1	2	3	4	5	6
y	10	13	7	22	28	19

   Using the least-squares line, find the predicted values corresponding to the six values of x and hence the six residuals. Plot each residual versus the corresponding value of x.

2. Consider the data of Exercises for Section 12.3. After fitting a straight line to the scatter plot of ABSORBANCE versus DOC, plot the residuals versus X = DOC. What do you conclude?
3. Fit a simple linear model predicting DIAVOL from SYSVOL to the data in Table 2.1.1 (see also Fig. 3.1.1). Plot the residuals versus X = SYSVOL. What problems with the model, if any, do you see?
4. Fit a simple linear model to the Olympic 1500 m winning times in Example 3.1.4. Plot the residuals versus X = TIME. What problems with the model, if any, do you see?

*5. Fit a quadratic model predicting NOX from CO to the data in Table 3.1.3, Example 3.1.3. Plot the residuals versus X = CO. What problems with the model, if any, do you see?

12.5 CORRELATION AND ASSOCIATION

12.5.1 Two Regression Lines

Throughout this chapter we have stressed the asymmetric nature of regression and the need to distinguish between response variables and explanatory variables. How much does this matter? Figure 12.5.1 plots the same data as Fig. 12.4.4. Each data point relates to a sexually abused child. The plot is constructed for predicting Y = teacher's rating (the teacher's rating of psychological damage) from X = parent's rating (the rating from the nonabusing parent). The flatter line, labeled "regression of Y on X," is the least-squares line for predicting teacher's ratings from parent's ratings. This line is

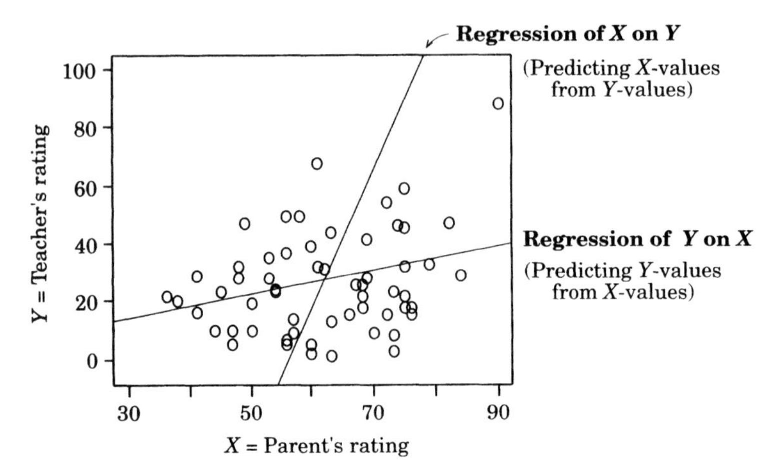

FIGURE 12.5.1 Two regression lines.

chosen to minimize *vertical* prediction errors. Both of these variables are random, so it also makes sense to look at things the other way round and determine the equation of the least-squares line for predicting parent's ratings (X) from teacher's ratings Y. We did this and drew that relationship on the same graph (labeled "regression of X on Y"). Because what we are trying to predict with the second line is the variable plotted against the horizontal axis,[8] this second line is chosen to minimize *horizontal* prediction errors.

We note that the least-squares line for predicting Y from X is quite different from the line predicting X from Y. This example emphasizes both the lack of symmetry in the way that regression methods treat data and the importance of choosing the correct variable to use as Y and the correct one to use as X. To be fair, we must point out that Fig. 12.5.1 depicts a rather extreme case, as the relationship between Y and X is rather weak (compare with the very strong linear relationship in Fig. 12.4.8). It turns out that the angle between the two lines grows smaller as any linear relationship in the data grows stronger. With an exact linear relationship (no scatter), both lines are identical.

It is clear that when we want a way of measuring or describing the closeness of a relationship between variables without singling out any variable to have a special role (i.e., a method that treats the variables symmetrically), we will not be able to use regression. What can we do? We start with a commonly quoted measure called the *correlation coefficient.*

12.5.2 The Correlation Coefficient

The *sample correlation coefficient, r*, is a measure of *how close* the points in the scatter plot of Y against X (or vice versa) come *to lying on a straight line.* For data $(x_1, y_1), (x_2, y_2), \ldots, (x_n, y_n)$, the correlation coefficient is

$$r = \frac{1}{n-1} \sum_{i=1}^{n} \left(\frac{x_i - \overline{x}}{s_X} \right) \left(\frac{y_i - \overline{y}}{s_Y} \right)$$

[8]Normally, when interested in predicting parent's ratings, we would use that variable for the vertical axis.

Inspection of the formula reveals that X and Y are treated symmetrically. If the x's and the y's are interchanged, we will still get the same answer. Also, the calculation uses the x's only through the standardized values $(x_i - \overline{x})/s_X$, that is, how many standard deviations each x_i is from the center of the x's (and similarly for the y's). Thus the resulting measure remains the same when we change measurement scales from, say, distance in miles to distance in kilometers, temperature in °F to °C, or weight in kilograms to grams or pounds.

An alternative formula for r that links correlation with regression is $r = \widehat{\beta}_1(s_X/s_Y)$, and it does not matter which variable we choose as the Y-variable in using this formula. If $\widehat{\beta}_1$ is very small, the slope of the least-squares line is almost 0 and the fitted line is almost flat. This means that there is very little trend in the data and very little linear association. This is reflected in the correlation also being very small.

The following properties of the sample correlation coefficient, which enable us to interpret r, are illustrated in Fig. 12.5.2.

(i) r always falls between -1 and $+1$.

(ii) *Positive* values of r occur when large values of X are associated with large values of Y so that Y gets larger as X gets larger (*uphill trend;* right column of Fig. 12.5.2). *Negative* values of r occur when large values of X are associated with small values of Y so that Y gets smaller as X gets smaller (*downhill trend;* left column of Fig. 12.5.2).

(iii) $r = +1$ if all points lie *exactly* on an uphill *line* [Fig. 12.5.2(i)].
$r = -1$ if all points lie *exactly* on a downhill *line* [Fig. 12.5.2(a)].

(iv) $r = 0$ when there is no uphill or downhill trend (or drift). Then X and Y are said to be *uncorrelated* [Fig. 12.5.2(e)].

(v) The value of $|r|$ (i.e., the size of r ignoring its sign) measures how close the points are to falling on a straight line. Starting from exact linear relationships when $|r| = 1$ and moving down Fig. 12.5.2, we see that as $|r|$ becomes smaller, the linear relationship looks weaker and weaker until at $r = 0$ [Fig. 12.5.2(e)] there is no longer any trend at all.

P-Values for Correlation Coefficients

When we asked Minitab for a correlation between the parent's rating and the teacher's rating of the psychological states of abused children (Example 12.4.1), we clicked the option that offered the inclusion of P-values. We got the following output:

```
Correlation of parent and teacher = 0.297, P-Value = 0.024
```

What does this P-value mean? In the same way that $\widehat{\beta}_1$ is an estimate of the population parameter β_1 underlying the linear model, the *sample* correlation coefficient r is also an estimate of a corresponding *population* correlation coefficient (conventionally denoted by ρ). Here ρ summarizes the true relationship between two variables in the population (or distribution) from which our data were sampled. The hypothesis addressed by the P-values printed for correlation coefficients by statistical packages is H_0: $\rho = 0$. There is a misconception that a very small P-value alongside a correlation coefficient tells you that the true coefficient is large. It doesn't! All a small P-value

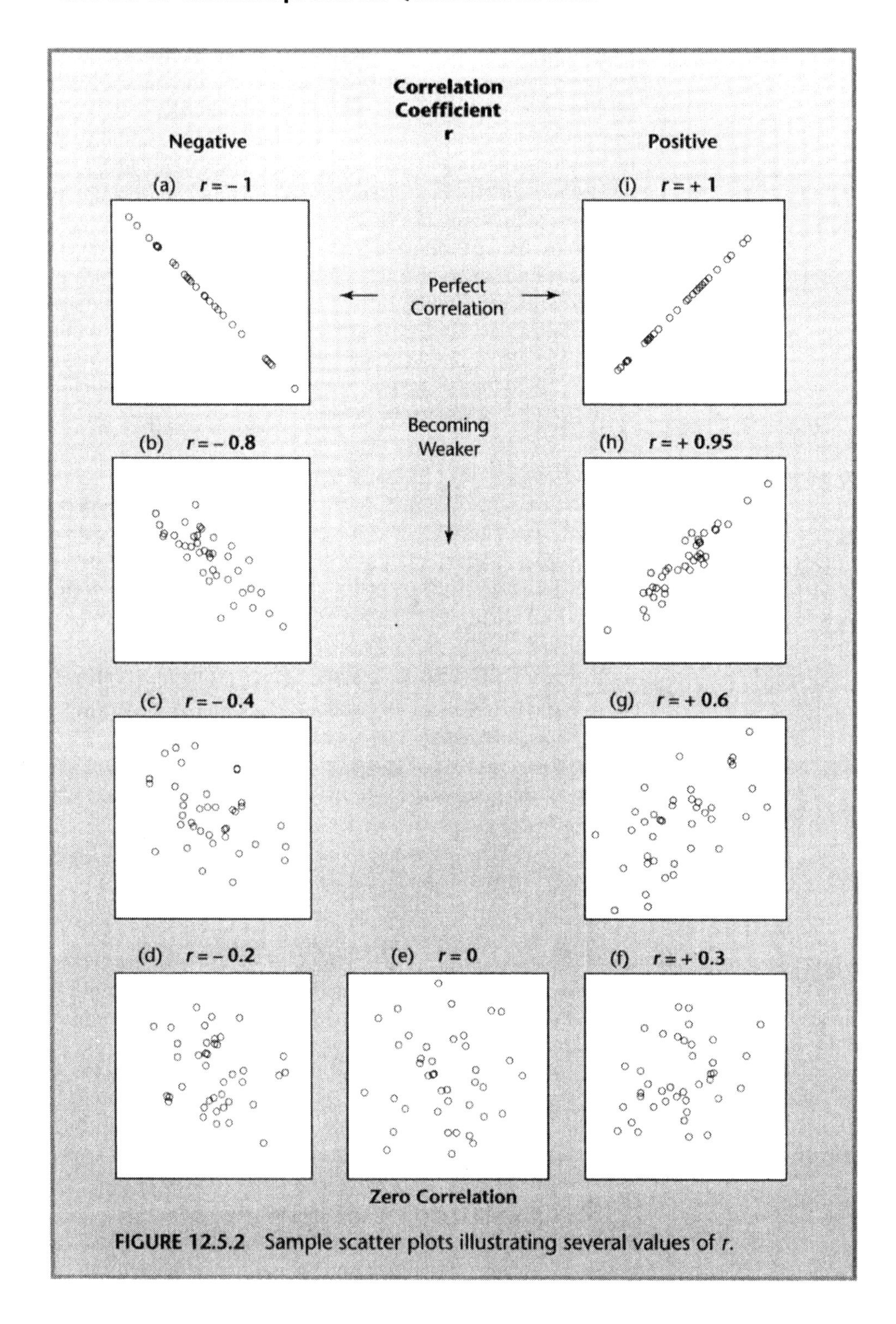

FIGURE 12.5.2 Sample scatter plots illustrating several values of r.

tells you is that you have strong evidence that the population correlation is not 0. To estimate how large the true correlation is, you need a confidence interval. Formal inference for correlation coefficients is outside the scope of this book, however.

Because of the relationship between correlation and regression, it is valid to test H_0: $\rho = 0$ by simply testing H_0: $\beta_1 = 0$, which we already know how to do. The preceding P-value is identical to the P-value Minitab prints for the slope in its regression output.

It should be emphasized that correlation coefficients should be calculated only when the trend is linear. Many researchers blindly calculate correlation coefficients for tens (or even hundreds) of pairs of variables without ever looking at a scatter plot. Figure 12.5.3 was designed to discourage this practice. Correlation coefficients should not be calculated for any of these plots because the primary impression is not that of a linear trend plus scatter. Each has an interesting story to tell that would be missed by anybody blindly calculating the correlation coefficient without first drawing a scatter plot. Figures 12.5.3(a)–(c) show patterns with $r = 0$ (i.e., no discernible upward or downward drift). The first is in fact an exact quadratic relationship. We could determine Y-values from X-values exactly and yet the two variables are

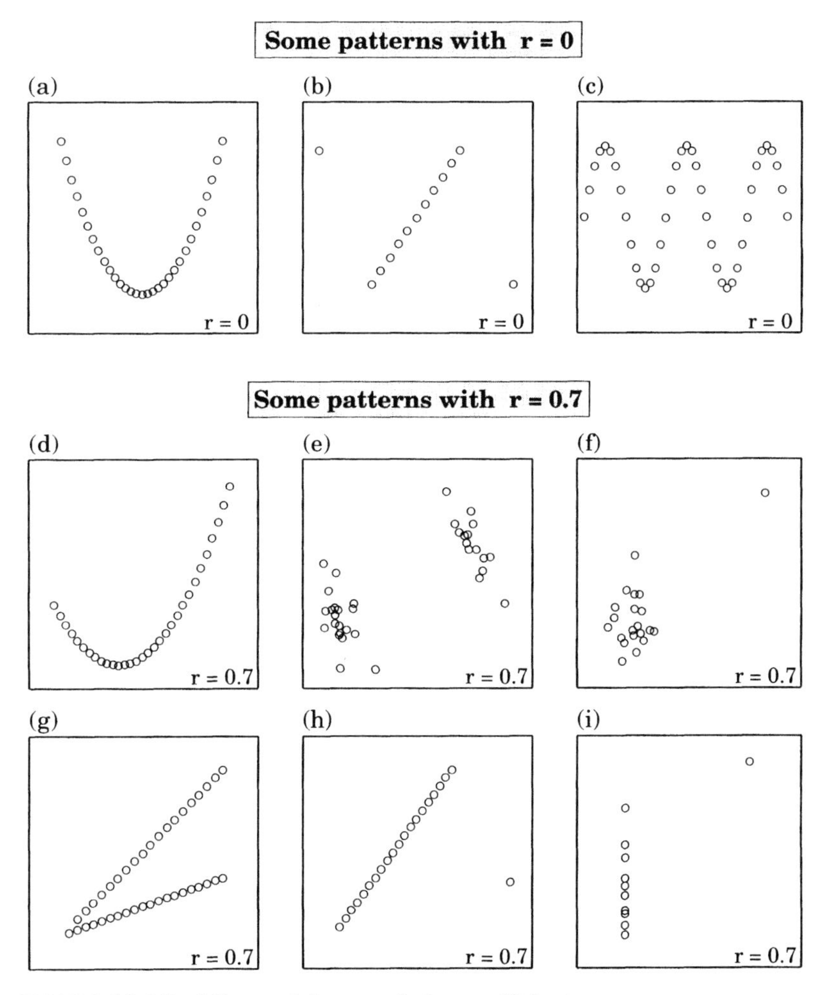

FIGURE 12.5.3 Misuse of the correlation coefficient.

"uncorrelated." Figure 12.5.3(c) is similar but comes from an exact sinusoidal relationship. Figure 12.5.3(b) could be part of a saw-tooth pattern or a linear relationship with two outliers.

Figures 12.5.3(d)-(i) all have correlation 0.7 but show very different features. Figure 12.5.3(d) is a portion of an exact quadratic. Figure 12.5.3(e) shows two clusters. Within each cluster, the trend is negative (downhill). The positive correlation coefficient is a result of the placing of the two clusters. Figure 12.5.3(f) is a single cluster, within which there is no correlation, and an outlier. Figure 12.5.3(g) shows two exact linear relationships, while Fig. 12.5.3(h) is a single line with an outlier. Finally, in Fig. 12.5.3(i) all impression of correlation is determined by a single point.

It is not easy to judge correlations from a scatter plot because changing the scale (such as stretching out one of the axes) can substantially affect the visual impression, even though r is unchanged. All plots in Fig. 12.5.2 had the variables standardized to have mean 0 and standard deviation 1. We will now give the sample correlation coefficients of the data used for some of the plots that you have seen in this chapter and Chapter 3: Fig. 12.2.1(a) $r = +0.97$; Fig. 12.2.1(b) $r = +0.94$; Fig. 3.1.5 $r = -0.94$; Fig. 3.1.6(a) $r = +0.96$; Fig. 12.5.1 $r = +0.30$; Fig. 12.1.1(a) $r = +0.83$; Fig. 3.1.13(a) $r = +0.82$; Fig. 3.1.8 $r = +0.5$; Fig. 12.4.1 $r = -0.91$; Fig. 12.4.8 (omitting obs. 35) $r = 0.998$.

Correlation and Causation

People falsely persist in thinking that a high correlation (i.e., a close to linear relationship) between two variables implies some sort of causal relationship, that is, that an increase in X actually causes a corresponding increase (or decrease) in Y. However,

Correlation does not necessarily imply causation.

This is just a restatement of our warning given earlier in Section 12.1.2 that simply finding a strong association between two variables does not demonstrate that the relationship is causal.

QUIZ ON SECTION 12.5

1. Describe a fundamental difference between the way regression treats data and the way correlation treats data.
2. What is the correlation coefficient intended to measure?
3. For what shape(s) of trend in a scatter plot does it make sense to calculate a correlation coefficient?
4. What is the meaning of a correlation coefficient of $r = +1$? $r = -1$? $r = 0$?

EXERCISES FOR SECTION 12.5

The data in Table 12.5.1 are a subset of a data set described in Seber [1984, pp. 120-123]. It consists of data on 27 boys aged 8-9 in the control group of a larger study. The variables are scores for auditory ASSOCIATION and auditory MEMORY given by a test called the Illinois Test of Psycholingual Ability (ITPA). We want to explore the relationship, if any, between these two variables.

TABLE 12.5.1 Audio Association and Auditory Memory

Association	20	44	26	19	28	37	42	16	31	41	46	20	48	24
Memory	38	39	42	40	48	43	36	33	40	51	50	38	54	49
Association	28	24	25	28	35	32	32	26	26	21	31	42	32	
Memory	43	32	42	15	51	36	47	38	33	31	41	44	36	

Source: Data courtesy of Peter Mullins, University of Auckland.

(a) Plot MEMORY versus ASSOCIATION. Does there seem to be a relationship?

(b) There is a clear outlier in these data. Is it visible when you look at the distribution of ASSOCIATION scores alone? in MEMORY scores alone?

(c) Calculate the correlation between ASSOCIATION score and MEMORY score both with and without the outlier.

(d) When people say that there is a "significant correlation," what do you think this statement means? What hypothesis is it related to?

(e) Omitting the outlier, check whether the correlation in (c) is "significant."

12.6 SUMMARY

12.6.1 Concepts

1. Relationships between quantitative variables should be explored using ***scatter plots.*** Usually the Y variable is continuous (or behaves as though it were in that there are few repeated values), and the X variable is discrete or continuous.

2. ***Regression*** singles out one variable (Y) as the response and uses the explanatory variable (X) to explain or predict its behavior. ***Correlation*** treats both variables symmetrically.

3. In practical problems, regression models may be fitted for any of the following reasons:
   - To understand a ***causal relationship*** better
   - To find relationships that may be ***causal***
   - To make ***predictions*** (Be cautious when predicting outside the range of the data.)
   - To ***test theories***
   - To ***estimate parameters*** in a theoretical model

4. In observational data, strong relationships are *not* necessarily causal. We can have reliable evidence of causation only from controlled experiments (see Section 12.1.2). Be aware of the possibility of ***lurking*** variables that may affect both X and Y.

5. Two important trend curves are the ***straight line*** and the ***exponential curve.***
   - A straight line changes by a *fixed amount* with each unit change in x.
   - An exponential curve changes by a *fixed percentage* with each unit change in x.

6. You should not let the questions you want to ask of your data be dictated by the tools you know how to use. You can always ask for help!

7. The two main approaches to summarizing trends in data are using *smoothers* and fitting mathematical curves.

8. The ***least-squares criterion*** for fitting a mathematical curve is to choose the values of the parameters (e.g., β_0 and β_1) to minimize the sum of squared prediction errors, $\sum(y_i - \widehat{y}_i)^2$.

12.6.2 Linear Relationships

1. We fit the linear relationship $\widehat{y} = \beta_0 + \beta_1 x$. The slope β_1 is the change in $\widehat{y}$ associated with a one-unit increase in x.

2. ***Least-squares estimates***
   - The least-squares estimates, $\widehat{\beta}_0$ and $\widehat{\beta}_1$, are chosen to minimize $\sum(y_i - \widehat{y}_i)^2$.
   - The ***least-squares regression line*** is $\widehat{y} = \widehat{\beta}_0 + \widehat{\beta}_1 x$.

3. ***Underlying model for statistical inference***
   Our theory assumes the model $Y_i = \beta_0 + \beta_1 x_i + U_i$, where the random errors, $U_1, U_2, \ldots, U_n$, are a random sample from a Normal(0, σ) distribution.
   This means that
   (i) the random errors are Normally distributed (with mean 0),
   (ii) the random errors all have the same standard deviation σ regardless of the value of x, and
   (iii) the random errors are all independent.

   These assumptions should be checked using residual plots (Section 12.4.4). The ith *residual* is $y_i - \widehat{y}_i$ = observed − predicted.

   An *outlier* is a data point with an unexpectedly large (positive or negative) residual.

4. ***Inferences for the intercept and slope*** are just as in Chapters 8 and 9, with confidence intervals being of the form *estimate* ± *t std errors* and test statistics of the form t_0 = (*estimate* − *hypothesized value*)/*standard error*.
   - We use ***df*** $= n - 2$.
   - To test for ***no linear association***, we test H_0: $\beta_1 = 0$.

5. ****Prediction***
   - The predicted value for a new Y at $X = x_p$ is $\widehat{y}_p = \widehat{\beta}_0 + \widehat{\beta}_1 x_p$.
   - The *confidence interval for the mean* estimates the *average* Y-value at $X = x_p$ (averaged over many repetitions of the experiment).
   - The *prediction interval* tries to predict the next *actual* Y-value.
   - The prediction interval is wider than the corresponding confidence interval for the mean.

6. ***The correlation coefficient r*** is a measure of linear association with $-1 \le r \le 1$.
   - If $r = 1$, then X and Y have a perfect positive linear relationship.
   - If $r = -1$, then X and Y have a perfect negative linear relationship.
   - If $r = 0$, then there is no linear relationship between X and Y.

   Correlation does not necessarily imply causation.

REVIEW EXERCISES 12

1. The following data come from some experiments carried out by an Auckland concrete manufacturer to determine in what way and to what extent the hardness of a batch of concrete depends on the amount of cement used in making it. Forty batches of concrete were made up with varying amounts of cement in the mix, and the hardness of each batch was measured after seven days. The data are given in Table 1 where CEMENT is measured in grams and HARDNESS in appropriate units.
   (a) Plot the data. What do you see?
   (b) Find the equation of the least-squares line, and superimpose it on your plot.

TABLE 1 Cement Hardness Data

Cement	Hardness	Cement	Hardness	Cement	Hardness
365	45.0	200	12.0	365	41.0
220	17.0	185	10.0	250	27.5
240	18.0	305	27.0	265	28.0
215	16.0	305	38.5	365	45.0
255	19.5	250	26.0	345	40.0
220	16.0	345	40.5	380	47.5
250	26.0	250	25.0	365	49.5
230	20.0	365	41.5	215	15.5
240	24.0	220	16.5	310	28.0
250	28.0	365	43.5	345	40.0
240	20.0	240	20.0	215	15.0
250	21.0	365	40.5	380	49.0
220	14.5	250	25.0	255	16.0
365	44.0				

Source: Data courtesy of Peter Mullins, University of Auckland.

(c) Why is the intercept not a meaningful quantity in this application?

(d) By how much would you estimate that the hardness would increase if the amount of cement used were increased by 20 g? (Give a single-number estimate.)

(e) Obtain a 95% confidence interval for the true slope. What does this tell you?

(f) Obtain a 95% prediction interval for the hardness of a new batch of concrete made in the same way with 275 g of cement.

(g) Would you expect batches of concrete made with 600 g of cement to be harder than those seen here? Justify your answer.

(h) Construct residual plots. Do you see anything that makes you doubt the applicability of the simple linear model to these data?

(i) The basic recipe for concrete is very simple. One mixes a small quantity of cement into a mixture of sand and stone chips, adds some water, and mixes the whole lot up. It is clear from the plot that batches of concrete with the same amount of cement can be quite different in their level of hardness. Why should this be so? Suggest some other possible causes of variability.

2. In 1993 several articles appeared in the press claiming that listening to Mozart raised one's IQ. The articles were based on a study published in a letter to the prestigious journal *Nature* (p. 611, 14 October 1993). In the study the performance on an IQ test of each subject was taken under three experimental conditions: after 10 minutes of listening to Mozart, after 10 minutes of listening to a relaxation tape, and after 10 minutes of silence. Thirty-six student volunteers were used, and each experienced all three conditions. Performance was "significantly higher" after listening to Mozart.

(a) What does this final phrase mean? How would a nonstatistical audience be likely to interpret it?

(b) Three different IQ tests were used. Why was this necessary?

The tests were a pattern-analysis test, a multiple-choice matrices test, and a multiple-choice paper-folding and -cutting test. The article stated the following: "For our sample, these three tasks correlated at the 0.01 level of significance. We were thus able to treat them as equal measures of abstract reasoning ability."

TABLE 2 Winning Times in Olympic Sprints (s)

	Men			Women		
Year	100 m	200 m	400 m	100 m	200 m	400 m
1896	12		54.2			
1900	11	22.2	49.4			
1904	11	21.6	49.2			
1908	10.8	22.6	50.0			
1912	10.8	21.7	48.2			
1920	10.8	22.0	49.6			
1924	10.6	21.6	47.6			
1928	10.8	21.8	47.8	12.2		
1932	10.3	21.2	46.2	11.9		
1936	10.3	20.7	46.5	11.5		
1948	10.3	21.1	46.2	11.9	24.4	
1952	10.4	20.7	45.9	11.5	23.7	
1956	10.5	20.6	46.7	11.5	23.4	
1960	10.2	20.5	44.9	11.0	24.0	
1964	10.0	20.3	45.1	11.4	23.0	52.0
1968	9.95	19.83	43.8	11.0	22.5	52.0
1972	10.14	20.00	44.6	11.07	22.40	51.08
1976	10.06	20.23	44.26	11.08	22.37	49.29
1980	10.25	20.19	44.60	11.60	22.03	48.88
1984	9.99	19.80	44.27	10.97	21.81	48.83
1988	9.92	19.75	43.87	10.54	21.34	48.65
1992	9.96	20.01	43.50	10.82	21.81	48.83
1996	9.84	19.32	43.49	10.94	22.12	48.25

(c) Critique the quotation.

(d) If you were conducting this study, how would you use the three different tests? In other words, what subjects would get which test when? Why would you do it this way?

(e) It was further claimed that the "pulse rates of the subjects did not change under any of the test conditions." What do you think this statement means?

3. Table 2 gives the winning times for the 100-m, 200-m, and 400-m running races for both men and women at Olympic Games up until 1996.

(a) It is well known that there has been a downward trend in the winning times for Olympic sprint races over the years. Do the trends appear linear? Is there any sign of a leveling-off in the winning times for any of these races over recent times?

(b) Compare graphically the winning times for a men's race and a corresponding women's race by plotting WINNING TIME versus YEAR with different plotting symbols or colors for men and for women. What do you see?

(c) Why must a linear model for Olympic winning times eventually fail?

(d) We will investigate the winning times for the men's 400 m.

(i) Plot the winning times versus year, and superimpose the least-squares regression line on the plot.

(ii) Obtain residual plots. Do you see anything that makes you doubt the applicability of the simple linear model to these data?

(iii) Reanalyze the data, omitting the data for the year 1896, and obtain a 95% prediction interval for the winning time of the men's 400 m in the Olympic Games in Sydney in the year 2000.

[Note: It is better to use something like *years since 1940* as the X-variable in your regression calculations in preference to *year* itself for computational reasons.]

(e) Repeat (d) for any other race that interests you.

(f) We can create a new variable, AVERAGE RUNNING SPEED = RACE DISTANCE/WINNING TIME. For each of the men's 100-m, 200-m, and 400-m races, plot AVERAGE RUNNING SPEED versus YEAR on the same graph using different plotting symbols to distinguish between the three types of race. What do you see?

(g) Repeat (f) for the women's 100-m, 200-m, and 400-m races.

TABLE 3 Pavement Condition Data

Age	14	16	12	18	15	13	13	12	11	15	15	15	12	18	13	13
PCI	61	56	75	56	54	57	55	60	68	67	59	59	58	49	54	62
Age	15	11	12	13	13	14	16	16	16	16	14	11	11	18	14	11
PCI	51	60	72	59	55	66	45	63	51	58	71	65	76	53	58	71

Source: Data courtesy of Prof. David Matthews, University of Waterloo.

TABLE 4 Olympic 100-meter Running Times (s)

React	Total	React	Total	React	Total	React	Total	React	Total
0.132	10.16	0.176	10.45	0.177	11.14	0.188	10.49	0.189	10.26
0.146	10.26	0.169	10.29	0.145	9.99	0.222	10.29	0.204	10.65
0.157	10.14	0.167	11.12	0.144	10.26	0.192	10.58	0.164	9.90
0.175	10.16	0.186	10.95	0.161	10.89	0.148	10.28	0.167	10.22
0.148	10.98	0.172	10.24	0.151	10.13	0.17	10.48	0.193	11.41
0.195	10.23	0.198	11.87	0.116	10.17	0.146	10.42	0.189	10.26

4. The data in Table 3 come from a study of road conditions in Canada. The variables given are the age of a portion of highway (in years) and PCI (pavement condition index), a composite measure of the quality of its surface that incorporates the effects of surface deterioration, potholes, cracks, undulations, and other engineering indicators of pavement deterioration.
   (a) Plot PCI versus AGE. What do you see?
   (b) Superimpose the least-squares regression line on your plot.
   (c) Investigate the fit of the simple linear model using residual plots.
   (d) Interpret both the *P*-value and the 95% confidence interval for the age effect (slope).
   (e) What does the interval for the intercept tell you about in this situation? Would you take this seriously here? Why or why not?
   (f) Obtain a 95% prediction interval for PCI for a road segment aged 20 years.
   (g) By looking at your plot in (a), do you think you could derive a useful rule for replacing highway based on age alone? Justify your answer.
   (h) One would expect road surface to deteriorate with age. What other variables would you expect to affect the quality of road surface?

5. How crucial is the start in the 100-m sprint at Olympic level? Twenty men's 100-m races were run at the 1996 Olympic Games at Atlanta (counting heats, semifinals, and the final), resulting in 168 finishes by the athletes. The data in Table 4 come from a sample of 30 of those finishes.[9] The variables are REACTION (reaction time in seconds, abbreviated to "react" in the table) and TOTAL (total time in seconds).
   (a) Plot TOTAL time versus REACTION time. What do you see?

   One might expect a relationship between TOTAL time and REACTION time since reaction time is incorporated in total time. Define RUNNING time[10] to be the difference: TOTAL time − REACTION time.

[9] The scatter plots for this sample closely resemble those for the whole data set.

[10] Using the full data set the authors investigated the effect of using running times rather than total times to determine the winners of the races. The identity of the winner would have changed in only 3 of the 20 races (the semifinals and finals were not affected). In those 3 races, the second-place runner would have won.

TABLE 5 Head Dimensions (cm) of First and Second Sons

1st	Length	191	195	181	183	176	208	189	197	188	192	179	183	174
1st	Breadth	155	149	148	153	144	157	150	159	152	150	158	147	150
2nd	Length	179	201	185	188	171	192	190	189	197	187	186	174	185
2nd	Breadth	145	152	149	149	142	152	149	152	159	151	148	147	152
1st	Length	190	188	163	195	186	181	175	192	174	176	197	190	
1st	Breadth	159	151	137	155	153	145	140	154	143	139	167	163	
2nd	Length	195	187	161	183	173	182	165	185	178	176	200	187	
2nd	Breadth	157	158	130	158	148	146	137	152	147	143	158	150	

Source: Frets [1921].

(b) Plot RUNNING time versus REACTION time. What do you see?

(c) Compute the correlation between **(i)** REACTION time and TOTAL time and **(ii)** REACTION time and RUNNING time. Compare these values. What is your impression from the analysis to date?

(d) Find the least-squares line for the regression of RUNNING time on REACTION time, and add it to your plot in (b).

(e) Test for a relationship between RUNNING time and REACTION time.

(f) Can you think of a possible explanation for the plot in (b)?

(g) What shape would you expect the plot of residuals versus reaction time to have for the regression in (d)? Obtain a plot of residuals versus X = REACTION and check out your expectations. Would you trust the least-squares analysis?

6. The data in Table 5 from Frets [1921] give the head dimensions in centimeters of the first and second adult sons for 25 families. This enables us to investigate one small aspect of familial resemblance.

(a) Plot the head length of the second son versus that of the first son. What do you see? Calculate the correlation coefficient. Repeat using the head breadths.

(b) For one of the plots in (a), the correlation coefficient fails to tell an important part of the story. Which one, and what is missing?

(c) Fit a simple linear regression model to predict the head length of the second son from that of the first.

(d) Interpret both the P-value and the 95% confidence interval for the slope.

(e) The P-value in (d) also tells us something about the correlation. What does it tell us?

*(f) Fit an appropriate model to the head breadths data.

(g) Suppose that instead of using lengths and breadths, you wanted to combine these variables to try to capture the ideas of "size" and "shape". What combination would you use for "size"? for "shape"?

(h) Explore the relationship between first and second sons in terms of your size variable. Repeat for your shape variable.

(i) Is there any evidence of differences in head length on average between first and second sons? Repeat using head breadths.

(j) How would you measure head breadths and head lengths? What difficulties would you anticipate?

7. The data in Table 6 originated from a car's journey log kept by a brother of Canadian statistics professor Jack Robinson over a 12-month period. The car was being used in Canada and the United States. Each time he filled his car with gas, he recorded these raw data:

- ODOMETER: number of kilometers the car had traveled to date
- DATE: the date

TABLE 6 Fuel Consumption Data

Obs.	Dist.	km/L	Days	Month	Obs.	Dist.	km/L	Days	Month	Obs.	Dist.	km/L	Days	Month
1	313	6.01	7	4	30	297	8.95	2	7	59	280	7.12	4	11
2	374	6.55	6	4	31	226	9.19	0	7	60	153	8.18	2	11
3	272	6.59	5	4	32	642	8.02	6	7	61	262	6.27	1	11
4	232	7.81	3	4	33	271	5.55	6	7	62	215	7.03	4	11
5	268	6.57	6	5	34	167	8.43	2	7	63	146	5.89	3	11
6	141	6.38	3	5	35	194	8.70	1	8	64	267	6.94	6	11
7	361	7.11	6	5	36	294	10.54	0	8	65	223	6.52	2	12
8	141	5.44	4	5	37	396	9.71	1	8	66	231	6.42	5	12
9	203	6.15	4	5	38	339	9.42	0	8	67	185	6.22	3	12
10	193	7.88	3	5	39	217	10.59	0	8	68	228	6.44	6	12
11	214	6.79	3	5	40	384	8.87	1	8	69	324	5.70	6	12
12	156	7.78	5	6	41	447	9.22	1	8	70	225	4.14	14	1
13	229	6.54	3	6	42	298	9.93	1	8	71	134	3.86	3	1
14	187	6.58	3	6	43	464	9.21	1	8	72	318	6.07	9	1
15	193	8.98	3	6	44	379	7.70	9	8	73	266	5.34	7	1
16	286	8.61	5	6	45	272	8.07	6	8	74	316	6.03	5	1
17	167	7.05	3	6	46	143	7.90	3	8	75	261	6.74	5	2
18	151	7.31	5	6	47	346	8.22	4	8	76	161	5.61	4	2
19	388	8.78	7	7	48	345	7.93	5	9	77	264	6.07	5	2
20	195	6.96	3	7	49	343	7.74	8	9	78	201	8.59	9	2
21	380	9.27	3	7	50	299	7.71	7	9	79	305	6.15	7	2
22	296	9.58	0	7	51	347	7.48	10	9	80	301	6.94	5	3
23	282	8.68	1	7	52	282	8.01	5	10	81	186	4.74	3	3
24	436	9.25	0	7	53	260	7.03	6	10	82	301	7.02	7	3
25	287	9.23	0	7	54	336	8.18	6	10	83	217	6.96	4	3
26	383	8.74	1	7	55	305	6.92	6	10	84	352	7.05	10	3
27	417	9.12	0	7	56	284	8.30	5	10	85	283	6.43	2	3
28	378	8.92	0	7	57	225	7.79	5	10	86	557	11.61	9	4
29	281	9.59	2	7	58	292	6.71	6	11					

Source: Data courtesy of Prof. Jack Robinson, University of Waterloo.

- LITERS: number of liters of fuel put into the car
- Some comments about servicing and so on

The data analyzed consist of variables derived from those preceding, namely

- DIST: distance traveled in kilometers since last fill-up
- KM.LTR: fuel consumption in kilometers traveled per liter
- DAYS: days since last fill-up
- MONTH: month of the year with values 1 = Jan, 2 = Feb, and so on
- LONG: for a long trip (prior fill-up less than two days before)

(a) Imagine you were keeping such a log. What types of errors would you be most likely to make?

(b) Explain how you would construct each variable given in the table from the information given in the log.

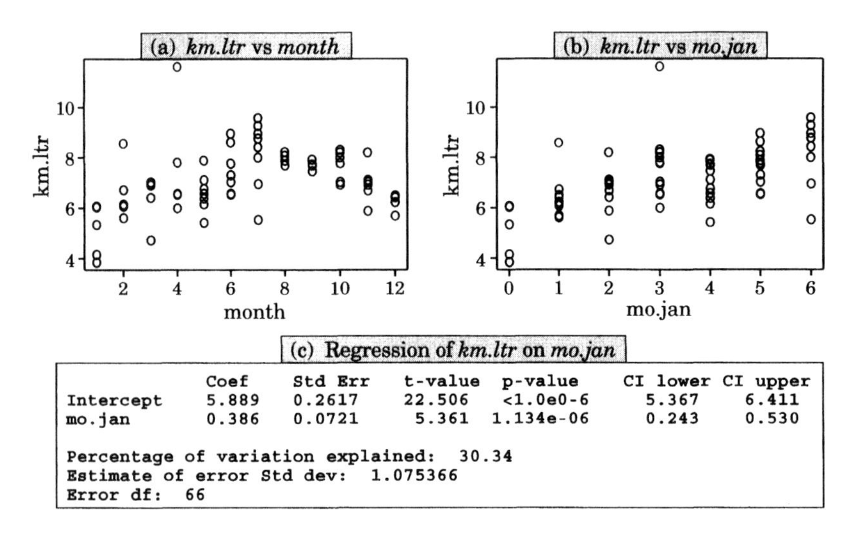

	Coef	Std Err	t-value	p-value	CI lower	CI upper
Intercept	5.889	0.2617	22.506	<1.0e0-6	5.367	6.411
mo.jan	0.386	0.0721	5.361	1.134e-06	0.243	0.530

```
Percentage of variation explained:  30.34
Estimate of error Std dev:  1.075366
Error df:  66
```

FIGURE 1 Fuel consumption data.

(c) Scan down the DAYS column. Do you notice anything, and can you think of an explanation?

(d) Figure 1(a) plots KM.LTR versus MONTH for short (i.e., not long) trips only. What do you see? Can you suggest a possible cause for the main pattern you see? Can you think of a possible cause for the worst outlier?

A new variable, MO.JAN, has been defined as the number of months away from January (MO.JAN is 0 for January, 1 for December and February, and so on).

(e) Figure 1(b) plots KM.LTR versus MO.JAN. Why might this be a reasonable way of looking at the data? (Hint: January is the coldest month of the winter in the Northern hemisphere.) What do you see?

Figure 1(c) gives computer output from a linear regression using MO.JAN to predict the response KM.LTR.

(f) Are you satisfied with the regression model we have fitted for predicting fuel consumption from "months away from January"? What other things would you want to do or check up on before using the model?

For parts (g)–(j) we will presume that the model is a good one.

(g) What quantity does the intercept tell you about? What does the confidence interval say about this quantity?

(h) What quantity does the slope tell you about? What do the data say about this quantity? (Interpret the confidence interval.)

(i) Is it clear that fuel consumption changes with month of the year? What tells you this?

*(j) Obtain a 95% prediction interval for a new measured fuel consumption for short trips in the month of June. Repeat for the month of August.

(k) Could you apply a one-way analysis of variance to the data in Fig. 1(b)? The F-test from the one-way ANOVA has $f_0 = 6.7$ and a vanishingly small P-value. What does this tell you? How is the regression model related to the model used in one-way ANOVA?

We tried a quadratic model regressing KM.LTR on both MO.JAN and MO.JAN^2. The line of the regression output relating to MO.JAN^2 gave coef = -0.059, se = 0.0415, t-value = -1.42, and P-value = 0.16.

(l) What does this information tell us?

(m) There were eight fill-ups on long trips (0 or 1 days since last fill) in July and nine in August. Find these data points and add them to Fig. 1(a). Does there seem to be a long-

TABLE 7 October Ozone Levels over Halley Bay Antarctica

Year	Column thickness	Year	Column thickness	Year	Column thickness	Year	Column thickness
1956	321	1967	323	1978	284	1989	164
1957	330	1968	301	1979	261	1990	179
1958	314	1969	282	1980	227	1991	155
1959	311	1970	282	1981	237	1992	142
1960	301	1971	299	1982	234	1993	111
1961	317	1972	304	1983	210	1994	124
1962	332	1973	289	1984	201	1995	129
1963	309	1974	274	1985	196	1996	139
1964	318	1975	308	1986	248		
1965	281	1976	283	1987	163		
1966	316	1977	251	1988	232		

Source: Data courtesy of Dr. J. D. Shanklin, British Antarctic Survey.

trip effect? Qualitatively, what sort of effect is it? What, if anything, is surprising about the augmented plot?

*8. The data in Table 7 consist of the mean October ozone column thickness (measured in Dobson units) measured at the British Antarctic Survey station at Halley Bay in Antarctica since 1956. It was dramatic data from Halley Bay that first alerted scientists to the recent development of severe ozone depletion, popularly known as the "ozone hole," in the atmosphere over Antarctica each spring.

Define TIME = *year* − 1956.

(a) Plot COLUMN THICKNESS versus TIME. What do you see?

(b) (i) Fit a quadratic curve in TIME to predict COLUMN THICKNESS.

(ii) Superimpose the quadratic curve on your graph from (a).

(iii) Using residual plots, check how well the quadratic model fits.

(iv) Use the quadratic model to obtain a 95% prediction interval for mean October column thickness of ozone in the year 2000.

(c) Why should you not trust the quadratic curve for predicting ozone levels before 1956?

9. Figure 2(a) plots student enrollment numbers in first-year statistics at the University of Auckland for 1980 to 1995. Statisticians hate graphs like this because they are so misleading. True,

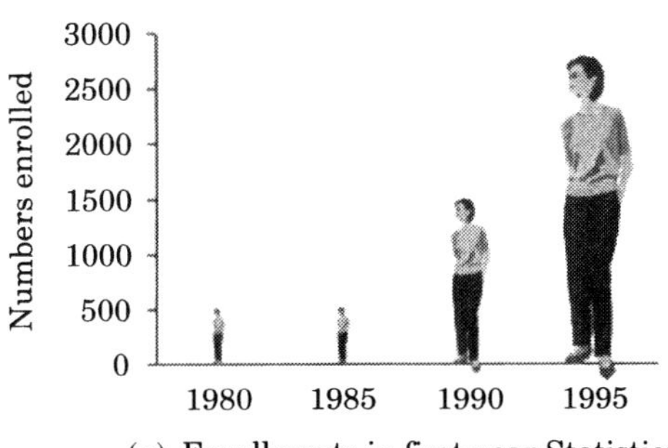

(a) Enrollments in first-year Statistics at Auckland

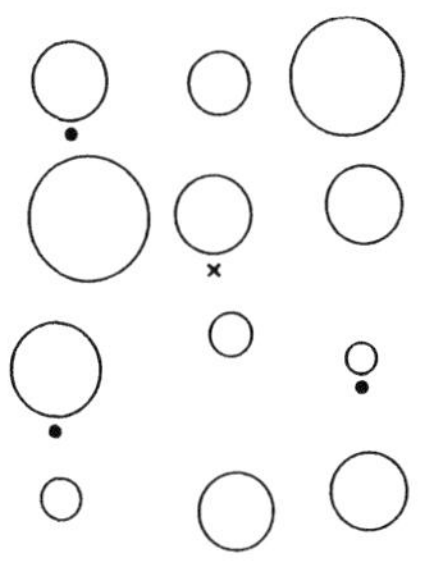

(b) A page of circles

FIGURE 2 Comparing sizes.

TABLE 8 Perceived Area of Circles Data

True area								
15.5	32.4	52.6	75.5	128.0	157.1	188.0	220.4	254.4
Student 1								
10 5	35 30	30 75	80 90	120 180	175 250	250 200	200 350	350 340
10 10	15 20	70 60	75 90	130 180	200 200	250 400	250 130	375 550
4 10	20 20	40 60	95 75	130 150	150 130	350 175	350 250	250 400
10 15	30 5	70 50	90 80	150 130	150 190	250 240	225 220	400 270
Student 4								
25 25	40 50	50 80	90 80	110 120	120 120	130 130	180 140	200 200
15 15	50 50	80 75	90 80	110 110	115 120	120 120	140 130	200 150
25 15	50 50	80 80	90 80	111 110	170 120	150 120	160 120	130 200
20 25	50 50	50 80	90 90	110 110	125 120	130 130	180 150	200 150
Student 8								
15 20	45 30	75 65	100 80	125 110	175 175	130 200	300 250	400 350
15 10	50 50	60 45	85 90	115 110	225 200	200 200	275 300	600 200
15 15	50 30	65 75	75 75	130 75	125 200	175 250	285 200	300 300
10 15	25 35	50 75	90 80	135 110	250 125	275 225	300 250	350 400
Student 10								
21 15	45 45	60 83	85 90	250 400	500 600	250 700	1000 945	1000 2000
15 10	56 45	80 63	96 87	125 225	600 350	500 460	1500 600	2000 2000
25 20	65 49	45 50	85 76	175 156	200 300	700 500	1020 800	200 800
15 20	35 20	50 75	90 87	425 110	500 400	1000 550	950 925	2000 4000

Source: Cleveland et al. [1982].

there has been substantial growth in statistics at Auckland recently, but not nearly as much as the picture suggests. For example, numbers almost doubled between 1990 and 1995, but the right-hand figure looks considerably more than twice as large. This is because when we enlarge the figure proportionately to double the height, we have a figure with four times the area, and roughly speaking, visual impression of size is proportional to area. But is it this simple?

Cleveland et al. [1982] conducted an experiment in which subjects were shown many different pages of circles like that in Fig. 2(b) (though much larger). One circle on each page, the *comparison circle,* was labeled with a cross. Three others, the *test circles,* were marked with a dot. The comparison circle was assigned the size 100. The subjects had to judge the size of each test circle relative to the comparison circle. Thus, if a subject allotted a score 300 to a circle, that meant the person had judged it to be three times as big as the comparison circle. Test circles in 9 sizes were used, and circles of each size appeared 8 times as test circles, resulting in 72 observations per subject.[11] Cleveland et al. [1982] used 14 scientists and 10 students. Table 8 gives the results for 4 of the students. For any given student, the 8 data points corresponding to a given true circle size are presented as a block 2 observations wide and 4 deep.

(a) On separate graphs, plot *perceived size* versus *true size* for each of the student subjects. On each graph, draw in the $y = x$ line. Why is this line relevant?

(b) What do you see? Comment both on general themes and features that are specific to particular students.

[11] We are only using the "nonmaplike stimulus" experimental condition in the paper.

(c) What conclusions should we draw about the wisdom of using areas to convey size in statistical graphs?

(d) Plot the logarithm of *perceived size* versus the logarithm of *true size* for Student 8. We will call this the log-log plot. What can you say about the shape of the plot?

*(e) Demonstrate mathematically that if perceived size is proportional to a power of the area (perceived size $= a\,\text{area}^b$), then that power is the slope of the log-log plot.

(f) Fit a linear model to the log-log data, and superimpose the least-squares line on your log-log plot from (d).

(g) Find a 95% confidence interval for the slope in (f). What does it tell you about the power of the area implicitly used by this student?

*(h) Find a 95% prediction interval for the perceived area of a circle with true area 300? (Use your linear model for the log-log data, and then convert the results back from a prediction for log(area) to a prediction for area.)

10. The data in Table 9, collected by McGill statistics professor Keith Worsley while on sabbatical leave in France, consists of YEAR, KM (distance traveled), and PRICE of 59 used Renault-5 cars advertised in a Toulouse newspaper one day in October 1985. Instead of YEAR we have worked with AGE calculated as (1985 − YEAR).

(a) Plot PRICE versus AGE and also log(PRICE) versus AGE. What do you notice? Does this suggest a model for the way that price changes as the car ages?

Table 10 gives some regression output from a computer program for a regression of log(PRICE) on AGE. The lower half of Table 10 gives the predicted value and 95% prediction intervals for log(PRICE) at each age up to 10 years.

TABLE 9 Renault-5 Car-Price Data

Obs.	Year	km[a]	Price[b]		Obs.	Year	km	Price		Obs.	Year	km	Price	
1	84	31	30	(h)[c]	21	82	38	20		41	76	22	7	(l)
2	81	86	13	(h)	22	78	104	10		42	80	28	30	
3	78	89	7.5		23	81	72	21		43	85	6	37	
4	84	21	27		24	85	2	48.5	(l)	44	85	9.5	45.5	
5	85	7.5	48		25	81	9	18	(l)	45	78	129	8	(h)
6	74	62	4		26	79	130	7	(h)	46	85	22	39.5	(h)
7	85	11	66		27	85	6.5	45		47	78	10	11	(l)
8	79	69	11		28	80	60	20		48	85	11	45	
9	82	80	34	(h)	29	85	11	45		49	84	31	32	(h)
10	79	125	12	(h)	30	75	100	6		50	76	8	13	(l)
11	76	15	5	(l)	31	82	26	32.5		51	81	88	18	(h)
12	80	72	20		32	82	60	26.5	(h)	52	75	100	6	
13	84	17	30		33	85	11	49		53	83	38	33.5	
14	83	42	46.5		34	83	50	33	(h)	54	79	38	10	
15	82	63	27	(h)	35	82	87	36	(h)	55	79	84	14	
16	78	10	11	(l)	36	85	16	40		56	81	60	19	
17	74	159	3		37	80	100	13	(h)	57	83	45	36	(h)
18	85	7	49.5		38	82	63	27	(h)	58	74	35	6	
19	85	7	55		39	84	21	38.5		59	78	75	8.5	
20	82	86	26	(h)	40	80	69	18						

[a]Distance traveled in thousands of kilometers.

[b]Price in thousands of French francs.

[c]Cars marked (h) have unusually high km for year; those with (l) have unusually low km.

Source: Data courtesy of Prof. Keith Worsley, McGill University.

TABLE 10 Regression of log(price) on AGE

	Coef	Std Err	t-value	p-value	CI lower	CI upper
Intercept	3.8511	0.0494	78.02	0	---	---
age	-0.2164	0.0095	-22.67	0	-0.24	-0.20

Percent of variation explained: 90.02
Estimate of error Std dev: 0.2433205
Error df: 57

Age	0	1	2	3	4	5	6	7	8	9	10
Predicted	3.85	3.63	3.42	3.20	2.99	2.77	----	2.34	2.12	1.90	1.69
Pred lower	3.35	3.14	2.93	2.71	2.49	2.28	----	1.84	1.62	1.40	1.18
Pred upper	4.35	4.13	3.91	3.69	3.48	3.26	----	2.83	2.62	2.40	2.19

(b) What does the value of the slope coefficient tell you?

(c) Calculate the missing confidence interval for the true intercept.

(d) Obtain the missing predicted value and 95% prediction interval for the log(PRICE) of a car with AGE = 6 years entering the market the following week.

Nobody is interested in the log(PRICE) of a car! To make the analysis understandable, we have to work out what the results say in terms of the prices themselves.

(e) What does the slope coefficient tell us about the relationship between the price of cars of a given age and the price of cars one year older? Apply the same sort of conversion to the confidence interval limits for the slope. What can you say on the basis of these calculations?

(f) Convert the predicted values for log(PRICE) in Table 10 into predicted values for actual PRICE. Apply the same type of conversion to the 95% prediction interval limits.

(g) Add these new predicted values and prediction intervals to your PRICE versus AGE graph. What do you notice?

(h) It is very tempting on the basis of the preceding analysis to make a statement like "Renault-5's lose about 20% of their value every year." What practical consideration, related to the way the data were collected, would make us hesitant to make such a statement?

A more thorough analysis would take into account the kilometers traveled by a car as well as its age.

(i) Plot KM versus AGE. What do you see?

(j) Add to the plot the line KM = 16 + 14 AGE, and also the line KM = 4 + 2.5 AGE.

Quite informally, using trial and error, we decided to classify those points falling above the former line as having unusually high kilometers for their age and those falling below the latter line as having unusually low kilometers.

(k) Does this classification seem reasonable? Would you prefer to have done it in some other way?

High-KM cars (according to our definition) are those labeled "(h)" in Table 9, and low-KM cars are labeled "(l)." Soon we are going to ask you to color high-KM points red and low-KM points blue in your plots in (a).

(l) What sort of pattern would you expect a strong kilometer effect to produce in the plots?

(m) Do the coloring. Is there an obvious kilometer effect?

(n) Keith bought a 1979 Renault-5 that had done 45,000 km for 16,000 francs. He thinks he probably paid too much for it. Plot this data point on your graph. Do you agree?

(o) What variables, other than age and kilometers traveled, would you expect to affect the value of a car? Is advertised price a good measure of the value of a car? Why or why not?

Further exercises, some involving large data sets, are available on our web site.

Appendixes

Statistical Tables

Appendix A1
RANDOM NUMBERS

This table contains blocks of random digits (see Section 1.5).

71578	81355	39007	60764	19852	87652	50354	22183	14935	09519	97820	86085
64215	70267	38804	00351	10065	27645	68107	21706	75781	04223	51761	05873
85586	70782	16377	09713	63471	93867	73473	22551	55024	09542	41617	83579
40922	52826	63529	38806	91458	67062	26370	30497	94358	69705	42120	88517
21169	11638	93363	00603	12371	39940	73570	56636	13663	47720	99518	97840
97248	78533	10849	86322	17818	95980	48813	92959	01602	62529	35575	24447
91623	07307	41478	61440	80083	73537	13436	99953	20202	99537	54308	78253
26508	84178	71632	65181	03772	61806	48375	16717	02711	56431	15463	48333
67108	16156	99133	67862	13572	23104	30784	06509	86187	41947	74514	85507
46400	89413	74507	49374	85117	99213	49557	98263	39623	38089	57890	71157
55333	97222	96104	65109	72010	50397	50996	63569	62204	40530	97192	73362
03521	10933	34889	13974	22627	89967	17772	37260	62576	01966	70333	12602
62387	33190	73580	15672	83265	94983	45401	59318	17077	13555	21212	89747
12965	44097	55481	22640	91748	57335	31097	24874	00267	84661	87853	61868
35560	83689	02626	32964	38791	02343	56275	25795	42090	54025	98377	70493
38905	93159	25252	29004	54972	73607	14151	57431	26736	88811	59533	33663
72329	44334	70842	93366	73428	80317	57771	45762	58234	73078	38946	36607
38868	89458	21217	89802	14210	84494	86517	25193	59752	42776	17976	02290
21633	28447	47136	39694	46548	54188	18153	02036	43884	84039	97730	17182
32215	44057	15706	55615	24632	77366	48561	17194	24541	32682	59173	17676
63669	73937	35964	17178	36804	50638	71972	73079	55746	69171	68684	59486
86383	50279	28234	09844	13578	28251	12708	24684	90170	80194	05569	32598
02687	70297	41900	36793	38314	03236	89243	88278	32991	39790	60135	93656
55269	99775	58393	41337	35439	57506	96195	34442	46408	26849	05553	25315
58841	72710	94082	43886	42436	28147	26488	86387	03710	24682	10587	01717
08382	43794	61510	16989	66863	09897	14697	00253	24678	41056	25886	13627
03316	27709	87263	60134	66341	43771	78054	60446	33196	17330	03423	11645
28247	07745	89325	69496	71694	88961	16079	17243	32843	98040	25866	50637
70111	40131	46657	39828	59286	72347	48453	02086	84718	69757	31580	37330
81281	10946	42393	09929	26565	41259	00055	75433	44947	97708	59872	63041
44647	09002	69021	93619	87392	86572	27661	00146	57180	93618	36382	05436
20837	85546	82953	14980	28104	25256	26545	98762	53874	52454	41547	84216
10973	04903	14270	93434	55394	46815	00642	92309	96336	99290	93273	13213
17508	59935	93902	02871	47830	91116	17804	57584	66385	88040	05182	43332
41921	48281	54456	48218	18232	03435	99630	18174	80514	00401	40216	67325
11362	35692	96337	90842	46843	62719	64049	17823	85899	11514	14557	48937
63394	27747	96239	22483	62553	45568	43041	96047	19285	73690	29780	88033
14168	49300	55727	25351	00206	07672	84073	01281	75177	92864	13769	41689
92814	04351	05957	09941	55423	03014	17425	36270	05099	03292	39526	01796
67234	45941	10747	25296	73782	34622	72824	06693	49142	97535	62240	54994

Table constructed by David Smith, University of Auckland.

Appendix A2
BINOMIAL DISTRIBUTION (INDIVIDUAL TERMS)

The tabulated value is $\text{pr}(X = x)$, where $X \sim \text{Binomial}(n, p)$.

For example, For $n = 5$, $p = 0.2$, $\text{pr}(X = 2) = 0.205$

For $n = 7$, $p = 0.3$, $\text{pr}(X = 5) = 0.025$

n	x	.01	.05	.10	.15	.20	.25	.30	.35	.40	.50	.60	.65	.70	.75	.80	.85	.90	.95	.99
2	0	.980	.902	.810	.723	.640	.563	.490	.422	.360	.250	.160	.122	.090	.063	.040	.023	.010	.003	
	1	.020	.095	.180	.255	.320	.375	.420	.455	.480	.500	.480	.455	.420	.375	.320	.255	.180	.095	.020
	2		.003	.010	.023	.040	.063	.090	.122	.160	.250	.360	.422	.490	.563	.640	.723	.810	.902	.980
3	0	.970	.857	.729	.614	.512	.422	.343	.275	.216	.125	.064	.043	.027	.016	.008	.003	.001		
	1	.029	.135	.243	.325	.384	.422	.441	.444	.432	.375	.288	.239	.189	.141	.096	.057	.027	.007	
	2		.007	.027	.057	.096	.141	.189	.239	.288	.375	.432	.444	.441	.422	.384	.325	.243	.135	.029
	3			.001	.003	.008	.016	.027	.043	.064	.125	.216	.275	.343	.422	.512	.614	.729	.857	.970
4	0	.961	.815	.656	.522	.410	.316	.240	.179	.130	.063	.026	.015	.008	.004	.002	.001			
	1	.039	.171	.292	.368	.410	.422	.412	.384	.346	.250	.154	.111	.076	.047	.026	.011	.004		
	2	.001	.014	.049	.098	.154	.211	.265	.311	.346	.375	.346	.311	.265	.211	.154	.098	.049	.014	.001
	3			.004	.011	.026	.047	.076	.111	.154	.250	.346	.384	.412	.422	.410	.368	.292	.171	.039
	4				.001	.002	.004	.008	.015	.026	.063	.130	.179	.240	.316	.410	.522	.656	.815	.961
5	0	.951	.774	.590	.444	.328	.237	.168	.116	.078	.031	.010	.005	.002	.001					
	1	.048	.204	.328	.392	.410	.396	.360	.312	.259	.156	.077	.049	.028	.015	.006	.002			
	2	.001	.021	.073	.138	.205	.264	.309	.336	.346	.313	.230	.181	.132	.088	.051	.024	.008	.001	
	3		.001	.008	.024	.051	.088	.132	.181	.230	.313	.346	.336	.309	.264	.205	.138	.073	.021	.001
	4				.002	.006	.015	.028	.049	.077	.156	.259	.312	.360	.396	.410	.392	.328	.204	.048
	5						.001	.002	.005	.010	.031	.078	.116	.168	.237	.328	.444	.590	.774	.951
6	0	.941	.735	.531	.377	.262	.178	.118	.075	.047	.016	.004	.002	.001						
	1	.057	.232	.354	.399	.393	.356	.303	.244	.187	.094	.037	.020	.010	.004	.002				
	2	.001	.031	.098	.176	.246	.297	.324	.328	.311	.234	.138	.095	.060	.033	.015	.005	.001		
	3		.002	.015	.041	.082	.132	.185	.235	.276	.313	.276	.235	.185	.132	.082	.041	.015	.002	
	4			.001	.005	.015	.033	.060	.095	.138	.234	.311	.328	.324	.297	.246	.176	.098	.031	.001
	5					.002	.004	.010	.020	.037	.094	.187	.244	.303	.356	.393	.399	.354	.232	.057
	6							.001	.002	.004	.016	.047	.075	.118	.178	.262	.377	.531	.735	.941
7	0	.932	.698	.478	.321	.210	.133	.082	.049	.028	.008	.002	.001							
	1	.066	.257	.372	.396	.367	.311	.247	.185	.131	.055	.017	.008	.004	.001					
	2	.002	.041	.124	.210	.275	.311	.318	.298	.261	.164	.077	.047	.025	.012	.004	.001			
	3		.004	.023	.062	.115	.173	.227	.268	.290	.273	.194	.144	.097	.058	.029	.011	.003		
	4			.003	.011	.029	.058	.097	.144	.194	.273	.290	.268	.227	.173	.115	.062	.023	.004	
	5				.001	.004	.012	.025	.047	.077	.164	.261	.298	.318	.311	.275	.210	.124	.041	.002
	6						.001	.004	.008	.017	.055	.131	.185	.247	.311	.367	.396	.372	.257	.066
	7								.001	.002	.008	.028	.049	.082	.133	.210	.321	.478	.698	.932

Table constructed by David Smith, University of Auckland.

Appendix A2
BINOMIAL DISTRIBUTION (INDIVIDUAL TERMS)

		p																		
n	*x*	.01	.05	.10	.15	.20	.25	.30	.35	.40	.50	.60	.65	.70	.75	.80	.85	.90	.95	.99
8	0	.923	.663	.430	.272	.168	.100	.058	.032	.017	.004	.001								
	1	.075	.279	.383	.385	.336	.267	.198	.137	.090	.031	.008	.003	.001						
	2	.003	.051	.149	.238	.294	.311	.296	.259	.209	.109	.041	.022	.010	.004	.001				
	3		.005	.033	.084	.147	.208	.254	.279	.279	.219	.124	.081	.047	.023	.009	.003			
	4			.005	.018	.046	.087	.136	.188	.232	.273	.232	.188	.136	.087	.046	.018	.005		
	5				.003	.009	.023	.047	.081	.124	.219	.279	.279	.254	.208	.147	.084	.033	.005	
	6					.001	.004	.010	.022	.041	.109	.209	.259	.296	.311	.294	.238	.149	.051	.003
	7							.001	.003	.008	.031	.090	.137	.198	.267	.336	.385	.383	.279	.075
	8									.001	.004	.017	.032	.058	.100	.168	.272	.430	.663	.923
9	0	.914	.630	.387	.232	.134	.075	.040	.021	.010	.002									
	1	.083	.299	.387	.368	.302	.225	.156	.100	.060	.018	.004	.001							
	2	.003	.063	.172	.260	.302	.300	.267	.216	.161	.070	.021	.010	.004	.001					
	3		.008	.045	.107	.176	.234	.267	.272	.251	.164	.074	.042	.021	.009	.003	.001			
	4		.001	.007	.028	.066	.117	.172	.219	.251	.246	.167	.118	.074	.039	.017	.005	.001		
	5			.001	.005	.017	.039	.074	.118	.167	.246	.251	.219	.172	.117	.066	.028	.007	.001	
	6				.001	.003	.009	.021	.042	.074	.164	.251	.272	.267	.234	.176	.107	.045	.008	
	7						.001	.004	.010	.021	.070	.161	.216	.267	.300	.302	.260	.172	.063	.003
	8								.001	.004	.018	.060	.100	.156	.225	.302	.368	.387	.299	.083
	9										.002	.010	.021	.040	.075	.134	.232	.387	.630	.914
10	0	.904	.599	.349	.197	.107	.056	.028	.013	.006	.001									
	1	.091	.315	.387	.347	.268	.188	.121	.072	.040	.010	.002	.001							
	2	.004	.075	.194	.276	.302	.282	.233	.176	.121	.044	.011	.004	.001						
	3		.010	.057	.130	.201	.250	.267	.252	.215	.117	.042	.021	.009	.003	.001				
	4		.001	.011	.040	.088	.146	.200	.238	.251	.205	.111	.069	.037	.016	.006	.001			
	5			.001	.008	.026	.058	.103	.154	.201	.246	.201	.154	.103	.058	.026	.008	.001		
	6				.001	.006	.016	.037	.069	.111	.205	.251	.238	.200	.146	.088	.040	.011	.001	
	7					.001	.003	.009	.021	.042	.117	.215	.252	.267	.250	.201	.130	.057	.010	
	8							.001	.004	.011	.044	.121	.176	.233	.282	.302	.276	.194	.075	.004
	9								.001	.002	.010	.040	.072	.121	.188	.268	.347	.387	.315	.091
	10										.001	.006	.013	.028	.056	.107	.197	.349	.599	.904
11	0	.895	.569	.314	.167	.086	.042	.020	.009	.004										
	1	.099	.329	.384	.325	.236	.155	.093	.052	.027	.005	.001								
	2	.005	.087	.213	.287	.295	.258	.200	.140	.089	.027	.005	.002	.001						
	3		.014	.071	.152	.221	.258	.257	.225	.177	.081	.023	.010	.004	.001					
	4		.001	.016	.054	.111	.172	.220	.243	.236	.161	.070	.038	.017	.006	.002				
	5			.002	.013	.039	.080	.132	.183	.221	.226	.147	.099	.057	.027	.010	.002			
	6				.002	.010	.027	.057	.099	.147	.226	.221	.183	.132	.080	.039	.013	.002		
	7					.002	.006	.017	.038	.070	.161	.236	.243	.220	.172	.111	.054	.016	.001	
	8						.001	.004	.010	.023	.081	.177	.225	.257	.258	.221	.152	.071	.014	
	9							.001	.002	.005	.027	.089	.140	.200	.258	.295	.287	.213	.087	.005
	10									.001	.005	.027	.052	.093	.155	.236	.325	.384	.329	.099
	11											.004	.009	.020	.042	.086	.167	.314	.569	.895

		p																		
n	*x*	.01	.05	.10	.15	.20	.25	.30	.35	.40	.50	.60	.65	.70	.75	.80	.85	.90	.95	.99
12	0	.886	.540	.282	.142	.069	.032	.014	.006	.002										
	1	.107	.341	.377	.301	.206	.127	.071	.037	.017	.003									
	2	.006	.099	.230	.292	.283	.232	.168	.109	.064	.016	.002	.001							
	3		.017	.085	.172	.236	.258	.240	.195	.142	.054	.012	.005	.001						
	4		.002	.021	.068	.133	.194	.231	.237	.213	.121	.042	.020	.008	.002	.001				
	5			.004	.019	.053	.103	.158	.204	.227	.193	.101	.059	.029	.011	.003	.001			
	6				.004	.016	.040	.079	.128	.177	.226	.177	.128	.079	.040	.016	.004			
	7				.001	.003	.011	.029	.059	.101	.193	.227	.204	.158	.103	.053	.019	.004		
	8					.001	.002	.008	.020	.042	.121	.213	.237	.231	.194	.133	.068	.021	.002	
	9							.001	.005	.012	.054	.142	.195	.240	.258	.236	.172	.085	.017	
	10								.001	.002	.016	.064	.109	.168	.232	.283	.292	.230	.099	.006
	11										.003	.017	.037	.071	.127	.206	.301	.377	.341	.107
	12											.002	.006	.014	.032	.069	.142	.282	.540	.886
15	0	.860	.463	.206	.087	.035	.013	.005	.002											
	1	.130	.366	.343	.231	.132	.067	.031	.013	.005										
	2	.009	.135	.267	.286	.231	.156	.092	.048	.022	.003									
	3		.031	.129	.218	.250	.225	.170	.111	.063	.014	.002								
	4		.005	.043	.116	.188	.225	.219	.179	.127	.042	.007	.002	.001						
	5		.001	.010	.045	.103	.165	.206	.212	.186	.092	.024	.010	.003	.001					
	6			.002	.013	.043	.092	.147	.191	.207	.153	.061	.030	.012	.003	.001				
	7				.003	.014	.039	.081	.132	.177	.196	.118	.071	.035	.013	.003	.001			
	8				.001	.003	.013	.035	.071	.118	.196	.177	.132	.081	.039	.014	.003			
	9					.001	.003	.012	.030	.061	.153	.207	.191	.147	.092	.043	.013	.002		
	10						.001	.003	.010	.024	.092	.186	.212	.206	.165	.103	.045	.010	.001	
	11							.001	.002	.007	.042	.127	.179	.219	.225	.188	.116	.043	.005	
	12									.002	.014	.063	.111	.170	.225	.250	.218	.129	.031	
	13										.003	.022	.048	.092	.156	.231	.286	.267	.135	.009
	14											.005	.013	.031	.067	.132	.231	.343	.366	.130
	15												.002	.005	.013	.035	.087	.206	.463	.860
20	0	.818	.358	.122	.039	.012	.003	.001												
	1	.165	.377	.270	.137	.058	.021	.007	.002											
	2	.016	.189	.285	.229	.137	.067	.028	.010	.003										
	3	.001	.060	.190	.243	.205	.134	.072	.032	.012	.001									
	4		.013	.090	.182	.218	.190	.130	.074	.035	.005									
	5		.002	.032	.103	.175	.202	.179	.127	.075	.015	.001								
	6			.009	.045	.109	.169	.192	.171	.124	.037	.005	.001							
	7			.002	.016	.055	.112	.164	.184	.166	.074	.015	.004	.001						
	8				.005	.022	.061	.114	.161	.180	.120	.035	.014	.004	.001					
	9				.001	.007	.027	.065	.116	.160	.160	.071	.034	.012	.003					
	10					.002	.010	.031	.069	.117	.176	.117	.069	.031	.010	.002				
	11						.003	.012	.034	.071	.160	.160	.116	.065	.027	.007	.001			
	12						.001	.004	.014	.035	.120	.180	.161	.114	.061	.022	.005			
	13							.001	.004	.015	.074	.166	.184	.164	.112	.055	.016	.002		
	14								.001	.005	.037	.124	.171	.192	.169	.109	.045	.009		
	15									.001	.015	.075	.127	.179	.202	.175	.103	.032	.002	
	16										.005	.035	.074	.130	.190	.218	.182	.090	.013	
	17										.001	.012	.032	.072	.134	.205	.243	.190	.060	.001
	18											.003	.010	.028	.067	.137	.229	.285	.189	.016
	19												.002	.007	.021	.058	.137	.270	.377	.165
	20													.001	.003	.012	.039	.122	.358	.818

Appendix A3
MINIMUM SAMPLE SIZES

Minimum sample sizes for large sample confidence intervals for a proportion based on the Normal approximation

$\hat{p}$	Minimum n			$\hat{p}$	Minimum n		
	15% rule	10% rule	5% rule		15% rule	10% rule	5% rule
0.01	2400	5400	21400	0.51	10	10	27
0.02	1200	2650	10400	0.52	10	10	27
0.03	767	1700	6700	0.53	11	11	28
0.04	550	1225	4875	0.54	11	11	28
0.05	440	960	3760	0.55	11	11	29
0.06	350	767	3033	0.56	11	11	30
0.07	286	643	2514	0.57	12	12	30
0.08	238	538	2125	0.58	12	12	31
0.09	211	456	1811	0.59	12	12	37
0.1	180	400	1570	0.60	13	13	40
0.11	155	345	1373	0.61	13	13	49
0.12	142	308	1208	0.62	13	16	58
0.13	123	269	1069	0.63	14	19	68
0.14	107	243	957	0.64	14	22	78
0.15	100	220	853	0.65	14	23	91
0.16	88	194	763	0.66	15	29	103
0.17	82	176	688	0.67	18	33	121
0.18	72	156	617	0.68	19	38	138
0.19	63	142	553	0.69	19	39	152
0.20	60	125	500	0.70	20	47	170
0.21	52	114	452	0.71	24	52	193
0.22	50	105	409	0.72	25	57	218
0.23	43	96	370	0.73	30	63	241
0.24	38	83	333	0.74	31	69	269
0.25	36	76	296	0.75	36	76	296
0.26	31	69	269	0.76	38	83	333
0.27	30	63	241	0.77	43	96	370
0.28	25	57	218	0.78	50	105	409
0.29	24	52	193	0.79	52	114	452
0.30	20	47	170	0.80	60	125	500
0.31	19	39	152	0.81	63	142	553
0.32	19	38	138	0.82	72	156	617
0.33	18	33	121	0.83	82	176	688
0.34	15	29	103	0.84	88	194	763
0.35	14	23	91	0.85	100	220	853
0.36	14	22	78	0.86	107	243	957
0.37	14	19	68	0.87	123	269	1069
0.38	13	16	58	0.88	142	308	1208
0.39	13	13	49	0.89	155	345	1373
0.40	13	13	40	0.90	180	400	1570
0.41	12	12	37	0.91	211	456	1811
0.42	12	12	31	0.92	238	538	2125
0.43	12	12	30	0.93	286	643	2514
0.44	11	11	30	0.94	350	767	3033
0.45	11	11	29	0.95	440	960	3760
0.46	11	11	28	0.96	550	1225	4875
0.47	11	11	28	0.97	767	1700	6700
0.48	10	10	27	0.98	1200	2650	10400
0.49	10	10	27	0.99	2400	5400	21400
0.50	10	10	26				

[For $\hat{p} < 0.01$ the 10% rule gives a count of 56 ($= n\hat{p}$).]

Source: Derived from Table 1 of Samuels and Lu [1992].

Appendix A4
STANDARD NORMAL DISTRIBUTION

Negative z-Values

For given z, the tabulated value is $prob = \text{pr}(Z \leq z)$ where $Z \sim \text{Normal}(0, 1)$.

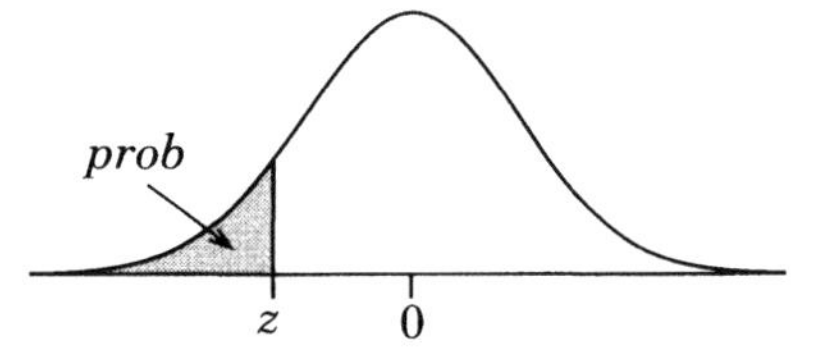

z	0	1	2	3	4	5	6	7	8	9
	Second decimal place of z									
−3.4	.000	.000	.000	.000	.000	.000	.000	.000	.000	.000
−3.3	.000	.000	.000	.000	.000	.000	.000	.000	.000	.000
−3.2	.001	.001	.001	.001	.001	.001	.001	.001	.001	.001
−3.1	.001	.001	.001	.001	.001	.001	.001	.001	.001	.001
−3.0	.001	.001	.001	.001	.001	.001	.001	.001	.001	.001
−2.9	.002	.002	.002	.002	.002	.002	.002	.001	.001	.001
−2.8	.003	.002	.002	.002	.002	.002	.002	.002	.002	.002
−2.7	.003	.003	.003	.003	.003	.003	.003	.003	.003	.003
−2.6	.005	.005	.004	.004	.004	.004	.004	.004	.004	.004
−2.5	.006	.006	.006	.006	.006	.005	.005	.005	.005	.005
−2.4	.008	.008	.008	.008	.007	.007	.007	.007	.007	.006
−2.3	.011	.010	.010	.010	.010	.009	.009	.009	.009	.008
−2.2	.014	.014	.013	.013	.013	.012	.012	.012	.011	.011
−2.1	.018	.017	.017	.017	.016	.016	.015	.015	.015	.014
−2.0	.023	.022	.022	.021	.021	.020	.020	.019	.019	.018
−1.9	.029	.028	.027	.027	.026	.026	.025	.024	.024	.023
−1.8	.036	.035	.034	.034	.033	.032	.031	.031	.030	.029
−1.7	.045	.044	.043	.042	.041	.040	.039	.038	.038	.037
−1.6	.055	.054	.053	.052	.051	.049	.048	.047	.046	.046
−1.5	.067	.066	.064	.063	.062	.061	.059	.058	.057	.056
−1.4	.081	.079	.078	.076	.075	.074	.072	.071	.069	.068
−1.3	.097	.095	.093	.092	.090	.089	.087	.085	.084	.082
−1.2	.115	.113	.111	.109	.107	.106	.104	.102	.100	.099
−1.1	.136	.133	.131	.129	.127	.125	.123	.121	.119	.117
−1.0	.159	.156	.154	.152	.149	.147	.145	.142	.140	.138
−0.9	.184	.181	.179	.176	.174	.171	.169	.166	.164	.161
−0.8	.212	.209	.206	.203	.200	.198	.195	.192	.189	.187
−0.7	.242	.239	.236	.233	.230	.227	.224	.221	.218	.215
−0.6	.274	.271	.268	.264	.261	.258	.255	.251	.248	.245
−0.5	.309	.305	.302	.298	.295	.291	.288	.284	.281	.278
−0.4	.345	.341	.337	.334	.330	.326	.323	.319	.316	.312
−0.3	.382	.378	.374	.371	.367	.363	.359	.356	.352	.348
−0.2	.421	.417	.413	.409	.405	.401	.397	.394	.390	.386
−0.1	.460	.456	.452	.448	.444	.440	.436	.433	.429	.425
−0.0	.500	.496	.492	.488	.484	.480	.476	.472	.468	.464

[For $z < -3.4$, $\text{pr}(Z \leq z) = 0.000$ to three decimal places.]

Table constructed by David Smith, University of Auckland.

Positive z-Values

For given z, the tabulated value is $prob = \text{pr}(Z \leq z)$ where $Z \sim \text{Normal}(0, 1)$.

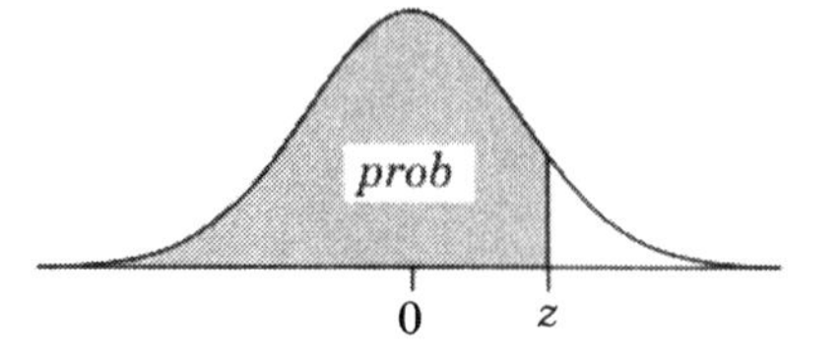

	Second decimal place of z									
z	0	1	2	3	4	5	6	7	8	9
0.0	.500	.504	.508	.512	.516	.520	.524	.528	.532	.536
0.1	.540	.544	.548	.552	.556	.560	.564	.567	.571	.575
0.2	.579	.583	.587	.591	.595	.599	.603	.606	.610	.614
0.3	.618	.622	.626	.629	.633	.637	.641	.644	.648	.652
0.4	.655	.659	.663	.666	.670	.674	.677	.681	.684	.688
0.5	.691	.695	.698	.702	.705	.709	.712	.716	.719	.722
0.6	.726	.729	.732	.736	.739	.742	.745	.749	.752	.755
0.7	.758	.761	.764	.767	.770	.773	.776	.779	.782	.785
0.8	.788	.791	.794	.797	.800	.802	.805	.808	.811	.813
0.9	.816	.819	.821	.824	.826	.829	.831	.834	.836	.839
1.0	.841	.844	.846	.848	.851	.853	.855	.858	.860	.862
1.1	.864	.867	.869	.871	.873	.875	.877	.879	.881	.883
1.2	.885	.887	.889	.891	.893	.894	.896	.898	.900	.901
1.3	.903	.905	.907	.908	.910	.911	.913	.915	.916	.918
1.4	.919	.921	.922	.924	.925	.926	.928	.929	.931	.932
1.5	.933	.934	.936	.937	.938	.939	.941	.942	.943	.944
1.6	.945	.946	.947	.948	.949	.951	.952	.953	.954	.954
1.7	.955	.956	.957	.958	.959	.960	.961	.962	.962	.963
1.8	.964	.965	.966	.966	.967	.968	.969	.969	.970	.971
1.9	.971	.972	.973	.973	.974	.974	.975	.976	.976	.977
2.0	.977	.978	.978	.979	.979	.980	.980	.981	.981	.982
2.1	.982	.983	.983	.983	.984	.984	.985	.985	.985	.986
2.2	.986	.986	.987	.987	.987	.988	.988	.988	.989	.989
2.3	.989	.990	.990	.990	.990	.991	.991	.991	.991	.992
2.4	.992	.992	.992	.992	.993	.993	.993	.993	.993	.994
2.5	.994	.994	.994	.994	.994	.995	.995	.995	.995	.995
2.6	.995	.995	.996	.996	.996	.996	.996	.996	.996	.996
2.7	.997	.997	.997	.997	.997	.997	.997	.997	.997	.997
2.8	.997	.998	.998	.998	.998	.998	.998	.998	.998	.998
2.9	.998	.998	.998	.998	.998	.998	.998	.999	.999	.999
3.0	.999	.999	.999	.999	.999	.999	.999	.999	.999	.999
3.1	.999	.999	.999	.999	.999	.999	.999	.999	.999	.999
3.2	.999	.999	.999	.999	.999	.999	.999	.999	.999	.999
3.3	1.00	1.00	1.00	1.00	1.00	1.00	1.00	1.00	1.00	1.00
3.4	1.00	1.00	1.00	1.00	1.00	1.00	1.00	1.00	1.00	1.00

[For $z > 3.4$, $\text{pr}(Z \leq z) = 1.000$ to three decimal places.]

Table constructed by David Smith, University of Auckland.

Appendix A5
ADDITIONAL STANDARD NORMAL TABLES

Z is Normal($\mu = 0, \sigma = 1$).

Appendix A5.1 Central Probabilities

For given *prob*, the tabulated value is z such that pr($-z \leq Z \leq z$) = *prob*.

For example, for *prob* = 0.9, we read z = 1.6449, telling us that 0.9 = pr($-1.6449 \leq Z \leq 1.6449$), which is interpreted for arbitrary Normal distributions as, "The probability of falling within 1.6449 standard deviations on either side of the mean is 0.9."

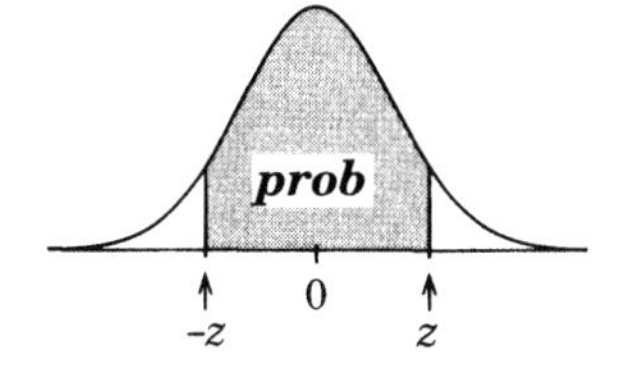

prob	0.50	0.70	0.75	0.80	0.85
z	0.6745	1.0364	1.1503	1.2816	1.4395
prob	0.90	0.95	0.98	0.99	0.999
z	1.6449	1.9600	2.3263	2.5758	3.2905

Appendix A5.2 Inverse Tail Probabilities

For given *prob*, the tabulated value is z such that pr($Z \geq z$) = *prob* and pr($Z \leq -z$) = *prob*.

For example, for *prob* = 0.05, we read $z(prob)$ = 1.6449, telling us that 0.05 = pr($Z \geq 1.6449$) and 0.05 = pr($Z \leq -1.6449$). 0.05 = pr($Z \geq 1.6449$) is interpreted for arbitrary Normal distributions as, "The probability of obtaining a value greater than 1.6449 standard deviations above the mean is 0.05."

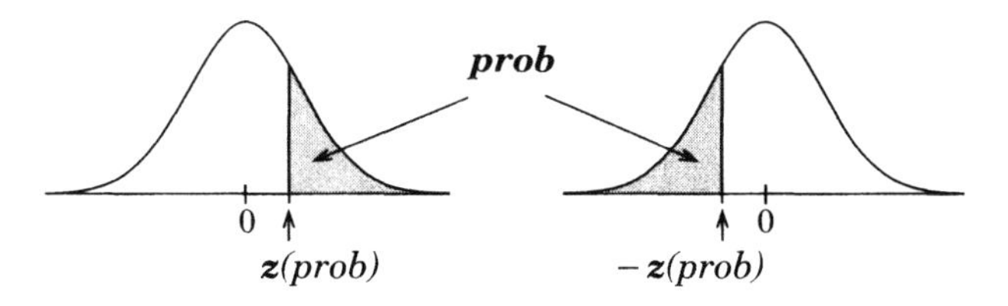

Fraction	(1/4)	(1/5)	(3/20)	(1/10)	(1/20)	(1/40)
prob	0.25	0.20	0.15	0.10	0.05	0.025
z(prob)	0.6745	0.8416	1.0364	1.2816	1.6449	1.9600
Fraction	(1/100)	(1/200)	(1/400)	(1/1000)	(1/2000)	(1/10,000)
prob	0.01	0.005	0.0025	0.001	0.0005	0.0001
z(prob)	2.3263	2.5758	2.8071	3.0902	3.2905	3.7195

Appendix A6
STUDENT'S *t*-DISTRIBUTION

For fixed *prob* and *df*, the tabulated value is the number $t = t_{df}(prob)$ such that for $T \sim$ Student(df), pr($T \geq t$) = *prob*.

For example, for $prob = 0.025$ and $df = 23$, $t_{23}(0.025) = 2.069$. A simple guideline when *df* is not in the table is to use the tabulated value immediately smaller than that required. When $df \geq 120$, use $df = \infty$.

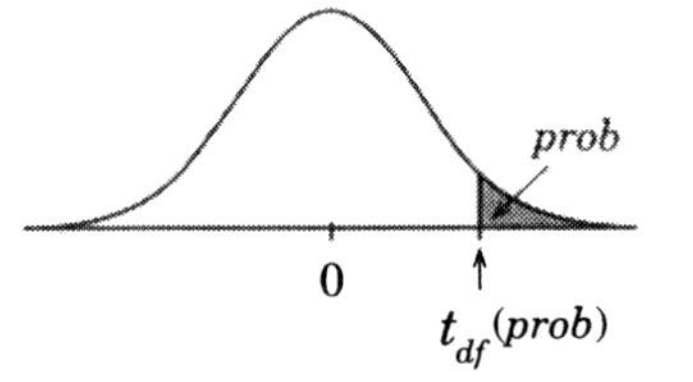

	prob									
df	.20	.15	.10	.05	.025	.01	.005	.001	.0005	.0001
3	0.978	1.250	1.638	2.353	3.182	4.541	5.841	10.21	12.92	22.20
4	0.941	1.190	1.533	2.132	2.776	3.747	4.604	7.173	8.610	13.03
5	0.920	1.156	1.476	2.015	2.571	3.365	4.032	5.893	6.869	9.678
6	0.906	1.134	1.440	1.943	2.447	3.143	3.707	5.208	5.959	8.025
7	0.896	1.119	1.415	1.895	2.365	2.998	3.499	4.785	5.408	7.063
8	0.889	1.108	1.397	1.860	2.306	2.896	3.355	4.501	5.041	6.442
9	0.883	1.100	1.383	1.833	2.262	2.821	3.250	4.297	4.781	6.010
10	0.879	1.093	1.372	1.812	2.228	2.764	3.169	4.144	4.587	5.694
11	0.876	1.088	1.363	1.796	2.201	2.718	3.106	4.025	4.437	5.453
12	0.873	1.083	1.356	1.782	2.179	2.681	3.055	3.930	4.318	5.263
13	0.870	1.079	1.350	1.771	2.160	2.650	3.012	3.852	4.221	5.111
14	0.868	1.076	1.345	1.761	2.145	2.624	2.977	3.787	4.140	4.985
15	0.866	1.074	1.341	1.753	2.131	2.602	2.947	3.733	4.073	4.880
16	0.865	1.071	1.337	1.746	2.120	2.583	2.921	3.686	4.015	4.791
17	0.863	1.069	1.333	1.740	2.110	2.567	2.898	3.646	3.965	4.714
18	0.862	1.067	1.330	1.734	2.101	2.552	2.878	3.610	3.922	4.648
19	0.861	1.066	1.328	1.729	2.093	2.539	2.861	3.579	3.883	4.590
20	0.860	1.064	1.325	1.725	2.086	2.528	2.845	3.552	3.849	4.539
21	0.859	1.063	1.323	1.721	2.080	2.518	2.831	3.527	3.819	4.493
22	0.858	1.061	1.321	1.717	2.074	2.508	2.819	3.505	3.792	4.452
23	0.858	1.060	1.319	1.714	2.069	2.500	2.807	3.485	3.768	4.415
24	0.857	1.059	1.318	1.711	2.064	2.492	2.797	3.467	3.745	4.382
25	0.856	1.058	1.316	1.708	2.060	2.485	2.787	3.450	3.725	4.352
26	0.856	1.058	1.315	1.706	2.056	2.479	2.779	3.435	3.707	4.324
27	0.855	1.057	1.314	1.703	2.052	2.473	2.771	3.421	3.690	4.299
28	0.855	1.056	1.313	1.701	2.048	2.467	2.763	3.408	3.674	4.275
29	0.854	1.055	1.311	1.699	2.045	2.462	2.756	3.396	3.659	4.254
30	0.854	1.055	1.310	1.697	2.042	2.457	2.750	3.385	3.646	4.234
31	0.853	1.054	1.309	1.696	2.040	2.453	2.744	3.375	3.633	4.215
32	0.853	1.054	1.309	1.694	2.037	2.449	2.738	3.365	3.622	4.198
33	0.853	1.053	1.308	1.692	2.035	2.445	2.733	3.356	3.611	4.182
34	0.852	1.052	1.307	1.691	2.032	2.441	2.728	3.348	3.601	4.167
35	0.852	1.052	1.306	1.690	2.030	2.438	2.724	3.340	3.591	4.153
36	0.852	1.052	1.306	1.688	2.028	2.434	2.719	3.333	3.582	4.140
37	0.851	1.051	1.305	1.687	2.026	2.431	2.715	3.326	3.574	4.127
38	0.851	1.051	1.304	1.686	2.024	2.429	2.712	3.319	3.566	4.116
39	0.851	1.050	1.304	1.685	2.023	2.426	2.708	3.313	3.558	4.105
40	0.851	1.050	1.303	1.684	2.021	2.423	2.704	3.307	3.551	4.094
45	0.850	1.049	1.301	1.679	2.014	2.412	2.690	3.281	3.520	4.049
50	0.849	1.047	1.299	1.676	2.009	2.403	2.678	3.261	3.496	4.014
60	0.848	1.045	1.296	1.671	2.000	2.390	2.660	3.232	3.460	3.962
80	0.846	1.043	1.292	1.664	1.990	2.374	2.639	3.195	3.416	3.899
100	0.845	1.042	1.290	1.660	1.984	2.364	2.626	3.174	3.390	3.861
∞	0.842	1.036	1.282	1.645	1.960	2.326	2.576	3.090	3.291	3.719

Table constructed by David Smith, University of Auckland.

Appendix A7
CHI-SQUARE DISTRIBUTION

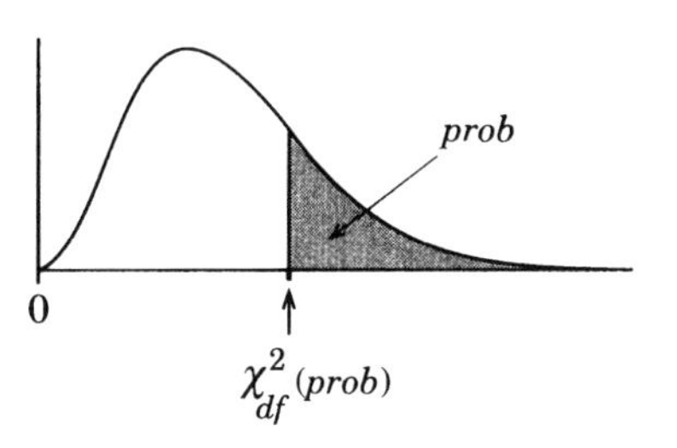

For fixed *prob* and *df*, the tabulated value is the number $\chi^2 = \chi^2_{df}(prob)$ such that for $X^2 \sim$ Chi-square(df), $\text{pr}(X^2 \geq \chi^2) = prob$. For example, for $prob = 0.10$ and $df = 13$, $\chi^2_{13}(0.10) = 19.81$.

	prob									
df	.975	.95	.20	.15	.10	.05	.025	.01	.005	.001
1	0.001	0.004	1.642	2.072	2.706	3.841	5.024	6.635	7.879	10.83
2	0.051	0.103	3.219	3.794	4.605	5.991	7.378	9.210	10.60	13.82
3	0.216	0.352	4.642	5.317	6.251	7.815	9.348	11.34	12.84	16.27
4	0.484	0.711	5.989	6.745	7.779	9.488	11.14	13.28	14.86	18.47
5	0.831	1.145	7.289	8.115	9.236	11.07	12.83	15.09	16.75	20.52
6	1.237	1.635	8.558	9.446	10.64	12.59	14.45	16.81	18.55	22.46
7	1.690	2.167	9.803	10.75	12.02	14.07	16.01	18.48	20.28	24.32
8	2.180	2.733	11.03	12.03	13.36	15.51	17.53	20.09	21.95	26.12
9	2.700	3.325	12.24	13.29	14.68	16.92	19.02	21.67	23.59	27.88
10	3.247	3.940	13.44	14.53	15.99	18.31	20.48	23.21	25.19	29.59
11	3.816	4.575	14.63	15.77	17.28	19.68	21.92	24.72	26.76	31.26
12	4.404	5.226	15.81	16.99	18.55	21.03	23.34	26.22	28.30	32.91
13	5.009	5.892	16.98	18.20	19.81	22.36	24.74	27.69	29.82	34.53
14	5.629	6.571	18.15	19.41	21.06	23.68	26.12	29.14	31.32	36.12
15	6.262	7.261	19.31	20.60	22.31	25.00	27.49	30.58	32.80	37.70
16	6.908	7.962	20.47	21.79	23.54	26.30	28.85	32.00	34.27	39.25
17	7.564	8.672	21.61	22.98	24.77	27.59	30.19	33.41	35.72	40.79
18	8.231	9.390	22.76	24.16	25.99	28.87	31.53	34.81	37.16	42.31
19	8.907	10.12	23.90	25.33	27.20	30.14	32.85	36.19	38.58	43.82
20	9.591	10.85	25.04	26.50	28.41	31.41	34.17	37.57	40.00	45.31
21	10.28	11.59	26.17	27.66	29.62	32.67	35.48	38.93	41.40	46.80
22	10.98	12.34	27.30	28.82	30.81	33.92	36.78	40.29	42.80	48.27
23	11.69	13.09	28.43	29.98	32.01	35.17	38.08	41.64	44.18	49.73
24	12.40	13.85	29.55	31.13	33.20	36.42	39.36	42.98	45.56	51.18
25	13.12	14.61	30.68	32.28	34.38	37.65	40.65	44.31	46.93	52.62
26	13.84	15.38	31.79	33.43	35.56	38.89	41.92	45.64	48.29	54.05
27	14.57	16.15	32.91	34.57	36.74	40.11	43.19	46.96	49.64	55.48
28	15.31	16.93	34.03	35.71	37.92	41.34	44.46	48.28	50.99	56.89
29	16.05	17.71	35.14	36.85	39.09	42.56	45.72	49.59	52.34	58.30
30	16.79	18.49	36.25	37.99	40.26	43.77	46.98	50.89	53.67	59.70
40	24.43	26.51	47.27	49.24	51.81	55.76	59.34	63.69	66.77	73.40
50	32.36	34.76	58.16	60.35	63.17	67.50	71.42	76.15	79.49	86.66
60	40.48	43.19	68.97	71.34	74.40	79.08	83.30	88.38	91.95	99.61
70	48.76	51.74	79.71	82.26	85.53	90.53	95.02	100.4	104.2	112.3
80	57.15	60.39	90.41	93.11	96.58	101.9	106.6	112.3	116.3	124.8
90	65.65	69.13	101.1	103.9	107.6	113.1	118.1	124.1	128.3	137.2
100	74.22	77.93	111.7	114.7	118.5	124.3	129.6	135.8	140.2	149.4

Table constructed by David Smith, University of Auckland.

References

Abbott, R. D., Yin Yin, M. A., Reed, D. M., and Yano, K. (1986). Risk of stroke in male cigarette smokers. *New England Journal of Medicine,* **315**, 717-720.

Agresti, A. (1990). *Categorical Data Analysis.* Wiley: New York.

Algina, J., Oshima, T. C., and Lin W.-Y. (1994). Type I error rates for Welch's and James's second-order test under nonnormality and inequality of variance when there are two groups. *Journal of Educational and Behavioral Statistics,* **19**, 275-291.

Allen, C. T., Schewe, C. D., and Wijk, G. (1980). More on self-perception theory's foot technique in the pre-call mail survey setting. *Journal of Marketing Research,* **17**, 498-502.

Altman, D. G. (1980). Misuse of statistics is unethical. *British Medical Journal,* **281** (6249), 1182-1184.

Andréasson, S., Allebeck, P., Engström, A., and Rydberg, U. (1987). Cannabis and schizophrenia: A longitudinal study of Swedish conscripts. *Lancet,* **2** (8574), 1483-1485.

Appel, V., Weinstein, S., and Weinstein, C. (1979). Brain activity and recall of TV advertising. *Journal of Advertising Research,* **19**, 7-18.

Atkinson, G., MacLaren, D., and Taylor, C. (1994). Blood lead levels of British competitive cyclists. *Ergonomics,* **37**, 43-48.

Atwood, E. L. (1956). Validity of mail survey data on bagged waterfowl. *Journal of Wildlife Management,* **20**, 1-16.

Banford, J. C., Brown, D. H., McConnell, A. A., McNeil, C. J., and Smith, W. E. (1982). Determination of thiol concentrations in haemolysate by resonance Raman spectrometry. *Analyst,* **107**, 195-199.

Barnard, G. A. (1988). Two-by-two (2×2) tables. In *Encyclopedia of Statistical Sciences, Vol. 9,* S. Kotz, N. L. Johnson, and C. B. Read (Eds.), pp. 367-371. Wiley: New York.

Barnett, V. (1978). The study of outliers: Purpose and model. *Applied Statistics,* **27**, 242-250.

Best, D. J., and Rayner, J.C.W. (1987). Welch's approximate solution to the Behrens-Fisher problem. *Technometrics,* **29**, 205-210.

Bhattacharyya, G. K., and Johnson, R. A. (1977). *Statistical Concepts and Methods.* Wiley: New York.

Bickel, P., and O'Connell, J. W. (1975). Is there a sex bias in graduate admissions? *Science,* **187**, 398-404.

Bohning, D. (1994). Better approximate confidence intervals for a binomial parameter. *Canadian Journal of Statistics,* **22**, 207-218.

Box, G.E.P., Hunter, W. G., and Hunter, J. S. (1978). *Statistics for Experimenters.* Wiley: New York.

Brook, R. J., Arnold, G. C., Pringle, R. M., and Hassard, T. M. (Eds.). (1986). *The Fascination of Statistics.* Marcel-Dekker: New York.

Brown, B. W., Jr. (1972). Statistics, scientific methods and smoking. In *Statistics: A Guide to the Unknown,* J. M. Tanur, F. Mosteller, W. H. Kruskal, R. F. Link, R. S. Pieters, and G. R. Rising (Eds.), pp. 40-51. Holden-Day: San Francisco.

Casler, C. (1964). The effects of hypnosis on GESP. *Journal of Parapsychology,* **28**, 126-134.

Cavendish, H. (1798). To determine the density of the earth. *Philosophical Transactions*, **88**, 405.

Chambers, J. M., Cleveland, W. S., Kleiner, B., and Tukey, P. A. (1983). *Graphical Methods for Data Analysis.* Wadsworth: Belmont, CA.

Childers, T. L., and Ferrell, O. C. (1979). Response rates and perceived length in mail surveys. *Journal of Marketing Research*, **16**, 429-431.

Cleveland, W. S., Harris, C. S., and McGill, R. (1982). Judgements of circle sizes on statistical maps. *Journal of the American Statistical Association*, **77**, 541-547.

Clogg, C. C., and Shockey, J. W. (1988). Multivariate analysis of discrete data. In *Handbook of Multivariate Experimental Psychology*, J. R. Nesselroade and R. B. Cattell (Eds.), pp. 337-365. Plenum Press: New York.

Cobb, G. W. (1987). Introductory textbooks: A framework for evaluation. *Journal of the American Statistical Association*, **82**, 321-339.

Cochran, W. G. (1954). Some methods of strengthening the common χ^2 tests. *Biometrics*, **10**, 417-451.

COMAP (Consortium for Mathematics and Its Applications). (1989). *Against All Odds: Inside Statistics* (Video). Annenberg/CPB Collection: Santa Barbara, CA.

Conover, W. J. (1980). *Practical Nonparametric Statistics*, 2nd ed. Wiley: New York.

Conover, W. J., Johnson, M. E., and Johnson, M. M. (1981). A comparative study of tests for homogeneity of variance, with applications to the outer continental shelf bidding data. *Technometrics*, **23**, 351-361.

D'Agostino, R. B. (1986a). Graphical analysis. In *Goodness-of-Fit Techniques*, Ch. 2, R. B. D'Agostino and M. A. Stephens (Eds.). Marcel Dekker: New York.

D'Agostino, R. B. (1986b). Test for the normal distribution. In *Goodness-of-Fit Techniques*, Ch. 9, R. B. D'Agostino and M. A. Stephens (Eds.). Marcell Dekker: New York.

D'Agostino, R. B., Belanger, A., and D'Agostino, R. B., Jr. (1990). A suggestion for using powerful and informative tests for normality. *American Statistician*, **44**, 316-321.

D'Agostino, R. B., Chase, W., and Belanger, A. (1988). The appropriateness of some common procedures for testing the equality of two independent binomial proportions. *American Statistician*, **42**, 198-202.

Dai, W. S., Gutai, J. P., Kuller, L. H., and Cauley, J. A. (1988). Cigarette smoking and serum sex hormones in men. *American Journal of Epidemiology*, **128**, 796-805.

Davidow, W. H., and Uttal, B. (1989). *Total Customer Service: The Ultimate Weapon.* Harper and Row: New York.

DeLotelle, R. S., and Epting, R. J. (1988). Selection of old trees for cavity excavation by red-cockaded woodpeckers. *Wildlife Society Bulletin*, **16**, 48-52.

Dodson, J. A., Tybout, A. M., and Sternthal, B. (1978). Impact of deals and deal retraction on brand switching. *Journal of Marketing Research*, **15**, 72-81.

Duncan, D. F. (1994). Drug law enforcement expenditures and drug-induced deaths. *Psychological Reports*, **75**, 57-58.

Efron, B., and Tibshirani, R. J. (1993). *An Introduction to the Bootstrap.* Monographs on Statistics and Applied Probability No. 57. Chapman and Hall: New York.

Ehrenberg, A.S.C. (1975). *Data Reduction.* Wiley: London.

Ehrenberg, A.S.C. (1977). Rudiments of numeracy. *Journal of the Royal Statistical Society, Series A*, **140**, 277-297.

Ehrenberg, A.S.C. (1981). The problem of numeracy. *American Statistician*, **35**, 67-71.

Emerson, J. D., and Simon, G. A. (1979). Another look at the sign test when ties are present: The problem of confidence intervals. *American Statistician*, **33**, 140-142.

Fairley, W. B., and Mosteller, F. (1974). A conversation about Collins. In *Statistics and Public Policy*, W. B. Fairley and F. Mosteller (Eds.), pp. 369-379. Addison-Wesley: Reading, MA.

Fienberg, S. E. (1971). Randomization and social affairs: The 1970 draft lottery. *Science*, **171**, 255-261.

Fisher, B., Bauer, M., Margolese, R., Poisson, R., Pilch, Y., Redmond, C., Fisher, E., Wolmark, N., Deutsch, M., Montague, E., Saffer, E., Wickerman, L., Lerner, H., Glass, A., Shibata, H., Deckers, P., Ketcham, A., Oishi, R., and Russell, I. (1985). Five-year results of a randomized clinical trial comparing total mastectomy and segmented mastectomy with or without radiation in the treatment of breast cancer. *New England Journal of Medicine,* **312**, 665-673.

Fleming, T. R. (1992). Evaluating therapeutic interventions: Some issues and experiences. *Statistical Science,* **7**, 428-456.

Freedman, D., Pisani, R., Purves, R., and Adhikari, A. (1991). *Statistics,* 2nd ed. Norton: New York.

Frets, G. P. (1921). Heredity of head form in man. *Genetica,* **3**, 193-384.

Freund, J. E., and Perles, B. M. (1987). A new look at quartiles of ungrouped data. *American Statistician,* **41**, 200-203.

Frigge, M., Hoaglin, D. C., and Iglewicz, B. (1989). Some implementations of the boxplot. A new look at quartiles of ungrouped data. *American Statistician,* **43**, 50-54.

Gastwirth, J. J. (1987). The statistical precision of medical screening procedures: Application to polygraph and AIDS antibodies test data. *Statistical Science,* **2**, 213-238.

Glaser, R. E. (1983). Levene's robust test of homogeneity of variances. In *Encyclopedia of Statistical Sciences, Vol. 4,* S. Kotz, N. L. Johnson, and C. B. Read (Eds.), pp. 608-610. Wiley: New York.

Grimshaw, J. J., and Jaffe, G. V. (1982). Statistics in relation to clinical trials in general practice. *Statistician,* **31**, 63-69.

Gunter, B. (1988). Subversive data analysis, part II: More graphics, including my favorite example. *Quality Progress,* November 1988, pp. 77-78.

Haedrich, R. L., and Merrett, N. R. (1988). Summary atlas of deep living demersal fishes in the North Atlantic basin. *Journal of Natural History,* **22**, 1325-1362.

Hahn, G. J., and Meeker, W. Q. (1991). *Statistical Intervals: A Guide for Practitioners.* Wiley: New York.

Halperin, S. A., Barreto, L., Eastwood, B. J., Law, B., and Roberts, E. A. (1994). Safety and immunogenicity of a five-component acellular pertussis vaccine with varying antigen quantities. *Archives of Pediatrics and Adolescent Medicine,* **148**, 1220-1224.

Hamill, R., Nisbitt, R. E., and Wilson, T. D. (1980). Insensitivity to sample bias: Generalizing from atypical cases. *Journal of Personality and Social Psychology,* **39**, 578-589.

Hansel, C.E.M. (1966). *ESP: A Scientific Evaluation.* Scribner: New York.

Hauck, W. W., and Anderson, S. (1986). A comparison of large-sample confidence interval methods for the difference of two binomial probabilities. *American Statistician,* **40**, 318-322.

Hicks, R. A., and Bautista, J. (1993). Snoring and nightmares. *Perceptual and Motor Skills,* **77**, 433-434.

Holt, J. D., and Prentice, R. L. (1974). Survival analysis in twin studies and matched pair experiments. *Biometrika,* **61**, 17-30.

Hoyle, R. H. (1993). Interpersonal attraction in the absence of explicit attitudinal information. *Social Cognition,* **11**, 309-320.

Hutchinson, T. P. (1987). *Road Accident Statistics.* Rumsby Scientific: Adelaide.

Jansen, J. H. (1985). Effect of questionnaire layout and size and issue involvement on response rates in mail surveys. *Perceptual and Motor Skills,* **61**, 139-142.

Johnson, N. L., and Leone, F. C. (1977). *Statistics and Experimental Design in Engineering and the Physical Sciences,* 2nd ed. Wiley: New York.

Jonakait, R. N. (1983). When blood is their argument: Probabilities in criminal cases, genetic markers, and, once again, Bayes' theorem. *University of Illinois Law Review,* **2**, 369-421.

Kahle, L. R. (1986). The nine nations of America and the value basis of market segmentation. *Journal of Marketing,* **50**, 37-47.

Kalton, G., and Schuman, H. (1982). The effect of the question on survey responses: A review. *Journal of the Royal Statistical Society, Series A,* **145**, 42-73.

Kaufman, L., and Rock, I. (1962). The moon illusion, I. *Science,* **136**, 953-961.

Kerrich, J. (1964). *An Experimental Introduction to the Theory of Probability.* University of Witwatersrand Press: Johannesburg.

Krummer, D. R., and Hoffmeister, J. R. (1978). Valuation consequences of cash tender offers. *Journal of Finance,* **33**, 505-516.

La Barbara, P. A., and MacLachlan, J. M. (1979). Response latency in telephone interviews. *Journal of Advertising Research,* **19**, 49–55.

Lendvoy, L. (1996). The mystery of "dumbing down" Nancy Drew. *STATS,* **16**, 23–25.

LeVay, S. (1991). A difference in hypothalamic structure between heterosexual and homosexual males. *Science,* **253**, 1034–1037.

Lorenzen, T. J. (1980). Determining statistical characteristics of a vehicle emissions audit procedure. *Technometrics,* **22**, 483–493.

Lucas, A., Morley, R., Cole, T. J., Lister, G., and Leeson-Payne, C. (1992). Breast milk and subsequent intelligence quotient in children born preterm. *Lancet,* **339** (8788), 261–264.

MacLachlan, J. M., and Siegel, M. H. (1980). Reducing the cost of T.V. commercials by use of time compressions. *Journal of Marketing Research,* **17**, 52–57.

Margolese, M. S. (1970). Homosexuality: A new endocrine correlate. *Hormones and Behaviour,* **1**, 151–155.

Margolin, B. H. (1988). Statistical aspects of using biologic markers. *Statistical Science,* **3**, 351–357.

Markowski, C. A., and Markowski, E. P. (1990). Conditions for the effectiveness of a preliminary test of variance. *American Statistician,* **44**, 322–326.

Mehta, C., and Patel, N. (1995). *StatXact 3 for Windows: User Manual.* CYTEL Software Corporation: Cambridge, MA.

Meier, P. (1986). Damned liars and expert witnesses. *Journal of the American Statistical Association,* **81**, 269–276.

Mitchell, R. J., Izatt, M. M., Sunderland, E., and Cartwright, R. A. (1976). Blood groups antigens, plasma protein and red cell isoenzyme polymorphisms in south-west Scotland. *Annals of Human Biology,* **3**, 157–171.

Moore, D. S. (1979). *Statistics: Concepts and Controversies.* W. H. Freeman: San Francisco.

Moore, D. S. (1986). Tests of chi-squared type. In *Goodness-of-Fit Techniques,* Ch. 3, R. B. D'Agostino and M. A. Stephens (Eds.). Marcell Dekker: New York.

Moore, D. S. (1991). *Statistics: Concepts and Controversies,* 3rd ed. W. H. Freeman: San Francisco.

Morgan, J. P., Chaganty, N. R., Dahiya, R. C., and Doviak, M. J. (1991). Let's make a deal: The player's dilemma (with discussion). *American Statistician,* **45**, 284–289.

Morton, R. H. (1990). Lotto, then and now. *New Zealand Statistician,* **25**, 15–21.

Neter, J. (1989). How accountants save money by sampling. In *Statistics: A Guide to the Unknown,* 3rd ed., J. M. Tanur, F. Mosteller, W. H. Kruskal, E. L. Lehmann, R. F. Link, R. S. Pieters, and G. R. Rising (Eds.), pp. 151–160. Holden-Day: San Francisco.

Newcombe, R. G. (1998a). Interval estimation for the difference between independent proportions. Comparison of eleven methods. *Statistics in Medicine,* **17**, 873–890. [Correction 1999, **18**, 1293.]

Newcombe, R. G. (1998b). Two-sided confidence intervals for the single proportion. Comparison of seven methods. *Statistics in Medicine,* **17**, 857–872.

Norušis, M. J. (1988). *SPSS^X Advanced Statistics Guide,* 2nd ed. McGraw-Hill: New York.

Pettit, A. N. (1982). Durbin-Watson test. In *Encyclopedia of Statistical Sciences, Vol. 2,* S. Kotz, N. L. Johnson, and C. B. Read (Eds.), pp. 426–428. Wiley: New York.

Phillips, D. P. (1972). Deathday and birthday: An unexpected connection. In *Statistics: A Guide to the Unknown,* J. M. Tanur, F. Mosteller, W. H. Kruskal, R. F. Link, R. S. Pieters, and G. R. Rising (Eds.), pp. 52–65. Holden-Day: San Francisco.

Pianka, E. R. (1994). Comparative ecology of *Varanus* in the Great Victorian Desert. *Australian Journal of Ecology,* **19**, 395–408.

Poole, D. I., and Pole, N. (1987). *The Maori Population to 2011: Demographic Change and Its Implications.* Technical Paper No. 1, NZ Demographic Society, Wellington.

Posten, H. O. (1979). The robustness of the one-sample *t*-test over the Pearson system. *Journal of Statistical Computation and Simulation,* **9**, 133–149.

Preston, R. H., Dwyer, F. R., and Rudelius, W. (1978). The effectiveness of bank premiums. *Journal of Marketing,* **42**, 96–101.

Proshaska, V. (1994). "I know I'll get an A": Confident overestimation of final course grades. *Teaching of Psychology,* **21**, 141-143.
Reid, D. D. (1989). Does inheritance matter in disease? The use of twin studies in medical research. In *Statistics: A Guide to the Unknown,* 3rd ed., J. M. Tanur, F. Mosteller, W. H. Kruskal, E. L. Lehmann, R. F. Link, R. S. Pieters, and G. R. Rising (Eds.), pp. 53-59. Holden-Day: San Francisco.
Resnick, R. B., Fink, M., and Freedman, A. M. (1970). A cyclazocine typology in opiate dependence. *American Journal of Psychiatry,* **126**, 1256-1260.
Roberts, G., Martyn, A. L., Dobson, A. J., and McCarthy, W. H. (1981). Tumour thickness and histological type in malignant melanoma in New South Wales, Australia, 1970-1976. *Pathology,* **13**, 763-770.
Rosa and Dorsey. (1907). A new determination of the ratio of the electromagnetic to the electrostatic unit of electricity. *Bulletin of the National Bureau of Standards,* **3**, 433-604.
Rosenblatt, J. R., and Filliben, J. J. (1971). Randomization and the draft lottery. *Science,* **171**, 306-308.
Roth, A. J. (1988). Welch tests. In *Encyclopedia of Statistical Sciences, Vol. 9,* S. Kotz, N. L. Johnson, and C. B. Read (Eds.), pp. 586-589. Wiley: New York.
Samuels, M. L., and Lu, T.-F. C. (1992). Sample size requirements for the back-of-the-envelope binomial confidence interval. *American Statistician,* **46**, 228-231.
Sanches, V. M., Alvarez-Guisasola, F., Cayla, J. A., and Alvarez, J. L. (1995). Predictive factors of mycobacterium tuberculosis infection and pulmonary tuberculosis in prisoners. *International Journal of Epidemiology,* **24**, 630-636.
Schumacher, M. (1972). Predicting subsequent conviction for individual male prison inmates. *New Zealand Statistician,* **8**, 26-34.
Scott, A. J., and Seber, G.A.F. (1983). Comparing populations from the same survey. *American Statistician,* **37**, 319-320.
Seber, G.A.F. (1982). *The Estimation of Animal Abundance and Related Parameters,* 2nd ed. Charles Griffin: London.
Seber, G.A.F. (1984). *Multivariate Observations.* Wiley: New York.
Shuptrine, F. K., and McVicker, D. D. (1981). Readability levels of magazine ads. *Journal of Advertising Research,* **21**, 45-51.
Sigelman, L. (1981). Question-order effects on presidential popularity. *Public Opinion Quarterly,* **45**, 199-207.
Smith, S. (1990). It sounds good, but .... *The Statistics Teacher Network,* December 1990, 1-2.
Snee, R. D. (1993). What's missing in statistical education? *American Statistician,* **47**, 149-154.
Solomon, H., and Stephens, M. A. (1988). Variance, sample. In *Encyclopedia of Statistical Sciences, Vol. 9,* S. Kotz, N. L. Johnson, and C. B. Read (Eds.), pp. 477-480. Wiley: New York.
Speed, T. P. (1977). *Negligible Probabilities and Nuclear Reactor Safety: Another Misuse of Probability.* Department of Mathematics, University of Western Australia: Perth.
Speed, T. P. (1985). Probabilistic risk assessment in the nuclear industry: WASH-1400 and beyond. In *Proceedings of the Berkeley Conference in Honor of Jerzy Neyman and Jack Kiefer, Vol. 1,* L. M. Le Cam and R. A. Olshen (Eds.), pp. 173-200. Wadsworth: Monterey.
Sprent, P. (1993). *Applied Nonparametric Statistical Methods.* Chapman and Hall: London and New York.
Stigler, S. M. (1977). Do robust estimators work with real data? (with discussion). *Annals of Statistics,* **5**, 1055-1098.
Stigler, S. M. (1986). *The History of Statistics: The Measurement of Uncertainty before 1900.* Belknap Press of Harvard University Press: Cambridge, MA.
Sudman, S., and Ferber, R. (1974). A comparison of alternative procedures for collecting consumer expenditure data for frequently purchased products. *Journal of Marketing Research,* **11**, 128-135.
Thornton, B., and Moore, S. (1993). Physical attractiveness contrast effect: Implications for self-esteem and evaluations of the social self. *Personality and Social Psychology Bulletin,* **19**, 474-480.

Tiku, M. L., Tan, W. Y., and Balakrishnan, N. (1986). *Robust Inference.* M. Dekker: New York.

Tjaden, P., and Thoennes, N. (1998). Prevalence, incidence, and consequences of violence against women: Findings from the National Violence Against Women Survey. A *Research in Brief* report, U.S. Department of Justice.

Trewin, D. (1977). Non-sampling errors in sample surveys. *CSIRO Division of Mathematics and Statistics Newsletter,* June 1977.

Trinkaus, J. (1994). Wearing baseball caps: An informal look. *Psychological Reports,* **74**, 585-586.

Tufte, E. R. (1983). *The Visual Display of Quantitative Information.* Graphics Press: Cheshire, CT.

Tufte, E. R. (1990). *Envisioning Information.* Graphics Press: Cheshire, CT.

U.S. Nuclear Regulatory Commission. (1975). *Reactor Safety Study WASH-1400.* NUREG-75/014. U.S. Nuclear Regulatory Commission: Washington, DC. [Also known as the Rasmussen Report.]

Vardeman, S. B. (1992). What about the other intervals? *American Statistician,* **46**, 193-197.

Vasil, L. (1987). Background paper: The prediction of violent reoffending. *NZ Department of Justice Report,* Wellington.

Ward, J. C., Russick, B., and Rudelius, W. (1985). A test of reducing callbacks and not-at-home bias in personal interviews by weighting at-home respondents. *Journal of Marketing Research,* **22**, 66-73.

Wasserstein, R. L., and Boyer, J. E., Jr. (1990). Probability and instant lottery games. *STATS,* **3**, 7-10.

Weiss, S. H., Goedert, J. J., Sarngadharan, M. G., Bodner, A. J., The AIDS Sereoepidemiology Working Group, Gallo, R. C., and Blattner, A. (1985). Screening test for HTLV-III (AIDS agent) antibodies. *Journal of the American Medical Association,* **253**, 221-225.

Whichelow, M. J., Golding, J. F., and Treasure, F. P. (1988). Comparison of some dietary habits of smokers and non-smokers. *British Journal of Addiction,* **83**, 295-304.

Wild, C. J. (1995). Continuous improvement of teaching: Case study in a large statistics course. *International Statistical Review,* **63**, 49-68.

Wild, C. J., and Seber, G.A.F. (1993). Comparing two proportions from the same survey. *American Statistician,* **47**, 178-181. [Correction 1994, **48**, 269.]

Wild, C. J., Triggs, C. M., and Pfannkuch, M. (1997). Assessment on a budget: Using traditional methods imaginatively. In *The Assessment Challenge in Statistical Education,* I. Gal and J. B. Garfield (Eds.), pp. 205-220. IOS Press: Amsterdam.

Winter, F. D., Snell, P. G., and Stray-Gundersen, J. (1989). Effects of 100% oxygen on performance of professional soccer players. *Journal of the American Medical Association,* **262**, 227-229.

Wright, T. A., and Bonett, D. G. (1991). The volunteer subject and job performance: Implications for research and management. *International Journal of Management,* **8**, 509-516.

Answers to Selected Problems

Chapter 1

Section 1.1

1. We could take all members of the population in the country at the time who were entitled to vote in national elections. In New Zealand this would exclude the young, the illegal immigrants, those in prisons, and people legally committed to mental hospitals. It would include anyone else legally resident in New Zealand whether or not they were citizens, and citizens living overseas. You might want to be more, or less, restrictive. In practice, one would probably sample from something like the electoral rolls—that subset of people who fit the eligibility criteria for voting and who have registered to do so.

2.–4. These are entirely for you to draw on your own experience. Similar stories are given in the chapter.

5. The survey was self-selective and therefore not necessarily representative of N.Z. women. The reporter compared the statement "92% of incest cases reported in the survey are European" with "81% of the population is European" and claimed unusually high abuse. The reporter should have compared the statement "92% of incest cases reported in the survey are European" with the fact that "91% of those responding to the survey were European." Far from the figure of "92% of incest cases" being an abnormally high proportion, it is almost exactly what you would expect if incest and ethnicity were unrelated.

Section 1.2

1. We would choose 10 rats at random to form the reward group. The remaining 10 form the punishment group. We might expect the first 10 rats caught to tend to be the slowest (and perhaps the most stupid!).

2. Product characteristics tend to vary over time, so it is possible that the two lengths chosen may differ in important ways from the other two. Another method would be to randomly choose two of the four rods available from each length for testing process 1 and the remainder for process 2. This would protect us from the possibility of systematic changes along the length of a rod. Alternatively we can use a systematic method: Choose rods 1 and 3 from each of the lengths for process 1, and rods 2 and 4 for process 2.

3. Important things are random selection and double blinding (cf. school milk story). The definition and measurement of intelligence is a thorny problem that has exercised psychologists for years. If one suspected that there would be an effect on a particular type of intelligence, that is the type that should be measured. Otherwise, you would probably just choose a standard broad-spectrum psychological test appropriate for the age group.

Section 1.3

Note: In each of these items there are other possible explanations besides the "obvious" one. Coming up with alternative possibilities is an important part of statistical thinking. Our answers are not the "right" answers. We have no particular expertise in the areas under discussion and just report a few ideas that occurred to us.

1. The "obvious" inference is that with the coming of electricity and the facilities it brings, there are more distractions, and birth rates fall. The observed differences in rates, however, may be related to economic issues, with people not yet having electricity tending to be poorer and less educated. There are probably rural/urban differences. Rural areas tend to be slower to get electricity, and there are practical reasons for rural people to want larger families.

2. The observed difference between the groups may be due to something that is associated with smoking but not caused by smoking. Perhaps people who choose to smoke tend not to work as hard on average, so that it is this and not the smoking that gives them the lower marks. Perhaps their average IQ is lower.

3. Not necessarily. Even long trips have a close-to-home portion. Most of our traveling experience is close to home, so we might expect more accidents close to home.

4. It could be the other way around. Maybe prolonged exposure to alcohol causes these chemical changes.

5. A host of factors may be at work here. Few would doubt that more men are involved in crime than women, so it is not surprising that more end up in jail. It is possible that mothers with dependent children are less likely to be sentenced to jail. The types of crime that men and women tend to commit may also differ.

6. A randomized experiment is not feasible. Subject choice affects the careers you can go into, for example. We cannot direct some randomly chosen people to head off in one direction and others in another. If the gender differences in mathematics scores disappear when we look just at people who also take physics, this would tend to support the "reinforcement" idea. A problem with this is that there may be gender differences in subject choice. Perhaps girls taking physics are better than boys on average because they are less likely to take the subject unless they know they are good at it. (This is easily investigated—how?) Any effect like this would bias the male-female comparison of mathematics scores confined to physics students.

Review Exercises 1

1. (a) One definition would be anyone not in full-time paid employment who wants a job and is actively seeking one. This is problematic in that it excludes the long-term unemployed who are so discouraged that they have given up on finding a job. (c) The definition of what it means to be unemployed differs between countries, as does the way the data are collected, for example, surveys (with many differences in survey design) versus official registers of unemployed people.

3. The experiences giving rise to anecdotal evidence are very small samples that may also have come from a very biased "sampling" mechanism.

5. (i) Big unrepresentative samples are still unrepresentative. (ii) This is fine so long as you see it as raising issues rather than representing popular opinion or experience. (iii) Pointing to other instances of unrepresentative research does not make your research any better.

7. (a) There will be people with Celtic names but with very little Celtic blood (e.g., a Celtic surname could have come from a sole Celtic male ancestor six generations ago). Similarly, there would be people who were predominantly Celtic but have non-Celtic names. (b) Including some non-Celts in with the Celtic group and some Celts in the non-Celtic group would make the observed differences between the groups *smaller* than they should be. (c) Psychiatric patients of the type that attended the particular hospital at that time. (d) Anything that was special about the types of patients admitted and the conditions that they were exposed to.

9. Fatalities per kilometer per car tell us how safe the road is for us to travel on. It tells us nothing, however, about the number of lives that we are likely to be able to save per dollar spent on highway improvement. Short stretches of road with many fatalities, like the bridge,

may be good to target in this regard. Fatalities per kilometer of highway may give a reasonable approximate measure.

11. Elevation, shading, steepness, and soil conditions are all factors to be considered. Laying out a grid of cells of equal area on a map of the area and randomly choosing cells to receive each of the three treatments is a reasonable way to proceed if practically feasible.

13. Different places use different definitions. Reporting rates may also differ for a variety of reasons. We have to be a little suspicious that people with a point to prove may be very selective in what they report and how they report it!

15. The paint surfaces that last longest at high temperatures may be different from the surfaces that last longest under normal operating conditions.

Chapter 2

Section 2.1

2. Income (continuous), religion (categorical), ethnicity (categorical), distance (continuous), transport (categorical), household number (discrete), smoking (ordinal), cigarettes (discrete), political party (categorical), and percentage income (continuous, though if the data are sufficiently rounded to a few significant figures, it could be treated as discrete).

3. Patient 69 did not continue to smoke, had surgery for symptoms within one year, and was alive at last follow-up.

4. Patient 201 was aged 42 at admission, had DIAVOL = 329, was not taking beta blockers, and had a cholesterol score of 39.

5. Patient 203 was aged 54 at admission, had occlusion score 0, was not taking beta blockers, and had no cholesterol score recorded.

6. Only five patients continued to smoke. Only two had surgery for symptoms after five years. Thirty-two were still alive at last follow-up.

Section 2.2

1. (a) South Africa has been by far the biggest producer, but its production figures have been decreasing a little over time. Trends include a gradual increase in world production since 1980. The major increases in production have been in the United States, Australia, Canada, and China. Figures for the USSR start in 1980, but given the large production occurring then, we imagine that they must have been producing substantial amounts before this (similarly for China). We use the original table for the following answers: (b) (1972) 65%; (1980) 55%; (1986) 40%. (c) (1972) 3.2%; (1980) 2.5%; (1986) 7.3%; (1991) 13.9%. (d) (1970s) Canada; (1980s) USSR; (1990s) United States.

Section 2.3.1

1. Spain stands out as a high-unemployment outlier.

2. Finland stands out as a high-unemployment outlier.

3. Plotting against the same scale allows us to compare unemployment patterns of members and nonmembers. We see that the EEC countries had higher unemployment on average than non-EEC members. The spreads are similar.

Section 2.3.2

1. (a) (i) 1.54. (ii) 0.67. (b) (i) 154. (ii) 67. (c) (i) 0.00154. (ii) 0.00067.

2. (a) 16|3 = 1.63 cm. (b) Round to two significant figures: 5|4 = 540 m. (c) Round to one decimal place: 16|3 = 16.3 m. (d) 166|3 = 1663 kg.

3. 5.4, 9.8, 10.1, 10.1, . . . , 25.6, 26.8.

4. We see a positively skewed shape (long tail toward large values) with a mode at about 145.

6. A stem-and-leaf plot for position 7 is given together with other plots in Fig. 3.2.5 in the main text. The body of the plot is bimodal, and there is a group of three much smaller observations.

Section 2.3.3

1. The endpoints of the intervals are slightly different from the stem-and-leaf plots, so that the frequencies are not identical. The shapes are very similar, however.

2. To compare them we would place one histogram above the other and use the same scale. The histogram for the males is shifted to the right, indicating their greater lengths. The one for the females is more peaked.

3. 20/40, or 50%.

Section 2.3.4

1. The shape is negatively skewed (long tail to the left).

2. The shape is bimodal with an outlier.

3. The stem-and-leaf plot is very similar in shape to the histogram. The overall shape is positively skewed (long tail to the right) with possibly two outliers.

Section 2.4.1

1. (a) 7.5. (b) 4. (c) 3.0.

2. Mean = 89.24; median = 89.75 (original data) or 90 (stem-and-leaf). Data from the stem-and-leaf plot are rounded. No.

3. Mean = 92.06 and median = 92 (either original data or rounded data from a stem-and-leaf plot). From the samples, the average length of male coyotes is greater than that for the female coyotes.

4. Mean = 48.43. Median = 48.

Section 2.4.2

1. (a) (1, 2, 8, 11.5, 14). (b) (1, 5, 10, 13, 20).

2. (6.8, 8.45, 8.75, 9.2, 10.4).

3. (36, 45, 48, 54, 59).

4. You would be tall and thin.

Section 2.4.3

1. There are eight of them, with diastolic volumes 133, 138, 193, 149, 124, 167, 156, 195. $\overline{x} = 156.875$, $s_X = 26.541$.

2. (a) $\overline{x} = 92.056$, $s_X = 6.696$. They are both larger for males. The male data set is slightly more variable. Although on *average* males are longer than females, when it comes to individual comparisons we could not be sure which one would be longer. (b) (1 sd) 29/43 = 0.67, (2 sd) 42/43 = 0.98. (c) Range = 105 − 78 = 27, IQR = 96 − 87 = 9. sd/IQR = 0.744, or almost exactly 75%.

Section 2.4.4

1. The data are positively skewed (long tail to the right) with two outliers (453, 589).

2. The data are positively skewed with one outlier (8.5).

Section 2.5.1

1. (a) 3.870. (d) 20.4%. (e) 65.2%. (f) 6.

2. They don't make much sense. For example, how would we define a median? Of the observations, 44.8% are smaller than 4, 20.4% are equal to 4, and 34.7% are bigger than 4.

Section 2.5.2

1. Mean = 227.419 and sd = 94.274. These answers are fairly close to the answers for the ungrouped data.

2. The shape looks like the right half of a bell. (a) 51/62. (b) 9/62. (c) 2/62.

3. The height of the rectangle above the interval 400–600 would be $3/(4 \times 62)$.

Section 2.6

2. The biggest changes are the reduction in the South African share and the growth in the U.S. and Australian shares between 1985 and 1991.

4. We used percentage of total downtime as our measure and ordered by descending percentage. The first two reasons (both die repairs) account for 70% of all downtime. It is here we look first.

Review Exercises 2

1. (a) Median = 5.8, IQR = 0.6. (c) Both show a central group with two large outliers. (d) The outliers are the only two cars with automatic transmissions.

3. (a) Home: mean = 20.227, sd = 6.510; away: mean = 14.364, sd = 5.104. Nearly six more goals scored on average for away games. The spread (sd) is also somewhat greater at home. (b) Use back-to-back stem-and-leaf plots. Yes. (c) Home: (Min, Q_1, Med, Q_3, Max) = (9, 16, 19, 23, 39). (d) We see the shift toward higher values for home games, the long lower whisker of the "away" box coming from the negative skew in the "away" stem-and-leaf plot. The unusual home-game score of 39 shows up in both the box plot and stem-and-leaf plot. (e) We need ways that let us know which pairs of home and away points belong together. We mention three ways. First, we can use a dot plot of the differences. This shows that the home − away differences are almost always positive (i.e., most teams got more goals at home than away). It also shows the variability in the differences. Second, we can use a linked dot plot. Here points that belong together are linked by lines. We can see that the home score is almost always bigger than the away score. This plot also displays the variability in home scores and in away scores. The variability in the differences is harder to see. Third, we can use a scatter plot (discussed in Chapter 3) to show a tendency for teams who get more away goals to also get more home goals. A point would lie on the $y = x$ line if the number of home and away goals were identical. Most points lie considerably above the line (more home goals than away).

5. (a) 31%. *(c) It depends on what is meant by "gain," as it can be measured by the actual change or the percentage change. Either of these can be plotted against the 1982 percentage using a scatter plot (Chapter 3). Both plots support the conjecture, though the latter does so more strongly. (d) Use a side-by-side bar graph with the two years together for each profession.

7. (a) The design seems reasonable. We could also have used BABA. To reduce the amount of change, ABBA or BAAB could have been used. (b) 738 caught with mesh size = 35, 787 caught with mesh size = 87. Slightly more caught with the larger mesh, whereas we would expect more to be caught with the smaller mesh (fewer escaping through the mesh). The most obvious explanation is that the boat just happened to come upon fewer fish when trawling the larger-mesh cod end. (c) Using the rule for estimating means and standard deviations from grouped data, we have: (35 mm) mean = 33.427, sd = 3.418; (87 mm) mean = 34.549, sd = 3.154. This tells us that the fish caught by the larger-mesh net are

slightly longer on average and slightly less variable in length. This is what we should expect because the larger mesh is failing to hold some of the smaller fish. The 1- and 2-sigma rules would lead us to expect that, for each mesh size, about 68% of the fish caught would have a length within *mean* ± *sd*, and about 95% would fall within *mean* ± 2*sd*. (In fact, these approximations work quite well with these data). (d) Relative frequency histograms let us compare the distributions of fish caught (e.g., the proportions caught within a given length range) by the two types of net in a way that is valid even if the numbers swimming into the net are quite different. (e) The 35-mm mesh caught a higher proportion of smaller fish, while the 87-mm mesh caught a higher proportion of larger fish, as expected. The 87-mm mesh would allow more of the smaller fish to escape. (f) We come to similar conclusions, though perhaps it is now easier to see the difference.

9. (a) Lizards and grasshoppers are very high on the diet list. (b) Percentages, as the total numbers of stomachs are different for the two species. (c) No, because some of the percentages are too small to show on a pie chart. (d) Depends on the purpose of the study. However, we suggest numbers, as they measure the number of "hits." (e) Lizards could be more abundant. Preference studies are not easy to carry out. However, the naive approach would be to put *Varanus eremius* in an enclosure with different prey species and see what happens.

11. There is an enormous range in life expectancy, with the highest being Morocco (69.5 years) and the lowest Malawi (36.2 years). They are fairly well spread out with a hint of small clusters of two or three countries about every five years. These clusters could be studied to see if the countries in them have any common features.

13. (a) The female life expectancy is three years longer than for males. Zero is special because it is the equality value (female life expectancy same as the male). Positive values correspond to a female life expectancy longer than the male. (b) The female life expectancy is 1.2 times as long as that for males (i.e., 20% longer). Unity is special because that is the equality value (female life expectancy same as the male). Values greater (resp. smaller) than 1 correspond to the female life expectancy longer (resp. shorter) than the male.

15. (a) Age structure of the population, availability of birth control methods, infant mortality rate, and so on. (b) The issues are complex. One reason for the higher death rates might be that the industrialized countries generally have higher life expectancies, which means there are more old people. This could apply to the Central and South American countries as well as the less industrialized Asian countries. Hong Kong, Japan, and Singapore, however, have lower death rates as well as lower infant mortality rates. The infant mortality rate will affect both the total death rate and the life expectancy, so that these factors are all interrelated. (c) Finland is different in two respects: It has the lowest male mortality rate, and this rate is less than its female mortality rate. (d) Not very informative, as the data set is small and the numbers quite variable. We have mean = 16.9 and median = 16.3. The median is more informative as it is less affected by the extreme values. (e) If the emphasis is on global improvement, we should look at the number of people experiencing economic growth rather than the number of countries. (f) $Increase = Migrants + Births - Deaths$.

Chapter 3

Section 3.1.1

1. (b), (e), (f), (g), (h): both random. (a), (c), (d), (i): X nonrandom and Y random.

Section 3.1.2

1. Compare your curves with those of fellow students. This exercise emphasizes the subjective nature of freehand curves.

2. (b) Predict X. (e) Symmetric comparison. (f) Choice is unclear, though you would probably want to predict Y. (g) Symmetric comparison, though it depends on why the data were collected. (h) Predict Y.

3. (a) The point (160, 160) is a highly influential observation and may be an outlier. If this point is ignored, the rest of the plot looks slightly curved, though a linear trend might adequately represent the data. (c) Most of the points lie below the line $y = x$. This indicates that the y-value is usually less than the x-value. This is not surprising; we might expect the blood pressure to be less the second time it is read, as the person interviewed will then be more relaxed.

4. (b) The two plots are quite deceptive. The plot for *Pride and Prejudice* seems to have an upward trend, while that for *Spy Hook* seems to have a curved trend initially heading downward. Both plots are strongly influenced by the percentages of "to" that are 7 or more. Ignoring these points changes the visual impression of each plot. These points, however, may be important indicators of style. (c) A combined plot does not show any trend, but it does reveal that the points for *Spy Hook* tend to lie to the right of the points for *Pride and Prejudice*. Thus *Spy Hook* tends to use "the" more frequently. There are other features that suggest that we can divide up the plot into regions where points from one of the books predominate. (d) Given a few more points from one of the two books, it might be possible to make an educated guess as to which book they come from. It would be interesting to have more data.

Section 3.1.3

These values are subjective and may differ slightly from your answers.

1. 180.

2. 1.4.

3. 215.7.

Section 3.2

1. (b) The three plots indicate that the two data sets have similar spreads and "shapes." The median of the control group is greater than the median of the TNT group. The control scores are higher on average than the TNT scores. (c) The three plots show the same features. It is tempting to read too much into such plots. You can assess a plot more readily by seeing what happens if you change one value slightly.

2. (b) Women clearly have lower waist-to-hip ratios than men, as expected. For the men, the two plots indicate symmetry and their spreads are similar. The older men appear to have slightly larger ratios on average than the younger men. For the women, the two plots indicate symmetry, and the spread for the older women is slightly greater than that for the younger women. Whether this is true for the whole population is not clear, as the difference is small. Older women have substantially greater ratios than younger women.

Section 3.3

1. The two-way table follows:

	SURG					
BETA	**0**	**1**	**2**	**3**	**4**	**Total**
1	7	1	0	0	0	8
2	25	4	4	2	2	37
Total	32	5	4	2	2	45

2. (a) 0.0289. (b) 0.1868. (c) 0.1689. (d) 0.2748. (e) 0.1100. (f) 0.1801. (g) 0.2583.

Review Exercises 3

1. (a) The plot shows a great deal of variability in the cost. The distribution has a long upper tail. (Min, Q_1, Med, Q_3, Max) = (412, 555.5, 701, 786.5, 1193). (b) Max/Min = 2.9 (to one decimal place). (c) There is no relationship between the fee charged and number of inspections performed. (Some of the highest charges come from cities that perform the fewest inspections.) The spread is greatest for cities with the lowest numbers of inspections.

3. (a) It is approximately linear. (b) Because of the trend, there is useful information in the data from the other years about 1979–1981 prices. It would appear from this trend that the 1981 prices are on the high side. (c) $6800. (d) $3600 to $8900. (e) Buy it! (f) Distance traveled, number of owners, condition of the truck, and so on.

5. (a) *Homicides*: A wide range of positively skewed ages, with the 20s being the predominant age group. *Accidents*: Ages between about 15 and the mid-30s. *Suicides*: Roughly uniform across all ages, with a drop in the 40-to-64 age group. Possibly three different groups here, namely 15 to 45, 45 to 64, and 65 to 80. The peak at 65 suggests possible retirement problems. *Self-defense*: A wide range of positively skewed ranges (2–56). (b) Have the four plots arranged side by side with the same scale to highlight the age differences in the plots. (c) The five-number summaries are: homicides (2, 22, 28, 38, 79); accidents (9, 17, 22, 29, 37); suicides (14, 27, 38, 65, 87); self-defense (21, 26, 34, 46, 56). The box plot for suicides does not reveal the bimodal nature of the data. No, as the median is a single measure of location. We need a measure for each of the two groups. (d) The combined stem-and-leaf plot will tend to be dominated by the homicide data, with perhaps a slight peak around 65 and a fatter upper tail. (e) No, as the gun laws are different.

7. (a) Figure (a) suggests that the measurements are not easy to make. From (b) we see that there are a lot of lighter perch and a few heavy pike. The weights for bream are less variable than for perch and pike. The perch data seem bimodal, and there is a hint of bimodality for both bream and pike. From (c), pike are longer on average, and bream are less variable in length. The weights of the few heavy pike are due to their length. From (d) the widths for perch are clearly bimodal, indicating the presence of two distinct groups. Figure (e) confirms (d). Figures (f), (g), and (h) indicate that pike tend to be longer with a smaller cross-sectional area than bream and perch. Figures (b) to (e) now suggest that all forms of the pike data are unimodal but positively skewed with a long tail to the right. Bream seem to be substantially higher and slightly narrower than perch for the same length. The bimodality of perch shows up most clearly in (h). The plots are curved in (f) and approximately linear in (h). (b) Height and length.

Chapter 4

Section 4.3

1. Spin the pointer a large number of times and count the proportion of spins that fall in the grey sector. Yes, only if the spinner is balanced and the angle of the grey sector is known.

2. Yes, by tossing the thumb tack a large number of times and calculating the proportion landing point down. No, as the outcomes are not equally likely. We cannot argue from symmetry conditions as we could with the pointer.

3. (a) F. (b) T. (c) F (previous outcomes do not affect the current outcome). (d) F (all sequences are equally likely).

Section 4.4.2

1. $A = \{(1,3), (2,2), (3,1)\}$.

2. $A = \{HHT, HHHHT, HHHHHHT, \ldots\}$, $\overline{A} = \{T, HT, HHHT, \ldots\}$
$\overline{A}$ = "odd (or zero) number of tosses before the first tail."

3. $S = \{TT, THT, HTT, THHT, HTHT, THHH, HHH, HTHH, HHTT, HHTH\}$, $A = \{THT, HTT, HHH\}$.

Section 4.4.3

1. $S = \{(T1), (T2), \ldots, (T6), (H1), (H2), \ldots, (H6)\}$.
   (i) $A = \{(T3), (H3)\}$, (ii) $B = \{(H1), (H2), (H3), (H4), (H5), (H6)\}$
   (iii) A and $B = \{(H3)\}$
   (iv) A or $B = \{(T3), (H1), (H2), (H3), (H4), (H5), (H6)\}$.

2. $S = \{(A, A), (A, B), (A, O), (A, AB), (B, A), (B, B), (B, O), (B, AB), (O, A), (O, B), (O, O), (O, AB), (AB, A), (AB, B), (AB, O), (AB, AB)\}$
   (a) $C = \{(A, A), (B, B), (O, O), (AB, AB)\}$.
   (b) $D = \{(A, A), (A, B), (A, O), (A, AB), (B, A), (O, A), (AB, A)\}$.
   (c) C and $D = \{(A, A)\}$.

Section 4.4.4

1. (i) $\frac{1}{9}$. (ii) $\frac{1}{2}$.

2. (a) (i) $\frac{363}{5584} = 0.0650$. (ii) $\frac{911}{5584} = 0.1631$. (b) $S = \{MW, MS, MP, FM, FS, FP\}$ with probability distribution 0.305, 0.214, 0.098, 0.217, 0.101, 0.065 (see Table 4.4.2). (c) $S = \{W, S, P\}$ with probability distribution 0.522, 0.315, 0.163. (d) It is easier to compute any probability we might be interested in.

3. (a) People may lose more than one job. (b) No, as there may be fewer females with jobs (which is likely). (c) We need to compare the proportions of those with jobs losing their jobs for both males and females.

Section 4.5.1

1. (a) 0.3. (b) 0.5. (c) 0.5.

2. (a) 0.415. (b) 0.211. (c) 0.895. (d) 0.105. (e) 0.376. (f) 0.141.

3. (a) 0.06. (b) 0.66. (c) 0.1.

4. You can use either Table 4.4.2 or, for more accuracy, Table 4.4.1 as follows: (a) $1 - \frac{1703}{5584} = 0.6950$. (b) $\frac{1703+1196+548+1210}{5584} = 0.8340$. (c) $\frac{1196+548}{5584} = 0.3123$.

Section 4.6.1

1. (a) 0.2794. (b) 0.416.

2. $\frac{5}{7}$

3. (a) (i) 0.1010. (ii) 0.2639. (b) (i) 0.2142. (ii) 0.3470. (iii) 0.3152. (c) (i) 0.1699. (ii) 0.1590. (iii) 0.1631.

4. (a) (i) 0.353. (ii) 0.5524. (iii) 0.5804. (iv) 0.195. (b) (i) Divorced (from the female divorced column). (ii) To get the conditional probabilities, we divide each entry in the column by the *same* total, 0.336. (iii) Never married, 0.7238. (c) (i) 0.7238. (ii) 0.401. (d) 0.6328.

5. 0.2633.

Section 4.6.2

1. H = "homeless," S = "schizophrenic." $\text{pr}(H) = 0.008$, $\text{pr}(S) = 0.01$, and $\text{pr}(S \mid H) = \frac{1}{3}$.

2. 0.2002.

3. $\frac{1}{15}$.

Section 4.6.3

1. 3.02%, 8.8%.

2. 0.0921, 0.5294.

3. 0.1402. The ratio seems to be about 4. Yes.

Section 4.7.1

1. 0.0084 (assuming independence). May not be independent; he may think he is safe, as she is on the pill!

2. 0.3836. As chosen at random.

Section 4.7.3

1. (a) 0.065625. (b) Hard to tell. Perhaps matching and curing may not be independent. Depends on what is meant by "matching."

2. Let A = "sirens not working," B = "visual signals not noticed," C = "batteries ran down," D = "power failure," and E = "routing switches shut off." If the sirens going off means that the visual warnings will be noticed (i.e., $\overline{A}$ implies $\overline{B}$), then A and B are not independent. Otherwise A and B are likely to be independent. We have the direct causal sequence $C \to D \to E$, so that any event that is independent of C is also independent of D and E. If the sirens go off, then C will not occur (i.e., $\overline{A}$ implies $\overline{C}$). The same is true if the visual warnings are noted. Thus A and B are indirectly related to C.

Review Exercises 4

1. (a) $S = \{\text{Yes, No}\}$. (b) $S = \{0, 1, 2, \ldots\}$. (c) $S = \{0, 1, 2, \text{unknown}\}$. (d) $S = [a, b]$ depends on where you live (e.g., [−20°C, 50°C]). (e) $S = \{0, 1, 2, \ldots\}$. (f) $S = \{b_1, b_2, \ldots, b_{10}\}$. (g) S = {bus, car, train, boat, two-wheeler, walk, other}. The "other" category would include combinations such as car, then bus.

3. 0.75.

5. (a) 0.6320. (b) 0.1671. (c) 0.6632. (d) 0.6761. (e) 0.3530. (f) We recommend the typeset large-page format.

7. (b) 50.11%. (c) (i) 38.91%. (ii) 7.26%. (iii) 2.10%. (iv) 9.82%.

9. Four per thousand.

11. (a) To see how the course-work score related to the distribution of final grades. (b) Set of all members of the class. (c) 0.101. (d) 0.08615. (e) 0.853. (f) 0.104 (g) 0.339. (h) 0.0451. (i) 0.133. (j) 0.2049, 0.7951. (k) 0.1916. (l) 0.4497.

13. (a) 0.4. (b) 0.4001. (c) 0.00075. (d) Recommend that no animal meat be consumed for at least a day before the test is taken.

15. (a) Let W and L denote a win and a loss, respectively, for player 1. Sample space is $S = \{WW, WLW, WLL, LWW, LWL, LL\}$. (b) (i) 0.5. (ii) 0.5. (iii) 0.25. (iv) 0.75. (v) 0.5. (c) Independent, as pr(A and B) = pr(A) × pr(B). (d) A and B are not mutually exclusive since they have an outcome, WW, in common.

17. (a) (i) $0.9^3 = 0.729$. (ii) $0.1^3 = 0.001$. (iii) $3 \times 0.1 \times 0.9^2 = 0.243$. (iv) $3 \times 0.1^2 \times 0.9 = 0.027$. (b) (i) $\frac{349}{350} \times \frac{99}{100} = 0.9872$. (ii) $\frac{1}{100} \times \frac{349}{350} = 0.009971$. (iii) $\frac{1}{350} \times \frac{1}{100} = 2.857 \times 10^{-5}$. (c) (i) $0.729 \times 0.9872 = 0.7197$. (ii) $(1 - 0.729) \times 0.9872 = 0.2675$.

19. (b) 0.04. (c) 0.028. (d) 0.04. (e) 0.25. *(f) 0.00877.

21. (a) (i) $\frac{97{,}473}{98{,}826} = 0.9863$. (ii) $\frac{89{,}099}{98{,}304} = 0.9064$. (b) (i) 20 to 25. (ii) 35 to 40. Young men are risk-takers. (c) (i) $0.84834 \times 0.90636 = 0.7689$. (ii) $0.15199 \times 0.90636 = 0.1375$. (iii) $0.15166 \times 0.09364 = 0.0142$. (d) Independent life times. No. Positive. Keep each other going! (e) 0.9766. (f) Assume survival rates haven't changed since the data were compiled. However, we can expect life expectancies, and hence the survival rates, to have increased a little. Still, the answers will give a reasonable approximation.

23. $\frac{2}{3}$ if always switch, $\frac{1}{3}$ if never switch.

25. (a) 0.9984. (b) (i) $0.9984^{1000} = 0.2016$. (ii) $0.9984^{10000} = 1.11 \times 10^{-7}$. (c) 0.9952 (the probability of a double reversal is negligible).

27. 17.36%.

Chapter 5

Section 5.2

1. 1.10, 0.11. Probabilities must lie between 0 and 1 and add to 1.

2. $x = 1, 2, 3$ with probabilities 0.88, 0.11, 0.01.

3. $x = 0, 1, 2, 3$ with probabilities $\frac{1}{8}, \frac{3}{8}, \frac{3}{8}, \frac{1}{8}$.

4. 0.844.

5. (b) 0.17. (c) 0.21. (d) 0.41. (e) 0.84. (f) 0.71. (g) 0.74. (h) 0, impossible value. (i) 0, impossible values of X.

6. $0.9^3 = 0.729$. As $\text{pr}(X \leq 6) = 0.469$ and $\text{pr}(X \leq 7) = 0.522$, she should try seven times.

Section 5.3

1. (a) 0.2668. (b) 0.3504. (c) 0.01059. (d) 0.9894. (e) 0.1493. (f) 0.3488. (g) 0.3488. (h) 0. (i) 1.

2. (a) Here the 20% refers to a conceptual population rather than an actual one, so that the coin-tossing model is a candidate. The urn model does not apply, as the 10 chosen cars are not a simple random sample of all the cars parking. We need to have the probability of overstaying remaining constant at 0.2. Here $n = 10$ and $p = 0.2$. The arrival of each car is like a binomial trial. We can use Binomial($n = 10, p = 0.2$). (b) Urn model. The 50 cars watched must be a simple random sample of the cars. Here $N = 10{,}000$; M, the total number of overstayers, is unknown; and $n = 50$. We can approximate by Binomial($n = 50, p$), where $p = M/N$, as $n/N < 0.1$. We may be able to use $p = 0.2$ from part (a). (c) Urn model. The 50 people dialed must be a simple random sample of subscribers. Here $N = 7400, M = 2730$, and $n = 50$. We can approximate by Binomial($n = 50, p = M/N = 0.3689$) as $n/N < 0.1$. (d) Urn model. Here N and M are large unknown numbers, $M/N = 0.45$, and $n = 100$. We can approximate by Binomial($n = 100, p = 0.45$) as we can expect $n/N < 0.1$. (e) Urn model. Here $N = 100$, $M = 12$ and $n = 7$. We can approximate by Binomial($n = 100, p = M/N = 0.12$) as $n/N = 7/100 < 0.1$. (f) Urn model. Here N and M are large unknown numbers, $M/N = 0.64$, and $n = 50$. We can approximate by Binomial($n = 50, p = 0.64$) as we can expect $n/N < 0.1$. (g) Urn model. Here $N = 188, M = 99$, and $n = 30$. We cannot use the Binomial as $n/N > 0.1$. (h) Coin-tossing model. Here $n = 10$ and $p = 1/6$. Can use Binomial($n = 10, p = 1/6$). (i) Urn model. Here $N = 52, M = 4$, and $n = 7$. We need random shuffling of the pack before dealing. We cannot use the Binomial as $n/N > 0.1$. (j) Urn model. Here N and M are large unknown numbers, $M/N = 0.1$, and $n = 30$. We can approximate by Binomial($n = 30, p = 0.1$) as we can expect $n/N < 0.1$. (k) Either model is a candidate. For an urn model, the 30% must refer to the actual population sampled with M and N unknown and $n = 20$. For the coin-tossing model, we must have p constant. Since we can expect $n/N < 0.1$, either model will lead to Binomial($n = 20, p = 0.3$). (l) Like (a), either model is a candidate. The urn model, however, will apply only if the 50 patients can be regarded as a simple random sample of patients, which is probably not the case. The coin-tossing model is dubious as p may not be constant.

Section 5.4.1

1. 5.0.

2. 0.6.

Section 5.4.2

1. $\sqrt{3.8} = 1.9494$.

2. $\sqrt{0.6875} = 0.8292$.

3. $\sqrt{0.42} = 0.6481$.

Section 5.4.3

(a) 6, 4. (b) 7, 2. (c) 7, 2. (d) 11, 6. (e) 19, 10. (f) −15, 10. (g) −11, 10. (h) −11, 10. (i) −30, 14.

Review Exercises 5

1. (a) Coin-tossing model. $X_1 \sim \text{Binomial}(n = 20, p = 0.2)$. (b) Urn model with $N = 1000$, $M = 100$, and $n = 20$. Can use the Binomial approximation, $X_2 \sim \text{Binomial}(n = 20, p = 0.1)$. (c) N/A. (d) Coin-tossing model. $X_4 \sim \text{Binomial}(n = 120, p = 0.6)$. (e) Urn model with $N = 120$, $M = 70$, and $n = 10$. Can use the Binomial approximation, $X_5 \sim \text{Binomial}(n = 10, p = 7/12)$. (f) Urn model with $N = 20$, $M = 9$, and $n = 15$. Cannot use a Binomial approximation. (g) Coin-tossing model. $X_7 \sim \text{Binomial}(n = 12, p = 1/6)$. (h) N/A. (i) Coin-tossing model. $X_9 \sim \text{Binomial}(n = 12, p = 1/36)$. (j) Urn model with $N = 98$, $M = 44$, and $n = 7$. Can use the Binomial approximation, $X_{10} \sim \text{Binomial}(n = 7, p = 44/98)$.

3. (a) $p = \frac{\text{sampled area}}{\text{population area}} = \frac{20\times100\times100}{2000\times2000} = \frac{1}{20}$. (b) The 420 animals may be regarded as 420 independent Binomial trials each with probability of success, where success means "found in the sample area" and failure means "found outside the sample area." The four assumptions are satisfied because the animals (trials) are independent. $X \sim \text{Binomial}(n = 420, p = \frac{1}{20})$. (c) For a single plot, $p = \frac{100\times100}{2000\times2000} = \frac{1}{400}$ and $W \sim \text{Binomial}(n = 420, p = \frac{1}{400})$. (d) Any two of the following: (i) Animals tend to exhibit social tendencies and so are not generally independent. (ii) Animals do not move randomly but usually have well-defined territories or "home ranges." (iii) The presence of observers may disturb the animals so that they move out of the area. (iv) Some animals may be missed. Deer, for example, are very hard to spot.

5. (a) $L \sim \text{Binomial}(n = 160, p = \frac{120}{80{,}000})$. (b) $E(L) = np = 0.24$, or about one every four years. (c) Each ship has the same probability of being lost. Probably not true, but it may be a reasonable approximation. Shipwrecks are all independent.

7. (a) $\text{Binomial}(n = 12, p = 0.18)$. (b) pr(no failures) $= \text{pr}(X = 0) = 0.09242$.

9. If X = number of rods that perform satisfactorily, we assume that $X \sim \text{Binomial}(n = 10, p = 0.80)$. Then $\text{pr}(X \leq 4) = 0.006369$.

11. (a) −0.39, 0.29. (b) 0.48. (c) 0.52. (d) $E(X) = 1.04$, $\text{sd}(X) = 3.32$.

13. (b) $E(X) = 0.1341$, or 13.41 cents. (c) $\text{sd}(X) = 2.2509$. (d) The expected redeemable value in cents is calculated by $E(X)$ − (postage cost) × pr(getting a voucher) = 10.74. Thus the expected cost of a box of Almond Delight is \$1.84 − \$0.11 = \$1.73. At \$1.60, the alternative brand is the better value.

15. Let X = number of defective rivets in the sample. (a) $X \sim \text{Binomial}(n = 8, p = 0.01)$. Then $\text{pr}(X \geq 2) = 1 - \text{pr}(X \leq 1) = 0.00269$ (b) $Y \sim \text{Binomial}(n = 8, p = 0.02)$. Then $\text{pr}(Y \leq 1) = 0.9897$.

17. Let X = number of attempts made. (a) $X = 1, 2, 3, 4$ with probabilities 0.1, 0.09, 0.081, 0.729. (b) $E(X) = 3.439$, $\text{sd}(X) = 1.0131$. (c) E(cost) = \$7000 × $E(X)$ = \$24,073. (d) pr(still childless) $= (0.9)^4 = 0.6561$. (g) $\frac{9400}{100{,}000} = 0.094$, $\frac{7896}{90{,}600} = 0.0872$, $\frac{6662}{82{,}704} = 0.0806$, $\frac{5647}{76{,}042} = 0.0743$.

19. If X is the number of incorrect identifications out of 91 calves, then $X \sim \text{Binomial}(n = 91, p)$. We can compute $\text{pr}(X \leq 1)$ for different p, namely, for p = 0, 0.02, 0.04, 0.06,

0.08 we obtain 1, 0.4545, 0.1167, 0.0244, 0.0045, respectively. By calculating a couple more points between 0.04 and 0.06 and plotting a graph, we estimate that the value of p that gives $\text{pr}(X \leq 1) = 0.1$ is 0.042.

21. If X is the number of mutations, then $X \sim \text{Binomial}(n = 20{,}000, p = 1/10{,}000)$. (a) $\text{pr}(X = 0) = 0.1353$. (b) $\text{pr}(X \geq 1) = 1 - \text{pr}(X = 0) = 0.8647$. (c) $\text{pr}(X \leq 3) = 0.8571$.

23. (a) $X \sim \text{Binomial}(n = 10, p = \frac{1}{50})$. The Binomial distribution is appropriate since the person either wins or does not win a bottle, sampling is with replacement so that the draws are (supposedly) independent, and the probability of winning is the same for each draw, namely, $\frac{1}{50}$. (b) (i) $\text{pr}(X = 0) = 0.8171$, $\text{pr}(X \leq 2) = 0.9991$. (ii) $\text{pr}(X \geq 3) = 1 - \text{pr}(X \leq 2) = 0.000864$. (c) Since the probability in (b)(ii) is so small, the fact that one person has won 3 bottles would lead us to suspect that the names were not properly stirred. (d) By (b)(ii), $\text{pr}(E_i) = \text{pr}(X \geq 3) \approx 0.00086 \neq 0$. If the event E_i has occurred for at least 3 people, then the event E_j cannot occur for the other people (since there are only 10 bottles to be won), so $\text{pr}(E_j) = 0$. This establishes that the outcome of E_i depends on the occurrence or otherwise of the other outcomes. Hence the E_i's are not independent. (e) Let Y = number of people who win 3 or more bottles. Assuming independence of the E_i's and regarding E_i as a binomial trial, we have $Y \sim \text{Binomial}(n = 50, p = 0.000864)$. Hence $\text{pr}(Y \geq 1) = 1 - \text{pr}(Y = 0) = 0.0423$. (f) Although the last probability in (e) is somewhat larger (slightly over 4 chances in 100), our opinion in (c) is not changed.

Chapter 6

Section 6.2.2

2. (a) 0.1908. (b) 0.4554. (c) 0.1908.

3. (a) 0.09121. (b) 0.5889. (c) 0.09121.

Section 6.2.3

2. (a) 249.4. (b) 298.9. (c) 245.5. (d) 274.4.

3. IQs are given to the nearest whole number. (a) 113. (b) 135. (c) 92.

Section 6.2.4

1. (a) [154.75, 170.65]. (b) [152.50, 172.90].

2. (a) [23.85, 30.75]. (b) [22.05, 32.55].

Section 6.3.1

1. (a) 280 is 0.875 sd's above the mean. (b) 250 is 1 sd below the mean. (c) 270 is 0.25 sd's above the mean.

2. (a) 80 is 1.3333 sd's below the mean. (b) 110 is 0.6667 sd's above the mean. (c) 90 is 0.6667 sd's below the mean.

Section 6.3.2

1. (a) $z = 1.9600$, $[\mu - z\sigma, \mu + z\sigma] = [234.6, 297.4]$. $z = 1.2816$, $[245.5, 286.5]$. (b) $z = 1.6449$, $\mu + z\sigma = 292.3$. $z = 2.3263$, 303.2. (c) $z = 1.6449$, $\mu - z\sigma = 239.7$. $z = 2.3263$, 228.8.

2. (a) $z = 1.96$, $[\mu - z\sigma, \mu + z\sigma] = [71, 129]$. $z = 1.2816$, $[81, 119]$. (b) $z = 1.6449$, $\mu + z\sigma = 125$. $z = 2.3263$, 135. (c) $z = 0.6745$, $\mu - z\sigma = 90$. $z = 1.2816$, 82.

Section 6.3.3

1. (a) (i) -2. (ii) 2. (iii) 0.5. (iv) -0.4. (b) The same numbers.

2. (a) 0.629. (b) 0.629. (c) 0.378. (d) 0.182. (e) 0.251.

3. (a) 0.309. (b) 0.067. (c) 0.3411.

Section 6.4.3

1. (a) 4, 3.61. (b) 1, 3.61. (c) 3.5, 6.56. (d) 6.5, 5.39. (e) -3.5, 5.39. (f) -1, 6.16. (g) 1, 5.83. (h) 9, 5.83. (h) 11.5, 6.56. (j) 9.5, 8.49. (k) 6.5, 17. (l) 29.5, 21.73.

2. Let D be the score for a depressed child, and N be the score for a child not depressed. We have $D \sim \text{Normal}(\mu = 11.2, \sigma = 6.8)$ and $N \sim \text{Normal}(\mu = 8.5, \sigma = 7.8)$. (a) We require $\text{pr}(D < N)$ or $\text{pr}(N - D \geq 0)$. Now $X = N - D \sim \text{Normal}(\mu = 8.5 - 11.2, \sigma = \sqrt{7.8^2 + 6.8^2})$, that is, $\text{Normal}(\mu = -2.7, \sigma = 10.3480)$. $\text{pr}(X > 0) = 1 - \text{pr}(X \leq 0) = 0.3971$. (b) Some of them may be depressed.

Review Exercises 6

1. (a) (i) $\text{pr}(X > 141) = 1 - \text{pr}(X \leq 141) = 0.06681$. (ii) $\text{pr}(120 < X < 132) = 0.4515$. (iii) $\text{pr}(X < 118.5) = 0.2266$. (b) -0.6. 120 is 0.6 sd's below the mean. (c) 1.4. 140 is 1.4 sd's above the mean. (d) $0.85 = \text{pr}(X \leq a)$, $a = 136.4$. (e) $0.9 = \text{pr}(X \leq b)$, $b = 138.8$. (f) $0.01 = \text{pr}(X \leq c)$, $c = 102.7$. (g) [109.6, 142.4]. (h) [117.6, 134.4]. (i) Total $= T \sim \text{Normal}(\mu = 20 \times 126, \sigma = \sqrt{20} \times 10)$, or Normal(2520, 44.72). $\overline{X} = \frac{T}{20} \sim \text{Normal}(\mu = 125, \sigma = \frac{10}{\sqrt{20}})$, or Normal(125, 2.24).

3. Let X be the distance reached. Then $X \sim \text{Normal}(\mu = 125, \sigma = 10)$. (a) $\text{pr}(X \geq 120) = 1 - \text{pr}(X < 120) = 0.6915$. (b) $\text{pr}(X \geq x) = 0.95$, $x = 108.55$, or about 109 cm.

5. Here $X \sim \text{Normal}(\mu_X = 266, \sigma_X = 16)$ (a) $\text{pr}(X > 349) = 1.0657 \times 10^{-7}$. (b) Looks like a wrong decision! (c) Let Y be the number of gestation periods lasting 349 or more days. Then $Y \sim \text{Binomial}(n = 10{,}000{,}000, p = 1.0657 \times 10^{-7})$ and $\text{pr}(Y \geq 1) = 1 - \text{pr}(Y = 0) = 0.6555$. It is a possibility. (This calculation relies on the Normal distribution giving a reliable probability in (a) for such a rare event.)

7. (a) 0.0401. (b) 0.0401. (c) $X \sim \text{Binomial}(n = 50, p = 0.0401)$. (d) The new cutoff level is 7.2184. (e) 0.0133.

9. (a) 11.5%. (b) $Y \sim \text{Binomial}(n = 100, p = 0.1151)$. (c) $P = (65-25)Y + (70-26.5)(100-Y) = 4350 - 3.5Y$. $E(P) = 4350 - 3.5E(Y) = 4350 - 3.5 \times 11.51 = 4310$, or \$43.10. *(d) 0.1208, or 12.1%.

11. (a) 0.1992. (b) 0.3830. (c) 0.2220. (d) Tend to be related.

13. Let X be the weight of an adult, M the weight of a male, and F the weight of a female. Then $X \sim \text{Normal}(\mu_X = 73, \sigma_X = 13)$, $M \sim \text{Normal}(\mu_M = 78, \sigma_M = 12)$ and $F \sim \text{Normal}(\mu_F = 68, \sigma_F = 12)$. (a) 0.5277. Both assumptions are doubtful because of sex differences in weight, and friends or acquaintances often take a lift together. (b) Normal(858, 39.800). 0.9275. (c) 0.6745. (d) 0.2386. *(e) Assume all men and there are m of them with total weight T. $T \sim \text{Normal}(m \times 78, \sqrt{m} \times 12)$. $\text{pr}(T \leq 800) = 0.9 \Rightarrow \text{pr}(Z \leq z) = 0.9$, where $z = \frac{800 - 78m}{12\sqrt{m}}$. Our target value of z is 1.28155. When $m = 8, z = 5.19$; $m = 9, z = 2.72$; and $m = 10, z = 0.53$. Limit of 9. (f) University students will be lighter, as people tend to gain weight with age.

15. (a) 0.6827. *(b) Let X be the number of weighings. Then $\text{pr}(X = 1) = \text{pr}(Y \geq 49.9) = 0.8413$. $\text{pr}(X = 101) = \text{pr}(Y < 49.9) = 0.1587$. Hence $E(X) = 1 \times 0.8413 + 101 \times 0.1587 = 16.87$, or about 17.

17. (a) (i) Normal(25, 0.35355). (ii) 0.00234. (b) $Y \sim$ Normal(μ = 49 × 0.05, σ = 7 × 0.02), or Normal(μ = 2.45, σ = 0.14). (c) 0.0127. (d) (i) Normal(2.45, 0.98). (ii) 0.2073. (e) M is much more variable than Y.

19. (a) 0.2119. Binomial(n = 64, p = 0.2119). (b) 6533 kg.

Chapter 7

Section 7.2.1

1. (a) 30. (b) 80. (c) 800.

2. Let X be the monthly profit. Then $\overline{X} \sim$ Normal($\mu = 10, \sigma = \frac{3.5}{\sqrt{6}} = 1.4289$). pr($\overline{X} > 8.5$) = 0.8531.

Section 7.2.2

1. 0.07865 (8%).

2. (a) 0.8807. (b) 0.2779.

3. (a) Although n is only 28, we shall assume we can use the Normal approximation. 0.0001. No, much higher for smokers, as the previous probability is small. (b) 0.2350. (c) No, as we are measuring different probabilities. In (a) the probability is associated with a sample mean, whereas in (b) the probability is associated with a single random observation, which will vary a lot more than a mean.

Section 7.2.3

1. (b) [5.07, 5.55]. (c) 4.09.

2. (a) [679.7, 910.5]. (b) No, as it does not lie in the two-standard-error interval.

3. Using a two-standard-error interval, [234.2, 265.8].

Section 7.3.1

1. (a) [0.29, 0.49]. (b) No, as 0.23 does not lie in the two-standard-error interval. 3.1.

2. Professionals: [0.30, 0.34]. University: [0.38, 0.42].

3. Using a two-standard-error interval: [0.445, 0.515]. No. The values of 0.5 to 0.515 are plausible values, as they lie in the interval.

4. (a) [0.167, 0.185]. (b) [0.19, 0.24]. (c) 55.8%. [0.53, 0.59].

Section 7.5

1. (a) 0.04314. (b) 3.96. Yes, as this number is much greater than 2. (c) [0.09, 0.26]. (d) There appears to be a drop-off in usage between grade 11 and grade 13.

2. (a) 1.36. No evidence, as less than 2. (b) [−0.08, 0.42]. Between −0.08 and 0.42.

3. $\widehat{p}_1 - \widehat{p}_2 = 0.31 - 0.20 = 0.11$, se($\widehat{p}_1 - \widehat{p}_2$) = 0.05291. (a) 2.08 standard errors apart. We suspect a difference. (b) [0.004, 0.216]. (c) Clients at a clinic at two different times. (d) Asked the same people twice.

4. (a) None: [559, 682]; 1–30: [627, 805]; 31–70: [680, 910]. (b) (i) The two-standard-error interval is [44, 305]. Some evidence of a difference. (ii) The two-standard-error interval is [−13, 203]. No evidence of a difference. (c) We can't use individual two-standard-error intervals for making comparisons, as the combined variation is not taken into account. (d) We have an observational study and not a controlled experiment. (e) No. The two-standard-error intervals refer to true means, not individual observations.

Review Exercises 7

1. (a) $\mu_Y = 609, \sigma_Y = 60.85$. (b) $\mu_W = 609, \sigma_W = 7 \times 23 = 161$. (c) $\mu_{\overline{X}} = 87, \sigma_{\overline{X}} = 3.83$. (d) Normal; central limit theorem. (e) More Normal looking, and narrower in distributional shape.

3. (a) $X \sim \text{Binomial}(n = 870, p = 0.79)$. (b) 29. Not random. (c) 1.95. (d) The first recommendation questions the randomness of the jury selection process, whereas the second questions whether a particular factor, race in this case, was involved in the teacher-hiring process. The first does not attempt to apportion blame to any factor in the event that the selection process was not random; the second does tend to attribute blame on racial discrimination as a primary factor in the event that the selection process was not random. The first mentions a specific group "social scientist." (e) No, we do not agree with the court. While the occurrence of gross statistical disparities is an indication that the selection of teachers is not random, the factors that are used as criteria in the hiring process may not necessarily be based on discrimination.
Reasons for: Past experience in other areas may have pointed to the practice of discrimination, and the strong opinion is an expression of the court's disapproval in law. Discrimination is hard to prove, so the opinion expressed tends to shift the burden to proving nondiscrimination instead.
Reasons against: In the absence of further evidence, it is dangerous to blame just one factor in the actual hiring process. While it may be true that discriminatory practices at other levels may have led to black teachers having lower qualifications, the use of qualifications as a hiring criterion is itself not discrimination. The problem has to be tackled at other levels.

5. (a) If $\widehat{P}$ is the sample proportion of voters who voted for her, then $\widehat{P} \sim \text{Normal}(\mu = p = 0.6, \sigma = \sqrt{\frac{0.6 \times 0.4}{n}})$. (b) 0.9981.

7. (a) E(X) = 0.05263, sd(X) = 0.9986. (b) (i) 2.6316, 7.0612. (ii) 0.05263, 0.14122. (c) (i) $\mu = n \times \frac{2}{38} = n(0.05263)$, $\sigma = \sqrt{n}(0.9986)$. (ii) $\mu = \frac{2}{38} = 0.05263$, $\sigma = \frac{0.9986}{\sqrt{n}}$. (d) 0.3547. (e) 0.0478, 0, 0.

9. (a) 0.024. (b) $\frac{(0.4)(0.6)}{\sqrt{200}} = 0.0170$. (c) Between the two. Use (a), as it is more conservative.

11. (a) 0.0154. (b) 0.0381. (c) 2.47. (d) 0.0495. (e) The ratios are $\frac{0.0495}{0.0316} = 1.57$ and $\frac{0.0495}{0.0381} = 1.30$. (f) It will be too small for looking at proportions calculated from subsets of the data and for looking at differences between proportions. The poll will appear to be more accurate than it really is in these situations.

13. (a) Interval is [0.378, 0.477]. (b) 2.55 standard errors apart, which is greater than 2, suggesting there is a difference. (c) No, as $\text{se}(\overline{x}_1 - \overline{x}_2) < \text{se}(\overline{x}_1) + \text{se}(\overline{x}_2)$. (d) Yes, as the intervals don't overlap and (c) implies that the two-standard-error interval will not contain zero. (e) Individuals tend to move around in groups who do similar things, like wearing their caps the same way.

15. (a) 5.1. Yes. (b) [6.2, 14.2]. (c) Babies who were preterm and had low birth weight. (d) Not necessarily. The desire to breast feed may be related to the mother's IQ. Mothers are not randomly selected for each group. (e) It rules out the query in (d).

17. (a) Answers to the three questions are: (i) Those in the "humane" group should score higher than those in the "inhumane" group. (ii) In the "humane" group, there should be a trend downward from typical to atypical. In the "inhumane" group, the reverse should hold. (iii) We expect the "control" group to be in between the "humane" and "inhumane" "no information" groups. (b) No difference has been demonstrated between any of the three in the "humane" group. No difference has been demonstrated between any of the three in the "inhumane" group. The "control" group scored lower than the "typical" and "no information" subgroups in the "humane" group and scored higher than the "typical" subgroup in the "inhumane" group. Groups 2 and 3 scored higher than groups 5, 6, and 7.

Group 4 scored higher than groups 5 and 7. The data support (i) in (a), do not support (ii), and partly support (iii). (c) Use larger samples with groups for each sex. (d) For example: Are there sex or age differences in the participants? What effect does the sex of the guard have?

***19.** (a) $\widehat{t}_{Can} = 0.1 \times 27 \times 10^6, \widehat{t}_{US} = 0.2 \times 250 \times 10^6, \widehat{t}_{Mex} = 0.6 \times 90 \times 10^6$. $\widehat{p} = 0.2907$. (c) $\text{sd}(\widehat{p}) = \sqrt{a_{Can}^2 \,\text{sd}(\widehat{p}_{Can})^2 + a_{US}^2 \text{sd}(\widehat{p}_{US})^2 + a_{Mex}^2 \text{sd}(\widehat{p}_{Mex})^2}$.
$\text{se}(\widehat{p}) = \sqrt{a_{Can}^2 \,\text{se}(\widehat{p}_{Can})^2 + a_{US}^2 \text{se}(\widehat{p}_{US})^2 + a_{Mex}^2 \text{se}(\widehat{p}_{Mex})^2} = 0.00944$.

Chapter 8

Section 8.2

1. [5.00, 5.62].

2. (a) [0.493, 0.516].

3. (a) (i) [559, 682]. (ii) [677, 913]. (b) (i) [697, 893]. (ii) [635, 955].

Section 8.3

1. (a) [0.647, 0.753]. (b) [0.617, 0.783].

2. [0.280, 0.440].

Section 8.4

1. (a) $df = 5$, [−0.496, 0.152]. (b) The two-standard-error interval [−0.424, 0.080] is narrower. (c) From either interval there is no evidence of a difference between the two means.

2. $df = 27$, [−229, 70].

Section 8.5

1. (a) Situation (c). (b) Situation (a). (c) Situation (a). (d) Situation (b). (e) Situation (a).

2. (a) [0.23, 0.37]. 23% to 37%. (b) [0.23, 0.35]. 23% to 35%. (c) [0.07, 0.17]. 7% to 17%. (d) [0.03, 0.13]. 3% to 13%. (e) [0.01, 0.11]. 1% to 11%.

3. (a) Situation (b), [0.29, 0.39] (i.e., 29% to 39%). (b) Situation (a), [−0.001, 0.08] (i.e., 0.1% to 8%). (c) Situation (a), [0.04, 0.12] (i.e., 4% to 12%). (d) Situation (c), [0.01, 0.11] (i.e., 1% to 11%). (e) The people in the UK, which spends the least on health care, seem happiest with their system.

Section 8.6

1. (a) 2401. (b) 1225. (c) 1225.

2. Using the 95% level of confidence: (a) 35. (b) 1193.

Review Exercises 8

1. (a) No. (b) $\bar{x} = 3.289$, $s_X = 0.04701$. (c) [3.26, 3.32]. Yes, as the value of 3.25 lies outside the interval. (d) The interval is now [3.27, 3.30]. If the multiplier did not change, the width of the confidence interval would halve. The multiplier also gets slightly smaller with the increase in df, so with more significant figures you will see that the width of the new interval for $n = 40$ is slightly less than half the width of the interval for $n = 10$. *(e) 38.

3. (a) A single piece of paper may look like it would take less time to answer. This should lead to a higher response rate. (b) (90% CI) [−0.014, 0.13]. The difference in the response rates could be anywhere between 1.4% in favor of the two-sheet version to 13% in favor of the one-sheet version. (95% CI) [−0.03, 0.15]. The difference in the response rates could be anywhere between 3% in favor of the two-sheet version to 15% in favor of the one-sheet version. (c) Since both intervals contain zero, the printing format may make no difference. We would be inclined to use the two-sided version in accordance with our intuition and the slight suggestion of an increased response rate given by the data. We note that both response rates were quite low.

***5.** (a) 1037. (b) 995. (c) 955. (d) Practical questions include how to sample students from the many schools in a city. There are also questions of definition such as how to handle students whose parents were never legally married but have now split up. (Whether to include them depends on the purpose of the survey. Are you interested in legalities or whether the students are living with both parents?)

7. (a) We would plot the data using box plots to compare groups (as these groups are quite large). We would also look at stem-and-leaf plots or histograms to check distributional shape. (b) [6.2, 14.2]. Breast-fed children have IQs that are between about 6 to 14 points higher on average than the IQs of non-breast-fed children. *(c) 1162. (d) The babies were all preterm and very small. The results may be special to this population. (e) It is an observational study in which mothers chose whether to breast feed. It does not demonstrate that the effect is causal. (f) [1.4, 19.0]. The interval has become more than twice as wide. (g) 0.329. There is approximately 1 chance in 3 that the bottle-fed baby will end up with a higher IQ. The confidence interval refers only to the difference between the means and says nothing about any other aspect of the distribution. In fact, there is substantial overlap between the IQ distributions for the two groups.

9. (a) The data suggest that ex-smokers have healthier eating patterns on average than smokers, both among manual workers and among nonmanual workers. Similarly, nonmanual workers seem to have healthier eating patterns on average than manual workers, for both smokers and ex-smokers. (b) Using only nonmanual workers, the following are 95% confidence intervals for the differences in the true proportions between ex-smokers and smokers for each category. Breakfast: [0.14, 0.25]. Brown bread: [0.12, 0.25]. Fresh fruit: [0.12, 0.24]. Fried food: [−0.17, −0.07]. The ex-smokers clearly do better than the smokers when it comes to both having breakfast and having a healthy breakfast.

11. (a) [0.461, 0.500]. (b) [0.043, 0.097]. (c) [0.050, 0.110]. (d) White: [0.539, 0.561]. Black: [0.09, 0.13]. Hispanic: [0.31, 0.39]. Asian: [0.32, 0.52]. Difference between Asian and Hispanic: [−0.04, 0.18].

13. 0.01057.

15. (a) [10.1, 11.8]. (b) [1.0, 3.6]. (c) [−0.98, 2.04].

***17.** (a) $m \geq z\sqrt{(\hat{p}_1(1-\hat{p}_1)+\hat{p}_2(1-\hat{p}_2))/n}$ or $n \geq (\frac{z}{m})^2 \times \{\hat{p}_1(1-\hat{p}_1)+\hat{p}_2(1-\hat{p}_2)\}$. (b) Both 0.5. (c) Take the proportions $\hat{p}_1$ and $\hat{p}_2$ equal to 0.5 to make the expressions in the inequality for n in (a) to be as large as possible. Then $n \geq (\frac{z}{m})^2 \times \{\frac{1}{2}\times\frac{1}{2}+\frac{1}{2}\times\frac{1}{2}\}$, that is, $n \geq \frac{1}{2}(\frac{z}{m})^2$. (d) $m \geq z\sqrt{(\hat{p}_1+\hat{p}_2-(\hat{p}_1-\hat{p}_2)^2)/n}$, that is, $n \geq (\frac{z}{m})^2 \times (\hat{p}_1+\hat{p}_2-(\hat{p}_1-\hat{p}_2)^2)$. Again setting $\hat{p}_1 = \hat{p}_2 = 0.5$ to maximize the right-hand side of the previous expression, we get $n \geq (\frac{z}{w})^2$.

Chapter 9

Section 9.2

1. (a) H_0: $\mu = 750$. (b) H_1: $\mu < 750$. (c) The sample mean $\overline{x}$ is too much smaller than 750 to be explained simply in terms of sampling variation.

2. (a) H_0: $\mu_{white} - \mu_{blue} = 0$. (b) H_1: $\mu_{white} - \mu_{blue} > 0$. (c) $\overline{x}_{white}$ is sufficiently larger than $\bar{x}_{blue}$ that it could not be explained simply in terms of sampling variation.

3. (a) H_0: $\mu_{French} - \mu_{math} = 0$. (b) H_1: $\mu_{French} - \mu_{math} > 0$. (c) $\overline{x}_{French}$ is sufficiently larger than $\overline{x}_{math}$ that it could not be explained simply in terms of sampling variation.

4. (a) H_0: $\mu = 10$. (b) H_1: $\mu \neq 10$. (c) $\bar{x}$ is too far from 10 (in either direction) to be explained simply in terms of sampling variation.

5. (a) H_0: $p = 0.5$. (b) H_1: $p \neq 0.5$. (c) $\hat{p}$ is too far from 0.5 to be explained simply in terms of sampling variation.

6. Same as for 5.

7. (a) H_0: $p_{giveaway} - p_{none} = 0$. (b) H_1: $p_{giveaway} - p_{none} < 0$, expecting people attracted by free gifts to be less loyal. (Could also argue for a "$\neq$" alternative.) (c) $\hat{p}_{giveaway}$ is sufficiently smaller than $\hat{p}_{none}$ that it could not be explained simply in terms of sampling variation.

8. (a) H_0: $\mu = 0$. (b) H_1: $\mu \neq 0$. (c) $\bar{x}$ is sufficiently far from zero that it could not be explained simply in terms of sampling variation.

Section 9.3

1. Let p be the proportion in the womb sucking their left thumbs. (a) H_0: $p = 0.1$ versus H_1: $p < 0.1$ (prior theory). $t_0 = -3.086$ and the (one-tailed) *P*-value is pr($Z \leq -3.086$) = 0.001. There is very strong evidence against H_0 in favor of H_1. (*Note*: The 10% rule gives n to be at least 960, which is not true.) (b) We assume that thumb-sucking behavior of fetuses relates to left- and right-handedness after birth (apart from some switching due to such things as "birth stress"). We also assume that Belfast left-handedness rates are 10% or more.

2. Let p be the probability of a person dying in the month before his or her birthday. (a) H_0: $p = \frac{1}{12}$ versus H_1: $p < \frac{1}{12}$ (prior theory). $t_0 = -3.327$ and the (one-tailed) *P*-value is pr($Z \leq -3.327$) = 0.0004. There is very strong evidence against H_0 in favor of H_1. (*Note*: The 10% rule gives n to be at least 960, which is not the case.) (b) Notable Americans. We assume that "ordinary" people have the same survival behavior as "notable" people. We also assume some sort of uniformity of the birth and death rates throughout the year. For example, if most births were in the summer and most deaths in the winter, for reasons that had nothing to do with "postponing" death, our estimate of p would be small.

3. Let μ be the mean nicotine content. H_0: $\mu = 18$, H_1: $\mu > 18$ (as the prior claim is one-sided). $t_0 = 2.0055$, and using $T \sim$ Student($df = 11$), the (one-tailed) *P*-value is pr($T \geq 2.00555$) = 0.035. We do have some evidence that the claim is false.

4. Let p_S be the proportion of smoking mothers with infants getting colic and p_{NS} be the proportion of nonsmoking mothers with infants getting colic. H_0: $p_S - p_{NS} = 0$ versus H_1: $p_S - p_{NS} \neq 0$. For independent proportions, $t_0 = 5$ and the (two-tailed) *P*-value is $2 \times$ pr($Z > 5$) = 0.0000. There is very strong evidence against H_0 (i.e., very strong evidence that smoking mothers are more likely to have colicky babies).

5. Let p_{ES} be the proportion knowing that Christ was resurrected on Easter Sunday and p_{GF} be the proportion knowing Christ was crucified on Good Friday. H_0: $p_{ES} - p_{GF} = 0$ versus H_1: $p_{ES} - p_{GF} \neq 0$ (no prior theory). For Situation (c), $t_0 = 1.945$ and the (two-tailed) *P*-value is $2 \times$ pr($Z > 1.945$) = 0.052. We do have some evidence against H_0 (i.e., evidence of a real difference).

6. Let p_B be the probability of accepting if claimed beneficial and p_{NB} be the probability of accepting if claimed not beneficial. H_0: $p_B - p_{NB} = 0$ versus H_1: $p_B - p_{NB} \neq 0$ (no prior theory). $t_0 = 5.368$, and the (two-tailed) *P*-value is $2 \times$ pr($Z > 5.368$) = 0.0000. There is very strong evidence against H_0, that is, very strong evidence that journals are more likely to accept articles claiming that intervention is beneficial. (*Note*: The 10% rule requires n_B to be at least 11 and n_{NB} to be at least 243, so that the large-sample theory is a little suspect.) Assume the 70 journals to get the "beneficial" version were selected at random and they independently made decisions; we do not want the situation where different journals are using the same referees to make their decisions.

7. Let μ_{HS} be the true mean length in hedge-sparrow nests and μ_{GW} be the true mean length in garden-warbler nests. H_0: $\mu_{HS} - \mu_{GW} = 0$ versus H_1: $\mu_{HS} - \mu_{GW} \neq 0$. $t_0 = 4.948$, and using $T \sim$ Student($df = 57$), the (two-tailed) *P*-value is $2 \times$ pr($T \geq 4.948$) $= 0.0000$. There is very strong evidence of a true difference in mean lengths. Bigger birds may select hedge-sparrow nests. There may be differences between habitats containing mainly hedge sparrows or mainly garden warblers.

Review Exercises 9

1. (a) Let p_T and p_C be the respective probabilities that a person will return a completed questionnaire with or without telephone contact. H_0: $p_T - p_C = 0$ versus H_1: $p_T - p_C \neq 0$. For independent proportions, $t_0 = 9.613$ and the (two-tailed) *P*-value is 0.0000. There is very strong evidence that phone calls make a difference. (b) A 95% CI for $p_T - p_C$ is [0.27, 0.41]. With 95% confidence, calls increase the response rate by between 27 and 41 percentage points. (c) Even though there is substantially less nonresponse in the treatment group, it is still quite high, so nonresponse bias would still be a worry. If only the people they contacted by phone were sent questionnaires, this could add further bias.

3. (a) Let p_{none} and p_{coup} be the respective probabilities that a customer buys the same brand next time without or with a coupon. H_0: $p_{none} - p_{coup} = 0$ versus H_1: $p_{none} - p_{coup} \neq 0$. For independent proportions, $t_0 = 19.566$ and the (two-tailed) *P*-value is 0.0000. There is very strong evidence that brand loyalty is lower. (b) A 95% CI for $p_{none} - p_{coup}$ is [0.34, 0.42]. With 95% confidence, brand loyalty is lower by between 34 and 42 percentage points when a coupon offer is involved. (c) Among new customers, the reduction in brand loyalty will probably be higher as new customers may have switched to the brand only during the coupon special. (d) The real issue here is whether such a promotion attracts sufficient new profits to be cost-effective. (It does not matter if only a low proportion of the customers who switched during the promotion stayed with the brand.) To test this, it would be better to look at sales trends before and after the promotion and analyze these to see if there has been a significant jump in sales. Why might you expect to see a short-term drop in sales immediately after a promotion?

5. (a) H_0: $p_{pay} - p_{control} = 0$ versus H_1: $p_{pay} - p_{control} \neq 0$ (though a case could perhaps be made for ">"). For independent proportions, $t_0 = 2.0816$ and the (two-tailed) *P*-value is pr($Z \geq 2.082$) $= 0.037$. There is evidence that payments make a difference. (b) The 95% CI for $p_{pay} - p_{control}$ is [0.007, 0.235]. There is likely to be a difference of between 0.7% and 24% more respondents with a \$5 payment. (c) Paying participants reduces the number of people you will be able to afford to survey. So one of the trade-offs is response rate versus sample size.

7. (a) H_0: $p_{before} - p_{during} = 0$ versus H_1: $p_{before} - p_{during} > 0$ (prior suspicion). For independent proportions, $t_0 = 2.7535$ and the (one-tailed) *P*-value is pr($Z \geq 2.7535$) $= 0.003$. There is strong evidence that the induced customers are less loyal. A 95% CI for $p_{before} - p_{during}$ is [0.029, 0.17], or between 3% and 17%. (b) The actual number of accounts retained and the value of the accounts to the bank; the cost of the promotion.

9. (a) H_0: $p_{smoke} - p_{nonsm} = 0$ versus H_1: $p_{smoke} - p_{nonsm} \neq 0$. For independent proportions, $t_0 = 5.294$ and the (two-tailed) *P*-value is 0.0000. There is strong evidence that smokers are more likely to have strokes. (b) A 95% CI for $p_{smoke} - p_{nonsm}$ is [0.015, 0.032], that is, the percentage of smokers having strokes is higher by between 1.5 and 3.2 percentage points.

11. (a) Let μ_{morn} and μ_{aft} be the respective population mean pH levels for morning and afternoon patients. H_0: $\mu_{morn} - \mu_{aft} = 0$ versus H_1: $\mu_{morn} - \mu_{aft} \neq 0$. $t_0 = 2.051$, and using Student($df = 48$), the (two-tailed) *P*-value is 0.046. We do have evidence of a difference. A 95% CI is [0.02, 2.00]. (b) Let p_{aft} and p_{morn} be the respective population proportions of morning and afternoon patients with a pH level below 2.5. H_0: $p_{aft} - p_{morn} = 0$ versus H_1: $p_{aft} - p_{morn} \neq 0$. For independent proportions, $t_0 = 2.169$ and the (two-tailed) *P*-value is 0.03. We have evidence of a true difference. A 95% CI for $p_{aft} - p_{morn}$ is [0.02, 0.40].

(c) It opens the possibility of bias. One would need to be assured that the allocation could not depend in any way on the metabolism of the patient.

13. (a) Substantial proportions of reoffenders in both groups. (b) Bias against the classification system as not paroling the "worst" prisoners should lower the reoffending rate in the high-risk group and make the rates in the two groups more similar. (c) Longer times would lead to higher proportions reoffending in both groups.

15. H_0: $p_{Asian} - p_{Hispan} = 0$ versus H_1: $p_{Asian} - p_{Hispan} \neq 0$. For independent proportions, $t_0 = 1.3019$ and the (two-tailed) *P*-value is 0.193. We have no evidence of a true difference. A 95% CI for $p_{Asian} - p_{Hispan}$ is $-0.035, 0.18]$.

17. (a) H_0: $p_{singpar} - p_{alone} = 0$ versus H_1: $p_{singpar} - p_{alone} \neq 0$. For independent proportions, $t_0 = 1.4986$ and the (two-tailed) *P*-value is 0.13. We have no evidence of a true difference. A 95% CI for $p_{singpar} - p_{alone}$ is $[-0.008, 0.060]$. (b) H_0: $p_{smoke} - p_{unhealthy} = 0$ versus H_1: $p_{smoke} - p_{unhealthy} \neq 0$. For situation (c) (Table 8.5.5), $t_0 = 2.6661$ and the (two-tailed) *P*-value is 0.008. We have strong evidence of a true difference. A 95% CI for $p_{smoke} - p_{unhealthy}$ is $[0.03, 0.19]$. (c) H_0: $p_{underw} - p_{overw} = 0$ versus H_1: $p_{underw} - p_{overw} \neq 0$. For situation (b) (Table 8.5.5), $t_0 = 1.6367$ and the (two-tailed) *P*-value is 0.10. We have only weak evidence of a true difference. A 95% CI for $p_{underw} - p_{overw}$ is $[-0.007, 0.081]$.

19. (a) $\frac{1}{3}$. (b) H_0: $p = \frac{1}{3}$, $t_0 = 2.7417$ and the (one-tailed) *P*-value is 0.003. (*Note:* The sample size is too small for the 10% rule.) (c) The *P*-value tells us that we have strong evidence that the identification rate is better than $\frac{1}{3}$. The magazine has got it wrong. (d) Possible differences in appearance can be compensated for by using blindfolds. There is the possibility of learning over the three attempts, so we could have more people and make only one identification each. Other ideas? (e) If you use H_0: $p = \frac{1}{2}$, the result is not significant.

Chapter 10

Section 10.1.2

1. (a) Plot the differences. Yes. (b) H_0: $\mu_{diff} = 0$ versus H_1: $\mu_{diff} \neq 0$. Assuming Normality, a paired-comparison *t*-test yields $t_0 = 1.25$ with a *P*-value of 0.22. There is no evidence of a difference between the head lengths. A 95% confidence interval for μ_{diff} is $[-1.23, 4.99]$. (c) How were the families selected? How were the measurements taken? Was a standard procedure strictly followed?

2. People would vary in how they administered the procedure. The size of any systematic difference between the two sets of calipers will vary with how the head measurement is taken and from what part of the head it is taken. As the cardboard calipers wear, they will tend to give bigger measurements.

Section 10.1.3

1. Let $\tilde{\mu}_{diff}$ be the population median of the differences. H_0: $\tilde{\mu}_{diff} = 0$ versus H_1: $\tilde{\mu}_{diff} \neq 0$. Using a sign test, the *P*-value is 0.84. There is no evidence of a difference. A sign 95% confidence interval for $\tilde{\mu}_{diff}$ is $[-3.80, 6.60]$.

2. Let $\tilde{\mu}$ be the population median. H_0: $\tilde{\mu} = 28$ versus H_1: $\tilde{\mu} \neq 28$. Using the sign test, the *P*-value is 0.18. There is no evidence that cyclozocine is an effective treatment. A sign 95% confidence interval for $\tilde{\mu}$ is $[27, 51]$.

Section 10.3

1. Using the *F*-test for equal means, $f_0 = 1.75$ with a *P*-value of 0.18. There is no evidence of racial differences. The dot plots indicate that the four samples have acceptably similar spreads (the standard deviations range from 0.49 to 0.84). A combined Normal probability plot looks reasonable, and the *W*-test is not significant.

2. Using the F-test for equal means, $f_0 = 7.90$ with a P-value of 0.001. There is very strong evidence of racial differences. Looking at the individual 95% confidence intervals for the four individual means, we see that Asian does not overlap with Caucasian or Native American, and Black does not overlap with Caucasian.

3. The dot plot shows a high outlier at 2.63. The numerator of the F-test measures how far apart the sample means are. Removing the outlier will reduce the mean of the Black group. This will move three of the means closer together, thus reducing the numerator. Removing the outlier will also substantially reduce the internal variation of the Black data, thus reducing the denominator. Since means are less sensitive than standard deviations to outliers, the F-ratio might be expected to increase, though we can't be sure!

4. We now find that $f_0 = 17.08$ and the P-value $= 0.000$, again indicating very strong evidence of racial differences. The individual 95% confidence intervals for Asian and Black no longer overlap, so that Asian is clearly different from the other three. The value of f_0 has increased, as suggested in 3.

5. The spread for Asian is much greater than that for the other three, which are quite similar. The standard deviation for Asian is 0.3933 and that for Caucasion is 0.1063, a ratio of nearly 4. The F-test is therefore of doubtful validity.

Review Exercises 10

1. (a) Let μ_{Gloo} and μ_{Cold} be the respective mean times for the Glooscap and Coldbrook schools. H_0: $\mu_{Gloo} - \mu_{Cold} = 0$ versus H_1: $\mu_{Gloo} - \mu_{Cold} \neq 0$. Using a Welch two-sample t-test, we have $t_0 = 2.73$ with P-value $= 0.012$. There is strong evidence of a difference between the two schools. A 95% confidence interval for the difference in the means is $[0.26, 1.91]$, that is, a difference of between about 0.3 and 1.9 seconds. The Normal theory methods appear to be applicable. *(b) Using medians, H_0: $\tilde{\mu}_{Gloo} - \tilde{\mu}_{Cold} = 0$ versus H_1: $\tilde{\mu}_{Gloo} - \tilde{\mu}_{Cold} \neq 0$. The Mann-Whitney (Wilcoxon) test gives a P-value of 0.036, which provides some evidence of a school difference. An approximate 95% confidence interval for the difference in the medians is $[0.16, 1.92]$. (c) The problem here is that we have an observational study, not an experiment, so that we cannot prove causality, namely, that the coach makes a difference. For example, the better runners might go to Glooscap. (How would you prove that coaching makes a difference?)

3. (a) From the dot plots we see that 56 in group 4 is an outlier. Also, group 5 has a larger mean and a larger spread than the other groups. (b) Using the F-test for equal means, we have $f_0 = 5.99$ and P-value $= 0.000$. There is very strong evidence of a difference in the group means. Without the outlier we have $f_0 = 7.55$ with P-value $= 0.000$. The conclusion is unchanged. If we leave out the outlier, the 95% confidence interval for the mean of group 5 does not overlap with the other four confidence intervals. The Normal probability plot is reasonable, and the maximum ratio of two standard deviations is (just) less than 2. (c) Without the outlier, Fisher's pairwise comparisons are: $1-2$: $[-1.88, 0.10]$, $1-3$: $[-2.56, 0.59]$, $1-4$: $[-1.82, 0.45]$, and $1-5$: $[-4.18, -1.72]$; $2-3$: $[-2.21, 1.13]$, $2-4$: $[-1.52, 1.03]$, and $2-5$: $[-3.86, -1.15]$; $3-4$: $[-1.12, 1.72]$, $3-5$: $[-3.46, -0.47]$, and $4-5$: $[-3.29, -1.24]$. The group 5 mean is clearly different from the other four means, which are not significantly different.

5. (a) Using a scatter plot, we see that poststerilization increases with presterilization. There is a definite trend upward. (b) We use a paired-comparison method. If $diff = pre - post$, we wish to test H_0: $\mu_{diff} = 0$ versus H_1: $\mu_{diff} \neq 0$. Using a t-test, $t_0 = 4.50$ with a (two-sided) P-value of 0.000. There is very strong evidence that sterilization makes a difference. A 95% confidence interval for μ_{diff} is $[82.9, 232.2]$. (c) A dot plot of the differences indicates an outlier. Removing it (donor number 16) and retesting gives us $t_0 = 5.62$ with a P-value of 0.000, so that there is no change in our conclusion. The confidence interval now becomes $[80.4, 179.6]$, which is a lot shorter. Various plots and the W-test indicate that the Normal theory seems to be applicable.

7. (a) We use the method of paired comparisons, as we have measurements on the same brand. Let *diff* = *high* − *low*. Then H_0: $\mu_{diff} = 0$ versus H_1: $\mu_{diff} > 0$. Using a one-sample *t*-test, $t_0 = 2.01$ with a (one-sided) *P*-value of 0.037. There is moderate evidence supporting the contention. The Normality assumption seems reasonable. (b) No, as two brands had negative differences. The test looks at the mean of the differences, not individual differences. (c) A scatter plot shows that there is a weak relationship that seems to be almost nonexistent for the seven observations closest to the origin. (d) Through randomization one can try and eliminate any systematic bias due to the order in which the ads are seen.

9. (a) We use a two-sample *t*-test. H_0: $\mu_{tiles} - \mu_{bedr} = 0$ versus H_1: $\mu_{tiles} - \mu_{bedr} \neq 0$. Using Welch's test, $t_0 = -0.69$ and *P*-value = 0.5. There is no evidence of a difference. For the tile data, the dot plot has the hint of an outlier. However, the Normal probability plot and *W*-test are supportive of the Normality assumption. For the bedrock data, the dot plot and Normal probability plot are unusual, and the *W*-test has a *P*-value of 0.03, indicating that the Normality assumption is not appropriate. (b) A Mann-Whitney (Wilcoxon) test has *P*-value = 1.000! The reason for this strange result is that it uses a t_0-statistic that takes the value of zero; this has a one-sided *P*-value of 0.5, which is doubled. An approximate 95% confidence interval for the difference in the medians is $[-10, 7]$. There is clearly no evidence of a difference. However, we need to be careful about the bedrock sample. The Mann-Whitney test is strictly a test to see if two independent samples come from the same distribution, and, although we don't have significance, the two samples are very different looking! (c) You would need to randomize the placing of the tiles and the selection of bedrock samples to avoid any systematic bias.

11. (a) From the dot plots, we see that the heterosexual data are slightly skewed while the homosexual data are strongly skewed, but in the opposite direction. The Normal probability plot and the *W*-test for the heterosexual data support the Normality assumption, while those for the homosexual data do not. (b) We use a two-sample *t*-test to test H_0: $\mu_{het} - \mu_{hom} = 0$ versus H_1: $\mu_{het} - \mu_{hom} \neq 0$. The Welch test gives $t_0 = 3.73$ with a *P*-value of 0.0008, giving very strong evidence for a difference. A 95% confidence interval for $\mu_{het} - \mu_{hom}$ is $[3.0, 10.3]$. (c) This is an observational study, so we cannot prove causality. The samples are not random, as a very high percentage (approximately 38%) of the heterosexual men died of AIDS.

13. (a) $\bar{x}_{87} = 101.59$, $s_{87} = 36.11$; $\bar{x}_{89} = 134.37$, $s_{89} = 76.89$; and $\bar{x}_{91} = 139.33$, $s_{91} = 66.19$. There is a substantial increase in the sample mean from November 87 to September 89 and August 91. There is also a substantial increase in the spread. From dot plots we see that this is possibly due to a few more expensive homes in 1989 and 1991. The box plots show similar trends, though the differences don't appear to be so obvious because of the compressed vertical scale. (b) The *F*-test yields $f_0 = 3.65$ with a *P*-value of 0.030, yielding moderate evidence against the three means being equal. The individual 95% confidence intervals indicate that 87 is very different from 89 and 91, which are similar. (c) Using Fisher's pairwise comparisons, we have 91 − 87: $[5.1, 70.3]$, 89 − 87: $[3.5, 62.1]$, 91 − 89: $[-29.0, 38.9]$. (d) We see that $s_{89} > 2s_{87}$. There are outliers present, and the 1989 data are clearly skewed. The Normal theory is therefore suspect. (e) Except for possible outliers, the dot plots indicate that much of the skewness seems to have been removed, and the spreads are more similar. (f) Using the *F*-test with the logged data, we get $f_0 = 4.00$ with a *P*-value of 0.022, so our conclusion is unchanged. The data are still skewed, as seen from the histogram of the residuals and the slight curvature in their Normal probability plot. The *W*-test for Normality has *P*-value < 0.01. However, the standard deviations are now similar. We again conclude that there is a significant increase in house prices from 1987 to 1989, and little change from 1989 to 1991. (g) An increase in a mean house price does not imply that all individual house prices go up; some will go down as well. Furthermore, any increase in the average may be due to just a few expensive houses being sold. These comments would apply to all houses. We would need to look at houses sold more than once or, if there are few in this category, compare houses of similar valuation.

15. (a) Select a random sample of 10 out of 20, and assign them to the standard treatment. (b) You could use a paired-comparison method based on the differences. (c) No, as we have two independent samples. (d) You can again use a paired-comparison method, though there are more complicated methods of analyzing design III. (e) To allow for any carry-over effect. (f) Design III. Any carry-over effect will be balanced out: half of the subjects will get treatment 1 first and the other half treatment 2 first. This is in contrast to design I, where you may not get 10 with each ordering.

17. (a) (i) One sample. Confidence interval (as we are not told what "effective" means). (ii) The percentages are approximately Normal with equal standard deviations. (iii) No placebo is used for a comparison. Also, some patients will have more headaches than others so that the (Binomial) percentages will have different standard deviations. (b) (i) Two independent samples. Confidence interval. (ii) The data set for each method is Normally distributed and the sets are independent. (We also need to use a completely randomized design.) (iii) Variability in the fertility, for example, of the plots, which may become confounded with the method difference. (c) (i) More than two independent samples. Confidence intervals. (ii) Assume that the numbers trapped for each color are Normally distributed and that the four standard deviations are all equal. Also, assume that the four samples are independent. (We need to have some randomized method, such as a randomized block design, for allocating the color to each board.) (iii) There may be a variation in the numbers of beetles in different parts of the field. (d) (i) Paired data. Hypothesis test. (ii) Differences Normally distributed with the same standard deviation. (iii) There may be a carry-over learning effect. The order of using the thread needs to be randomized so that half the students use the right-hand thread first, and the other half use the left-hand thread first.

Chapter 11

Section 11.1

1. (a) H_0: $p_0 = p_1 = \cdots = p_{10} = \frac{1}{10}$. Each $E_i = \frac{110}{10} = 11$. $x_0^2 = 20.36$ with $df = 9$ and P-value $= 0.016$. There is clear evidence that digits are not random. (b) We don't have a simple random sample of telephone numbers. However, the sampling method is not unreasonable, particularly if the (same) number used for each page is a random number.

2. (a) H_0: $p_A = \frac{1}{4}$, $p_{AB} = \frac{1}{2}$, and $p_B = \frac{1}{4}$. (b) $x_0^2 = 0.93$ with $df = 2$ and P-value $= 0.63$. No evidence against the law.

Section 11.2

1. Testing for independence yields $x_0^2 = 39.2$ with $df = 12$ and P-value $= 0.0001$. We very strongly reject independence. Deleting the Pacific group, $x_0^2 = 20.2$ with $df = 8$ and P-value$= 0.0095$ (i.e., still reject independence very strongly).

2. (a) Situation (2). (b) A test for homogeneity yields $x_0^2 = 55.08$, $df = 4$, and P-value ≈ 0. We very strongly reject homogeneity.

Review Exercises 11

1. H_0: $p_A = p_B = p_C = \frac{1}{3}$. $x_0^2 = 6.26$ with $df = 2$ and P-value $= 0.044$. There is some evidence that the three candidates are not equally preferred.

3. (a) A test for homogeneity yields $x_0^2 = 29.55$ with $df = 2$ and P-value ≈ 0. There is very strong evidence that the three rates are different. (b) Using independent proportions, a 95% confidence interval for $p_{high} - p_{med}$ is $[-0.11, 0.19]$. The corresponding interval for $p_{med} - p_{low}$ is $[0.21, 0.51]$. The interval for $p_{high} - p_{low}$ will have even greater values. We conclude that there is a difference between the low and the medium or high rates, but no difference between the medium and high rates. (c) No, the number of reoffenders will be greater than those reconvicted, as some won't get caught.

5. (a) A test for homogeneity yields $x_0^2 = 146.2$ with $df = 30$ and P-value ≈ 0. There is very strong evidence of a difference. (b) From the row proportions we see that, except for the medical students, the schools are similar with respect to status 3–6. Medicine has greater numbers in status 1.

7. (a) Incomplete information. For class discussion. (b) (i) Smoke, sleep prone, and breast feed. (ii) Socioeconomic status and season. Those in (i). (c) In what follows, the cases and controls come from independent populations, so that the corresponding proportions are independent.
Smoke versus crib death rate: $x_0^2 = 35.1$ with $df = 1$ and P-value ≈ 0. There is very strong evidence that smoking is related to crib death. A 95% confidence interval for the difference $p_{cases} - p_{controls}$ is $[0.20, 0.38]$. Smoking is clearly a major factor.
Sleep prone versus crib death rate: $x_0^2 = 36.1$ with $df = 1$ and P-value ≈ 0. There is very strong evidence that sleeping in the prone position is related to crib deaths. A 95% confidence interval for the difference $p_{cases} - p_{controls}$ is $[0.21, 0.39]$. Sleeping in a prone position is clearly a major factor.
Breast feed versus crib death rate: $x_0^2 = 24.1$ with $df = 1$ and P-value ≈ 0. There is very strong evidence that not breast feeding is related to crib deaths. A 95% confidence interval for the difference $p_{controls} - p_{cases}$ is $[0.11, 0.26]$. Not breast feeding is clearly a major factor.
Mother's smoking in the last two weeks versus crib death rate: We have already established a smoking relationship. Here $x_0^2 = 42.0$ with $df = 1$ and P-value ≈ 0. We would like to know, however, if incidence is related to the number of cigarettes smoked. From the table of column proportions we see that the proportion of cases increases with the number of cigarettes smoked.
Socioeconomic status versus crib death rate: $x_0^2 = 9.3$ with $df = 2$ and P-value $= 0.01$. There is strong evidence relating crib deaths to socioeconomic status. From the table of column proportions we see that the cases group are weighted more toward the lower V and VI classes.
Season versus crib death rate: $x_0^2 = 10.26$ with $df = 5$ and P-value $= 0.068$. There is weak evidence relating crib deaths to season.
(d) (i) $x_0^2 = 13.94$ with $df = 5$ giving P-value $= 0.016$. There is evidence that cot deaths do not occur uniformly throughout the year. The low point is January–February (N.Z. summer), and the high point is August–July (N.Z. winter). (ii) $x_0^2 = 4.18$ with $df = 5$ and P-value $= 0.52$. There is no evidence of a monthly variation. (e) We are using collapsed tables, which ignores interactions and hidden variables. We cannot conclude causality from this evidence alone.

9. (a) A test for homogeneity yields $x_0^2 = 48.7$ with $df = 35$ and P-value $= 0.062$. There is weak evidence of regional differences. From a table of column proportions we see that the first four values seem to be more popular than the second four. (b) Segments that are as similar as possible. To use a common strategy.

11. $x_0^2 = 63.9$ with $df = 2$ and P-value ≈ 0. There is very strong evidence of differences between the locations.

13. (a) The percentages answering yes are: principals (48%), teachers (34%), 15–17 (23%), and 12–14 (10%). Teachers and principals have much higher percentages than the students. Assuming that principals tend to be older than teachers, the percentage seems to increase with age. (b) $x_0^2 = 13.9$, with $df = 2$ and P-value $= 0.001$. There is very strong evidence of a difference between teachers and principals. (c) $x_0^2 = 42.0$, with $df = 2$ and P-value $\approx$ 0. There is very strong evidence of a difference between the two groups of students.

15. A test for independence yields $x_0^2 = 30.2$ with $df = 1$ and P-value ≈ 0. There is very strong evidence of a relationship.

17. A test for independence yields $x_0^2 = 30.1$ with $df = 1$ and P-value ≈ 0. There is very strong evidence of a relationship. The Chi-square test is approximately the square of the test statistic from problem 9.

Chapter 12

Section 12.1.3

(a) We measure V for different values of I. We can then construct a scatter plot using the pairs of values of I and V using I for the x-axis and V for the y-axis. A straight line. It should pass through the origin.

(b) With a straight-line trend drawn through the origin, an estimate of r is the slope of the fitted line.

Section 12.2

(a) No, as a statistical relationship does not necessarily imply a causal relationship. The plot of budget versus deaths is not very helpful, as there does not seem to be a clear statistical relationship after 9000 deaths.

(b) A straight line is not an adequate fit. Another possible trend is an exponential curve, or possibly a quadratic curve.

(c) There does not seem to be any relationship between budget and deaths other than that they both tend to increase with time.

Section 12.3

(a) Yes, there is a clear upward trend.

(b) The least-squares line is $\hat{y} = 0.795 + 16.1x$.

(c) The nature of the trend is not so clear now. The least-squares line is $\hat{y} = -0.0049 + 0.0343x$. To superimpose this on our previous plot, we need to express x in terms of y, and then interchange x and y. Thus the superimposed line is $y = 29.15x + 0.143$.

(d) It depends on the purpose of the study. If we wish to see how optical absorbance varies with DOC, then we would use (c). This appears to be the more likely use. We might, however, want to predict DOC from a measure of optical absorbance. We would then use (a).

Section 12.4.2

1. (a) A strange-looking plot! Cyclist 1, the cyclist with 8 hours of training, could be an outlier. If this point is ignored, there appears to be an increasing trend that looks curved. On the other hand, cyclist 10, with a reading of 0.87, could be an outlier. If this point is ignored, the plot looks somewhat flat. Clearly, these two points have a big effect on what we see. (b) H_0: $\beta_1 = 0$ versus H_1: $\beta_1 \neq 0$. Using a t-test, $t_0 = 2.35$ and P-value $= 0.047$, which indicates some evidence of a trend. (c) Omitting cyclist 10, $t_0 = 1.38$ with P-value $= 0.211$, which suggests there is no evidence of a trend. (d) No, as no good reason is given for why this cyclist should be omitted. As we mentioned in (a), the picture would be very different if we omitted cyclist 1. (e) The data will depend on the training habits of each cyclist, for example, when and where they cycle. Other factors that could affect their exposure to lead are where they live and work. More information is needed about each cyclist.

2. The confidence interval is [6.44, 25.71].

Section 12.4.3

The prediction intervals follow: (a) When $x = 0.5$ the interval is [0.170, 0.339] with width 0.17. (b) When $x = 6$ the interval is [2.583, 2.764] with width 0.18. The second interval is slightly wider, as 6 is further from $\overline{x}$ than 0.5 (see Fig. 12.4.8).

Section 12.4.4

1. The residuals are: $1, 1, -8, 4, 7, -5$.
2. The least-squares line is $\hat{y} = -0.0049 + 0.0343x$. The residual plot is very similar to the original scatter plot. The spread does not seem to be constant. However, there may be two outliers, which need closer investigation.
3. The least-squares line is $\hat{y} = 74.713 + 1.060x$. The residual plot suggests the possibility of a slight fan effect, though if the largest positive residual is ignored, the plot looks satisfactory.
4. The residual plot is curved, indicating that a straight-line model is not appropriate.
5. From the residual plot the quadratic model seems to fit well. However, there is clearly an outlier and a hint of a reducing spread with increasing CO, though the points are sparse with large values of CO.

Section 12.5

(a) From the scatter plot there seems to be an upward trend that is roughly linear.

(b) It is visible only with the MEMORY scores.

(c) $r = 0.457$ with the outlier and $r = 0.542$ without the outlier.

(d) It means that the correlation is significantly different from zero and relates to H_0: $\rho = 0$, the hypothesis of no (linear) relationship.

(e) Without the outlier P-value$= 0.004$, indicating the existence of a relationship.

Review Exercises 12

1. (a) Plotting HARDNESS versus CEMENT, we get an approximate linear relationship. (b) The least-squares line is $\hat{y} = -24.1 + 0.186x$. (c) Once the amount of cement gets below a certain positive level c, say, the mix will not harden at all, that is, $y = 0$ for all values X below c. The regression line is therefore not relevant to the physical situation below $x = c$, and certainly not at $x = 0$, which gives a negative hardness of -24.1! (d) 3.72. (e) [0.170, 0.203]. It indicates that there is a positive trend, as it lies to the right of zero. (f) At $x = 275$ the 95% prediction interval is [20.85, 33.58]. (g) Yes, but we can't be sure, as the model may change as the amount of cement increases, particularly with such a large value of x. (h) The residual plot looks a little strange because of replicated x-values. However, there is no obvious departure from a simple linear model. The Normal probability plot is closely linear, and the W-test for Normality is not significant. (i) The variability will depend on the degree of mixing of all the ingredients, namely, sand, chips, cement, and water. The composition of batches of the sand-chip mix may vary considerably. There may also be variability in the quality of the cement from bag to bag.
3. (a) For the three men's races, the plots are surprisingly linear apart from initial outliers for the 100-m and 400-m races. There is no sign of leveling off for 100 m and 200 m, though for 400 m there appears to be some leveling off as the most recent points are above the general trend. The plots for the women's races are less well defined but, except for recent points, are still approximately linear for 100 m and 200 m. In these last two, the trend is increasing for the last three points, indicating some leveling off. The plot for 400 m is unusual, with a change in the slope of the downward trend perhaps indicating some leveling off. (b) The women's times have fallen more sharply than the men's times. (c) A descending straight line must eventually cut the x-axis, giving a zero time! (d) (i) The least-squares line is $\hat{y} = 196 - 0.0769x$. (ii) The residual plot shows that there are too many positive residuals at the right-hand end, suggesting that the linear model is breaking down at this end. The Normal probability plot shows some curvature. (iii) Without the outlier, the line is $\hat{y} = 176 - 0.0666x$. The Normal probability plot and the W-test do not indicate any

non-Normality. The 95% prediction interval for the year 2000 is [41.37, 44.31]. (f) For the men's races, the speeds for 100 m and 200 m are similar, especially after 1940. The three plots are reasonably linear with less evidence of a leveling off. (g) For women's races, the speed for 200 m in recent games has increased relative to that for 100 m so that they are now similar. There has only been a slight increase in speed for 400 m in the last few games.

5. (a) The TOTAL time seems to increase with the REACTION time. Most of the appearance of upward trend, however, is provided by the two highest points in the plot. Without them there is little visual evidence of a trend. Also, with or without these two points, the spread is not constant. (b) The plot is almost the same as before. (c) (i) $r = 0.358$. (ii) $r = 0.312$. There is little difference between using TOTAL time and RUNNING time. (d) $\widehat{y} = 9.32 + 5.87x$. (e) Using either a test for zero ρ or an (equivalent) test for zero slope, we get P-value $= 0.094$, which indicates little evidence of a relationship. (f) One interpretation of the scatter plot is that there is more than one trend superimposed here (maybe three?). This could, for example, be due to the performance differences for the heats, semifinals, and the final, particularly with the better runners. (g) Fan shaped. No, as the spread is not constant. A more careful analysis of the original data is needed.

7. (a) The tank will not be completely empty each time. There may also be a slight variation in the way it is "completely filled." There is more than one month involved after some fill-ups, for example, from the last few days of March through the first few days of April. Unless the car is always filled at the same time of the day, there is the problem of determining the number of days. (b) DAYS are measured from each date up to and including the day before the next date. The amount used to fill the car should be the same each time as it is going from "empty" to "full." These numbers should be checked and perhaps the average used to convert km to km per liter. (c) There are several zeros, which means that there is more than one fill-up on some days because of a long journey. (d) The plot is curved. There are better consumption rates in the summer months. For the outlier the distance is recorded as 557, which looks far too large, suggesting a mistake or a fill-up not recorded. (e) Such pairs of months might have similar driving conditions in terms of weather. Apart from apparent outliers, there seems to be an increasing trend from left to right, which appears to be related to weather conditions. (f) We would want to see if there are any differences within the pairs of months that are linked together, and check overall for outliers. We would also look at a residual plot, as the trend does not look particularly linear. (g) The intercept of the straight line is the mean for the month of January ($x = 0$) and is estimated by 5.889. The 95% confidence interval is [5.37, 6.41]. However, this interval does not contain two of the four observations at $x = 0$. This confirms what we see visually, namely, that a straight line does not fit the data too well at this point. (h) The slope of the straight line, with an estimate of 0.386, gives the increase in consumption as we move one month further away from January. The 95% confidence interval is [0.243, 0.530], which says that the consumption increases by between 0.24 and 0.53 km per liter for each month distant from January. (i) Yes, the fuel consumption changes with month of the year because the confidence interval for the slope does not contain the value 0. Also, when testing for zero slope, P-value $= 0.000$. *(j) For both June and August ($x = 5$) a 95% prediction interval is [5.64, 10.00]. (k) Yes, provided the standard deviations are not too different. To achieve this some outliers may need to be removed, for example, the largest values for $x = 1$ and $x = 3$, and possibly the smallest value for $x = 6$. The F-test indicates that the means are clearly not all equal. The null hypothesis for zero slope in the regression model is the same as the equal-means null hypothesis for the one-way ANOVA. However, the alternative hypothesis is more restrictive for the regression model in that it specifies that the means all lie on a line rather than just being not all equal, as in the case of the ANOVA model. (l) It means that there is no evidence against the null hypothesis that the quadratic coefficient is 0 (i.e., we don't need a quadratic model). (m) Yes, there seems to be a long-trip effect. There seems to be a decrease in consumption for long trips, as we might expect.

9. (a) The line $y = x$ corresponds to the perceived size being the same as the true size. (b) The students vary a great deal in their perception of size. Students 1 and 8 are inclined to overestimate the sizes of the larger circles. Student 4 tended to overestimate the sizes of smaller circles and underestimate the sizes of larger circles. Student 10 strongly overestimated the sizes of larger circles. (c) Using areas is not a good idea, as there is substantial variation in the perception of relative size. (d) The trend looks reasonably linear. *(e) Taking logs, we get $\log(\textit{perceived size}) = \log a + b \log(\textit{area})$, which is a straight line with slope b. (f) The fitted line is $\hat{y} = -0.184 + 1.06x$. (g) A 95% confidence interval for the slope is [1.000, 1.127]. This implies that the power is greater than 1. *(h) A 95% prediction interval for $\log y$ at $\log 300 = 5.704$ is [5.388, 6.378]. For y it is [exp(5.388), exp(6.378)], or [218.70, 588.75].

Index

Special needs literacy
Resources for group time

National curriculum

Level 3

activities based on:

'The owl who was afraid of the dark' & other texts

Kate Grant

Contents

Author Kate Grant
Editor Clare Gallaher
Assistant editor Roanne Davis
Series designer Paul Cheshire
Designer Micky Pledge
Illustrations Jane Cope

Designed using Adobe Pagemaker

Published by Scholastic Ltd, Villiers House, Clarendon Avenue, Leamington Spa, Warwickshire CV32 5PR

1 2 3 4 5 6 7 8 9 0   0 1 2 3 4 5 6 7 8

British Library Cataloguing-in-Publication Data
A catalogue record for this book is available from the British Library.

ISBN 0-439-01711-4

Acknowledgements

The publishers gratefully acknowledge permission to reproduce the following copyright material:

- **Egmont Children's Books** for *The Owl Who Was Afraid of the Dark* by Jill Tomlinson, illustrated by Susan Hellard © 1968, The Estate of Jill Tomlinson (1968, Methuen).
- **Faber & Faber Ltd** for *Tales of Polly and the Hungry Wolf* by Catherine Storr, illustrated by Jill Bennett © 1982, Catherine Storr (1982, Puffin).
- **Orion Children's Books** for text and 'David and Goliath Rap' from *Gynormous!* by Adrian Mitchell, illustrated by Sally Gardner © 1996, Adrian Mitchell (1996, Orion Children's Publishing).
- **Usborne Publishing Ltd** for the use of text and illustrations from *Understanding Your Senses: Science for Beginners* by Rebecca Treays © 1998, Usborne Publishing Ltd (1998, Usborne).

Every effort has been made to trace copyright holders for the works reproduced in this book, and the publishers apologize for any inadvertent omissions.

Introduction

Why do I need special needs resources for group time?

The National Literacy Strategy *Framework for Teaching* requires that all children's needs are catered for in the daily Literacy Hour. The class teacher can address individual needs through targeted questions and direct teaching in the first part of the hour when the whole class is taught. However, there is a real necessity for purposeful activities for children with special needs during the group/independent work section of the hour, when the teacher is working elsewhere with guided reading/writing groups. This book is geared to allowing children with special needs to work at an appropriate level on relevant objectives under the guidance of a classroom assistant or other adult.

How does the book work?

Each of the six chapters in this book contains five photocopiable lessons, designed to fit into the 20-minute group/independent work slot in the Literacy Hour and to be used by a classroom assistant or other additional adult. The lessons are based on particular texts, and children working on the activities will need to have access to them. The three story books are:

- *The Owl Who Was Afraid of the Dark* by Jill Tomlinson (Mammoth)
- *Tales of Polly and the Hungry Wolf* by Catherine Storr (Puffin)
- *Gynormous!* by Adrian Mitchell (Dolphin Paperbacks).

The non-fiction texts and poems are provided as photocopiable pages. One story or extract is the focus for each week's lessons; the activities are at text, sentence and word level and include writing frames and homework tasks.

How does it fit into the Literacy Hour planning?

The genre for the lessons is chosen from the range for Year 5, but the learning objectives are from previous years or terms, since children with special needs are still developing and practising skills from an earlier stage. In this way, if the class is studying traditional stories, myths and legends, for example, the chapter on 'Don Quixote and the windmill giants' (Week 3) could be selected for the group working with the classroom assistant.

The objectives are clearly indicated in the grid at the start of each chapter, allowing for simple group target-setting. The same fiction text could be used for guided reading with the group in the previous week if required, although this is not essential.

How flexible is it?

Each chapter in this book can be used independently – there is no need to follow any particular order. Although the texts have been selected to appeal to Year 5 children reading Level 3 texts, they may also be suitable for older children with learning difficulties who are working at this level. The materials can also be used outside the Literacy Hour for additional support if required.

Will classroom assistants need any training to use the book?

Everything that is needed to carry out the group-time lesson is explained in the straightforward notes for each day. There are also useful hints provided in the next few pages to help classroom assistants support children with reading, writing and spelling.

What is the range of the texts?

For Year 5/Level 3, the texts include:

- stories by significant children's writers *(The Owl Who Was Afraid of the Dark, Tales of Polly and the Hungry Wolf)*

- a legend ('Don Quixote and the windmill giants' from *Gynormous!*)
- narrative poetry ('David and Goliath rap' from *Gynormous!*)
- explanations (non-fiction text)
- letters to complain or protest.

What objectives are covered?

Learning objectives include:

- **Text level:** patterns of rhyme and verse, structure in poems, writing poems, comprehension, story planning, settings, narrative order, a book review, features of legends, story themes, characters, writing alternative sequels, factual research, structure of explanatory texts, information in diagrams, presenting a point of view in writing
- **Sentence level:** powerful verbs, adjectives, devices for presenting texts, adverbs, speech marks, sentence types, verb tenses, plurals, pronouns
- **Word level:** spelling by analogy, high-frequency words, suffixes, using a dictionary and thesaurus, apostrophes, antonyms, synonyms, vocabulary extension.

High-frequency words/onset and rime

The following lists show the high-frequency words and onset and rime spelling patterns that can be focused on when you are using the texts in the activities.

The Owl Who Was Afraid of the Dark		Tales of Polly and the Hungry Wolf		Gynormous!	
baby	asked	thought	sure	many	turned
garden	-ould	about	-aw	high	-ool
Explanations		**Narrative poetry and performance poetry**		**Letters to complain or protest**	
think	only	once	upon	heard	birthday
more	-art	first	-ear	also	-ass

Supported reading

Before choosing a week's lessons, make sure the text being used is at a suitable reading level (most children in the group should be able to read with no more than one mistake in every ten words.)

Fiction books: the lessons which use the story books as their basis assume that the children have already read the relevant chapters in the books. If the children have not already done so, introduce the story in the following ways:

- Look at the cover and read the title together and the names of the author and illustrator.
- Look through the story together, helping the children to have ideas about it from the illustrations. (Don't give away the ending!)
- Tell them a few key words and help them to find them in the text.

Now read the story together. The children can take it in turns to read a page aloud individually, or they could read some pages together as a group, or read silently to themselves. When they are reading aloud, see if they are changing the expression in their voices and noticing speech marks and other punctuation. Check that they have understood the parts they have read silently by asking a few key questions at the end of those pages.

Non-fiction/information texts: the lessons based on the non-fiction texts do not assume that children will have read the texts before. Although Day 1 for each week explains how to introduce and read the texts, it is helpful to bear in mind the following general points about introducing non-fiction texts:

- Ask the children to read some of the captions to the pictures and the headings for the different sections.
- Help the children to think of some things that the text might help them to learn about.

Poems: the children do not need to have read the poem ('David and Goliath rap' from *Gynormous!*) before you begin the lessons for Week 5. The notes for Day 1 explain how to introduce the poem but, before reading, you could also help the children to consider how a poem looks different from a story or information text. Look at the lengths of the lines and how lines are arranged into verses, and notice any rhymes.

Helping children with reading

Ask the children to tell you how they can work out a word they are finding difficult to read. They could:

- go back to the beginning of the sentence
- re-read the sentence and say the first sound of the word they cannot read, thinking of a word that fits
- leave out the hard word, read a few more words, then have another try
- look for a little word they know inside the hard word, for example *l*_**and**_*ed*
- look for a pattern they know, for example *old* in *sc*_**old**_

Remind the children of things they should ask themselves as they read:

- Does it make sense?
- Does it sound right?
- Does it look right?

Helping children with writing

The aim should be to enable the children to write as much as possible on their own, and your help will be geared to developing their skills. The writing frames provided in this book give the children a 'scaffold' so that they are not faced with a blank sheet of paper, but you will still need to discuss with them some ideas for what they might want to write. It is a good idea not to let them have pencils until you have finished a preliminary discussion. Have scrap paper or a dry-wipe board and pen available for trying out spellings.

If children ask you how to spell a common (or high-frequency) word that they will need to use fairly often, teach them to spell it by getting them to write it a few times using the Look–Say–Cover–Write–Check method (see page 6).

Another way to help is to draw boxes for the letters in the word (if it is one that can be 'sounded out' fairly easily). Ask what they can hear at the beginning, middle and end of the word, then help them to put in the missing letters.

For example, if a child wants to write *rocket*, ask him or her to say the word. Once the child can hear the *r*, the *k* in the middle and the *t* at the end, you can help him or her to put the letters in the correct boxes.

r			k		t

Next, help the child to say the word again, emphasizing each sound so that he or she can hear more letters. You may need to explain that there is another letter that sounds the same as *k*. The *e* sounds like *i* so you may need to help with this letter, too. Eventually the whole word will appear in the boxes.

r	o	c	k	e	t

Encourage the children to keep on reading their sentences as they are writing them, so that they don't leave words out. Remind them to put in capital letters, full stops and so on, and make sure they read their writing through carefully when they have finished.

Supporting spelling (Look–Say–Cover–Write–Check)

If you train children to learn a new word by this method every time they want to remember a spelling, they will nearly always be successful.

> Have a good LOOK at the word.
> SAY the word and spell out the letters.
> COVER the word so that you can't see it.
> WRITE the word down.
> Uncover the word to CHECK if it's right.

The children should repeat the process as often as necessary until they get the word right three times running. Even if the spelling is forgotten later, they will probably have most of the letters right, and relearning will be much quicker the second time.

One reason this method works is because it uses three different ways to learn the same thing. You use your:

- **eyes** to look at the word
- **ears** to hear the letter names as you say them
- **hand** to write the word, and get the feel of its pattern and shape.

This gives the brain several chances to store the word in the memory.

It can help children's handwriting development as well as their spelling if they write the words in joined writing, so that they are learning the spelling patterns as a continuous movement.

Making spelling resources

Some of the activities in this book refer to the use of a word wheel and flick book. These can be made very easily and children really do enjoy using them.

Making a word wheel to learn spelling patterns

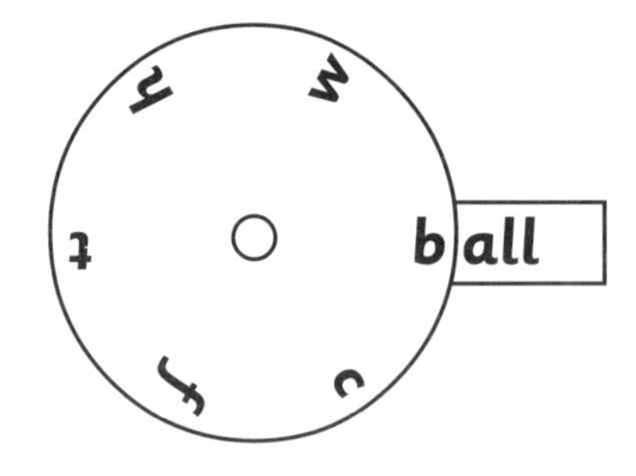

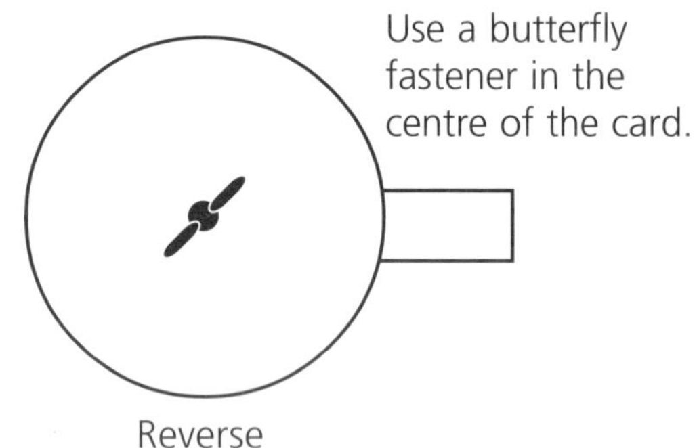

Making a flick book to learn spelling patterns

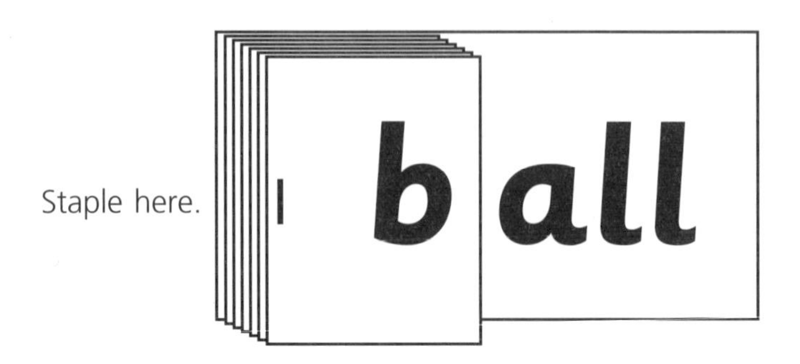

Write a different letter on each small piece of paper to make words, eg *wall, fall, tall* and so on.

Week 1 Dark is exciting

(from *The Owl Who Was Afraid of the Dark* by Jill Tomlinson)

Introduction

Jill Tomlinson is a popular children's author who has written a delightful series of books about young animals who are getting to grips with the process of growing up. Children easily identify with their dilemmas and struggles. *The Owl Who Was Afraid of the Dark* is an amusing book about a baby barn owl named Plop who has not come to terms with one important aspect of his life as an owl – he is afraid of the dark.

The text for Week 1 is the first chapter ('Dark is exciting') where Plop meets a boy who tries to convince him that dark is exciting by telling him about fireworks and Bonfire Night.

If the children enjoy this story, encourage them to try reading other chapters in the book for themselves.

Week 1 Objectives

	Word level	Sentence level	Text level
Day 1 Reading		To notice a range of devices for presenting texts Y3 T1 (9)	To identify the main characteristics of the key characters Y4 T1 (2)
Day 2 Comprehension sequencing			To retell main points of a story in sequence Y3 T3 (1)
Day 3 Spelling: high-frequency words, onset and rime	To recall high-frequency words Y4 T1–3 (1) Spelling by analogy with known words Y4 T1–3 (3)		
Day 4 Adverbs Speech marks	To use suffixes to generate new words from root words Y3 T3 (14)	To identify adverbs and understand their functions in sentences Y4 T1 (4) To use speech marks appropriately in writing Y3 T3 (4)	
Day 5 Writing composition			To plan main points as a structure for story writing Y3 T2 (6), Y4 T1 (10)
Homework task 1 Speech marks		To use speech marks appropriately in writing Y3 T3 (4)	
Homework task 2 Comprehension		To use awareness of grammar, phonic and graphic knowledge, word recognition and context when reading Y3 T2 (1)	

Reading

The children should already have read this chapter with their teacher in guided reading (see page 4).

The characters

Discuss how the story begins (a description of Plop, the main character). Make sure they are familiar with the word *character*. Help the children to write a list of the characters in this chapter, starting with the main characters (Plop, Mr and Mrs Barn Owl, the boy). Ask them to think of a few phrases to describe one of the characters (working in pairs) and find the parts of the text which show what their character is like. The children should share their ideas with the group.

Text layout

- Ask the children to find three different kinds of typeface used on page 10 (*italic*, CAPITAL or UPPER-CASE LETTERS and regular print). Make sure they know the term *italic*. Talk about why they are used. (To make certain words more important.)
- Practise reading sentences with words in italic and capital letters, stressing the important words. Then read the same sentences as if they had no important words, to hear the differences.

Week 1 **The Owl Who Was Afraid of the Dark**

Day 1

Special Needs Literacy

Comprehension

Ask the children to retell the main points of the story. Remind them that this means putting the story into their own words, and encourage them to be brief. Let them take turns to say a sentence and to try to link it to the previous one (using phrases such as *next, but then, after that, later on, finally* and so on).

Quiz (oral or written answers)

1. What did the boy think Plop was?
A. A Catherine wheel.
2. Describe Plop's eyes.
A. Enormous and round.
3. What colours other than black can the dark be?
A. Blue, grey and silver.
4. When did Plop say, 'Actually, I'm a barn owl'?
A. When he met the little boy.
5. Who said, 'Your landing will improve with practice'?
A. Plop's mother.
6. What kind of knees did Plop have?
A. Knackety.

Quiz (written answers)

1. Where did Plop live?
A. At the top of a tree.
2. Who did Plop meet when he fell off his branch?
A. A little boy.
3. What did Plop say when his mother asked him what he had found out?
A. 'The little boy says DARK IS EXCITING.'
4. True or false? Plop was a tawny owl.
A. False.
5. Why did Plop's father say he sounded like a tawny owl?
A. Plop said, 'Ooooh!'
6. What was the boy doing when Plop saw him?
A. Collecting sticks.

Week 1 **The Owl Who Was Afraid of the Dark**

Day 2

Special Needs Literacy

Sight vocabulary

Ask the children to find *baby* (page 6), and to learn to spell the word using the Look–Say–Cover–Write–Check method (see page 6) until they can write it from memory. Do the same with *garden* (page 12) and *asked* (page 10). The children should write the words in their spelling books to take home for further practice.

Help them to think of a sentence with all three words in it and to write it down, using a coloured pencil or felt-tipped pen to highlight *baby, garden* and *asked*.

Spelling by analogy

Spelling pattern: *-ould*
(would, could, should, wouldn't, couldn't, shouldn't)

Ask the children to find *would* (page 11) and to look carefully at the spelling. As they may already know how to spell *wood*, you may need to explain the difference between *would* and *wood*. They should learn to spell *would* using the Look–Say–Cover–Write–Check method (see page 6) until they can write it from memory.

Encourage them to think of words that rhyme with *would.* Dictate the words for the children to write in their spelling books.

Dictate the following sentences, for the children to write. (Encourage them to write without looking at their word lists.) Remind them about capital letters and full stops, if necessary.

*I **would** help you, if I **could**.*
*You **shouldn't** do that.*
*I **wouldn't** if I were you.*

The children should underline all the *-ould* words in their completed sentences with a coloured pencil or felt-tipped pen.

They could make word wheels or flick books with the spelling pattern to take home (see page 6).

Adverbs

Ask the children to find *beautifully* (page 14). Can they find other words on page 14 that end with the same two letters? *(cautiously, quietly.)* Help them to work out which words are the roots for each of the three words *(beautiful, cautious, quiet)* and to notice the two letters *-ly* in the ending (or suffix).

Check if the children know what kind of words end in *-ly.* (Adverbs – that is, words that tell the reader how something was done.)

- Help them to make a list of adverbs, starting with the three from the story, then adding adverbs made from the following words: *slow, quick, kind, helpful, sudden, loud, soft.*
- Encourage them to write three or four sentences, each containing two of the adverbs.

Speech marks

Ask the children to find the speech marks on page 7. They should mark the beginning and end of the speech by putting one finger from each hand on each speech mark. Some children may find it helpful to think of the shape of the marks (66 and 99).

Help them to notice where in the line the opening speech marks (the 66) are. (At the beginning.) Look on other pages to check if this is always the case. (Yes.)

Which word often follows straight after the closing speech marks? *(said.)* Can the children find other words that are used instead of *said* on pages 10, 11, 14 and 15? *(asked, begged, laughed, called, squeaked.)*

Read the following sentence to the children and help them to rewrite it using speech marks.

Plop told his mother that he wanted to be a day bird, but Mrs Barn Owl replied that owls are night birds.

Answer: Plop told his mother, "I want to be a day bird."
Mrs Barn Owl replied, "Owls are night birds."

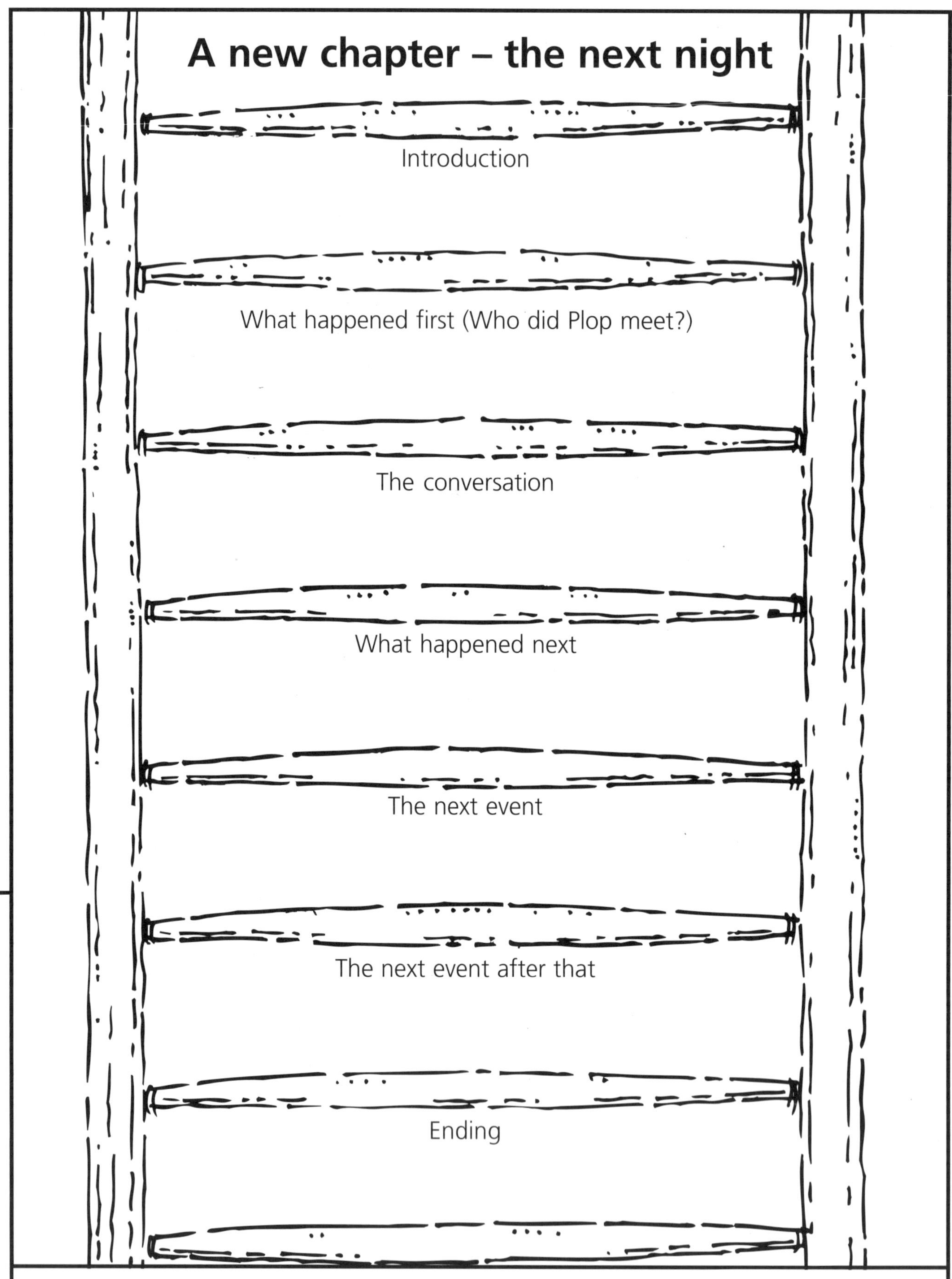

Week 1 **The Owl Who Was Afraid of the Dark**

Day 5

Writing composition:

- Ask the children to use the 'ladder' writing frame to plan another story for Plop. On each rung they should write a few key words, not whole sentences. They can then use their completed story plans to write their stories; each rung on the ladder with its key words can be the basis of a paragraph. (They may need to finish their writing outside the Literacy Hour.)
- See notes on page 5 ('Helping children with writing').

Speech marks

All the speech marks have been lost from this extract from *The Owl Who Was Afraid of the Dark*. Can you put them all back in?

What are fireworks? asked Plop. I don't think owls have them – not Barn Owls, anyway.

Don't you? said the little boy. Oh, you poor thing. Well, there are rockets, and flying saucers, and volcanoes, and golden rain, and sparklers, and…

But what *are* they? begged Plop. Do you eat them?

NO! laughed the little boy. Daddy sets fire to their tails and they *whoosh* into the air and fill the sky with coloured stars – well, the rockets, that is. I'm allowed to hold the sparklers.

What about the volcanoes? And the golden rain? What do they do?

Oh, they sort of burst into showers of stars. The golden rain *pours* – well, like rain.

Homework

Comprehension

Read this extract from *The Owl Who Was Afraid of the Dark*, then answer the questions.

When the very last firework had faded away, Mr Barn Owl turned to Plop.

"Well son," he said. "I'm off hunting now. Would you like to come?"

Plop looked at the darkness all around them. It seemed even blacker after the bright fireworks. "Er – not this time, thank you, Daddy. I can't see; I've got stars in my eyes."

"I see," said his father. "In that case I shall have to go by myself." He floated off into the darkness like a great white moth.

Plop turned in distress to his mother.

"I *wanted* to go with him. I *want* to like the dark. It's just that I don't seem to be able to."

"You will be able to, Plop. I'm quite sure about that."

"I'm not sure," said Plop.

"Well I *am*," his mother said. "Now come on. You'd better have your rest. You were awake half the day."

1 Why did Plop not want to go hunting?

____

2 What did Mr Barn Owl look like when he flew off?

____

3 What had the Barn Owls been doing before this chapter started?

____

4 What does **in distress** mean? (Use a dictionary or ask someone if you don't know.)

____

5 Which words are written in italic print?

____

6 Find three words that have apostrophes in them (for example, **won't**).

____

Homework

Week 2 Songs my mother taught me

(from *Tales of Polly and the Hungry Wolf* by Catherine Storr)

Introduction

Catherine Storr's series of books about Polly, the clever little girl, and the wolf who is always trying to gobble her up, have become modern-day classics. The stories appeal to children because they use humour cleverly to deal with their fears and because the child always wins the arguments and ultimately outwits the wolf.

In this amusing chapter ('Songs my mother taught me') the wolf tries to convince Polly that she has learned the wrong words to the songs and nursery rhymes she knows. The wolf's versions of the songs are all about eating people, particularly children.

If the children enjoy this story, encourage them to try reading other chapters for themselves.

Week 2 Objectives

	Word level	Sentence level	Text level
Day 1 Reading	To collect and use new words from reading Y3 T3 (12)	To use awareness of grammar to decipher new words Y3 T1–3 (1)	To explore narrative order Y4 T1 (4)
Day 2 Comprehension			To investigate how settings are built up from small details Y4 T1 (1)
Day 3 Spelling: high-frequency words, onset and rime	To recall high-frequency words Y4 T1–3 (1 & 4) Spelling by analogy with known words Y4 T1–3 (3 & 4)		
Day 4 Adjectives		To understand the function of adjectives in sentences Y3 T2 (2) To understand how grammar is changed when sentence type is altered Y4 T3 (3)	
Day 5 Writing composition			To write a book review based on evaluation of plot and character Y3 T3 (14)
Homework task 1 Cloze	To collect and use new words from reading Y2 T3 (12)		
Homework task 2 Vocabulary		To use awareness of grammar, phonic and graphic knowledge, word recognition and context when reading Y3 T2 (1)	

Day 1 Week 2 **Tales of Polly and the Hungry Wolf**

Reading

The children should already have read this chapter with their teacher in guided reading (see page 4).

Developing the story

Remind the children of the story about Polly and the wolf. Help them to understand how the author develops this particular part of the story by asking them:

- What is the **introduction**? (The setting – Polly upstairs in her house, by the window, playing her recorder, in the evening, a big moon in the sky.)
- What is the **problem?** (The wolf wants to eat Polly.)
- What is the **conflict?** (The disagreement between Polly and the wolf about the words of the songs.)
- What is the **crisis?** (When the wolf said he would wait until it was darker so he could eat Polly.)
- What is the **resolution?** or **How was the problem sorted out?** (Polly's version of 'Boys and girls come out to play' scared the wolf and he ran off before he could be eaten.)

Using dictionaries

Look with the children at these difficult words from the story, but don't tell them what they mean.

caterwauling (page 38) – screaming like a cat
yodelled (page 40) – high-pitched singing
declaimed (page 41) – recited
agape (page 44) – wide open
imploring (page 47) – begging
barbarous (page 47) – cruel
refrain (page 47) – stop
melodious (page 47) – tuneful
infinitely obliged (page 48) – very grateful

Ask them to:

- choose three of the words, find them in the story and read the sentences again (working in pairs)
- discuss what they think the words mean (continuing to work in pairs)
- look up the words in a dictionary to see if they were right
- rewrite the sentences from the book using simpler words in place of the words they have been studying.

Day 2 Week 2 **Tales of Polly and the Hungry Wolf**

Story settings

Read the first page of the story (page 38) again with the children.

- What is the reader told about the setting in the first sentence? (Time of day – early evening.)

Ask the children to look through the page again and find all the other details which help the reader to have a picture of where and when the story took place.

Comprehension

Quiz (oral or written answers)

1. What was the first song the wolf heard?
A. I had a little nut tree.
2. What is the real name for the 'Onions and Kidneys' song?
A. 'Oranges and lemons'.
3. When did the story take place?
A. Evening.
4. What happened to the nursery rhyme character Humpty Dumpty in the wolf's song?
A. He was eaten.
5. Why did Polly mention the moon being bright?
A. To make the wolf think the boys and girls would come to eat him.

Quiz (written answers)

1. Where was Polly playing her recorder?
A. Upstairs by the window.
2. Who ate a lot of lamb in the wolf's song?
A. Mary.
3. When do boys and girls eat wolf, according to Polly?
A. At midnight feasts.
4. True or false? The wolf was kind and generous.
A. False.
5. Where did the story take place?
A. Polly's house.

Sight vocabulary

Ask the children to find *thought* (page 40), and to learn to spell the word using the Look–Say–Cover–Write–Check method (see page 6) until they can write it from memory. Do the same with *about* (page 41) and *sure* (page 41). The children should write the words in their spelling books to take home for further practice.

Help them to think of a sentence with all three words in it and to write it down, using a coloured pencil or felt-tipped pen to highlight *thought, about* and *sure*.

Spelling by analogy

Spelling pattern: *-aw*
(law, saw, paw, claw, jaw, raw, straw)

Ask the children to find *law* (page 44) and to look carefully at the spelling. They should learn to spell the word using the Look–Say–Cover–Write–Check method (see page 6) until they can write it from memory.

Encourage them to think of words that rhyme with *law*. Dictate the words for the children to write in their spelling books. (Common exceptions to the *-aw* pattern are *for, door* and *more*; if the children are not sure of any of these three words, they could revise them using Look–Say–Cover–Write–Check.)

Dictate the following sentences, for the children to write. (Encourage them to write without looking at their word lists.) Remind them about capital letters and full stops, if necessary.

*I **saw** a wolf with **straw** in his **jaws**.*
*The wolf pulled at the **raw** meat with his **claws**.*

The children should underline all the *-aw* words in their completed sentences with a coloured pencil or felt-tipped pen.

They could make word wheels or flick books with the spelling pattern to take home (see page 6).

Week 2 **Tales of Polly and the Hungry Wolf**

Day 3

Special Needs Literacy

Adjectives

Ask the children to find *huge* (page 38). What is being described as *huge*? (The moon.) Can they remember what 'describing words' are called? (Adjectives.) Help them to find a few more adjectives in the story and write them down.

page 38	*huge, yellow, early, easy, horrible*
page 39	*disappointing, silver, golden, little, difficult, revolting*
page 40	*new, great, right, poor, old, fat, big, good, raw*
page 41	*interesting, white, horrible, anxious*
page 42	*real, solid, miserable, clever*
page 43	*stupid, boiling, short, soothing, beautiful, lovely, frightening, boring, difficult*

Ask the children to rewrite three sentences from the book, using their own adjectives (which must mean nearly the same as those in the book, for example *huge/large*). They could use a thesaurus or dictionary to find alternative words.

Questions and statements

Look at page 43 with the children to find examples of sentences which are questions. Make sure they notice the question marks and speech marks.

The children should rewrite each question (there are four) to make it into a statement. Explain that for some questions there are no 'right answers'.

1. Question: 'I haven't jumped into a pot of boiling water, have I?'
   Statement: 'I haven't jumped into a pot of boiling water.'
2. Question: 'Who did?'
   Statement: 'My mother did.' *(Answers may vary.)*
3. Question: 'Are you going to play again?'
   Statement: 'You are going to play again.'
4. Question: 'What's that tune?'
   Statement: 'I don't know what that tune is' or 'That tune is…' *(Answers may vary.)*

Week 2 **Tales of Polly and the Hungry Wolf**

Day 4

Special Needs Literacy

Book review

I have just read a chapter from

written by

illustrated by

published by

I thought that the plot was because

The character I liked best was because

My favourite part was

I didn't like

I would give it a score of

Boring										Brilliant
0	1	2	3	4	5	6	7	8	9	10

Reviewed by

Writing composition:

- Discuss the story with the children. Help them to express their opinions and to give their reasons, to prepare them for writing the book review. Encourage them to think of something to write for each section.
- See notes on page 5 ('Helping children with writing').

Tales of Polly and the Hungry Wolf

Songs my mother taught me

Find the missing words and fill in the lines.

Polly was practising her ________________ one evening, when the wolf appeared. He ________________ the tunes she was playing, but he had learned ________________ words to the songs from his ________________. Nearly all the wolf's songs were about ________________ children. He tried to get ________________ to lean out of the window, so that he ________________ catch her, but she was too ________________ for him. Polly quickly made up ________________ words to 'Girls and boys come ________________ to play' – in her song the children ate the wolf! This ________________ worked. The wolf was so scared that he ________________ off down the road.

could	**recorder**	**eating**	**clever**
new	**trick**	**ran**	**mother**
different	**out**	**knew**	**Polly**

Homework

Change the words

Rewrite these sentences using simple words to replace the circled words. If you can't remember what the words mean, you could use a dictionary to find out.

1 Please refrain from caterwauling.

2 I am imploring you to play your recorder in a melodious way.

3 I will be infinitely obliged if you would stop being so barbarous to your little sister.

Homework

Week 3 Don Quixote and the windmill giants

(from *Gynormous!* by Adrian Mitchell)

Introduction

This famous story is from the traditional stories, myths, legends and fables genre. It is about the elderly nobleman, Don Quixote, a legendary character with a romantic soul, who wanted to be a knight. He went to find adventure and rescue maidens in distress and entered into everything he did with great enthusiasm and vanity, but very little common sense. His faithful servant, Sancho Panza, was the opposite. He was a very sensible, practical man who thought his master was mad, but was too polite to say so. Sancho Panza was often exasperated by Don Quixote, but remained his loyal servant. So although Don Quixote is the master, Sancho Panza, his faithful servant, is clearly the one with the brains.

Week 3 Objectives

	Word level	Sentence level	Text level
Day 1 Reading		To use awareness of grammar to decipher new words Y3 T1–3 (1)	To identify the features of legends Y5 T2 (1) To identify typical story themes Y3 T2 (2)
Day 2 Characters Comprehension	To explore opposites Y3 T2 (24)		To identify and discuss main characters Y3 T2 (3) To identify the main characteristics of the key characters Y4 T1 (2)
Day 3 Spelling: high-frequency words, onset and rime	To recall high-frequency words Y4 T1–3 (1 & 4) Spelling by analogy with known words Y4 T1–3 (3 & 4)		
Day 4 Punctuation Vocabulary	To collect new words, infer the meaning and use dictionaries to check Y3 T2 (17–19) To use the apostrophe to spell shortened forms of words Y3 T2 (15)		
Day 5 Writing composition			To write own versions of legends using structure identified in reading Y5 T2 (11) To write alternative sequels to traditional stories Y3 T2 (10)
Homework task 1 Vocabulary	To collect new words, infer the meaning and use dictionaries to check Y3 T2 (17–19)		To retell main points of story in sequence Y3 T3 (1)
Homework task 2 Sequencing			

Reading

Ask if any of the children have heard of Don Quixote. Read the introduction to 'Don Quixote and the windmill giants' (page 24), which explains the background to the story, then read the story together (see 'Supported reading' on page 4).

After reading

- Do the children think this is a true story? Why not?
- Check that they have heard the word *legend*, and ask if they know what makes a story a legend. Explain that it is often a story about a character who may have lived long ago, but stories of his or her life are exaggerated, to make them more fantastic.
- Do they know of any other legends (for example, King Arthur and the knights of the round table or Robin Hood)?
- Help the children to make a list of the difficult vocabulary in the story. (This will be used later in the week, as a dictionary activity.) You will probably want to include *monstrous, advance, engage, ferocious, spurs, brutes, lance, glory*.

Characters

Opposites

Talk about the differences between Don Quixote and Sancho Panza. Who is the wise (or clever) character in this story? Who is foolish (or stupid)?

Ask each child in turn to think of a word or phrase to describe Don Quixote. They could get ideas from the illustrations as well as the story. Write them in a list. Do the same with Sancho Panza.

Compare the two lists. Can the children see any opposites in the two lists? (For example, Don Quixote is thin, Sancho Panza is fat.)

Help the children to write their own lists of opposites. Start with the ones from the story and add more until they reach ten words in each list.

Comprehension

Quiz (oral or written answers)

1. What did a knight wear on his shoes to prick the horse?
A. Spurs.
2. How did Don Quixote explain his mistake?
A. He said his enemies turned the giants into windmills.
3. Why were the giants waving their arms at Don Quixote?
A. The wind made the sails turn.
4. What is a lance and how is it used?
A. It is a weapon like a pole with a sharp point used for charging enemies.
5. Did Sancho Panza think Don Quixote was wise or foolish?
A. Foolish.

Quiz (written answers)

1. Spell *knight.*
2. What does *Don* mean?
A. Sir/lord/gentleman.
3. How many windmills were there?
A. Thirty or forty.
4. What part of the windmills looked like arms?
A. The sails.
5. Who was Rosinante?
A. The horse.

Sight vocabulary

Ask the children to find *many* (page 26), and to learn to spell the word using the Look–Say–Cover–Write–Check method (see page 6) until they can write it from memory. Do the same with *high* (page 26) and *turned* (page 29). The children should write the words in their spelling books to take home for further practice.

Help them to think of a sentence with all three words in it and to write it down, using a coloured pencil or felt-tipped pen to highlight *many, high* and *turned*.

Spelling by analogy

Spelling pattern: *-ool*
(fool, cool, pool, tool, stool, school)

Ask the children to find *fool* (page 29) and to look carefully at the spelling. They should learn to spell the word using the Look–Say–Cover–Write–Check method (see page 6) until they can write it from memory.

Encourage them to think of words that rhyme with *fool.* (Notice that *rule* follows a different pattern.) Dictate the words for the children to write in their spelling books.

Dictate the following sentences, for the children to write. (Encourage them to write without looking at their word lists.) Before they start, ask them to think about how sentences begin and end (capital letters and full stops or question/exclamation marks.)

*Keep the rules in **school** unless you are a **fool**.*
*Does your **school** have a swimming **pool**?*

The children should underline all the *-ool* words in their completed sentences with a coloured pencil or felt-tipped pen and circle the odd one out *(rule*).

They could make word wheels or flick books with the spelling pattern to take home (see page 6).

Week 3 **Gynormous!**

Day 3

Special Needs Literacy

Punctuation

Look with the children at the first sentence of the second paragraph on page 25:

'We're in luck,' said Don Quixote.

Can they find the comma and the apostrophe? Ask the children if they know the difference between a comma and an apostrophe.

- The comma divides a long sentence into shorter sections and always comes before the closing speech mark, when the character carries on talking. Commas are always written on the line, like full stops.
- The apostrophe here is to show two words have been shortened into one word – *we are* becomes *we're*. This is called a contraction. Apostrophes are always above the letters.

Ask the children to find more examples of apostrophes, where two words have been contracted into one, in the story.

Try reading a few sentences as if the words had not been shortened. Does it sound different? Which do they prefer?

The children can write a list of the words they have found:
we'll, aren't, they're, don't, I'm, can't, didn't, it's

Next to each word they can write the two words it is made from:
we'll, ***we will*** *aren't,* ***are not*** and so on

Vocabulary

Look with the children at the difficult words you found on Day 1:
monstrous, advance, engage, ferocious, spurs, brutes, lance, glory.

- Working in pairs, the children can choose three of the words, find them in the story and read the sentences again.
- Help the children to write what they think each word means, then to look it up in the dictionary and write down the definition.
- The children could make up and write sentences of their own, each containing one of the words they have been studying.

Week 3 **Gynormous!**

Day 4

Special Needs Literacy

Another adventure

How the story starts

Setting

Characters

The adventure or problem

The main event

The resolution (how the problem is solved or what happens in the end)

The link to the next story

Writing composition:

- Help the children to use the story frame to plan another adventure about Don Quixote and Sancho Panza. Before they begin, remind them that Don Quixote is foolish, but thinks he is wise, and Sancho Panza is his clever servant. They can then use their completed story plans to write their stories. (They may need to finish their writing outside the Literacy Hour.)
- See notes on page 5 ('Helping children with writing').

Gynormous!

Vocabulary

Can you remember what these words mean?
Use a dictionary, or ask someone, if you get stuck.

advance **lance** **ferocious** **spurs** **glory**

1 meaning: a long, sharp spear used by a knight **word:**
2 meaning: sharp spikes worn on boots which are dug into a horse to make it gallop faster **word:**
3 meaning: wild and fierce **word:**
4 meaning: victory and fame **word:**
5 meaning: to go forward **word:**

Homework

Gynormous!

Sort the sentences

These sentences from the story have become mixed up.
Put a number in front of each sentence to show the correct order.

○ Don Quixote said, "Do you take me for a fool?"

○ "I will do battle with those giants," said Don Quixote.

○ At this moment a wind arose, and the great sails of the windmills began to turn.

○ Don Quixote's lance broke and he tumbled to the ground.

○ Don Quixote and Sancho Panza were riding along when they saw thirty or forty windmills.

○ Then the two of them rode away in search of high adventure.

○ Sancho Panza didn't think they were giants, but Don Quixote was determined to fight them.

○ Don Quixote told Sancho Panza to run away and say his prayers.

Can you find the missing words?

"I'm a ____________________ and you can't frighten me."

"Of course I ____________________ a windmill when I see one."

Did you notice anything about the spelling?

____

____

____

Homework

Week 4 Explanations

Taste & smell (from *Understanding Your Senses: Science for Beginners* by Rebecca Treays)

Introduction

Reading for information requires a different approach from story reading. Although both fiction and non-fiction can be read for pure enjoyment, information texts are also for finding out. The KWL grid, which the children will use on Day 1, will help them to focus more clearly on what they already know and on what they want to find out from their reading.

Non-fiction can be divided into different categories, and the text used for this week is 'explanation'. Many science texts fall into this category. This extract about taste and smell is one that makes a lively introduction to the senses, as part of a science topic ('life processes and living things'), or it can be used as an opportunity for revision.

(*Note:* there is one technical term which the children will need to know before reading the text – *receptors*; explain that these are the parts of your tongue that respond to food and send electrical messages to your brain so that you know what the taste is.)

Week 4 Objectives

	Word level	Sentence level	Text level
Day 1 Reading		To use awareness of grammar to decipher new words Y3 T1–3 (1)	To prepare for factual research Y4 T2 (16)
Day 2 Comprehension			To locate key words and mark extracts Y4 T2 (17 & 18) To discuss purpose and structure of explanatory texts Y4 T2 (20)
Day 3 Spelling: high-frequency words, onset and rime	To recall high-frequency words Y4 T1–3 (1 & 4) Spelling by analogy with known words Y4 T1–3 (3 & 4)		
Day 4 Verb tenses Plurals		To investigate verb tenses Y4 T1 (2) To extend knowledge of pluralisation Y3 T2 (4), Y4 T3 (1)	
Day 5 Writing composition			To present information in labelled diagrams Y4 T2 (23) To make a record of information Y3 T1 (22)
Homework task 1 Comprehension		To use grammar and context when reading Y3 T2 (1)	To present information in labelled diagrams Y4 T2 (23)
Homework task 2 Wordsearch	To collect and use new words from reading Y3 T3 (12)		

Reading

Before reading

Before looking at the extract on photocopiable pages 46 and 47, begin the KWL grid (see below). The children will need an A4 size version of the grid.

What do I already know?	What do I want to find out?	What have I learned?

K = What do I already **k**now?
W = What do I **w**ant to find out?
L = What have I **l**earned?

To help the children to complete the first two columns of the KWL grid:

- Ask, 'What do you already know about your senses of taste and smell? Encourage them to write at least three pieces of information in the first column.
- Help them to think of three questions about taste and smell. They should then write their questions in the second column.

Give the children a few minutes to look at the extract and enjoy the illustrations, before reading it. Now read the text together (see 'Non-fiction/ information texts' on page 5).

After reading

The children can complete the third column of the KWL grid by writing two facts they have learned from the text.

Week 4 **Explanations**

Day 1

Special Needs Literacy

Finding key words

Ask the children to re-read one section of the text (choosing from 'Tongue-tastic', 'Yum or yuk' and 'Sweet tooth'). Working in pairs, they should:

- discuss which are the key words (the most important ones) which tell the reader the main idea
- highlight or underline the key words
- explain to the group why they chose those words – the others can disagree, but they must be able to say why, and to give alternative key words.

Looking at the structure

Discuss with the children how the extract is organized.

- Why are there pictures as well as words?
- Can they find the introduction, diagrams, captions and headings?

What kind of text is it?

- A story?
- Instructions about how to do something?
- A report about something that has happened?

Make sure the children understand that the text is explaining how something works and is called an explanation text.

Comprehension

Quiz (oral or written answers)

1. Where are your taste receptors?
A. On your tongue.
2. What does a putrid smell mean?
A. Something that smells rotten or bad.
3. What is missing: salty, bitter, sour, _______
A. Sweet.
4. How can you smell food in your mouth?
A. Your mouth and nose are linked.
5. Why do we need a sense of taste?
A. So that we know if something is safe to eat.

Quiz (written answers)

1. How many different kinds of taste are there?
A. Four.
2. What is the name for the water in your mouth?
A. Saliva.
3. True or false? Sweet food gives you an energy boost.
A. True.
4. Name something that smells fragrant.
A. Roses. *(Answers may vary, if children are using their own knowledge.)*
5. Which part of your tongue tastes bitter food?
A. The back.

Week 4 **Explanations**

Day 2

Special Needs Literacy

Sight vocabulary

Ask the children to find *think* (in 'Tongue-tastic'), and to learn to spell the word using the Look–Say–Cover–Write–Check method (see page 6) until they can write it from memory. Do the same with *more* (in 'Tongue-tastic') and *only* (in the first paragraph of 'Sugar and spice'). The children should write the words in their spelling books to take home for further practice.

Help them to think of a sentence with all three words in it and to write it down, using a coloured pencil or felt-tipped pen to highlight *think, more* and *only*.

Spelling by analogy

Spelling pattern: *-art*
(part, cart, dart, smart, start, heart, depart)

Ask the children to find *part* (in the first label of the diagram in 'Sugar and spice') and to look carefully at the spelling. They should learn to spell the word using the Look–Say–Cover–Write–Check method (see page 6) until they can write it from memory.

Encourage them to think of words that rhyme with *part*. (Notice that *heart* has an extra letter – *e*.) Dictate the words for the children to write in their spelling books.

Dictate the following sentences, for the children to write. (Encourage them to write without looking at their word lists.) Before they start, ask them to think about how sentences begin and end (capital letters and full stops.)

*If you are **smart**, you won't **depart** in that **cart**.*
***Part** of her **heart started** to go bump.*

The children should underline all the *-art* words in their completed sentences with a coloured pencil or felt-tipped pen.

They could make word wheels or flick books with the spelling pattern to take home (see page 6).

Week 4 **Explanations**
Day 3

Verb tenses

Read the introduction to the extract again, as a group. Ask the children if they can remember what a verb is (an 'action' word).

Help them to underline all the verbs in the introduction *(eat, drink, need, live, comes, be, had, allow, recognize)*. The verbs *be* and *had* are difficult ones – you could mention that they are 'helpful' verbs.

Are these verbs in the past or present time? It may help if you give an example of a sentence from a story written in the past tense (such as *The princess wanted to get her golden ball back so she asked the frog*). Non-fiction usually has verbs in the present tense. Explain that *tense* means time, when it is used about verbs.

Help the children to find and underline the verbs in the 'Yum or yuk' paragraph. What tense are the verbs in? (Present.)

Ask them to look carefully at the extract and find sections that are in the past tense. ('Sweet tooth' and 'The smelly T-shirt test'.) Why are they in the past tense? (Because they are about things that have already happened.)

Plurals

Ask the children if they can remember the special word that means 'more than one' of something. Remind them that the word we use is *plural* and it could refer to just two or millions.

Help the children to find all the plural words in the 'Tongue-tastic' paragraphs: *scientists, flavours, tastes, things, oranges, grapefruits, crisps.* What would the singular (one) of each of these words be? What is the difference in the spelling? (You add *s* for the plural.)

Dictate the following words for the children to write in another list: *child, man, foot, tooth.* The children should now write the plural of each word beside the first word (for example, *child, children*). Ask them if they always have to add *s* to write a plural word.

Week 4 **Explanations**
Day 4

Understanding your computer

A computer has several different parts. The part that looks like a television screen is called the **monitor**. The **keyboard** has letters and numbers on it. When you press a letter it appears on the monitor.

There is usually a big box called the **hard drive**, which is where the computer's memory is stored. If you do some work on the computer, you can store it in the memory. There is a slot for the **floppy disk**, and sometimes a slot where you can put a **CD-ROM**, called the **CD-ROM drive**.

A computer also has a **mouse**. This can be moved around and has a **button** you can click to make something happen on the monitor.

The **printer** is linked to the hard drive, so that you can make a copy of your work on paper. There are **wires** connecting all the parts together.

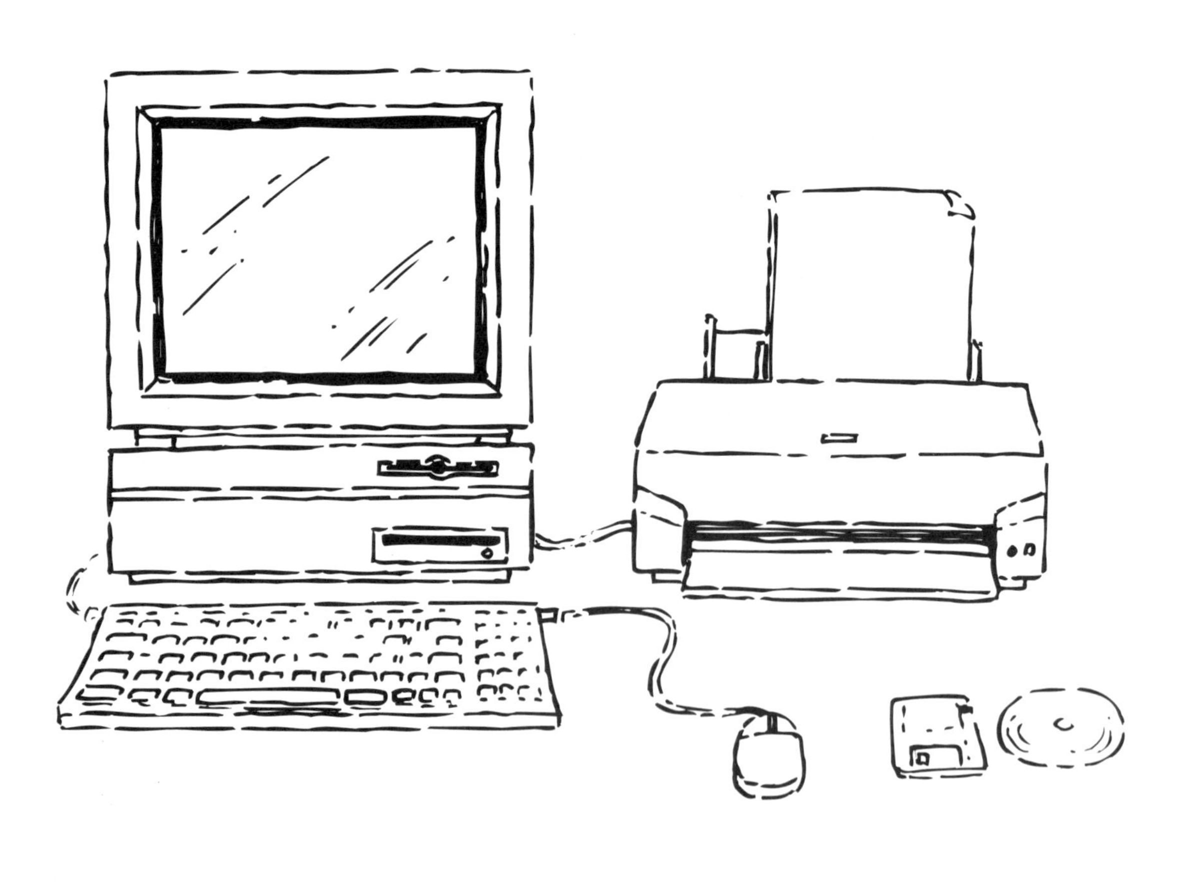

Week 4 Explanations

Day 5

Writing composition:

Ask the children to read the explanation, then add labels to the diagram to show the different parts of the computer. Make sure they include all the words in bold.

Explanations

Taste and smell

Can you remember what you read about taste and smell?
Put a circle around the correct answers.

1 Where are your taste receptors?

on your tongue **in your nose** **in your throat**

2 What does **putrid** mean?

sweet **rotten** **salty**

3 How many different kinds of taste are there?

three **four** **hundreds**

4 What is the name for the water in your mouth?

receptors **chemicals** **saliva**

5 Which of these things smells fragrant?

wood **flowers** **books**

6 Sweet food gives you an energy boost.

true **false**

On the back of this sheet, draw a diagram of a tongue and label it to show which parts respond to the four kinds of taste: sweet, salty, sour and bitter.

Homework

Wordsearch

b	i	t	t	e	r	y	o
o	i	b	r	a	i	n	o
i	s	s	m	e	l	l	y
r	e	c	e	p	t	o	r
h	s	e	c	t	i	o	n
s	a	l	i	v	a	s	a
t	a	s	t	e	s	e	a
n	e	e	n	e	r	g	y

Find each of these words in the wordsearch square. Draw a line through each word as you find it.

smelly **energy** **tastes** **bitter**
section **saliva** **brain** **receptor**

Week 5 Narrative poetry and performance poetry

David and Goliath rap (from *Gynormous!* by Adrian Mitchell)

Introduction

Many children, particularly boys, are fans of 'rap' songs and enjoy performing them (with appropriate actions). This modern rap poem by a well-known children's writer, Adrian Mitchell, is based on the famous Bible story of the young shepherd boy David who, despite his youth and size, fought with the giant Goliath and won, using his 'wicked little sling'. It is a very amusing retelling of a traditional story. The strongly repetitive rhythm and its fast-paced storyline make it an ideal choice for a poem to be read aloud. It combines graphic language, humorous dialogue and entertaining illustrations in a way which will appeal to children. The theme of a small child overcoming a powerful giant is one which children will relish.

Week 5 Objectives

	Word level	Sentence level	Text level
Day 1 Reading	To use phonic/spelling, graphic, grammatical and contextual knowledge as a cue when reading Y4 T2 (1)		To identify patterns of rhyme and verse and read aloud effectively Y4 T2 (7)
Day 2 Syllables Comprehension			To understand how the use of expressive/descriptive language can build tension Y4 T2 (4) To clap out and count the syllables in lines of poetry Y4 T3 (5)
Day 3 Spelling: sight vocabulary, onset and rime	To recall high frequency words Y4 T1–3 (3 & 4) Spelling by analogy with known words Y4 T–3 (3 & 4)		
Day 4 Adjectives and verbs		To revise adjectives and link to expressive language in poetry Y4 T2 (1) To identify the use of powerful verbs Y4 T1 (3)	
Day 5 Writing composition			To write poetry based on structure/style of poems read Y4 T2 (11)
Homework task 1 Review			To write a review Y3 T3 (14)
Homework task 2 Poem structure			To identify patterns of rhyme and verse Y4 T2 (7)

Reading

Before reading

Ask the children if they know the story of David and Goliath – tell them briefly that it is from the Bible and tells the story of how the Philistines were fighting the Israelites. King Saul of Israel was worried about a Philistine giant called Goliath. The Israelite soldiers were too scared to fight him. Then a young shepherd boy called David came and offered to take Goliath on.

Reading the poem

Read the poem on photocopiable pages 44 and 45 in the following way:

- Read the first four verses aloud, so that the children can hear the strong rhythm.
- Ask them to read the next three verses silently, trying to hear the rhythm in their heads.
- Ask for volunteers to read a verse or two, then read the first seven verses together as a group, emphasizing the regular beat of the rhythm.
- Read the rest of the poem. The children can read from verse 8, either in a group or pairs, taking it in turns to read aloud, or they can read to themselves silently. Ask a few questions to check that they have understood what they have read.

After reading

Ask the children:

- How many lines are there in each verse? (Four.)
- Why are all the verses the same length? (To make the poem very regular.)
- Can you hear any rhymes? What is the pattern? (The rhymes are at the ends of pairs of lines – tell the children that this pattern of rhyme is called 'rhyming couplets'.)
- Do all rhyming words have to have the same spelling pattern? (No – in pairs, the children could look at a few verses of 'David and Goliath rap' and find, then highlight, the rhyming words on their copies of the poems.)
- One line in the verse beginning *David takes a staff and he goes to look*… has a rhyme inside the line. Can you find it? *(Goliath replieth.)*

If there is time, each child could read aloud their favourite few verses, either alone or in pairs.

Syllables

Check that the children remember what syllables are. Quickly revise by counting the number of claps while saying the names *David* (two syllables) and *Goliath* (three syllables).

Read the first verse aloud as a group. Half the group should read the first line, while the other half count the syllables in the whole line (12 syllables). Do the same with the other lines in verse 1 (line 2 has 12 syllables; line 3 has 14 syllables; line 4 has 12 syllables).

Choose any other verse and do the same (most of the verses have lines of between 11 and 14 syllables).

Comprehension

Read verses 8, 9 and 10. Ask each child to tell you something that Goliath says or does in these three verses that makes the reader think that he will kill David. Which is the line where the tension and fear disappears? *(But David just smiles as he takes his stand.)*

Quiz (oral or written answers)

1. Where is this story from?
A. The Bible.
2. How had David demonstrated his strength?
A. He had killed a lion and a bear.
3. Which army was Goliath in?
A. The Philistines.
4. Why do you think Goliath was surprised?
A. Because David was so small.
5. How did Goliath's mother find out he had died?
A. In a newspaper (the *Palestine Post*).

Quiz (written answers)

1. How many pebbles did David collect?
A. Five.
2. What kind of a poem is it?
A. A rap.
3. Who is the king of the Israelites?
A. Saul.
4. What rhymes with Goliath?
A. Replieth.
5. How tall was Goliath?
A. Eleven feet.

Sight vocabulary

Ask the children to find *once* (verse 1), and to learn to spell the word using the Look–Say–Cover–Write–Check method (see page 6) until they can write it from memory. Do the same with *upon* (verse 1) and *first* (verse 2). The children should write the words in their spelling books to take home for further practice.

Help them to think of a sentence with all three words in it and to write it down, using a coloured pencil or felt-tipped pen to highlight *once, upon* and *first.*

Spelling by analogy

Spelling pattern: *-ear*
(fear, spear, hear, clear, gear, near, tear, year)
Ask the children to find *fear* and *spear* in the poem (verse 10) and to look carefully at the spellings. What little word can they see in both *fear* and *spear*? *(ear.)* They should learn to spell *ear* using the Look–Say–Cover–Write–Check method (see page 6) until they can write it from memory.

Encourage them to think of words that rhyme with *ear*. (If they suggest *beer, deer, cheer* and so on, explain that some words that rhyme with *ear* are spelled differently.) Point out that the way to remember whether to use *here* or *hear* is to think of what you do with your ears (*ear* and *hear* are spelled the same). Dictate the words for the children to write in their spelling books.

Dictate the following sentences, for the children to write. (Decide whether you want them to have their word lists for reference or not.) Remind them about capital letters, question marks and full stops, if necessary.

*Can you **hear** a **spear** coming **near**?*
*It is **clear** that I have no **fear** in **Year** 5.*

The children should underline all the *-ear* words in their completed sentences with a coloured pencil or felt-tipped pen.

They could make word wheels or flick books with the spelling pattern to take home (see page 6).

Week 5 **Narrative poetry and performance poetry**

Day 3

Special Needs Literacy

Adjectives

Ask the children if they can remember what an adjective is. (If not, remind them that it is a word that describes or tells us more about something.)

- What is the adjective in verse 1 that describes the war? *(terrible.)*
- How is David's sling described (verse 6)? *(wicked, little.)*

Help the children to find four other adjectives in the poem and write them down – they could search through a few verses each or work in pairs.

Verbs

Read the first line of verse 2 together:

They shouted at each other, they hollered and cursed.

Ask the children:

- Which words are the verbs? (*shouted, hollered, cursed*.)
- Why do you think the poet used three verbs that mean nearly the same? Would the effect have been the same if he had just said, *They shouted*?
- Can you think of other verbs for *shouted* that could have been used? (The children could use a thesaurus or dictionary to find alternative words, for example *cried, called, yelled, roared, screamed, shrieked, thundered.*)

They should rewrite the first line from verse 2, substituting different verbs for the ones used.

- Make sure they notice that *cursed* rhymes with *first*. Their new word for *cursed* would mean the second line would need to change too.

Ask them to read aloud their new lines, then the original one. Which do they prefer, and why?

Week 5 **Narrative poetry and performance poetry**

Day 4

Special Needs Literacy

Goldilocks rap

One day a girl called Goldilocks
Came to a door – do you think she knocks?
The house belonged to the three fine bears,
They liked their beds and they liked their chairs.

Here are some rhymes that may be useful:

walk	**talk**		
bear	**chair**	**there**	**dare**
hot	**not**	**got**	**lot**
wood	**good**	**could**	**should**
bed	**said**	**red**	

Week 5 **Narrative poetry and performance poetry**

Day 5

Writing composition:

- Explain to the children that they are going to write a poem like 'David and Goliath rap'. They can either continue the Goldilocks rap or choose their own subject. They can write as many lines as they like. They may wish to work in pairs.
- They should try to write in verses of four lines, if they can. It may help if they think of some rhymes first, suitable for the subject of the poem, and note them down so they can refer to them as they write. A rhyming dictionary may also be useful.

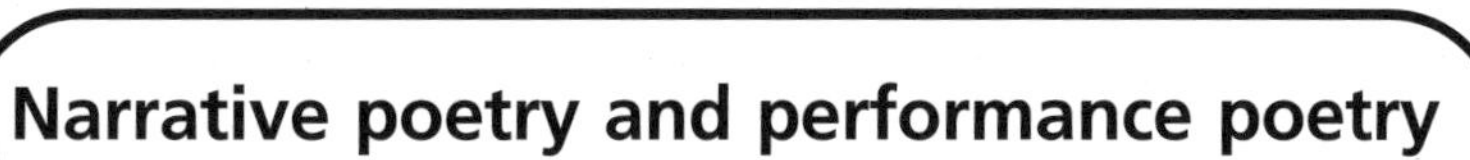

Poem review

I have just read **David and Goliath rap**

written by **Adrian Mitchell**

It was about

I thought that the poem was

because

The part I liked best was

I didn't like

I would recommend this poem to

I would give it a score of

Boring										Brilliant
0	1	2	3	4	5	6	7	8	9	10

Reviewed by

Homework

David and Goliath rap

The computer has taken out all the line breaks in this poem. Can you write it out properly, remembering to start each new line with a capital letter? There should be four lines to each verse.

"I'm just a shepherd boy, but I guard my flock with a wicked little sling that I load with rock. Every one of the slings I've slung were aces – I can hit a wolf's nose at a thousand paces. I smashed up a lion and demolished a bear and I'll do just the same to that hulk over there." David grinned and the King gave a sigh and said: "You've got the guts boy, give it a try." David takes a staff and he goes to look for five smooth pebbles from a nearby brook then he calls out: "Goliath, you're in for a shock." And Goliath replieth: "I'm ready to rock."

____

____

____

____

____

____

____

____

____

____

____

____

Homework

Week 6 Letters to complain or protest

Introduction

In an age of telephones and e-mails, children (and adults) do not have many opportunities for letter writing. However, in Year 5, children are expected to learn about different styles of letter writing during the Literacy Hour. Both these extracts are concerned with writing to complain or protest. The ability to write a letter of complaint is a useful life skill. This is a letter style which children may be familiar with from TV shows about holidays which turned into disasters!

The first extract is a letter from Anna Brown, a ten-year-old girl, to the manager of a theme park, complaining about several disappointments during her birthday visit. The second is a short newspaper article about a proposed threat to a wood near the children's school. On Day 5, the children respond to this article by writing their own letters of protest, using a writing frame, to a local paper in an attempt to save the wood.

Week 6 Objectives

	Word level	Sentence level	Text level
Day 1 Reading Synonyms	To understand the purpose and organisation of the thesaurus and to find synonyms Y3 T1 (16)	To use awareness of grammar, phonic and graphic knowledge, word recognition and context when reading Y3 T2 (1)	To understand how arguments are presented Y4 T3 (17) To investigate how style and vocabulary are used to convince the reader Y4 T3 (18)
Day 2 Spelling: high-frequency words, onset and rime	To recall high frequency-words Y4 T1–3 (1 & 4) Spelling by analogy with known words Y4 T1–3 (3 & 4)		
Day 3 Comprehension Pronouns		To identify and understand functions of pronouns Y3 T3 (2)	To read letters, to understand form and layout Y3 T3 (16)
Day 4 Reading			To investigate how style and vocabulary are used to convince the reader Y4 T3 (18) To summarise the key ideas Y4 T3 (24)
Day 5 Writing composition (letter)			To present a point of view using a writing frame Y4 T3 (21 & 22)
Homework task 1 Comprehension		To use context when reading Y3 T3 (1)	
Homework task 2 Cloze		To predict from the text Y3 T3 (1)	

Week 6 **Letters to complain or protest**

Day 1

Reading

Letter of complaint

Read the introduction to the letter (see the box on photocopiable page 43) to the children. Discuss any similar experiences they have had. Have they or their families ever written to complain about something?

Now read the letter, either aloud in a group, taking it in turns to read a paragraph, or ask the children to read it to themselves silently. Ask a few questions to check that they have understood what they have read.

After reading

Ask the children:

- How can you tell this is a letter? (Make sure they notice the layout – the addresses, date, *Dear…* and *Yours sincerely*.) Why doesn't Anna end with *Best wishes* or *Love from*?

Make sure the children understand the terms *formal* and *informal*.

- When would a formal letter be written? (To someone you don't know well, or someone at their place of work.)

Ask the children to find and highlight the words Anna uses to show she is making a list. *(First, Next, Also, Finally.)* Then help them to find and highlight the words that show how Anna felt about her ruined day out *(annoying, disappointed, unhappy, very upset, appalled.)*

Vocabulary

Anna could have written that the theme park was *awful*, but she found more interesting words to use. Using dictionaries or thesauruses, help the children to find and make a list of other, similar words which could have been used in the letter. (For example, *irritating, horrified, cross, angry, furious, terrible, shocked, dismayed, dissatisfied, let down, difficult, unpleasant, disagreeable*.)

Special Needs Literacy

Week 6 **Letters to complain or protest**

Day 2

Sight vocabulary

Ask the children to find *heard* in the letter on photocopiable page 43, and to learn to spell the word using the Look–Say–Cover–Write–Check method (see page 6) until they can write it from memory. Do the same with *also* and *birthday*. The children should write the words in their spelling books to take home for further practice.

Help them to think of a sentence with all three words in it and to write it down, using a coloured pencil or felt-tipped pen to highlight *heard, also* and *birthday*.

Spelling by analogy

Spelling pattern: *-ass**
(glass, class, pass, grass, lass, mass, brass)

Ask the children to find *glass* in the letter on photocopiable page 43 and to look carefully at the spelling. They should learn to spell the word using the Look–Say–Cover–Write–Check method (see page 6) until they can write it from memory.

Encourage them to think of words that rhyme with *glass*. Dictate the words for the children to write in their spelling books.

Dictate the following sentences, for the children to write. (Encourage them to write without looking at their word lists.) Remind them about capital letters and full stops, if necessary.

*There was broken **glass** on the **grass**.*
*All the **class** had to **pass** the test.*

The children should underline all the *-ass* words in their completed sentences with a coloured pencil or felt-tipped pen.

They could make word wheels or flick books with the spelling pattern to take home (see page 6).

* Depending on the children's regional accent, these words may not all rhyme.

Special Needs Literacy

Comprehension

Quiz (oral or written answers)

1. What was wrong with the picnic area?
A. The only tables which were free were broken or dirty.
2. Which ride was not working?
A. The watersplash.
3. What is Anna's address?
A. 12 Mountview Road, Arnsdale BL2 5JD
4. Why was Melanie upset?
A. She couldn't use the playground.
5. True or false? The ice-creams were good value.
A. False.

Quiz (written answers)

1. How many children went to Funland?
A. Six.
2. When did the trip take place?
A. 15 May 2000.
3. How many things did Anna complain about?
A. Four.
4. Who was appalled at the prices?
A. Anna's dad.
5. Where is Funland?
A. Westlingsea.

Pronouns

Ask the children to read the first paragraph of Anna's letter again (see photocopiable page 43) and to find the word she uses instead of her own name. *(I.)* If the children do not know already, tell them the name for this sort of word is 'pronoun'. Pronouns are used instead of nouns (or names for people, things and so on).

Read the following sentence, with no pronouns in it, and ask the children to improve it:

Anna told the manager that Anna and Anna's family and Anna's friends had a dreadful time at Funland.

Check that they use *she* and *her* instead of repeating *Anna*.

Ask them to find and highlight all the pronouns in Anna's letter. They should count the number of times each pronoun is used to find which are the two most frequent. (*I* – 8, *my* – 8, *you* – 5, *we* – 8, *our* – 2.)

Help them to choose a sentence from the letter and rewrite it, using nouns instead of pronouns.

The children can read their sentences aloud to the group, to hear how odd the text sounds without pronouns.

Week 6 **Letters to complain or protest**

Day 3

Key ideas

Read the introduction to the newspaper article (see photocopiable page 48) with the children, as a group.

Now read the article itself, while the children follow the text in their copies. Some of the vocabulary is difficult to understand, so after you have read it once through first, read it again, one sentence at a time, with the children joining in as much as possible. Discuss what each sentence means, asking the children what they think the tricky words mean and explaining where necessary.

Ask the children to put each sentence into their own words. (They could imagine they are explaining it to younger children.) Help them to highlight key words, such as *Beechnut Wood, threat, Safeco, car park, protesting.*

Read the article again, as a group, looking for words which the writer has used to persuade readers that the supermarket's plan is not a good idea. (For example, *beauty spot, destroyed, giant supermarket chain, national heritage, appalling.*)

Week 6 **Letters to complain or protest**

Day 4

A letter of protest to the newspaper

Ms N Fairview
Editor
Arnsdale Post
High Street
Arnsdale
BL5 4KS

Arnsdale Primary School
Rowling Road
Arnsdale
BL4 8NJ

Date

Dear Ms Fairview

I am about the plan to

First, the children at Arnsdale Primary School

Next, I live very close to Beechnut Wood and

Also, my friends and I

Finally,

Yours sincerely

Week 6 **Letters to complain or protest**

Day 5

Writing composition:

- Explain that Class 5's teacher, Mrs Jackson, has suggested that her class writes a letter to the Arnsdale Post, protesting about the plan to destroy trees to build a car park. The children should now imagine they are in Class 5 and write a letter giving their own points of view. They should write at least three reasons why Beechnut Wood should be saved.
- Remind them to date the letter and to add their name.

Letters to complain or protest

Threat to Beechnut Wood

Can you remember? Draw a circle around the correct answers.
(Some questions may have more than one answer.)

1 Which of these trees grow in the wood?

beech **oak** **sycamore**

2 Who wants to destroy the trees?

the council **Safeco** **Class 5**

3 What is the teacher's name?

Mr Jackson **Mrs Jones** **Mrs Jackson**

4 What does the teacher want Class 5 to do next term?

collect leaves **pond-dipping** **draw trees**

5 What is the name of the newspaper?

Arnsdale News **Daily Post** **Arnsdale Post**

6 How far is the wood from the school?

one hour **five minutes** **ten minutes**

7 Which of these animals live in the wood?

hares **squirrels** **hedgehogs**

8 Who is protesting?

local people **Safeco** **the school**

9 Who wrote to the paper?

Jenny King **Class 5** **Mrs Jackson**

Homework

Letters to complain or protest

Threat to Beechnut Wood

Find the missing words and write them on the lines. Check that it makes sense!

Beechnut Wood is ten __________________ walk from Arnsdale Primary __________________. Classes visit the wood regularly to __________________ the animals and plants there. It has __________________, oak and birch trees, and the wildlife includes squirrels, __________________ and hedgehogs. Mrs Jackson wants to take Class 5 there to go __________________-dipping next term. __________________ of the children live __________________ the wood and love to play there with __________________ friends. The children of Class 5 __________________ furious! They have just __________________ an article in their local __________________, the Arnsdale Post.

minutes'	**newspaper**	**rabbits**	**are**
their	**study**	**near**	**Many**
School	**pond**	**beech**	**read**

Homework

Funland

Ms K Jones
Manager
Funland Theme Park
Westlingsea
CC4 9PZ

12 Mountview Road
Arnsdale
BL2 5JD

15 May 2000

Dear Ms Jones,

I visited the Funland Theme Park on Sunday 12 May with my family and four friends for my tenth birthday. We had heard that Funland was the best theme park in our area and we were really looking forward to our visit. I am writing to you to complain about several things which spoiled our day.

First, the new watersplash ride was not open. This was especially annoying as you had been advertising it on TV the week before. My friends and I were very disappointed.

Next, when we went to the picnic park for lunch we could not find a table to sit at, as the only empty ones were either broken or dirty. My mum and dad were very unhappy about this.

Last Sunday, Anna visited the Funland Theme Park for her birthday treat with her mum and dad, her younger sister, Melanie, and four friends. She had really been looking forward to the day out, but it turned out to be very disappointing. Anna was so upset and angry that she wrote to the Manager to complain.
This is her letter.

Also, we were unable to play in the playground because there was broken glass around the climbing frame. My sister was very upset.

Finally, when we went to buy ice-creams and drinks, my dad was appalled by the high prices we had to pay.

My birthday, which should have been an enjoyable and happy time, was ruined. Funland is an expensive day out and I do not think it was worth the money we paid.

I would like you to refund some of the cost to make up for the ruined day. I hope that you are going to improve your services so that I can tell my friends that Funland is worth visiting again. I look forward to hearing from you.

Yours sincerely,

Anna Brown

David and Goliath rap

Once upon a time there was a terrible war –
Though no one ever told me what the war was for –
But the army of the Philistine was hot for a fight,
And so was the army of the Israelite.

They shouted at each other, they hollered and cursed
But nobody wanted to charge in first
When a Philistine giant stepped from the front rank,
With a ton of brass armour going clank clank clank.

He was eleven feet high and so was his spear.
He yelled: "My name's Goliath, and I've come here
To ask you to send me a man to fight –
Come on – feed me an Israelite."

But none of the Israelites fancied a tussle
With eleven foot of blood bone brass and muscle
Till a kid called David says: "I know about giants,
They can always be beaten by a little bit of science."

The King of Israel says: "Forget it son,
He weighs ten ton, you'll be out in round one."
David says: "I can take him, OK?"
King Saul says: "Who are you, anyway?"

"I'm just a shepherd boy, but I guard my flock
With a wicked little sling that I load with rock.
Every one of the slings I've slung were aces –
I can hit a wolf's nose at a thousand paces.

"I smashed up a lion and demolished a bear
And I'll do just the same to that hulk over there."
David grinned and the King gave a sigh
And said: "You've got the guts, boy, give it a try."

David takes a staff and he goes to look
For five smooth pebbles from a nearby brook
Then he calls out: "Goliath, you're in for a shock."
And Goliath replieth: "I'm ready to rock."

When Goliath sees David he cackles: "Little boy,
I'm going to shake you and break you like a baby's toy,
I'll crack your every bone from your skull to your toes
And I'll feed what's left to the jackals and crows."

All the Israelite army shudders with fear
When Goliath gives a roar and lifts his mighty spear
And the shadow of the giant falls cold on the land –
But David just smiles as he takes his stand

And he takes one pebble and pops it in his sling
And he slings that sling in a circling ring
And he slings that sling and he lets go – now!
And there's a pebble sunk deep into Goliath's brow.

For a moment there's a look of terrible surprise
Which lights up both of that giant's eyes,
Then the lights go out and Goliath is dead
With his face in the dust and a stone in his head.

So that's a happy ending for old King Saul
And the army of Israel and most of all
For little David with his cunning sling –
And that shepherd boy grew up to be King.

But when Goliath's mother read the *Palestine Post*
And learned how her loving son gave up the ghost,
Her great tears flooded the wilderness
And the whole world shook at the grief of a giantess.

Adrian Mitchell

TASTE & SMELL

You eat and drink because you need energy to live, and energy comes from food. But it would be really boring if your food had no taste or smell. Your receptors allow you to recognize a whole range of tastes, from chocolate to chili con carne.

The picture above shows which parts of the tongue respond to which tastes.

TONGUE-TASTIC

Scientists think that all flavours are made up of four basic tastes: sweet, salty, sour and bitter. The things you eat are a mixture of these tastes. So, for example, oranges are sweet and sour, grapefruits less sweet and more sour, and crisps are salty and a bit sweet.

Receptors on your tongue respond to chemicals in your food dissolved in saliva. Different parts of your tongue respond to each of the four basic tastes.

Section of tongue

YUM OR YUK

The main purpose of your sense of taste is to tell you whether something is safe to eat. Dirt, muddy water and most poisonous plants taste horrid. So your immediate reaction is to spit them out. Most foods that are good for us don't taste nasty.

SWEET TOOTH

Many of us have a sweet tooth. This is because thousands of years ago, sweet things were extremely rare. But they were also extremely important, as they gave people a much needed energy boost.

So our ancestors developed a "taste" for sweet things, to make sure they would eat them whenever they found them.

Anthology

SUGAR AND SPICE

If you find it hard to believe that all the flavours of all the different foods come from only four tastes, you'd be right. This is because flavours aren't only made up of tastes, but also of smells.

This is why, if your nose is blocked with a really bad cold, you can feel as if you are eating cotton wool or cardboard.

The difference between taste and smell is that taste receptors respond to dissolved chemicals, while smell receptors respond to chemicals in the air.

You use these parts of your body to smell.

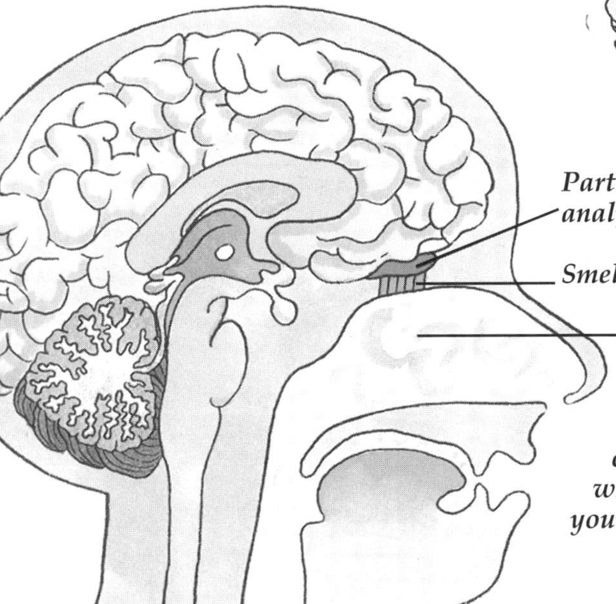

Scientists don't understand exactly how smell receptors work. But, like tastes, they think there are probably only four basic smells: fragrant (like roses), fresh (like pine), spicy (like cinnamon) and putrid (like rotten eggs).

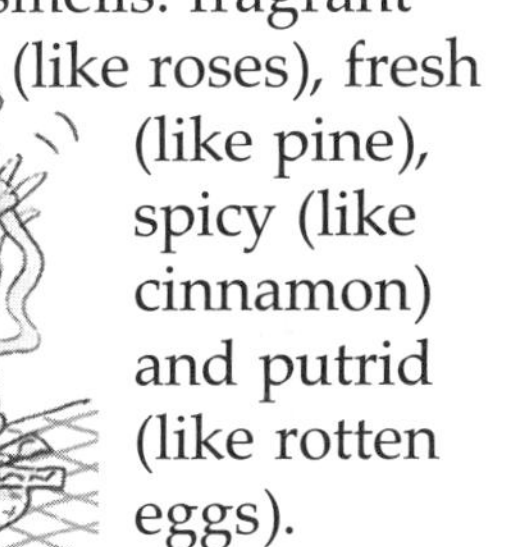

THE SMELLY T-SHIRT TEST

In order to test people's sense of smell, a group of men and women were asked to wear the same T-shirt for 24 hours without washing. The T-shirts were then sealed in plastic bags. Each person was asked to take a sniff of three bags: one containing their own T-shirt and two belonging to strangers – one man's and one woman's. About 75% of people could identify their own T-shirt and tell the difference between the man's and the woman's T-shirts.

Anthology

Beechnut Wood

Beechnut Wood is ten minutes' walk from Arnsdale Primary School. Classes visit the wood regularly to study the animals and plants there. It has beech, oak and birch trees, and the wildlife includes squirrels, rabbits and hedgehogs. Mrs Jackson wants to take Class 5 there to go pond-dipping next term. Many of the children live near the wood and love to play there with their friends. It is a great place for hide-and-seek games and for picnics in the summer.

ARNSDALE POST

Threat to Beechnut Wood

A local beauty spot, Beechnut Wood, is under threat from a giant supermarket chain. Safeco is seeking planning permission to build a new store adjoining the Arnsdale by-pass. Part of Beechnut Wood would be destroyed to make way for a car park. Local residents are protesting about the Safeco bid. "This proposal is appalling," said Jenny King. "Beechnut Wood is part of our national heritage. Some of these trees are hundreds of years old."

Anthology